ISCHEMIA AND REPERFUSION OF VARIOUS ORGANS: INJURY MECHANISMS, METHODS OF PREVENTION AND TREATMENT

Ischemia and Reperfusion of Various Organs: Injury Mechanisms, Methods of Prevention and Treatment

M. V. Bilenko

Nova Science Publishers, Inc.
Huntington, New York

Senior Editors: Susan Boriotti and Donna Dennis
Senior Coordinator: Tatiana Shohov
Office Manager: Annette Hellinger
Graphics: Wanda Serrano
Book Production: Cathy DeGregory, Matthew Kozlowski, Jonathan Rose and Jennifer Vogt
Circulation: Ave Maria Gonzalez, Ron Hedges and Andre Tillman

Library of Congress Cataloging-in-Publication Data

Bilenko, M.V.
Ischemia and reperfusion of various organs: injury mechanisms, methods of prevention and treatment / M.V.Bilenko
p. cm.
Includes bibliographical references and index.
ISBN 1-56072-800-0.
1. Ischemia—Pathophysiology. 2. Reperfusion injury—Pathophysiology. 3. Free radicals (Chemistry)—Pathophysiology. 4. Membrane lipids—Peroxidation.

RB144 .B554 2000
616.07—dc21

00-025271

227 Main Street, Suite 100
Huntington, New York 11743
Tele. 631-424-6682 Fax 631-425-5933
E Mail Novascil@aol.com

Printed in the United States of America

CONTENTS

PREFACE

Throughout the world, cardiovascular diseases associated with ischemic and reperfusion organ injuries are among the main factors of human mortality. Hence, the analysis of mechanisms responsible for these injuries and the development of approaches to their prevention and treatment present a pressing problem for specialists in biology and medicine.

This book is based on firsthand experience gained by the author and her research team during many years of experimental studies, supported by the analysis of abundant literature published between 1970 and 1999. It deals with the comparative assessment of ischemic and reperfusion injuries and their mechanisms in vitally important organs differing in their sensitivity to oxygen deficiency, metabolic activity, and provision with radical-generating and antioxidant system. A distinctive feature of the monograph, compared to previous studies in this field, is that attention is focused on the role of free-radical reactions, their place among other damaging factors (disturbances of energy metabolism, intensification of lipolysis and phospholipolysis, etc.), a detailed analysis of mechanisms responsible for the formation of free radicals during ischemia (the roles of electron transport pathways, radical-generating enzymes, macrophages, endothelial cells, etc,), and a general review of modern methods of antiradical and antiischemic therapy. None of the books published to date is characterized by such an approach to the problem.

Professor Marianna V. Bilenko is head of the Antiischemic and Antiradical Drug Department of the Institute of Biomedical Chemistry (Russian Academy of Medical Sciences, Moscow, Russia).

She is the author of three monographs and more than 280 scientific papers, and the most of them have been published in reviewed journals. She is a far-famed distinguished scientist in the field of transplantology, membranology, ischemia and reperfusion of organs, free-radical lipid peroxidation; 25 Cand. Sci. dissertations were prepared under her supervision. Prof. Bilenko is a member of several scientific societies, including the European Section of the International Society for Heart Research. As an invited lecturer, she participated in more than 30 international scientific conferences and workshops.

Prof. Bilenko and her colleaques were the first who have demonstrated the role of free radicals in the pathogenesis of ischemic and reperfusion injuries. This pioneering work was awarded a Diploma of Discovery (No. 393, priority publication in 1973) under the name *The Phenomenon of Intensification of Lipid Peroxidation in Ischemic (Myocardial and Renal) Tissues*.

INTRODUCTION

Prevention and treatment of acute ischemia of organs is one of the most pressing problems in medicine. This is true not only because of the steadily worsening statistics of cardiovascular diseases worldwide, but also due to the increasing number of surgical operations requiring temporary cessation of blood flow (operations on the aorta, main arteries, "dry" organs), as well as the development and introduction into clinical practice of organ transplantation, in which acute total ischemia of an organ during the period between its removal from the donor and implantation to the recipient is inevitable and has a decisive effect on the results of transplantation.

Moreover, ischemic injuries are in the background of various pathological processes, such as inflammation, different types of stress and shock, traumas, and toxic injuries.

Our knowledge about the mechanisms of injury in ischemic organs has undergone drastic revision over the past 15–20 years.

First, the very concept of acute ischemia has been revised. Today, this term implies not only the impaired supply of arterial blood to the organ, with the venous blood outflow remaining unaffected; acute ischemia is now understood as a severe impairment (partial ischemia) or complete block (total ischemia) of all the three main functions of local circulation, namely, oxygen supply to the tissue, supply of oxidation substrates, and the removal of metabolites from the tissue. It is the interruption of these three functions (primarily of oxygen supply) that accounts for the severity of processes leading to marked structural and functional deterioration of the organ or, in extreme cases, its death.

Second, new possibilities of pharmacological and surgical intervention into the ischemic processes and the increasing number of cases associated with spontaneous or induced (by drug therapy or surgery) reperfusion of ischemic organs have led to the discovery that the latter process is responsible for organ damage as well as ischema per se and provided the basis for understanding ischemia as an integrated two-phase process in which injuries inflicted during acute ischemia are aggravated and often become irreversible after the restoration of blood flow in the organ. A special role also belongs to studies that provided evidence for the central role of O_2 in reperfusion injuries and allowed their authors to formulate the concept of the "oxygen paradox" [Irvin and O'Brien].

Third, abundant new data on the role of membrane structures in organs have been obtained. Membranes have been found to play a major role not only in ensuring compartmentalization of cells and cell organelles, i.e., in the maintenance of electrolyte, water, and other types of homeostasis in the cell (barrier function), but also in providing conditions for the normal functioning of their constituent protein molecules, including membrane-dependent enzymes, transport and regulatory proteins, and receptors (structural, or matrix, function). A new scientific field—medical membranology—has developed, with its hypotheses, theories, and experimental evidence that membrane damage is a primary factor in a number of pathological states.

Fourth, progress in research on this problem helped to partially elucidate mechanisms responsible for membrane damage. Ample evidence has been accumulated to support the role of the qualitative composition and physicochemical properties of the lipid membrane component in maintaining protein–protein and protein–lipid interactions and the normal structural and functional state of membranes. On this basis, new knowledge has been acquired on the role in membrane damage of such factors as products of membrane lipid hydrolysis (lipolysis and phospholipolysis) and products of arachidonic acid metabolism arriving from the cyclooxygenase and lipoxygenase pathways. The key role of Ca^{2+} ions was also demonstrated, and the "calcium paradox" was described [Hearse D. J., Humphrey S. M. and Bullock G. R; Nayler W. G., Ferrari R., Poole-Wilson P. A. et al.].

Fifth, the new concepts concerning the role of free-radical lipid peroxidation (LPO) processes in membrane damage have been formulated. Initially, the "peroxide hypothesis" of cell death was proposed and corroborated only as applied to different pathological states progressing against the background of normal or excess oxygen content in the tissues, such as radiation damage, malignant growth, hyperoxybaria, E vitamin deficiency, etc. [Vladimirov Yu. A. and Archakov A. I.].

The role of lipid peroxidation processes (LPO) in the pathogenesis of membrane damage in the states characterized by oxygen deficiency (hypoxia, anoxia, partial and, especially, total ischemia) for many years remained unknown or even was negated, as it was erroneously assumed that oxygen almost completely disappears from the tissue during its local or total ischemia, which inevitably leads to the cessation of free-radical reactions.

As long ago as between 1973 and 1976, studies performed in our laboratory demonstrated that, even upon exposure to total ischemia, tissues retain residual amounts of oxygen for a fairly long period of time, and oxygen content during the postischemic period significantly exceeds the requirement for oxygen in the organ and the antioxidant potential of the latter (the state of "relative hyperoxia" develops) [Bilenko M. V., Shlifer A. Y. and Epstein I. M.; Bilenko M. V. (1978a)], and this creates favorable conditions for LPO both during and after ischemia. Direct evidence for LPO intensification during total ischemia and after reperfusion of kidneys and the heart was obtained, and the role of these processes in the pathogenesis of structural and functional organ damage was confirmed [Alessenko A. V., Bilenko M. V. and Shelenkova L. N.; Bilenko M. V., Alessenko A. V., Burlakova E. B. et al. (1991); Bilenko M. V. (1989b); Bilenko M. V., Burlakova E. B., Alessenko A. V. and Shelenkova L. N.].

Since that time, abundant experimental and clinical data supporting the above findings have been accumulated, and this problem has been analyzed in many articles, reviews, and special monographs and discussed at international and local conferences.

Today, it is apparent that a major role in LPO belongs not so much to the absolute concentration of oxygen in tissues as to the relationship between oxygen content and the activity (integrity) of the systems producing reactive oxygen species and those protecting tissues against oxygen (antioxidant systems), as it is precisely this relationship that determines the onset and progression (branching) of free-radical chain reactions.

Most of specialists have already accepted that the formation of reactive oxygen species (ROS) and their involvement in the initiation of oxidation chain reactions do occur and are even enhanced upon oxygen deficiency in the organ, as well as under conditions of its excess. In fact, the time has proven the validity of the concept concerning the role of free-radical reactions in pathogenesis of ischemic and reperfusion injuries.

However, many key questions still remain to be answered unequivocally. They are as follows. Whether free-radical reactions are intensified and their products accumulate during the period of oxygen deficiency or only after the restoration of blood flow in the organ, and how important is their role in pathogenesis of structural and functional injuries of membranes and the entire organ caused by ischemia and reperfusion? What are the mechanisms providing for intensification of free-radical reactions during ischemia and after reperfusion, and what are the major sources for free oxygen radical production? What are the timing and cause-and-effect relationships of free-radical processes and other membrane-modifying mechanisms?

Sixth, during the last five to eight years, new data have been obtained on the role of NO, endothelial cells, and phagocytes; effects of oppositely acting vasoactive agents; concepts of necrosis and apoptosis as two variants of cell death; progress in genetic engineering and experiments with transgenic mice, which made possible selective tests for the contribution of each radical-producing or radical-scavenging factor (enzyme) to pathogenesis of ischemic injury; the discovery of new synthetic antioxidants, enzyme system blockers, cell receptors, etc. All these data provide a new means for theoretical analysis and allow us to broaden our views on the key role of free-radical reactions in ischemic and reperfusion organ injuries.

In this book, our experimental results and abundant published data (including publications of 1999) are used as the basis for analyzing pathogenesis of acute ischemic and reperfusion injuries from the membranological standpoint. Attention is mostly concentrated on the interaction between free-radical reactions and other factors responsible for ishemic and reperfusion injuries, such as disturbances of energy metabolism; acidosis; damage to the structure and barrier function of cell membranes; free-radical damage of lipids, proteins, and DNA; activation of phospholipolysis and arachidonic acid metabolism; and the impairment of protective (antioxidant) systems and activation of systems generating free radicals.

The problem of mechanisms and factors responsible for ischemic and reperfusion injuries is usually analyzed using an example of a certain organ. This book, by contrast, is an attempt to analyze it by comparing the response to ischemia and reperfusion of several vitally important organs (the kidney, liver, heart, brain, skeletal muscles) differing in sensitivity to oxygen deficiency, the development of collateral blood flow, tissue metabolism, and provision with free-radical-generating and antioxidant systems.

An attempt is also made to analyse the role of macrophages and endothelial cells in free radical production, to reveal a relationship between the accumulation of products of disturbed lipid metabolism and local (the no-reflow phenomenon) or generalized (ischemic shock) postischemic hemodynamic disorders in organs and the entire organism; and to revise the data on methods for preventing and treating ischemic injuries by means of membrane-stabilizing drugs and inhibitors of free-radical processes.

The book is supplied with numerous references to both classical and recent publications, and many of references are to the studies performed by Russian specialists, which are insufficiently known in the West.

Experimental research that provided the basis for this monograph was initiated in the Laboratory of Antiischemic Drugs of the Drug Research Institute (USSR Ministry of Medical and Microbiological Industry) and continued in the Antiischemic and Antiradical Drug Department of the Institute of Biomedical Chemistry (Russian Academy of Medical Sciences), headed by prof. M. V. Bilenko.

It is my great pleasure to thank all my colleagues in the laboratory and specialists from other Moscow Institutes, who were involved in this research. Their names are indicated in references.

Mechanisms responsible for ischemic and reperfusion injuries of organs have biochemical, biophysical, pathomorphological, and many other aspects. In this monograph, the problem is dealt with "at the interface" of corresponding disciplines. Hopefully, emphasis on practical aspects will make this monograph useful for specialists in many fields, including experimental surgeons, transplantologists, and other specialists in clinical medicine dealing with prevention and treatment of ischemic and reperfusion injuries, reconstructive surgery, and resuscitation.

Studies on the problem of ischemic and reperfusion injuries develop rapidly, and the entire spectrum of corresponding experimental and clinical data is difficult to review comprehensively. The purpose of this book is to help in gaining better understanding of this problem and to attract more attention to this interesting and important field of biomedical research.

All kinds of comments and criticism are welcome.

M. V. Bilenko

ABBREVIATIONS

ACA - artifical circulation apparatus
AA - arachidonic acid
AD - aminopyrine N-demethylase
ADN - adenine nucleotides (ATP + ADP + AMP)
ADP - adenosine diphosphoric acid
AsA - ascorbic acid
AMI - acute myocardial infarction
AMP - adenosine monophosphoric acid
AO - antioxidant
AOA - total antioxidant activity
ARA - total antiradical activity
ASA - acetylsalicylic acid
ATP - adenosine triphosphoric acid
ATA - atmospheres absolute
BHT - butylated hydroxytoluene (ionol)
BLM - bilayer lipid membrane
CA - catecholamines
CAT - catalase
cAMP - cyclic adenosine monophosphoric acid
CCO - cytochrome oxidase
CDI - circulation distribution index
CK - creatine kinase
CL - chemiluminescence
COC - critical oxygen concentration
CP - creatine phosphate
CP - creatine phosphokinase
CPR - coronary perfusion rate
Cr - creatine
cyt. P-450 - cytochrom P-45
DC - diene conjugates
DFO - desferrioxamine, Desferal
DMSO - dimethyl sulfoxide
E - epinephrine
EC - endothelial cells
EDP - end diastolic pressure
EDRF - endothelium derived relaxing factor
EGM - enriched gas mixture
EDTA - ethylenediaminetetraacetic acid
EPR - electron paramagnetic resonance
ER - endoplasmic reticulum
ESR - electron spin resonance
FA - fatty acids
FR - free radicals

16:0	- palmitic acid
16:1	- 8-hexadecenoic acid
18:0	- stearic acid
18:1	- oleic acid
18:2	- linolic acid
18:3	- linolenic acid
20:4	- arachidonic acid (AA)
20:5	- eicosapentaenoic acid
20:6	- eicosahexanoic acid
22:4	- docosatetranoic acid
22:6	- docosahexaenoic acid
FA	- fatty acid
FAD	- flavine adenine dinucleotide
FFA	- free fatty acids
GSH	- glutathione
GSHpx	- glutathione peroxidase
GSSG	- oxidized glutathione
G-6-PDH	- glucose-6-phosphate dehydrogenase
HbO2	- oxyhemoglobin
HGM	- hypoxic gas mixture
HMPS	- hexose monophosphate shunt
HP	- hydroperoxides
HPLC	- high-performance liquid chromatography
1-HTMP	- 1-Tetramethylpiperidine
HUVEC	- human umbilical vein endothelial cells
i. m.	- intramuscularly
i. p.	- intraperitoneally
IC	- ischemic contracture
IMP	- inosine 5-monophosphate
KH	- Krebs-Henseleit solution
LAC	- left circumflex coronary artery
LAD	- left arterior descending coronary artery
LBF	- local blood flow
LCA	- left coronary artery
LDH	- lactate dehydrogenase
LDL	- low density lipoproteins
LDL_{ox}	- oxidized LDL
LMWI	- low molecular wight iron
LP	- lipid peroxides
LPO	- lipid peroxidation
LT	- leukotriene
LysoPL	- lysophospholipid
LysoPLase	- lysophospholipase
Mb	- myoglobin
MCA	- middle cerebral artery
MDA	- malonic dialdehyde
MetHb	- methemoglobin
MPO	- myeloperoxidase
Mt	- mitochondria
N	- nuclei
NE	- norepinephrine
NMR	- nuclear-magnetic resonance
PB	- phenobarbital
PC	- phosphatidylcholine
Pcox	- oxidized phosphatidylcholine
PEA	- phosphatidylethanolamine

per os	- perorally
PI	- phosphatidylinosite
PL	- phospholipids
PL-ase	- phospholipase
PMN	- polimononuclear
PS	- phosphatidylserine
PSE	- proliferation-stimulation extract
PUFA	- polyunsaturated fatty acids
RBC	- red blood cell
ROS	- reactive oxygen species
RPR	- renal perfusion rate
RVR	- renal vascular resistance
SB	- Schiff bases
SDH	- succinate dehydrogenase
SM	- sphignomyelin
SMP	- submitochondrial particles
SOD	- superoxide dismutase
SR	- sarcoplasmic reticulum
TBA	- thiobarbituric acid
TBARS	- TBA-reactive substances
TC	- triene conjugates
TLC	- thin-layer chromotography
TNF-α	- tumor necrosis factor α
VPR	- volume perfusion rate
W. T.	- wet tissue
XDH	- xanthine dehydrogenase
XO	- xanthine oxidase

Chapter 1

CIRCULATORY AND OXYGEN-SUPPLY DISORDERS DURING ACUTE ISCHEMIA AND AFTER REPERFUSION OF ORGANS

1.1 CIRCULATORY AND OXYGEN-SUPPLY DISORDERS IN ORGANS AFFECTED BY DIFFERENT TYPES OF ACUTE ISCHEMIA

Acute ischemia of organs is caused either by general hemodynamic disorders (cardiovascular insufficiency, arterial hypotension, stress and shock of different etiology, apparent death with subsequent resuscitation, etc.) or by local circulatory disorders caused by the restriction or cessation of blood flow through the organ's main arteries.

Local ischemia occurs more often and, depending on the affected area, may be global (involving the entire organ) or regional (involving a part of the organ); as regards the degree of circulatory disorder in the organ, ischemia may be partial (incomplete) or total (complete).

Since it is the extent to which an organ becomes isolated from blood supply that determines the severity of ischemic and postischemic damage, a significant (if not the main) role belongs to the model of organ ischemia created for analyzing the ischemic injuries themselves and evaluating the contribution of each mechanism to the latter.

Some of the most commonly used clinical and experimental models of acute local ischemia are listed in Table 1. It can be seen that the models for inducing acute local ischemia in different organs are of principally the same type. More importantly, each model determines whether the organ will be affected by *partial ischemia,* varying in severity and always maintaining reduced blood and metabolite inflow and outflow (because of ligation of accompanying veins or reduced inflow), or by *total ischemia,* when the blood and metabolite inflow via both the main and collateral vessels is blocked completely for some time and that the outflow also ceases. Such ischemia is always severe. The degree of severity depends on organ-specific differences in collateral circulation, which, in turn, depends on the number of anastomoses inside and outside the organ.

In the kidneys, internal anastomoses are scarce and lie mainly in the medullary (rather than cortical) layer, and the external collateral circulation is either via additional renal arteries or anastomoses with vessels of the perirenal fat, adrenal gland, diaphragm, and ureter; in other words, the severity of ischemia depends on how complete is the cessation of external collateral circulation, including that through the ureteral artery.

In the liver, the blood supply is dual (the flow through the portal vein accounts for 70–80% of the total circulation in the liver, and that through the hepatic artery, for 20-30% [Seleznev S. A]), and the organ has fairly well-developed internal (arterioportal, arteriovenous, portovenous, sinusoid-to-sinusoid) and external anastomoses (with vessels of the diaphragm, spleen, retroperitoneal space, and others, including biliary duct vessels). As a consequence, total liver ischemia occurs only when both afferent vessels and the biliary duct are occluded and the organ is isolated from the external collaterals.

Table 1. Experimental and clinical models of ischemia of organs.

Severity of ischemia	Basic principle of model	Artery occlusion in organs					Corresponding clinical situation	Reperfusion
		Kidney	Liver	Brain	Heart	Limb muscles		
Partial ischemia: Mild or moderate	Cessation of blood flow through the main artery with continuing collateral circulation and venous return.	Abdominal aorta proximal to the renal arteries	Hepatic artery	Middle cerebral artery, common carotid artery	LAD or LAC	Iliac or femoral artery	AMI, insult, reconstructive surgery	Discontinuance of occlusion
Severe, up to subtotal	Cessation of blood flow through the main artery with reduced collateral circulation and venous return	Abdominal aorta proximal and distal to renal arteries; renal artery; same plus separation of the kidney from the perirenal fat; renal artery and vein; same plus abdominal aorta.	Same plus portal vein; same plus hepatic veins (calls for portocaval shunting)	Two carotids; same plus arterial hypotension	Same in animals with weak collateral circulation	Same plus ligation of the abdominal aorta branches near the bifurcation	Vast AMI, vast insult	Discontinuance of occlusion; extracorporeal circulation; extracorporeal perfusion; cross circulation
Complete (total) ischemia	Cessation of arterial and collateral circulation; lack of venous return	Entire renal pedicle plus separation of the kidney from the perirenal fat	Hepatic artery, portal and hepatic veins, plus separation of the liver from the surrounding fat (calls for portocaval shunting); lobar vessels if the liver is lobate	Two carotids and two vertebral arteries or two carotids if (Willis) circle is not closed; cerebral compression; intracranial hypertension	Thoracic aorta (calls for cardio-pulmonary shunting)	Tourniqueting of the limb; compression of the limb muscles	Operations on the "dry" organs	Same plus transplantation; extracorporeal perfusion; decompression
		Storage of the organ in vitro					Organ plantation	Extracorporeal perfusion, transplantation

In the brain, all the four main blood-supplying arteries (two carotid and two vertebral) are connected by anastomoses to form the arterial (Willis') circle at the base of the brain, which ensures abundant internal collateral blood flow whose intensity sharply decreases when the anterior or one of the posterior anastomosing arteries is blocked so that the circle is not closed. Such a situation may arise as a congenital abnormality, and in the animal kingdom it occurs normally in a rodent, the Mongolian gerbil. The external collateral circulation in the brain mainly occurs through the anastomosing branches of the internal and external carotids.

In the heart, internal collateral circulation is through numerous anastomoses between branches of the coronary arteries, coronary arteries proper, and the Thebesian–Vieussens' vessels, whereas external circulation proceeds via anastomoses with branches of the mediastinal and bronchial arteries. The anastomoses are located primarily in the outer (subepicardial) layer of the myocardium, while vessels in the median and especially subendocardial layers are practically terminal in nature, which leads to anatomical insufficiency of the collateral circulation through them. Abundant collateral blood flow, both internal and external, is also characteristic of limb muscles.

In addition to the reduction of collateral circulation, the severity of ischemia also depends on whether an organ's main veins are functional or occluded, as the absence of valves in the veins of many organs (liver, heart, kidneys, etc.) may lead, especially when the arterial flow discontinues, to the backward blood flow (regurgitation) into the organ through the main veins [see Gugushvili L. L.].

Since each model of experimental partial and total ischemia has its counterpart in an almost equivalent clinical situation (see Table 1), the analysis of mechanisms of ischemia and methods for its correction is possible for every clinical case, but with due account for the type of model used for this purpose.

Along with methods for inducing organ ischemia *in situ*, the most widespread approaches include experimental models involving ischemia of isolated organs under conditions of their perfusion and reperfusion with saline solutions *in vitro*, such as the models of total ischemia resulting from temporary cessation of perfusion (stop-flow), low-volume perfusion (hypoperfusion), and anoxic (hypoxic) perfusion. The model of so-called anoxic perfusion of the entire organ or its slice or section does not reproduce the true anoxic state but merely gives rise to the state of hypoxia, as it is impossible to reduce pO_2 in the perfusate below 6 mm Hg under atmospheric conditions [Kolchinskaya A. Z.]. Moreover, similarly to many other types of hypoxia *in vivo*, this model only creates the deficiency of O_2, but not of substrates, in the organ and flushes the toxic metabolites out of it.

The hypoperfusion models (constriction of the afferent vessel or a decrease in perfusion pressure) and the stop-flow model are convenient for inducing desired degrees of partial or total (and at the same time global) ischemia of the heart, liver, kidney, and even brain: they allow researchers to use small animals and, in experiments with perfusion of the heart, to evaluate the degree of ischemic and reperfusion injury under conditions of partial (Fallen's modification of Langendorf's model) or full load on the myocardium (the "working heart" model) [Neely J. R. and Rovetto M. J.] and, under milder conditions (Langendorf's coronary perfusion model), of the cardiac function after a long period of ischemia, without any risk of its decompensation. Recently, numerous cultivated cell types (vascular endothelial, neurons, liver, kidney, etc.) have been used as models for studying ischemic and reperfusion injuries.

Let us now briefly consider the main types of ischemia.

Acute partial ischemia differs from hypoxia by circulatory disorders in the organ and reduced supply of not only oxygen, but also of oxidation substrates, with reduced metabolite outflow. Although oxygen deficiency in this case is a factor determining the severity of ischemia, organ transition from the open to a partially closed state is responsible for substrate, antioxidant, hormonal, and other deficiencies; moreover, it creates conditions for the accumulation in tissues of substances resulting from carbohydrate and lipid metabolism (lipid peroxidation and other products), which produce a toxic effect on the tissue. The severity of partial ischemia varies widely, which depends primarily on the experimental or clinical model of its initiation (see Table 1) and the degree of development of collateral circulation in the organ, which is species- and organ-specific and varies individually. Species-specific differences in collateral circulation, which determine the severity of partial ischemia in the same model, have been observed in most organs. It is known that the internal and external collateral blood flows in the heart are especially well-developed in dogs [Biro G. P. and Beresford-Kroeger D.; Vasdev S. C., Biro G. P., Narbaiz R. and Kako K. J.]; to a lesser extent, in cats, rabbits, rats, and guinea pigs; and the lest developed in pigs and primates [Lavallee M. and Vatner S. F.; Shaikh N. A. and Downar E. (1981); White F. C. and Bloor C. M.], with the anatomical structure of coronary vessels in baboons and humans being remarkably similar. Hence, upon ligation of the coronary artery branch in dogs, blood flow in the ischemic zone remained at the level of 16–40% of the initial [Biro G. P. and Beresford-Kroeger D.; Darsee J.R., Kloner R.A. and Braunwald E.; Jugdutt B.I., Grover Ch., Hutchins G.M. et al.; Vasdev S. C., Biro G. P., Narbaiz R. and Kako K. J.]. The collateral blood flow in dogs was 21.4% of the initial in the center of ischemic cardiac tissue, 52% at its periphery, and 127% in the intact tissue [Biro G. P. and Beresford-Kroeger D.]. By contrast, ligation of the left anterior descending coronary artery (LAD) in cats and rats reduced blood flow to 9–29% of the initial; when the same vessel was ligated in pigs, the residual blood flow was only 4–2% [Fujiwara H., Ashraf M., Sato S. and Millard R. W.] and, at the point of ramification of the first diagonal branch, as low as 1% [Shaikh N. A. and Downar E. (1981)]. LAD ligation in baboons reduced blood flow to 3–5% of the initial level [Heyndrickx G. R, Amano J., Patrick Th. A. et al.]. In the brain, collateral circulation is also better manifested in dogs, then follow cats and rabbits, and humans are the last [Shafranova V. P. and Ryasina T. V.]. The residual blood flow in ischemic tissue after ligation of the middle cephalic artery in cats was 20–30% of the initial [Flamm E. S., Seligman M. L. and Demopoulos H. B.], while in apes it was about 20% [Garcia J. H., Conger K. A., Morawetz R. et al.].

Bilateral ligation of the carotids in rats reduced blood flow through the cerebral hemispheres to 18.7% of the initial [Gasteva S.V., Raize T.E., and Sharagina L.M.; Zamuruev O. N. (1979)], whereas in the diencephalon and midbrain this percentage was two to three times higher, and circulation in the cerebellum and medulla remained the same as before ligation.

In addition to anatomical (preceding) insufficiency of collateral circulation, partial ischemia may be accompanied by functional insufficiency, which is aggravated as the duration of ischemic period increases. The phenomenon of increasing *functional insufficiency of collateral circulation,* which promotes the increase in the size of an ischemic focus, has been observed after occlusion of the middle cephalic artery in cats [Demopoulos H.B., Flamm E.S., Pietronigro D.D. and Seligman H.L.], occlusion and stenosis of a coronary artery branch in dogs [Gorman M.W. and Sparks H.V.; Grayson J., Irvine M., Parratt J.R. et al.], and occlusion of a coronary artery branch in cats [Kogan A. Kh., Losev N. I., Kudrin A. N. et al.]. However, some authors found that the collateral circulation remained unchanged [Lavallee M. and Vatner S. F.] or even improved, rather than deteriorated, as the time after induced acute myocardial infarction (AMI) increased [Olbinskaya L.I. and Litvitsky P.F.; Marshall R. and Parratt J.].

The gradual time-dependent reduction of collateral circulation during partial ischemia of organs seems to be due to vasoconstriction of the collateral vessels and local rheological changes in the blood [Demopoulos H.B., Flamm E.S., Pietronigro D.D. and Seligman H.L.].

A distinguishing feature of partial ischemia, both global or local, is the nonuniformity of circulatory disorders in different parts, layers, or structural units of the organ, due to which ischemic and reperfusion injuries of different severity are observed in them.

It is known that upon the reduction of blood flow through major vessels, circulatory disorders (and severity of ischemic injuries) in the kidney are more pronounced in the cortical, rather than medullary layer [Bilenko M. V. (1978a)]; in the heart of animals with well-developed collateral circulation, they are more pronounced in the subendocardial, rather than subepicardial layer [Allison T.B. and Holsinger V.W.; Biro G. P. and Beresford-Kroeger D.; Kloner R.A., Ganote C.E, Whalen D.A. et al.; Quinn P.J.]; and in the liver, they are most severe in the center of a lobe, around the central venule [Bradford B.U., Marotto M., Lemasters J.J. et al.]. The difference in blood flow through different layers of the myocardium during AMI is so obvious that partial ischemia, especially in the dog heart, is often severe in the subendocardial layer and moderate or even mild in the subepicardial one [Allison T.B. and Holsinger V.W.].

Differences in blood flow through the outer and inner myocardial layers also increase with the duration of partial ischemia [Darsee J.R., Kloner R.A. and Braunwald E.; Gavin J.B., Thomson R.W., Humphrey S.M. et al.].

Thus, partial ischemia varies in severity depending on the model of its initiation as well as on organ-specific, species-specific, and individual differences in the collaterals; it is unstable and changes (more often for the worse) as the ischemic period increases, and is also heterogeneous in different parts of the organ. Consequently, according to different authors, the tolerable period of partial ischemia varies widely, and the method itself is not very appropriate for the comparative analysis of the nature and molecular mechanisms of ischemic and reperfusion injuries in various organs.

Acute total (complete) ischemia is characterized by the complete cessation of circulation in the organ, so that the latter is converted into a closed system. Experimental models of total ischemia differ markedly from those of partial ischemia.

The model of total ischemia offers the following possibilities:

(1) To block the delivery of oxygen and oxidation substrates to the organ and outflow of metabolites from it, i.e., to reproduce the entire complex (triad) of adverse factors that not only determine the severity of ischemia, but also affect the mechanisms of ischemic injuries, as the products of carbohydrate and lipid metabolism, accumulating in the organ during total ischemia, make a major contribution to their pathogenesis.

(2) To create ischemia of controlled severity (with respect to time, temperature and other conditions).

(3) To ensure the uniformity of circulatory disorders in functionally different parts of the organ and, hence, to create identical conditions in the latter for revealing the respective mechanisms of ischemic and reperfusion injuries.

(4) The model involving isolated organ storage with subsequent transplantation, perfusion *in vitro*, or extracorporeal perfusion makes it possible to exclude (or minimize) its exposure to the effects of emotional and pain stress, hyperadrenalemia, arterial hypotension, and other extrinsic factors during ischemia and reperfusion; in other words, such a model provides the possibility of revealing the exact mechanism of the damaging effect exerted on the organ by ischemia and reperfusion proper. The model of ischemia created in an organ immediately after its dissection from the cadaver is less consistent, as the period of agony is accompanied by generalized ischemia of uncertain duration and severity, which aggravates the subsequent ischemic injuries. This model is unsuitable for organs sensitive to ischemia, such as the brain and the liver.

(5) To create identical conditions in different organs, which is practically impossible in the case of partial ischemia, and to compare the nature, severity and mechanisms of ischemic and reperfusion injuries in organs characterized by different sensitivity to ischemia.

As demonstrated on the total ischemia models used to evaluate a number of different parameters in rats, dogs, pigs, and humans, the maximal (critical) period of tolerance (reversibility) is 30–60 min for hepatic ischemia [Marubayashi S., Dohi K., Ezaki H. et al. (1982)], 60 min for cardiac ischemia [Przyklenk K., Vivaldi M.T., Schoen F.J. et al.; Schaper J., Mulch J., Winkler B. et al.], 60–90 min for renal ischemia [see Bilenko M. V. (1978a)], 5–10 min for cerebral ischemia [Tel'pukhov V.I., Trenin S.O., Knyazeva G.D. and Lapkina T.I.], and 6–9 h for ischemia of limb muscles [Oksman T.H., Dalin M.V. and Kovanov V.V.; Savel'ev V.S., Chekareva G.A., Mishnev O.D. et al.].

The partial tolerance period usually exceeds the above values by 30–60 min [Bilenko M. V. (1978a); Marubayashi S., Dohi K., Ezaki H. et al. (1982); Nördlinger B., Douvin D., Javaudin L. et al.], depending on individual variations. Currently, many specialists believe that the model of total ischemia of organs (especially the heart) *in vitro* has certain advantages over the *in vivo* model [Jennings R.B., Reimer K.A., Hill M.L. and Mayer S.E.; Nayler W.G. and Elz J.S.; Schaper J., Mulch J., Winkler B. et al.], while the hypoxic perfusion model is not quite adequate [Nayler W.G. and Elz J.S.].

Since oxygen deficiency plays a key role in the pathogenesis of ischemic injuries and it will be shown later that one of the principal mechanisms of these injuries is intensification of lipid peroxidation (LPO) processes, which are oxygen-dependent, special attention should be given to *the amount of "residual" O_2 in organs with different types of ischemia,* usually determined from the value of pO_2. Judging from the parameters of residual blood flow in the models of partial ischemia, pO_2 even in the center of ischemic cerebral tissue is no less than 4–8 mm Hg [Flamm E. S., Seligman M. L. and Demopoulos H. B.], whereas in the heart it is at least 4–10 mm Hg [Meerson F.Z., Saltykova Y.A., Didenko V.V. et al.; Olbinskaya L.I. and Litvitsky P.F.; Shargorodsky B.M., Rastorguev B.P. and Raiskina M.E.].

Total ischemia leads to a rapid (within 1.5–4 min) decrease of pO_2 in the organ [Olbinskaya L.I. and Litvitsky P.F.; Sorokovoy V.I., Petrenko Yu.U., Mochenova G.M. et al.; Shargorodsky B.M., Rastorguev B.P. and Raiskina M.E.], and it has long been believed that pO_2 drops all the way to zero.

In our experiments with a kidney stored at 20°C *in vitro* and occlusion of the renal artery in dogs *in vivo* [Bilenko M. V. (1978a)], the pattern of variations in pO_2 was as follows (Fig. 1): rapid decrease (in about 1 min), slow decrease (up to 5 min), and the period of remaining at the residual level (about 2–5 mm Hg) for 1.5–2 h. Periods of rapid and slow pO_2 decrease, although of greater duration than in the kidney, were also observed during total ischemia of limb muscles in rats, after their tourniqueting or isolated storage at 20°C, residual pO_2 (4–2 mm Hg) remaining even after 1–6 h of observation [Arkhipenko Yu. V., Bilenko M. V., Dobrina S. K. et al.].

Fig. 1: pO_2 in the cortex (1) and medulla (2) of the dog kidney after renal artery occlusion (a) and during isolated kidney storage at 20°C (b)

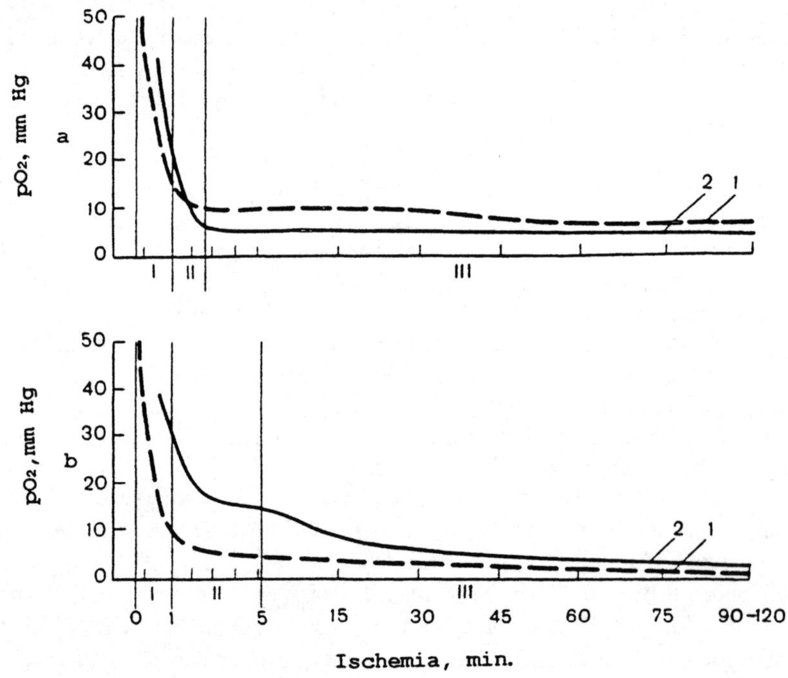

I - period of rapid decrease in pO_2; II - period of slow decrease in pO_2; III -period of very slow decrease in pO_2 (stabilization). pO_2 was measured using an LP-7E polarograph with platinum electrodes implanted to depths of 5 mm (cortex) and 15 mm (medulla)

In our study on the pattern of pO_2 decrease during total ischemia of the isolated dog heart stored at 20 and 37°C [Otverchenko V.N.], pO2 in the left ventricle dropped from 23 ± 0.8 to 4–7 mm Hg during 10 min of storage and then remained virtually unchanged, almost irrespective of storage temperature (Fig. 2). Thus, pO2 never dropped to zero during our experiments with total ischemia of the kidney, limb muscles and the heart, the residual pO_2 level being 2–7 mm Hg.

The fact that residual amounts of O_2 remain in totally ischemic organs is consistent with other findings as well [Oda T.]. In the latter study, anoxic cardiac arrest in dogs, resulting from occlusion of the ascending aorta and aortopulmonary shunting, was accompanied by pO_2 decrease to 50% of the initial within the first 2 min; thereafter, pO_2 continued to decrease for another 3–5 min but never went below 20% of the initial level during 120 min of observations (Fig. 3). Apparently, the decrease of pO_2 had not received sufficient attention in the past, and the residual pO_2 level had been underestimated. Recently, it has been speculated that a biological system under atmospheric pressure can never be truly anoxic [Hess M.L. and Manson N.H.].

Fig. 2: pO_2 in an isolated dog heart after 10min-4h storage at 20°C or 37 °C and after 10-60 min of reperfusion (mean ± SEM).

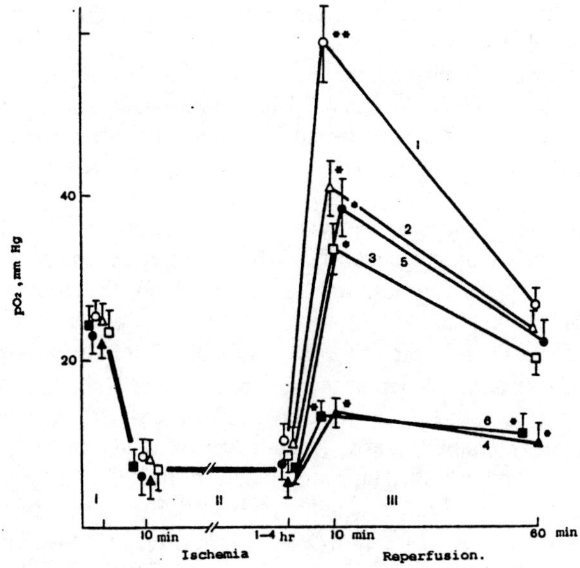

The mean value of pO_2 for different storage periods and temperature are represented by the thick line. Ischemia at 20°C (1-4): 1 - 1 h; 2 - 2 h; 3 - 4 h; 4 - 8 h. Ischemia at 37°C (5-6): 5 - 1 h; 6 - 2 h. *-$p < 0.05$; **-$p < 0.01$ versus initial level. pO_2 was measured using electrodes and a KSP-6 recorder. The initial values were measured on the heart *in situ.* The reperfusion was performed by heart extracorporeal connection to dog recipient cervical vessels.

Fig. 3: pO_2 in the dog heart during its anoxic arrest (% of the initial level) [Oda T.]

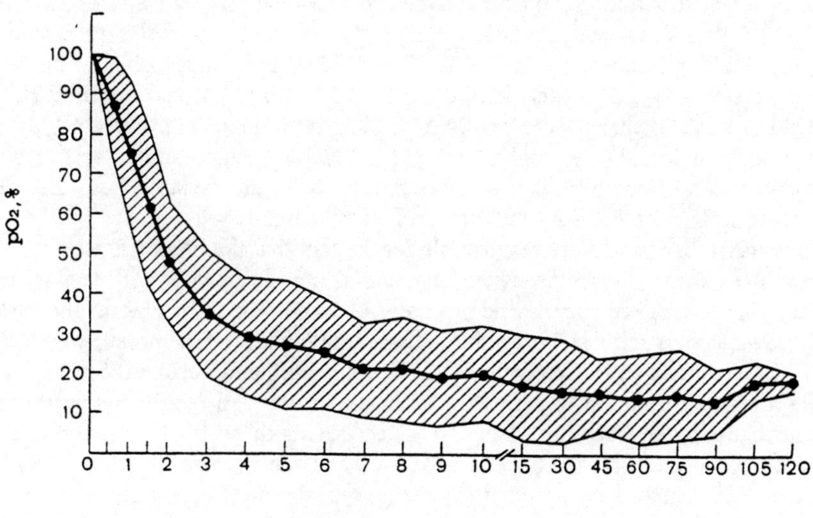

It is known that the value of pO_2 in tissues is a resultant of two quantities: the rate of oxygen supply and the rate at which oxygen is expended in O_2-dependent processes within the cell (oxidation of substrates in mitochondria, LPO processes, etc.), so that the persistence of residual pO_2 in a totally ischemic organ can be explained by the following processes:

(1) Continuing dissociation of HbO_2 within capillaries and O_2 diffusion across the capillary wall into the intercellular fluid and across the plasma membrane into the cell. The release of O_2 from HbO_2 is facilitated by the decreasing affinity between O_2 and Hb during ischemia [Grigoryants R. A., Meitina R. A. and Gasilin V. S.]; acidosis promotes this effect (the Bohr effect) [Wolbarsht M.L. and Fridovich I.; Borovskikh N.A., Filev L.V. and Luk'yanov Yu.V.].

(2) Oxygen release from intracellular depots, such as O_2 bound to myoglobin (Mb) in muscle cells [Araki R., Tamura M. and Yamazaki J.; Kreuzer F.] and tissues of mollusks [Karnaukhov V. N.]; to carotenoids and lipid granules, as in tissues of mollusks (and, possibly, other animals); and to liver cytochrome P-450, which is a hypothetical O_2 carrier [Archakov A. I. (1975); Longmuir I.S., Young A. and Mailman K.].

During cardiac ischemia, Mb deoxygenation by 50% leads to a decrease in O_2 consumption rate, this decrease becoming twofold upon Mb deoxygenation by 70% [Araki R., Tamura M. and Yamazaki J.].

(3) A sharp decrease in oxygen consumption for the most O_2-demanding oxidative processes in mitochondria after the "critical oxygen concentration" (COC) is reached. The upper and lower COC levels—the minimal amounts of O_2 at which the mitochondrial respiration (and ATP synthesis) either continue at the same rate or discontinue altogether—vary from 8–10 to 1 mm Hg [Araki R., Tamura M. and Yamazaki J.; Brezis M., Rosen S., Silva P. et al.; Chance B.; Jones D.P. and Mason H.S.]; they differ markedly from one organ to another and are dynamic quantities. Changes in the behavior of the mitochondrial electron transfer chain and activation of succinate dehydrogenase (SDH) (both phenomena being common during ischemia; see Chapter 3) may result in the reduction of COC. There is evidence [Marshall R.S., Koch C.J. and Rauth A.M.] that at least 0.02–0.04% O_2 remained in a cell suspension *in vitro* after respiration had stopped.

Since other O_2-dependent processes require much more less oxygen than respiration [Lukyanova L.D., Balmukhanov B.S. and Ugolev A.T.] (see Chapter 9), the residual (subcritical) O_2 content may indeed remain at the same level for a long period of time.

(4) More rapid O_2 diffusion from capillary lumina and extracellular fluid into the cell because of lower membrane resistance to O_2 diffusion.

The O_2 diffusion coefficients in tissues under atmospheric pressure are known to vary from $0.6 \cdot 10^{-5}$ to $2.3 \cdot 10^{-5}$ $cm^2 \cdot s^{-1}$; that is, O_2 diffuses within an hour to a depth of 0.013 mm [Berezovsky V.A. and Sushko B.S.; Evans N.T.S., Naylor PFO, and Rowlinson G.].

The existing diffusion limitations are determined by membranes, and O_2 diffusion coefficients vary in different organs, different structural units of the same organ [Bilenko M. V., Shlifer A. Y. and Epstein I. M.; Evans N.T.S., Naylor PFO, and Rowlinson G.], and apparently in different structural units of a membrane, as O_2 solubility in nonpolar media at 22 and 37°C is known to be 3.5 to 5 times as high, respectively, as in polar media [Lawrence J.H., Loomis W.F., Tobias C.A. and Turpin F.H.; Merck Index]. Membrane permeability for O_2 changes drastically depending on changes in the electrolytic composition of the cell and the composition and physicochemical properties of the membrane itself, especially its lipid bilayer [Alferov Yu.M. and Berezovskiy V.A.; Berezovsky V.A. and Sushko B.S.; Burlakova E.B. and Khrapova N.G.; Kreuzer F.; Min'ko A.A., Petunin Yu.I. and Sushko B.S.].

These and possibly other factors are responsible for the fact that the barrier function of cell membrane structures (at least the plasma membrane of the myocardial muscle in rats) with respect to O_2 markedly decreases [Epshtein I.M.]. This process may be promoted by oxidation of membrane phospholipids because the less ordered the molecules and the shorter their chains, the higher the membrane permeability for gases [see Berezovsky V.A. and Sushko B.S.; Min'ko A.A., Petunin Yu.I. and Sushko B.S.];

(5) The nonuniform distribution of residual O_2 content in the cell due to the appearance of oxygen concentration sites and oxygen-free sites. The O_2 content seems to be higher, firstly and naturally, near capillaries; secondly, near membranes, since it is known that the vectored motion of molecular oxygen along radially oriented membranes is advantageous both thermodynamically [Skulachev V.P.] and biologically, for this is how cytochrome P-450-dependent oxidative processes seem to be supplied with O_2 [Jones D.P. and Mason H.S.]; thirdly, in hydrophobic (nonpolar) membrane portions where O_2 solubility is higher. The heterogeneous (mosaic-like) distribution of residual O_2 in cells and membranes during ischemia of an organ [Demopoulos H.B., Flamm E.S., Pietronigro D.D. and Seligman H.L.; Halliwell B. and Gutteridge J.M.C. (1984)] may correspond to the heterogeneity of metabolic processes occurring in the ischemic organ and intensify LPO processes by creating locally increased O_2 concentrations, which exceed those determined from an averaged pO_2 value (see Chapter 10).

The fact that residual O_2 can be found (at least temporarily) even in a totally ischemic organ had led us in 1973, to the hypothesis of the possible involvement of LPO processes in the pathogenesis of ischemic injuries. Arguments in its favor will be discussed below (see Chapter 5).

1.2 CIRCULATORY AND OXYGEN-SUPPLY DISORDERS IN ORGANS DURING THE REPERFUSION PERIOD

Reperfusion of an ischemic organ or reoxygenation of a hypoxic one is virtually never accompanied by rapid and complete restoration of its circulation. Changes in local hemodynamics after reperfusion of the ischemic organ may take the form of reactive hyperemia followed by either the recovery of normal circulation or the reduction of blood flow below the initial level (posthyperemic ischemia or delayed incomplete blood flow recovery, also known as the no-reflow phenomenon); the third alternative is incomplete restoration of blood flow immediately after reperfusion (the primary no-reflow phenomenon). The direction, extent, and reversibility of these changes depend on a number of factors, including the type of organ, the extent and duration of ischemia, the integrity of neural connections between the organ and the entire organism, and the temperature at which an isolated organ is stored.

The dependence of changes in blood flow during **reperfusion of the kidney** on the extent and duration of ischemia was studied in our laboratory using three experimental models: (1) occlusion of the renal artery in dogs (acute partial ischemia), (2) the same combined with separation of the kidney from the perirenal fat (acute subtotal ischemia), and (3) isolated dog kidney storage at 20°C (acute total ischemia) [Bilenko M. V. (1978a); Bilenko M.V., Dmitrieva I.S. and Kotlyarov P.M.; Kovanov V. V., Bilenko M. V., Amosov I. S. et al.].

When partial ischemia (the first model) was used, brief 30-s and 1-min ischemia was not accompanied by significant circulatory changes during the reperfusion period. As the ischemic period increased to 3–15 min, the state of reactive hyperemia developed; after 30–60 min of ischemia, reactive hyperemia was less pronounced and was followed after 3–5 min by the no-reflow phenomenon; ischemia lasting for 120 min was not accompanied by reactive hyperemia but was immediately followed by the no-reflow phenomenon. In subtotal ischemia (the second model), reactive hyperemia changed to the no-reflow phenomenon after a shorter period of occlusion (Fig. 4).

Studies on the nature and reversibility of hemodynamic changes during reperfusion in dog kidneys exposed to total ischemia (the third model) under moderately hypothermic conditions (20°C) and subsequently perfused extracorporeally by connecting them to the femoral vessels of a recipient dog (Fig. 5) showed that short periods of ischemia (5–10 min) under such experimental conditions are not followed by substantial circulatory changes in the organ; one-hour ischemia is accompanied by a reversible decrease in blood flow and a reversible increase in renal vascular resistance (RVR); two- and three-hour ischemia results in a sharper and irreversible (within the reperfusion period under study) reduction of circulation and an increase in RVR. Reactive hyperemia during a 15-min to 4-h reperfusion period was not observed in these experiments. The degree of functional degradation corresponded to the severity of postischemic hemodynamic changes in the kidney [Bilenko M. V. (1978a); Kovanov V. V., Bilenko M. V., Amosov I. S. et al.]. The relationship between the intensity and duration of the no-reflow phenomenon in the kidney, on the one hand, and the extent and duration of its ischemia, on the other, was revealed by other researchers as well [Mason J., Welsch J. and Takabatake T.; Riley A.L.; Wolgast M., Karlberg L., Källskog Ö. et al.].

Fig. 4: Renal perfusion rate (RPR) in the dog kidney reperfused 1-3 min or 10 min after different time of complete in vivo ischemia (% of the initial level)*

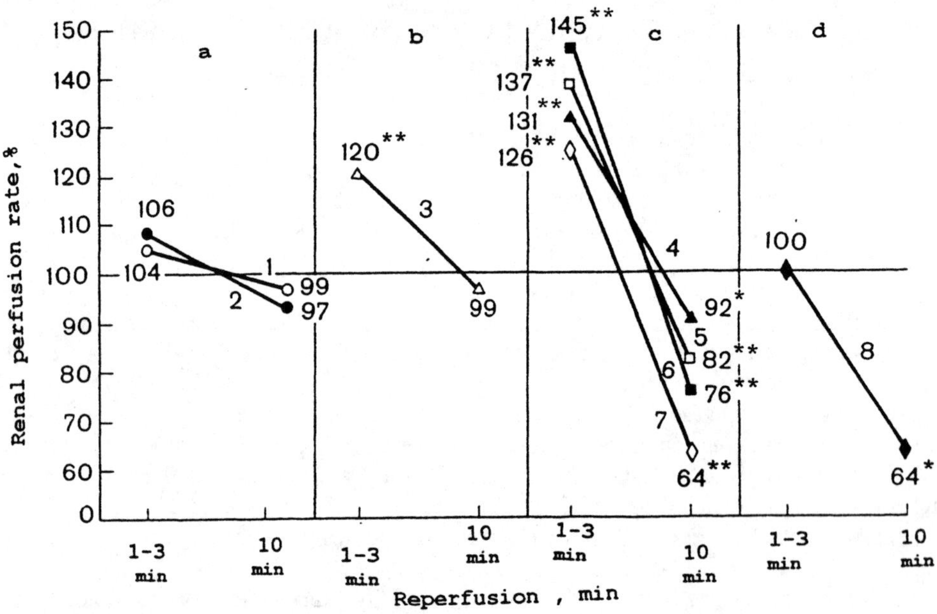

Ischemia at 37°C: 1 - 30 s; 2 - 1 min; 3 - 3 min; 4 - 5 min; 5 - 15 min; 6 - 30 min; 7 - 60 min; 8 - 120 min; a - no changes; b - reactive hyperemia; c - reactive hyperemia followed by the no-reflow phenomenon, d - primary no-reflow phenomenon. *-$p < 0.05$; **-$p < 0.01$ versus initial level. RPR was measured using a Statham electromagnetic fluometer (in ml •min^{-1}per 100 g of wet weight). *note: here and in most next figures, percentage of the initial level is given as mean values, the significant differences being determinated by Student's test using mean ± SEM of absolute values.

A distinguishing feature of the no-reflow phenomenon in ischemic kidneys is the redistribution of blood from the cortical into the medullary layer. In our experiments with reperfusion of dog kidneys after different ischemic periods, this redistribution was indicated by a lower amplitude of the cortical wave on the rheonephrogram, reduced turgor of the kidney [Bilenko M. V. (1978a)], decreasing pO_2 in the cortical layer without any perceptible change in medullary pO_2, and a lower circulation distribution index (CDI) (Fig. 6) [Bilenko M. V., Shlifer A. Y. and Epstein I. M.].

Blood shunting through the juxtamedullary bypass during reperfusion was also observed when the microangionephrographic, histological, isotope, injection, corrosion, and other techniques were used [Bilenko M. V. (1978a)]. Microangiography combined with histological examination [Bilenko M. V. (1978a); Kovanov V. V., Bilenko M. V., Amosov I. S. et al.] showed that the sites of incomplete restoration of renal circulation were primarily associated with distal portions of interlobar arteries at the beginning of the afferent arteriole, whereas the subcapsular region, medulla, and kidney pelvis were characterized by an increased number of vessels and straight arteries filled with blood. Prolongation of the ischemic period to 2–3 h may restrict not only cortical, but also juxtamedullary circulation [Mason J., Welsch J. and Takabatake T.; Mason J., Forhorst J. and Welsch J.; Solez K., Idbura T., Silva C.B. et al.; Wolgast M., Karlberg L., Källskog Ö. et al.]. The nature and pattern of hemodynamic changes during reperfusion of ischemic kidneys were described in our previous publications [Bilenko M. V. (1978a); Bilenko M.V., Dmitrieva I.S. and Kotlyarov P.M.].

Fig. 5: Renal perfusion rate (RPR, a) and renal vascular resistance (RVR, b) in the dog kidneys reperfused 15min or 3-4 h after different time of complete in vitro ischemia (isolated storage at 20°C) (% of the initial level).

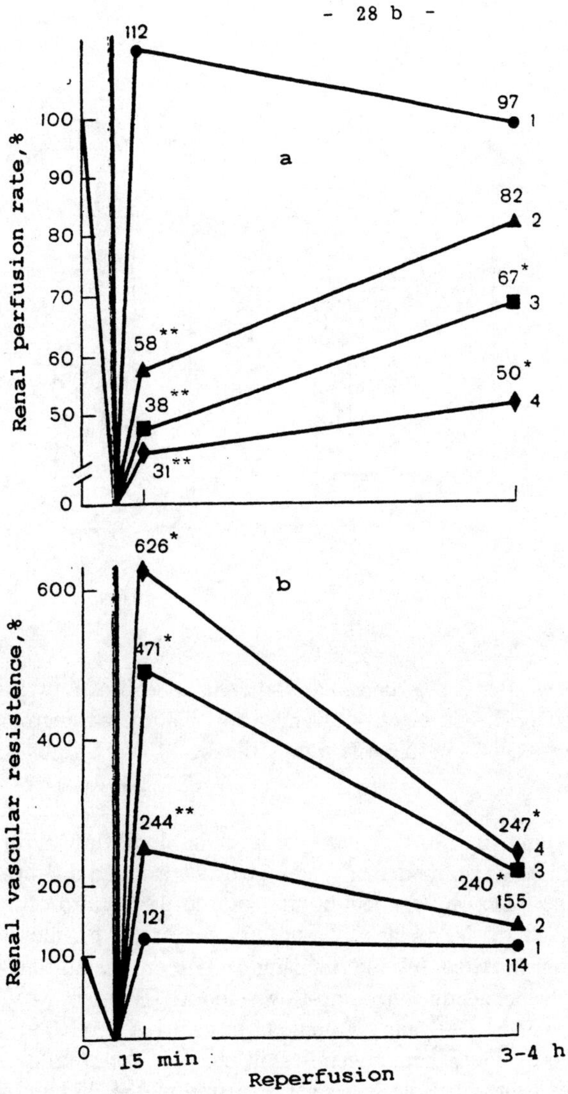

Ischemia at 20°C: 1 - 10 min; 2 - 1 h; 3 – 2 h; 4 - 3 h; *-p<0.05; ** -p<0.01 versus initial level. The RPR was measured using a bubble photodiode flowmeter (in ml•min-¹ per 100 g of wet weight). The RVR was calculated using the Poiseuille formula.

Fig. 6: pO_2 in the cortex (a) and medulla (b) and circulation distribution index (CDI, c) in the dog kidneys reperfused during 6h after 2 h of complete in vitro ischemia (isolated storage at 20°C) (mean ± SEM)

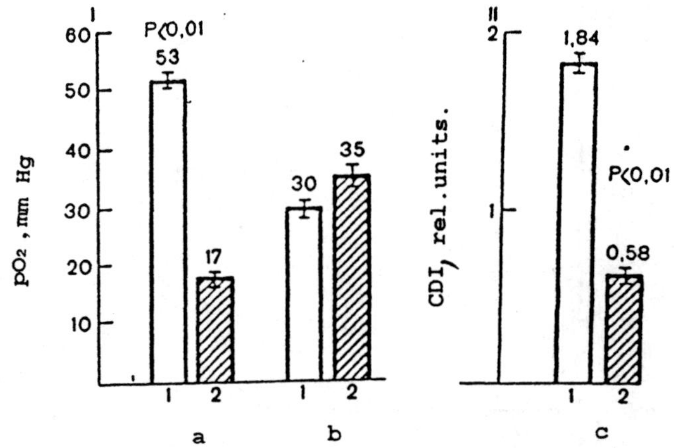

1 - intact kidney; 2 - kidney after 2 h of complete ischemia and 6h of reperfusion; P-1 versus 2. pO_2 was measured using Cu/Mg-Cd electrodes and an M-95 microampermeter. The CDI was calculated using pO_2 in cortex: pO_2 in medulla ratio [Bilenko M. V., Shlifer A. Y. and Epstein I. M.].

In our experiments with **reperfusion of the heart,** changes in blood flow depending on the extent and duration of ischemia were studied on two models: (1) storage of the isolated dog heart at 20° and 37°C with its subsequent extracorporeal perfusion by connecting the aorta to the carotid of a recipient dog [Bilenko M.V., Otverchenko V.M., Astaf'ev V.V. et al.], and (2) perfusion of the isolated rat heart with Krebs–Henseleit (KH) solution *in vitro* by the Langendorf–Fallen method, which was discontinued temporarily and followed by reperfusion (the stop-flow model) [Bilenko M.V., Bulgakov V.G. and Morgunov A.A. (1987); Bilenko M.V., Bulgakov V.G. and Morgunov A.A. (1989); Morgunov A.A.].

Analyzing the parameters of local hemodynamics in the dog heart after its total ischemia and extracorporeal perfusion, it was found that the coronary perfusion rate (CPR) and pO_2 in the heart during the reperfusion period depend on storage temperature and the duration of ischemia (Fig. 7). After 10-min reperfusion, CPR as well as pO_2 (see Fig. 2) reached a peak in the heart kept at 20°C for 1 h (reactive hyperemia); after 2-h and 4-h storage at 20°C and 1-h storage at 37°C, both CPR and pO_2 were lower; and after storage for 8 h at 20°C or 2 h at 37°, their values were extremely low. In all experimental series except the last two, pO_2 and CPR decreased almost in a similar manner after 60 min of reperfusion.

Fig. 7: Coronary perfusion rate CPR in the dog hearts reperfused 10 or 60 min after different time of complete heart ischemia isolated storage at 20°C or 37°C (% of the initial level).

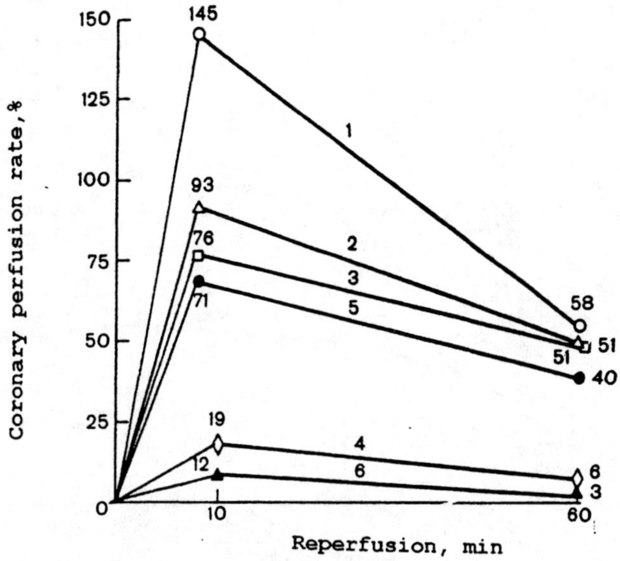

Ischemia at 20°C (1-4): 1-1h; 2-2h; 3-4h;4-8h. Ischemia at 37°C (5-6): 5-1h;6-2h. Significant differences after 10 min of reperfusion: P1 <0.05, 1-versus 2-6; P2<0.05, 2-versus 4 and 6; P3<0.01, 3-versus4; P4<0.05, 5-versus 6.
CPR was measured using a Statham electromagnetic fluometer, in ml•min^{-1} per 100 g of wet weight.

According to published data, CPR in the "nonworking" pig heart under conditions of aortopulmonary shunting and artificial coronary at pressures of 50 and 100 mm Hg was 150 ± 30 and 426 ± 30 ml/100 g per min, respectively [Engelman R.M., Adler S., Gouge T.H. et al.]; in the nonworking dog heart perfused *in situ* with the donor blood, CPR was 55 ml/100 g per min [Trubetskoy A.V. and Lysenko L.T.]. In other words, the "initial" CPR was higher than that observed in some our experiments after 10-min cardiac reperfusion and in all experiments after 60 min of cardiac reperfusion; that is, the delayed or primary no-reflow phenomenon, dependent on the duration of ischemia and storage temperature, was always observed in the ischemic and reperfused dog heart.

Comparison of pO_2 in the reperfused heart with the initial level (see Fig. 2) showed that in most experiments (series 1–3 and 5) it was significantly higher ($p<0.05$) after 10 min of reperfusion and, notwithstanding a low rate of coronary blood flow, remained close to the initial after 60 min of reperfusion (see Fig. 7). The increase of pO_2 and its maintenance at a relatively high value in the reperfused heart against the background of low CPR are indicative of reduced O_2 consumption by the cardiac tissue and, hence, the excessive (or relatively excessive) amount of O_2 in the organ. The state of absolute or relative hyperoxia in ischemic and reperfused organs during the reperfusion period will be discussed in Chapter 6.

The analysis of CPR in the isolated rat heart *in vitro* using the stop-flow model (Fig. 8) showed that 30- and 60-min total cardiac ischemia at 20°C and 30-min ischemia at 37°C lead to short-term reactive hyperemia; 120-min ischemia at 20°C results in a moderate no-reflow phenomenon; and 60- and 120-min ischemia at 37°C is followed by the primary no-reflow phenomenon, which becomes more pronounced as the duration of ischemia increases (CPR by the end of reperfusion decreased to 82 and 43% of the initial, respectively; $P<0.05$).

Fig. 8: Coronary perfusion rate (CPR, (a) and myocardial function (P_{max}, b) in the isolated rat hearts reperfused after different periods of stop-flow ischemia (at 20°C and 37°C) (mean ± SEM, %).

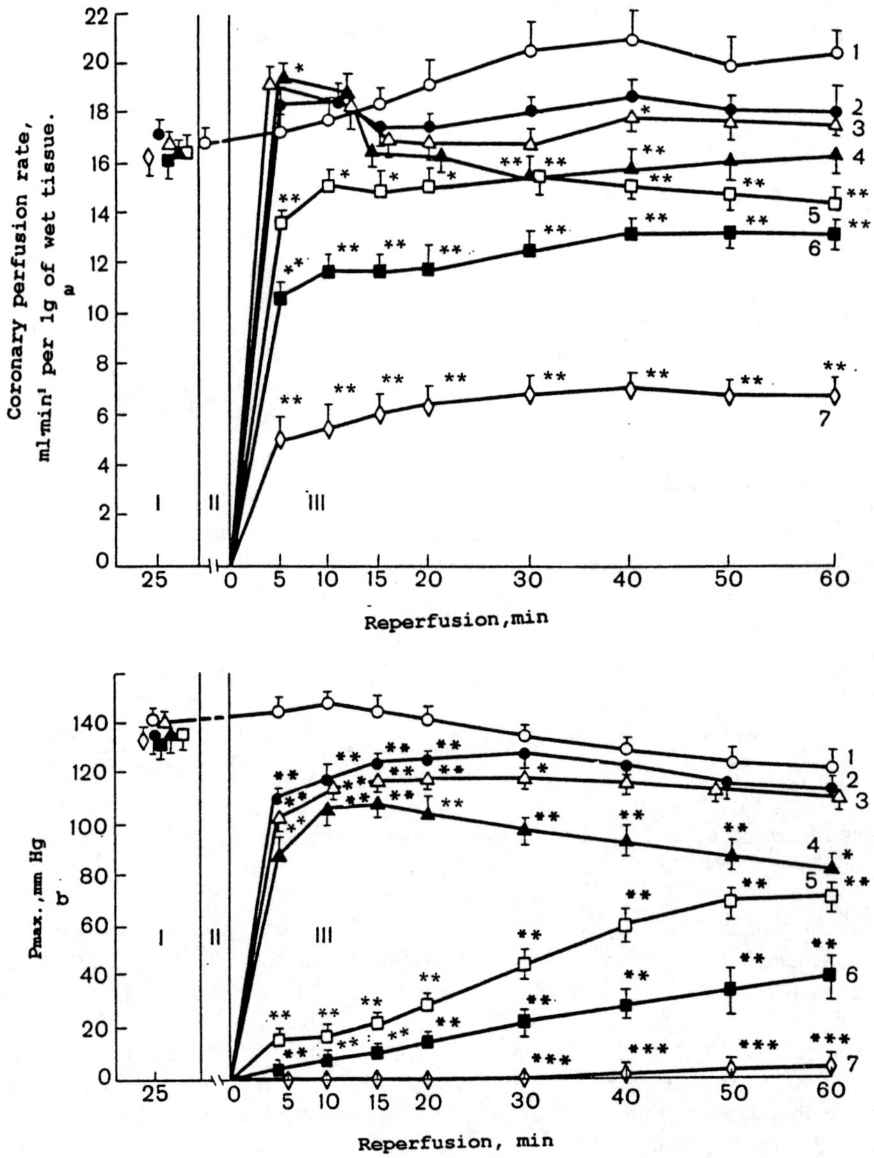

1 - without ischemia (control); ischemia at 20°C (2-4): 2 - 30 min; 3 - 60 min; 4 - 120 min; ischemia at 37°C (5-7): 5 – 30 min; 6 - 60 min; 7 - 120 min. I - initial (stabilizing) perfusion (25 min); II – stop-flow (30 – 120min, see upper); III - reperfusion (60 min). *-$p < 0.05$; ** -$p < 0.05$; versus 1 (control). The perfusion was performed using a KH solution, pH 7.4, 37°C, oxygenated with carbogen (95% O_2, 5% CO_2) by the Langendorf- Fallen method [see Neely J. R. and Rovetto M. J.]. The CPR was measured by the efflux from the pulmonary artery; P_{max} was determined in the left ventricle by the isovolume technique using a Statham LP-1400 blood pressure monitor.

The recovery of cardiac function (P_{max} in the left ventricle) was satisfactory when the heart was stored at 20°C for 30 and 60 min, slow when the heart was stored at 20°C for 120 min and at 37°C for 30 and 60 min, and failed when proceeded against the background of a pronounced no-reflow phenomenon in the heart stored at 37°C for 120 min.

Thus, hemodynamic changes during reperfusion of totally ischemic dog and rat hearts follow the same pattern and depend on the extent and duration of ischemia; they are also similar to those in ischemic kidneys. The local hemodynamic changes taking place in the dog heart after longer periods of ischemia, as compared to experiments with rats, seem to be due differences either in reperfusion conditions or in susceptibility of the dog and rat hearts to ischemia.

Artificial reperfusion of the pig heart with blood pumped after 30 and 60 min of its intermittent total ischemia *in situ* caused reactive hyperemia, while after 90 min of intermittent or, especially, continuous ischemia, reactive hyperemia was replaced by the no-reflow phenomenon [Engelman R.M., Adler S., Gouge T.H. et al.].

Circulatory disorders due to reperfusion have been studied in detail on the model of coronary vessel occlusion during myocardial infarction. When using such a model of partial ischemia in dogs and other animals with comparable collateral circulation (rats and cats), a slight reactive hyperemia occurred after 20 min of ischemia [Allison T.B. and Holsinger V.W.] or was absent. After 2–6 h of ischemia, reactive hyperemia was weak or virtually nil [Higginson L.A.J., Beanlands D.S., Nair R.C. et al.; Parker P.E., Bashour F.A., Downey H.F. et al.], although the blood flow rate was reduced by one-third after 1-h reperfusion and by factors of 2 and 3.5–5 after 2–4-h reperfusion (in the subepicardial and subendocardial layers, respectively), compared to that during the first minutes of reperfusion [Parker P.E., Bashour F.A., Downey H.F. et al.]. In other words, "latent" reactive hyperemia was observed.

The no-reflow phenomenon in this model of myocardial ischemia in dogs was not revealed after 5–40 min of ischemia but became fairly pronounced 90–120 min after the onset of ischemia [Allison T.B. and Holsinger V.W.; Kloner R.A., Ganote C.E, Whalen D.A. et al.; Parker P.E., Bashour F.A., Downey H.F. et al.]. In rats with temporary LAD occlusion, the no-reflow phenomenon manifested itself after 30–60 min [Kogan A. Kh., Miyeegombyn A., Losev N. I. and Kudrun A. N.] or 90–120 min of ischemia [Litvitsky P.F., Yakunin G.A., Lavrova L.A. et al.; Olbinskaya L.I. and Litvitsky P.F.], and experiments with injections of stained latex particles into vessels revealed three phases in its development: (1) reduction of the noninjected area, (2) significant expansion of the noninjected area (the true no-reflow phenomenon), and (3) secondary reduction of the noninjected area [Kogan A. Kh., Miyeegombyn A., Losev N. I. and Kudrun A. N.].

It is quite obvious that these phases correspond to the periods of reactive hyperemia, the no-reflow phenomenon itself, and its subsequent reversion, observed in experiments on other animals and in other organs [Camilleri J.P., Joseph D., Fabiani J.N. et al.; Eriksson E, Replagle R.L. and Geagov S.].

In the pig heart, where the collateral circulation is less pronounced [White F. C. and Bloor C. M.], the ischemic period necessary to cause the no-reflow phenomenon was shorter, 30–40 min [Klein M.H., Nebendahl K., Lindert S. et al.]. On the other hand, in experiments with LAD occlusion in baboons, reactive hyperemia of the heart was observed even after 1 h of ischemia, and the no-reflow phenomenon developed only after 3 h, accompanied by significant deterioration of the cardiac function [Heyndrickx G. R, Amano J., Patrick Th. A. et al.]. The authors of the latter study regard the absence of reactive hyperemia as a diagnostic trait correlating with severe necrotic changes in the heart and, hence, indicating that prognosis is unfavorable.

Reactive hyperemia after occlusion of a coronary artery branch in animals with well-developed collateral circulation is most pronounced in the subepicardial layer of the myocardium, whereas the no-reflow phenomenon manifests itself in the subendocardial layer [Kloner R.A., Ganote C.E, Whalen D.A. et al.; Parker P.E., Bashour F.A., Downey H.F. et al.]. Marked local hemodynamic disorders have also been revealed during reperfusion of AMI foci in humans, and, on this basis, it was recommended to perform operations of aortocoronary shunting no later than 6 h after the onset of AMI [Iwakura K., Ito H. and Takiuchi S. et al.; Phillips S.J., Kungtahworn C., Zeff R.H. et al.].

The extent and duration of ischemia affect reperfusion-dependent hemodynamic disorders in other organs as well.

Reperfusion of the brain 1 h after transorbital ligation of the middle cerebral artery in primates was accompanied by reactive hyperemia (the blood flow rate increasing 2.0–2.2-fold) without the no-reflow phenomenon [Garcia J. H., Conger K. A., Morawetz R. et al.]. After 15-min total ischemia of the rabbit brain and its reperfusion, most of the organ (95%) remained without blood supply [Ames A., Wright R.Z., Kowada M. et al.]; after 60-min bilateral occlusion of the carotids and pronounced hypotension, the blood flow through most of brain regions was only 50% of the initial [Workshop on: Oxygen free-radicals in shock]. In newborn lamb, asphyxia with subsequent resuscitation resulted in reactive hyperemia of the brain followed by the phase of hypoperfusion [Rosenberg A. A.]. The recovery of circulation in the rat brain after 60-min occlusion of both carotid arteries was accompanied by increased blood flow through the hemispheres, diencephalon, and midbrain, whereas blood flow through the cerebellum and medulla (unaffected during ischemia) decreased. According to the author, this was due to blood redistribution between the coronary and vertebral arteries [Zamuruev O. N. (1979)]. The increase of pO_2 after reperfusion was also found in the ischemic rat brain [Raguzin A.V.].

Reperfusion of the liver after its partial and, especially, total ischemia resulted in the reduced volume rate of hepatic circulation [Shumakov V.I., Galperin E.I. and Nekhludova E.A.], the increased tone of vascular wall, and the constriction of small hepatic veins accompanied by "venous outflow block. " In addition, the portal flow increased, whereas blood flow through the hepatic artery decreased. Thus, the blood became redistributed between the vessels supplying the organ with substrates (portal vein) and oxygen (hepatic artery).

The proportion of hepatic parenchyma with unrestored circulation in the early reperfusion period increased in parallel with the duration of ischemia and equaled 21, 32, and 31% after 10, 30, and 50 min of ischemia, respectively [Sergeeva V.S., Altuckova L.K. and Kremlin S.M.].

Experiments on **reperfusion of limbs** showed that dog limb muscles retain the capacity for reactive hyperemia (determined from an increase in pO_2) after the periods of ischemia as long as 1, 2, and 4 h; in the latter two cases, reactive hyperemia was followed by the no-reflow phenomenon [Eriksson E, Replagle R.L. and Geagov S.].

The no-reflow phenomenon after reactive hyperemia and drastic microcirculatory changes in ischemic striated muscles during the reperfusion period, dependent on the extent and duration of ischemia, have also been observed by many other researchers [Braunwald E. and Kloner R.A.; Fox K.A.A., Saffitz J.E. and Corr P.B.; Hurn P.D., Koehler R.C., Blizzard K.K. and Traystman R.J.; Menger M.D., Steiner D. and Messmer K.]. The corresponding data have also been obtained on other organs and tissues, such as intestine [Kaminski P.M. and Proctor K.G.] and skin [Kucukcelebi A. and Ozcan M.].

Thus, hemodynamic changes in different organs during the postischemic period are of basically the same type and occur stepwise: a short ischemic period leads to reactive hyperemia during reperfusion; a longer period, to reactive hyperemia followed by the no-reflow phenomenon; and critically long ischemia results in the no-reflow phenomenon not preceded by reactive hyperemia.

These most commonly observed phases may also be accompanied by the fourth phase of postischemic vascular reaction, namely, paralytic dilatation of blood vessels. The latter is observed when the duration of ischemia significantly exceeds the critical period. Such a response, observed in virtually every organ, indicates that changes occurring after prolonged ischemia and reperfusion are irreversible [Bilenko M. V. (1978a)].

As *the no-reflow phenomenon is the primary and most typical manifestation of severe postischemic disturbances of local hemodynamics in the organ,* it is of particular interest to analyze vascular changes responsible for its development.

There is conclusive evidence that the no-reflow phenomenon is caused by microcirculatory disorders resulting from a number of factors: the narrowing of the lumen because of endothelial cell damage, swelling, and shedding [Bilenko M.V., Bulgakov V.G. and Morgunov A.A. (1987); Ramazanova L.I. and Bardakhchyan E.A.; Kempski O. and Behmanesh S.; Kloner R.A., Ganote C.E, Whalen D.A. et al.; Nishigaya K., Yoshida Y., Sasuga M.et al.]; compression of small vessels by extracellular edema, which increases during the reperfusion period [Ames A., Wright R.Z., Kowada M. et al.; Camilleri J.P., Joseph D., Fabiani J.N. et al.; Kloner R.A., Ganote C.E, Whalen D.A. et al.; Nayler W.G. and Elz J.S.]; in the heart and skeletal muscles, increasing ischemic contracture [Camilleri J.P., Joseph D., Fabiani J.N. et al.; Gavin J.B., Thomson R.W., Humphrey S.M. et al.; Humphrey S.M., Gavin J.B. and Herdson P.B. (1980)]; neutrophil adhesion to the endothelium, increased accumulation of blood cells (particularly blood platelets and leukocytes) in microvessels, and agglutination of red blood cells [Alexandrov P.N., Shinkarenko U.S., Khugaeva U.K. and Morozov S.E.; Muller J.M., Vollmar B. and Menger M.D.; Del Zoppo G.J., Schmid-Schonbein G.W., Mori E. et al.; Eriksson E, Replagle R.L. and Geagov S.; Korthuis R.J. and Gute D.C.; Mullane K.M., Read N., Salmon J.A. and Moncada S.; Romson J., Hook B., Kunkel S. et al.]. The latter process contributes to reperfusion injury by physically obstructing microcirculation and by leading to the release of vasoactive and cytotoxic products. Blood platelets that have passed through the ischemic kidney [Bilenko M. V. (1978a)], brain [Dougherty V.H., Levy D.E. and Weksler B.B.], and heart [Litvitsky P.F., Yakunin G.A., Lavrova L.A. et al.; Olbinskaya L.I. and Litvitsky P.F.] have been found to exhibit an increased tendency for aggregation and adhesion.

At the same time, it is known that blood vessels at the sites of most severe edema are often dilated rather than constricted; in the ischemic heart, the vascular lumen may be blocked when muscles are relaxed, and vice versa [see Gavin J.B., Thomson R.W., Humphrey S.M. et al.]; microcirculatory changes can only affect the overall throughput capacity of an organ to an extent that is usually unnoticeable in ischemic organs. Apparently, all the aforementioned mechanical factors play a supplementary role, whereas the increased tone of small resistive arteries, i.e., the vasoconstrictive reaction (vasospasm) is of primary significance.

Evidence that changes responsible for the no-reflow phenomenon in the early postischemic period are related to vasoconstriction and associated with small resistive vessels (terminal arteries, arterioles) has been obtained in histological [Ramazanova L.I. and Bardakhchyan E.A.], microangiographic [Bilenko M. V. (1978a); Kovanov V. V., Bilenko M. V., Amosov I. S. et al.; Shinkarenko V.S. and Morozov S.E.], fluorescent [Gavin J.B., Thomson R.W., Humphrey S.M. et al.; Hellstrom H.R.; Humphrey S.M., Gavin J.B. and Herdson P.B. (1982)], and other studies. The functional nature of these changes is also borne out by their reversibility, which has been repeatedly demonstrated in the course of drug treatment.

In the liver, vasoconstriction of the sinusoidal sphincters (which selectively reduces portal blood inflow) may be accompanied by that of the venous sphincters (venous outflow block). Thus, during some generalized postischemic disorders (such as ischemic shock), the liver may become a blood depot (see Chapter 8).

Today, it is known that endothelial cells play a key role in vascular tone regulation, synthesizing numerous vasoactive substances, including kininase II, the enzyme converting inactive angiotensin I into angiotensin II and inactivating bradykinin [Davies M.G. and Hagen P.; Ryan J.W., Ryan U.S., Schultz D.R. et al.; Ryan J.W. and Ryan U.S.]; prostaglandins with vasodilatory (PgJ_2, PgE_2) and vasoconstrictive properties (TXA_2, $PgF_2\alpha$) [Gerlach E., Nees S. and Becker B.F.; Davies M.G. and Hagen P.]; and endothelins (ET-1, ET-2, ET-3), which are extremely active vasoconstrictor peptides [Vitola J.V., Forman M.B., Holsinger J.P.et al; Zaugg C. E., Hornstein P. S., Zhu P. et al; Kelly R.F., Hursey T.L., Schaer G.L. et al.; Lerman A., Hildebrand F.L., Marqulies K.B. et al.; Yanagisawa M., Kurihara H., Kimura S. et al.].

Endothelial cells can also release the endothelium-derived relaxing factor (EDRF), a vasodilator whose activity is mediated by nitric oxide (NO) or other nitrosyl-containing compounds (see Chapter 10).

Ischemia and reperfusion of the vascular endothelium lead to the enhancement of vasomotor activity, which is explained by PgJ_2 synthetase inhibition and a decrease in PgJ_2 production against a background of retained TXA_2 synthetase activity and TXA_2 production [Katusic Z.S. and Vanhoutte P.M.; Palace G.P., Del-Vecchio P.J., Horgan M.J. et al.; Rubanyi G.M. and Vanhoutte P.M.], and by increased production of endothelins and sensitization to them [Davies M.G. and Hagen P.; Neubauer S., Ertl G., Pulzer F. et al.]. Moreover, in experiments on ischemia and reperfusion of the cat heart [Lefer A.M. and Lefer D.J. (1991)], rat brain [Prado R., Watson B.D. and Wester P.], rat liver [Scommotau S., Uhlmann D., Loffler B.M. et al.], and rat kidney [Cristol J.P., Thiemermann C., Mitchell J.A. et al.], specialists revealed the decrease or cessation of EDRF production in the endothelium and the decrease of endothelin/NO ratio [Scommotau S., Uhlmann D., Loffler B.M. et al.; Pannen B.H., Al-Adili F., Bauer M., et al.]. The role of EDRF and ET/NO balance in restoring blood flow in these organs during the reperfusion period was demonstrated. Inactivation of EDRF may also be induced by reactive oxygen species (ROS) [Chimowitz M.I., Furlan A.J., Nayak S. and Sila C.A.; Rubanyi G.M. and Vanhoutte P.M.], the formation of which drastically increases during ischemia and reperfusion of different organs and tissues (see Chapter 5), including the vascular endothelium (see Chapter 10).

Ischemic endothelial cells, activated leukocytes, macrophages, and cardiac myocytes, are capable of producing cytokines, in particular, the tumor necrosis factor alpha (TNF-α) [Frolkis J., Gurevitch J., Yuhas Y.; Gurevitch J., Frolkis I., Yuhas Y. et al.; Murota S., Nishida M. and Morita J.], and the platelet-activating factor (PAF). The latter is a vasoconstrictor involved in reperfusion disorders of microcirculation, especially in liver capillaries, portal and hepatic veins [Wang H.G., Shibamoto T. and Koyama S.]. PAF activity is apparently mediated by the increase of reactive O_2 forms [Park T.S., Gonzales E.R. and Gidday J.M.] and promotes adhesion of blood cells (polymorphonuclear leukocytes) to the capillary endothelium [Milhoan K.A., Lane T.A. and Bloor C.M.; Nose P.S.; Petzelbauer P., Bender J.R., Wilson J. and Pober J.S.]. Lipid peroxidation (LPO) products formed in the ischemic organ and involved in the pathogenesis of the no-reflow phenomenon also have vasoactive (vasoconstrictive) properties (see Chapter 8).

An endothelium-related increase of microvascular tone may also be associated with the self-regulatory response of the tissue to the decrease in its metabolic requirements. This aspect and biological rationale of the no-reflow phenomenon were analyzed in detail in our previous work [Bilenko M. V. (1978a)].

According to [Hellstrom H.R.], transition from vasodilatory to vasoconstrictive response or, in other words, from reactive hyperemia to the no-reflow phenomenon is determined by the vasoconstrictor–vasodilator balance that exists in blood vessels and is apparently responsible for heterogeneity of circulation in a normal organ. The author of this work suggested that the ischemia-induced myocardial spasm gives rise to a vicious circle, as it leads to more pronounced ischemia, which in turn, aggravates the spasm. Depending on the degree of ischemic injury, the vasoconstrictive response may prevail over reactive hyperemia or, on the contrary, reactive hyperemia may prevent spasm in resistive vessels.

1.3 Conclusions

(1) The data discussed above show that it is important to make distinction between the states of hypoxia and ischemia as well as between partial and total ischemia; one should be aware of the fact that the severity and reversibility of ischemic and reperfusion injuries depend not only on the conditions under which ischemia occurs (*in vivo* or *in vitro*, in the laboratory or clinical environment), but also on the degree of circulatory disorders, i.e., the reduction of O_2 and oxidative substrate inflow and metabolite outflow from the organ. Among the most essential factors are the duration of ischemia and temperature at which it develops; if the extent, duration and conditions of ischemia are the same, the key factor is the type of organ, as organs differ in their sensitivity to ischemia, at least by the metabolic criterion. As estimated using an easily standardizable model of total ischemia, they can be arranged in the following descending series: brain > liver > kidney and heart > skeletal muscles > endocrine organs > skin (see Chapter 3).

(2) It is necessary to understand differences between numerous models of ischemia initiation, most of which ensure only partial (global or local) ischemia the severity of which broadly varies depending on animal species, individual pattern of collaterals, organ under study, its layer, and the time elapsed from the onset of ischemia (appearance of functional insufficiency of collaterals). This makes the models of partial ischemia poorly suitable for comparative studies on the nature and mechanisms of ischemic and reperfusion injuries in different organs and also makes it difficult to extrapolate the results of animal experiments to clinical cases. The models of total ischemia are characterized by greater uniformity of injuries and lesser dependence on animal species and development of collaterals.

On this basis, the published data on ischemic and reperfusion changes in organs are discussed in the following chapters with due regard to the model of ischemia used by the author of a given study, whereas our own experiments on comparing membrane changes in different organs and elucidating some of their mechanisms were performed on the model of total ischemia because, as shown above, it has obvious advantages over the partial ischemia model.

Variants of the model used in our studies are listed in Table 2. Most experiments were carried out on Wistar rats which, as shown in our laboratory, are sensitive to acute hypoxia [Khachatur'yan M.L., Gukasov V.M., Komarov P.G. et al.] and demonstrate small individual differences in this respect. All operations were performed under anesthesia.

(3) Another important fact is that even totally ischemic organs retain residual amounts of molecular oxygen, which is due to its facilitated diffusion from capillaries and various depots and to a sharp decrease of its expenditure for oxidative processes in mitochondria. This finding had led to a hypothesis, later proved, that LPO processes are involved in the pathogenesis of ischemic and reperfusion injuries (see Chapter 5).

(4) The analysis of circulatory disorders in organs during the early postischemic period revealed several phases in their development and showed that each stage depends on the type of organ, the extent of ischemia, and its duration. The main manifestation of postischemic hemodynamic disorders is incomplete recovery of circulation, or the no-reflow phenomenon, which may be observed either immediately after reperfusion (the primary no-reflow phenomenon) or after a period of reactive hyperemia. The no-reflow phenomenon may be reversible (spontaneously or after administration of drugs) and irreversible; it may

also be followed by the phase of paralytic dilatation. In most organs, this phenomenon manifests itself differently in each functional organ unit and may have a mosaic pattern within the same unit. It is now evident that endothelial cell–leukocyte interactions (including activation of both these cell types along with the release of ROS and vasoactive mediators, an increase in the vascular tone and microvascular permeability, and the reduction of small-vessel and microvascular flow) play the most important role in the no-reflow phenomenon.

(5) The hemodynamic disorders developing stepwise during the postischemic period lead to changes of oxygen regime in the organ, which also occur in phases. However, the absolute O_2 excess in the organ (or one of its layers) during reactive hyperemia is not necessarily followed by its deficiency even at the no-reflow phase. Hence, either the same absolute excess of O_2 or, more often, its level close to initial (relative excess) is maintained in the organ. These findings provided the basis for our concept of "relative local hyperoxia" [Bilenko M. V. (1978a)]. We have explained the role of this state in damage caused by reoxygenation and reperfusion, including the no-reflow phenomenon itself, and proposed a method for preventing reperfusion injuries by creating temporary moderate hypoxemia during the early postischemic period (for details, see Chapter 6).

Table 2: Models of total ischemia and reperfusion of organs used in our experiments

Organ	Animal	Conditions	Ischemia model	Mode of reperfusion
Kidneys	Wistar rats August rats	*in vivo*	Occlusion of entire renal pedicle + separation of kidney from perirenal fat	Discontinuance of occlusion
	Mongrel dogs	*in vitro*	Storage of isolated kidney at 20°C	Extracorporeal perfusion with connection to femoral vessels of recipient dog
Liver	Wistar rats August rats	*in vivo*	Occlusion of vascular pedicle of central and left lobes	Discontinuance of occlusion*
Heart	Wistar rats	*in vitro*	Storage of isolated heart at 20 and 37°C	Heterotopic transplantation onto abdominal vessels of recipient rats
		in vitro	Stop-flow model of total global ischemia at 20 and 37°C	Reperfusion by Langendorf-Fallen or Neely models
	Mongrel dogs	*in vitro*	Storage of isolated heart at 20 and 37°C	Extracorporeal perfusion with connection to cervical vessels of recipient dog
Brain	Wistar rats	*in vivo*	Occlusion of two carotid and two sub-clavian arteries under operating microscope	Discontinuance of occlusion
Limb muscles	Wistar and August rats	*in vivo*	Tourniqueting the upper third of femur	Removal of tourniquet

* To determine the survival rate, the nonischemic (right and caudate) lobes were resected.

Chapter 2

STRUCTURAL AND FUNCTIONAL MEMBRANE DISORDERS DURING ISCHEMIA AND AFTER REPERFUSION OF ORGANS

There is no doubt at present that cell and intracellular membranes play an extremely important role in the life activity of the cell and affect its metabolic processes, division, motility, excitability, and conduction of nerve impulses; they also perform such functions as reception, intercellular contacts, homeostasis, and others.

It is also obvious that biological membranes are structures whose integrity and functional capabilities are practically indivisible; they change, therefore, in parallel and can be studied only together.

Structural concepts about most biological membranes are based on the Singer-Nicholson model of the liquid-crystal (mosaic) membrane (Fig. 9), although membrane structure and functions vary depending on the organ and, to a lesser extent, on the animal species concerned. The widest variations are observed in plasma membranes (multiple-layer, microvillous, electrically excitable, glycocalyx-containing, and others) and membranes of the smooth endoplasmic reticulum (ER), which largely determine the specialized cell function (detoxication of xenobiotics in the liver, intestinal secretion of enzymes, transport of ions, electromechanical conjugation in muscles, etc.). More homogeneous are membranes of mitochondria, lysosomes, peroxisomes, and nuclei; they have similar structures and perform similar functions irrespective of organ.

The major components of membranes are lipids and proteins, and minor components are glycoproteins and glycolipids. The latter are found predominantly in plasma membranes where they perform receptor and immune functions as well as regulate intercellular interactions.

The lipid - protein ratio varies in different membranes; they are arranged asymmetrically and are in constant motion (see Fig. 9).

According to their mobility, lipids can be divided into three groups: (1) highly mobile liquid lipids of the bilayer; (2) moderately mobile (immobilized) lipids of the annular monolayer, associated with proteins in the lipoprotein complex, which includes 22-50 molecules per protein molecule (20-30% of the total amount of lipids [Boldyrev A.A. and Tverdislov V.A.; Boldyrev A.A. (1986)]); (3) low-mobility (immobil) ordered lipids, forming clusters. In membranes of the sarcoplasmic reticulum (SR) of skeletal muscles, these groups of lipids account for 65, 36, and 2%, respectively [Azizova O.A., Maksina A.G., Klaan N.K. et al.; Kozlov Yu. P., Kagan V. E. and Arkhipenko Yu. V.].

Fig. 9: Model of mosaic membrane and mobility of membrane proteins and lipids, according to [Boldyrev A.A. (1985); Boldyrev A.A. (1986)] with modifications

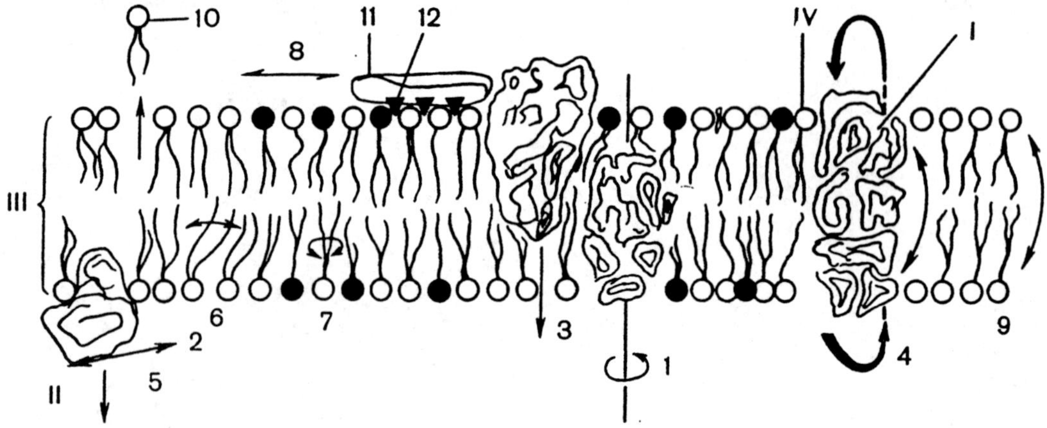

I - integral proteins; II - peripheral proteins; III - lipid bilayer; IV - annular lipids; 1 - axial rotation of proteins on membrane surface; 2 - lateral diffusion of proteins (primarily peripheral proteins); 3 – in-out motion of proteins (flip-flop, integral proteins); 4 - rotation of proteins in a perpendicular to membrane surface (transport, integral proteins); 5 - escape of protein from membrane; 6 - segmental mobility of lipids; 7 - rotation of lipids about their own axis; 8 - lateral diffusion of lipids; 9 - flip-flop motions of lipids; 10 - escape of lipids from membrane; 11 - glycocalyx; 12 - calcium depot.

The mobility of proteins and the strength binding them to the lipids in a membrane bilayer depend on how deeply they are embedded in the membrane.

As has been established in recent years (EPR, NMR, luminescent spectroscopy, scanning calorimetry, de- and relipidization of membranes, and other techniques), a change in composition, mobility, charge, conformation, etc., of one of its components brings about structural and functional changes of the membrane as a whole.

There is direct evidence that the function of membrane proteins (especially lipid-dependent ones) is markedly affected by the following parameters characterizing the state of membrane lipids [see reviews Boldyrev A.A. and Mel'gunov V.J.; Boldyrev A.A. (1986); Burlakova E.B., Dzhailbova M.I. and Gvakhariya V.O.; Burlakova E.B., Zaslavsky Yu.A. and Shishkina L.N.; Burlakova E.B. and Khrapova N.G.; Coleman R.; Quinn P.J.; Sandermann H.; Spector A.A. and Yorek M.A.]: (1) quantitative and qualitative composition of lipids, which has to do with their specific allosteric (activating or inhibiting) action on proteins; (2) polarity, microviscosity (fluidity), and heterogeneity of the lipid bilayer, which depend on the phospholipid (PL) composition, relative amount of cholesterol, and degree of nonsaturation of fatty acids (FA); (3) the presence in the lipid bilayer of lipid metabolites of lipase, phospholipase, lipoxygenase, or peroxide nature, influencing the hydrophobic volume, microviscosity, clustering, and other indicators of the physicochemical state of lipids and also exerting direct damaging effect on the three-dimensional structure (conformation) and functional properties of membrane proteins (see Chapters 4 and 5).

The lipid-dependent proteins of membranes first of all include integral proteins performing enzymic, metabolic, transport, regulatory, receptor, or other functions. These are Na, K-ATPase, Ca-ATPase, adenylate cyclase, cytochrome oxidase, and many other enzymes of the mitochondrial respiratory chain, glucose-6-phosphatase, pyruvate oxidase, NADH-cytochrome c-reductase, cytochrome P-450, 5-nucleotidase, hexokinase, acyl-CoA: cholesterol acyltransferase, and so on [Burlakova E.B., Dzhailbova M.I. and Gvakhariya V.O.; Sandermann H.; Zakim D.].

Changes in conformation properties and, consequently, accessibility of active sites for the substrate, mobility, sequential arrangement, and possibility of formation of specialized polyenzymic complexes (ensembles), as well as other protein properties, are strongly affected by changes in the properties of annular lipids [Kagan V. E., Azizova O. A., Arkhipenko Yu. V. et al.; Ritov V.B., Murzakhmetova H.K. and Kozlov Yu.P.], particularly lipids with affinity for a particular protein.

It has been shown that phosphatidylserine (PS) is such a lipid of preference for Na, K-ATPase; cardiolipin and phosphatidylethanolamine (PEA) - for enzymes of the mitochondrial respiratory chain; phosphatidylinosite (PI), for adenylate cyclase of the plasma membrane, and lysophosphatidylcholine (lysoPC), for Ca-ATPase [Boldyrev A.A. and Tverdislov V.A.; Edlund A., Sahlin K. and Wennmalm A.; Lenaz G.].

Protein composition and conformation properties in turn affect membrane lipids; more particularly, as lipids are screened by proteins, they become less susceptible to phospholipases, LPO, and other pathological factors.

Ischemia and reperfusion are accompanied by early and drastic changes in the composition, physicochemical properties, and structure of membranes, and the disruption of most of their functions. We shall now turn to some of these changes, most important for ischemic and reperfusion injury of organs.

2.1 Disturbances of the Barrier Function of Membranes

The barrier function is one of the most important membrane functions. Inherent in all membranes without exception, in some membranes (mitochondrial and nuclear) it is somewhat less important than the structural (catalytic, protein-fixing) function, in others (plasma, ER and SR membranes)- it is more or less equal to the latter; yet in other membranes (lysosomes, peroxisomes), the barrier function is predominant.

Owing to the barrier function the cell content keeps apart from the extracellular fluid; it also provides for separation and relative independence of the organelle content from the intracellular fluid and one organelle from another - that is, it provides for the cell's homeostasis and compartmentalization.

Increasing permeability of membranes to electrolytes and water is the earliest disorder due to ischemia and reperfusion of organs. As is known, the content of such electrolytes as Na^+, K^+, Cl^-, H^+, Ca^{2+}, Mg^{2+}, and other minor ones in the extracellular fluid, cytoplasm, and cell organelles is substantially different under physiological conditions (Fig. 10), which gives rise to an electrochemical gradient (diffusion potential) over the membrane surface and leads to a constant passive diffusion of ions along the concentration gradient into and out of the cell plus ion movement between the cytoplasm and organelles.

Ionic asymmetry is a dynamic phenomenon for it depends on the cell's functional state. Maintenance of ionic asymmetry in all the cells of the body and its restoration upon completion of a phase of functional activity in excitable cells are possible by virtue of the following membrane properties: (a) the initially low permeability of the hydrophobic layer of lipids to water-soluble substances (ions, water, etc.); (b) the existence of selective transport systems by diffusion along the concentration gradient through membrane pores or hydrophilic protein channels (passive and assisted diffusion); (c) existence of active (ATP-dependent) transport systems versus the concentration gradient, by means of transport enzymes (Na, K-ATPase, H-ATPase, and Ca-ATPase of plasma and SR membranes), Na^+/Ca^{2+} and H^+/Ca^{2+} ion-exchange mechanisms (see Fig. 10); (d) the presence of protein, lipoprotein and other complexes displaying affinity to electrolytes and serving as intracellular depots. The chief pathways and mechanisms of electrolyte transport in the cell have been studied in detail in [Boldyrev A.A. (1985); Verkhatskii A.N. and Pidoplichkov I.; Orlov O.N. (1981); Ritov V.B., Murzakhmetova H.K. and Kozlov Yu.P.; Lee C.O.].

Fig. 10: Schematic representation of cell ionic asymmetry and electrolyte transport systems, according to [Boldyrev A.A. (1985); Orlov O.N. (1981); Rasmussen H. and Goodman D.B.P.] with modifications

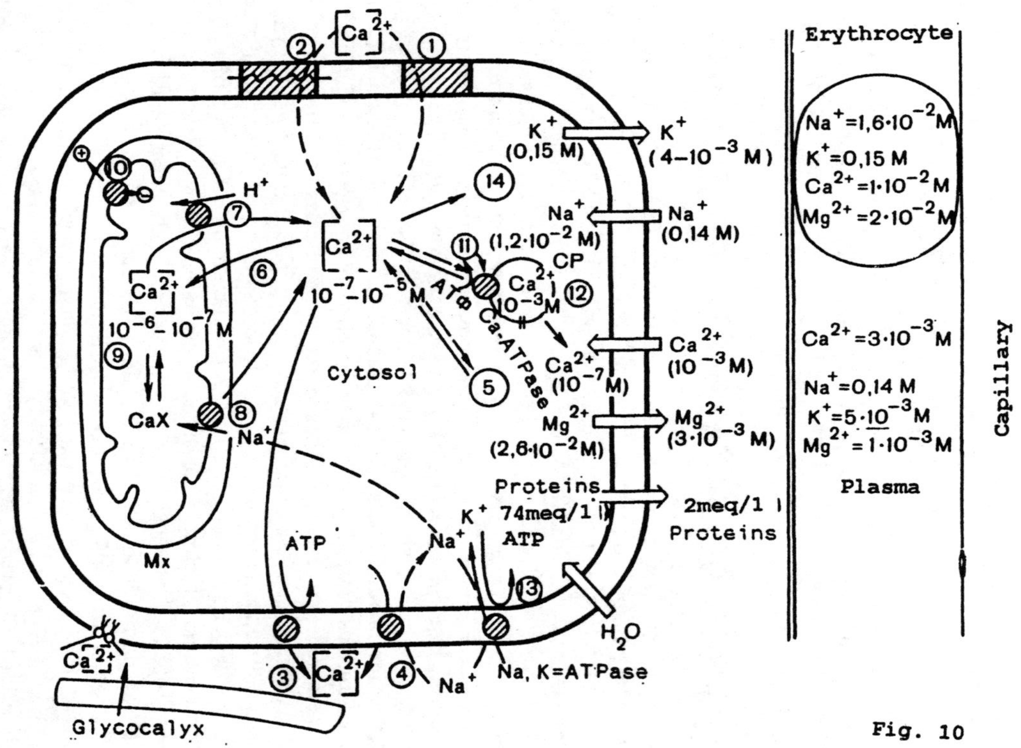

Fig. 10

1 - potential-independent Ca^{2+} channels; 2 - potential-dependent Ca^{2+} channels; 3 - Ca-ATPase; 4 - Na^+/Ca^{2+} exchange; 5 - Ca^{2+}- binding buffers (proteins, nucleic acids, etc.): 6 - Ca^{2+} accumulation in mitochondria; 7 - H^+/Ca^{2+} exchange on inner mitochondrial membrane; 8 - Na^+/Ca^{2+} exchange in mitochondria; 9 - reversible transition of Ca^{2+} from a difficulty soluble form into a membrane-bound one; 10 - systems for potential generation on inner mitochondrial membrane; 11 - Ca-ATPase of SR membranes; 12 - two Ca^{2+}-binding SR proteins; 13 - plasma membrane Na, K-ATPase; 14 - secretory granules

2.1.1 Disturbances of the Membrane Barrier Function with Respect to Univalent Cations and Anions

Disturbances of the membrane barrier function with respect to univalent cations and anions (K^+, Na^+, Cl^-, H^+) during ischemia and reperfusion are first of all due to damage of the plasma membrane which regulates their content with the help of Na, K-ATPase and Na^+/Ca^{2+}-exchange system.

Ischemia leads to early redistribution of univalent ions between the cell and extracellular fluid. Selective electrodes were used to demonstrate that during cardiac ischemia in pigs and dogs the extracellular K^+ content started to increase already within the first seconds and after 4-5 min exceeded the initial level four-fold. The rapid increase phase gave way to a plateau (about 10-15 min) followed by a slow increase phase. After 60-70 min of ischemia, the extracellular K^+ content exeeded the initial level 6 times. The rapid increase and plateau phases were reversible, whereas that of slow increase was irreversible [Hill J.L. and Gettes L.S.; Hirche H., Bissig R., Friedrich R. et al.]. The escape of only one percent of intracellular ion across the cell membrane is quite sufficient for the extracellular K^+ content to double. However, a higher increase of this content causes irreversible changes in the cardiac function [Kapelko V. I., Gorina M. S. and Novikova N. A.].

Rapid (within the first 2-3 min) depolarization of membranes, manifesting itself in a drop of the membrane or transepithelial potential, was also observed during hepatic and renal ischemia [Sarbash V.I., Rymarchuk V.I. and Malenkov A.G.; Hasselgren P., Jehnische E., Fornander J. and Hellman A.; Molitoris B.A., Wilson P.D., Schrier R.W. and Simon F.R.].

In the kidneys, the transmembrane Na^+ and K^+ gradient equalized, and Na^+ diffusion into the cell stopped after 20-30 min of their total ischemia [Shumakov V.I., Onishchenko N.A. and Shtengol'd E.S.].

In the early anoxic or ischemic period, the diffusion of univalent ions was due to enhanced passive K^+ escape through the membrane [Reeves W.B. and Shah S.V.], and was not connected with the disturbed function of Na, K-ATPase, but later it was accompanied by disturbance of this function and severe sarcolemma damage [Shine K.I.]. Decreasing activity of Na, K-ATPase during ischemia was also reported by other workers [Chien K.R., Reeves J.P., Buja L.M. et al.; Ishiharajima S., Aida T., Nakagawa R. et al.; Takahashi K. and Kako K.J.].

Pronounced barrier function disturbances with respect to K^+ and Na^+ arise during ischemia in intracellular membranes as well. Decreasing K^+ concentration, increasing Na^+ content, and lower K^+/Na^+ ratio have been observed in mitochondria of the liver, the heart, and the kidneys (after 30 min of ischemia) as well as in mitochondria of the brain (after 5 min of ischemia) [Ozawa K., Seta K., Araki H. and Handa H.].

After reperfusion of an ischemic organ, the redistribution of univalent ions (accumulation of Na^+, Cl^- ions and loss of K^+) was more pronounced than during ischemia [Olbinskaya L.I. and Litvitsky P.F.].

In an ischemic, and especially reperfused, organ with increasing metabolic acidosis (see Chapter 3), membrane permeability to H^+ ions also increases, which causes them to leave the cell and acidulate the extracellular fluid [Hirche H., Bissig R., Friedrich R. et al.] and the blood, and when a large organ is reperfused, generalized metabolic acidosis sets in.

2.1.2 Disturbance of the Membrane Barrier Function with Respect to Water

Disturbance of the membrane barrier function with respect to water and the cell losing its capacity to regulate the water balance are among an organ earliest and most easily detectable ischemic and reperfusion injuries. It has been established that quantities of total, free and bound water increase during histological and electron microscope examinations of ischemic and, especially, reperfused organs, appearing as a thickening and loosening of membranes, with water inclusions (bubbles) in them, in the cytoplasm, and in organelles, and as an enlargement and deformation of organelles and the cell.

Edema (swelling) of mitochondria and clarification of their matrix are the earliest manifestations of ischemic changes in the myocardium, liver, kidneys, and brain, preceding those in other organelles, and clearly visible already within the first 5-15 min of ischemia [Sorokovoy V.I. and Vladimirov Yu.A.]. In an ischemic heart, the swelling and clarification are followed by vacuolation of mitochondria (after 10-20 min), ruptures of the outer membrane (after 40-60 min), pronounced edema, straightening of the inner membrane with loss of cristae, its multiple ruptures (after 2-3 h), and complete disintegration of mitochondria (after 6-12 h) [Kloner R.A., Fishtein M.C., Hare C.M. et al.]. Thickening, loosening, and vacuolation during ischemia are also observed in SR membranes of the myocardium, ER membranes of the liver, and plasma membranes [Farber J.L., Chien K.R and Mittnacht S.; Jennings R.B., Reimer K.A. and Steenbergen C.Jr. (1985); Kloner R.A., Fishtein M.C., Hare C.M. et al.]; within the first 15-20 min of total ischemia, nuclei of the tubular epithelium in the kidneys [Shumakov V.I., Onishchenko N.A. and Shtengol'd E.S.] and myocardium [Sharov V.G., Severin V.V. and Mogilevsky G.M.; Tranum-Jensen J., Janse M.I., Fidett I.W. et al.] undergo swelling and vacuolation.

The degree of edematization during the postischemic period increases with severity and duration of ischemia (see Fig. 33), the increase being particularly marked in the case of an organ's irreversible ischemic injury (up to 4-9% after 3-6 h of ischemia of dog limbs and up to 36% after 9-12 h) [Savel'ev V.S., Chekareva G.A., Mishnev O.D. et al.].

Factors upsetting the cellular water balance in an ischemic organ are rather numerous. One of them is a rise in *intracellular osmotic pressure* due to ion redistribution between the extracellular fluid and the cell, and also accumulation in the cell of medium- and low-molecular glycogen metabolites, hydrolysates of adenine nucleotides, inorganic phosphorus (P_i), etc. (see Chapter 3). It has been found that after 35-50 min of local myocardial ischemia in pigs, osmotic pressure in the cell rises from 6 to 35-40 osm/kg [Tranum-Jensen J., Janse M.I., Fidett I.W. et al.]. The role of the osmotic factor in damaging the water homeostasis is corroborated by the protective effect observed when solutions close in osmolarity to the intracellular fluid are used prior to reperfusion [Tranum-Jensen J., Janse M.I., Fidett I.W. et al.] or long-term storage of organs [see Shumakov V.I., Onishchenko N.A. and Kirpatovsky V.I.].

The influx of water and edema of the cell are especially promoted by increasing intracellular Na^+ concentration, in view of the fact that this ion is highly hydrophilic.

Another important factor responsible for enhanced water permeability of membranes during ischemia and reperfusion is the changing physicochemical properties of membrane lipids as well as formation of peroxide channels in the lipid clusters (see Chapter 5). The resulting edema aggravates the damage to membranes, causing their rupture, with outflow of organelle contents from the cell (osmotic shock phenomena).

2.1.3 Disturbances of the Membrane Barrier Function with Respect to Bivalent Cations

Disturbances of the membrane barrier function with respect to bivalent cations, particularly $Ca^{2+}$and Mg^{2+}, are currently believed to be the severest manifestations of ischemic and reperfusion damage to membranes and, in turn, affect substantially the structure and functions of the cell.

Systems ensuring cellular homeostasis in terms of bivalent ions (Ca^{2+}) are located in the plasma membrane, SR membranes, and inner mitochondrial membrane (see Fig. 10) with the result that their damage upsets Ca^{2+} homeostasis in all cell particles of ischemic organs. Since rhythmic variations in Ca^{2+} distribution between extracellular fluid and the cell, as well as between cytosol and intracellular organelles, are essential for the functioning of muscle cells, derengement of the membrane barrier function with respect to Ca^{2+} during acute ischemia and reperfusion have been studied primarily on cardiac and skeletal muscle ischemia models.

The studies pointed to two key factors injuring Ca^{2+} homeostasis in the cell: (1) Ca^{2+} accumulation in the cytoplasm and (2) Ca^{2+} accumulation in the mitochondrial matrix [Nayler W.G. and Elz J.S.; Poole-Wilson P.A., Harding D.P., Bourdillon P.D.V. et al.].

Enlargement of the cytoplasmic Ca^{2+} pool is due primarily to ion escape from such intracellular depots as mitochondria and SR vesicles. The outflow of Ca^{2+} from mitochondria into the cytoplasm during ischemia of different organs was revealed by electron microscopy already within the first 5-20 min [Sorokovoy V.I. and Vladimirov Yu.A.]. In the early ischemic period, the mitochondria were found to gradually lose their capacity for accumulating Ca^{2+} [Dhalla N.S., Das P.K. and Sharma G.P.; Lee S.L., Balasabramanian V. and Dhalla W.S.]. These phenomenon seem to be determined not so much by ATP deficiency, since the ATP content is still high enough at this time, as by the transmembrane drop of the electric potential normally arising by virtue of electron transport, associated with electron transfer along the respiratory chain, and oxidation [Lehninger A.L., Reynafarje B., Vercesi A. et al.].

The failure of SR membranes to uptake Ca^{2+} from the cytoplasm and the resulting Ca^{2+} accumulation in the cytosol also belong to early ischemic manifestations and depend on ischemia extent [Hess M.L., Krause S.M. and Greenfild L.J.; Lee S.L., Balasabramanian V. and Dhalla W.S.; Toba R., Katagiri T. and Takeyama U.]. In spite of the fact that the total volume of SR vesicles and cisternae is much smaller that the overall capacity of mitochondria, it is quite sufficient not only in skeletal muscles but also in the heart (pigeons and guinea pigs), to fully provide for contraction and relaxation processes (efflux and deposition of Ca^{2+}) [Levitsky D. O., Benevolensky D. S. and Levchenko T. C.], and the speed of Ca^{2+} transfer into SR vesicles and its following release into the cytoplasm is very high [Kozlov Yu. P., Kagan V. E. and Arkhipenko Yu. V.]. Together, mitochondria and SR vesicles accumulate about 98% of total Ca^{2+} present in the cell [Orlov O.N. (1984)] so that the failure of organelles to accumulate Ca^{2+} will inevitably lead to a marked increase in cytoplasmic Ca^{2+} content.

During longer and more sever ischemia deterioration of the plasma membrane barrier function with respect to Ca^{2+} is also observed, accompanied by Ca^{2+} influx into the cytoplasm from the glycocalyx or extracellular fluid along the concentration gradient [Nayler W.G. and Poole-Wilson P.A.; Nayler W.G. and Elz J.S.; Poole-Wilson P.A., Harding D.P., Bourdillon P.D.V. et al.]. Other workers, however, consider the influx rate of extracellular Ca^{2+} into the cell during total ischemia to be negligible or even refute it [Sharov V.G., Severin V.V. and Mogilevsky G.M.].

Reperfusion of a long ischemic organ or reoxygenation of a hypoxic one enhances the permeability of the plasma membrane to Ca^{2+} almost instantly, followed by extensive Ca^{2+} transport from the extracellular fluid into the cell [Meerson F.Z.; Olbinskaya L.I. and Litvitsky P.F.; Farber J.L., Chien K.R and Mittnacht S.; Nakanishi T., Nishioka K. and Jarmakani J.M.; Nayler W.G. and Elz J.S.; Poole-Wilson P.A., Harding D.P., Bourdillon P.D.V. et al.].

Reperfusion accompanied by pronounced of the sarcolemma damage and increasing concentration of intracellular Ca^{2+} gives rise to another important manifestation of disturbed Ca^{2+} homeostasis, namely, a higher Ca^{2+} content in mitochondria, due to excessive Ca^{2+} accumulation both in the cytosol and in the mitochondria, which is more preferable than ATP production when the mitochondria are energized [Lehninger A.; Lehninger A.L., Reynafarje B., Vercesi A. et al.]. Excessive Ca^{2+} accumulation in mitochondria which, according to most authorities [Dhalla N.S., Das P.K. and Sharma G.P.; Kreuzer F.; Miyamae M., Camacho S.A., Weiner M.W. and Figueredo V.M.; Peng C.F., Murphy M.L. and Straub K.O.; Wilson D.R., Arnold P.E., Burke Th.J. and Schrier R.W.], is the main cause of the uncoupling of the processes of respiration and phosphorylation in them as of the drop of high-energy phosphates' synthesis i.e. is the reason why mitochondria lose their main, energy-synthesing function (see Chapter 3).

The phenomenon of excessive Ca^{2+} transport from the blood (or perfusate) into the cell across the plasma membrane and then into mitochondria during organ reperfusion, has come to be known as the "calcium paradox" [Hearse D. J., Humphrey S. M. and Bullock G. R]. It is closely related to the oxygen paradox (see Chapter 5) and is among the most important signs of reperfusion injury. The close similarity of the two phenomena mentioned above stems from the following observations: both paradoxes arise only after prolonged ischemia or anoxia, they are not so pronounced under hypothermic conditions, and both can be either inhibited or enhanced when the O_2 or Ca^{2+} level in the reperfusion solution or blood drops or rises [Levin R. M., Leggett R., Whitbeck C. and Horan P.; Hearse D. J., Humphrey S. M. and Bullock G. R].

The effect of excessive Ca^{2+} accumulation in an ischemic organ cytoplasm and mitochondria, particularly pronounced during the reperfusion period, on the severity of reperfusion injuries has been established from the well known anti-ischemic action of Ca^{2+} antagonists [Farber J.L., Chien K.R and Mittnacht S.; Higgins A.J. and Blackburn K.J.; Nayler W.G. and Poole-Wilson P.A.]; Ca^{2+} chelating agents [Ogilby J.D. and Apstein C.S.]; as well as from findings that an ischemic heart function improves, when reperfused with solutions with a low Ca^{2+} content [Chappell S.P., Lewis M.J. and Henderson A.H.; Ferrari R., Albertini A., Curello S. et al.; Shine K.I.], and deteriorates, when the perfusate Ca^{2+} content is high [Yano J., Riggs Th.R., Milam D.F. and Alexander J.C.] (see Chapter 11).

Among the major and obvious damaging factors associated with increased Ca^{2+} in the cytoplasm during ischemia and in the cytoplasm plus mitochondrial matrix during reperfusion are: (1) disruption of the mitochondria energy-synthesizing function; (2) activation of phospholipases (and, consequently, of the AA metabolic cascade); (3) activation of proteases; and (4) aggregation of thrombocytes [Das D.K., Engelman R.M., Rauson J.A. et al. (1987); Nayler W.G. and Elz J.S.]. Evidence of Ca^{2+} involvement in these and some other mechanisms will be discussed more precisely in Chapters 4, 5 and 10.

The Ca^{2+} pathways from the extracellular fluid (or glycocalyx) into the cell, i.e. the type and location of membrane damage leading to the loss of the barrier function to Ca^{2+}, are key factors in understanding the essence of ischemic and reperfusion membrane injuries. Ealier hypotheses about extracellular Ca^{2+} entering the cytoplasm through natural channels [Jennings R.B., Reimer K.A. and Steenbergen C.Jr. (1985)], ruptures, or microdefects in membranes [Sharov V.G., Severin V.V. and Mogilevsky G.M.; Ganote C.F.] are now doubtful [Crake T. and Poole-Wilson P.A.; Poole-Wilson P.A., Harding D.P., Bourdillon P.D.V. et al.]. Direct methods have been proved that the cytoplasm Ca^{2+} content of rat myocardial cells increases already after 15 minutes of total myocardial ischemia, when structural changes in membranes are still absent [Steenbergen C., Murphy E., Levy L. and London R.E.]. Increased membrane permeability or formation of peroxide channels (clusters) are more probable pathways of Ca^{2+} inflow (see Chapter 5). Ischemia and reperfusion of heart are also followed by the increase of intracellular Mg^{2+} [Krause S.M. and Rozanski D.].

2.1.4 Disturbances of the Membrane Barrier Function with Respect to High-Molecular Compounds

Disturbances of the membrane barrier function with respect to high-molecular compounds are later manifestations of ischemic and reperfusion injuries of organs.

Under physiological conditions, cell and intracellular membranes are virtually impermeable to high-molecular compounds, which ensures the normal functioning and retention of the individual protein, enzyme, receptor, and other membrane-bound enzymes of functionally different cells and intracellular organelles.

The transport of high-molecular compounds across a membrane under physiological conditions is ensured by various mechanisms, such as pinocytosis, exocytosis, RNA transfer from the nucleus through selective "pores" closed for other substances and ions, ATP energy transfer across the mitochondrial membrane, assisted by creatine phosphate, and others.

Ischemia and especially reperfusion of ischemic organs are accompanied by increasing membrane permeability for both exogenous (diagnostic) and endogenous high-molecular compounds, including insulin, colloidal lanthanum, sulfacyl, Krebs cycle intermediates, myoglobin, proteins, and so on [Gatsura V.V. and Pichugin V.V.; Sharov V.G., Jennings R.B. and Hawkins H.K.; Hearse D.J., Humphrey S.M. and Chain E.B.].

The escape of membrane-bound protein (enzyme) molecules from an organelle into the cytosol or from an organelle and cytosol into the blood (perfusate) is indicative of serious barrier dysfunction of a respective intracellular membrane or the intracellular and cell membranes together; it is also indicative of the membrane-bound protein escape from the membrane itself, which is possible only in the case of disturbed structural function of membrane lipids, disodered protein-lipid interactions, or membrane fragmentation (break-up). Since the binding of peripheral proteins to membranes is weaker, their escape is possible even if the membrane structure is more or less intact. The escape of structural (especially integral) proteins is possible only when the membrane is more badly damaged or fragmented, which is why this event is often considered an indicator of irreversible damage. This is achieved *in vitro* by means of chaotropic agents or mechanical damage; in pathology, this may be due to changes in the composition and physicochemical properties of the annular lipids, the detergent-like effect of lipolysis, phospholipolysis and LPO products on the membrane (see Chapters 4 and 5), and other factors.

Thus, membrane enzyme escape from the cell always points to disruption of both structural and barrier functions of membranes and serves as a diagnostic and prognostic criterion of ischemic damage severity.

The enzyme escape from membranes and their diffusion into the cytosol is easily detected by histoenzymochemical and immunofluorescent techniques. By using them in total ischemia and especially after reperfusion of the kidneys, we observed, already after 25-30 minutes of ischemia and later, diffusion of the enzymes in the cell and their migration from epithelial cells into the lumen of renal tubules [Bilenko M. V. (1978a)].

During hypoxic perfusion, partial ischemia, or reperfusion of various ischemic organs (heart, liver, kidneys), many enzymes were observed to escape from the cytosol or subcellular organelles into the perfusate or blood: lactate dehydrogenase (LDH) [Gulak P. V., Dudchenko A. M., Zaysev V. V. et al.; Bradford B.U., Marotto M., Lemasters J.J. et al.; Higgins A.J. and Blackburn K.J.], aspartate aminotransferase, transaminase, and glutamyl dehydrogenase [Higgins A.J. and Blackburn K.J.; Lohse W., Otto G., Pahlig H. et al.], creatine phosphokinase (CPK) and its isoenzyme [Golikov A.P., Avilova O.A.,

Polumiskov V.Yu et al.; Olbinskaya L.I. and Litvitsky P.F.; Frolkis R.A.; Hearse D.J.; Hearse D. J., Humphrey S. M. and Bullock G. R]; the escape into blood of subcellular (mitochondrial) enzymes occurs after longer ischemic periods and testifies to cell necrosis [Kamiike W., Fujikawa M., Koseki M. et al.].

Estimation of the severity of ischemic damage, based on MB CPK escape into the blood, is widely used in AMI patients as well.

The exact nature of membrane damage responsible for disruption of the barrier function with respect to high-molecular compounds is not yet fully understood. There is evidence that plasma membrane permeability in cardiac tissue to enzymes increases after 60 min of ischemia, with the membranes sustaining morphologically identifiable ruptures [Jennings R.B., Reimer K.A. and Steenbergen C.Jr. (1985)]. According to other investigators, however, the increasing permeability of plasma and mitochondrial membranes to colloidal lanthanum in an ischemic rat heart was associated with simultaneous barrier dysfunction in regard of Ca^{2+} [Sharov V.G., Jennings R.B. and Hawkins H.K.], while LDH escape from an ischemic rat heart into the perfusate and that of myoglobin and MB CPK from the AMI focus into the blood in dogs [Golikov A.P., Avilova O.A., Polumiskov V.Yu et al.; Higgins A.J. and Blackburn K.J.] was successfully decreased by administering agents blocking Ca^{2+}-channels and calmodulin, also with the antioxidant BHT. At the same time, CPK escape from a rabbit cardiac tissue strip was not associated with the Ca^{2+} and Ba^{2+} permeability of the sarcolemma [Nakanishi T., Nishioka K. and Jarmakani J.M.].

It is possible that the escape of high-molecular compounds, just as of electrolytes, is due to formation of peroxide channels (clusters) in the membrane, but here more study is needed.

The increasing permeability during ischemia of lysosomal membranes is of particular importance because their labilization and escape of lysosomal enzymes into the cytosol not only characterize the severity of ischemic injury, but also exert a marked damaging effect on the cell.

Lysosomal membranes perform mainly barrier functions and accommodate more than 60 enzymes, of which the most important are lipolytic (lipases, triacylglyceride lipases, phospholipases A_1 and A_2, sphingomyelinases, etc.) and proteolytic ones (cathepsins, acid phosphatase, etc.) i.e., enzymes capable of damaging both the lipid and protein components of subcellular and cell membranes [see Demurov E. A. and Ignatova V. A.; Lizenko E. I.; Grinde B.]. As was believed until recently, lysosomal enzyme escape is one of the most important factors of ischemic organ injury. At present, electron microscopy and cytochemical studies have revealed that the disturbance of lysosomal membrane permeability (at first to water, with vacuolation of the organelle and subsequent fragmentation of the membrane, then to enzymes themselves, with their escape into the cytosol) in ischemia take place later than in other membranes [Decker R.S. and Wildenthal R.; Kloner R.A., Ganote C.E, Whalen D.A. et al.] (see Chapter 10).

Reperfusion of ischemic organs aggravates the damage to lysosomal membranes, but damage in lysosomes occurs after a much longer ischemic period than in other membranes [Decker R.S., Poole A.R., Crie J.S. et al.; Decker R.S. and Wildenthal R.].

None the less, an increase of free (non-precipitable) activity of various lysosomal enzymes (acid phosphatase, β-galactosidase, E-cathepsin, etc.) is a concomitant of ischemic and reperfusion periods [Bilenko M.V. (1982); Sharapov V.I., Grek O.R., Dolgov A.V.; Sharov V.G., Jennings R.B. and Hawkins H.K.; Decker R.S., Poole A.R., Crie J.S. et al.; Decker R.S. and Wildenthal R.].

Higher lysosomal membrane resistance to ischemia, as compared to other membranes, has fostered the belief that lysosomal enzymes become involved in ischemic damage of cells only at the stage where reversible changes become irreversible [Meerson F.Z.; Katagiri Y., Sasai N., Nakamura M. et al.], or even after cell death [Decker R.S., Poole A.R., Crie J.S. et al.]. That this is true follows from the well known facts that the action of lysosomal lipases, and especially lysosomal phospholipases, is optimal at pH 5.0-3.5 (see Chapter 4); such pH values set up only after extremely long ischemia and only locally, in some negatively charged parts of the membrane [Beckman J.R., Owens K., Knauer T.E. et al.].

During the reperfusion period, lysosomes may take part in salvaging of reversibly damaged cells or complete the destruction of lethally damaged mitochondria or whole cells [Decker R.S. and Wildenthal R.].

Ischemia also damages peroxisomal membranes leading to the liberation of enzymes generating H_2O_2 [Panchenko L.F., Gerasimov A.M. and Antonenkov V.D.]. The role of H_2O_2 in initiating LPO processes during ischemia is considered in Chapter 10.

Thus, membrane barrier dysfunction provides a constant sign of an organ's ischemic injury. The severily of the barrier dysfunction depends on the severity and duration of the ischemia. Reperfusion of the organ aggravates the situation with membrane damage progressing in the following sequence: disturbed permeability to univalent ions → disturbed permeability to bivalent ions → disturbed barrier dysfunction in regard of high-molecular compounds and proteins.

2.2 CHANGES IN THE LIPID COMPOSITION OF MEMBRANES

The phospholipid (PL) composition of membranes, particularly the content of the two major PL, namely, phosphatidylcholine (PC) and phosphatidylethanolamine (PEA), does not differ significantly in organs of different animals and in different subcellular membranes of the same organ; it is more or less constant. PC prevails in all membranes (38-59% of the total PL content), its amount being 1.5-2.5 times greater than that of PEA (15-40%). Cardiolipin is present mainly in inner mitochondrial membranes (10-20%). The percentage of other PL (PI, PS, SM, etc.) is low, and the lysoPL content is minimal (0.4-1.1%) [Shaikh N. A. and Downar E. (1981); Vasdev S. C., Biro G. P., Narbaiz R. and Kako K. J.; Weglicki W.B., Owens K., Urchel C.W. et al.]. The total PL content in mitochondrial, microsomal and nuclear lipids in the liver is 92-94%, neutral fats accounting for 0-3% and cholesterol, for 5-6%. The latter content in the myelin sheath and plasma membrane is much higher [Kagava Y.]; the cholesterol/PL ratio in plasma membranes of different organs varies from 0.24-0.26 to 1.0-1.3.

The fatty acid composition of membrane PL is also rather uniform in mammal tissues. In the cardiac membranes of various animals, the total content of monoenoic (18:1) and polyenoic FA (18:2, 20:4, 22:6) exceeds that of saturated FA (16:0, 18:0) 1.5-2.5 times, whereas the content of polyenoic FA, (polyunsaturated, PUFA) alone is 45-55%, i.e. 1.5-2 times higher than that of saturated acids [Shaikh N. A. and Downar E. (1981)].

The greatest amount of unsaturated FA (more than 90%) is contained in cardiolipin [Kagava Y.], which makes it the most readily oxidizable lipid.

A roughly similar PL and FA content ratio has been found in rat livers in our laboratory (Table 3) [Dudnik L.].

Table 3: Phospholipid and fatty acid content in subcellular membrane of intact rat liver. (mean ± SEM) [Dudnik L.].

Content	Membranes		
	Mitochondrial	Microsomal	Nuclear
Total PL[1]	77.3 ± 2.9	63.8 ± 5.5	42.8 ± 1.7
Individual PL[2]:			
PC	50.1 ± 2.4	54.8 ± 1.0	47.4 ± 1.1
PEA	33.0 ± 0.3	24.0 ± 1.7	24.9 ± 1.3
cardiolipin	6.9 ± 1.3	1.2 ± 0.4	5.6 ± 0.9
SM	2.3 ± 0.6	4.6 ± 1.2	3.5 ± 1.4
PI	4.8 ± 1.5	9.0 ± 3.7	10.8 ± 1.1
PS	3.0 ± 2.3	3.6 ± 1.1	2.8 ± 0.5
lysoPL	1.4 ± 1.0	1.3 ± 0.6	5.1 ± 1.3
Fatty acids[3]:			
unsaturated			
16:0	12.2 ± 0.5	20.6 ± 3.1	18.9 ± 2.1
18:0	21.3 ± 2.2	23.9 ± 1.2	16.4 ± 2.7
total	33	45	35
monoenoic			
16:1	3.3 ± 0.5	1.0 ± 0.6	2.1 ± 1.2
18:1	12.3 ± 0.7	11.2 ± 1.2	17.8 ± 2.8
total	15.6	12.2	19.9
polyenoic			
18:2	20.0 ± 1.1	18.7 ± 2.0	20.6 ± 2.5
20:4	31.1 ± 1.1	24.9 ± 2.3	24.2 ± 1.1
total	51.1	43.6	44.8
monoenoic + polyenoic	66.7	55.8	64.8

[1] Total PL content was determined as a percentage of lipid content.

[2] Individual PL content was determined as a percentage of total PL content.

[3] Individual fatty acid content was determined as a percentage of FA total content.

Unlike PL in neutral lipids (triacylglycerides) in the heart of dogs and rats, saturated 16:0, 18:0 and monoenoic 16:1, 18:1 FA constitute the greatest amount (70-85%), while AA accounts for as little as 0.4-1.6% [Van der Vusse G.I., Roemen Th.A., Prinzen F.W. et al.].

Unsaturated FA occupy primarily position 2 (β) (95%), and saturated FA, position 1 (α) (94-97%). AA is present chiefly in PEA and occupies the β position exclusively [Kagava Y.].

Along with ordinary PL, membranes in some organs (myocardium, brain, skeletal muscles) [Swoboda G., Fritzsche J. and Hasselbach W.] contain plasmalogens differing from PL by having one long FA tail and one long aliphatic chain tail linked to the second (β) hydroxyl of glycerol. They are present mainly in PEA (about 65% of the total amount in the SR of skeletal muscles), which, together with the presence of AA, determine a greater degree of PEA nonsaturation in comparison with PC.

Lipids are arranged asymmetrically in the membrane: the outer monolayer contains primarily PC, SM, cholesterol, and glycolipids, the inner monolayer - PEA, PS, PI, aminophosphates. Such an arrangement exposes the different lipids to the different effects of endogenous and exogenous factors (PL, ROS, LPO products, ets.).

The physicochemical properties of membranes and their main characteristic, fluidity, are dependent on composition and the ratio between individual PL, neutral lipids, and cholesterol. PL containing PUFA are less ordered because they are curved at double bonds and, therefore, more fluid. The presence in membranes of saturated FA and especially cholesterol exerts a compacting effect [Skulachev V.P.].

The charge of membranes is influenced by changing amounts of negatively charged PL, such as PS, PI, cardiolipin, and phosphatidic acid in them. A drop of the negative charge lowers the Ca^{2+} fixing capacity of the plasma membrane (glycocalyx).

During ischemia and after reperfusion of organs, the composition of membrane PL undergoes drastic changes and is studied in sufficient detail for most organs.

Two-hour ***liver ischemia*** led to a marked decrease in cardiolipin content (from 13 to 10.1%) and a moderate drop in PEA level but did not affect the PC content in mitochondrial membranes in rat liver [Boime I., Smith E.S. and Hunter F.E.]. After 3 h of ischemia, the total PC content also decreased (by 40-50%), which correlated with the appearance of irreversible changes in the organ, a 25- to 50-fold decrease in Ca^{2+} transport, diminishing activity of glucose-6-phosphatase, and lower cytochrome P-450 content [Chien K.R., Abrams J., Serroni A. et al.].

Experiments conducted in our laboratory [Dudnik L. B., Bilenko M. V. Alessenko A. V. et al. (1980); Dudnik L. B., Bilenko M. V. Alessenko A. V. et al. (1981); Bilenko M.V. (1978b); Alessenko A.V., Bilenko M.V., Dudnik L.B. et al.] on a total ischemia model of the left lobe of the liver in August rats species revealed substantial changes in total lipid, individual PL and FA content in mitochondrial, microsomal and nuclear membranes, and in the homogenate. The total lipid content in the homogenate, mitochondrial and microsomal fractions was found to decrease significantly after 2 h of ischemia. The same applies to the PL content in the mitochondrial fraction and the total neutral lipids content in the homogenate and mitochondrial, microsomal and nuclear fractions (Fig. 11). The total neutral lipids content dropped significantly in mitochondrial and microsomal fractions already within the first 15 min of ischemia. In other words, degradation of neutral lipids in this time interval was more pronounced, than PL destruction. This phenomenon is also indirectly corroborated by the fact that the total PL as percentage of total lipids did not decrease within the first 15 min but, on the contrary, moderately increased (by 17% in mitochondria, by 33% in microsomes, by 53% in nuclei); in the homogenate it increased by 16 and 23%, within the first 15 and 30 min, respectively, i.e. the PL percentage of total lipids was higher at the expense of drop in other lipid components, most probably neutral lipids.

Fig. 11: Content of total lipids (1), phospholipids (2) and neutral lipids (3) in the homogenate (a), mitochondria (b), microsomes (c) and nuclei (d) of complete in vivo ischemic rat liver (mg • g^{-1} of wet tissue) (mean ± SEM) [Dudnik L. B., Bilenko M. V. Alessenko A. V. et al. (1980)]

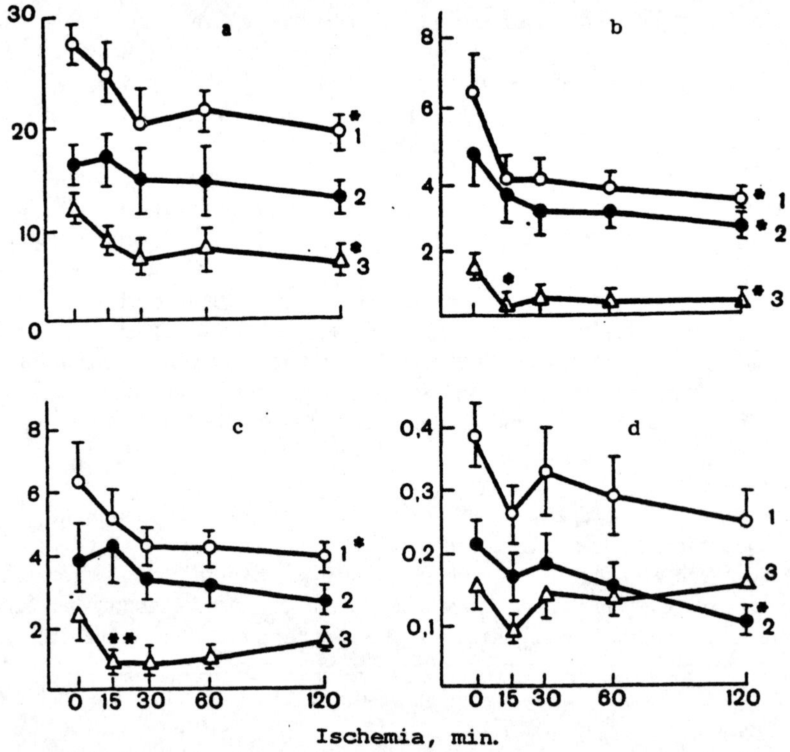

*-p < 0.05; **-p < 0.01 versus initial level. Subcellular fractions were separated by differential centrifugation; lipids were extracted according to [Bligh E.G. and Dyer W.J.]; total PL content was determined from the amount of P_i; neutral lipids content was determined by two-dimensional thin- layer chromatography [see Dudnik L. B., Bilenko M. V. Alessenko A. V. et al. (1980)].

The lipid content in all examined fractions, including the microsomal one, in which the protein level also went down significantly after 2 h of ischemia, decreased to a greater extent than that of proteins, i.e. the lipid/protein ratio in all fractions decreased to 68-72% of the initial level after 2 h of ischemia (Fig. 12). The PL/ protein ratio also decreased significantly in the mitochondrial and microsomal fractions.

The most drastic decrease of individual PL in the ischemic liver was observed in PC, PEA, and especially cardiolipin. The changes were most pronounced in the mitochondrial and microsomal fractions (Fig. 13). The PEA and PC contents in both fractions especially decreased between the 1st and 2nd hours of ischemia dropping after 2 h, to 60 and 66% of the initial level in the mitochondrial fraction and to 82 and 65% in the microsomal fraction. The cardiolipin content in the mitochondrial fraction decreased much earlier- within the first 15 min, after 60 and 120 min it dropped to 38 and 39% of the initial level. Such a decrease was registered not only per gram of protein, but also as a percentage of total lipids, i.e. compared to other lipids, cardiolipin underwent selective degradation during ischemia.

Fig. 12: Lipid/protein (a) and phospholipid/protein (b) ratios in the homogenate (1) mitochondria (2), microsomes(3) and nuclei(4) of ischemic rat liver (g per g of protein, rel. units) (mean ± SEM).

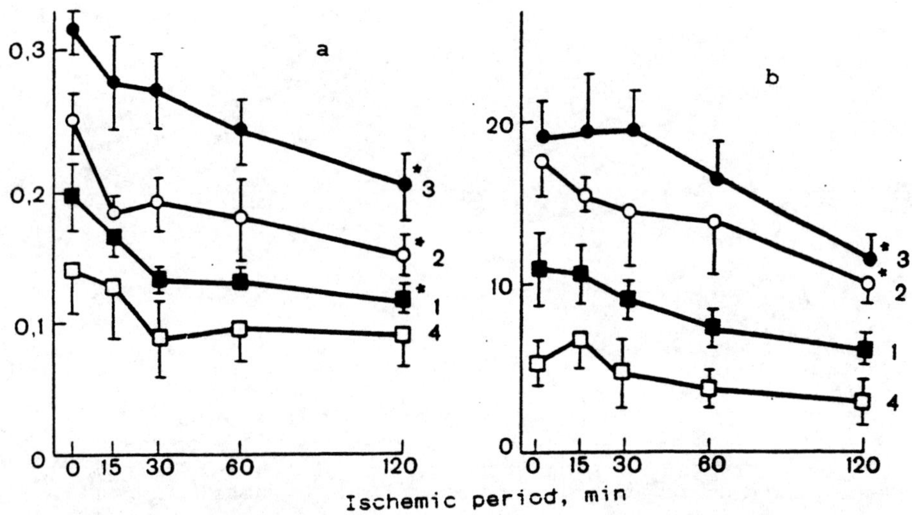

*-p<0.05 versus initial level. Protein was determined by the buret method.

Analysis of the FA composition of PL has revealed a tendency toward a diminution of the total content of all FA (in milligrams per gram of protein). After 2 h of ischemia the content of saturated FA (16:0 and 18:0) went down to 61-78% of the initial level, while that of mono- and diunsaturated FA (16:1, 18:1, 18:2) went down to 53-70%. A significant sharp decrease was observed in AA (20:4), its content after 2 h of ischemia, dropping to 67% of the initial level in the nuclear fraction, 40% in the homogenate, 48% in the mitochondrial fraction, and 58% in the microsomal fraction (see Fig. 13). The AA percentage of total FA also diminished significantly, which is indicative of the changes in this acid content being more marked than in others. It decreased in all the examined fractions, especially between 30 and 60 min of ischemia.

Comparing the changes in the content of different PL and AA showed (see Fig. 13) that in the mitochondrial fraction changes in total PL, PEA and PC follow the same pattern; the drop in AA content was rather pronounced and in cardiolipin-even more so. In the microsomal fraction, the same pattern is observed in the changes undergone by total PL and PC, while the decrease in PEA content was less pronounced than in total PL and PC, and in AA was comparatively more pronounced. Contents of all lipid components change earlier and more drastically in mitochondrial fragments, as opposed to microsomal ones (see Fig. 13).

Fig. 13: Content of total and individual phospholipids (PL) and arachidonic acid (AA) in mitochondrial (a) and microsomal (b) membranes of complete ischemic rat liver (% of initial level) [Dudnik L. B., Bilenko M. V. Alessenko A. V. et al. (1981)].

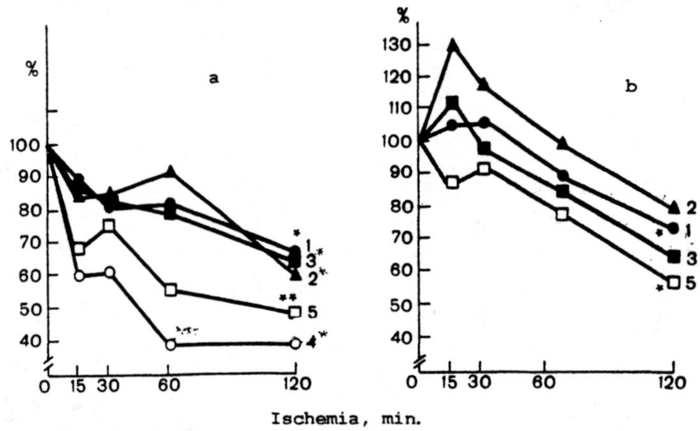

1 - total PL; 2 - PEA; 3 - PC; 4 - cardiolipin; 5 - AA. *-p<0.05;**-p<0.01; versus initial level. Individual PL were determined by two-dimensional thin-layer chromatography; PC was identified using Dragendorf's reagent. The FA composition was determined by gas-liquid chromatography (see Dudnik L. B., Bilenko M. V. Alessenko A. V. et al. (1981)).

The lysoPL content in intact hepatic tissue was 4-7 times higher in the nuclear fraction than in the homogenate and mitochondrial and microsomal membranes. During ischemia, total PL percentage doubled in the homogenate after 60 min of ischemia ($P < 0.05$), the increase being especially marked in the microsomal fraction where the initial level was exceeded three-fold after 2 h of ischemia ($P < 0.05$) (Fig. 14).

Fig. 14: Content of lysophospholipids in homogenate(1) and microsomes (2) of ischemic rat liver (% of lipid P_i content) (mean ± SEM).

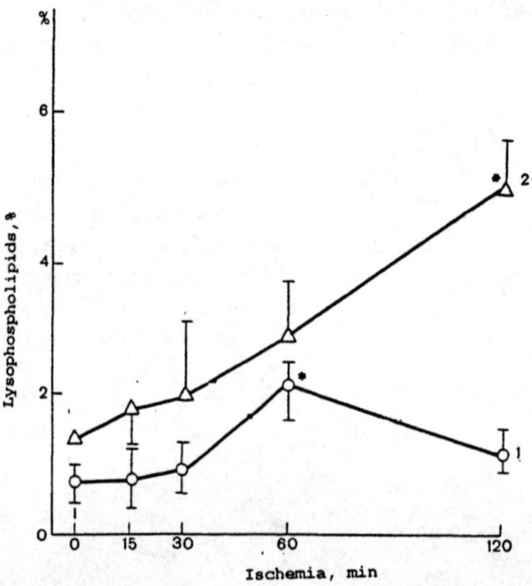

*-p<0.05 versus initial level.

The obtained results bring to light several important regularities.

Firstly, the total lipids and PL content in the homogenate and subcellular fractions decreased during ischemia both in overall wet tissue weight and the amount of protein, i.e. both lipids and PL (especially in mitochondrial and microsomal membranes) deteriorated more readily (possibly earlier) than membrane proteins. As already pointed out, the changing quantitative ratios between membrane lipids and proteins is a major factor disrupting the normal functioning of membrane proteins.

Secondly, the results indirectly point to the fact that the neutral lipids content in the homogenate and subcellular membranes of ischemic liver seems to decrease more rapidly within the first 15-30 min of ischemia, than PL. Such degradation, which occurs earlier and only temporarily (after 15-30 min the quantity of neutral lipids practically did not change for up to 2 h of ischemia) (see Fig. 11), is perhaps due to lipase activation. This is possibly connected with early accumulation in the ischemic organ of epinephrine in an amount exceeding that of norepinephrine (see Chapter 10), which activates intracellular lipases.

Thirdly, the most unsaturated PL and FA, such as cardiolipin, PEA, PC, and AA, were the first to undergo degradation. This, naturally, leads to a relative increase of saturated lipids in membranes, followed by a change in the phase transition temperature, lower membrane fluidity, and deterioration of the conditions for the functioning of membrane (lipid-dependent) proteins.

Changes in the physicochemical properties of biological membranes in ischemic liver were corroborated by the results of experiments using spin probe 1 (derivative of benzo- -carboline) [Dudnik L.], according to which hepatic ischemia brought about changes in the probe's rotational correlation time and effective rotational energy, indicative of the examined membranes adversely affected microviscosity.

Since the probe used in the experiments may be embedded both in the membrane's hydrophobic bilayer, away from its proteins, and in the annular lipid layer immobilized by integral proteins, it fails to pinpoint the site of changes in the lipids' physicochemical properties.

Fourthly, lysoPL accumulation was observed only in two out of the four examined fractions, namely, in the homogenate and microsomal fraction. In both cases, it was identified late in the ischemic period - after 1 and 2 h, respectively, which suggests that the phospholipase mechanism gets involved in the pathogenesis of lipid degradation later than other possible mechanisms. The role of this process will be discussed in detail in Chapter 4.

Evidence of substantial changes in the lipid composition of ischemic liver was later obtained also by other workers [Finkelstein S.D., Gilfor D. and Farber J.L.; Petrovich D.R., Finkelstein S., Waring A. et al.]. It was established that after 3 h of rat liver subtotal ischemia (ligation of portal vein and hepatic lobe artery) the total PL level, the contents of individual PL and esterified FA in PL go down by 28-30%, while the neutral lipids and cholesterol content remains the same and the cholesterol/PL ratio increases by 22% with a corresponding decrease in the fluidity of smooth ER membranes.

Ischemia of the heart was also accompanied by substantial changes in membrane lipids composition; these changes, however, were studied primarily on a model of partial ischemia induced by occlusion of the LAD.

It was found that after 10 and 20 min of myocardial ischemia in dogs the mitochondrial membrane cardiolipin content decreased by 16 and 25%, respectively, and the PC content decreased by 23% [Ekholm R., Kerstell I., Olson R.E. et al.].

In experiments on rats [Chien K.R., Pfau R.G. and Farber J.L.; Chien K.R., Reeves J.P., Buja L.M. et al.], no changes in the PL composition were observed within the first 10 min after LAD occlusion, after 3 h the total PL content dropped only by 15% (at the expense of PC and PEA), but after 12 h of ischemia the PL decrease was already by 33% (PC by 25% and PEA by 45%). The percentage decrease in total PL in myocardial SR membranes after 12 h of ischemia was close to that in the homogenate. The loss of membrane PL after 12 h of ischemia was combined with a decrease in the capacity of cardiac SR to uptake Ca^{2+} (up to 40% of the control) as well as with a sharp increase (by a factor of 6-7) in the passive permeability of SR membranes to Ca^{2+}. Moreover, SR membrane dysfunction was due to changes in their lipid component, because they were not accompanied by quantitative or qualitative changes in membrane proteins [Chien K.R., Pfau R.G. and Farber J.L.]. Nor did we observe any changes in the total PL content in cardiac SR membranes during the first 2 h in experiments carried out on a model of total canine heart ischemia (see Fig. 40).

During reperfusion the total and individual PL decrease in dogs with occluded LAD depended on the duration of ischemia; the decrease was more pronounced in sarcolemma than in mitochondrial membranes and was combined with ultrastructural changes in the mitochondria [Vasdev S. C., Biro G. P., Narbaiz R. and Kako K. J.].

No total and individual PL changes were observed 30-60 min after LAD occlusion in pigs, while reperfusion of the heart with blood already after 15 min caused their significant decrease to 91-80% of the initial level; this low level persisted for all the 60 min of reperfusion [Das D.K., Engelman R.M., Rauson J.A. et al. (1986); Shaikh N. A. and Downar E. (1981)]. The PL decrease and negatively charged PC and PI in the myocardial sarcolemma, along with the drop of the total PL, PC and PEA levels in the sarcolemma of the inner myocardial layer after LAD occlusion coincided with diminishing activity of Na, K-ATPase in the sarcolemma [Takahashi K. and Kako K.J.], while the total PL and PEA decrease coincided with diminishing activity of sarcolemmal Ca-ATPase [Chien K.R., Reeves J.P., Buja L.M. et al.]; arachidonic acid accumulation in isolated rabbit hearts was significantely time related to morphological changes in sarcolemma membranes in both endothelial cells and myocytes [Schrijvers A.H., De Groot M.J., Heijnen V.V. et al.].

Ischemia of the kidneys also leads to changes in the membrane lipid composition, qualitatively similar to those in other organs, even though less pronounced and occurring later in time.

During total rat kidneys ischemia (isolation storage at 37°C), no decrease in the content of total and most of individual PL was observed in mitochondrial membranes within 24 h, the lysoPL level did not increase, but cardiolipin decreased by 50 and 70% after 4 and 24 h, respectively, the content of esterified FA decreased, and the free fatty acids (FFA) content (mainly 18:2 and 20:4) increased towards 4th h of storage. These changes were accompanied by decreasing mitochondrial functions [Smith D.S., Rehncrona S. and Siesjö. Bo.K.]. The same model (rat kidneys total ischemia) demonstrated that changes in total and individual PL in the cortical layer homogenate were almost similar to those in ischemic liver (total PL and all individual PL levels dropping at the same rate, esterified FA decreasing, cholesterol/PL ratio increasing) but occurring much later (after 6-8 rahter than 3 h) and were accompanied by accumulation of FFA (10-fold after 20 h of ischemia) [Finkelstein S.D., Gilfor D. and Farber J.L.]. Clamping of the renal pedicle for 50 min revealed an increase of membrane lysoPC content and a drop of PC, SM, PS and phosphatidic acid [Molitoris B.A., Wilson P.D., Schrier R.W. and Simon F.R.].

In hypothermically stored dog kidneys, the drop of PL in lipids of the homogenate and mitochondrial fraction after 24 h was small and did not correlate with the kidney functional capacity after their transplantation [Southard J.H., Ametani M.S., Lutz M.F. and Belser F.O.].

Brain ischemia leades to changes in PL composition, largely dependent on the severity of ischemia and less pronounced than in other ischemic organs.

In total ischemia of the Mongolian gerbil brain, total or grave partial ischemia of the rat brain for 30-60 min, and also in moderate partial rat brain ischemia for 1-5 h, the total and individual PL contents did not change or decreased insignificantly [Gasteva S.V., Raize T.E., and Sharagina L.M.; Zamuruev O. N. (1984); Enseleit W.H., Domer F.K., Jarott D.M. et al.; Rehncrona S., Smith D.S., Akesson B. et al.; Yoshida S., Inoh S., Asano T. et al.].

In long partial ischemia (ligation of two rat carotids for 6 h), the FA composition did not change in homogenate lipids, in the mitochondrial fraction total FA content increased to 82%, and in the myelin the total FA dropped to 63% due to a more or less equal decrease in saturated and unsaturated acids, while hydrolysis of these acids resulted in growth of FFA [Kuwashima J., Nakamura K., Fujitani B. et al.].

In contrast to these data, in total guinea pig brain ischemia (decapitation model) the contents of total PL, PEA, PEA-plasmalogens, and PC decreased by 20, 20, 33, and 15% respectively, already within the first minute of ischemia, but did not perceptibly change over the next 4 min [Majewska M.D., Strosznajder J. and Lazarewicz J.]. In a totally ischemic Mongolian gerbil brain (ligation of two carotids), a significant decreases was observed within the first 3-5 min only in the PC contents (to 85%), PI content (to 77%) and PEA-plasmalogen content (to 86%), as well as in the AA level in PC, PI, and PEA-plasmalogen. The diglycerides content increased moderately within the first 30 s (to 116%, $P < 0.05$), the triglycerides content remained the same; the lysoPC content also increased moderately (to 141%) [De Medio G.E., Goracci G. and Horrocks L.A. et al.].

Thus, as has already been convincingly demonstrated, the qualitative and quantitative composition of lipids changes during ischemia of all organs, the changes being most pronounced during prolonged total ischemia; they greatly depend on the kind of organ; appearing earlier in the brain (the first 1-5 min), then in the liver (2-3 h), and much later and least pronounced, in the kidneys (after 6-8 h). According to our data, in total liver ischemia, the PL level drops more sharply in microsomal and mitochondrial membranes than in nuclei [Dudnik L. B., Bilenko M. V. Alessenko A. V. et al. (1980)].

During partial ischemia of the heart and brain, the total and individual lipid content diminishes to a small extent (goes unobserved for up to 3-5 h of ischemia), however, changes occur early in the reperfusion period. According to our results and data in the literature, mainly dropping in ischemia is the level of PL containing a high percentage of PUFA (PEA, PC, cardiolipin) or esterified PUFA (AA), although some investigators have failed to observe a selective decrease in any individual PL [Finkelstein S.D., Gilfor D. and Farber J.L.; Smith D.S., Rehncrona S. and Siesjö. Bo.K.].

2.3 Changes in Membrane Protein Activity

The functional activity of membrane proteins also changes considerably in the course of ischemia and after reperfusion of organs. In the early ischemic stage, the functional activity of some proteins (e. g., transport proteins, glycolytic enzymes, phospholipases, xanthine oxidases, etc.) intensifies for a while, but as the ischemic period becomes longer, the activity of both temporarily "activated" and most of the other proteins decreases, either succumbling to reversible inhibition, or irreversible inactivation.

A large body of literature has been concerned with this issue, especially the decreasing activity of sarcoplasmic Na, K-ATPase enzymes in the mitochondrial respiratory chain, glycolytic enzymes, ATP-ADP translocase (see Chapter 3), and adenylate cyclase (see Chapter 4).

Therefore, we shall analyze here only our own results and some published data concerning the effect of ischemia on the activity of Ca-ATPase in myocytes and the cytochrome P-450-dependent enzymic system in the hepatic ER, both being typical representatives of lipid-dependent enzymes.

2.3.1 Ca-ATPase of SR Membranes in Myocytes

Ca-ATPase of SR membranes in myocytes accounts for about 70 to 76% of all proteins in these membranes [Ritov V.B.] and is substantially dependent on the composition and properties of the annular lipid environment, forming together a single lipoprotein complex [Kozlov Yu. P., Kagan V. E. and Arkhipenko Yu. V.].

In our laboratory, experiments were conducted to elucidate the effect of ischemia and reperfusion on the enzymic system of Ca^{2+} transport and Ca-ATPase activity in SR membranes of skeletal muscles in rats and the myocardium in dogs.

It was shown that total ischemia of August rat skeletal muscles, (tourniqueting model) [Arkhipenko Yu. V., Bilenko M. V., Dobrina S. K. et al.; Bilenko M.V. (1978b)] leads to decrease of both parameters of the enzymic Ca^{2+} transport system (Ca^{2+}/ATP ratio and Ca-ATPase activity) [Arkhipenko Yu. V., Bilenko M. V., Dobrina S. K. et al.; Bilenko M.V. (1978b)] (Fig. 15).

The tendency toward a decrease in the Ca^{2+}/ATP ratio, showing how effectively the Ca^{2+} transport system works, was observed from the first hour of ischemia, and after 6 h of ischemia this value went down to 36% of the initial level. At the same time, Ca-ATPase activity within the first 3 h was virtually the same and diminished only between the 3rd and 6th h of ischemia (to 62% of the initial level). The decrease in Ca^{2+}/ATP with unchanged hydrolytic function of Ca-ATPase means that it takes more ATP to transfer a single Ca^{2+} molecule, which is indirectly indicative of possible multiple transfer of the same Ca^{2+} molecule as a consequence of its "leakage" from SR vesicles into the cytosol, i.e. early disturbance of the SR membranes' passive permeability in skeletal muscles for Ca^{2+} and later damage to Ca-ATPase itself. Reperfusion of skeletal muscles resulted in the further deterioration of both parameters of the enzymic Ca^{2+} transport system: it led to the SR membranes' failure to accumulate Ca^{2+} (Ca^{2+}/ATP dropping to zero) and markedly inhibited ATPase activity (to 42% of the initial level).

Fig. 15: Ca^{2+} /ATP (1) and Ca-ATPase activity (2) in SR membranes of rat skeletal muscles after 1 and 3h complete ischemia and 4 h of reperfusion (mean ± SEM).

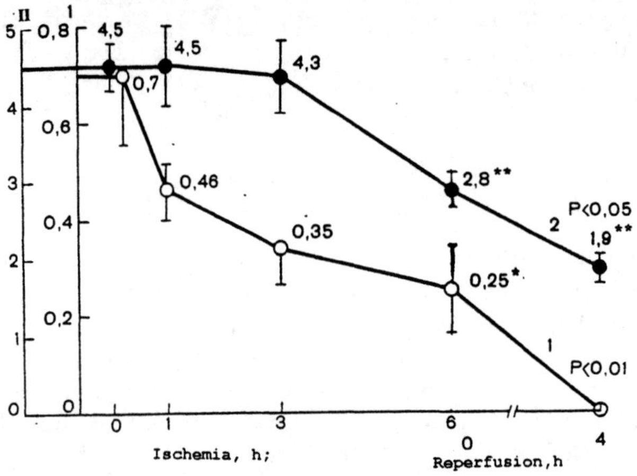

1 - Ca^{2+}/ATP (in relative units); 2 - Ca-ATPase activity (nmol • min-¹ per mg of protein). *-p < 0.05; **-p < 0.01 versus intial level. P - significant differences between parameters before and after reperfusion. SR membranes were separated by differential centrifugation. The Ca^{2+}/ATP ratio and Ca-ATPase activity were determined *in vitro* pH-metrically from acidification of the medium in the course Ca^{2+}- and Mg^{2+}-dependent ATP hydrolysis [Arkhipenko Yu. V., Bilenko M. V., Dobrina S. K. et al.]; protein in SR membranes was measured by the buret method.

In experiments with myocardial ischemia in dogs it was found [Bulgakov V.G. and Bilenko M.V.] that the SR membrane capacity for Ca^{2+} uptake decreased significantly already within the first 15-30 min of ischemia and continued to drop to the second hour of ischemia, falling by that time to 16% of the initial level, whereas the Ca-ATPase activity after 60 min of ischemia showed only a slight tendency for decrease, but dropped sharply towards the 2nd hour of ischemia to 27% of the initial level (Fig. 16).

Fig. 16: Parameters of Ca^{2+} transport enzyme system in SR membranes of canine myocardium after different time of complete ischemia in vitro(isolated storage at 37°C) (mean ± SEM) [Bulgakov V.G. and Bilenko M.V.].

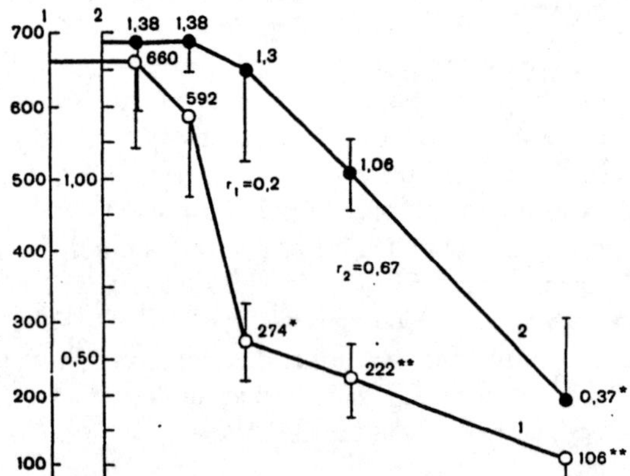

1 - Ca^{2+} uptake (nmol•mg-¹of protein for 10 min); 2 - Ca-ATPase activity (μmol P_i per mg of protein for 5 min); r_1, r_2 - see text. *-p<0.05; **p-<0.01 versus initial level. Methods of SR membrane separation, Ca^{2+} uptake determination and Ca-ATPase activity, see [Bilenko M.V., Otverchenko V.M., Astaf'ev V.V. et al.].

No correlation existed between Ca^{2+} uptake and Ca-ATPase activity in SR membranes of the myocardium during the first 30 min of ischemia (correlation factor $r_1 = 0.2$), while during the next 60-120 min there was a strong correlation ($r_2 = 0.67$, $P < 0.05$).

These findings suggest that the early inhibition of the Ca^{2+}-accumulating capacity of myocardial SR membranes, just as SR membranes in skeletal muscles, is mainly due to enhanced SR membranes' passive permeability to Ca^{2+}, "leakage" of Ca^{2+} oxalate precipitates into the cytosol and, as a result, lower effectiveness of the work performed by Ca-ATPase, and later on - due to lower activity of the enzyme itself.

These experiments revealed that, at different for each organ early moments in the ischemic period, the SR membranes of skeletal muscles and the myocardium retain the Ca-ATPase capacity for ATP hydrolysis, with a diminished Ca^{2+}-accumulating capacity of SR, uncoupling of ATP hydrolysis from Ca^{2+} transport was later corroborated on an ischemic heart model by other authors as well [Hess M.L., Krause S.M. and Greenfild L.J.; Hess M.L., Manson N.H. and Okabe E.].

At the same time, it is important to point to evidence [Rapundalo S.T., Briggs F.N. and Feher J.J.] that the Ca^{2+}-transporting capacity of SR membranes, isolated from the homogenate of an ischemic heart or from a fraction obtained after brief centrifugation, was higher than in an SR membrane fraction obtained in the usual manner, which, according to the authors, is due to the fact that centrifugation separates not all but only badly damaged membranes. In other words, estimation of the damage extent suffered by SR membranes *in vitro* may register much graver changes than those occurring in an organ *in vivo*.

We also studied the changes in the enzyme system of Ca^{2+} transport after 1 h of storage an isolated dog heart under less severe ischemic conditions (at 20°C) with subsequent reperfusion by extracorporeal connection of the cardiac vessels to the cervical ones of a recipient dog for 4 h [Bilenko M.V., Otverchenko V.M., Astaf'ev V.V. et al.]. The experiments showed (Fig. 17) that after 1 h of ischemia at 20°C the decrease in the Ca^{2+} uptake function of SR was less pronounced than after 1 h of ischemia at 37°C (61% versus 34%), and the activity of Ca-ATPase even tended to grow (at 37°C it displayed a tendency to go down). After 4 h of reperfusion, both parameters decreased (especially Ca^{2+} uptake); that is, the enzyme system of Ca^{2+} transport in myocardial SR, just as in skeletal muscle SR, sustained both ischemic and reperfusion damage with passive permeability of membranes to Ca^{2+}, first increasing, then considerably decreasing in Ca-ATP-ase activity. These changes started much earlier, in cardiac SR membranes (after 30-60 min at 37°C), than in the skeletal muscle SR membranes (after 3 and 6 h).

Persistence of high Ca-ATPase activity over a rather long period of time, with the earlier increased passive Ca^{2+} permeability of SR membranes in skeletal and myocardial muscles, must inevitably lead to a greater load on the Ca-ATPase, have to "pump" the escaped Ca^{2+} ions or Ca^{2+} precipitates into SR vesicles, which is bound to cause additional ATP waste. Indeed, such a situation is most likely to arise, because it is precisely within the first 3 h of skeletal muscle ischemia and the first 30 min of myocardial ischemia, i.e., during the period when the ATPase function is still retained, the ATP content in the organs diminished rather quickly (see Chapter 3), but still seems quite sufficient for Ca-ATPase action. Excess of transported Ca^{2+} may deplete ATP at the site of its utilization. However, the action of Ca-ATPases in SR membranes is ensured predominantly by ATP of creatine kinase origin, rapidly depleted during ischemia (see Chapter 3). Consequently, it is hard to tell at present whether ATPase retains (or even raises) its functional activity during ischemia or whether it becomes functionally active only when ATP is added into the medium *in vitro*.

Proceeding from the above data, one may present the Ca^{2+}-transport dysfunctioning of SR membranes during ischemia in the following sequence: ischemia $\rightarrow$ disordered passive membrane permeability to Ca^{2+} $\rightarrow$ (intensified Ca-ATPase action and rapid ATP consumption?) $\rightarrow$ decrease in Ca-ATPase activity. The stage in parentheses is hypothetical, inferred from the conducted experiments. During reperfusion of ischemic organs, the sequence of occurring changes is similar to that given for ischemia, but each stage is faster. The hypothetical sequence of disorders of the enzyme system of Ca^{2+} transport in SR membranes places emphasis on the disordered passive permeability of SR membranes (of their lipid bilayer) to Ca^{2+}, which precedes ATP deficiency, and clarifies the role of these changes in the ischemic and reperfusion damage mechanism of Ca^{2+} transport systems.

Fig. 17: Parameters of Ca^{2+} transport enzyme system in SR membranes of canine myocardium after 1h complete ischemia in vitro(isolated storage at 20°C) followed by 4h of reperfusion (mean ± SEM)

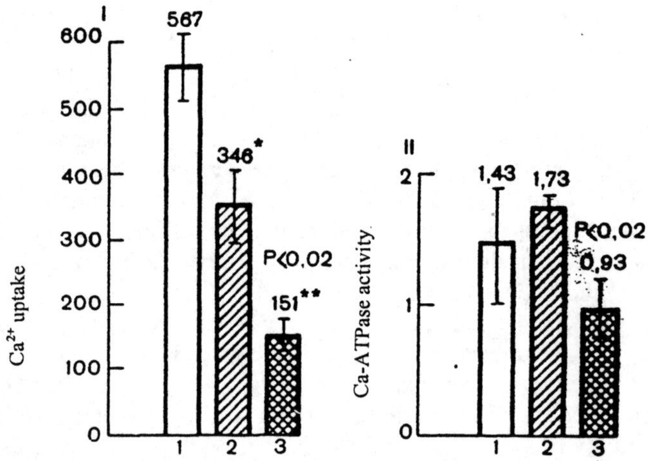

I - Ca^{2+} uptake (nmol•mg-¹ of protein for 10 min); II - Ca-ATPase activity(µmol Pi per mg of protein for 5 min); 1 - initial level; 2 - 1 h of ischemia; 3 - 4 h of reperfusion; -p<0.05; **-p<0.01 versus initial level, P-2 versus 3. Methods are the same as in Fig. 15 and 16.

Findings about Ca-ATPase activity going down later than the Ca^{2+} uptake capacity of SR membranes and their enhanced passive permeability to Ca^{2+} during total ischemia agree with the results obtained during *in vitro* hypoxic perfusion of the rat heart [Lee S.L., Balasabramanian V. and Dhalla W.S.], on a LAD occlusion model in the canine heart *in vivo* [Toba R., Katagiri T. and Takeyama U.], and on other membrane-damaging models [Geimonen E.R., Rubtsov A.M. and Boldyrev A.A.].

Expansion of the intracellular Ca^{2+} pool is also essential for development of one of the earliest ischemic and reperfusion injuries in the heart, namely, *ischemic contracture of the miocardium, or the "stone heart" phenomenon* [Hearse D.J.; Lakatta E.G., Nayler W.G. and Poole-Wilson P.A.; Lewis M.J., Grey A.C. and Henderson A.H.; Nayler W. G., Ferrari R., Poole-Wilson P. A. et al.], though other workers believe that ATP content plays a more importante role here [Lewis M.J., Grey A.C. and Henderson A.H.].

In order to elucidate the contribution of the ATP- and Ca^{2+}-dependent mechanisms to the pathogenesis of the heart ischemic contracture (IC), a study was made into the dynamics of rigidity (compressibility) of the myocardium, changes in ATP content, and Ca^{2+} uptake in SR membranes after 60 and 120 min of total global ischemia of an isolated canine heart stored at 37°C and after 30 min of cardiac reperfusion [Otverchenko V.N.; Otverchenko V.N., Bulgakov V.G., Volkova I.A. and Bilenko M.V.]. SR membranes were separated from the subendocardial (inner third of the left ventriclar wall) and subepicardial (outer third of same) layers of the miocardium separately.

As seen from Figure 18, after 30 min of ischemia the compressibility of the inner myocardial layer was 3 times lower than initially, significantly differing from that of the outer layer, i.e., contracture changes during this ischemic period occurred only in the myocardium inner layer. After 2 h, the compressibility was equally low in both myocardial layers, i.e., the development of total myocardial contracture was observed.

Fig. 18: Myocardial compressibility (a), Ca^{2+} uptake by SR membranes (b), and ATP content (c) in the outer (1) and inner (2) layers of canine myocardium after different periods of complete heart ischemia in vitro(isolated storage at 37°C) (mean ± SEM).

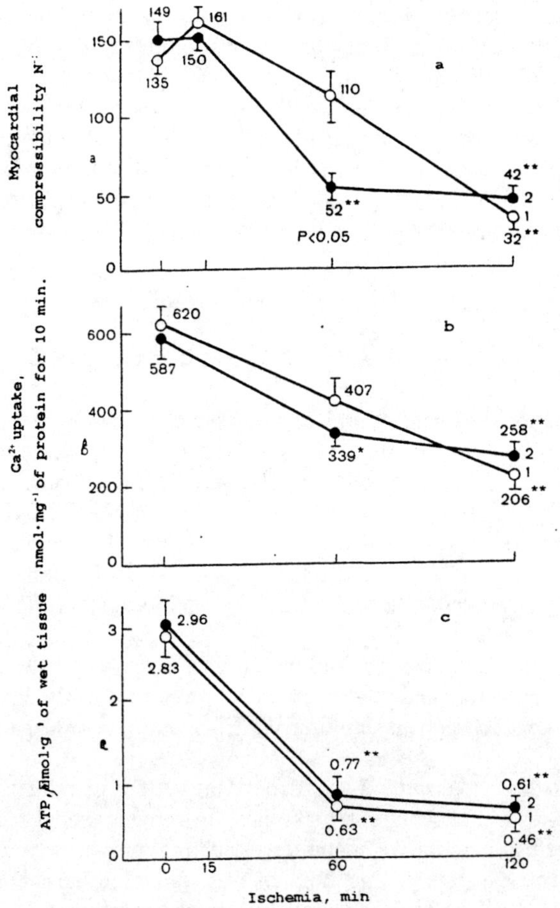

-p<0.05;**-p<0.01 versus initial level; P-1 versus 2. Myocardium compressibility was determined using a method [Seelye R.N., Nevalainen T.J., Gavin J.B. et al.], in modification [Otverchenko V.N., Bulgakov V.G., Volkova I.A. and Bilenko M.V.]. Ca^{2+} uptake was determined as in [Bilenko M.V., Otverchenko V.M., Astaf'ev V.V. et al.]; ATP content was determined by the hexokinase method.

The SR membrane Ca^{2+} uptake after 1 h decreased more steeply in the inner myocardial layer ($P < 0.05$, as compared to initial level) than in the outer one, whereas ATP content decreased in both layers at the same rate. Correlation analysis showed that the outer and inner myocardial layers' compressibility is in a stronger inverse correlation with the SR Ca^{2+}-accumulating function ($r_1 = -0.72$ for outer, $r_2 = -0.80$ for inner layer, $P < 0.05$-0.01) than with the ATP content in each layer ($r_1 = -0.45$ and $r_2 = -0.61$, respectively, $P < 0.05$).

Hearts reperfused after 1 h of ischemia regained their contractile function, while compressibility of the outer myocardial layer and the Ca^{2+} uptake capacity of its SR membrane tended to increase after 30 min of reperfusion, and the ATP content increased significantly. Rigidity in the inner layer continued to increase rather than go down, while other parameters remained without change. Two-hour reperfusion of an ischemic heart did not restored its myocardial contractile function, compressibility, and ATP content remained low or kept decreasing, and the left ventricular wall passive elasticity (stretchability) was negligible [Otverchenko V.N., Bulgakov V.G., Volkova I.A. and Bilenko M.V.].

These findings suggest a number of important conclusions.

Firstly, heart ischemic contracture (IC) may develop not only during reperfusion, but also during total cardiac ischemia.

Secondly, ischemic contracture changes in the inner myocardial layer occur earlier than in the outer one, and after reperfusion they remain the same or continue to worsen, whereas in the outer layer they may disappear after reperfusion, i.e., outer layer contracture changes are reversible.

Thirdly, the leading cause of IC injuries is not only ATP deficiency [Lewis M.J., Grey A.C. and Henderson A.H.], but the SR decreasing Ca^{2+}-accumulating capacity as well. As has already been shown (see Fig. 16), the SR membranes' Ca^{2+} uptake changes after 1 h of ischemia with the same Ca-ATPase activity or, in other words, these changes are due to SR membranes disodered passive permeability to Ca^{2+}; it may be assumed, therefore, that the leading role in the pathogenesis of the myocardium ischemic and reperfusion contracture damage is played precisely by the disturbed passive ionic permeability of the membranes' lipid bilayer.

2.3.2 Enzymes of the Polyfunctional System of Monoxygenases

Enzymes of the polyfunctional system of monoxygenases also belong to lipid-dependent membrane enzymes. They are located in hepatic ER membranes and specialize in detoxication of xenobiotics and endogenous toxic products. These enzymes are also involved in the biosynthesis of steroids, hormones, and lipids [Archakov A. I. (1975); Archakov A. I. (1983); Lyakhovich V.V. and Tsyrlov I.B.; Metelitsa D.I.]. The main protein components of this system are cytochrome P-450 (terminal oxidase) and NADPH-cytochrome P-450 reductase. The system of monoxygenases is labile; its functional activity varies depending on substrate supply and, according to our findings, on biological (diurnal) rhythm [Kagan V. E., Kozlov Yu. P., Bilenko M. V. et al.].

Changes in the quantitative composition and functional activity of the monoxygenase system's components depend on the composition and physicochemical properties of the lipid environment; they also depend on the decomposition and subsequent 'build-up' of ER membranes [Kagan V. E., Kotelevtsev S. V. and Kozlov Yu. P.].

The known properties of the cytochrome P-450-dependent system of monoxygenases and its key role in ensuring the hepatic cell's essential functions make it a convenient model for evaluating the effect of acute liver ischemia and reperfusion on the ER membrane structural properties and barrier function.

The monoxygenase system of hepatic ER membranes was studied in our laboratory on a model of total ischemia (occlusion of the left lobe of the Wistar rat liver) with reperfusion [Bilenko N.V., Kagan V.E., Velikhanova D.M. and Komarov P.G.; Velikhanova D.M., Bilenko M.V. and Kagan V.E.; Velikhanova D.M.; Komarov P. G., Velikhanova D. M. and Bilenko M. V.].

The experiments showed (Fig. 19) that the cytochrome P-450 content decreased moderately but significantly already after 15 min of ischemia (to 67% of the initial level), stayed at more or less the same level for 1 h, then dropped sharply within the next hour (to 39% of the initial level). During this interval a significant decrease in total protein content (to 64% of the initial level) was also observed. The activity of NADPH-cytochrome c reductase decreased insignificantly throughout the ischemic period (from 15 min to 2 h) to 61-70% of the initial level.

During the one-hour liver ischemia, the content of cytochrome b_5 in ER membranes remained virtually the same, and aminopyrine-N-demethylase (AD) activity fell sharply to 37% of the initial level.

The above results suggest that the sensitivity to ischemia of the examined enzymes of the monoxygenase system in hepatic microsomes decreases in the following order: AD > cytochrome P-450 > NADPH-cytochrome c reductase > cytochrome b_5.

The early AD inactivation (and, possibly, its decreasing content) during hepatic ischemia may be ascribed to the enzyme asymmetrical (peripheral) arrangement in the membrane, owing to which it is particularly sensitive to changing protein-lipid interactions. The stability of cytochrome b_5 seems due to the great amounts of nonpolar amino acids it contains and strong hydrophobic bonds to PL, whereas that of NADPH-cytochrome c reductase may be due to the location of its active site (flavoproteid) in the hydrophilic portion of the molecule and its ability to function in the aqueous phase even after complete (or partial) membrane breakdown.

Fig. 19: Content and activity of some components of the monoxygenase system in ER membranes after 15-20 min of ischemia of rat liver (% of the initial level)

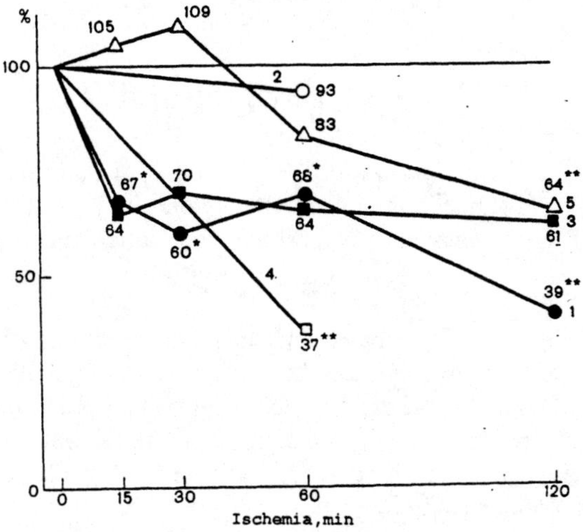

1 – cyt. P-450; 2 - cyt. b_5; 3 - NADPH-cytochrome c reductase; 4 - aminopyrine-N-demethylase; 5 - proteins. *-p<0.05; **-p<0.01 versus initial level. Hepatic microsomes were separated by differential centrifugation. Methods of determining cytochrome P-450 content, activity of aminopyrine-N-demethylase and NADPH-cytochrome c reductase, see [Velikhanova D.M., Bilenko M.V. and Kagan V.E.]; the protein was determined by the buret method.

The high sensitivity of enzymes of the cytochrome P-450-dependent system of monoxygenases in the liver and of cytochrome P-450 itself to ischemia is consistent with earlier [Ferrero M.E., Orsi R., Bernelli-Zazzera A.] and later [Voronov G.G. and Lukienko P.I. (1982); Voronov G.G., Lukienko P.I. and Bushma M.I.; Grek O.R., Sharapov V.I., Litvinov V.S. et al.] findings.

The decrease in cytochrome P-450 content during ischemia may be due not only to damage of the protein molecule itself or its escape from the membrane, but also to conversion of cytochrome P-450 into cytochrome P-420 [Voronov G.G., Lukienko P.I. and Bushma M.I.].

An analysis of reperfusion damage in enzymes of the monoxygenase system in the liver ER, which we performed after the "critical" (1 h) period of the organ ischemia and 4 h of its reperfusion, revealed a significant decrease in the content or activity of all examined enzymes, including those more stable ones, that, unlike others, were still unchanged after 1 h of ischemia (e. g., NADPH-cytochrome c reductase and cytochrome b_5).

Changes in the activity of enzymes of the monoxygenases system in the ER during hepatic ischemia were also studied in experiments using an inducer of this system, namely, phenobarbital (PB) [Velikhanova D.M., Bilenko M.V. and Kagan V.E.]. Its preoperative administration in a 80 mg/kg daily dose for 3 days raised the cytochrome P-450 content and activity of AD 2.2- and 2.3-fold, respectively ($P < 0.01$), while the activity of NADPH-cytochrome c reductase increased at a slower rate (1.2-fold, $P < 0.01$), and the cytochrome b_5 content did not change significantly. The changes in all examined enzymes (except for NADPH-cytochrome c reductase) during ischemia and after reperfusion in PB-treated animals were similar to those in PB-untreated animals (see Fig. 37 below), which differs from other data [Ferrero M.E., Orsi R., Bernelli-Zazzera A.] and suggests that the tolerance of the microsomal monoxygenase system to ischemia and reperfusion damage does not vary with changes in its functional state. This is most probably due to the fact that the major role in this system degradation is played by protein-lipid interactions and the lipid composition which is not affected by PB.

At the same time, the absolute cytochrome P-450 content, AD activity and NADPH-cytochrome c reductase in PB-induced animals both after ischemia and after 4 h of reperfusion were significantly higher than in non-induced animals (see Fig. 37), which corresponded to a higher survival rate in PB-induced rats (47% versus 15% in the control). This observation suggests that the pre-ischemic increase in the membrane enzyme content or activity, if their tolerance to ischemia remaining the same (or diminishing), may provide one of the ways for anti-ischemic preventive treatment of an organ.

2.4 Conclusions

As can be seen from the above findings, there is currently a body of experimental data attesting to early and substantial damage of subcellular and cell membranes in the course of ischemia, as well as to the role of membrane damage in the loss of structural and functional integrity by ischemic and reperfused organs.

Here are some key points.

(1) The barrier function, one of the basic and universal functions of cell membrane structures, is disturbed early in the course of organ ischemia. As a result, some organelles and the cell as a whole lose the capacity to maintain individuality of their composition, which is an absolute prerequisite for normal functioning. The membranes' barrier function is disturbed first in relation to univalent ions and water, then with respect to bivalent ions (mainly calcium) and, if damage is severe (often morphologically evident), to high-molecular compounds (proteins, intermediates, myoglobin, etc.).

The earliest changes are observed in mitochondrial membranes, then in ER (SR) membranes and nuclei, later on - in the plasma membrane, and the latest - in lysosomal membranes.

(2) During ischemia of an organ, the total membrane PL content decreases and their composition considerably changes, the degree and onset of which depend on the severity and duration of ischemia, sensitivity of the affected organ and kind of membrane.

The sharpest drop is observed in PL containing PUFA (PEA, PC, cardiolipin), with a corresponding decrease in the level of PL-associated PUFA. Selective delipidization of membranes leads to an increase of the cholesterol/PL ratio, alters the membrane physicochemical properties and lowers their fluidity (microviscosity), which apparently, is a major factor in the disturbance of the membranes' barrier and structural functions.

(3) During ischemia, the functional activity of membrane proteins changes (possibly increases for a short period, then drops down to complete inhibition). This is due to a number of mechanisms (see Chapters 3-5) of which the decreasing total and individual PL content and altered physicochemical properties of membrane lipids are more important, because the barrier dysfunction of the lipid bilayer precedes a decrease in the functioning of lipid-dependent proteins. Indeed, correlations between the decreasing level of membrane lipids and dysfunction of some lipid-dependent membrane proteins (Na, K-ATPase and Ca-ATPase of the sarcolemma, cytochrome P-450 and glucose-6-phosphatase in the liver's ER, mitochondrial enzymes, etc.), have been clearly demonstrated in ischemia.

(4) Reperfusion of long ischemic organs, or reoxygenation of organs after prolonged hypoxia aggravate virtually all membrane injuries: it promotes K^+ escape from the cell and influx of Na^+ and water into it, considerably intensifies Ca^{2+} influx to the cell, enhances degradation of membrane PL and further delipidization of membranes. It also enhances membrane permeability to proteins (enzymes), which cross the cell membrane into the blood or perfusate.

The degree of reperfusion damage to membranes depends on the duration of ischemia; the irreversibility of ischemic injury in an organ is characterized, according to most authors, by two main phenomena: (1) the loss by cell membranes of their barrier function against Ca^{2+} and substantial Ca^{2+} influx into the cell [Chien K.R., Abrams J., Serroni A. et al.; Hearse D.J.; Jennings R.B. and Reimer K.A. (1981); Nayler W. G., Ferrari R., Poole-Wilson P. A. et al.; Nayler W.G., Ferrari R., Poole-Wilson P.A. et al.; Nayler W.G. and Poole-Wilson P.A.; Poole-Wilson P.A., Harding D.P., Bourdillon P.D.V. et al.]; and (2) Ca^{2+} accumulation in mitochondria leading to their failure to restore their function [Farber J.L., Chien K.R and Mittnacht S.; Vogt M.T. and Farber E.] (see also Chapter 3).

The less commonly accepted criteria of membrane damage irreversibility include also such changes as loss by membranes of their barrier function against high-molecular proteins, the cell's inability to regulate homeostasis with respect to K^+, water balance, and volume [Hirche H., Bissig R., Friedrich R. et al.], pronounced degradation of membrane PL and delipidization of membranes [Farber J.L., Chien K.R and Mittnacht S.].

However, all the described kinds of membrane injury may be local rather than extensive (disseminated), which must be taken into account since local injury leads to membrane disintegration in only some organelles or cells, which can be compensated by intensification of reparative processes in intact organelles, whereas in the case of extensive damage, compensating responses either do not take place at all or end in decompensation and death of the remaining mildly damaged (or undamaged) components. Since quantitative analysis of irreversible changes in membranes is difficult and not always possible by morphological, chemical, or biophysical indices, the membrane failure to restore their specialized functions in the postischemic period still remains, in our opinion, the principal criterion of membrane damage irreversibility.

(5) The described changes in membranes, leading, after long period of ischemia, to irreversible changes in the cell and its death after short period of ischemia, are adaptive and have biologically sound. Moderate escape of K^+ from the myocardiocyte into extracellular space brings about partial depolarization of the sarcolemma and reduces the electrogenic and non- electrogenic Ca^{2+} influx into the cell, which lowers contractility and save energy [see Kapelko V. I., Gorina M. S. and Novikova N. A.]; moderate PL degradation, lysoPL and FFA accumulation without their escape from membranes facilitate the action of lipid-dependent enzymes (see Chapter 4); the enhanced cell membrane permeability to H^+ tends to depress intracellular acidosis [Gatsura V.V. and Sidorenko A.F.], etc. After ischemia of long duration these changes assume a pathologic aspect, frequently ending in cell death.

(6) The most challenging issue here is elucidation of the molecular mechanisms of membrane damage. The following mechanisms are of the greatest importance: (1) energy deficiency; (2) hydrolysis of membrane lipids with endogenous lipases and phospholipases (including lysosomal ones); (3) chemical modification of membrane lipids as a result of intensification of LPO processes.

Each of these mechanisms is discussed in Chapters 3, 4 and 5.

Chapter 3

DISORDERS OF ADENINE NUCLEOTIDE METABOLISM DURING ACUTE ISCHEMIA AND AFTER REPERFUSION OF ORGANS

Most processes providing for the functional activity of organs and maintaining the structural integrity and functions of cell and intracellular membranes are energy-dependent. Hence, the normal state of systems responsible for the synthesis of high-energy phosphates in the cell, their transport to target sites, and energy consumption at these sites is prerequisite to the structural and functional health of all cells and organs.

The most universal energy sources in mammalian cells are high-energy phosphates such as adenine nucleotides (ADN), primarily adenosine triphosphate (ATP). The latter is capable of donating phosphate or pyrophosphate groups during hydrolysis (ortho- and pyrophosphate pathways, respectively), thus releasing chemical energy which is easily converted into other forms of energy (thermal, mechanical, etc.) necessary for cell functioning. Losing these groups one by one, ATP is transformed into adenosine diphosphate (ADP) and monophosphate (AMP), which also are energy-rich.

Less universal energy sources include high-energy triphosphates with nitrogenous bases other than adenine, creatine phosphate, reduced forms of NAD, NADP and flavin adenine dinucleotide (FAD), and other phosphorylated compounds, both energy-rich (1,3-diphosphoglycerate, phospho(enol)pyruvate, acetyl phosphate, arginine phosphate, acetyl-CoA, etc.) and with lower energy values (glucose-1-phosphate, fructose-6-phosphate, glucose-6-phosphate, glycerol-2-phosphate, etc.). Energy contents of all these compounds, as well as energy stored in the form of substrates (glucose, glycogen, FFA, etc.), may be converted into the energy of ATP. Therefore, in addition to the amount of ATP, the contents of substances involved in its resynthesis (especially ADP, AMP, and nucleosides) also determine energy supply to the cell and serve as indicators of intracellular energy accumulation and consumption in tissues.

The study of synthesis, transport, and consumption of high-energy phosphates in biological systems is one of the cardinal problems in biology, extensively analyzed and discussed in scientific literature. This chapter will only deal with a few aspects of this problem that are important for understanding the role of disorders of adenine nucleotide metabolism in ischemic and reperfusion injuries of various organs.

3.1 DISORDERS OF ADENINE NUCLEOTIDE PRODUCTION, TRANSPORT, AND UTILIZATION

3.1.1 Disturbances of ATP Production Pathways

Disturbances of ATP production pathways are one of the earliest signs of ischemia that, at the same time, persists for a rather long period of time after the reperfusion of organs. There is convincing evidence that different processes occurring in the cell consume ATP of different origin, i.e., ATP origination and utilization is distinctly compartmentalized [Nunnally R.L. and Hollis D.P.], the question as to which pathways of ATP synthesis are damaged and which are temporarily intensified to compensate damage or can be intensified with the aid of drugs acquires particular importance.

In terms of origin and consumption, there are several categories of ATP differing in their roles under normal physiological conditions and during ischemia (Table 4).

Respiratory ATP, produced at inner mitochondrial membranes via oxidative phosphorylation of carbohydrate, lipid, and protein breakdown products (Fig. 20), is the most effective form of energy accumulation and prevails over other categories of ATP in most organs under normal conditions.

Fig. 20: Changes in the cell respiration pathways during ischemia, according to [Lehninger A.; Musil J.] with modifications.

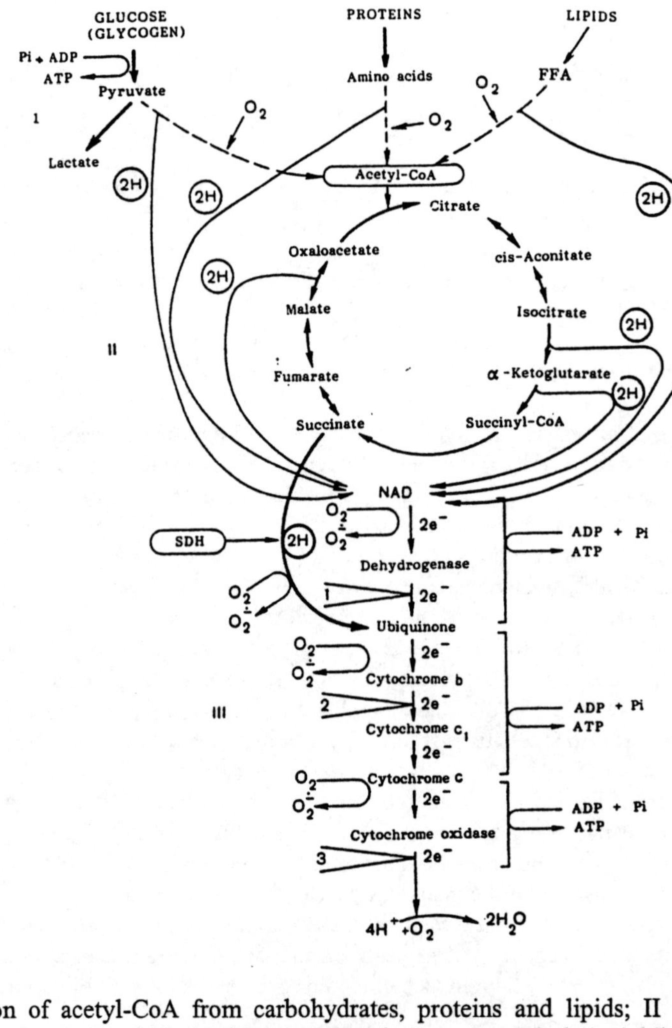

I - formation of acetyl-CoA from carbohydrates, proteins and lipids; II - Krebs cycle; III - electron-transfer chain and oxidative phosphorylation; sites of chain inhibition (1-3) with: 1 - rotenone; 2 - antimycine A; 3 - cyanide. Pathways enhanced during ischemia are marked by solid lines; pathways weakened during ischemia are marked by dashed lines

Table 4: Main pathways of high-energy phosphate production and sites of their consumption in organs under physiological conditions and during ischemia.

Pathway	High-energy phosphate production					
	Site	Mechanism	Resulting compounds	Physiological conditions	Ischemia	High-energy phosphate consumption
Aerobic	Inner MX membrane	Respiratory phosphory-lation	ATP (respira- Tory and creatine kinase's)	Myocardium, red skeletal muscles, renal cortex, cerebral cortex, liver	During partial ischemia decreased, during total ischemia stoped.	Na, K-ATPase, MX and plasma membranes, Ca-ATPase of SR membrane
	Cytosol	Adenylate kinase reaction	ATP (adenylate kinase)	All organs	Temporarily increased, then stoped	The same
	ER membranes	Oxidation of glucose in HMPS	Other high-energy compounds	Renal medulla, liver breast blood and tissue phagocytes	The same	On local synthetic reactions, Na, K-ATPase of plasma membranes, Ca-ATPase in conducting myocardiocytes
Anaerobic	Cytosol (ER membranes)	Glycolysis	ATP (glycolytic)	White skeletal muscles, renal medulla blood and tissue phagocytes, endocrine glands	The same	Ca-ATPase of SR, NA, K-ATPase in conducting myocardiocytes
	MX, Krebs cycle	Substrate phosphorylation of α-ketoglutarate	ATP (substrate)	Most organs, more intensive in the kidneys, liver, heart	Increased, especially in the presence of substrates	Electrolyte transport
	Inner MX membrane, Krebs cycle	Reduction of fumarate to succinate	ATP (respiratory)	The same	The same	The same
	All membranes	Dephosphory- lation of membrane proteins	ATP, other high-energy compounds	All organs	Temporarily increased	For local ATPases

At the same time, the respiratory pathway of ATP synthesis is the most sensitive to O_2 deficiency: it suffers severe damage during hypoxia or partial ischemia and is blocked during total ischemia of organs.

The basic factors responsible for disturbances of oxidative phosphorylation during hypoxia or ischemia of organs include a pO_2 decrease below the critical level (see Chapter 1), a decrease in the contents of oxidation substrates or in the ability of the tissue to extract them from blood, the increased level of reduction in the respiratory chain and higher NADH/NAD and NADPH/NADP ratios, structural damage to mitochondrial membranes, and reduced activities of mitochondrial enzymes.

It has been found that the affinity of cytochrome oxidase, the terminal oxidase of the respiratory chain, to oxygen is very high ($K_m = 10^{-7}$–10^{-6} M), and that the earliest changes in the respiratory chain (its block) affect its initial NAD-dependent component [Korneev A. A., Komissarova I. A. and Nartsissov Ya. R.; Lukyanova L.D., Popova O.A., Romanova B.E. et al.; Lukyanova L.D., Balmukhanov B.S. and Ugolev A.T.; Lukacova N., Gottlieb M. and Marsala J.], which inhibits oxidation of NAD-dependent substrates and stimulates succinate oxidation in mitochondria via the succinate dehydrogenase (SDH)-dependent pathway. Different resistance of NAD- and SDH-dependent links of the respiratory chain apparently account for differences in sensitivity to O_2 deficiency among low- and highresistant animals [Belousova V.V., Dudchenko A.M. and Luk'yanova L.D.; Dudchenko A.M., Chernobaeva G.N., Belousova V.V. et al.].

The SDH-dependent pathway of electron transfer, compared to NAD-dependent, is more tolerant to ischemia, and many intermediates of the Krebs cycle enter terminal oxidation through it, bypassing the NAD-dependent link [Klimovitskaya P. L.; Kondrashova M. N., Mayevsky E. I., Balayan G. V. et al.; Lukyanova L.D., Popova O.A., Romanova B.E. et al.; Lukyanova L.D., Balmukhanov B.S. and Ugolev A.T.; Lukacova N., Gottlieb M. and Marsala J.; Rouslin W.]. Moreover, although the number of phosphorylation sites is reduced from three to two, thermodynamic advantages of succinate provide for a better energy supply to tissues [Kondrashova M. N., Mayevsky E. I., Balayan G. V. et al.; Kondrashova M. U. and Maevsky E. I.].

Hence, the importance of the SDH-dependent pathway of respiratory ATP synthesis significantly increases during hypoxia or ischemia, and its directed intensification through the use of succinic acid or other substrates of the Krebs cycle (glutamate, fumarate, etc.) as the oxidation substrate is one of the effective methods for protecting organs against ischemia [Gatsura V.V. and Pichugin V.V.; Kondrashova M. N., Mayevsky E. I., Balayan G. V. et al.; Kondrashova M. U. and Maevsky E. I.; Lukyanova L.D., Balmukhanov B.S. and Ugolev A.T.].

Drastic changes during ischemia also occur at the substrate stage of the pathways for oxidative ATP production, which involves acetyl-CoA synthesis from various substrates and its entry into the Krebs cycle (see Fig. 20). The amount of pyruvate available for acetyl-CoA synthesis decreases owing to intensified glycolysis, and that of FAA, to their retarded (or terminated) β-oxidation. Combined with temporary intensification of protein catabolism, these factors lead to the reduction or cessation of acetyl CoA supply to the Krebs cycle [Shumakov V.I., Onishchenko N.A. and Shtengol'd E.S.].

These changes have a strong effect on the production of respiratory ATP because in some tissues (myocardium, red skeletal and smooth vascular muscles) FFA are normally the preferable substrate for acetyl-CoA synthesis [Olbinskaya L.I. and Litvitsky P.F.; Saks V.A. (1983)], and the energy value of this compound differs depending on the substrate for its synthesis, being maximal in the case of its formation through FFA oxidation [Kondrashova M. U. and Maevsky E. I.].

Respiratory ATP production in the early postischemic period is also strongly inhibited because of mitochondrial membrane damage and reduced activities of the membrane-dependent enzyme ensembles, which appear during ischemia and become most pronounced in the early reperfusion period, combined with the excessive Ca^{2+} accumulation in mitochondria after ischemia (see Chapter 2). Hence, respiratory ATP deficiency persists for a long time after ischemia.

The second aerobic pathway of energy production—*glucose oxidation in the hexose monophosphate shunt (HMPS)*—does not lead to ATP formation. Instead, energy is accumulated in the form of compounds reducing NADP to NADPH to produce pentoses and several low-energy phosphorylated compounds consumed at the site of their formation primarily for the maintenance of biosynthetic processes, e. g., the synthesis of nucleic acids, nucleotides, steroids, and FA, the functioning of Na, K-ATPase and Ca-ATPase, in conducting cardiomyocytes. The regulatory factor in this pathway is the decrease of the NADPH/NADP ratio, and the key enzyme is glucose-6-phosphate dehydrogenase (G-6-PDH). During ischemia (LAD occlusion in dogs), G-6-PDH activity was found to increase by a factor of 23–24 by day 10 of occlusion [Gudbjarnason S., Cowan Ch., Braasch V. and Bing R.J.], which appears to reflect both the intensification of HMPS reactions and, to a greater extent, the accumulation in the AMI focus of blood phagocytes in which HMPS is more resistant to hypoxia than in other cells (see Chapter 10).

During total ischemia, the block of this energy synthesis pathway also depends on pO_2 decrease to the critical level, which leads to NADPH and pentose deficiency, inhibition of structural and biosynthetic processes in the cell, and disturbances of K^+ transport. Reorientation of energy synthesis toward the HMPS pathway is one of the methods for enhancing reparative processes during partial ischemia or in the early postischemic period.

Glycolytic ATP is produced as the result of anaerobic glucose (glycogen) oxidation via the Embden-Meyerhof pathway. The latter is thermodynamically inefficient, and its role under normal conditions is substantial in only a few tissues, including kidney medulla, white skeletal muscles, blood cells, and endocrine glands. Glycolytic enzymes are apparently located on ER membranes, and ATP synthesized in the heart is mainly used to support Ca-ATPase functioning in SR membranes [Humphrey S.M., Gavin J.B. and Herdson P.B. (1980)] and, possibly, to supply energy to the cardiac conduction system [Meerson F.Z.]. According to published data, glycolytic ATP also provides energy for Na, K-ATPase in the liver [Cascarano J. and Seidman I.] and smooth muscle cells [Soloviev A.I. and Bershtein S.A.]; moreover, this ATP produced from endogenous glycogen maintains membrane functions (in particular, membrane potential) more effectively than that produced from exogenous glucose [Cowan J.C. and Vaughan Williams E.M.].

In the course of ischemia, the dynamics of glycolysis has the following three stages: (1) initial rapid intensification, (2) subsequent decrease, and (3) cessation. The first stage, at which the rate of glycolysis in the heart and the liver increases approximately 20-fold [Hasselgren P., Jehnische E., Fornander J. and Hellman A.], is a compensatory process alternative to oxidative phosphorylation. It is conditioned by the removal of the inhibitory effect of ATP, which is superceded by the activating influence of its degradation products, and the local release of catecholamines (CA), cyclic AMP (cAMP) formation, and Ca^{2+} accumulation in the cytosol [Jennings R.B., Reimer K.A., Hill M.L. and Mayer S.E.]. Another regulatory factor of glycolysis intensification during ischemia is a decrease in creatine phosphate (CP) content, which stimulates an increase in ADP and AMP concentrations and the activity of phosphofructose [Lyzlova S.N., Stefanov B.E. and Taame N.; Saks V.A. (1983)].

At this stage, glycolysis is intensified owing to activation of the glycolytic enzyme system [Dolgova N. P.]; however, even during the period of its highest activity, when glycolysis accounts for up to 80% of the total energy supply, it satisfies the requirements of basal metabolism by no more than one- to two-thirds [Shumakov V.I., Onishchenko N.A. and Shtengol'd E.S.; Shumakov V.I., Onishchenko N.A. and Kirpatovsky V.I.; Cunarro J.A., Schults S.E., Johnson W.A. and Weiner M.W.] and at the same time leads to lactate accumulation in tissues and the consequent sharp decrease of their pH.

The time of transition from this to the subsequent stages of glycolysis is organ-specific and depends on the severity of ischemia and temperature at which it occurs. At 37°C, this time is usually 30–40 min, although it may vary from 20 to 90 min [Dolgova N. P.; Onishchenko N.A., Kirpatovsky V.I., Artamonov S.D. et al.; Shumakov V.I., Onishchenko N.A. and Shtengol'd E.S.; Shumakov V.I., Onishchenko N.A. and Kirpatovsky V.I.; Jennings R.B., Reimer K.A., Hill M.L. and Mayer S.E.].

In our experiments on isolated dog kidneys kept at 20°C [Bilenko M. V. (1978a); Klimovitskaya P. L.], the intensification of glycolysis (estimated from lactate accumulation) was observed for the first 15–60 min (stage 1); thereafter, its activity decreased (60–120 min, stage 2) and eventually disappeared after 120 min (stage 3) (Fig. 21).

Glycolysis retardation and cessation depend on several factors of which the most important are as follows:

(1) Decreasing activity of polyenzymatic glycolytic systems, primarily phosphofructokinase (the key enzyme), phosphoglucomutase, and hexokinase. Until recently, this had been attributed to substrate inhibition by the end products of hydrolysis, the accumulation of reduced NADH and NADPH, and acidosis [Hochachka P.W.; Jennings R.B., Reimer K.A., Hill M.L. and Mayer S.E.]. However, the probable association of glycolytic enzyme systems with ER membranes suggests that their damage may result from structural and functional disturbances in these membranes.

(2) A decrease in the tissue content of ATP, the substance necessary at early stages of glycolysis. Jennings *et al.* [Jennings R.B., Reimer K.A., Hill M.L. and Mayer S.E.] concluded that this factor is crucial because the cessation of glycolysis after 90 min of cardiac ischemia occurred at an ATP-dependent stage (fructose-6-phosphate conversion into fructose-1,6-diphosphate) and coincided with tissue ATP decrease

below the "critical" level. Experiments showed that glycolysis ceases completely when the ATP content decreases to 1–2% of the initial.

(3) Depleted reserves of glycolysis substrates (glucose, glycogen) and cofactors [Bilenko M. V. (1978a); Shumakov V.I., Onishchenko N.A. and Kirpatovsky V.I.]. Apparently, this is not a crucial factor because even after 40 and 90 min of total ischemia, the dog heart still retains 62 and 49% of glycogen, respectively [Shumakov V.I., Onishchenko N.A. and Kirpatovsky V.I.], and, according to our experimental data, dog kidneys exposed to ischemia for two hours retain about 30% of glycogen (see Fig. 21). The cessation of glycolysis against the background of relatively high contents of substrates in the heart was also observed by other researchers [Dolgova N. P.; Jennings R.B., Reimer K.A., Hill M.L. and Mayer S.E.].

Fig. 21: Lactate (a) and glycogen (b) contents in the canine renal cortex after different time of complete ischemia in vitro (isolated storage at 20°C) (1) and after 6 h of reperfusion (2) (% of the initial level).

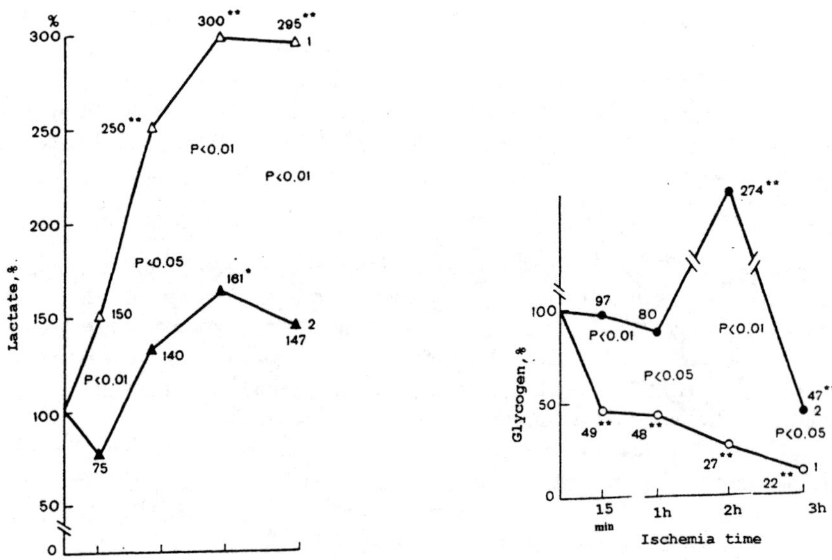

*-P < 0.05; **-P < 0.01, versus initial level; P – 1 versus 2. The reperfusion was performed by extracorporeal connection kidney vessels to femoral vessels of recipient dog [Kovanov V. V., Bilenko M. V., Amosov I. S. et al.].

In the early reperfusion period, glycolytic ATP synthesis is restored and even proceeds more intensely for a certain time. In our experiments (see Fig. 19), lactate content in the cortex of ischemic kidneys decreased after 6 h of their reperfusion; however, its return to values approaching the initial level was only observed after ischemia lasting for 15–60 min. After 2 and 3 h of ischemia, lactate content in the renal cortex remained markedly excessive.

Along with continuing glycolysis, organs in the postischemic period may demonstrate active energy storage in the form of glycogen (glyconeogenesis), which is untypical for normal organs. Figure 21 shows that the content of glycogen in the cortex of dog kidneys after 2 h of ischemia and 6 h of reperfusion was 10 times that during ischemia and almost 3 times higher than the initial level. Excessive glycogen accumulation during the postischemic period was also observed in the dog intestine [Saburova L.M. and Berezov T.T.] and the rat heart [Meerson F.Z.] but not in dog skeletal muscles [Selivanova L.P. and

Saburova L.M.]. Such an excessive accumulation may be explained by the ability of CP to switch metabolic pathways from glycolysis to glyconeogenesis [Lyzlova S.N., Stefanov B.E. and Taame N.].

During ischemia and the early reperfusion period, tissues also gradually lose the ability *to ATP synthesis in the adenylate kinase reaction.* The latter, described by the formula AMP + ATP ADP, is short-term and energy-dependent. Moreover, two additional anaerobic pathways of ATP formation in mitochondria, less common for tissues of higher animals, become intensified during ischemia: (1) anaerobic *substrate phosphorylation* through α-ketoglutarate decarboxylation with the formation of GTP and then ATP [Randall H.M. and Cohen J.J.] and (2) anaerobic formation of respiratory ATP through *fumarate reduction to succinate* with NADPH oxidation and partial restoration of the respiratory chain [Runeberg L. and Pakarinen A.].

These pathways of anaerobic ATP synthesis were analyzed in experiments with mitochondria, submitochondrial particles, homogenates, and tissue sections of the kidney, liver, and heart [see Gronow G.H.J. and Cohen J.J.]. Reactions of the first pathway lead to the production of high-energy phosphates at a rate of about one-half that in the case of glycolysis; the efficiency of the second pathway reaches 2–6% of the maximal aerobic α-ketoglutarate oxidation. This energy appears to be expended for ion transport, which follows from the fact that K^+ reaccumulation in liver tissue sections under anaerobic conditions proceeds in the presence of oxalacetate and α-ketoglutarate but not glucose [Cascarano J. and Seidman I.].

Reactions of these pathways intensified by introducing corresponding substrates into an anoxic (more correctly, hypoxic; see Chapter 1) perfusate proved to be sufficient for maintaining certain functions of an isolated rat kidney [Gronow G.H.J. and Cohen J.J.], but only their combination with intensified glycolysis provided for the maintenance of contraction of the isolated rat heart [Cascarano J., Chick W.L. and Seidmann I.].

Ischemia also stimulates reactions of the *anaerobic pathway of ADP phosphorylation by high-energy phosphate groups of membrane phosphoproteins* [Vale M.G., Osorio V.R. and Carvalho A.P.] *or phospholipids* (diphosphoinositide, triphosphoinositide) [Kislev G. V., Pavlinova Z. I. and Chetverikov D. A.]. Performing a compensatory function in the early ischemic period and apparently providing an additional amount of ATP energy for membrane functions (primarily cation transport), such dephosphorylation (energy depletion) of membranes eventually leads to membrane damage (see below). It appears that under ATP and CP deficiencies, membranes may be depleted of energy through the degradation (consumption) of ATP bound to the sarcolemma of cardiomyocytes in the form of the Ca^{2+}–ATP–protein complex, and the addition of exogenous CP considerably retards this degradation, thus protecting the membranes [Preobrazhensky A.N., Dzhavadov S.A. and Saks V.A.].

Thus, ischemia causes drastic changes in virtually all biosynthetic pathways leading to the formation of high-energy compounds (see Table 4): the rates of ATP synthesis via the respiratory pathway and the adenylate kinase reaction decrease or cease completely, and the same applies to the synthesis of high-energy compounds in the HMPS; glycolytic ATP synthesis temporarily becomes more intense and then ceases as well; some tissues demonstrate a certain increase in respiratory ATP production in mitochondria via fumarate reduction to succinate and in the substrate phosphorylation of α-ketoglutarate; dephosphorylation of membrane structures in ischemic organs is initially a compensatory process but subsequently has a damaging effect.

During the reperfusion period, the time and extent of recovely of the damaged energy formation pathways depend on the severity and duration of ischemia, which determine the degree of postischemic energy deficiency.

3.1.2 Disturbances in ATP Transport Pathways

Disturbances in ATP transport pathways is another sign of early and severe ischemia.

As follows from the previous section and Table 4, most of synthesized ATP is consumed at the site of its formation (by the "neighborhood" principle). The only exception is respiratory ATP, which is synthesized in mitochondria and consumed by energy-dependent systems outside these organelles. Its transport from mitochondria to the sites of ATP consumption in the myocardium and other excitable tissues, particularly the brain, is implemented by ATP–ADP translocase, adenylate kinase, and the creatine kinase enzyme system comprising four isozymes (MM, BB, MB, KK) [Saks V.A. (1983); Saks V.A. (1985)].

The essence of this process is that exchanging ATP for equimolar amounts of ADP, the carrier (ATP-ADP translocase) does not transport its molecules through both mitochondrial membranes but leaves them on the outer surface of the inner membrane where ATP reacts with creatine (Cr). This reaction is catalyzed by creatine kinase (CK) isozyme KK, which is specific to mitochondria. The resulting creatine phosphate (CP) freely diffuses through the outer mitochondrial membrane into the cytoplasm and, near consumers (myofibrils exhibiting ATPase activity, Na, K-ATPase in the sarcolemma, and Ca-ATPase in the SR), reacts with ADP, producing ATP and liberating Cr, which returns into mitochondria (Fig. 22).

ATP formation from CP is catalyzed by CK isozymes MM, BB, and MB [Levitsky D. O., Levchenko T. S., Saks V. A. et al.; Saks V.A. (1983); Saks V.A. (1985)], which ensures strictly directed transport of respiratory ATP and its differential consumption. It is the need for directed ATP consumption, rather than diffusion limitations (the rate of ATP diffusion in the cell is low), that probably accounts for the existence of the CK system in excitable cells [Saks V.A. (1983)]. There is also cytoplasmic CK, which similarly provides for the directed transport of glycolytic ATP to the sites occupied by Na, K-ATPase in the plasma membrane, Ca-ATPase in SR membranes, and myosin ATPase in myofibrils (see Fig. 22), and regulates the intensity of glycolysis under ischemic or anoxic conditions [Saks V.A. (1983)].

According to certain data, the CK system and intracellular energy transport, in addition to cells of excitable tissues, are also typical (altough to a lesser extent) for other cell types, including leukocytes, adipose cells, synaptosomes and nuclei of cardiac and muscle cells [Lyzlova S.N., Stefanov B.E. and Taame N.]; in other words, this is a universal system of the organism.

Fig. 22: Phosphocreatine pathway of intracellular energy transport in cardiocytes [Saks V.A. (1983)]

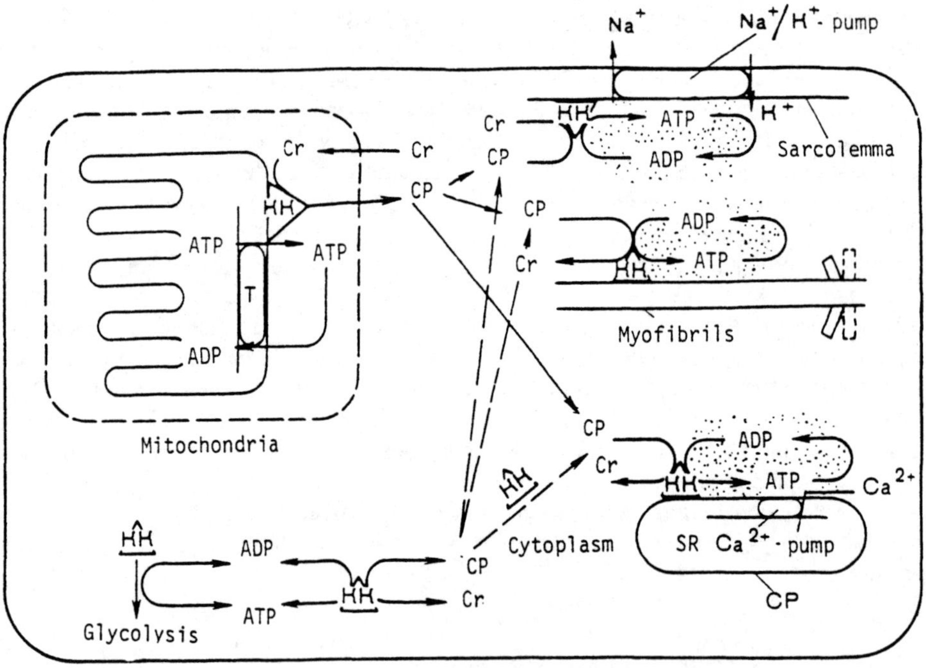

Cr-creatine, CK-creatine kinase, CP-creatine phosphate, KK-mitochondrial isoenzyme of CK, T-ATP-ADP translocase, a-inner and b-outer mitochondrial membrane.

During ischemia, the content of CP decreases much earlier than that of ATP (see 3.2). This is not due to the expenditure of CP as an independent energy source, as was believed earlier; the cause is rapid CP consumption for ATP synthesis [Saks V.A. (1983)] and termination of its subsequent transport from mitochondria. The latter results from the low resistance of ATP–ADP translocase to hypoxia and ischemia

[Farber J.L., Chien K.R and Mittnacht S.]. In the heart, ischemia lasting for 15 min leads to a 80% decrease in ATP–ADP translocase activity [Asimakis G.K. and Conti V.R.]. The activity of the CK enzyme system also decreases rapidly [Krasnova E. I., Minin P. P. and Zavodskaya I. S.; Asimakis G.K. and Conti V.R.; Takeo S. and Sakanashi M.], and during partial ischemia and in the early reperfusion period, corresponding enzymes (especially the soluble isozyme MB) pass through damaged mitochondrial and cell membranes into the extracellular fluid and blood (see Chapter 2). Ischemia also leads to a decrease in the content of Cr, which operates as a shuttle in energy transport, because of its high consumption, impaired or blocked transport from blood, and outflow in blood [Davydov V. V., Tverdokhleb V. P. and Yakushev V. S.]

ATP transport systems often become damaged earlier than ATP production pathways, which results in the uncoupling of ATP synthesis and consumption sites. This fact partially explains why irreversible injuries of organs can appear against the background of a relatively high ADN level (see below).

3.1.3 Disturbances of ATP Consumption by Tissues

Disturbances of ATP consumption by tissues also arise both during ischemia and in the early postischemic period.

Energy demand of organs under normal conditions depends on a number of factors.

(1) The amount of energy consumed by an organ is determined by its functional state. An actively working organ needs the amount of energy corresponding to its *functional consumption*, and a resting organ, to the *basal consumption*. The ratio between these parameters varies from one organ to another, and basal consumption may be regarded as practically equal to the amount of energy necessary for maintaining the structural and functional integrity of biological membranes as a prerequisite to cell homeostasis.

In the kidneys, functional consumption (tubular secretion, reabsorption of K^+, Na^+, Cl^-, protein, and other substances) accounts for 81%, and basal consumption, for only 19% of the total amount of expended energy [Shumakov V.I., Onishchenko N.A. and Shtengol'd E.S.].

In the heart, basal consumption reaches 10–30% of the total energy supply, the rest being expended to maintain cardiac contraction [Opie L.H.]. This is in agreement with calculations [Artamonov S. D. and Onishchenko N. A.] showing that the intact rat heart expends a greater part of energy for mechanical work (39%) and the maintenance of cyclic Ca^{2+} fluxes associated with it (45.5%), whereas expenditures for other purposes are much lower: the maintenance of K^+ and Na^+ gradients, 5–10%; protein and lipid synthesis *de novo*, 1–2%; and metabolic regulation, less than 3%. According to [Bershtein S.A., Gurevich M.I. and Soloviev A.N.], electrogenic ionic pumps in excitable tissues consume less than 20% of the total energy supply. Note that the guinea pig hearts with initially higher functional activity and the hypertrophic dog hearts proved to be more sensitive to ischemia than control organs [Mogilevskii G.M., Sharov V.G., Severin V.V. and Semenovskii M.L.; Veksler V.I., Khatkerich A.N. and Kapel'ko V.I.].

The energy budget of reticulocytes is as follows: synthetic processes (globin and heme synthesis), 28.5%; ion transport, 25.6% (Ca^{2+}, 2.6%; $Na^+ + K^+$, 23%), and proteolysis (degradation of mitochondria), 18.5% [Siems W., Dubiel W., Dumdey R. et al.].

(2) Energy demand for basal consumption depends on the rate of metabolic processes in the organ. By this criterion, organs form the following decreasing series: kidneys > gastrointestinal tract > heart > liver > skeletal muscles in dogs; kidney > heart > liver > brain > lungs > skin in rats; and heart > kidneys > brain > liver > skin > skeletal muscles in man [see Bershtein S.A., Gurevich M.I. and Soloviev A.N.].

These series do not exactly correspond to the gradient of organ sensitivity to ischemia (the criterion being the duration of reversible ischemia, see Chapter 1), which indicates that energy deficiency is not the only factor determining the severity of ischemic injury in organs.

(3) Requirements for energy within the same organ differ depending on the metabolic activity of its structural components. Thus, renal cortex requires more energy than medulla; gray matter of the brain, more than white matter; the inner (subendocardial) layer of cardiac tissue, more than the outer (subepicardial) layer; red skeletal muscles, more than white muscles; parenchymal cells of all organs, more than connective tissue cells; and functionally active structural units require more energy than resting units. The latter relation may account for the facts that organs consuming small amounts of O_2 during the preischemic period are more tolerant (resistant) to ischemia, and drugs reducing the metabolic activity of organs before ischemia produce the anti-ischemic effect [Shumakov V.I. and Onishchenko N.A.,; Artamonov S. D. and Onishchenko N. A.].

The *amount of energy consumed by an ischemic organ* broadly varies and largely depends on how quickly and completely the organ ceases to perform specialized functions and switches from the functional to basal energy consumption. At the same time, the above data on energy demand for basal consumption in cell and membrane structures only apply to intact organs. Membrane damage in the course of ischemia should inevitably lead to higher energy expenditures for ion transport (see Chapter 2), biosynthesis, and other processes; in other words, the calculated energy demand appears to be underestimated, and the experimental assessment of this parameter in ischemic organs is as yet methodologically difficult.

The rate of energy consumption by organs decreases with an increase in the duration of ischemia. This follows from lower ATP utilization (see below), which largely results from a decrease in the intensity of synthetic processes in the organ and the activity of ATPase during ischemia. The decrease of activity after AMI was observed for Na, K-ATPase in the sarcolemma, mitochondrial ATPase, and myosin ATPase [Davydov V. V., Tverdokhleb V. P. and Yakushev V. S.; Olbinskaya L.I. and Litvitsky P.F.] during total ischemia of the heart and skeletal muscles, for Mg-ATPase and Ca-ATPase in the SR [Arkhipenko Yu. V., Bilenko M. V., Dobrina S. K. et al.; Bulgakov V.G. and Bilenko M.V.; Meerson F.Z.; Shkolovoy V.V.] and myosin ATPase [Maudal A.K. and Miller J.]. In proximal tubule segments of the rat kidney, after 45 min of ischemia and 15 min of reperfusion, the energy distributed functional pump activity was reduced to 61% of the control energy. For non-transport purposes, the energy was reduced to only 78% of the control [Gaudio K.M., Thulin G., Ardito T. et al.]. During severe ischemic injuries, the activity of energy consumption systems continues to decrease in the postischemic period or remains low for a long period of time [Arkhipenko Yu. V., Bilenko M. V., Dobrina S. K. et al.; Olbinskaya L.I. and Litvitsky P.F.]. In any case, however, their activity during ischemia decreases later and is recovered in the postischemic period more rapidly than the activity of ATP-producing systems. Such a prevalence of energy consumption over the synthesis of high-energy phosphates leads to earlier depletion of intracellular ADN reserves during ischemia and their retarded restoration (or no restoration at all) after ischemia. As the dynamics of ADN stored in cells illustratively characterizes the severity of ischemic and reperfusion injuries, it deserves more detailed consideration.

3.2 Changes in Contents and Composition of Adenine Nucleotides and Their Critical Levels

3.2.1 Adenine Nucleotide Content

Adenine nucleotide content decreases for covering energy expenditures by various cell structures involves ATP hydrolysis to ADP and then to AMP, which breaks down to nucleosides and purine bases.

During the reperfusion period, ADN and ATP reserves of the cell, which are especially important in emergency situations, are restored via ADP, AMP, or CP rephosphorylation to ATP in the systems of energy supply described above (see 3.1). Naturally, this is possible only provided the enzyme systems of ATP synthesis in mitochondria remain intact, the processes of respiration and phosphorylation are still coupled, and the aforementioned ATP precursors are available in sufficient amounts.

ATP synthesis takes more time when cells are depleted of AMP. Resynthesis of AMP proceeds through *de novo* formation of purine nucleosides from precursors, condensation of the purine base, and phosphorylation of the purine nucleoside. This is only possible when the enzyme systems of purine nucleoside phosphorylation are intact, and the organ contains necessary precursors.

The rate and pattern of decrease in ADN reserves during ischemia and the rate of their restoration in the postischemic period reflect the degree of discrepancy between the synthesis and consumption of high-energy phosphates and characterize the state of energy supply systems, i.e., they serve as integral indicators of ADN metabolism. Both processes, of course, have organ specificity and depend on the severity and duration of ischemia.

During ischemia of the kidneys, ADN contents and composition change at an extremely fast rate. In our experiments, [Bilenko M. V. (1978a); Klimovitskaya P. L.], ATP content in the dog kidney cortex became significantly lower as soon as after 15 min of total ischemia and continued to gradually decrease on the first, second, and third hours (Fig. 23) in spite of the fact that the isolated kidneys were exposed to ischemia *in vitro* at 20°C. The dynamics of total ADN was close to that of ATP, the content of ADP moderately increased after 1–3 h, whereas that of AMP remained above the initial level for a long period of time. In the medulla of dog kidneys, where glycolysis is the main form of high-energy phosphate synthesis, ATP content decreased later and at a slower rate than in the cortex.

Total ischemia of the kidneys *in vivo* (at 37°C) was accompanied by an earlier and more abrupt decrease in ATP content. In the cortex of dog kidneys, the latter decreased to 30–40% of the initial after the first 30–60 s and to 15–20% after 2–5 min; in dog and rat kidneys, ATP content decreased after 10–20 min to 10–20% of the initial [Bore P.J., Paratheofanis I. and Sells R.A.; Cunarro J.A., Schults S.E., Johnson W.A. and Weiner M.W.; Vogt M.T. and Farber E.]. In experiments performed in our laboratory, the exposure of rat kidneys *in vivo* to total ischemia at 37°C resulted in a more drastic decrease in ATP content and a less pronounced increase in AMP content than the exposure of dog kidneys to ischemia at 20°C *in vitro* for the same period of time (Fig. 24) [Bilenko M.V., Volkova I.A. and Danilov S.P.; Volkova I.A.].

The degree and time of ATP resynthesis in the kidneys during the reperfusion period also depend on the duration and conditions of ischemia. The contents of ATP and total ADN in the cortex of dog kidneys after 15 min of ischemia at 20°C and 6 h of reperfusion were excessive; after 1–3 h of ischemia and 6 h of reperfusion, both parameters increased but did not reach the initial level (see Fig. 23).

Fig. 23: ATP, ADN, ADP and AMP contents in the canine renal cortex after different time of complete ischemia in vitro isolated storage at 20°C, 1) and after 6 h of reperfusion (2) (% of the initial level).

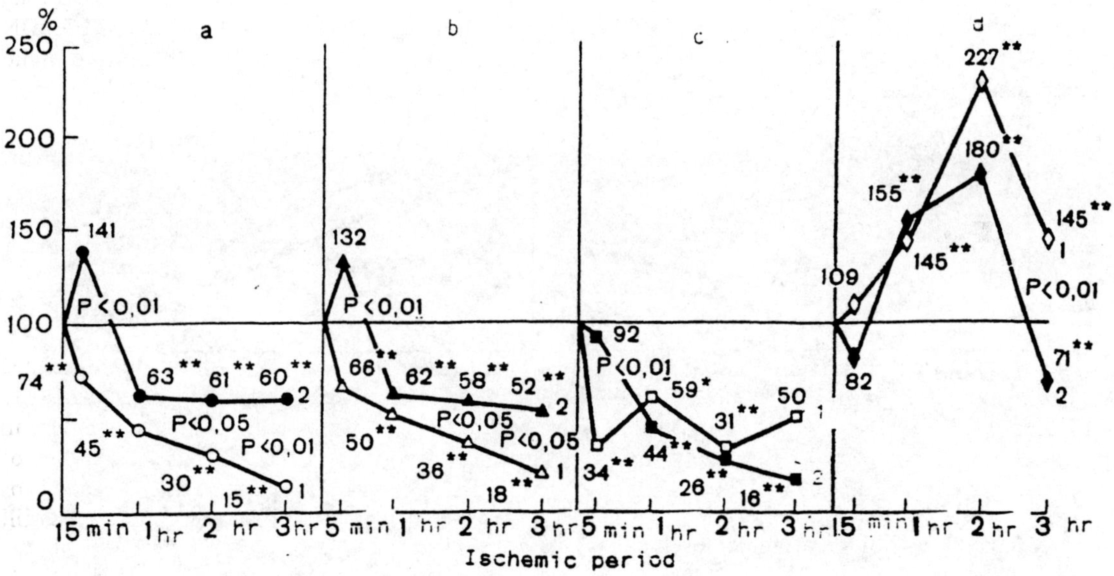

*-p<0.05; **-p<0.01 versus initial level; P-1 versus 2; The ATP content was determined by the hexokinase method; the ADP and AMP contents were determined by the pyruvate kinase method.

The rate of ATP resynthesis during reperfusion of rat kidneys after 2 h of total ischemia at 37°C was even lower (see Fig. 24): ATP content slightly exceeded that during ischemia only after 24 h of reperfusion and even after 48–72 h reached only 26% of the initial level. Unlike ATP, total ADN did not increase for the first 24 h but continued to decrease because of AMP degradation. The shift of total ADN metabolism toward degradation, rather than rephosphorylation, is a typical sign of the severity of postischemic changes in the kidneys [Bilenko M.V., Volkova I.A. and Danilov S.P.]. Slow restoration of the ADN pool in this organ after long-term exposure to total ischemia at 37°C has also been observed by other researchers [Bore P.J., Paratheofanis I. and Sells R.A.].

Changes of ADN composition ***during ischemia in the liver*** are similar to those in the kidneys. During total ischemia of the dog liver *in vivo* at 37°C, ATP content after 30, 50, and 90 min decreased to 9.6, 5.3, and 2.4% of the initial, whereas AMP content at the same time increased by factors of 9, 4.4, and 2.6, respectively [Kono Y., Ozawa K., Tanaka J. et al.]. In experiments on total ischemia of the rat liver, ATP content also sharply decreased within the first 15 min, continued to decrease during the following interval (15–30 min), and did not change significantly during the later ischemic period. Note that both ATP and total ADN pools were recovered within the first 2–3 h of reperfusion only in cases when ischemia lasted for 15 and 30 min, and this never occurred after 60 or 120 min of ischemia [Marubayashi S., Dohi K., Ezaki H. et al. (1982)].

Fig. 24: ATP(1), ADN (2), ADP (3) and AMP (4) contents in the rat renal cortex after 2 h of complete ischemia in vivo (occlusion two renal vessels plus separation kidney from perirenal fat) and different time of reperfusion (% of the initial level).

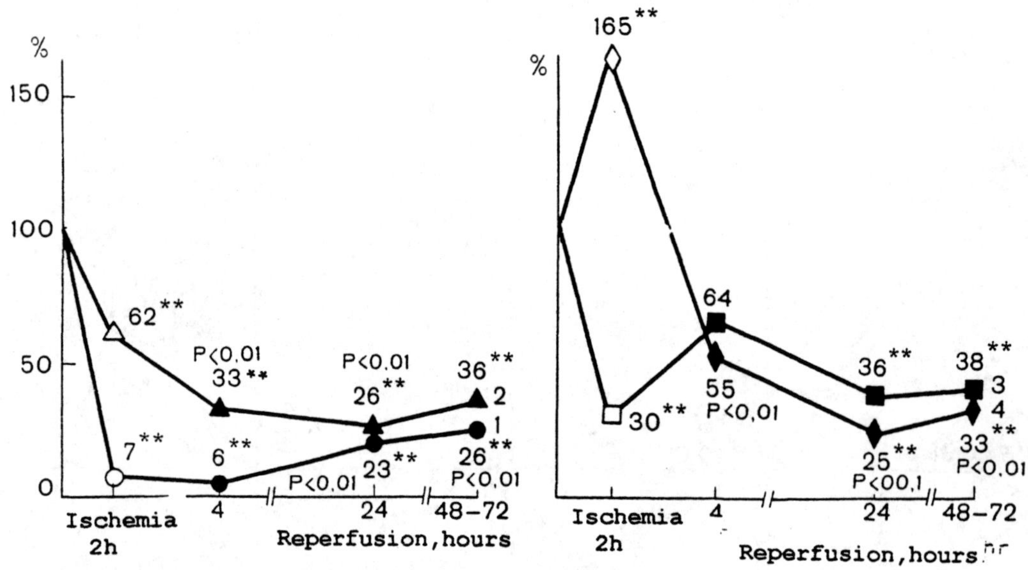

*-$p<0.05$; **-$p<0.01$ versus initial level; P-1 versus 2.

During ischemia of the brain, changes in ADN content and composition occur earlier than in other ischemic organs.

Severe partial ischemia in rats (ligation of the carotids and bleeding) led after 5 min to a rapid decrease in both CP and ATP contents (to 7.9 and 17%, respectively) and a marked increase in AMP content (by a factor of 42) [Nordström C.H. and Siesjö.Bo.K.]. During hypoxic perfusion of the dog brain (pO_2 30 and 40 mm Hg), CP concentration sharply decreased after 15–30 min, whereas ATP remained almost unchanged: after 30 min, CP and ATP contents after two different pO_2 were 0.6–4% and 73–90%, respectively [Kintner D., Fitzpatrick J.H., Louie J.A et al.]. Moderate partial ischemia of the rat brain (occlusion of the carotids) was accompanied by less pronounced changes: CP and ATP contents equaled 32 and 40% after 2 h of ischemia and simultaneously decreased to 5.8 and 10%, respectively, after 4 h; AMP content became 8–5.5 times higher [Kuwashima J., Nakamura K., Fujitani B. et al.]; after reperfusion for 1 h, ATP content was restored if the brain remained ischemic for 3 h but this did not occur after 5 h of ischemia [Yamamoto M., Shima T., Uozumi T. et al.].

During ischemia of the heart, the dynamics of ADN content and composition, compared to those in other organs, have certain peculiarities conditioned by a specific mode of ATP transport in the myocardium (see above), a high CP content in this tissue, and a relatively early decrease in the activities of ATP–ADP translocase and enzymes of the CK system, which results in the segregation of ATP stored in mitochondria from the sites of its consumption. CP content is the first to decrease during myocardial ischemia and the first to be restored during its reperfusion, whereas the reduction of ATP pool and its restoration after cardiac reperfusion occur later.

In experiments, a decrease in CP content directly correlated with the impairment of blood flow through the coronary artery. As the blood flow rate became 50, 70, and 90% lower, CP content decreased to 50, 20 and 25% of the initial level, respectively, whereas ATP and glycogen contents decreased only after the blood flow rate was reduced by 70–90%, remaining virtually unchanged after its reduction by 30–50% [Frolkis R.A.; Frolkis R.A., Voronkov G.S., Dubur G.Ya. et al.]. Reduction by 31% in coronary blood flow, within 5-10 min, led to subendocardial CP and ATP depletion by 47% and 63%, respectively [Arai A.E., Pantely G.A., Anselone C.G. et al.].

When a branch of the dog coronary artery was occluded, CP content in the myocardium decreased to 42–17% of the initial 30 s–3 min after this operation [Jennings R.B. and Reimer K.A. (1981); Jennings R.B., Reimer K.A., Hill M.L. and Mayer S.E.]; during anoxic perfusion of the isolated heart, this content became one-half lower after 5 s, whereas ATP content during this period remained unchanged.

In the dog heart exposed to partial ischemia for 15 min, . CP content in the AMI focus returned to the initial level rapidly (during 1.5 h), whereas the total recovery of the ATP pool took 7 days [Kloner R.A., De Boeh L.W.V, Darsel J, R. et al.].

As shown in our laboratory [Bilenko M.V., Otverchenko V.M., Astaf'ev V.V. et al.; Volkova I.A.] (Fig. 25), ATP content of an isolated dog heart kept at 37°C decreased at the highest rate during the first 15 min (by 42%) and twice slower during the next 15 min (by 18%); within the following 30 min, this rate slowed down again by one-half (10% decrease per 15 min), and remained very low after 1 h (3–1.5% decrease per 15 min). The dynamics of total ADN was similar to that of ATP, while ADP only decreased in the period between the first and second hours of ischemia and thereafter remained significantly below the initial level; the energy charge (EC) became lower after 30 min, and the AMP content showed a slight tendency toward increase within the first 2 h.

Storage of an isolated heart at a lower temperature (20°C) was more favorable for the maintenance of ADN level (Fig. 26): after 1 h, ATP and total ADN contents were twice as high, ADP content, 1.6 times, and AMP content, 2.7 times as high as those during the same period at 37°C. In the early reperfusion period after 1 h of total cardiac ischemia at 20 and 37°C, ATP and ADP remained at the same level as during ischemia but total ADN content continued to decrease because of AMP degradation. Similar but more pronounced changes were also observed after reperfusion of a heart exposed to total ischemia for 2 h at 37°C.

Moderate (up to 20°C) hypothermy of the heart was shown to result in an approximately similar decrease in the rates of ATP synthesis (glycolytic ATP) and energy consumption (estimated from ATPase activity), i.e., this treatment helps to reduce consumption of the deposited (reserve) high-energy phosphates.

During partial cardiac ischemia (LAD occlusion) in rats, ATP content after 10, 20, 40 and 120 min decreased to 48, 42, 36 and 25% of the initial, respectively [Olbinskaya L.I. and Litvitsky P.F.]; that is, the rate of decrease during the first 40 min was almost the same as in experiments with total cardiac ischemia in dogs (see Fig. 25), and after 120 min the amount of ATP in the AMI focus was four times greater than in our experiments with total ischemia (25% vs. 8%). ATP recovery to the initial level in the early reperfusion period was only observed if ischemia lasted for 10 min.

Fig. 25: Adenine nucleotide content (a) (% of the initial level) and energy charge (b) (in relative units) in the dog heart after different periods of complete ischemia in vitro (isolated storage at 37°C).

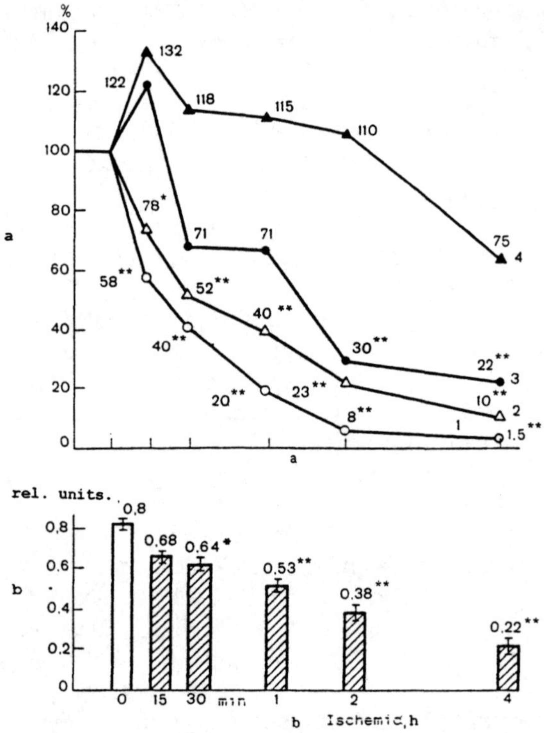

a: 1 - ATP; 2 - ADN; 3 - ADP; 4 - AMP; b: 0 - initial level. *-p<0.05; **-p<0.01 versus initial level. The energy charge (EG) was calculated according to Atkinson.

According to Jennings *et al.* [Jennings R.B., Reimer K.A., Hill M.L. and Mayer S.E.], ATP content in the AMI focus after 15, 30 and 40 min of LAD occlusion in dogs decreased to 38, 14.7 and 6% of the initial level, which approximately resembles the dynamics of ATP in the dog heart during a longer period of total ischemia *in vitro* (34, 7.8 and 1.8% after 60, 90 and 120 min). A more pronounced decrease of ATP content in AMI *in vivo*, compared to that in total ischemia of the heart *in vitro* (the decrease of ATP level after 15 min of partial ischemia *in vivo* equals that after 60 min *in vitro*), is apparently associated with the fact that the heart functioning was instantly arrested by K-cardioplegia and, hence, ATP expenditure for the maintenance of cardiac contraction was reduced.

A decrease in both CP and ATP contents in the inner myocardial layer during total cardiac ischemia in rats was more drastic than in the middle and outer layers [Fiolet J.W., Baartscheer A., Schumacher C.A. et al.]. The inhibition of mitochondrial ATPase activity sharply decelerated ATP depletion [Jennings R.B., Reimer K.A. and Steenbergen C. (1991)].

Fig. 26: ATP (a), ADP (b), ADN (c) and AMP (d) contents in dog heart, after different time of complete ischemia in vitro before(I) and after 4 h of reperfusion (II) (% of the initial level).

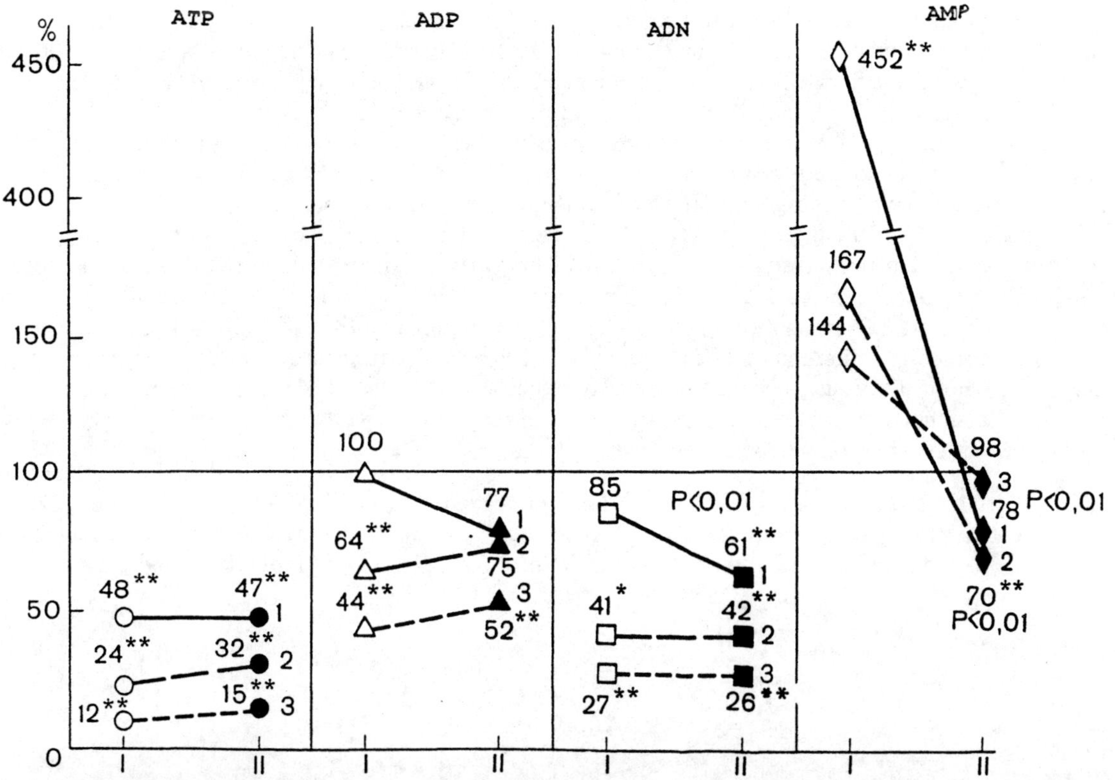

1 - 1 h isolated heart storage at 20°C (the myocardial function was restored and stable); 2 - 1 h of isolated heart storage at 37°C (the myocardial function was restored but was not stable); 3 - 2 h of isolated heart storage at 37°C (the myocardial function was not restored). *-p<0.05; **-p<0.01 versus initial level; P-significant differences before versus after reperfusion.

During ischemia of skeletal muscles, the decrease in ADN follows almost the same pattern as in the myocardium but is much slower. When tourniquets were applied to human limbs, CP content decreased to 52% of the initial after 30 min and to 40% after 60–90 min, whereas ATP remained at the same level [Haljamäe H. and Enger E.]. In experiments with this model of limb ischemia in dogs, most of CP was consumed within 1 h, whereas ATP remained at the same level for 2 h, decreased by 40% one hour later, and was exhaused after 8 h. In total limb ischemia in rats (the tourniquet ischemic model), total ADN content decreased after 2–6 h but its recovery in the reperfusion period only occurred after 2 h of ischemia and was not achieved after 4–6 h [Idström J.P., Soussi B., Elander A. and Bylund-Fellenius A.C.].

In the case of subtotal ischemia of dog limbs, ATP content decreased after 10–16 h only to 70% of the initial, AMP content became 4.8 times higher, and total ADN content remained relatively stable but sharply decreased (to 55%) 2–3 h after the blood flow was restored, as a result of further degradation of AMP [Klyachkin M. L. and Zakharova N. B.]. As a result of storage of an isolated dog limb at 18–20°C for 3 and 6 h, ATP content decreased to 24 and 10% of the initial and, in both cases, demonstrated a tendency toward recovery after 3 h of reperfusion; after 9 h of storage, ATP did not return to the initial level, and AMP degraded completely during the reperfusion period [Saburova L.M. and Berezov T.T.].

Thus, the content of stored ADN in all organs markedly decreases during ischemia. This primarily concerns the amount of CP and ATP in the early ischemic period. Disturbances of the energy utilization systems result in retarded ATP consumption; the period of ischemia after which ATP consumption at 37°C sharply decelerates is 60 min for the heart (our results, see above), 30 min for the liver [Marubayashi S., Dohi K., Ezaki H. et al. (1982)], 15 min for the lung [De Leyn P.R., Lerut T.E., Schreinmakers H.H., et al.], and 5–10 min for the kidneys [Bore P.J., Paratheofanis I. and Sells R.A.].

3.2.2 Adenine Nucleotide Degradation

Adenine nucleotides degradation to nucleosides and purine bases has certain organ-specific features (Fig. 27). During ischemia of **the heart and brain**, AMP is catabolized to adenosine in the interstitial fluid near the plasma membrane, where 5'-nucleotidase is located, whereas catabolism of adenosine to inosine and inosine to hypoxanthine occurs in the vascular endothelium and pericytes [Jennings R.B., Reimer K.A., Hill M.L. and Mayer S.E.; Van Belle H., Goossens F. and Wynantsi J.].

In **the kidneys**, AMP immediately degrades to inosine via inosine 5'-monophosphate (IMP), and degradation proceeds to the formation of hypoxanthine, while xanthine may also be produced in a parallel catabolic reaction involving 5-guanine nucleotides [Buhl M.R. and Jörgensen S.].

Total ischemia of **the liver** proved to result in the accumulation in tissues of all metabolites, including allantoin, which indicates that the entire chain of enzymes is involved in AMP degradation. Intense catabolism of ADN is promoted by the accumulation of CA [Kuzmin A. I., Shulzhenko V. S. and Medvedev O. S.]. AMP catabolism does not depend on the depletion of ATP and ADP pools but begins virtually from the very first minutes of ischemia. This has been determined in numerous experiments from the accumulation of inosine, hypoxanthine and xanthine in the ischemic organ itself and, if the organ is partially ischemic or undergoes hypoxic perfusion, from the sharp increase in the contents of these catabolites in the outflowing blood or perfusate.

Fig. 27: Metabolic pathways of formation and degradation of adenine nucleotides during ischemia of organs [Lehninger A.; Melikyants A.G. and Kucheryanu V.G.; Buhl M.R. and Jörgensen S.; Jennings R.B. and Reimer K.A. (1981)] with modifications

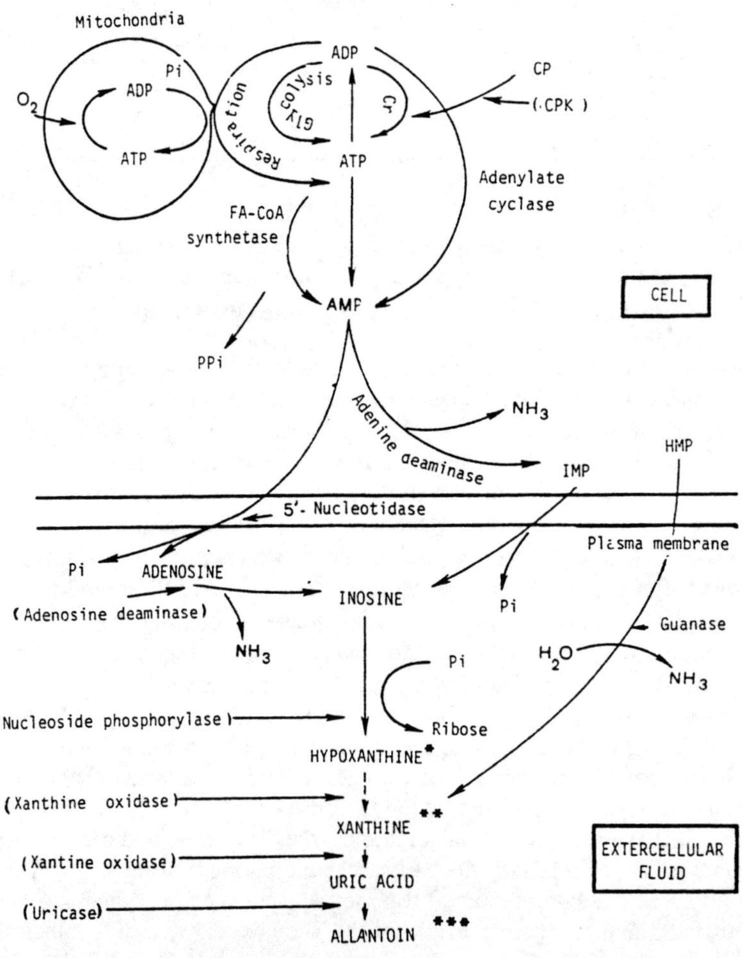

The end products of ATP degradation are marked as follows: * - in the kidneys; ** - in the heart and brain: *** - in the liver.

In the brain, the accumulation of adenosine and inosine was found to occur as early as 5 s after the onset of ischemia [Berne R.M., Belardinelli L., Winn H.R. et al.]. In the heart, 45 s after LAD occlusion in dogs, the subsequent reperfusion caused an increase in adenosine, inosine, hypoxanthine and xanthine contents in the outflowing blood; 5 min after LAD occlusion, reperfusion was accompanied by an almost instant release of inosine and hypoxanthine whose contents exceeded the initial level by factors of 400 and 130, respectively [Fox A.C., Reed G.E., Meilman H. et al.].

A sharp increase in the content of ADN degradation products was observed in the course of hypoxic perfusion of the heart in rabbits [Takeo S. and Sakanashi M.], total cardiac ischemia in dogs [Jennings R.B., Reimer K.A., Hill M.L. and Mayer S.E.], total and partial ischemia and hypoxic perfusion of the heart in rabbits and rats [Van Belle H., Goossens F. and Wynantsi J.], total ischemia and reperfusion of the kidneys [Warnick C.T. and Lazarus H.M.], total ischemia, reperfusion, and hypoxic perfusion of the liver [Nishida T., Koseki M., Kamiike W. et al.].

In many experiments, the release of AMP degradation products from ischemic organs was observed for a long time after their reperfusion, which is consistent with our data on the decrease of ADN level during the period after long-term ischemia of organs.

The outflow of ADN degradation products from organs is one of major factors responsible for the reduced synthesis of AMP *de novo* and, consequently, for reduced ATP resynthesis during the postischemic period [Berne R.M., Belardinelli L., Winn H.R. et al.; Fox A.C., Reed G.E., Meilman H. et al.; Takeo S. and Sakanashi M.]. This explains the rapid postischemic recovery of CP content (which incicates that mitochondria are intact) but not of ATP content [Swain J.L., Sabina R.L., McHall Ph. A. et al.; Takeo S. and Sakanashi M.], more rapid recovery of ATP in the rat brain perfused after total ischemia, compared to that after partial ischemia [Rehncrona S., Mela L., Siesjљ B.K. et al.], and the protective anti-ischemic action of ribose, adenosine, other ADN precursors and inhibitors of adenosine catabolism [Palombo J.D., Bowers J.L., Clouse M.E. et al.; Abd-Elfattah A.S., Maddox R.P., Jessen M.E. et al.; Kondo S.; Mandel L.J., Tanako T., Soltoff S.P. and Murdaugh S.; Newman G.C., Hospod F.E., Trowbridge S.D. et al.; Van Waarde A. Avison M.J. Thulin G. et al.].

3.2.3 Criteria of Reversibility of Ischemic Injuries

Criteria of reversibility of ischemic injuries, mainly based on the evaluation of *the content and composition of ADN* in ischemic organs, due to the major role of ADN in cell activity and their marked depletion during ischemia. However, it was found that ATP content, which rapidly diminishes during ischemia, does not correlate with the ability of the kidneys [Bore P.J., Paratheofanis I. and Sells R.A.] and the liver [Marubayashi S., Dohi K., Ezaki H. et al. (1982)] to recover their functions during the reperfusion period. ATP content as a parameter appeared to be more informative in case of the ischemic heart, in which the rate of ATP decrease is slower. For a long time, specialists distinguished several critical levels of ATP in the heart: (1) the level ensuring the absence of contracture and the adequate restoration of cardiac function (over 50% of the initial level) [Alanen K., Lipasti J. and Nevalainen T.; Takeo S. and Sakanashi M.]; (2) the level ensuring the reversibility of structural damage in the myocardium (35–40%) [Jennings R.B., Reimer K.A., Hill M.L. and Mayer S.E.]; and (3) the level ensuring the recovery of CP content in the myocardium, which corresponds to 2 μmol/g wet tissue weight [Bretschneider H.J., Hübner G., Knoll D. et al.]. In our experiments, however, cardiac contractions were restored even after 1 h of ischemia at 20 and 37°C, i.e., at an ATP content below the critical level (see Fig. 26). Other researchers also revealed no correlation between ATP content in the heart and the cardiac function or irreversible structural damage to the organ [Neely J.R. and Grotyohann L.W.; Watts J.A., Maiorano L.J and Maiorano P.C.]. Note that Irreversible changes in the AMI focus were observed against the background of relatively high residual ATP content (7–8 μmol/g dry weight) [Jennings R.B. and Reimer K.A. (1992)]. It was shown [Lindsay T.F., Liauw S., Romaschin A.D. and Walker P.M.] that after 4 and 5h of sceketal muscle ischemia, the ATP content was equal to 4% and 18% of the control level, respectively, the necrosis of the muscle tissue was in the same respect, 21 and 51% of the control. Thus, the decrease of ATP during ischemia reveals a very informative index gauging tissue injury.

Taking into account the compartmentalization of ATP, its selective local consumption (see above), and the small proportion of total ATP that can be utilized by myofibrils for contraction (only 5.6%) [Geisbuchler T., Altschuld R.A., Trewyn R.W. et al.], it becomes apparent that the total ATP content in an ischemic organ (heart) can hardly be regarded as a prognostic parameter in tests for various specific functions of the heart, including its contraction.

Some specialists attribute greater prognostic value to the total ADN content, as this parameter proved to closely correlate with ischemic kidney function (but not with the cardiac function) [Kuzmin A. I., Shulzhenko V. S. and Medvedev O. S.], or to the dynamics of AMP contents in tissues and, namely, its change from an increase to a sharp decrease, which adequately reflects the severity of ischemic injuries in the kidneys, intestines, and limb muscles [Saburova L.M. and Berezov T.T.].

Apparently, the most reliable prognostic indicator of the integrity of an organ is its capacity for ADN (especially ATP) resynthesis in the early postischemic period, rather than ADF or the contents of individual adenine nucleotides in the end of the ischemic period. The existence of a relationship between the rate of ATP resynthesis after ischemia and the structural–functional state of the organ was confirmed in studies on reperfusion of the kidneys [Takenaka M., Tatsukawa Y., Dohi K. et al.; Tatsukawa Y., Dohi J., Yamada K. et al.; Trifillis A.L., Kahug M.W., Cowley Kh. et al.], the heart [Ichigara K., Abiko Y. and Neely J.R.; Houston R.J., Heerschap A., Skotnicki S.H. et al.; Mori S., Tanaka A., Kitai T. et al.], the liver [Marubayashi S., Dohi K., Ezaki H. et al. (1982)], the skeletal muscles [Rubin B.B., Liauw S., Tittley J. et al.], the brain [Tatsukawa Y., Dohi J., Yamada K. et al.], and the small intestine [Hirata Y., Taguchi T., Nakao M. et al.], though in some studies intestine proved to be an exception to this rule [Canada A.T., Coleman L.R. Jr., Fabian M.A. and Bollinger R.R.].

In our experiments [Bilenko M.V., Volkova I.A. and Gushcha I.A.; Bilenko M.V., Gushcha I.A., Managadze L.G. et al.; Volkova I.A.; Bilenko M.V. (1980); Gushcha I.A.; Gushcha I.A., Borzov D.B., Volkova I.A. and Bilenko M.V.], Wistar rats after 2 h of renal ischemia received Furosemide (10 mg/kg), the saluretic inhibiting the most energy-demanding function of Na^+ and Cl^- reabsorption in renal tubules, or sodium succinate (60 mg/kg), which intensifies respiratory ATP synthesis via the SDH-dependent pathway (see above), simultaneously inhibiting Na^+ reabsorption in renal tubules [Fyodorova Z.P., Fyodorov V.I. and Shmyreva T.I.]. The results showed that the rate of ATP and ADN recovery during the postischemic period highly correlated with the function of ischemic kidneys evaluated by criteria of animal survival rate and nitrogen release. We also observed a strong correlation between ATP and total ADN resynthesis during the postischemic period, on the one hand, and the functional state of the kidneys, on the other, when animals with ischemic kidneys inhaled a hypoxic gas mixture (see Chapter 7).

There are data on successful attempts to use other metabolic parameters as prognostic criteria, namely: the degree of lactate accumulation in the heart [Neely J.R. and Grotyohann L.W.]; the rate of membrane potential and energy charge restoration during the postischemic period [Oredsson S., Arlock P., Plate G. and Qvarfordt P.; Kono Y., Ozawa K., Tanaka J. et al.]; the absence of a decrease in the total ADN level continuing after reperfusion due to AMP degradation or, as a concomitant of such degradation, the release of ADN degradation products from the heart, kidneys, liver, and skeletal muscles into the venous blood or perfusate [Idström J.P., Soussi B., Elander A. and Bylund-Fellenius A.C.; , Pisarenko O.I., Studieva I.M., Shul'zhenko V.S. et al.]. The ratio of the total ADN degradation products to ADN can also be used as a sensitive index of kidney ischemia severity [Maessen J.G., van der Vusse G.J., Vork M. et al.].

These parameters, reflecting the capacity of organs for resynthesizing high-energy phosphates or, on the contrary, their degradation continuing in the reperfusion period, actually characterize the integrity of the organ membrane systems because the latter form the basis of membrane-dependent ATP-producing enzyme complexes.

3.3 Effects of Disorders of Adenine Nucleotide Metabolism on Functions and Structure of Membranes

As O_2 deficiency and the consequent disorders in the metabolism of readily utilizable ADN are the initial (trigger) factors responsible for ischemic injuries in organs, it is important to analyze mechanisms of membrane damage resulting from these disorders and accompanying phenomena.

The principal membrane-damading factors associated with disorders of AND metabolism are as follows.

3.3.1 Energy Deficiency

The first factor is energy deficiency in the ischemic organ, which develops because the amount of synthesized ATP is insufficient for maintaining its metabolism at the level adequate for either functional consumption (hypoxia, partial ischemia) or even basal consumption (total ischemia).

An acute energy deficiency remains to be a serious membrane-damaging factor during the reperfusion period, when organs require energy in amounts sufficient for functional consumption but respiratory ATP resynthesis is negligible because of damaged mitochondrial membranes, uncoupling of respiration processes from phosphorylation, Ca^{2+} accumulation in mitochondria, lack of precursors, etc.

The damaging action of energy deficiency on the function and structure of membranes is realized in three basic ways: (1) by disturbing the functioning of ATP-dependent ion pumps, (2) by impairing processes necessary for the replacement of membrane structural components, and (3) by causing dephosphorylation of membrane proteins.

The impairment of membrane ATPase functioning because of energy deficiency had been regarded as one of the main causes of membrane damage in ischemic organs. Recent studies showed, however, that the decrease and even complete loss of ATPase activity during ischemia occur while the ATP level is sufficiently high and are largely associated with deterioration of the physicochemical properties of membranes, qualitative changes in lipid environment, protein–lipid interactions, and some other factors, rather than with energy deficiency proper (see Chapters 2, 4 and 5).

According to [Geisbuchler T., Altschuld R.A., Trewyn R.W. et al.], myocytes with ATP content below 0.1 nmol/mg protein (less than 0.4% of normal total amount), retained the ability to maintain the barrier function of the sarcolemma, although they were even incapable of anaerobic glycolysis.

Synthetic processes, particularly resynthesis of proteins, phospholipids, and other structural components of membranes, actually suffer from a substantial energy deficiency. Many studies showed that protein synthesis in ischemic organs sharply decreases after 5 min and ceases after 60 min [Shumakov V.I., Onishchenko N.A. and Kirpatovsky V.I.]. The decreasing intensity and the following complete cessation of reparative processes during ischemia have an adverse effect on the structure and functions of membranes because of enhanced protein catabolism (see 3.1), phospholipolysis of membrane PL (see Chapters 2 and 4), chemical modification of lipids, and protein damage due to peroxidation (see Chapter 5). Active reparative processes are especially important after ischemia, but the postischemic energy deficiency may result in even worse membrane damage instead of repair. A favorable result of our experinments, in which postischemic ATP deficiency in the kidneys was reduced by inhibiting tubular electrolyte reabsorption with the aid of furosemide [Bilenko M.V., Volkova I.A. and Gushcha I.A.; Bilenko M.V., Gushcha I.A., Managadze L.G. et al.; Volkova I.A.; Bilenko M.V. (1980)], was apparently achieved owing to the enhancement of reparative processes in renal membrane structures.

Dephosphorylation (energy depletion) of membrane proteins is another serious factor of membrane damage. It is directly related to the decrease of ATP content in the cell and the resulting energy deficiency [Gazitt Y., Ohad I. and Logter A.]. Phosphorylation–dephosphorylation processes are known to be effected by the phosphoprotein phosphatase enzyme system comprising 4–5 isoenzymes [Parsadanian H.K., Ter-Tatevosian L.P., Aslanian J.G., Adunz G.T.]. Membrane dephosphorylation upon energy deficiency leads to changes in protein electric charge and conformation, alters protein–lipid interactions, and promotes local PL uncovering, making them more sensitive to the hydrolytic action of phospholipases C and A_2 [Gazitt Y., Ohad I. and Logter A.]. The phosphorylated state is especially important for "channel" proteins participating in passive Ca^{2+} transport in myocytes because this state provides for keeping the channels open to allow Ca^{2+} influx into the cell. Dephosphorylation of these proteins under conditions of energy deficiency closes the channels, which leads to lower vascular motility and tone. The introduction of ATP into the medium containing SR membranes from the rabbit myocardial and skeletal muscles, completely protected Ca^{2+}ATPase activity from damaging effect of $^{\bullet}OH$ generating system [Xu K.Y., Zweier J.L. and Becker L.C.]. Authors suggested that $^{\bullet}OH$ denatures the Ca^{2+}ATPase of SR by directly attacking the ATP binding sites; therefore the depletion of ATP during ischemia may enhance the toxic effect of $^{\bullet}OH$ on Ca^{2+}ATPase of SR. The effect of other oxidants (H_2O_2, $HOCl^-$) on high affinity ATP-binding sites in myocardiocyte sarcolemma was shown in [Musat S. and Dhalla N.S.].

The damaging effect of energy deficiency depends on the degree of its expression, that is, on the discrepancy between the rates of synthesis of high-energy phosphates and their consumption in organs. Any measures to increase the rate of ATP synthesis or to reduce the rate of its consumption during the ischemic or reperfusion period alleviate energy deficiency and improve the structural and functional state of membranes and ischemic organs.

3.3.2 Metabolic Acidosis

The second membrane-damaging factor is metabolic acidosis always accompanying disturbances of ADN metabolism.

The main cause of acidosis upon O_2 deficiency is the formation of H^+ ions during certain processes occurring in the cell. These are glycolysis described by the formula: glucose → 2 lactates + $2H^+$ [Gevers W.], the accumulation of incompletely oxidized products of disturbed carbohydrate and lipid (fatty acid) metabolism, hydrolysis of ATP and other high-energy phosphates, reduction of NAD to NADH and NADP to NADPH. Among other factors, an important role belongs to a decreased H^+ utilization because of reduced ATP synthesis [Johnston D.G. and Alberti K.G.].

The degree of metabolic acidosis during ischemia of organs depends on the ischemic model. During hypoxic perfusion or partial ischemia, when the blood (perfusate) outflow from the organ proceeds at a usual or reduced rate, metabolic acidosis is usually moderate. During total ischemia and the absence of H^+ outflow, acidosis increases along with the duration of ischemia, and pH factor under these conditions *in vivo* may decrease to 6.9–5.7 [Cobbe S.M. and Poole-Wilson P.A.; Poole-Wilson P.A.; Whorton A.R., Montgomery M.E. and Kent R.S.].

Membrane structures and cell functions respond to acidosis depending on its degree.

Moderate acidosis, when pH factor decreases by 0.5 (pH 7.2–6.8) exerts a protective effect by reducing the activity of CA-stimulated adenylate cyclase, inhibiting cyclic AMP (cAMP) formation and activities of phospholipases, stimulating Ca-ATPase in cardiac SR membranes against the background of high Ca^{2+} content, and intensifying succinate oxidation in mitochondria of the liver and skeletal muscles in rats [Agureev A.P., Altukhov N.D., Mokhova E.N. and Savelyev I.A.; Benevolensky D.S. and Levitsky D.O.; Volkova I.A.]. The protective effect of moderate acidosis was also observed during hypoxia or ischemia of tissue sections or cell suspensions of the kidneys [Kleinman J.G., Brown W.W., Ware R.A. et al.; Weinberg J.M.; Martin L.F., Peter A.O., Fehr D.M. et al.], the myocardium [Lakatta E.G., Nayler W.G. and Poole-Wilson P.A.], cultured cells [see Gatsura V.V. and Sidorenko A.F.].

Pronounced acidosis with pH being reduced by more than 0.5–0.6 (up to pH 6.8–6.2) and *acute acidosis* (when pH is decreased by more than 1.5) produce a deleterious effect on cell metabolism and membrane structures of organs in the following ways:

(1) By inhibiting glycolytic enzymes such as phosphofructokinase and others [Reimer K.A., Jennings R.B. and Hill M.L.], which further aggravates energy deficiency. In dogs, a decrease in ATP content to the critical level (4 μmol/1 g wet tissue) under conditions of total cardiac ischemia and the use of various cardioplegic solutions was found to occur at pH 6.1–5.6 [Preusse C.S., Gebhard M.M. and Bretschneider B.H.].

(2) By enhancing the hydrolytic effect of some phospholipases (see Chapter 4).

(3) By activating LPO processes (see Chapter 10).

Intensified phospholipolysis and lipid peroxidation apparently contribute to the membrane-damaging action of acidosis and, owing to the vasoactive properties of their products (see Chapter 8), lead to an increase in vascular resistance and deterioration of the myocardial function observed during acidosis [Cobbe S.M. and Poole-Wilson P.A.; Fisher D.J.];

(4) By damaging membranes directly, e. g., by altering the conformation properties of membrane proteins. This effect was regarded as responsible for the inhibition of several enzyme systems during acidosis, such as creatine kinase [see Gatsura V.V. and Sidorenko A.F.] and myosin ATPase [Kiselev V. D.; Poole-Wilson P.A.], and for priming the enzymatic receptors of lysosomes resulting in activation of these organelles [Pokrovsky A.A.]. Acute acidosis inhibits the interaction between lipid chains, i.e., directly affects the physicochemical properties of the lipid bilayer.

3.3.3 Accumulation of Products of Disturbed Metabolism

The third membrane-damaging factor is the accumulation of products of disturbed metabolism of oxidation substrates (carbohydrates, lipids, and proteins) in ischemic organs in the course of alteration (reorientation) of ADN synthesis pathways.

Disorders of carbohydrate metabolism or, more precisely, its marked intensification along the anaerobic glycolysis pathway result in a sharp decrease in the content of synthesized ATP and, moreove, lead to the accumulation of incompletely oxidized intermediate products and the end product (lactate), giving rise to intracellular metabolic acidosis. The inverse relation between lactate accumulation in the ischemic rat heart and the degree of restoration of its contractility during the reperfusion period was demonstrated unambiguously [Neely J.R. and Grotyohann L.W.].

The opinion that the accumulating products of anaerobic glycolysis generally exert a more pronounced damaging effect on the structure and function of membranes than the developing energy deficiency is supported by many specialists [Rovetto M.J., Lamberton W.F. and Neely J.R.; Steenbergen C., Murphy E., Levy L. and London R.E.].

The damaging effect of lactate manifests itself either independently, by inhibiting glycolysis [Rovetto M.J., Lamberton W.F. and Neely J.R.], or, which is more likely, through the enhancement of tissue acidosis [Neely J.R. and Grotyohann L.W.].

Disorders of lipid metabolism—inhibition or cessation of lipid β-oxidation in mitochondria, which produces acetyl-CoA subsequently utilized in the Krebs cycle—lead to the accumulation of FFA and their intermediate incompletely oxidized products (acyl-CoA, acyl carnitine, β-hydroxy fatty acids). The latter substances *per se* are toxic to membranes, and FFA are also precursors of prostaglandins and leukotrienes, which are highly active with respect to membranes (see Chapter 4). Products of incomplete FFA β-oxidation also contribute to tissue acidosis.

Disorders of protein metabolism, namely, enhanced protein catabolism during ischemia and proteolysis during severe ischemia, result in direct damage to structural protein components of membranes, accompanied by the accumulation of highly toxic medium- and high-molecular protein derivatives (see Chapter 8). The degree of their accumulation correlates with the severity of ischemia. It is obvious that under conditions of total ischemia, when such metabolites are not washed out of the organ, their toxic effect on membranes is especially strong.

3.3.4 Accumulation of Reduced Carrier Molecules

The fourth factor membrane – damaging factor in organs is the accumulation of reduced carrier molecules (NADP, NADPH, etc.) during ischemia. It leads to the cessation of electron transfer along the mitochondrial respiratory chain, reduced thermostability of membranes, their lower resistance against lysis by phospholipases and proteases, electron "leakage, " and the formation of "reactive" O_2 forms (see Chapter 10).

3.3.5 Accumulation of Products of ADN Catabolism

Finally, the fifth membrane-damaging factor is the accumulation of products of ADN catabolism in ischemic organs. As shown above, this process is especially intense during ischemia and continues in the early reperfusion period. It leads to the accumulation of ADP and AMP, which are chelators of Fe^{2+} ions enhancing their prooxidant function, on one hand, and of hypoxanthine and xanthine, which are substrates for the hypoxanthine-xanthine oxidase reaction, on the other. Such conditions are prerequisite to the formation of reactive O_2 forms. Both effects are especially obvious during total ischemia accompanied by zero outflow of nucleotide degradation products from the organ and play an important role in triggering LPO processes in the ischemic organ (see Chapter 10).

However, nucleoside outflow from organs during partial ischemia or reperfusion impairs and retards ATP resynthesis, i.e., aggravates energy deficiency and membrane damage. It is still unclear which of the two effects is more deleterious: accumulation of catabolites in the organ or, on the contrary, their washing out of the organ.

Thus, disorders of ADN metabolism during ischemia of organs play a key role in membrane damage. Metabolic disorders, however, should be regarded merely as the initial factor as they trigger other, apparently more powerful membrane-damaging processes, including phospholipolysis, AA metabolism along different pathways, and LPO. Their damaging effects on membranes are analyzed in Chapters 4 and 5.

3.4 Conclusions

The reduction or complete block of oxygen supply to organs lead to rapid changes in the pathways of synthesis, transport, and consumption of high-energy phosphates readily utilized by tissues. This causes the depletion or exhaustion of ADN reserves (particularly CP and ATP) and the accumulation of products of incomplete or disturbed metabolism and molecular degradation. In general, there processes result in structural and functional membrane damage and the loss of structure, function, and viability of the ischemic organ.

Basic manifestations of impaired energy supply to organs during ischemia are as follows.

(1) Alteration (reorientation) of ADN metabolic pathways, which involves a sharp reduction or complete cessation (in total ischemia) of energy production by aerobic oxidation of carbohydrates, lipids, and proteins in mitochondria (in the adenylate kinase reaction) and oxidation of glucose in the HMPS and, conversely, temporary intensification of anaerobic glucose phosphorylation in reactions of the glycolytic pathway and the enhancement of atypical for normal tissues anaerobic pathways of fumarate reduction to succinate and substrate phosphorylation in the Krebs cycle.

When respiration and oxidative phosphorylation are partially retained in mitochondria during ischemia, respiration changes over from the NAD- to SDH-dependent pathway.

During the reperfusion period, the ADN synthesis pathways are restored slowly and incompletely, which leads to long-term energy deficiency and the formation of toxic products of incomplete or untypical metabolism in the early postischemic period.

(2) In certain organs (especially muscles and the brain), severe disturbances of ATP energy transport from mitochondria to the sites of its consumption in myofibrils and membranes. This is associated with the high sensitivity of the carrier enzyme (ATP–ADP translocase) and enzymes of the creatine kinase system to ischemia and leads to the uncoupling of the sites of ATP synthesis and consumption.

(3) Possible selective disturbances of cell functions receiving energy from ATP produced in metabolic pathways that are most sensitive to ischemia and suffer greater damage. This phenomenon is associated with the established compartmentalization of ATP.

(4) Early damage to energy consumption systems, primarily membrane-dependent ATPase, which disturbs the barrier function of membranes while ADN content remains sufficiently high. These systems, however, become damaged later than those synthesizing high-energy phosphates. The discrepancy between ATP synthesis and ATP consumption, apparent within the very first minutes of ischemia, intensifies ADN catabolism, reduces CP and ATP contents, increases AMP content, and leads to the accumulation of products of subsequent ADN degradation, such as nucleosides and purine bases. The extent and duration of recovering ATP and total ADN pools and normalizing the ratio between individual components of the adenylate pool during the postischemic period depend on the severity of ischemia and availability of precursors, i.e., the possibility of AMP synthesis *de novo*.

(5) ATP and total ADN contents in the ischemic organ do not always correlate with organ ability to recover its functions, which appears to result from ATP compartmentalization. The ability of an organ to resynthesize ATP and the absence of ADN degradation after ischemia are more reliable prognostic parameters.

(6) The deficiency of O_2 and the resulting disturbances of ATP synthesis, transport and consumption pathways have a membrane-damaging effect. The basic mechanisms of its realization include energy deficiency; accumulation of intermediate and toxic products of carbohydrate, lipid, and protein metabolism; pronounced acidosis; accumulation of reduced carrier molecules in tissues and the respiratory chain; and accumulation of ADN degradation products.

In addition, these factors initiate other, more powerful membrane-damaging processes, such as phospholipolysis and lipid peroxidation discussed in Chapters 4 and 5.

Chapter 4

INTENSIFICATION OF HYDROLYSIS OF MEMBRANE PHOSPHOLIPIDS AND DISORDERS OF FREE FATTY ACID METABOLISM DURING ACUTE ISCHEMIA AND AFTER REPERFUSION OF ORGANS

4.1 INTENSIFICATION OF MEMBRANE PHOSPHOLIPIDS HYDROLYSIS

It is known that neutral lipids, or triacylglycerides, in the cell are located in the cytosol, phospholipids occur primarily in membranes, and FFA are found both in the cytosol and in membranes. FFA after oxidation are used as a substrate in the Krebs cycle (see Fig. 20); under physiological conditions, their pool is mainly restored owing to supply with blood and, to a lesser extent, to their release during hydrolysis of triacylglycerides (catalysed by lipases) and membrane phospholipids (catalysed by phospholipases A_1 and A_2 and plasmalogenases) and subsequent hydrolysis of lysophospholipids (catalysed by lysophospholipases). The processes of phospholipid and lysophospholipid hydrolysis *in vivo* exist in a certain equilibrium with one another and are reversible.

FFA supply with blood sharply decreases during partial ischemia and ceases during total ischemia so that the cell receives FFA owing to the processes of intracellular lipolysis and phospholipolysis.

The contribution of intracellular lipolysis to FFA accumulation in most tissues (except adipose) is small because triacylglycerides contain no more than 10–13.7% of esterified fatty acids in the cell [Van der Vusse G.I., Roemen Th.A., Prinzen F.W. et al.; Weglicki W.B., Owens K., Urchel C.W. et al.].

Phospholipids, by contrast, account for about 85.7–90% of the total intracellular fatty acids [Van der Vusse G.I., Roemen Th.A., Prinzen F.W. et al.; Weglicki W.B., Owens K., Urchel C.W. et al.], which makes them the most likely source of FFA during ischemia. For example, when lipolysis and phospholipolysis were stimulated by acidosis (at pH 5.0 and 4.0), phospholipids provided the source of FFA in 71.1% of cases, whereas triacylglycerides, in only 28.9% of cases [Weglicki W.B., Owens K., Urchel C.W. et al.].

Phospholipids are hydrolyzed by four phospholipases (Fig. 28) of which the most important role in mammalian cells belongs to phospholipases A_1 and A_2 hydrolysing FFA in the α– and β–positions, respectively.

Phospholipases A_1 and A_2 have been found virtually in all organs and subcellular organelles [see Waite M.; Wolf R.A. and Gross R.W.] but their properties and pH optima depend on their location.

The most extensive data concerns mitochondrial *phospholipase A_2* (PLA_2), which plays the key role in mitochondrial membrane damage during anoxia [see Sorokovoy V.I. and Vladimirov Yu.A.]. This enzyme is Ca^{2+}-dependent, is inhibited by EDTA and activated by polyunsaturated FA (PUFA), its optimal pH ranges from 8.0 to 9.5, and its affinity for phospholipids decreases in the order: PEA > PS > PC > cardiolipin > phosphatidic acid > PI. As shown on the rat liver model, the inner and outer mitochondrial membranes contain different phospholipases A_2: the enzyme in the inner membrane is Ca^{2+}-dependent, is stimulated by increasing pH, and makes a greater contribution to the overall activity, whereas the enzyme in the outer membrane is Cu^{2+}-dependent; both phospholipases are inhibited by EDTA [Gorbataya O.N., Ziyatdinova R.Kh., Akhmedjanov R. et al.]. There is evidence that the Ca^{2+}-dependent phospholipase A_2 in the inner mitochondrial membrane in the rat liver is activated by calcium only on the side of mitochondrial matrix [Severina E.P. and Evtodienko Yu.V.], i.e., its activation is promoted by Ca^{2+} accumulation inside mitochondria.

Fig. 28: Hydrolysis of phospholipids by phospholipases A_1, A_2, C and D

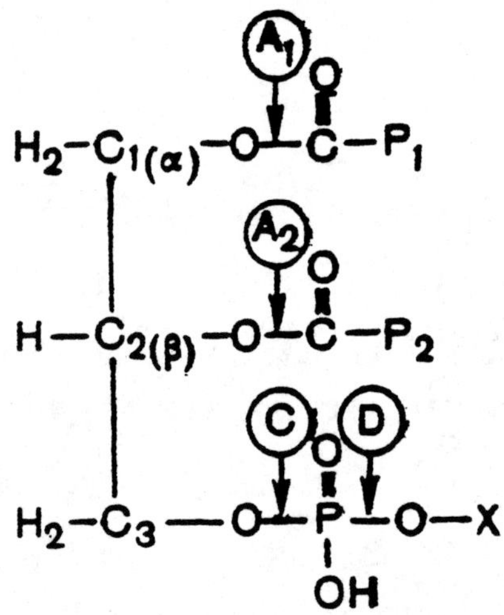

P_1 and P_2 - acyls of fatty acids; X - choline, ethanolamine, etc. α- and- β- first and second positions of phospholipases A1 and A2 action.

Mitochondrial *phospholipase A_2* in the rabbit liver is more active than in the heart, which was regarded as evidence that the phospholipase-dependent pathway of membrane damage is less important for the heart, compared to other organs [Corr P.B., Gross R.W. and Sobel B.E.]. At least three subtypes of this enzyme were identified in the heart: Ca^{2+}-dependent low-molecular PLA_2, cytoplasmic high-molecular PLA_2, and plasmalogen-specific PLA_2 [Van der Vusse G.J., Reneman R.S. and van Bilsen M.]. The identity of the latter to plasmalogenase (see below) remains to be established. Several types of myocardial PLA_2 and their different roles in ischemic and reperfusion-induced injuries were discussed in [De Windt L.J., Reneman R.S., Van der Vusse G.J. and Van Bilsen M.].

Phospholipase A_1 (PL A_1) in mitochondrial membranes, the plasma membrane, and membranes of the Golgi complex is Ca^{2+}-dependent, thermally stable, and its pH optimum is in the neutral to alkaline range (pH 7.5–9.0) [Bazan N.G. and Rodriquez de Furco E.B.; Waite M.].

Phospholipases from the plasma membrane of cardiomyocytes demonstrate optimal activity at lower values of pH factor than enzymes from other organs: phospholipase A_1 from the sarcolemma of the hamster heart has a bimodal optimum of activity, at pH 6.0 and 9.0, whereas that from the dog heart, at pH 7.0.

The activity of lysosomal phospholipases A_1 and A_2 is optimal at low pH values (4.0 and 6.5); they demonstrate a high thermostability and are not activated by Ca^{2+}. The most active phospholipases are those from the soluble lysosomal fraction: their activity is 12 times higher than that of phospholipases from lysosomal membranes, 40 and 20 times higher than the activity of phospholipases from mitochondrial and microsomal membranes, and 55 times higher than the total phospholipase activity of the homogenate [Mellors A. and Tappel A.L.]. The authors emphasize that, although protein content in rat liver lysosomes accounts for less than 1% of the total protein in the organ, the hydrolytic activity of lysosomal phospholipases is equal to the sum of mitochondrial and microsomal phospholipase activities.

The involvement of *lysophospholipases 1 and 2* in the second stage of phospholipid hydrolysis is better manifested in the intestines; then follow the kidneys, lungs, liver, and brain; lysophospholipases of the liver and heart are Ca^{2+}-independent or even inhibited by these ions, and their pH optimum is in the acidic range (pH 6.5).

Membrane plasmalogens are hydrolyzed by *plasmalogenase*. This enzyme is present primarily in the plasma membrane and in mitochondrial and microsomal membranes of oligodendroglial cells of the cerebral white matter and is absent from the cytosol and lysosomes. It is thermolabile, Mg^{2+}-dependent, and inhibited by Ca^{2+}. The activation of plasmalogenase is independent of that of phospholipase, precedes the activation of lysosomal enzymes, and mainly results in liberation of PUFA (20:4 and 22:4) [Horrocks L.A., Spanner S., Mozzi R. et al.].

A certain role in the hydrolysis of membrane phospholipids may belong to *phospholipase C,* which represents a group of substrate-dependent enzymes, each hydrolyzing a particular PL, including PC, PEA, and PI [Waite M.; Wolf R.A. and Gross R.W.]. The hydrolysis of PI by phospholipase C is accompanied by the release of a large amount of AA.

Recently, the role of *sphingomyelase* and *"sphingomyelin signaling pathway"* in the early ischemic/reperfusion injury of mouse renal cortex was assessed [Zager R. A., Iwata M., Conrad D. S. et al.]. The results showed that the major toxic factor consists in in an increase in ceramide content, which induces kidney apoptosis (see Chapter 9), although the content of sphingomyelin (SM) does not decrease [Zager R. A., Iwata M., Conrad D. S. et al.; Halat G., Lukacova N. Chavko M. and Marsala J.].

The activity of phospholipases increases markedly during ischemia, and their key role in ischemic/reperfusion injuries of different organs is confirmed by the following observations.

First, an increase in activities of endogenous phospholipases was observed in direct experiments on incubating ischemic tissue extracts with exogenous substrates (PC, PEA, etc.) *in vitro*. Endogenous phospholipase activity with respect to PEA increased by a factor of 2 after 1 min of total cerebral ischemia in a Mongolian gerbil and by a factor of 1.4 30 s after decapitation of rats [Edgar A.D., Strosznajder J., and Horrocks L.A.]; the activity of plasmalogenase increased by a factor of 2.3 20 min after decapitation [Horrocks L.A., Spanner S., Mozzi R. et al.]. Activities of phospholipases A_1, A_2, and lysophospholipase remained unchanged during the first 60 min of total cerebral ischemia in dogs (compression method) but increased considerably after 120 min of ischemia. After longer exposure to ischemia, activities of the three enzymes decreased again, but lysophospholipase retained a higher level of activity for a much longer period of time than phospholipases [Hirashima Y., Koshu K., Kamiyama K et al.]. PLA_2 from the mitochondrial and cytosolic fractions of the homogenate prepared from the rat renal cortex after 1 h of ischemia and 1 h of reperfusion proved to be activated, whereas there was no enhancement of its activity in corresponding fractions of rat liver ischemic for 1 h and reperfused for up to 24 h [Terao Y., Shibata O., Goto S. et al.]. PLA_2 activation in mitochondrial and microsomal fractions of rat kidney was observed as soon as after 45 min of ischemia; after repesfusion, activities of PLA_2 from both these sources and of cytosolic PLA_2 (with respect to PC and PEA) were enhanced [Nakamura H., Nemenoff R.A., Gronich J.H. and Bonventre J.V.].

Second, the induction of endogenous phospholipases proved to result in changes similar to those observed during ischemia *in vivo*. This series of observations includes experiments in which activation of endogenous phospholipase of mitochondrial membranes in the rat liver by Ca^{2+} ions caused degradation of PL; moreover, the loss of PEA was much higher, as compared to PC (60 and 20%, respectively), depended on the Ca^{2+} dose, and was accompanied by damage to enzymes, manifested in lower glucose-6-phosphatase activity and decreased cytochrome P-450 content [Chien K.R., Cherman S.C., Mittnacht S. et al.].

Third, there is evidence that the addition of exogenous phospholipases to intact or hypoxic (ischemic) tissues leads to changes similar to those occurring during ischemia. The treatment of skeletal muscle SR membranes with phospholipase A from the cobra venom *in vitro* enhanced their Ca^{2+} permeability and reduced ATP-dependent Ca^{2+} transport [Fiehn W. and Hasselbach W.]. A similar effect was also observed in cardiac SR membranes [Swoboda G., Fritzsche J. and Hasselbach W.]. Phospholipase A_2 from the giant hornet venom in small doses led to the uncoupling of respiration and phosphorylation in mitochondria of the rat liver, while its large doses additionally inhibited the rate of O_2 consumption [Muksimov F.A., Almatov K.T. and Tuichibaev M.U.]. Low concentrations of phospholipase A_2 increased the activity of succinate oxidase, cytochrome C oxidase, and NADH oxidase, whereas high concentrations had an inhibitory effect on these enzymes [Muksimov F.A., Almatov K.T. and Tuichibaev M.U.].

ATPases of the plasma membrane are especially sensitive to exogenous phospholipases. Incubation of rat brain synaptic membranes and synaptosomes with phospholipase A reduced total ATPase and Mg^{2+}-dependent ATPase activities by 35-44% and the activity of Na, K-ATPase by 70–80% [Flyorov M.A., Borovitskaya A.E. and Tolstukhina T.I.]. Exogenous PLA_2 seriously aggravated damage to hypoxic renal tubular cells [Nguyen V.D., Cieslinski D.A. and Humes H.D.].

Fourth, some findings provide evidence for an increase in the contents of two main products resulting from activation of phospholipases in the ischemic organ, namely: (1) FFA, with a pronounced decrease (partial ischemia) or complete cessation (total ischemia) of their supply to cells from the blood; and (2) products of phospholipid hydrolysis, or lysophospholipids. These observations directly indicate that phospholipolysis is intensified during ischemia. Below, the corresponding data will be discussed in more detail.

Fifth, numerous selective and nonselective inhibitors of PLA_2 and PLC proved to reduce the severity of ischemic / reperfusion injuries in rat cerebral cortex [Phillis J.W. and O'Regan M.H.; Sonnino R.E., Pigatt L., Schrama A. et al.; Estevez A.Y. and Phillis J.W.].

The question as to *which phospholipases are activated during ischemia* was partially answered in the course of studies on the fatty acid composition of lysoPL.

It was established that lysoPL in mitochondria of the intact rabbit heart contain up to 95% FFA, but after 1 h of total cardiac ischemia the content of FFA, especially AA, in mitochondrial lysoPC decreased considerably [Dagys A. I., Toleykis A. I. and Praskevičius A. K.]. This finding suggested that the activity of phospholipase A_1 prevails under physiological conditions, whereas that of phospholipase A_2 prevails during ischemia. At the same time, the FA composition of lysoPC and lysoPEA in the pig heart 40 min after LAD occlusion was the same as before ischemia [Shaikh N. A. and Downar E. (1981)], which means that ischemia activates phospholipases nonspecific with respect to the type of PL [Shaikh N. A. and Downar E. (1981)], or several phospholipases simultaneously [Vasdev S. C., Biro G. P., Narbaiz R. and Kako K. J.], or both phospholipases and lysophospholipases [Weglicki W.B., Owens K., Urchel C.W. et al.].

Lysosomal phospholipases are also activated during ischemia. According to some data, chloroquinone (a lysosome inhibitor) added to the perfusate prevented an increase in FFA content during normoxic and hypoxic perfusion of the isolated rat heart [Victor T., La Cock C. and Zochner H.]. It was also reported [Weglicki W.B., Owens K., Urchel C.W. et al.] that FFA (primarily unsaturated), lysoPC, and lysoPEA accumulated in rat and dog heart homogenates incubated at pH 4.0 and 5.0, i.e., under conditions optimal for the activity of lysosomal phospholipases. However, low-volume perfusion of the dog heart for 80 min led a much greater increase in FFA (especially 18:2 and 20:4) than incubation at low pH. As ischemia *in vivo* does not lead to such a decrease in pH, it is apparent that neutral nonlysosomal (rather than acidic lysosomal) phospholipases play the key role in PL degradation during ischemia. Microsomal lipids, owing to their close location, may be more vulnerable to lysosomal phospholipases than other cell membranes.

Activation of *intracellular lipases,* in addition to phospholipases, during ischemia is confirmed by the aforementioned data on a decrease in the content of neutral lipids in the liver during the early period of exposure to total ischemia (see Chapter 2). However, the role of local lipolysis in ischemic organs is less important than that of local phospholipolysis because during partial cardiac ischemia in dogs (LAD occlusion) [Van der Vusse G.I., Roemen Th.A., Prinzen F.W. et al.] or low-volume perfusion of the rat heart [Lochner A., Kotze J.C.N., Benade A.J.S. et al.], the level of triacylglycerides in cells increased instead of decreasing, and lipolysis inhibition by nicotinic acid did not prevent FFA accumulation [Victor T., La Cock C. and Zochner H.]. Since phospholipases retain a high activity during hypothermia and are more resistant to low temperatures than lipases [Giacomini K.M., Swezey S.E. and Giacomini J.C. et al.], they still participate in ischemic membrane damage when organs are preserved at low temperature.

4.2 Increasing Contents of Free Fatty Acids and Intermediate Products of Their Oxidation, and Their Damaging Effect on Membranes

Intensification of phospholipolysis and disturbed β–oxidation of FFA in mitochondria lead to intracellular accumulation of certain substances. Of the latter, the most toxic for membranes are FFA proper, acyl-CoA, acyl carnitine, incompletely oxidized FFA (β–hydroxy FFA), and their compounds with CoA and carnitine (Fig. 29).

4.2.1 The Increase of FFA Content and their Damaging Effect on Membranes

The increase of FFA content and their damaging effect on membranes have been observed during ischemia of most organs.

During ischemia of the liver in rats (complete cessation of blood flow), FFA content in mitochondria remained unchanged during 30 min of ischemia but increased 6.5-fold after 2 h, with a corresponding decrease in the respiratory function of mitochondria [Boime I., Smith E.S. and Hunter F.E.]. A similar accumulation of FFA was observed when a liver was stored for 13 h at 24°C; that is, hypothermia retarded FFA accumulation but did not block it. A toxic effect on the respiratory function of mitochondria was also exerted by exogenous FFA added to the medium in which mitochondria were incubated. In this case, albumin neutralized the effect of both endogenous and exogenous FFA and restored the respiratory function of mitochondria. In the hepatic microsomal fraction, FFA content after 30 min and 2 h of ischemia exceeded the initial levels by factors of 1.9 and 4.2, respectively, and the accumulation of unsaturated FA (18:2 and 20:4) in both mitochondria and microsomes was much more pronounced than that of saturated FA.

Fig. 29: FFA β-oxidation pathways during ischemia according to [Corr P.B., Lee B.I. and Sobel B.E.; Moore K.H., Koen A.E. and Hull F.E.] with modifications.

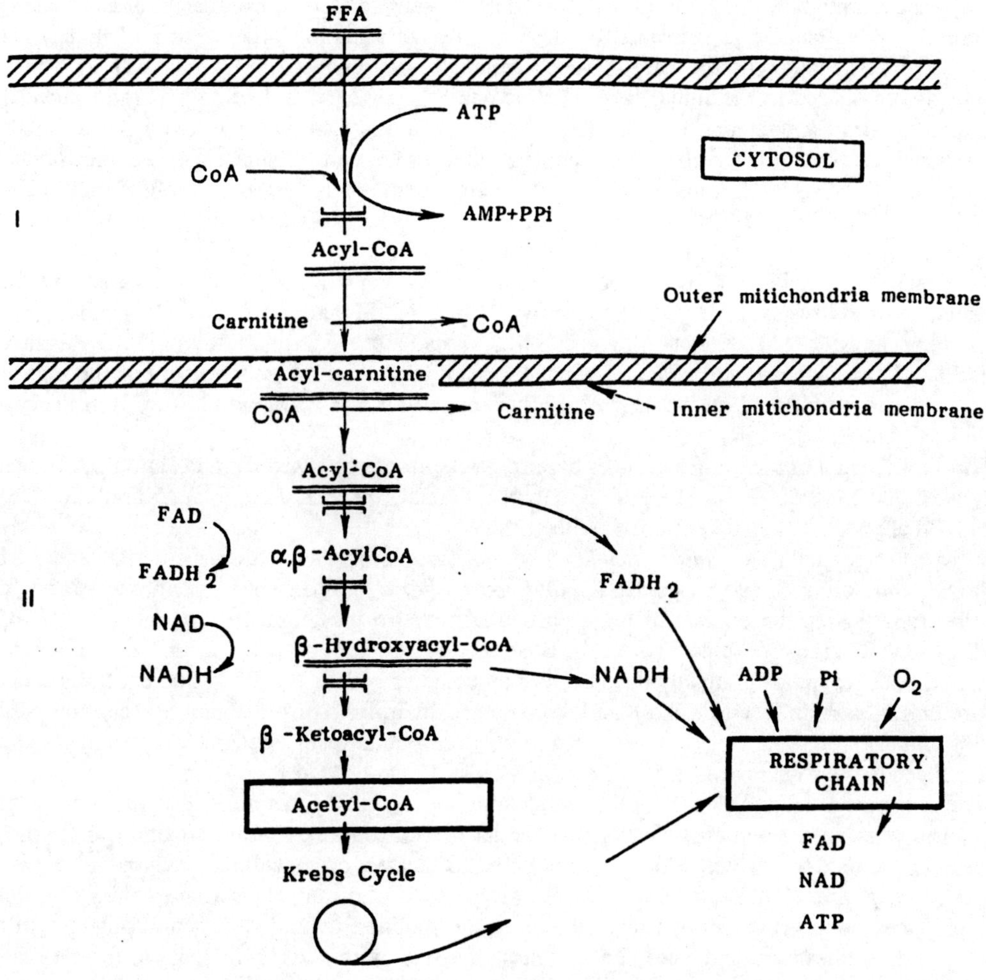

I - FFA activation stage; II - FFA β-oxidation stage. The sites of blocks during ischemia are indicated as. The intermediate products accumulating during ischemia are marked by double underlines.

Accumulation of FFA *during ischemia of the heart* has been reported by many researchers [see Opie I.H.; Van der Vusse G.J., Reneman R.S. and van Bilsen M.].

Low-volume perfusion of the dog heart with blood for 30 min led to a substantial increase in FFA, especially unsaturated: 18:2 and 20:4 increased by factors of 1.6 and 5.3, respectively [Weglicki W.B., Owens K., Urchel C.W. et al.]. The total FFA content and especially that of unsaturated FA were found to increase after partial LAD occlusion in dogs for 120 min: 18:1, 18:2, and 20:4 increased by factors of 6, 7.3, and 11, respectively [Van der Vusse G.I., Roemen Th.A., Prinzen F.W. et al.]. According to this report, the increase in FFA content was maximal in the subendocardial layer, moderate in the median layer, and minimal in the subepicardial layer (5.4-, 3.4-, and 1.6-fold) and inversely proportional to the rate of blood flow through the same layers, which was reduced by factors of 9, 5, and 2, respectively. In other words, the lower the blood flow rate, the more FFA accumulated in the myocardium.

A marked increase in FFA content, especially in that of unsaturated FA (16:1, 18:2, and 20:4), also occurred during hypoxic perfusion of the isolated rat heart [Victor T., La Cock C. and Zochner H.]. Albumin added to the perfusate prevented such an increase, whereas the addition of FA to the perfusate led to FFA accumulation in the extracellular medium but not in the cell. An increase in the intracellular contents of total FFA (1.9-fold) and of individual fatty acids (1.3- to 4.5-fold) was only observed when PC alone or together with phospholipase A_2 was added to the perfusate. Under such conditions, mitochondria lost their capacity for oxidative phosphorylation and respiration [Victor T., La Cock C. and Zochner H.].

Ischemia of the brain was also accompanied by an increase in FFA. Within the first 5 min of total cerebral ischemia in guinea pigs (decapitation model) [Majewska M.D., Strosznajder J. and Lazarewicz J.], FFA content in mitochondria increased by 33.7%, whereas after total (compression model) and severe partial ischemia (ligation of the carotids and hypotension), the increase in rats was 12- and 10-fold, respectively, compared to the initial level [Rehncrona S., Folbergrova J., Smith D. and Siesjö B.K.]. A linear increase in FFA content during ischemia of the brain was also reported by other specialists: during total ischemia in Mongolian gerbils, FFA content after 1, 15, and 30 min exceeded the initial level by factors of 1.7, 8, and 11 [Yoshida S., Inoh S., Asano T. et al.]; after 30 s, 1, 3, and 5 min, FFA content reached 112, 132, 159, and 224% of the initial, respectively [De Medio G.E., Goracci G. and Horrocks L.A. et al.].

In experiments with all models of ischemia, FFA content primarily increased owing to the accumulation of PUFA, especially AA [Lukacova N., Gottlieb M. and Marsala J.; De Medio G.E., Goracci G. and Horrocks L.A. et al.; Kubota M, Nakane M., Narita K. et al.; Saluja I., Song D., O'Regan M.H. and Phillis J.W.; Umemura A.], and although after 30 min of total cerebral ischemia in rats the increase in the FFA content did not depend on the degree of FA saturation, it was maximal (370-fold) in the case of AA [Yoshida S., Abe K., Busto R. et al.].

During ischemia of the brain and other organs, AA is mainly released from PC and PEA [Lukacova N., Gottlieb M. and Marsala J.; De Medio G.E., Goracci G. and Horrocks L.A. et al.; Kubota M, Nakane M., Narita K. et al.; Marion J. and Wolfe L.S.; Umemura A.].

FFA also accumulate *during ischemia of the kidneys* [Finkelstein S.D., Gilfor D. and Farber J.L.; Smith D.S., Rehncrona S. and Siesjö. Bo.K.] *and other organs*; in other words, this process is universal.

After reperfusion, the content of FFA, particularly AA in the brain, rapidly decreased, [Yoshida S., Inoh S., Asano T. et al.; Yoshida S., Abe K., Busto R. et al.], although in the heart it continued to increase against the background of gradually decreasing total contents of PL, PC, PEA, and cardiolipin [Vasdev S. C., Biro G. P., Narbaiz R. and Kako K. J.]. Apparently, hydrolysis of membrane PL continues during the reperfusion period as well, but FFA are more readily reacylated to PL, become converted into acyl-CoA, and are expended in LPO processes [Yoshida S., Abe K., Busto R. et al.].

The damaging effect of high FFA concentrations on the function and structure of biological membranes was mainly demonstrated on the model of isolated heart with exogenous FA added to the perfusate or the medium in which tissue sections, homogenates, or subcellular fractions of the heart were incubated. Thus, the addition of exogenous FA (16:0, 18:2, and other long-chain acids) to the perfusate either produced no effect or very weakly influenced the contractility and slow action potential of the heart in rabbits, rats, guinea pigs, and chicks under conditions of its normal oxygenation, but further inhibited the cardiac function and action potential and promoted the release of enzymes to the perfusate during hypoxic or low-volume perfusion of the heart [Harada H., Azuma J., Hasegawa H. et al.; see also Chapter 8].

Exogenous FFA added *in vitro* to the incubation medium for various membranes of the liver, heart, brain, and limb muscles suppressed the mitochondrial function and uncoupling of oxidative phosphorylation and respiration [Boime I., Smith E.S. and Hunter F.E.; Katz A.M. and Messineo F.C.; Moknova E.N., Starkov A.A. and Bobylova V.A.], inhibited Na, K-ATPase and adenylate cyclase activities in plasma membranes [Karli J.N., Karikas G.A., Hatzipavlou P.K. et al.; Leoni S., Luly P, Mangiantini M.T. et al.], and adversely affected Ca^{2+} transport in SR membranes [Kagan V. E., Churakova T. D., Karagodin V. P. et al.; Vashchenko V.I., Esyrev O.V. and Uspanova Zh.K.].

FFA *in vitro* also inhibited glycolytic enzyme systems and promoted gluconeogenesis [Lea M.A. and Weber G.; Tsutsumi E. and Takenaka F.]. The effect of FFA on the glycolytic metabolic pathway under normoxic and hypoxic conditions differs: in the former case it is manifested in inhibition, whereas in the latter, in intensification of glycolysis and aggravation of energy deficiency (toxic effect) [Cowan J.C. and Vaughan Williams E.M.]. It is the excess of FFA during the postischemic period that may cause the intensified gluconeogenesis observed in reperfused organs (see Chapter 3).

4.2.2 The Increase of Intermediate FFA Oxidation Product Content and their Damaging Effect on Membranes

The increase of intermediate FFA oxidation product content and their damaging effect on membranes are mainly observed during partial ischemia or hypoxia and during the reperfusion period. This is explained by the fact that FFA transport in mitochondria and β–oxidation of FFA are retained during partial ischemia of organs and are partially restored during their reperfusion (see Fig. 29). However, because of O_2 deficiency, structural damage of mitochondria, accumulation of reduced pyridine nucleotides, and higher NADH/NAD ratio, these processes are recovered incompletely, which leads to the accumulation of intermediate FFA oxidation products such as acyl-CoA, acyl carnitine, and β–hydroxy FA that are free or bound to CoA or carnitine.

The accumulation of intermediate products of oxidative FFA metabolism is especially important for the partially ischemic or reperfused heart in which FFA are the preferable oxidation substrate.

Acyl-CoA is accumulated predominantly in mitochondria because almost 90% of CoA content in the tissue are located in the mitochondrial matrix. After LAD occlusion in dogs, acyl-CoA content in the ischemic tissue doubled within 15 min and increased 2.3-to 4.5-fold after 30–60 min. At the same time, the mitochondrial function deteriorated, and the activity of ATP-ADP translocase decreased (to 34% of the initial after 60 min of occlusion) [Shug A.L., Shrago E., Bittas N. et al.; Sugiyama S., Miyazaki Y., Kotaka K. and Ozawa T.]. The mixture of various FFA injected to dogs intravenously promoted acyl-CoA accumulation in the ischemic tissue and affected the mitochondrial function even more adversely [Sugiyama S., Miyazaki Y., Kotaka K. and Ozawa T.].

Reperfusion of an AMI focus for 40 min after 30-min ischemia resulted in the partial removal of excessive acyl-CoA from the heart and in the significant recovery of mitochondrial functioning. After reperfusion against the background of lipid emulsion injections to animals, both decrease in CoA content and the recovery of mitochondrial functioning were much less pronounced [Kotaka K., Miyzaki Y. Ogawa K. et al.]. Acyl-CoA contents and the functioning of mitochondria during cardiac ischemia and reperfusion demonstrated a strong inverse correlation with one another ($r = -0.96$, $P < 0.01$), which suggested that acyl-CoA accumulation may be an important factor responsible for dysfuction of mitochondria and the ensuing irreversibility of their damage [Kotaka K., Miyzaki Y. Ogawa K. et al.; Sugiyama S., Miyazaki Y., Kotaka K. and Ozawa T.]. A significant (twofold) increase in the acyl-CoA content of the heart was also revealed on the model of hypoxic perfusion of the rat heart [Neely J.R., Rovetto M.J. and Whitmer J.T.; Shug A.L., Thomsen J.H. and Folts J.D.].

Accumulation of acyl carnitine in the heart is less pronounced than that of acyl-CoA; it occurs in the cytoplasm near the sarcolemma and has been observed on all models of ischemia [Toleikis A.I., Vavzhinkova Kh. and Tutterova M.; Neely J.R., Rovetto M.J. and Whitmer J.T.; Shug A.L., Thomsen J.H. and Folts J.D.].

FFA added to the perfusate during hypoxic perfusion of the heart accelerated the accumulation of acyl carnitine, which was consistent with the ultrastructural changes occurring in mitochondria, while introduction of various forms of exogenous carnitine improved lipid metabolism and alleviated ischemic membrane damage: it improved electrophysiological parameters during cardiac reperfusion, stimulated the function of ATP–ADP translocase, raised ATP and CP levels, lowered acyl-CoA content, and enhanced the mitochondrial function [Kotaka K., Miyzaki Y. Ogawa K. et al.; Liedtke A.J., Nellis S.H. and Whitesell L.F.; Sugiyama S., Miyazaki Y., Kotaka K. and Ozawa T.].

The anti-ischemic effect of carnitine seems to result from the removal of excessive acyl-CoA from the tissue, although the possibility of its acetylcholine-like activity, antilipolytic and vasodilative effects should not be ruled out.

Accumulation of β–hydroxy FA and their compounds with CoA and carnitine usually occurs at the same time and under the same conditions as that of acyl-CoA and acyl carnitine.

After 2, 5, and 10 min of low-volume perfusion of an isolated rabbit heart (1-3 ml/min) and 20 min of high-volume reperfusion with palmitic acid added to the perfusate, the content of β–hydroxypalmitate increased markedly (by factors of 5, 7, and 10, respectively), whereas the content of β–hydroxystearate increased at half that rate [Moore K.H., Koen A.E. and Hull F.E.]. These β–hydroxy FA quickly passed from mitochondria into the cytosol (after 2 and 5-10 min, their level in the cytosol rose to 38-44 and 50-54%, respectively), which was due to their binding with carnitine to produce β–hydroxyacyl and β–hydroxyacetyl carnitines.

The toxic effect of intermediate FFA oxidation products was mainly studied with reference to acyl-CoA. As shown on isolated cardiac and hepatic membranes, acyl-CoA is a specific inhibitor of ATP–ADP translocase in the inner mitochondrial membrane [Shug A.L., Shrago E., Bittas N. et al.].

In experiments *in vitro*, acyl-CoA also inhibited the mitochondrial Ca^{2+} transport in the rat heart and liver, together with mitochondrial respiration and phosphorylation [Beckman J.R., Owens K., Knauer T.E. et al.; Lochner A., Kotze J.C.N., Benade A.J.S. et al.]. The effect of excessive amounts of long-chain acyl-CoA may also manifest itself in the inhibition of some other oxidases and enzymes of FA and PL metabolism. All FFA metabolites also inhibit Na, K-ATPase in the cardiac sarcolemma, the toxic effect increasing in the following order: FFA < acyl-CoA < acyl carnitine [Wood J.M., Rush B. and Pitts B.J.].

Along with the abundant data on the toxic effect of FFA metabolites during ischemia, there is also evidence against their involvement in the pathogenesis of ischemic injuries. For instance, the addition of glucose to the perfusate during hypoxic low-volume perfusion of the rat heart led to an increase in acyl-CoA level in the tissue, but it also enhanced the activity of ATP–ADP translocase [Lochner A., Van Niekerk I. and Kotze J.C.N. (1981)]; during perfusion *in vitro*, the cardiac function in rats was not restored completely after ischemia at 37°C, but it did recover after ischemia under conditions of K^+-cardioplegia or hypothermia, although the accumulation of acyl-CoA and acyl carnitine in all three cases was approximately the same [Ichigara K., Abiko Y. and Neely J.R.].

4.3 Accumulation of Lysophospholipids and Their Damaging Effect on Membranes

Phospholipid hydrolysis during ischemia, producing FFA, also results in the accumulation of membrane lysoPL.

The first evidence for this was obtained in experiments [Sobel B.E., Corr P.B., Robinson A.K. et al.] that revealed an increase in the total lysoPC and lysoPEA contents of the rabbit heart up to 162 and 169% of the initial 5–15 and 30–60 min after LAD occlusion *in vivo* and up to 153 and 129% after 3–5 and 5–15 min of low- volume perfusion of the organ *in vitro*. Subsequent detailed studies revealed the accumulation of lysoPC and lysoPEA in ischemic tissues after 10 min of LAD occlusion in cats [Corr P.B., Gross R.W. and Sobel B.E.], after cardiac reperfusion with a Krebs–Henseleit solution [Snyder D.W., Grafford W.A., Glashow J.L. et al.], after LAD occlusion in pigs [Shaikh N. A. and Downar E. (1981)], in the outflowing perfusate during low-volume perfusion of the rabbit heart [Sobel B.E., Corr P.B., Robinson A.K. et al.], and in the totally ischemic rat liver (our experiments) [Dudnik L. B., Bilenko M. V. Alessenko A. V. et al. (1980); Dudnik L.; Dudnik L. B., Bilenko M. V. Alessenko A. V. et al. (1981); Bilenko M.V. (1978b)]. An increase in lysoPL content was even revealed in cases when preparative lysoPL extraction was performed by a modified method under conditions preventing the conversion of membrane plasmalogens into lysoPL [Corr P.B., Gross R.W. and Sobel B.E.; Shaikh N. A. and Downar E. (1981); Shaikh N.A. and Downar E. (1985)].

In some experiments, however, either no accumulation of lysoPL was observed during ischemia or their content even decreased in locally ischemic myocardial tissue [Vasdev S. C., Biro G. P., Narbaiz R. and Kako K. J.] or in a totally ischemic liver [Chien K.R., Abrams J., Serroni A. et al.]. Such a discrepancy in results may be explained by analyzing lysoPL metabolic pathways. Under normal conditions only a small amount of emerging lysoPL undergoes either reacylation or transacylation (both are possible if there is enough energy), or, otherwise, further hydrolysis to glycerophospholipids (glyceroPL). Synthesis of lysoPL *de novo* is also possible. During ischemia, virtually no reacylation and transacylation of lysoPL takes place because the cell is depleted of energy. In addition, reacylation is prevented because of its inhibition by accumulating acyl carnitine and reactive oxygen species (ROS) or acyl-CoA lysotransferase inhibition by excessive glyceroPL. As a consequence, further hydrolysis of lysoPL is the preferable or even the only possible pathway of lysoPL metabolism during ischemia, but even this pathway may be disturbed if pH decreases to 6.5 or acyl carnitine accumulates during ischemia [see Corr P.B., Gross R.W. and Sobel B.E.].

Thus, lysoPL accumulation in membranes during ischemia depends both on activation of phospholipases A_1 and A_2 and on the lack of balance between the deacylation and reacylation processes [Corr P.B., Lee B.I. and Sobel B.E.; Das D.K., Engelman R.M., Rauson J.A. et al. (1986); Lochner A., Kotze J.C.N., Benade A.J.S. et al.].

The involvement in PL hydrolysis of both phospholipases A_1 and A_2, which metabolize PL to glyceroPL without an intermediate stage of lysoPL formation, and lysoPL metabolism along one or several possible pathways may result in the absence of lysoPL accumulation, despite active hydrolysis of membrane PL.

The toxic effect of lysoPL was confirmed in experiments with adding exogenous lysoPL during perfusion of the heart or a slice of cardiac tissue *in vitro*.

Exogenous lysoPC and lysoPEA in amounts equal to those accumulating during ischemia *in vivo* (10-21 μmol/g dry weight) disturbed electrophysiological parameters of isolated conducting cardiomyocytes (Purkinje fibers) from the dog, sheep, and cat heart and of an isolated papillary muscle of the rabbit heart [Arnsdorf M.F. and Sawicki G.J.; Clarkson C.W. and Ten Eick R.E.; Corr P.B., Lee B.I. and Sobel B.E.]; a similar effect on the isolated hamster heart was observed at doses ranging from 0.01 to 0.05 μmol [Man R.Y.K., Wong T. and Choy P.C.], and on the rat heart reperfused after 15 min of ischemia, at doses from 2.5 to 10 μmol [Lochner A., Van Niekerk I. and Kotze J.C.N. (1985)].

Acidosis (pH 6.7) markedly enhanced the damaging effect of lysoPL [Corr P.B., Lee B.I. and Sobel B.E.], which is of particular interest taking into account that a similar decrease in pH also occurs during ischemia (see Chapter 3).

The toxic effect of lysoPL is similar to that of acyl carnitine; moreover, toxic effects of both substances are additive. Exogenous lysoPL may cause arrhythmia [Man R.Y.K., Wong T. and Choy P.C.] and contractural changes in the heart [Bergmann S.R., Ferguson T.B. and Sobel B.E.], act as vasopressors [Takumura A., Fukuzuwa K., Akamatsu Y. et al.], and inhibit Na^+ / Ca^{2+} exchange in canine myocardial sarcolemma [Bersohn M.M., Philipson K.D. and Weiss R.S.].

The toxic effect of lysoPL was manifested most markedly with respect to the plasma membrane into which labeled ^{14}C-lysoPC were mainly incorporated during perfusion of conducting cardiomyocytes or ventricular cardiac muscles in dogs [Saffitz J.E., Corr P.B., Lee B.I. et al.]. Note that lysoPL primarily inhibited Na, K-ATPase and adenylate cyclase [Karli J.N., Karikas G.A., Hatzipavlou P.K. et al.].

Despite an increased lysoPL content during ischemia and the toxic effect exerted by exogenous lysoPL on membrane structures *in vitro*, their role in ischemic damage *in vivo* is not universally recognized. As demonstrated using the modified method of lysoPL extraction [Shaikh N. A. and Downar E. (1981)], the amount of lysoPL accumulating in the heart during AMI is negligibly small (less than 0.6% of the total PL content) and can hardly be responsible for electrophysiological and other ischemic damage to the myocardium. At the same time, even such a small amount of lysoPL is quite sufficient for exerting a local toxic effect when concentrated at a certain site of a cell or subcellular membrane. Indeed, autoradiographic studies showed that an adverse effect on electrophysiological parameters of conducting cardiomyocytes and ventricular muscles of the myocardium arises when only 1% of myocardial PL is replaced by exogenous lysoPL; moreover, the decrease of pH to 6.7 did not lead to an increase in the proportion of emerging lysoPL but enhanced their adverse effect on electrophysiological parameters [Gross R.W., Corr P.B., Lee B.I. et al.].

4.4 Increasing Content of Arachidonic Acid Metabolism Products and their Protective and Damaging Effects on Membranes

As shown above, among FFA accumulating in ischemic tissues due to membrane phospholipid hydrolysis, the key role belongs to arachidonic (eicosatetraenoic) acid (AA, 20:4) [Boime I., Smith E.S. and Hunter F.E.; De Medio G.E., Goracci G. and Horrocks L.A. et al.; Van der Vusse G.I., Roemen Th.A., Prinzen F.W. et al.; Victor T., La Cock C. and Zochner H.; Weglicki W.B., Owens K., Urchel C.W. et al.].

Most of AA (95-98%) is included in membrane PL. In the latter, AA is in β–position and, hence, is liberated from PL molecules primarily by phospholipase A_2 and plasmalogenase, whereas lysophospholipase A_2 liberates AA from lysoPL. In thrombocytes and the vascular endothelium, phospholipase C and PI- and PC-specific phospholipases may also be involved in AA release.

Substances especially rich in AA include PI, PEA, PI, and PS. In thrombocytes, AA accounts for one-third to one-sixth of the total FA content. There is a large amount of AA in PI of the brain, renal cortex, and thrombocytes [see Varfolomeev S.D. and Mezch A.T.]. Linoleic acid (18:2) may also be a source of AA in the cell.

AA may be metabolized by the cyclooxygenase and lipoxygenase pathways (Fig. 30). The contents of AA metabolites produced in both pathways markedly increases during acute ischemia of organs, and their role in pathogenesis of ischemic and reperfusion injuries is now being intensively studied.

Fig. 30 : Arachidonic acid meabolism pathways (AA cascade).

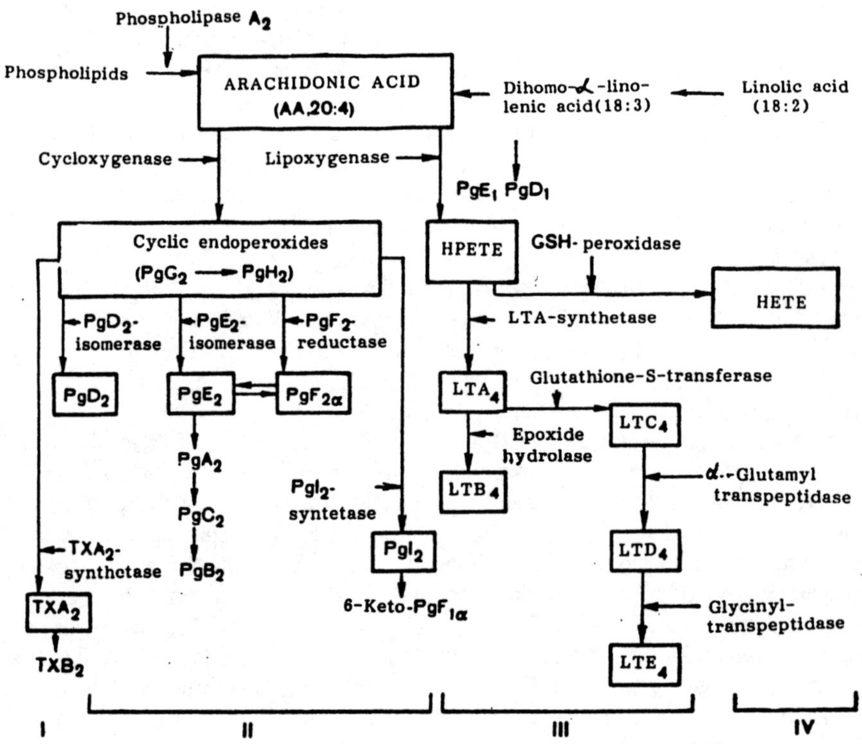

I - Thromboxanes (TX); II - Prostaglandins (Pg); III - Leukotrienes (LT); IV - Hydroxy acids

4.4.1 The Cyclooxygenase Pathway of AA Metabolism

The cyclooxygenase pathway of AA metabolism produces two main groups of compounds: prostaglandins (PgI_2, PgE_2, PgF_2, and PgD_2) and thromboxanes (TXA_2). Common precursors of both groups are represented by unstable and short-lived cyclic endoperoxides (PgG_2 and PgH_2) whose formation is catalysed by the enzyme with peroxidase properties, namely, cyclooxygenase of fatty acids. This enzyme is nonspecific and occurs in microsomal membranes of virtually all organs, especially in endothelial cells, thrombocytes and testes. Conversely, enzymes involved in biosynthesis of prostaglandins and found primarily in microsomal or plasma membranes are organ-specific. Prostacyclin (PgI_2) is produced largely in the vascular endothelium and smooth muscles of the vascular wall, the adrenal cortex, and smooth muscles of non-vascular origin (bronchi, uterus, urinary bladder, vesicular glands) [Smith W.L., De Witt D.L. and Day J.S.]. It is unstable and is easily hydrolysed to a more stable but inactive metabolite (6-keto-$PgF_{1\alpha}$), which is used to determine prostacyclin in tissues. Roles of other prostaglandins are less important, and the sites of their synthesis are less known: PgD_2 appears to be synthesized mainly in microvessels of the brain; $PGF_{2\alpha}$, in vesicular glands (probably in the kidneys); and PgE_2, in vesicular glands, the uterus, and smooth muscle cells.

TXA_2 is mainly synthesized in activated thrombocytes and, to a lesser extent, in blood vessel walls. However, TXA_2 synthesis was also observed in the lungs and organs associated with the metabolism and accumulation of thrombocytes. The half-life of TXA_2 is about 30 s, and the activity of its stable metabolite (TXB_2) is only one-thousandth that of TXA_2. In addition to AA, substrates for cyclooxygenases and precursors of prostaglandins and thromboxanes also include dihomo–α–linoleic acid (yielding monoenoic Pg with one double bond) and eicosapentaenoic acid (20:5) (yielding trienoic Pg with three double bonds). The biological properties of monoenoic and trienoic prostaglandins are less known than those of dienoic prostaglandins.

All cyclooxygenase products of AA metabolism are vasoactive. Cyclic endoperoxides display diverse vasoactivity, which appears to be due to the instability of these compounds and their ability to rapidly convert into different prostaglandins with opposite effects on blood vessels.

PgE_2 and PgD_2 also produce diverse effects on blood vessels. When introduced *in vivo*, PgE_2 acts mainly as a vasodilator, while *in vitro* it acts as a vasoconstrictor. PgE_2 acts differently on smooth muscles in different organs: it improves the tonicity of the uterus but causes relaxation of tracheal and gastric muscles in rats and rabbits.

However, the tone of vascular muscles under normal and pathological conditions is not determined by these minor prostaglandins: the main factor is the ratio between prostacyclin and thromboxane, the two substances producing opposite effects on the vascular tone virtually irrespective of the type of vessel, mode of administration, or experimental conditions (*in vitro* or *in vivo*).

The vasodilative action of PgI_2 on coronary, renal and femoral vessels and its overall hypotensive effect are 5 to 10 times stronger than those of PgE_2 [Markov Kh.M., Kozlov A.V. and Pinelis V.G.; Belo S.E. and Talesnik J.]. TXA_2, on the contrary, is a universal and the most potent vasoconstrictor [Lefer A.M. (1985)].

In addition to their vasoactive effect, cyclooxygenase metabolites of AA also influence the reaction of thrombocyte aggregation: PgH_2 and PgG_2 promote aggregation, PgI_2 inhibits this process and causes deaggregation of thrombocytes, and TXA_2 drastically accelerates their aggregation. PgI_2 and TXA_2 also act as antagonists with respect to adenylate cyclase, and TXA_2 enhances the permeability of lysosomal membranes [Vermylen J. and Deckmyn H.].

Since the effects of PgI_2 and TXA_2 are antagonistic, the PgI_2/TXA_2 ratio acquires special importance [Dusting G.J. and Vane J.R.; Moncada S. and Vane J.R.].

The role of cyclooxygenase metabolites of AA in ischemic and reperfusion injuries in organs is revealed from the following data.

(1) It was found that in all cases of hypoxia and partial or total cardiac ischemia, PgI_2 level in the tissue proper and the outflowing blood or perfusate increases [Evers A.S., Murphree S., Saffitz J.E. et al.; Prosdocimi M., Finesso M., Gorio A. et al.]; after reperfusion of ischemic myocardium and during and after cerebral hypoxia [Gabrielyan E.S. and Amroyan E.A.; Govorova L.V., Kucherenko A.G., Tashchaev Sh.S. et al.], prostaglandins of group E accumulate at a faster rate, whereas the content of prostaglandins possessing vasoconstrictive properties (TXA_2, $PgF_{2\alpha}$) increases later, sometimes only after stimulation, and this increase is less pronounced than that of prostaglandins with vasodilative properties [Panchenko E.P., Smirnov I.E. and Zyabkina A.G.; Evers A.S., Murphree S., Saffitz J.E. et al.; Sakay K., Ito T. and Ogawa K.]. This fact may be attributed to the adaptive role of cyclooxygenase metabolites of AA during ischemia [Meerson F.Z.]. The early expression of inducible cyclooxygenase was found also in infarcted human brains [Sairanen T., Ristimaki A., Karjlainen-Lindsberg M.L. et al.]; the release of FFA was discoved after 5 h ischemia and reperfusion of rabbit aortic endothelial cells [Mackenzie R.J., Furnival C.M., Wood C.B. et al.].

(2) It was shown that the synthesis of prostaglandins and thromboxanes during cardiac ischemia takes place precisely in the ischemic tissue [McCluskey E.R., Corr P.B., Lee B.J., et al.], and after additional stimulation it becomes more pronounced in ischemic, rather than intact, heart [Evers A.S., Murphree S., Saffitz J.E. et al.]. In addition to the ischemic tissue proper, this synthesis may also occur in blood cells accumulated in vessels and vascular smooth muscle cells, but not in the vascular endothelium [Madden M.C., Vender R.L. and Friedman M.].

(3) The exogenous PgI_2 administered to dogs reduces the infarct area and, being added to the perfusate, increases the coronary perfusion rate and improves contraction of the ischemic heart [Lupinetti F.M.,

Starnes V.A., Laws K.A. et al.]. A similar effect was produced by iloprost, a stable PgI_2 analogue, during AMI in rats [Darius H., Osborne J.A., Reibel D.K. et al.]. However, PgI_2 did not demonstrate any direct cytoprotective effect during AMI in rabbits [Edlund A., Sahlin K. and Wennmalm A.]. Administration of PgE_2, PgE_1, and PgI_2 (or its analogue) also reduced the infarct area and improved the collateral circulation during local heart ischemia in dogs and cats [Jugdutt B.I., Hutchins G.M., Bulkle B.H. et al. (1979); Jugdutt B.I., Hutchins G.M., Bulkle B.H. et al. (1981); Ogletvee M.L., Lefer A.M., Smith G.B. and Nicolaou K.C.]. The vasodilative effect of PgE_2 on an isolated strip of the ductus arteriosus *in vitro* was manifest at a low pO_2 in the perfusate and decreased or disappeared when pO_2 in the perfusate returned to normal [Coceani F., Bishai I., White E. et al.].

The data presented above provide convincing evidence that the cyclooxygenase pathway of AA metabolism is intensified both during ischemia and after reperfusion of organs, and this intensification is apparently both adaptive and destructive. The insufficiency of the adaptive reaction follows from the protective effect of administering exogenous prostaglandins with vasodilative properties, observed in most experiments.

4.4.2 The Lipoxygenase Pathway of AA Metabolism

The lipoxygenase pathway of AA metabolism involves a group of lipoxygenases present in the cytosol or ER membranes of blood cells (thrombocytes, reticulocytes, neutrophiles), macrophages, and mast cells; they can also be found in seminal vesicles, the heart, and the liver [Borgeat P. and Sirois P.; Evers A.S., Murphree S., Saffitz J.E. et al.; Hamberg M., Hedqvist P. and Radergranb K.; Samuelsson B., Hammarstrom S., Murphy R.S. and Borgeat P.].

It is of particular importance that AA metabolism along the lipoxygenase pathway intensely proceeds in vascular smooth muscle and endothelial cells [Samuelsson B., Hammarstrom S., Murphy R.S. and Borgeat P.].

AA oxidation by lipoxygenases yields AA hydroperoxides (HPETE), which are either converted into hydroxy acids (HETE) by GSH peroxidase or, with the aid of leukotriene A synthetase, give rise to a group of leukotrienes (LT): unstable and virtually inactive leukotriene A_4 (LTA_4), B_4 (LTB_4) and C_4 (LTC_4), the latter being metabolized into highly active products such as leukotrienes D_4 and E_4 (see Fig. 30).

All products of the lipoxygenase reaction demonstrate extremely high biological activity and pronounced vasoactive properties. HETE and LTB_4 drastically enhance the chemotaxis of neutrophils, eosinophils, monocytes, and macrophages, the chemotactic activity of hydroxy acids decreasing in the order: 5HETE > 8HETE > 11HETE = 12HETE > 15HETE.

Hydroperoxides and hydroxy acids resulting from AA metabolism mediated by 15-lipoxygenase (15HPETE and 15HETE) proved to exert a pronounced vasoconstrictive effect on veins and arteries (coronary, renal, splenic, femoral, cerebral) in dogs [Koide T., Noda Y., Nata S. et al.; Van Diest M.I., Verbeuren T.J. and Herman A.G.], coronary arteries in cats [Trachte G.J., Lefer A.M., Aharony D. and Smith J.B.], and the aorta in rabbits [Asano M. and Hidaka H.]. The vasoconstrictive action on arteries was not mediated by endothelium (the effect persisted after de-endothelization of the vessel wall) but was suppressed by Ca^{2+} inhibitors [Van Diest M.I., Verbeuren T.J. and Herman A.G.] (see Chapter 8).

Leukotrienes C_4, D_4 and E_4 (sulfide peptides) display primarily bronchospastic, vasoconstrictive and neuron-exciting action [see Sagach V.F.; Ezra D., Boyd L.M., Feuerstein G. and Godlstein R.E.; Lefer A.M. (1985); Lewis R.A. and Austen K.F.].

The constrictive effect of these leukotrienes (especially LTC_4 and LTD_4) with respect to smooth muscles of the small intestine and bronchi is 10^3–10^4 times that of histamine [Piper P.J., Letts L.G., Samhoun M.N. et al.]; sulfide peptides were found to act as vasoconstrictors with respect to arterioles, arteries, and veins in different organs, this action increasing as follows: $LTC_4 > LTD_4 > LTE_4$ [Lefer A.M. (1985)].

In addition to their local and generalized vasoconstrictive effect, LTC_4, LTD_4, and LTE_4 almost immediately enhance the permeability of microvessels. Apparently, both leukocytes and the endothelium of blood vessels possess specific receptors and serve as target cells for leukotrienes [Lewis R.A. and Austen K.F.].

Lipoxygenase metabolites of AA appear to play all very important role in ischemic and reperfusion injuries in organs.

It was demonstrated that concentrations of leukotrienes C_4, D_4 and E_4 increase during myocardial ischemia and upon the traumatic shock [Lefer A.M. (1985)]. Inhibition of AA metabolism along both pathways, in contrast to that of the cyclooxygenase pathway alone, reduced the infarct area in dogs [Jolly S.R. and Lucchesi B.R.]. Stimulation of both AA metabolic pathways during isolated perfusion of the rabbit heart with temporary occlusion of the left circumflex coronary artery (LAC) was accompanied by an increase in PgE_2, PgI_2, TXA_2 and LTC_4 levels in the effluent and an increase in the resistance of coronary vessels. The latter did not respond to treatment with either a TXA_2 inhibitor or indometacin but could be prevented completely using the local inhibitor of leukotriene receptors (FPα–55712) [Evers A.S., Murphree S., Saffitz J.E. et al.]. The pronounced vasoconstrictive effect of LTC_4 and profuse infiltration of the heart with leukocytes (within the first day), mononuclears, and fibroblasts capable of LTC_4 production (on day 4) suggest that leukotrienes are responsible for the coronaroconstriction and myocardial injury during infarction [Evers A.S., Murphree S., Saffitz J.E. et al.].

AA and especially its lipoxygenase metabolites have also been ascribed a major role in ischemic edema of the brain [Chan P.H. and Fishman R.A.].

A damaging effect may also be exerted by hydroxy acids (in particular, 12HETE) whose production in the infarct area of canine myocardium was ten times that in the adjacent intact tissues [Mullane K.M., Read N., Salmon J.A. and Moncada S.].

Intensification of AA oxygenation along the lipoxygenase pathway is associated with cell excitation (unspecific stimulation, activation of phagocytosis) and the increased accumulation of intracellular Ca^{2+}, i.e., with the phenomena always occurring during ischemia.

The relationship between cyclooxygenase and lipoxygenase pathways of AA metabolism is both direct and reciprocal.

On the one hand, both pathways have a common precursor, AA, and, hence, are simultaneously intensified during PL hydrolysis and simultaneously suppressed by phospholipase A_2 inhibitors, Ca^{2+} antagonists, α–adrenolytic substances, steroids, albumin, etc. There are also compounds inhibiting both cyclooxygenase and lipoxygenase, such as eicosa-5,8,11,14-tetraenoic acid, BW-755C, phenidone, FPp-755C, AHR-6293, and others.

The existence of a direct relationship between the cyclooxygenase and lipoxygenase pathways of AA metabolism is also confirmed by the data that vasoconstriction and the enhancement of vascular permeability under the effect of leukotrienes introduced into the lungs may be mediated by the release of TXA_2 and other prostaglandins from the vessel wall [Piper P.J., Letts L.G., Samhoun M.N. et al.; Bodr K.F., Baylis C., Pfeffer S.M. et al.]; PgE, on the contrary, enhances the ability of leukotrienes to increase vascular permeability [Lewis R.A. and Austen K.F.].

On the other hand, selective inhibition of the cyclooxygenase pathway by indometacin resulted in intensification of the lipoxygenase pathway; the addition of indometacin or acetylsalicylic acid to AA hydroxyperoxides or hydroxy acids (15HPETE and HETE) enhanced the vasoconstrictive effect of the latter [Van Diest M.I., Verbeuren T.J. and Herman A.G.]; and, vice versa, intensification of the cyclooxygenase pathway and TXA_2, PgI_2, and PgF_2 production by adding bradykinin to the perfusate during the rabbit heart reperfusion after temporary LAD occlusion reduced the yield of leukotrienes [Evers A.S., Murphree S., Saffitz J.E. et al.].

AA hydroxyperoxides (HPETE) resulting from AA metabolism along the lipoxygenase pathway (see Fig. 30) inhibit PgI_2 synthetase and suppress PgI_2 synthesis in the walls of various vessels and in an endothelial cell culture [Moncada S. and Vane J.R.; Okuma M., Takayama H. and Uchino H.].

4.5 Possible Factors Promoting Phospholipid Hydrolysis, and Mechanisms of the Damaging Effect Exerted by Its Products on Membranes

4.5.1 Factors Responsible for Activation of Phospholipases and Intensification of Membrane Phospholipid Hydrolysis during Ischemia of Organs

Factors responsible for activation of phospholipases and intensification of membrane phospholipid hydrolysis during ischemia of organs have been widely discussed in scientific literature but still remain to be studied in detail. The most important of them are as follows.

(1) Increasing content of intracellular Ca^{2+}, which is an activator (cofactor) of most intracellular phospholipases (except for lysosomal enzymes). This is confirmed by the fact that Ca^{2+} transport inhibitors effectively suppress phospholipase activities and prevent PF degradation [Chien K.R., Abrams J., Serroni A. et al.; Grinwald P.M, and Nayler W.G.]. These activities may also be suppressed by calmodulin inducers, which is evidence that the stimulating effect of Ca^{2+} on phospholipase may result from calmodulin binding [Azizova N.N., Muratova U.Z., Gagel'gans A.I. and Saatov T.S.; Wong P.Y.K. and Cheung W.Y.]. However, the relationship between activation of phospholipases and Ca^{2+} concentration in the cell implies a priori that the involvement of the phospholipase mechanism in the pathogenesis of membrane damage is secondary because, as shown in Chapter 2, Ca^{2+} accumulation in the cell results from an increase in the passive membrane permeability for this ion, which should be conditioned by some other mechanisms operating at earlier stages.

(2) Decreasing intracellular pH, which is especially important for activation of lysosomal phospholipases (see above), although other phospholipases (e. g., phospholipases A in the plasma membrane and lysophospholipases) may also function in a more acidic medium.

(3) Increasing intracellular catecholamine (CA) content or changes in the ratio of its components, which is especially manifested during partial ischemia of an organ (heart) and in the reperfusion period, i.e., in situations accompanied by stress and possible supply of epinephrine (E) from the adrenals into the organ with blood. During total ischemia, however, the content of norepinephrine (NE) in the organ also increases (see Chapter 10).

(4) Disturbances of ADN metabolism pathways and energy deficiency in the tissue, causing dephosphorylation of membranes and disturbing their structural organization (see Chapter 3), which facilitates the interaction between phospholipases and membrane PL [Nguyen V.D., Cieslinski D.A. and Humes H.D.; Shaikh N. A. and Downar E. (1981)].

(5) Intensification of LPO processes, which is one of the most important factors responsible for activation of phospholipases during ischemia and after reperfusion of organs. Exogenous oxidation of membrane phospholipids or activation of LPO *in vivo* [Kondakova I. V., Borunob E. V. and Antipova E. V.; Lukacova N., Gottlieb M. and Marsala J.; Gogvadze V.G., Bruetovetskii N.N. and Zhukova A.A.; Sevanian A., Muakkassah-Kelly S. and Montestruque S.; Weglicki W.B., Dickens B.F. and Tong-Mak I.; Chan P.H., Yurko M. and Fishman R.A.; Okayasu T., Curtis M.T. and Farber J.L.] stimulate hydrolytic activity of PLA_2, whereas antioxidants counteract this process [Musat S. and Dhalla N.S.]. Numerous studies showed that PLA_2 preferentially hydrolyzes oxidized FA [Kagan V. E., Shvedova A. A. and Novikov K. N.; Sevanian A., Muakkassah-Kelly S. and Montestruque S.; Sevanian A. and Kim E.], which is evidence that PLA_2, like proteolytic enzymes, has a membrane-protecting effect, or serves as a secondary antioxidant defense because it excises oxidized FA from the membrane, preventing their participation in free radical progradation reactions [Sevanian A.; Davies K.J.A., Wiese A.G., Sevanian A. and Kim E.H.; Rashba-Step J., Tatoyan A., Duncan R. et al.]. The interrelationship between phospholipase and mechanisms of lipid peroxidation is also discussed in Chapter 10.

(6) PLA_2 activation resulting from an increase in membrane lysoPL, which gives rise to the situation of a vicious circle [see Hashizume H., Hoque A.N., Magishi K. et al.].

(7) An important factor responsible for the accumulation of phospholipolysis products during ischemia is disturbed PL biosynthesis, particularly deacylation reacylation processes [Corr P.B., Gross R.W. and Sobel B.E.; De Windt L.J., Reneman R.S., Van der Vusse G.J. and Van Bilsen M.; Van der Vusse G.J., Reneman R.S. and van Bilsen M.]. The prevalence of accumulation of phospholipolysis products (FFA) is typical for total ischemia.

4.5.2 Factors Responsible for Intensification of Cyclooxygenase and Lipoxygenase Pathways of AA Metabolism during Ischemia of Organs

Factors responsible for intensification of cyclooxygenase and lipoxygenase pathways of AA metabolism during ischemia of organs are even less known than those involved in phospholipid hydrolysis, because their role in ischemic damage is a relatively recent discovery.

The main reason why these pathways of AA metabolism are stimulated appears to be the accumulation of AA in the tissue, because the rate of phospholipid hydrolysis correlates with that of prostaglandin production, while phospholipase inhibitors drastically reduce both AA hydrolysis and its metabolism along the two pathways [Otani H., Engelman R.M., Ronson J.A. et al.].

Another equally important factor providing for intensification of AA metabolism is the accumulation of blood cells in microvessels of organs during ischemia and after reperfusion (upon the development of no-reflow phenomenon, see Chapter 1). Thrombocytes are especially important: they are very rich in cyclooxygenase, thromboxan synthetase, and lipoxygenase, and these enzymes are activated upon aggregation and adhesion of blood cells. This factor may partially account for the higher rate of prostaglandin production during the reperfusion period, observed in [Otani H., Engelman R.M., Ronson J.A. et al.].

4.5.3 Mechanisms of Membrane Damage by Phospholipid Hydrolysis Products and AA Metabolites during Ischemia of Organs

Mechanisms of membrane damage by phospholipid hydrolysis products and AA metabolites during ischemia of organs have been studied mainly on model systems involving the use of exogenous products. As corresponding results were described in a number of reviews [Brecher P.; Corr P.B., Gross R.W. and Sobel B.E.; Hashizume H., Hoque A.N., Magishi K. et al.; Katz A.M. and Messineo F.C.], they will be considered very briefly.

All long-chain FFA, their incompletely oxidized metabolites, CoA, and carnitine derivatives are water-soluble amphiphiles, i.e., their molecules contain both hydrophilic and hydrophobic groups. The effect of these substances on membranes depends on their concentration.

At low concentrations, amphiphiles exist in aqueous solutions as monomers that can incorporate into the hydrophobic layer of the membrane and, by changing its physicochemical properties, stabilize the membrane and reduce its permeability. The alteration of physicochemical properties of membrane lipids (especially annular) by low doses of amphiphiles, combined with the possible removal of Ca^{2+} from the sites of its binding with negatively charged heads of PL, can lead to changes in the function of membrane-bound (lipid-dependent) proteins, which may be manifested both in increase and in decrease of enzyme activities.

At high concentrations, amphiphiles in the aqueous medium aggregate into micelles, and their effect on membranes is almost invariably damaging (detergent-like). This effect arises because micelles penetrate into the membrane, which destroys the bilayer, "expels" individual phospholipid molecules or membrane-bound Ca^{2+} from the latter, and renders the membrane highly permeable (Fig. 31). When micelles are embedded into annular lipids, they selectively damage the function of the protein molecule and displace protein from the membrane [De Leiris J. and Opie L.H.]. As already mentioned, proteins most sensitive to amphiphiles include Na, K-ATPase, Ca-ATPase, and ATP–ADP translocase.

However, the membrane-damaging effect of phospholipolysis products at high doses may be due to the detergent-like properties of monomers, as well as of micelles. This primarily applies to acyl carnitine [Corr P.B., Lee B.I. and Sobel B.E.] and lysoPL [Bergmann S.R., Ferguson T.B. and Sobel B.E.; Gross R.W., Corr P.B., Lee B.I. et al.; Man R.Y.K., Wong T. and Choy P.C.]. Because of the irregular (cuneiform) shape of their molecules, acyl carnitine and lysoPL may cause membrane bending and even rupture [Lucy J.A.] (see Fig. 31). Moreover, their effect on membranes is additive [Corr P.B., Lee B.I. and Sobel B.E.].

Cyclooxygenase and lipoxygenase metabolites of AA act predominantly through specific cell receptors. The possibility and mechanisms of their direct action on membranes are virtually unstudied, although it is known that the ability of TXA_2 to aggregate thrombocytes [Avdonin P.V., Men'shikov M.Yu., Orlov S.N. et al.; Panchenko E.P., Smirnov I.E. and Zyabkina A.G.], as well as the vasoconstrictive and bronchospastic effects of leukotrienes, at least LTC_4 and LTD_4, along with their ability to disturb capillary permeability [Weiss E.B. and Mullick P.C.], depend on intracellular Ca^{2+} accumulation resulting from its intensified influx from the extracellular space.

Fig. 31: Possible mechanisms of membrane damage by high dose of amphiphiles (FFA, incompletely oxidized metabolites, FFA derivatives, lysoPL, etc.) acording to [Lucy J.A.]

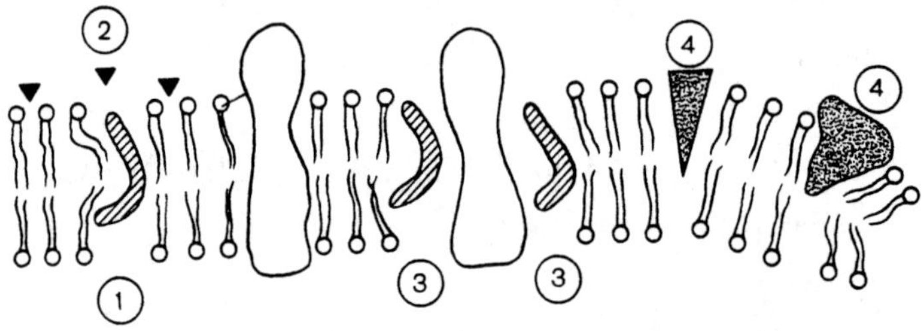

1 - Penetration into the membrane; 2 - pushing of Ca^{2+} from the membrane; 3 - rupture (damage) of annular lipids; 4 - curving of the membrane by lysoPL.

It was also shown that the direct Ca^{2+}-dependent effect on vascular smooth muscles is also exerted by products of AA oxidation by lipoxygenase, namely, its HP and hydroxy acids (15HPETE and 15HETE), and hydroxy acids are capable of damaging lysosomal membranes, thereby facilitating the release of lysosomal enzymes [Van Diest M.I., Verbeuren T.J. and Herman A.G.].

An equally important factor of membrane damage during hydrolysis of membrane PL is that the pool of total and individual PL (PEA, PI, PC, etc.) in cell membranes becomes depleted [Halat G., Lukacova N. Chavko M. and Marsala J.; Oubidar M., Boquillon M., Maril Ch. et al.; Van der Vusse G.J., Reneman R.S. and van Bilsen M.]. When SR membranes from skeletal muscles were incubated with phospholipase A_2, PC and PEA contents sharply decreased with a corresponding increase in lysoPC and lysoPEA [Swoboda G., Fritzsche J. and Hasselbach W.]. Albumin added to the incubation medium promoted further extraction of PL, lysoPC, and PUFA from membranes. These changes affected the activity of Ca-ATPase: it increased during incubation without albumin but drastically decreased when the latter was added. According to the authors of this work [Swoboda G., Fritzsche J. and Hasselbach W.], this effect resulted from the removal of FA from the membrane, and the reversibility of changes depended on the critical PL level: at 0.4 and 0.2 μmol of remaining PL, Ca-ATPase could be reactivated by oleinic acid to 70 and 30% of the initial activity, respectively.

A decrease in the content of cardiolipin, a tetra-acyl phospholipid, is the factor having an extremely adverse effect of mitochondrial membrane functioning. Cardiolipin, which is very rich in unsaturated bonds, is located in the inner mitochondrial membrane and is a likely target for peroxidation and hydrolysis by LPA_2. During ischemia and reperfusion of organs, an earlier and more drastic decrease in cardiolipin content, compared to PC and PEA, was revealed in the heart [Vasdev S. C., Biro G. P., Narbaiz R. and Kako K. J.], kidney [Smith M.W., Collan Y., Kahng M.W. and Trump B.F.], liver [Okayasu T., Curtis M.T. and Farber J.L.], and skeletal muscles [Soussi B., Idström J.P., Schersten T. and Bylund-Fellenius A.C.]. The reduced content of cardiolipin results in mitochondrial dysfunction and especially affects the activity of cytochrome C oxidase [Okayasu T., Curtis M.T. and Farber J.L.; Soussi B., Idström J.P., Schersten T. and Bylund-Fellenius A.C.].

The degree and role of the decrease in membrane PL (delipidization) in ischemic and reperfusion injuries were discussed in Chapter 2; however, the involvement of the phospholipase and lipid peroxidation mechanisms requires further study. This issue will be reviewed in Chapter 5.

4.6 Conclusions

The data discussed above provide evidence that an important role in the pathogenesis of membrane damage during acute ischemia of organs and after their reperfusion belongs to the accumulation of products of PL hydrolysis and their subsequent metabolism by different pathways.

Activation of phospholipolysis under ischemia is promoted by the existence of its anaerobic pathway; its relationship with intracellular Ca^{2+} accumulation, which considerably increases during ischemia (see Chapter 2); its dependence on acidosis typical for ischemia; and, what is most important, by the fact that it is facilitated in membranes already damaged by energy deficiency (see Chapter 3) or lipid peroxidation (see Chapter 5), which completes this vicious circle and leads to further membrane damage.

In recent studies, researchers elucidated and analyzed in detail the involvement of cyclooxygenase and lipoxygenase metabolites of AA in ischemic and reperfusion injuries. These results significantly improved our knowledge about the mechanism of damaging action of PL hydrolysis products and demonstrated its universal nature in different organs and models of ischemia.

The basic conclusions summarizing these data are as follows.

(1) During ischemia, both lipolysis (hydrolysis of neutral lipids in the cytosol) and phospholipolysis (hydrolysis of membrane PL) in the cell are intensified. The latter process in more important, especially in total ischemia.

(2) Lipolysis and phospholipolysis actively proceed both during ischemia and in the reperfusion period. The ambiguity of the results obtained in attempts to evaluate their intensity on the basis of an increase in FFA level (see above) appears to stem from underestimating FFA consumption LPO, which are intensified during ischemic and postischemic periods (see Chapter 5), FFA (AA) metabolism by the cyclooxygenase and lipoxygenase pathways, accelerated conversion of FFA into acyl-CoA and acyl carnitine after ischemia, and their reacylation to PL.

(3) Membrane damage during intracellular hydrolysis of lipids and phospholipids is inflicted by FFA themselves and products of their incomplete oxidation, such as acyl-CoA, acyl carnitine, and β–hydroxy FA and lysoPL. The effect of these compounds is implemented either via alteration of the physicochemical properties of the lipid bilayer in membranes or through direct impairment of lipid-dependent enzyme functioning. An equally important role belongs to changes in the qualitative and quantitative composition of membrane PL and delipidization of membranes.

(4) Since lipolysis and especially phospholipolysis lead to the maximal release of AA, a rather important (if not the leading) role in the damaging effect of these processes belongs to products of AA metabolism along the cyclooxygenase pathway (particularly prostaglandins and thromboxanes with vasoconstrictive and thrombocyte-aggregating properties) and the lipoxygenase pathway (particularly hydroxy acids and leukotrienes C_4 and D_4). These metabolites appear to be especially important for the pathogenesis of collateral circulation disorders during partial ischemia and the no-reflow phenomenon in the early postischemic period.

Identification of selective inhibitors for these metabolites and the elucidation of the mechanism of their membrane-damaging action will open up new possibilities for preventing and treating ischemic injuries.

(5) The damaging effect of intracellular lipolysis and phospholipolysis products depends on their levels, which is why the extent of their removal from organs acquires special importance: effluent from organ remains unchanged during hypoxia, decreases during partial ischemia, and ceased to zero during total ischemia. Hence, the damaging effect of PL hydrolysis is more pronounced during total ischemia.

Chapter 5

INTENSIFICATION OF LIPID PEROXIDATION DURING ACUTE ISCHEMIA AND AFTER REPERFUSION OF ORGANS

It is now evident that free radicals and LPO products play a major role in the regulation of membrane lipid metabolism and changes in the physicochemical properties and permeability of biological membranes under physiological conditions and in many pathological states [Burlakova E.B., Alessenko A.V., Molochkina E.M. et al.; Burlakova E.B. and Khrapova N.G.; Vladimirov Yu. A. and Archakov A. I.; Hess M.L. and Manson N.H.].

Studies on the role of free radicals and LPO in pathogenesis of ischemic and reperfusion injuries began later than those on other pathologies because these processes have been known as oxygen-dependent, and this provided the basis for a long-standing concept (which eventually proved to be erroneous) that they cannot occur under oxygen deficiency. Some specialists even believed that "the peroxide hypothesis adequately explains many cases of cell death (*except ischemia*)" [Vladimirov Yu. A. and Archakov A. I.].

Evidence for the accumulation of LPO products and a decrease in antioxidant activity *in vivo* or, in other words, for LPO intensification during ischemia and after reperfusion of organs (namely, kidneys and the heart) was obtained for the first time in our studies performed in cooperation with A. V. Alessenko and E. B. Burlakova. Our results were first presented at the First All-Union Symposium on Acute Ischemia of Organs (Moscow, November 1973) and than on the Bilateral Symposium on Experimental and Clinical Surgery (Smolenice, Czechoslovakia, October 1974). In the next years (1975–1978) these data were many times published in Russia and abroad [Alessenko A.V., Bilenko M.V., Burlakova E.B. et al.; Alessenko A. V., Bilenko M. V. and Shelenkova L. N.; Bilenko M. V. (1978a); Bilenko M.V. (1978b); Bilenko M.V., Alessenko A.V., Burlakova E.B. et al. (1975); Bilenko M.V., Alessenko A.V. and Shelenkova L.N.]. Simultaneously and independently, A. Kh. Kogan, A. N. Kudrin, and S. M. Nikolaev revealed the increase of chemiluminescence in the zone of ischemia and developing myocardial infarction. These both studies were deemed pioneering, as was confirmed after long expertise in 1990 by the USSR Certificate of Discovery entered into the State Register of Discoveries (Diploma No. 399, priority of 1973) [Bilenko M. V., Alessenko A. V., Burlakova E. B. et al. (a)].

To date, we have already studied changes in LPO intensity in far greater detail and not only in the heart and kidney, but also in other organs differing in sensitivity to ischemia, including the brain, liver, and skeletal muscles. Numerous observations on LPO intensification during ischemia and reperfusion have also been published by other authors, and the number of such publications is progressively increasing each year. This chapter deals with the analysis of these processes, whereas the causes and possible mechanisms of ROS formation and LPO intensification during ischemia will be discussed in Chapter 10.

Here, it is necessary to discuss what is meant by "LPO intensification. " The point is that corresponding processes are usually quantified by assessing the accumulation of LPO products in membrane structures or the homogenate of a particular organ, and the content of these products, just as that of any other cell metabolites, is an integral parameter depending on the rates of two opposite processes, namely, product formation and product elimination (by way of metabolism, consumption, hydrolysis of phospholipases, or flushing out of the organ).

An increase in the content of LPO products in an organ is indicative of their more rapid formation (i.e., of true, or absolute LPO acceleration) only if the rate of metabolism and elimination of these products remains stationary. However, in many pathological states, including ischemia, metabolism and consumption of these products slow down, whereas their elimination from the organ either decelerates (partial ischemia) or ceases (total ischemia) and is resumed only after reperfusion. Under these conditions, the content of LPO products may increase as a result of either an actual increase in peroxidation rate or (which is more likely) more intensive accumulation of LPO products. In any event, this increase suggests that the balance between the formation and consumption (elimination) of LPO products is upset, and the production of lipid metabolites prevails over their consumption. This is the relative aspect that should be taken into account while using the term " LPO intensification, " which actually implies an increase in the content of LPO products.

5.1 METHODS FOR DETERMINING ROS AND LPO PRODUCTS IN ISCHEMIC AND REPERFUSED ORGANS

Determining the content of LPO products accumulating in membranes *in vivo* is difficult for a number of reasons: (1) most of these substances (lipid radicals, hydroperoxides) are unstable (the more reactive the product, the lower its stability); (2) their amount is extremely low, and very sensitive methods are needed to detect them; (3) LPO products may be flushed out of an organ during reperfusion; and (4) both the formation and degradation of LPO products may occur in the course of tissue treatment and lipid extraction, causing an increase or decrease in their content in the sample. The latter is most relevant to determining the content of LPO products in ischemic organs because they may be formed *de novo* when the homogenate or extracted lipids are exposed to atmospheric oxygen. Consequently, in order to positively identify the LPO products formed during ischemia, rather than subsequent treatment of material, all manipulations (homogenization, lipid extraction, and determination of LPO products) should be performed under conditions preventing or minimizing the exposure of lipids to the air.

Our experience in working with ischemic tissues and special control tests showed that tissues should be homogenized after rapid freezing in liquid nitrogen or in a chamber blown with argon or highly purified nitrogen, and lipids should be extracted at a low temperature (0–2°C), with the extraction mixture supplemented with BHT (10^{-5} M) or bubbled with highly purified nitrogen.

The correct use of adequate methods (preferably, more than one) for determining LPO products is a necessary condition for properly assessing the content of LPO products in ischemic organs.

Toxic metabolites accumulating in tissues owing to intensified free-radical reactions and LPO can be conventionally divided into reactive oxygen species, intermediate LPO products of free-radical nature, and the primary, secondary, and end products of LPO. Their origin and sequence of formation are described in Fig. 81 (Chapter 10).

Reactive oxygen species (ROS), which include superoxide ion radical $O_2^{\cdot-}$, hydrogen peroxide H_2O_2, and hydroxyl radical $^{\cdot}OH$, are the main inducers of membrane phospholipid oxidation via the nonenzymatic pathway. The increase of their content in organs indicates the existence of prerequisites for LPO initiation or the progression of LPO.

Intermediate LPO products include alkyl ($R^{\cdot}$), alkoxyl ($RO^{\cdot}$), and peroxyl ($ROO^{\cdot}$) radicals. They are formed at the first (initiation, $R^{\cdot}$), second (propagation, $RO_2^{\cdot}$), and third (multiplication, or branching; $R^{\cdot}$, $RO^{\cdot}$, $RO_2^{\cdot}$) stages of LPO. All the oxygen-centered ($O_2^{\cdot-}$, $^{\cdot}OH$, $RO^{\cdot}$, $ROO^{\cdot}$) and carbon-centered ($R^{\cdot}$) free radicals are highly reactive and have low steady-state concentrations. Hence, they are mainly detected by methods of electron spin resonance (ESR) or electron paramagnetic resonance (EPR) in modifications without or with spin trap agents, such as *tert*-nitrosobutane (*t*NB), phenyl-*tert*-butylnitrone (PBN), α-4-pyridyl-1-oxide N-*tert*-butylnitrone (4-POBN), 5.5-dimethylpyrroline N-oxide (DMPO), or others (OXANOH, PMA, CLA-phenyl, etc.). In the latter modification, free radicals react with the double bond of spin trap molecules to form radical adducts, which are more stable than the initial free radicals. This technique is now widely used [Arroyo C.M., Kramer J.H., Leiboff R.H. et al.; Li X.Y., McCay P.B., Zughaib M. et al.; Togashi H., Shinzawa H., Yong H. et al.; Vartanyan L.S., Rashba Yu.E., Nayler L.G. et al. (1990); Weixiong H., Aneman A., Nilsson U. and Lundgren O.; Wu W. and Qiu F.; Zweier J.L., Duke S.S., Kuppusamy P. et al.], although it is not absolutely specific and also measures free radicals of proteins, amino acids, and nucleic acids.

The oxygen-centered $ROO^{\cdot}$ (peroxyl) radical can also be determined by the chemiluminescence method, i.e., by measuring quanta of visible light emitted during $ROO^{\cdot}$ recombination in the following reactions:

$$ROO^{\bullet} + ROO^{\bullet} \rightarrow ROH + RO + {}^{1}O_2 \text{ or}$$

$$ROO^{\bullet} + ROO^{\bullet} \rightarrow ROH + RO^* + O_2,$$

where ${}^{1}O_2$ is singlet oxygen, and RO* is the excited triplet carbonyl group (reaction constant is 10^6–10^8 l/mole • s) [Cadenas E., Varsavsky A.I., Boveris A. et al.]. Chemiluminescence intensity is proportional to the square of peroxide radical concentration.

However, the intensities of chemiluminescence and LPO do not always demonstrate a strict quantitative correlation [Vladimirov Yu. A. and Archakov A. I.; Vladimirov Yu.A., Olenev V.I., Suslova T.B. et al.; Vladimirov Yu.A., Suslova T.B. and Olenev V.I.]. Moreover, chemiluminescence can only be measured after transferring the tissue to aerobic conditions, and, hence, its intensification cannot serve as a direct indication of LPO intensification *in vivo* during ischemia. It can only provide evidence that the ischemic tissue is susceptible (less resistant) to the initiation of free-radical processes (for example, in the postischemic period).

Primary LPO products—hydroperoxides, endoperoxides, etc. —are products of enzymatic (see Fig. 30) and nonenzymatic oxidation (see Fig. 81). They are determined in extracted lipids by measuring the contents of lipid peroxides (LP), hydroperoxides (HP), and diene conjugates (DC). The latter are the product of rearrangement of double bonds in fatty acids containing more than two such bonds. LP are usually measured by iodometric titration [Burlakova E.B., Alessenko A.V., Molochkina E.M. et al.; Vladimirov Yu. A. and Archakov A. I.] with the equivalence point determined amperometrically [Neyfakh E.A. and Kagan V.E.]; HP are determined polarographically [Danilov V. S., Sitkovsky M. V., Kagan V. E. and Kozlov Yu. P.]; and DC and ketodienes, by UV spectrophotometry at 232–235 and 272–280 nm [Bolland J.L. and Koch H.J.], either after reducing LPO products to the stable hydroxyl form [Romaschin A.D., Rebeyka I., Wilson G.I. et al.] or without this procedure. Lipid purification by high performance liquid chromatography (HPLC) and isolation of the phospholipid fraction before measurements provides for a significant increase in specificity of these methods [Hegstad A.C., Strand H. and Ytrehus K.].

Secondary LPO products, which result from oxidative degradation of phospholipid hydroperoxides, include triene conjugates (TC) and numerous (more than 18 [Vladimirov Yu. A. and Archakov A. I.]) short- and long-chain carbonyl compounds: epoxides, epoxy alcohols, ketoacids, alkanals, alkenals, malonic dialdehyde (MDA), 4-hydroxyalkenal, 4-hydroxy-2-nonenal, etc. [see Cheeseman K.H.].

TC are determined by UV spectrophotometry at 270–280 nm, although several unsaturated ketone-containing fatty acids have a similar absorbance range. The degree of lipid oxidation can also be estimated from the ratio of products absorbing UV light of different wavelengths, e. g., from the 232/215 and 232/206-nm absorbance ratio (oxidation index), with absorbance by unoxidized lipids being maximal at 215 nm. To determine 4-hydroxy-2-nonenal, different variants of chromatography and an immunohistochemical method are used.

MDA is generally regarded as a typical component of lipid peroxidation products reacting with 2-thiobarbituric acid (TBA) with the formation of TBA-reactive substances (TBARS), which are registered spectrophotometrically by absorbance at 532–535 nm. This method is relatively simple and does not require lipid extraction but has many disadvantages: it is insufficiently specific, as TBA reacts not only with MDA, but also with many other compounds (aromatic aldehydes, amino acids, carbohydrates, etc.); MDA is only formed after oxidation of linolenic (18:3) and arachidonic (20:4) acids, but not of oleic acid (18:1), linolic acid (18:2), or cholesterol; and it is rapidly metabolized in mitochondria or consumed in reactions with compounds containing sulfhydryl or amino groups to form end products. The yield of MDA is very low, and, hence, the formation of TBARS is mainly accounted for by MDA formed *in vitro,* in the course of reaction with TBA, rather than *in vivo*.

Carbonyl compounds are formed during the oxidative destruction of primary products (lipid hydroperoxides). As corresponding reactions are oxygen-dependent, these compounds are mainly revealed during partial ischemia or in the early reperfusion period, rather than in cases of total ischemia (see section 5.3).

Several authors have expressed an opinion that MDA determination by means of the TBA assay (at least in cases when only this method is used) is only expedient in experimental models, rather than in biological systems [Vladimirov Yu. A. and Archakov A. I.; Kozlov Yu. P., Danilov V. S. and Kagan V. E.; Nohl H. and Hegner D.]. This is especially true of ischemic tissues because the TBA assay is performed in an acidic medium, at high temperature (100°C) and, necessarily, in aerobic conditions, in which the results of measurements are almost ten times as high as those obtained under anaerobic conditions [Kostyuk V. A. and Potapovich A. I.].

Because of these shortcomings, numerous modifications of the TBA assay have been developed. For example, it was proposed to conduct the reaction in the presence of high concentrations of Fe^{2+} and BHT, an antioxidant (10^{-2} M), or in two variants: in the presence of Fe^{2+} (to determine alkadienols) and EDTA (to determine MDA) [Kikugawa K., Kojima T., Yamaki S. and Kosugi H.]. In addition, a HPLC-based TBA assay is used [Bird R.P. and Draper H.H.; Ceconi C., Cargnoni A., Pasini E. et al.; Grune T., Sommerburg O., Petras T. and Siems W.G.; Kinoshita T., Mira O.S., Matsumoto K. and Nakamura T.; Kikugawa K., Kojima T., Yamaki S. and Kosugi H.; Ohta H., Sweeney E.A., Masamune A. et al.], which is more sensitive because excludes the formation of non-lipid-related MDA-like TBARS.

End products of LPO are often determined spectrofluorometrically by measuring the content of Schiff bases resulting from irreversible covalent binding of MDA or other carbonyl products to peptides, proteins, PEA, PS, or other compounds containing sulfhydryl or amino groups. An alternative approach is to measure the content of gaseous carbohydrates (ethane and pentane) in the exhaled air [Lawrence G.D. and Cohen G.; Andreoni K.A., Kazui M., Cameron D.E. et al.; Kazui M., Andreoni K.A., Norris E.J. et al.]. This method is noninvasive and, hence, very convenient for clinical studies, but it only allows evaluation of general changes in the organism as a whole.

Methods for inducing Fe^{2+}–ascorbate- and NADPH-dependent *LPO* in damaged organs (including ischemic) *in vitro* have the same advantages and disadvantages as the chemiluminescent method. Therefore, they are poorly suitable for estimating LPO intensity during ischemia, at least in some organs (e. g., the liver; see below).

Other important characteristics of LPO processes include *tissue contents of natural antioxidants or prooxidants* expended or produced in the course of LPO (vitamin E, coenzymes Q, ascorbic acid, GSH, GSSG, etc.) and *the activity of individual antioxidant enzymes and overall antiradical (ARA) or antioxidant (AOA) activity of lipids*. These parameters indicate whether the observed accumulation of LPO products is absolute or relative and, in cases of natural antioxidants consumed only in LPO reactions, provide indirect evidence for LPO intensification (see Chapter 10). However, to prove that LPO processes are actually intensified, it is necessary to combine these indirect approaches with direct methods for determining LPO products.

More detailed data on methods for assessing the contents of ROS and LPO products can be found elsewhere [Burlakova E.B., Alessenko A.V., Molochkina E.M. et al.; Burlakova E.B., Kruglyakova K.E. and Shishkina L.N.; Burlakova E.B. and Khrapova N.G.; Vladimirov Yu. A. and Archakov A. I.; Bird R.P. and Draper H.H.; Kennedy Ch.H., Maples K.R. and Mason R.P.; Slater T.F.].

The development of sufficiently sensitive methods made it possible to demonstrate that lipids of every intact organ contain traces of LPO products [Kozlov Yu. P., Danilov V. S. and Kagan V. E.; Neyfakh E.A. and Kagan V.E.], although in different amounts.

The experimental results obtained in our laboratory [Bilenko M.V., Churakova T.D., Arkhangelskaya S.A. and Komarov P.G.] show that organs of intact rats can be arranged in the following series: with respect to the content of primary LPO products (HP and DC), brain < limb muscles < heart and kidneys < liver < blood; with respect to MDA content, limb muscles < brain < heart < liver < kidneys; and with respect to antioxidant activity, liver and kidneys < heart and limb muscles (Fig. 32). Thus, the contents of primary and secondary LPO products were minimal in the brain and limb muscles, more or less similar in the heart and kidneys, and maximal in the liver and circulating blood; moreover, the relation between the content of LPO products and AOA of lipids roughly resembled an inverse correlation. Parameters of LPO (DC and TC contents) and activities of antioxidant systems in the heart and liver of Wistar rats dependent on animal resistance to hypoxia [Khachatur'yan M.L., Gukasov V.M., Komarov P.G. et al.].

Fig. 32: LPO product contents (a, b, c) and AOA (d) in the blood and intact Wistar rat organs (mean ± SEM) [Bilenko M.V., Churakova T.D., Arkhangelskaya S.A. and Komarov P.G.].

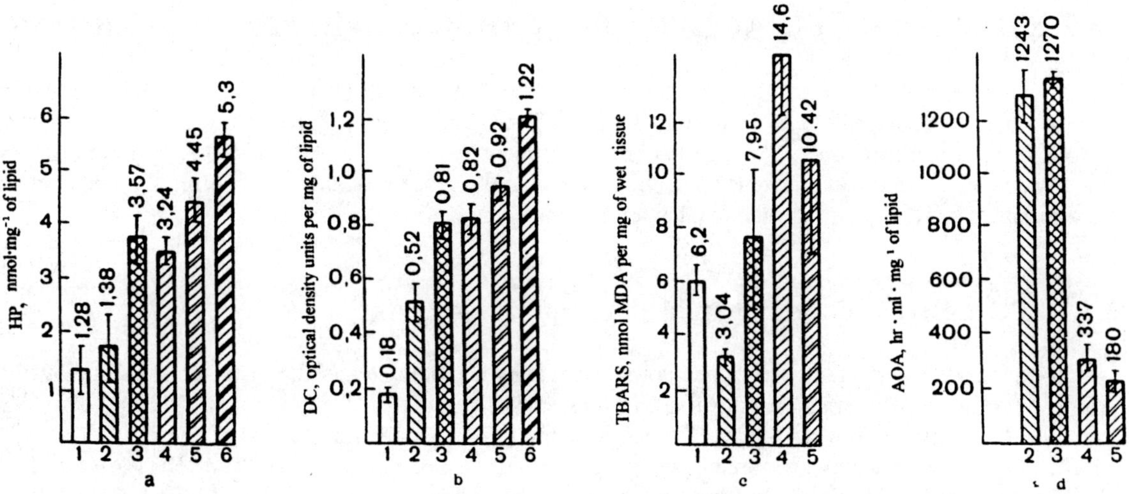

1 - brain; 2 - limb muscles; 3 - heart; 4 - kidneys; 5 - liver; 6 - blood. Lipids were isolated from homogenate by the Folch method [Folch B.J., Lees M. and Stanley H.S.] in modification [Bligh E.G. and Dyer W.J.]; HP was determined by polarographic method [Danilov V. S., Sitkovsky M. V., Kagan V. E. and Kozlov Yu. P.]; DC - by spectrophotometric method [Bolland J.L. and Koch H.J.]; TBARS - by TBA assay[Mihara M., Uchiyama M. and Fukuzawa K.]; AOA - on a methyl oleate model [Burlakova E.B., Alessenko A.V., Molochkina E.M. et al.].

In our experiments, the contents of primary products (HP, DC) in blood plasma of rats [Bilenko M.V., Churakova T.D., Arkhangelskaya S.A. and Komarov P.G.] and healthy humans [Churakova T.D., Bilenko M.V., Pokazeeva Z.T. et al.] were 2.4 to 1.7 times higher than in erythrocytes, whereas those of secondary products were approximately the same. The liver of intact rats contained more LP in mitochondrial than in microsomal membranes; lipid AOA in both mitochondrial and microsomal membranes was equally high, whereas in nuclear membranes it was low or absent [Dudnik L. B., Bilenko M. V. Alessenko A. V. et al. (1981)].

In the rat liver, the levels of ARA in mitochondrial and microsomal membranes proved to be high, exceeding ARA of nuclear membranes by one order of magnitude. The contents of natural antioxidants and membrane lipid viscosity were higher in mitochondrial than in microsomal fractions, which was regarded as evidence that mitochondrial membranes are better protected against peroxidation than other cell membranes [51].

The cortical, medullary, and papillary regions of normal rat kidney were analyzed for a series of parameters related to oxidative stress [Gonzalez-Flecha B., Evelson P., Sterin-Speziale N. and Boveris A.]. The results showed that the rates of oxygen uptake and MDA production in the cortex exceeded those in the medulla by a factor of 1.5 and those in the papilla by factors of 2.0 (oxygen) and 7.3 (MDA). The activities of SOD, Cat, and glutathione peroxidase (GP) in the cortex were 1.5, 1.6, and 1.2 times as high as in the medulla and 6.3, 9.8, and 1.2 as high as in the papilla, respectively. The contents of α-tocopherol in the cortex and medulla were similar, being less that in the papilla by a factor of 1.5.

Susceptibility of intracellular membranes to LPO induction *in vitro* is organ-dependent. In limb muscles of mongrel rats, the rate of NADH-dependent LPO in microsomal membranes was 4.5 times that in mitochondrial membranes [Koshkin V. V.]; in the heart, this rate was similar in both membrane types, but the effect of peroxidation inducers decreased in the series: ascorbate > NADH > NADPH [see Free radicals and biological stabilizators]. On the other hand, the results concerning LPO-inducing effects of ascorbate and NADPH on ER membranes of the rat liver are controversial [Voronov G.G., Lukienko P.I.

and Bushma M.I.; Sharapov V.I., Grek O.R., Dolgov A.V.; Sharapov V.I., Sokirchenko I.A., Greek O.R. et al.]. The above facts may be useful for interpreting the data on the dynamics of LPO products obtained by different methods in experiments on different organs exposed to ischemia and reperfusion.

5.2 Estimating LPO Activity in Ischemic and Reperfused Organs

In our laboratory, LPO activity (the content of LPO products and changes in AOA) was studied in organs differing in sensitivity to ischemia—kidneys, liver, brain, heart, and limb muscles—in different periods of total ischemia and after reperfusion. Experiments were performed on Wistar and August rats and, in some cases (renal and cardiac ischemia), on mongrel dogs (for methods, see Table 2). The data on the assessment of ROS are presented in Chapter 10.

5.2.1 LPO Activity in the Ischemic and Reperfused Kidney

In experiments with rats, renal ischemia was created by occluding either the aorta above and below the point where the renal arteries branch off or, in most cases, vascular peduncles of both kidneys dissected from the surrounding fat (the model of total ischemia). The pioneering results of these studies were described as mentioned above in numerous publications [Alessenko A.V., Bilenko M.V., Burlakova E.B. et al.; Alessenko A. V., Bilenko M. V. and Shelenkova L. N.; Bilenko M. V. (1978a); Bilenko M.V. (1982); Bilenko M.V. (1978b); Bilenko M.V., Alessenko A.V., Burlakova E.B. et al. (1975); Shelenkova L.N.; Bilenko M.V., Alessenko A.V. and Shelenkova L.N.; Bilenko M.V. (1980)].

As shown in Figure 33, the contents of LP and DC 60 and 120 min after the onset of ischemia sharply increased (by a factor of 5.5–7.4, data statistically significant), whereas lipid AOA significantly decreased and, after 1.5 and 2 h of ischemia, even became "negative, " i.e., lipids from the ischemic kidneys began to accelerate methyloleate oxidation, instead of inhibiting it, which provided evidence for the prevalence of prooxidants over antioxidants. Kidney reperfusion for 4 h after 60 min of ischemia resulted in normalization of all the parameters; however, when the kidneys remained ischemic for 90 and 120 min, LP content remained high, exceeding the initial level by a factor of 3.4 to 3.8, with AOA values remaining negative, as during ischemia.

A high LP content and low AOA in 90- and 120-min ischemic kidneys during the reperfusion period correlated with inadequate restoration of blood flow through the cortex, blood shunting via the juxtamedullary pathway (pale cortex, hyperemic medullary substance), an increase in kidney weight, and a sharp impairment of the integral kidney function as estimated from animal survival rate (100% mortality after 1.5 and 2 h of ischemia) (Fig. 33, curves 4b and 5b). Thus, exposure to ischemia for the period of time sufficient for the accumulation of LPO products and a decrease in AOA to negative values (without normalization of both these parameters after 4-h reperfusion) caused irreversible changes in the kidneys, such as disturbed hemodynamics, pronounced edema, and markedly impaired integral function.

The major role of LPO in pathogenesis of ischemic and reperfusion injuries in the kidneys is additionally confirmed by the protective action of antioxidants (α-tocopherol, BHT, diludin, 6-mercaptopurine) [Alessenko A. V., Bilenko M. V. and Shelenkova L. N.; Bilenko M.V., Shelenkova L.N., Dubur G.Ya. and Velena A.H.; Shelenkova L.N. and Bilenko M.V.]. Corresponding observations are described in Chapter 11.

A similar but slightly less pronounced LPO intensification was observed in dog kidneys isolated and stored for 2 h at 20°C: LP increased by a factor of 6.8, and AOA decreased far below the zero level. Extracorporeal reperfusion of ischemic kidneys for 4 h, performed by connecting them to blood vessels of a healthy recipient dog, did not result in normalization of these parameters but even impaired them to a certain extent [Alessenko A.V., Bilenko M.V., Burlakova E.B. et al.].

Fig. 33: AOA level(I), LP content, kidney weight, and rat survival rate(II) after different time of complete rat kidney ischemia in vivo and after 4 h of kidney reperfusion.

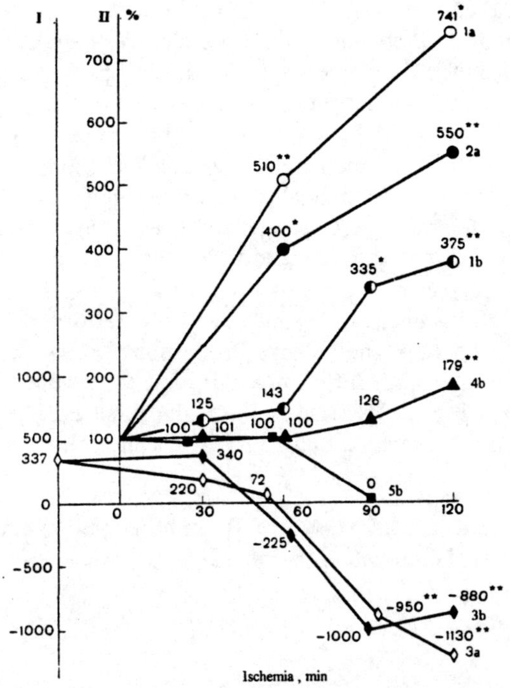

I - AOA (hr • ml • mg^{-1} of lipid); II - all other indexes (% of initial level); 1 - LP; 2 - DC; 3 - AOA; 4 - kidney weight; 5 - rat survival rate; a - after ischemia; b - after ischemia and reperfusion. *-$p < 0.05$; **-$p < 0.01$ versus initial level. The LP content was determined by iodometric titration [Neyfakh E.A. and Kagan V.E.]. Other methods are the same as in Fig. 32.

Our results were subsequently confirmed by other authors. Thus, an increase in the content of primary LPO products (DC) in rabbit kidneys was observed after 60 and 120 min of total ischemia, whereas the end products (Schiff bases) accumulated in the reperfusion period, along with the further increase of DC [Green C.J., Healing G., Simpkin S. et al. (1986b)]. The accumulation of LPO products in preserved dog and rabbit kidneys was also described in [Green C.J., Healing G., Lunec J. et al.; Green C.J., Healing G., Simpkin S. et al. (1986c); Southard J.H., Senzig K.A., Hoffmann R. et al.]. The time-dependent increase of oxygen- and carbon-centered lipid radicals, determined by the ESR method with PBN used as a spin trap agent, was detected in blood outflowing from 15- and 60-min ischemic rabbit kidneys within early period after reperfusion [Defraigne J.O., Detry O., Pincemail J. et al.; Pincemail J., Detry O., Philippart C. et al.]. In isolated human renal tubular cells, MDA was found to increase after 60-min anoxia *in vitro* and then, more significantly, after 5-min reoxygenation; the content of 4-hydroxynonenal, a long-chain aldehydic LPO product, also exceeded the initial level by a factor of up to 3, but this occurred only after 30- and 60-min reoxygenation, i.e., later than in the case of MDA [Grune T., Sommerburg O., Petras T. and Siems W.G.].

5.2.2 LPO Activity in the Ischemic and Reperfused Liver

The formation of ROS (O_2 radicals) in the liver by the cytochrome P-450-dependent system of microsomal monooxygenases is a continuous physiological process (see Chapter 10). Moreover, the liver is highly sensitive to O_2 deficiency. This is why studying the role of LPO processes in ischemic and reperfusion injury of the liver was of particular interest.

In experiments performed in our laboratory, LPO activity was studied in lipids of the total homogenate and of mitochondrial, microsomal and nuclear membrane fractions during ischemia (occlusion of hilar peduncles) or ischemia and reperfusion of the median or median and left lobes of the rat liver.

As shown at the first stage of these studies [Dudnik L. B., Bilenko M. V. Alessenko A. V. et al. (1980); Dudnik L.; Dudnik L. B., Bilenko M. V. Alessenko A. V. et al. (1981); Alessenko A.V., Bilenko M.V., Dudnik L.B. et al.] (Fig. 34), LP content in the ischemic liver homogenate of August rats reached a peak as soon as after 15 min of ischemia, exceeding the initial level by a factor of 13; in mitochondrial and microsomal fractions, the corresponding increase was nine- and sevenfold, respectively. 7-fold. After longer exposure to ischemia, LP content slightly decreased but, even after 2-h ischemia, exceeded the initial levels by a factor of 3–5.5 in all membrane preparations. Lipid AOA of the homogenate and microsomal and nuclear fractions decreased gradually, acquiring significantly lower (negative) values after 30 min (nuclear fraction) and 60 min (homogenate and microsomal fraction). Changes in AOA of mitochondrial lipids were more complex: it markedly decreased 15 min after the onset of ischemia, demonstrated a tendency toward recovery after 30 min, and decreased again after 60–120 min. The dynamics of this type may be accounted for by an attempt to restore AOA in the functionally more important organelles by redistributing antioxidants from other parts of the cell, with subsequent decompensation of this process. The possibility of antioxidant redistribution in the organism has been described previously [Burlakova E.B., Alessenko A.V., Molochkina E.M. et al.; Vladimirov Yu.A. and Archakov A.I.], and such redistribution within individual organs and cells in certain pathological situations is even more likely.

Fig. 34: LP content (a) and AOA level (b) in homogenate and subcellular fraction of complete ischemic rat liver (mean ± SEM).

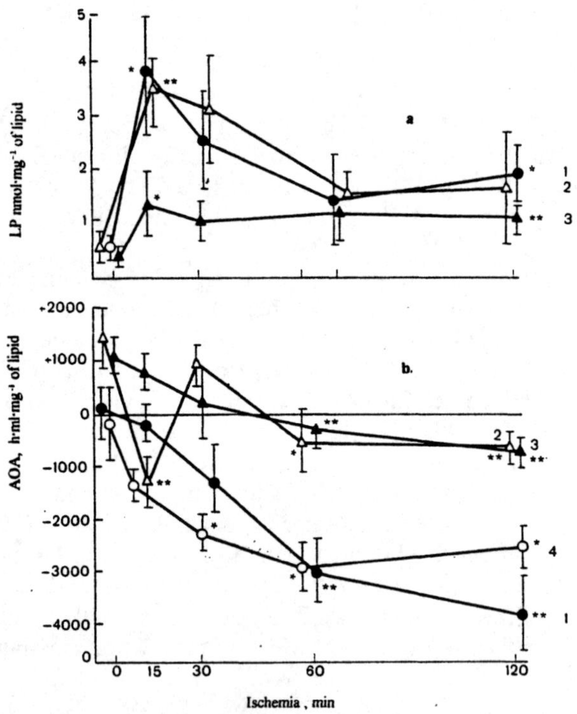

1 - homogenate; 2 - mitochondria; 3 - microsomes; 4 - nuclei. *-$p<0.05$; **-$p<0.01$ versus initial level. Methods are the same as in Figs. 32 and 33.

Figure 34 shows that the accumulation of LP in the mitochondrial fraction was more pronounced, and the decrease of AOA began earlier than in the microsomal fraction. The peak of LPO product accumulation in these membranes (15 min) coincided in time or preceded the onset of decrease in LPO-sensitive PL (PEA and PC) and AA (see Fig. 13) started to decrease, which indicated the role of LPO in the degradation of membrane lipids during ischemia [Alessenko A.V., Bilenko M.V., Burlakova E.V. and Dudnik L.B.; Dudnik L. B., Bilenko M. V. Alessenko A. V. et al. (1981)].

With respect to the time and degree of LPO intensification and decrease in LPO-dependent lipid content during ischemia, membrane fractions can be arranged in the following descending series: mitochondrial > microsomal > nuclear membranes. These observations were later confirmed by other authors, who showed that MDA accumulation in subcellular fractions of rat liver exposed to 1–3-h ischemia or incubated at 37°C extracorporeally decreases in the series: mitochondria > microsomes > nuclei > lysosomes [Dzhafarov A. I. (1981)]. Moreover, in different experimental models, MDA accumulation proved to be more pronounced in the inner than the outer mitochondrial membrane [Dzhafarov A. I., Magomedov N. M., Kulieva E. M. et al.].

At the second stage of our experiments on the same model of Wistar rat liver ischemia (with and without subsequent reperfusion), the content of LPO products in microsomal membrane lipids was compared with the functional state of membranes [Bilenko N.V., Kagan V.E., Velikhanova D.M. and Komarov P.G.; Velikhanova D.M., Bilenko M.V. and Kagan V.E.; Velikhanova D.M.; Komarov P. G., Velikhanova D. M. and Bilenko M. V.; Bilenko M.V., Bulgakov V.G. and Velikhanova D.M.; Bilenko M.V. (1980)].

Microsomal membranes were chosen because of sufficiently high LPO activity in them during ischemia and the possibility of assessing the specific function of these membranes by the state of their cytochrome P-450-dependent monooxygenase system whose components are highly sensitive to ischemia (see Chapter 2). Moreover, this system and NADPH-dependent LPO in microsomal membranes of the liver share the same electron transport chain are responsible for active ROS production [Archakov A. I. (1975); Kagan V. E., Kotelevtsev S. V. and Kozlov Yu. P.]. There is evidence that LPO is involved in enzyme degradation and membrane "dismantling" after the cessation of the cytochrome P-450-dependent monooxygenase system functioning *in vitro* and *in vivo* [Archakov A. I. (1975); Kagan V. E., Kotelevtsev S. V. and Kozlov Yu. P.]. According to our findings, rhythmic diurnal changes in the activity of this system and the dynamics of LPO are in antiphase and demonstrate a strong inverse correlation [Kagan V. E., Kozlov Yu. P., Bilenko M. V. et al.].

Figure 35 shows that the content of both primary (HP) and end (SB) products of LPO in microsomal membranes was already increased after 15 min of hepatic ischemia, as in experiments performed at the first stage of these studies. A sharp increase was observed during the period from the first to the second hour of ischemia, especially in HP content (by a factor of 6.8, compared to 2.7 in the case of SB). The accumulation of LPO products occurred against the background of a decrease in the content of cytochrome P-450, which is the most ischemia-sensitive component of the liver monooxygenase system (see Chapter 2). The contents of LPO products and cytochrome P-450 demonstrated a strong inverse correlation with one another ($r_1 = -0.94$, $r_2 = -0.91$, $P < 0.01$), and this may be interpreted as evidence for the involvement of LPO products in degradation of the cytochrome P-450-dependent monooxygenase system during ischemia [Velikhanova D.M., Bilenko M.V. and Kagan V.E.].

The role of LPO in modifying the functioning of this system in the ER of liver cells during ischemia also follows from the fact of high sensitivity of cytochrome P-450 and, conversely, resistance of NADPH-dependent cytochrome c reductase and cytochrome b_5 to LPO induced *in vitro* [Kagan V. E., Kotelevtsev S. V. and Kozlov Yu. P.], which is similar to the situation observed during ischemia.

The decrease in cytochrome P-450 content during ischemia may depend not only on its degradation, but also on its conversion into an inactive form, namely, cytochrome P-420 [Voronov G.G., Lukienko P.I. and Bushma M.I.]. The probable involvement of LPO in this conversion is indirectly confirmed supported by the fact that the antioxidant benzo[*a*]pyrene inhibited NADPH-dependent LPO *in vitro* and, at the same time, protected cytochrome P-450 against transformation into cytochrome P-420 [Gamrekeli D. V., Savov V. M., Stepanova L. I., et al.].

The toxic effect of LPO products on the cytochrome P-450-dependent monooxygenase system of liver cells may also be accounted for by the secondary carbonyl products of LPO and, in particular, 4-hydroxyalkenals mainly represented by 4-hydroxynonenal. It was shown that this compound is produced in the course of Fe^{2+}–NADPH-, CCl_4- or $BrCCl_3$-induced LPO in microsomes [Benedetti A., Comporti M. and Esterbauer M.; Esterbauer H., Cheeseman K.H., Dianzani M.U. et al.]; actively binds to protein SH groups [Cadenas E., Müller A., Brigelius R. et al.]; inhibits cytochrome P-450, aminopyridine methylase, and Ca^{2+} transport in liver microsomes [Benedetti A., Fulceri R. and Comparti M.]; and lyses isolated hepatocytes [Benedetti A., Comporti M. and Esterbauer M.].

Fig. 35: LPO products and cytochrome P-450 contents in the microsomal fraction of complete ischemic in vivo rat liver (% of the initial level) [Velikhanova D.M., Bilenko M.V. and Kagan V.E.]

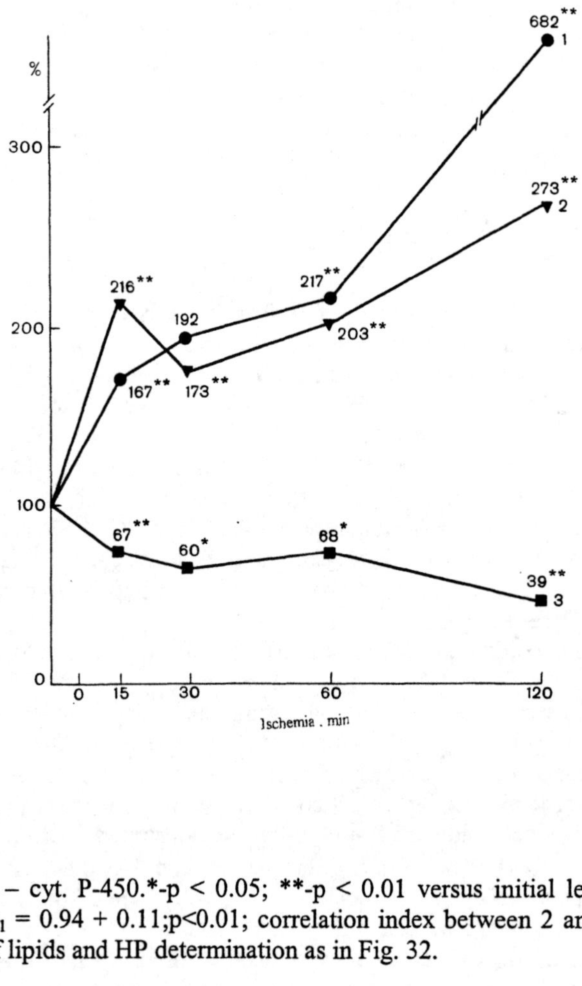

1 - HP; 2 - SB; 3 – cyt. P-450.*-$p < 0.05$; **-$p < 0.01$ versus initial level. Correlation index between 1 and 3: $r_1 = 0.94 + 0.11$;$p<0.01$; correlation index between 2 and 3: $r_2 = 0.91 + 0.04$, $p<0.01$. Isolation of lipids and HP determination as in Fig. 32.

In our experiments, the relationship between the dynamics of LPO product accumulation in microsomal membranes and their functional state was also analyzed after 4-h reperfusion of rat liver exposed to 1-h ischemia (Figs. 36, 37). The results showed that reperfusion of the ischemic liver led to further accumulation of HP (a 37% increase) and SB (a 42% increase) combined with a decrease in cytochrome P-450 content (by 27%), cytochrome b_5 content (by 33%), and the activity of NADPH-dependent cytochrome c reductase (by 26%).

Treatment of rats with BHT (an antioxidant), BHT combined with phenobarbital (PB; an inducer of the monooxygenase system), or PB alone had no significant effect on HP content in lipids and SB content in microsomes of the intact liver but moderately or significantly inhibited the accumulation of both LPO products in the course of 1-h ischemia and after 4-h reperfusion of the liver (Fig. 36).

In our experiments, BHT had no significant influence on cytochrome P-450 content and activity of aminopyrine N-demethylase (AD) in the liver of intact animals (Fig. 37), in contrast to its inducing action at multiple or higher doses [Shuliakovskaya T.S., Arshinov V.Yu, Pakhomov V.Ju. et al.], but displayed a moderate protective effect on AD during ischemia and on both cytochrome P-450 and AD in the course of reperfusion. The more pronounced effect of BHT on both parameters in the reperfusion period may also be attributed to the fact that BHT did not completely inhibited the accumulation of LPO products during liver ischemia but prevented their additional increment in the course of reperfusion (see Fig. 36).

Fig. 36: HP (a) and SB contents (b) in microsomal fraction of rat liver after 1 h of complete ischemia and 4 h of reperfusion in experiment with BHT (ionol) PB and BHT+PB (mean ±SEM).

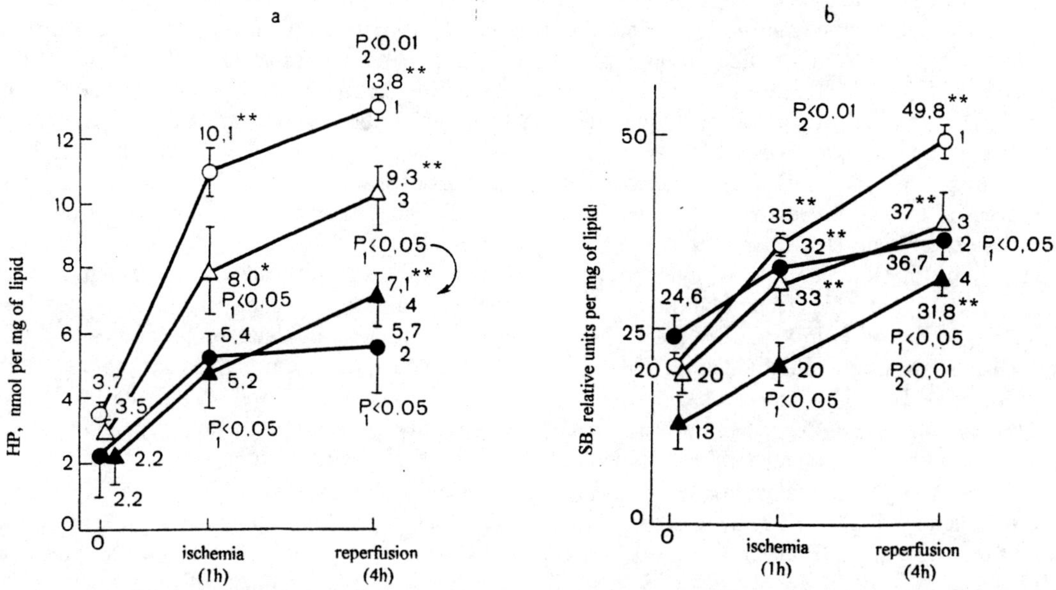

1 - without drug (control); 2 - with BHT; 3 - with PB; 4 - with BHT + PB. *-p < 0.05; **-p < 0.01 versus initial level. P_1 - versus control; P_2 – significant differrences after versus before reperfusion. BHT (ionol) was administered i. p. in dose 240 mg/kg in twin 80 24 h bifore ischemia; PB - 80 mg/kg once a day for 3 days. The initial values were obtained on intact rats receiving the preparations in the same doses.

Fig. 37: Cytochrome P-450 content (a) and aminopyrine N-demethylase activity (b) in the microsomal fraction of rat liver after 1 h of complete ischemia and 4 h of reperfusion in experiment with BHT, PB and BHT+PB (% of the initial level in the control series)

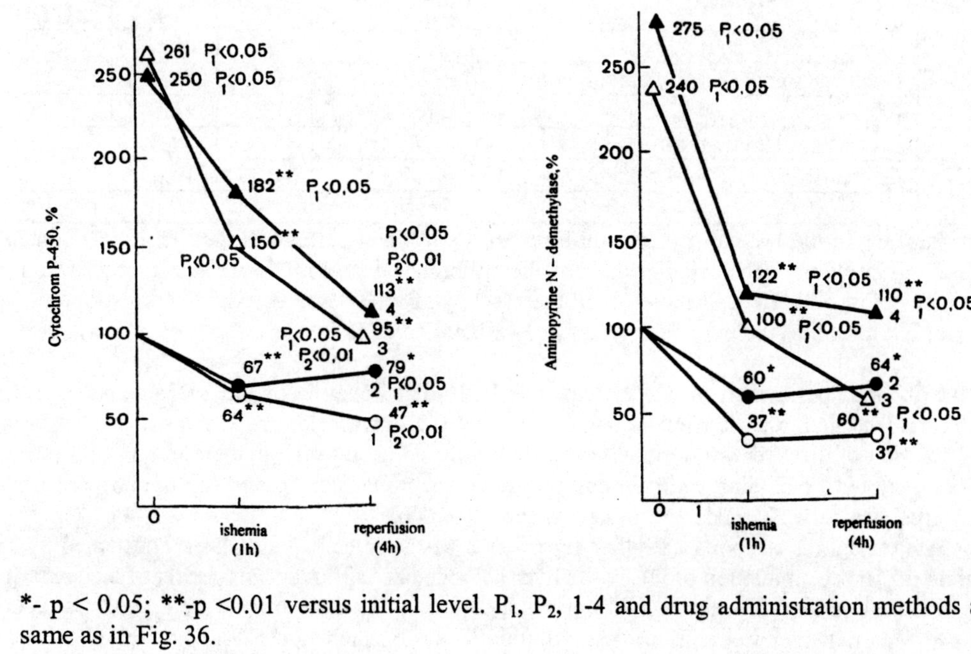

*- p < 0.05; **-p <0.01 versus initial level. P_1, P_2, 1-4 and drug administration methods are the same as in Fig. 36.

BHT combined with PB or PB alone did not prevent the decrease in cytochrome P-450 content and AD activity during ischemia and reperfusion of the liver. However, as the initial values of both parameters were relatively high, the absolute content of cytochrome P-450 and AD activity remained elevated, suggesting that BHT and its combination with PB may be promising as agents protecting the liver against ischemia. The same conclusion follows from the data on rat survival after 1-h ischemia of the liver lobe remaining after resection of other lobes: after combined treatment with BHT and PB, survival rate increased to 47% vs. 15% in the control. It is appropriate to note that in 1983, we published the first paper describing the method of assessing the integral liver function by the survival rate of rats in which hilar peduncles of one or two liver lobes were clamped, while the remaining lobes served as a temporary pathway of portal blood shunting and then, after reperfusion of ischemic lobes, were removed [Bilenko N.V., Kagan V.E., Velikhanova D.M. and Komarov P.G.]. Since then, this method has been used in numerous experiments (see Chapter 9).

Assuming that the decrease of cytochrome P-450 content during ischemia and reperfusion depends, fully or partially, on the accumulation of LPO products in microsomal membranes, it is possible to calculate cytochrome P-450 stability under the effect of LPO products in standard conditions and during antioxidant therapy and LPO induction. The results of such calculations, summarized in Table 5 [Bilenko N.V., Kagan V.E., Velikhanova D.M. and Komarov P.G.], show that 21 and 26 HP molecules per a degraded cytochrome P-450 molecule were formed during ischemia and reperfusion, respectively. This is close to the calculated number of HP molecules necessary for breaking down one cytochrome P-450 molecule *in vitro* (15–20 molecules) [Kagan V. E., Kotelevtsev S. V. and Kozlov Yu. P.]. In the series with BHT treatment, HP formation decreased to 13 molecules during ischemia and zero molecules during reperfusion, as neither HP accumulation nor cytochrome P-450 degradation was detected. In experiments with PB and PB + BHT, HP increment was small, whereas the absolute content of induced cytochrome decreased significantly; hence, only 2.3 to 5.4 HP molecules per destroyed cytochrome P-450 molecule were formed during ischemia and reperfusion of the liver. It remains unknown, however, whether these results reflect increased sensitivity of cytochrome P-450 to LPO products after its induction or the main role in degrading induced and noninduced cytochrome P-450 belongs to different membrane-modifying factors.

Table 5: Ratio between increasing HP and decreasing cyt. P-450 contents in the microsomal fraction of ischemic and reperfused rat liver.

Time	Parameters	Without drug	With drugs		
			BHT	PB	BHT+PB
Ischemia (1h)	Increase in HP content	6.4	3.2	4.5	3.0
	Decrease in cyt. P-450 content	-0.30	-0.24	-0.91	-0.55
	K	21.0	13.3	4.9	5.4
Reperfusion (4h)	Increase in HP content	3.7	0.3	1.3	1.9
	Decrease in cyt. P-450 content	-0.14	-0.09	-0.45	-0.57
	K	26.4	-	2.3	3.3

NOTE: The increase in the HP content (in nmol•mg^{-1} of lipid) and decrease in the cyt. P-450 content (in nmol•mg^{-1} of protein) were calculated from the difference between the contents before and after ischemia or before and after reperfusion; HP increase: cyt. P-450 decrease ratio K corresponding to the number of HP molecules per one destroyed cyt. P-450 molecule.

Thus, the above experimental results indicate that both total ischemia of the liver and its reperfusion after long (critical) ischemic periods are accompanied by LPO intensification. The latter processes in the microsomal fraction of the liver are definitely responsible for damage to the cytochrome P-450-dependent monooxygenase system, including its main component (cytochrome itself), and the impairment of Ca^{2+}-ATPase activity [Watanabe N., Onda M., Genga A. and Asano G.].

LPO activation during ischemia and after reperfusion of the liver has also been confirmed by many other researchers. The accumulation of DC in the liver homogenate within a short period of ischemia (up to 30 min) and increase in the MDA content during a longer period (up to 60 min) were demonstrated on the model of severe partial liver ischemia in dogs [Shilina M.K., Ivashchenko S.P., Shrainberg A.I. et al.]. MDA content was also shown to increase in isolated rat liver after long-term storage [Panchenko L.F.,

Gerasimov A.M., Nozdrachova L.I., Kariakina G.A.] and in microsomal fractions incubated for 45–60 min under anoxic conditions [Sorokovoy V.I., Petrenko Yu.U., Mochenova G.M. et al.].

In the isolated rat liver incubated at 37°C, MDA content increased significantly after 1 h of ischemia and exceeded the initial level by a factor of 4 after 6 h. At 20°C, HP, DC, and MDA contents demonstrated a significant increase only after 2–4 h of ischemia; subsequently, the content of primary LPO products decreased, reaching the initial level after 6 h of incubation, which coincided with the peak of MDA content in the tissue [Kozlov A. V., Shinkarenko L. I., Vladimirov Yu. A. and Azizova O. A.].

Under conditions of acute hypoxia (the rats were exposed to asphyxia for 6.5 min), lipid HP content in the liver increased by a factor of 1.8, whereas ARA decreased to 64% of that in the control [Konvai V. D., Lukoshkin A. V. and Smirnova V. B.].

LPO intensification (an increase in the contents of DC and Schiff bases) was also revealed in the liver of sacrificed rabbits stored for a long time at 4–8°C [Lankin V. Z., Tikhaze A. K., Voronina N. V. et al.], in rat liver 15 min after vena porta occlusion [Lee K.T., Tsai L.Y. and Sheen P.C.], and in reperfused liver [Konvai V. D., Lukoshkin A. V. and Smirnova V. B.; Silver E.M. and Szabo S.]. In some experiments, however, no MDA accumulation and only a slight increment of DC were observed during liver ischemia [Marubayashi S., Dohi K., Ezaki H. et al. (1982)], but the contents of both these LPO products sharply increased in the reperfusion period. Using the ESR method with spin trapping, a time-dependent increase in generation of free oxygen radicals (DMPO-OOH$^\bullet$ and DMPO-OH) was revealed in the effluent after 6-min reperfusion of the rat liver exposed to total ischemia for 10, 30, 60, and 90 min [Togashi H., Shinzawa H., Yong H. et al.], as well as in the liver itself after 30- and 60-min ischemia and after 40-min reperfusion [Wu W. and Qiu F.]. Pigs subjected to 2-h warm total ischemia of the liver with subsequent reperfusion demonstrated the growth of ethane content in the exhaled air [Kazui M., Andreoni K.A., Norris E.J. et al.].

In contrast to the data obtained by measuring LPO products directly, the results of experiments on LPO induction in the liver already exposed to ischemia are controversial. In our studies on NADPH-dependent LPO induction in the liver after different periods of ischemia or after reperfusion, no significant change from the control was revealed [Bilenko N.V., Kagan V.E., Velikhanova D.M. and Komarov P.G.; Velikhanova D.M., Bilenko M.V. and Kagan V.E.].

In experiments of other scientists, NADPH-dependent LPO induction in the liver was absent after 30–120 min of its ischemia [Sharapov V.I., Grek O.R., Dolgov A.V.] and proved to be inhibited much stronger than Fe^{2+}–ascorbate-dependent LPO after 60–180 min of ischemia [Voronov G.G., Lukienko P.I. and Bushma M.I.]. Spontaneous MDA accumulation in the hepatic tissue during ischemia was more pronounced as compared to induction of ascorbate-dependent LPO [Sharapov V.I., Grek O.R., Dolgov A.V.], and that in liver mitochondria incubated under anoxic conditions, as compared to induction of Fe^{2+}-dependent LPO [Sorokovoy V.I., Petrenko Yu.U., Mochenova G.M. et al.].

Apparently, the potential of tissues for responding to artificial LPO induction is reduced or exhausted after long-term ischemia, which is evidence for extensive structural damage of membranes [Sharapov V.I., Grek O.R., Dolgov A.V.]. Consequently, it is necessary to distinguish between *the accumulation of endogenous LPO products during ischemia* and *the potential of the ischemic organ for secondary LPO intensification in vitro* (see section 5.1).

Thus, LPO intensification during ischemia and reperfusion of the liver has been confirmed for virtually all the models of ischemia, namely, total ischemia of liver lobes, partial ischemia or hypoxia of the liver, and storage of the isolated liver at normal temperature and under hypothermic conditions. This process, as lipid depletion of membranes, is especially manifested in mitochondrial membranes, which are the first to suffer damage during ischemia [Sorokovoy V.I. and Vladimirov Yu.A.].

5.2.3 LPO Activity in the Ischemic and Reperfused Brain

It is well-known that the brain is not only highly sensitive to O_2 deficiency, but also is strongly predisposed to induction of free-radical reactions, and LPO processes induced *in vitro* have a strong deleterious effect on membrane structures of the brain. Thus, the intensity of chemiluminescence induced by 100% O_2 bubbling through the rat brain homogenate and the accumulation of MDA in it after incubation for 3 h at 37°C proved to be 20 times higher than in the liver homogenate [Cadenas E., Varsavsky A.I., Boveris A. et al.].

Incubation of rat brain slices with a complex of agents inducing formation of free O_2 radicals (xanthine + xanthine oxidase + ADP + Fe^{2+}) [Chan P.H. and Fishman R.A.] led to MDA accumulation (71, 155 and 231% increase after 30, 60 and 90 min, respectively, compared to the control) and enhanced edema of the brain tissue by 18–27%. After 30 min of incubation, MDA accumulation was maximal in myelin (176% increase); then followed mitochondria and synaptosomes (155 and 133%) and, finally, nuclei and microsomes (85 and 73%). MDA accumulation after 30 min of incubation was accompanied by a decrease in PL content in synaptosomal membranes; after 60 min, the concentration of free fatty acids (FFA; 18:1, 18:2, 20:4, 22:6) increased by a factor of 2–4. The most pronounced inhibitory effect (according to the MDA accumulation test) was observed after adding chlorpromazine and BHT to the incubation medium. In addition, severe edema and lysis of rat brain mitochondria resulted from Fe^{2+}–ascorbate-induced LPO *in vitro* [Kubo K., Yoshitake I., Kumada Y. et al.].

Apparently, LPO induction in the brain is easily accomplished owing to a number of factors facilitating this process. Thus, the brain is characterized by the increased contents of PL [Cadenas E., Varsavsky A.I., Boveris A. et al.], oleic acid, and plasmalogens [Cadenas E., Varsavsky A.I., Boveris A. et al.]; a high lipid/protein ratio (2.5–10 times that in limb muscles) [Boldyrev A.A. and Tverdislov V.A.]; low content of vitamin A (10–20 times as low as in the liver); extremely low activity of GSH peroxidase (in the cytosol and mitochondria, 30 and 15 times as low as in the liver) [De Marchena O., Guarnieri M. and McKhann G.; Sies H., Roch O.R., Martin O.E. and Boveris A.]; virtually complete absence of catalase [Sailer N.]; high Fe^{2+} content; and low concentrations of transferrin and ceruloplasmin in the cerebrospinal fluid [Halliwell B. and Gutteridge J.M.C. (1985)].

Insufficiency of protective antioxidant systems of the brain against the background of ample blood supply, high oxygen consumption, and high CA content [Halliwell B. and Gutteridge J.M.C. (1985)] appears to be partially compensated by the presence of a relatively great amount of ascorbic acid, a natural antioxidant [Demopoulos H.B., Flamm E.S., Pietronigro D.D. and Seligman H.L.], and systems capable of increasing its content almost tenfold, compared to that in blood [Halliwell B. and Gutteridge J.M.C. (1985)].

Low tolerance of the brain to ischemia, the high intensity of induced free-radical processes, and the damaging effect of LPO induced *in vitro* on the brain strongly suggest the involvement of LPO processes in pathogenesis of ischemic and reperfusion injury of the brain.

In our laboratory, LPO intensity in the brain was determined from the accumulation of primary (DC) and secondary (TC and MDA) LPO products. Experiments were performed on the model of total rat brain ischemia involved temporary bilateral ligation of carotid and subclavian arteries with the use of microsurgical technique and a special dissecting microscope proposed previously [Tel'pukhov V.I., Trenin S.O., Knyazeva G.D. and Lapkina T.I.] and allowed for subsequent reperfusion of the brain [Bilenko M.V., Tel'pukhov V.I., Churakova T.D. and Komorav P.G.; Churakova T.D., Bilenko M.V., Tel'pukhov V.I. and Komarov P.G.].

Cerebral ischemia was regarded as total when blood pressure in the distal sections of the carotids dropped to zero, the basal line on the ECG appeared after 20–30 s, rats stopped breathing after 4 min, and the contrast substance ceased entering the cervical and cerebral vessels located distally from the occlusion point [Tel'pukhov V.I., Trenin S.O., Knyazeva G.D. and Lapkina T.I.]. As LPO processes were easy to induce in the brain, special precautions were taken to avoid exposure of material to the air during processing.

Total cerebral ischemia was accompanied by the accumulation of all the LPO products examined (Fig. 38). An increase in DC content became significant after 5 min of ischemia; that in TC and MDA contents reached a significant level 5–15 min after the onset of ischemia and persisted throughout the experiment; and the contents of secondary LPO products reached a peak after 30-60 min of ischemia, exceeding the initial value by a factor of 1.8–2.5.

Fig. 38: LPO product content in rat brain after different time of complete ischemia in vivo (% of the initial level)

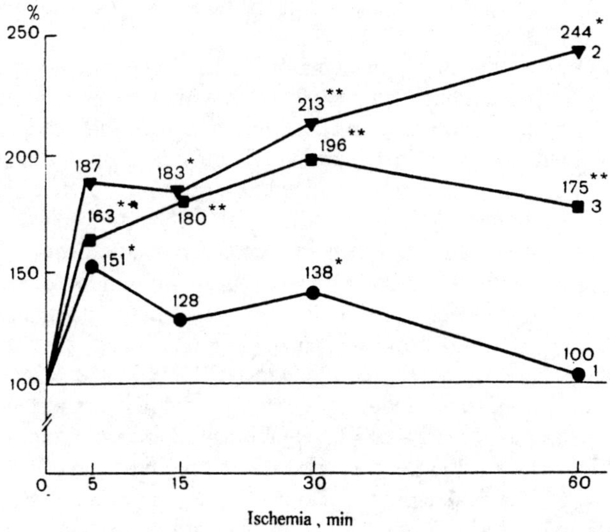

1 - DC; 2 - TC; 3 -TBARS; *-p<0.05, **-p<0.01 versus initial level. Methods are the same as in Fig. 32.

Reperfusion of the brain after 30-min ischemia (Fig. 39) was accompanied by a significant increase in DC content (by 33% after 20 min) and a tendency toward the accumulation of secondary LPO products (MDA after 5 min and TC after 20 min). After reperfusion for 60 min, the content of LPO products decreased, approaching the level observed in the ischemic brain; however, the contents of secondary products still demonstrated a significant excess over the initial value.

Fig. 39: LPO product content in rat brain after 30 min of complete in vivo ischemia and after 5, 20 and 60min of brain reperfusion (% of the initial level).

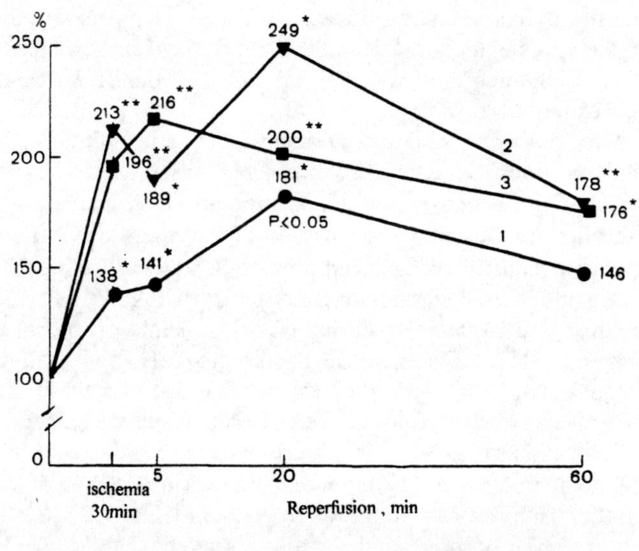

1 - DC; 2 – TC; 3 – TBARS; *-p<0.05; **-p<0.01 versus initial level. P-significant differences after versus before reperfusion. Methods are the same as in Fig 38.

Antioxidant treatment (BHT) inhibited DC and MDA accumulation in the totally ischemic brain both in the course of 30-min ischemia and after 20-min reperfusion; at the same time, it prolonged the period of tolerance to cerebral ischemia as estimated from animal survival rate (see Chapter 11). This fact provided additional evidence for the role of LPO processes in pathogenesis of ischemic and reperfusion brain damage.

The dynamics of LPO processes in the ischemic brain has been studied by many researchers in on different models, and this problem has been extensively discussed in the literature over the past few years.

In experiments with guinea pig brain, animal decapitation was followed after 5 min by a two- to threefold increase in HP content and a less pronounced (1.3-fold) increase in DC content in the microsomal fraction, although MDA and fatty acid contents remained unchanged [Strosznaider J. and Dabrowitecki Z.]. After NADPH-dependent LPO induction in microsomes of the ischemic brain *in vitro,* the increase of the above LPO parameters was more drastic than in nonischemic brain microsomes, whereas the response to Fe^{2+}–ascorbate-dependent LPO was similar in both cases. In retaining the potential for LPO induction *in vitro,* the ischemic brain proved to differ from the ischemic liver both quantitatively and qualitatively: in the liver, this potential decreased, and the organ retained residual capacity for ascorbate-dependent (rather than NADPH-dependent) LPO [Voronov G.G. and Lukienko P.I. (1983)] (see above).

Subsequent experiments on the same model of cerebral ischemia (decapitation) revealed the twofold increase of MDA and the decrease of water- and fat-soluble antioxidants in microsomal membranes after 5 min [Majewska M.D., Strosznajder J. and Lazarewicz J.]. The accumulation of lipid HP in the rat brain was also observed 6.5 min after mechanical asphyxia [Konvai V. D.], and that of HP, DC, and MDA (against the background of decreasing ARA), 5 min after clinical death caused by bleeding [Semchenko V.V., Poluektov L.V. and Konvai V.D.]; in another instance, LPO products accumulated during severe partial ischemia, 60 min after ligation of both carotid arteries and bleeding [Zamuruev O. N. (1984)] or during local ischemia 10 or 60 min after middle cerebral artery occlusion in rats [Gumuslu S., Serteser M., Ozben T., et al.]. In all these case, the high LPO contents persisted for 30–60 min during the resuscitation or reperfusion period.

Partial cerebral ischemia created by ligating the carotids led to the accumulation of DC and Schiff bases in the 3-h ischemic or 30-min ischemic and 1-h reperfused rat brain [Plotnikov M.B., Kobzeva E.A. and Plotnikova T.M.] and of Schiff bases in the outflowing venous blood [Ryasina T.V., Korshunova T.S., Polyanskii N.B. and Muranov K.O.]. In spontaneously hypertensive rats, exposure to the same model of cerebral ischemia for 1, 3, and 5 h produced an increase in MDA content up to 130, 135 and 160%, respectively; reperfusion for 1 and 3 h after 3-h ischemia caused further significant rise of MDA level in the brain tissue (by a factor of 1.5–1.6) and in blood [Yamamoto M., Shima T., Uozumi T. et al.]; note that α-tocopherol administered in a dose of 20 mg/kg 30 min before occlusion of the carotids did not prevent MDA accumulation during ischemia but significantly reduced its repeated increase during reperfusion. There are also data on the accumulation of DC in the brain during focal compression ischemia [Plotnikov M.B., Kobzeva E.A. and Plotnikova T.M.] and of DC and Schiff bases during cerebral hypoxia [Goroshinskaya I.A. and Mogil'nitskaya L.V.].

At the same time, some researchers did not reveal any increase in MDA content after cerebral ischemia [MacMillan V.; Yoshida S., Inoh S., Asano T. et al.] but observed it only after 15-30 min of brain reperfusion [Yoshida S., Inoh S., Asano T. et al.]. Although no MDA increase occurred after 30-min total ischemia and during the first 15–30 min of reperfusion of the brain, this period was accompanied by a decrease in the levels of α-tocopherol and reduced forms of CoQ_9 and CoQ_{10} [Yoshida S., Abe K., Busto R. et al.] and rapid consumption of polyunsaturated fatty acids (PUFA).

Several authors estimated LPO intensity during partial ischemia of the cat brain and spinal cord by indirect criteria [Demopoulos H.B., Flamm E.S., Pietronigro D.D. and Seligman H.L.; Flamm E. S., Demopoulos H. B., Seligman M. L. et al.]. In their experiments, the content of ascorbic acid after 1, 5 and 24 h of partial cerebral ischemia decreased by 21, 28 and 34%, while the spinal cord compression for 1, 3, and 24 h resulted in its reduction by 40, 48 and 60%, respectively. A decrease in ascorbic acid content (to 48%), GHS (to 44%), total thiols (to 42%), protein thiols (to 38%) and α-tocopherol (to 26%) in hippocamp was found after asphyxia and 10min of reperfusion [Katz L.M., Callaway C.W., Kagan V.E. and Kochanek P.M.]; decreased in ascorbic acid was also observed when LPO in liposomes was induced *in vitro* [Demopoulos H.B., Flamm E.S., Pietronigro D.D. and Seligman H.L.].

There is an opinion that, as LPO products themselves are difficult to detect, conclusive evidence for activation for free-radical reactions during ischemia may be obtained using indirect criteria, such as the decrease of cholesterol and ascorbic acid levels and the asymmetrical reduction of fatty acid contents in PL, i.e., prevailing loss of PUFA (20:4, 22:6) and retention of saturated FA, which occur in a certain sequence during ischemia. According to some authors, an important indicator of LPO intensification during cerebral ischemia is a decrease in GSH content, which was observed after total or partial ischemia and reperfusion of the rat brain [Rehncrona S., Folbergrova J., Smith D. and Siesjö B.K.]. Although no reciprocal increase in GSSG content or changes the FA and PL composition were detected, these authors did not rule out the possibility of local free-radical reactions in subcellular structures of the brain.

The involvement of LPO in ischemic and reperfusion injuries of the brain is also confirmed by the protective effect of various antioxidants, such as 3-dihydropyridine derivatives tested on the model of severe cerebral hypoxia in rats or barbiturates used in experiments on the model of local and total cerebral ischemia in different animals, including monkeys [Smith D.S., Rehncrona S. and Siesjö. Bo.K.] (see Chapter 11).

Thus, our data and other experimental evidence obtained by both direct and indirect methods testify that LPO processes are intensified during total and partial cerebral ischemia and repeatedly activated in the reperfusion period.

5.2.4 LPO Activity in the Ischemic and Reperfused Heart

The importance of studies on the role of LPO processes in cardiac ischemia is determined by not only the extremely wide occurrence of the disease, steadily increasing number of operations on the open heart, and organization of major heart transplantation centers, but also the increasing level of emotional and other types of stress resulting in the release of CA, which initiate LPO (see Chapter 10) and are targeted toward the heart.

As already noted (see Chapter 1), only the model of total ischemia ensures regular and strictly time-dependent cessation of blood flow in the heart, which is characterized by considerable species-specific and individual variability of the collateral circulation. At present, the model of total cardiac ischemia created by cardiopulmonary shunting *in situ* and isolated heart storage *in vitro* are widely used for evaluating ischemic and reperfusion injuries in experimental studies.

In our laboratory, evidence for the accumulation of LPO products in heart (a twofold increase) was first obtained in 1976 in experiments on storing the isolated rat heart for 2 h and the dog heart for 3 h at 37°C [Bilenko M. V., Burlakova E. B., Alessenko A. V. and Shelenkova L. N.]. Thereafter, LPO processes were investigated not only during ischemia, but also after reperfusion of the heart, using different methods for determining primary and secondary LPO products and antioxidant activity in lipids of the total homogenate and SR membranes [Bilenko M.V., Bulgakov V. and Otverchenko V.; Bilenko M.V. (1992b); Bilenko M.V., Otverchenko V.M., Astaf'ev V.V. et al.; Bilenko M.V. and Otverchenko V.N.; Bulgakov V.G. and Bilenko M.V.; Bulgakov V.G., Volkova I.A., Astafyev V.V. and Bilenko M.V.; Bulgakov V.G.; Bilenko M.V., Bulgakov V.G. and Velikhanova D.M.; Bilenko M.V. (1980)].

All experiments were carried out on mongrel dogs using the model of total ischemia of the heart, which was dissected under anesthesia and artificial respiration and then stored for different periods of time (up to 2 h) at 37 or 20°C. Reperfusion was performed extracorporeally by connecting cardiac vessels to the cervical vessels of heparinized recipient dogs under superficial morphine–hexenal anesthesia (the model of heart heterotransplantation to the neck).

Determination of LPO products in lipids extracted from the heart after different periods of storage at 37°C (Fig. 40) showed that the content of primary products (HP and DC) increased significantly as soon as after 30 min of ischemia and remained high throughout the 2-h period of ischemia. Lipid AOA also began to decrease after 30 min, acquired a negative value after 1 h, and decreased to a minimum after 2 h of ischemia (data statistically significant). The accumulation of primary LPO products and decrease in lipid AOA demonstrated an inverse correlation ($r = -0.69$, $P < 0.05$).

In SR membrane lipids, MDA content significantly exceeded the initial level after 30 minutes and remained high throughout the ischemic period; the contents of DC and TC did not demonstrate any significant changes (data not shown). The accumulation of primary LPO products in the homogenate lipids and the absence of changes in their content in the SR membrane lipids were probably due either to the increasing hydrophily of LP and their resulting release from membranes into the cytosol or to more rapid metabolism of primary LPO products in membrane structures than in the homogenate. The actual situation is as yet unclear.

Fig. 40: LPO products, total PL, AOA and Ca^{2+} enzyme transport system indexes in the homogenate and SR membranes of canine heart after different time of complete ischemia in vitro at 37°C.

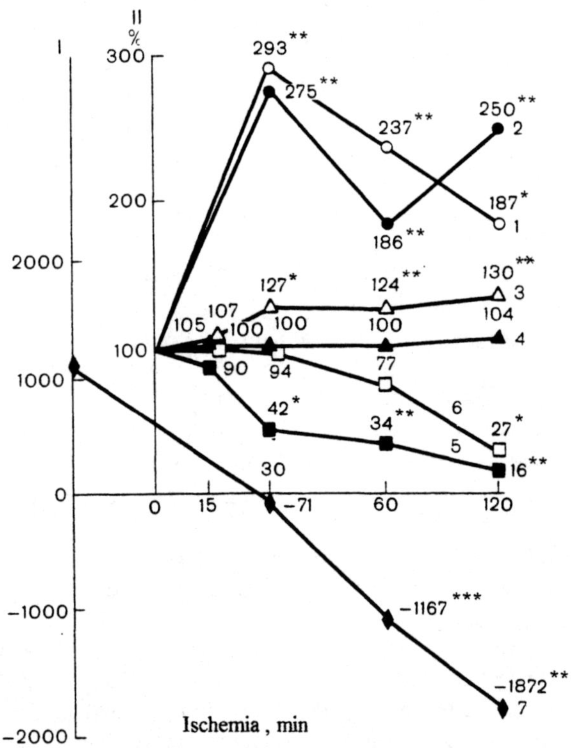

I - AOA (hr • ml • mg^{-1} of lipid); II - other parameters (% of the initial level); 1 - HP in homogenate; 2 - DC in homogenate; 3 - TBARS in SR membranes; 4 - total PL in SR membrane; 5 - Ca^{2+} uptake in SR membranes; 6 - Ca-ATPase activity in SR membranes; 7 - AOA in homogenate; *-$p<0.05$, **-$p<0.01$ versus initial level. The methods are the same as in Figs. 16 and 32.

Comparison of changes in the content of LPO products with the functional activity of SR membranes determined from the parameters of enzymatic Ca^{2+} transport showed that the accumulation of primary LPO products in the homogenate lipids and that of MDA in SR membrane lipids, clearly manifested after 30 min of ischemia, was accompanied by a marked drop in Ca^{2+} uptake by SR membranes, whereas Ca-ATPase still retained its activity at this time (see Fig. 40); in other words, the capacity of Ca-ATPase for ATP hydrolysis became "uncoupled" from that for Ca^{2+} transport. A sharp decrease in the hydrolytic activity of Ca-ATPase in the ischemic isolated heart was only observed after 2 h of storage. We also revealed similar uncoupling in SR membranes of ischemic limb muscles (see Chapter 2), and this result was later corroborated by other researchers [Hess M.L., Krause S.M. and Greenfild L.J.; Hess M.L., Manson N.H. and Okabe E.]. The possible mechanisms of the involvement of LPO products in this phenomenon will be discussed below. It should only be pointed out that early damage of Ca^{2+} transport during ischemia, which is indicative of primary disturbances of passive membrane permeability for Ca^{2+} and the resulting low efficiency of the Ca^{2+} pump, did not result from lipid depletion of membranes because the overall PL content in SR membranes during the ischemic period remained virtually unchanged (see Fig. 40).

When the isolated heart was stored at 20°C, the accumulation of primary LPO products (HP and DC) and changes in Ca^{2+} uptake by SR membranes were significantly less pronounced than at 37°C (Fig. 41). Reperfusion of the heart after storage at 20° was accompanied by either a tendency toward the additional accumulation of LPO products (DC in the homogenate) or their actual significant increase (DC in SR membranes) and by a further decrease in AOA of homogenate lipids, down to negative values (data not shown). LPO activation in the reperfusion period was accompanied by a significant decrease in the ability of SR to take up Ca^{2+}, followed by a significant decrease in the activity of Ca-ATPase (see Fig. 41). The role of LPO in ischemic and reperfusion damage of the heart was also demonstrated in our numerous experiments on antioxidant therapy (see Chapter 11) and inhibition of LPO processes by blood hypooxygenation in the reperfusion period (see Chapter 6).

LPO intensification during cardiac ischemia was also revealed by other researchers who worked with different models of cardiac ischemia.

In experiments on the *total ischemia* of the dog heart *in vivo*, the content of DC (reduced in advance to the stable hydroxyl form) exceeded the initial level by a factor of 4 after 45 min of ischemia, by a factor of 20 after 5 min of reperfusion, and returned to the initial level after 30 min of reperfusion; moreover, most of these peroxidized metabolites (at least 5 min after reperfusion) was in the PUFA fraction (18:2, 20:4, 22:6) [Romaschin A.D., Rebeyka I., Wilson G.I. et al.]. In the isolated rat heart, 15-min ischemia resulted in the accumulation of MDA in sarcolemma, which was accompanied by a decrease in Na, K-ATPase activity [Kim E.A., Danilenko M.P., Murzakhmetova M.K. et al.]; 25-min ischemia led to an increase of 4-hydroxy-2-nonenal (HNE) content in membranes of mitochondria and concomitant inhibition of their respiratory function. The analysis of MDA by a modified HPLC method showed that the increase of its content in the rat heart during the first 5–30 min of ischemia occurred more rapidly than the decrease of adenine nucleotide contents [Tavazzi B., Di Pierro D., Bartolini M. et al.]. The isolated rat heart also demonstrated intensification of chemiluminescence, which reached a peak after 2–3 h of storage.

Experiments on the isolated rabbit heart revealed the accumulation of DC during ischemia and after reperfusion [del Nido P.J., Nakamura H., Mickle D.A. et al.] and of MDA in the homogenate and DC membranes of mitochondria and lysosomes after reperfusion [Ambrosio G., Flaherty J.T., Duilio C. et al.]. In HPLC-purified phospholipids from the rat heart, by contrast, DC content increased during 60-min ischemia but remained unchanged after reperfusion for 2–10 min [Hegstad A.C., Strand H. and Ytrehus K.]. The increase of oxygen-centered, carbon-centered, and nitrogen-centered (flavosemiquinones) radicals was observed during total ischemia and/or after reperfusion of the rabbit heart [Togashi H., Shinzawa H., Yong H. et al.; Zoli M., Ferraguti F., Zini I. et al.].

Fig. 41: HP(I) and DC(II) contents in homogenate, DC content(III), Ca^{2+} uptake(IV) and Ca^{2+}-ATPase activity(V) in SR membrane in dog heart after 1h complete ischemia in vitro (1, 2) and after 4h of reperfusion(3) (% of the initial level).

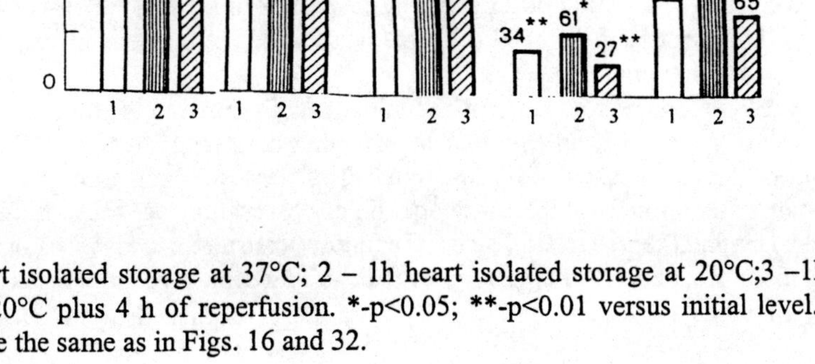

1 – 1h heart isolated storage at 37°C; 2 – 1h heart isolated storage at 20°C;3 –1h heart isolated storage at 20°C plus 4 h of reperfusion. *-p<0.05; **-p<0.01 versus initial level. P- 3 versus 2. Methods are the same as in Figs. 16 and 32.

The dynamics of LPO processes was studied in greatest detail using the model of *experimental AMI with LAD or LAC occlusion*. In this model, coronary circulation preserved in a reduced form and, consequently, constant influx of substrates and O_2 may support high LPO activity over a longer period of time. On the other hand, the maintenance of venous blood outflow from the heart reduces the degree of acidosis, the severity of ischemic injury, and the accumulation of LPO products in the organ. The main feature, however, is that this model (except for experiments under anesthesia) is associated with pain and stress reactions that lead to the active formation of CA and their influx into the organ. This makes it difficult to estimate the role of LPO *per se* during ischemic injury, and, hence, corresponding effects can only be assessed in combination.

In experiments on LAD occlusion in rats, spontaneous and H_2O_2-induced chemiluminescence of lipids increased 2 h after the operation, continued to increase on the third day, and persisted until the 14th day [Kogan A. Kh., Kudrin A. V. and Nikolaev S. M. (1976); Kogan A. Kh. and Mednykh A. Ya; Kogan A. Kh., Kudrin A. V. and Nikolaev S. M. (1979); Kudrin A. N., Kogan A. Kh., Korolev V. V. et al.]. Preventive treatment with antioxidants (selenite, Na, α-tocopherol, or their combination) resulted in lower chemiluminescence, improved cardiac function, and reduced size of ischemic necrosis. Later, the same model was used to demonstrate that an increase in chemiluminescence and the content of secondary LPO products (Schiff bases) occurs after a shorter exposure to ischemia (10–40 min after the occlusion) and in the early reperfusion period [Kogan A. Kh., Kudrin A. N., Kakturskii L. V. and Losev N. I.; Litvitsky P.F., Kogan A.Kh., Kudrin A.N. et al.; Olbinskaya L.I. and Litvitsky P.F.].

Experiments with rabbits showed that MDA content in the cytoplasm, mitochondrial membranes of the heart, and blood increases on the first and fifth days after LAD occlusion [Yakushev V.S. and Lifshits R.I.]. In experiments on rats, HP content in the ischemic tissue was found to increase 3.5-fold 24 h after LAD occlusion of the LAD, the increase in Schiff bases being 13.3-fold. In this case, the accumulation of LPO products was accompanied by lower activity of antioxidant enzymes (SOD and GSH peroxidases I and II) [Meerson F.Z., Kagan V.E., Azkhipenko Yu.V. et al.]. There is evidence [Lukyanova L.O., Kudrin A.N. and Kogan A.Kh.] that the contents of MDA and Schiff bases 2 h after LAD occlusion in rats exceeded the initial values by factors of 7 and 7.8, respectively. The peak of MDA content 3 h after coronary occlusion was also reported elsewhere [Röth E., Török B., Zsoldos T. and Matkovics B.]. In experiments on dogs, the content of free radicals in the AMI focus, estimated by the EPR method, increased earlier than MDA content (15 and 45 min after LAD occlusion, by factors of 1.5 and 1.4, respectively); moreover, the free radical content of venous blood outflowing from the AMI focus increased fourfold after 5 min, i.e., stronger and earlier than in the ischemic tissue itself [Rao P.S., Cohen M.V. and Mueller H.S.].

The generation of ROS (DMPO-OH) and oxygen- and carbon-centered lipid free radicals ($RO^{\bullet}$, $ROO^{\bullet}$, $O^{\bullet}$) in the dog heart exposed to regional ischemia (AMI) after 20–90 min of LAD [Arroyo C.M., Kramer J.H., Leiboff R.H. et al.; Timoshin A.A., Tskitishvili O.V., Serebryakova L.I. et al.] and after 15 min of LAD and 3-min heart reperfusion [Li X.Y., McCay P.B., Zughaib M. et al.] was revealed by the ESR method with PBN and MNP used as spin trapping agents. DC content in the venous effluent from the ischemic region increased in parallel with the duration of ischemia (from 8- to 21-fold after 5–90 min) [Arroyo C.M., Kramer J.H., Leiboff R.H. et al.].

The release of LPO products in the blood and the correlation between their content and severity of myocardial damage were repeatedly observed clinically in patients with AMI using the chemiluminescent method [Gorgoshidze M.L. and Megreladze I.I.] and by measuring the accumulation of primary and secondary LPO products (HP, DC, MDA, Schiff bases) [Volkova L.P., Bondarenko M.I. and Severin N.F.; Lankin V. Z., Zakirova A. N., Akhmetova B. Kh. et al.; Panchenko E.P., Smirnov I.E. and Zyabkina A.G.; Ageta M., Yamasaki S., Ueki K., et al.; Jayakumari N., Ambikakumari V., Balakrishnan K.G. and Iyer K.S.; Loeper J., Goy J., Rozensztajn L. et al.]. An increase in MDA content in urine was revealed in patients after aortocoronary shunting.

In our laboratory, LPO products were determined in patients admitted to the hospital not later than 2–4 h after the onset of primary vast AMI not complicated by other diseases [Churakova T.D., Bilenko M.V., Pokazeeva Z.T. et al.]. The results showed that the content of these products in blood and plasma of the patients upon admittance and throughout the 15-day period of observation was much higher than in healthy individuals and patients with coronary heart disease. The most persistent increase was observed in DC and TC contents, being more pronounced in erythrocytes than in blood plasma. In patients with complicated AMI, the contents of these LPO products were 1.4–1.5 times higher than in patients without any complications. Additional treatment with the antioxidant Dibunol (BHT) reduced the level of LPO products and had a favorable effect on the course of recovery from AMI (see Chapter 11).

LPO intensity in t he heart has also been studied on experimental models of *total hypoxia in vivo or hypoxic perfusion in vitro.* However, LPO activation in these models was less pronounced and, judging from MDA content, either did not manifest itself at all [Varskeviciene Z.Z., Cerniauskiene R.C. and Grybauskas P.S.] or was observed only during the reoxygenation period [Meerson F.Z.; Gauduel Y. and Duvelleroy M.A.; Guarnieri C., Flamigni F. and Caldarera C.M.]. Nevertheless, O_2 radical scavengers (mannitol, histidine), antioxidant enzymes (SOD, catalase, GSH peroxidase), and natural or synthetic antioxidants produced a protective effect even on these models of O_2 deficiency.

One of the models of ischemic damage in the myocardium also involved *injection of epinephrine, norepinephrine, and adrenomimetic drugs (isoproterenol, methoxyamine, metraminol, and others)* in high doses causing numerous small necroses in the myocardium [Serov R.A., Terekhova T.V., Makarova L.D. et al.; Shlyapnikov V.N. and Bogdanova A.A.]. A similar necrogenic effect on the heart is also exerted by high doses of coronaroconstrictors, such as angiotensin, histamine, and vasopressin.

LPO intensity in the heart after treatment with the above drugs was rather high. Spontaneous lipid chemiluminescence in the rat heart increased 2–12 h after administration of epinephrine at a necrogenic dose [Kogan A. Kh., Kudrin A. V. and Nikolaev S. M. (1979)]; a high dose of isoproterenol led to the accumulation of MDA in the homogenate, mitochondria, and microsomes of the rat heart [Chernysheva G.V., Stoida L.V., Amakantova G.G. et al.]; intracardiac administration of epinephrine to rats reduced SOD and antioxidant activities [Demurov E. A., Mil'chakov V. I. and Gerasimov A. M.]; vasopressin injections to rabbits caused an increase in the contents of DC and MDA in mitochondrial and SR membranes of the heart [Tsyomik V.A.]. The necrogenic effect of large doses of isopropyl norepinephrine on the rat heart was more pronounced in E-deficient animals and insignificant in rats receiving vitamin E over a long period of time [Singal P.K., Kapur N., Dhillon K.S. et al.], and damage caused by exogenously introduced CA was alleviated by antioxidant drugs [Pronchuk N.F., Gurevich M.I. and Meerson F.Z.; Shimkovich M.V. and Petkov O.].

Another model close to CA-induced necrosis of the heart is the *stress model,* in which blood supply to the heart becomes markedly impaired and the content of LPO products increases significantly, although to a smaller extent than during local ischemia [Meerson F.Z., Kagan V.E., Azkhipenko Yu.V. et al.; Meerson F.Z.].

The data discussed above show that LPO processes are intensified in all models of cardiac ischemia and reperfusion, although to different degrees. In the total ischemia and AMI models, LPO proved to be intensified both during ischemia itself and in the course of subsequent reperfusion. The accumulation of LPO products was accompanied by disturbances of the barrier function of membranes and reduced activities of membrane-bound enzymes, and antioxidant drugs produced a protective effect. Under conditions of hypoxia or hypoxic perfusion of the heart, MDA and SB accumulation and the release of MDA to the perfusate occurred only during reperfusion, although the protective effect of antioxidants was also observed. In the model of hypoxic perfusion, heart injury is less pronounced than during partial and especially total ischemia, and LPO products can be flushed out of the heart. These factors may account for less pronounced changes in LPO intensity or the less successful detection of the resulting LPO products in experiments with this model.

LPO activation was also observed upon action of CA and vasopressors and in conditions of CA release under stress. In these cases, LPO intensity may depend on both ischemia spreading in the heart and LPO-inducing properties of epinephrine itself (see Chapter 10).

5.2.5 Assessment of LPO in Ischemic and Reperfused Limb Muscles

Striated (skeletal) limb muscles belong to tissues highly resistant to ischemia.

It is known that the critical period of their exposure to total ischemia at 36–37°C is 6 h, and changes occurring in them become irreversible only after 9 h of ischemia [Oksman T.H., Dalin M.V. and Kovanov V.V.; Savel'ev V.S., Chekareva G.A., Mishnev O.D. et al.]. The much higher resistance of limb muscles to ischemia, compared to that of other tissues, indicates the importance of studies on the time and degree of LPO intensification and its role in ischemic and reperfusion injury of this tissue.

In our laboratory, the intensity of LPO in ischemic and reperfused limb muscles was studied in experiments on applying tourniquets on the upper one-third of the hind limbs of August or Wistar rats for 1-6 h [Arkhipenko Yu. V., Bilenko M. V., Dobrina S. K. et al.; Bilenko M.V. (1982); Bilenko M.V., Churakova T.D., Arkhangelskaya S.A. and Komarov P.G.; Kagan V. E., Churakova T. D., Karagodin V. P. et al.; Bilenko M.V. (1980)].

The results (Fig. 42) showed that LP content in the homogenate and SR membrane lipids was very low in intact muscles but increased significantly after 3–6 h of ischemia. The initial HP content in SR membrane lipids exceeded that of LP in homogenate lipids; it increased significantly (threefold) as soon as after 1 h of ischemia are remained at the same level until the sixth hour of the ischemic period.

Fig. 42: AOA (I), DC(II), LP and HP (III) in homogenate and SR membrane of rat limb muscles after 1, 3, and 6h of complete (tourniquet) ischemia (mean ± SEM).

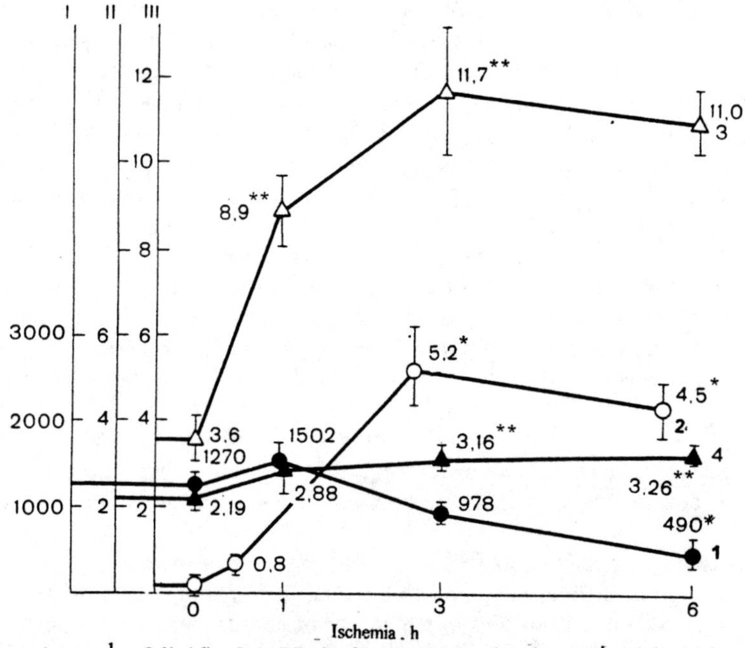

1 - AOA (h• ml• mg^{-1} of lipid); 2 - LP in homogenate (nml• mg^{-1} of lipids); 3 - HP in SR membrane (nml• mg^{-1} of lipid); 4 - DC in SR membrane optical density units per mg of lipid). *-p<0.05; **-p<0.01 versus initial level. LPO product and AOA determination methods are the same as in Fig. 32; the LP method as in Fig. 33.

DC content in SR membrane lipids increased after 3 h of ischemia. The decrease of AOA began during the same period of time but became statistically significant only after 6 h. In contrast to other organs, however, AOA values remained positive, i.e., antioxidants in limb muscles quantitatively prevailed over prooxidants.

Limb reperfusion after 6 h of ischemia (Fig. 43) was followed by repeated and rapid (within the first 10 min) significant accumulation of HP in the tissue, with MDA content remaining high for 1 h. After 12 h of reperfusion, both these parameters returned to the initial level. DC content in limb muscle homogenate demonstrated no significant changes.

Comparison of changes in the content of LPO products with those in the enzymatic system of Ca^{2+} transport in SR membranes of ischemic limb muscles (Fig. 39) showed that the accumulation of LPO products in SR membrane lipids preceded changes in the Ca^{2+} transport system, as the Ca^{2+}/ATP ratio demonstrated a tendency to decrease after 1–3 h, significantly decreased not earlier than after 6 h, and Ca-ATPase activity also decreased after 6 h. The increase of HP content and the reduction of the Ca^{2+}/ATP ratio demonstrated a strong inverse correlation with one another ($r = -0.73$, $P < 0.01$).

During 4-h reperfusion of limb muscles exposed to 6-h ischemia, the content of DC remained high, that of HP increased, whereas the Ca^{2+}/ATP ratio dropped to zero, and Ca-ATPase activity decreased further (data statistically significant) (Fig. 44).

Fig. 43: LPO product content in homogenate of 6h ischemic rat limb muscles (tourniquet method) after different time of reperfusion (% of the initial level).

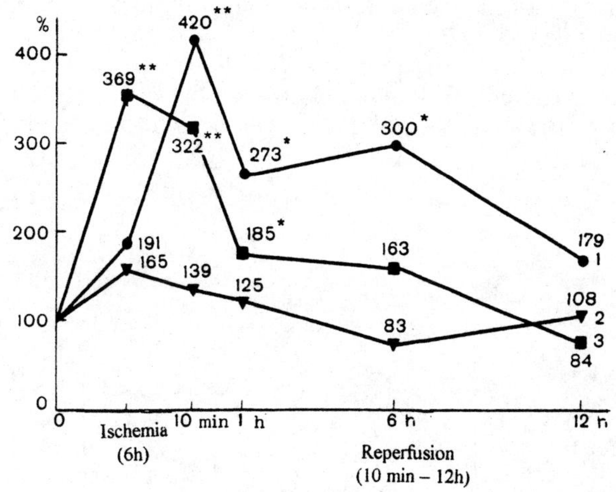

1 - HP; 2 - DC; 3 - TBARS; *-p<0.05, **-p<0.01 versus initial level. Methods of LPO product determination see in Fig. 32.

Fig. 44: LPO product content and Ca^{2+} transport enzyme system indexes in SR membranes of rat limb muscles after 1, 3 and 6h of complete ischemia and 6 h of ischemia + 4 h of reperfusion (% of the initial level).

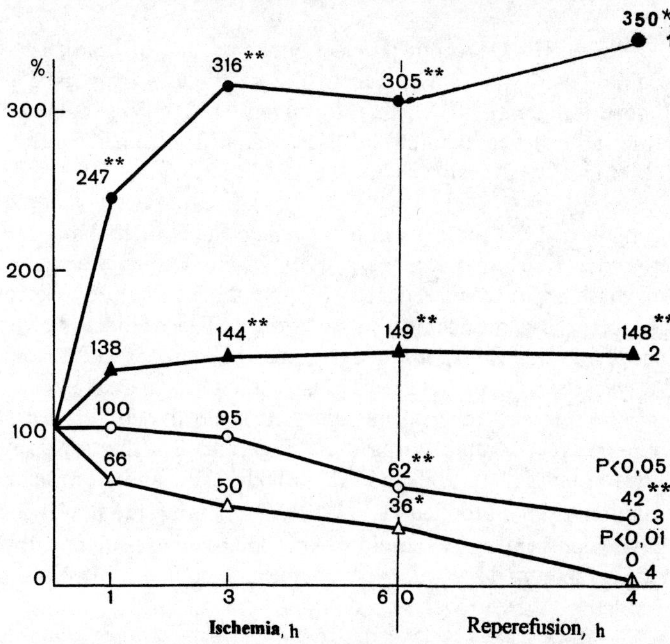

1 - HP; 2 - DC; 3 - Ca-ATPase activity; 4 - Ca^{2+}/ATP ratio; *-p<0.05, **-p<0.01 versus initial level. P - significant differences before versus after reperfusion.

The involvement of LPO in structural and functional damage of SR membranes in limb muscles was also demonstrated in experiments with preventive administration of natural (α-tocopherol) and synthetic

(BHT) antioxidants to rats. As shown in Fig. 45, both antioxidants significantly reduced the accumulation of LPO products and, at the same time, slightly improved the efficiency of the enzymatic Ca^{2+} transport system, as determined from the Ca^{2+}/ATP ratio. Since Ca-ATPase activity remained unchanged during the same period of ischemia, the effect of antioxidants on it did not manifest itself (data not shown). A similar protective effect on Ca^{2+} uptake in SR membranes of 3-h ischemic and 19-h reperfused rat limb muscles was observed after treatment with a combination of antioxidant enzymes, SOD and catalase [Cronenwett J.L., Lee K.R., Shlafer M. and Zelenock G.B.].

Fig. 45: Effect of α-tocopherol and BHT(ionol) the AO on the LPO product content and Ca^{2+} /ATP ratio in SR membranes of rat limb muscles after 3 h of ischemia (mean ± SEM)

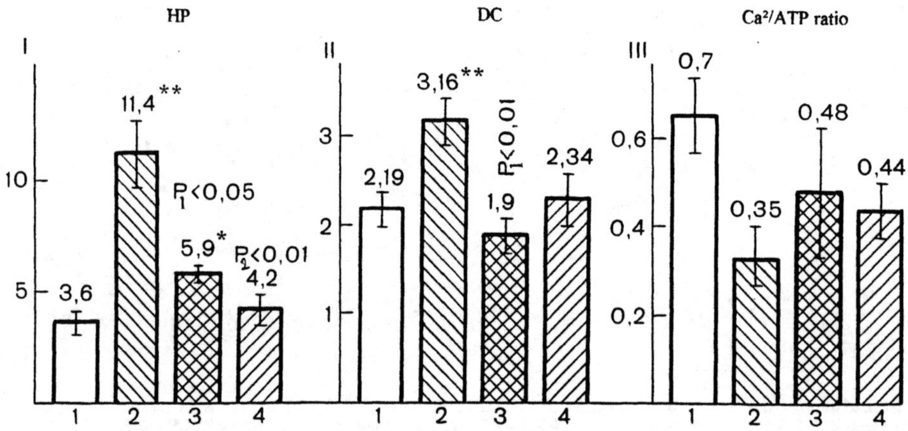

I - HP (nmol• mg^{-1} of lipid); II - DC (optical density units per mg of lipid); III - Ca^{2+}/ATP ratio (relative units); 1 - initial level; 2-4 - 3 h of ischemia: 2 - without AO drugs; 3 - with BHT (120 mg/kg inp. 12 h before ischemia); 4 - with α-tocopherol (50 mg/kg i. p. for 3 days before ischemia); *-p<0.05, **-p<0.01 versus initial level. P1 - significant differences with and without AO; P2 -4 versus 2. Methods are the same as in Figs. 15 and 32.

An increase in the content of LPO products in homogenate lipids of limbs ischemic for 6 h was also accompanied by an increase in lysosomal enzyme activity in the soluble fraction (Fig. 46). Preventive administration of BHT in these experiments reduced HP content and inhibited the release of cathepsin E (completely) and acid phosphatase (significantly) in the cytosol. These findings provide evidence for the role of peroxidation processes in destabilization of lysosomal membranes and the earlier involvement of LPO (compared to that of lysosomes) in pathogenesis of ischemic tissue damage.

The role of LPO in ischemic and reperfusion injury of skeletal muscles has also been studied on models of the gastrocnemius muscle in rats [Kawasaki S., Sugiyama S., Ishiguro N. et al.; Seyama A.] and the gracilis muscle in dogs [Lindsay T., Walker P.M., Mickle D.A. and Romaschin A.D.]. The results demonstrated the accumulation of LPO products (HP, DC, MDA) during the first 1–3 h of reperfusion after 2–5-h ischemia, which was prevented by administering long-acting polyoxyethylene-modified SOD [Kawasaki S., Sugiyama S., Ishiguro N. et al.] and allopurinol or a combination of SOD and catalase [Seyama A.].

Fig. 46: HP content(III), acid phosphatase non-precipitable ("free") activity (I) and cathepsin E (II) after 6 h rat limb muscles ischemia in experiments without and with BHT(ionol) (mean ± SEM).

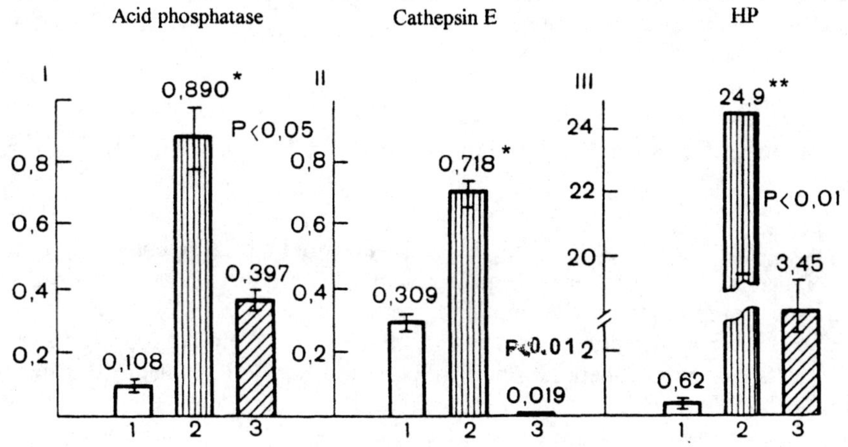

I – Acid phosphate (nmol• min^{-1} per mg of protein); II - cathepsin E (μg• min^{-1} per mg of protein); III - HP (nmol• mg^{-1} of lipid); 1 - initial level; 2 - 6 h of ischemia without BHT; 3 - 6 h of ischemia with BHT (120 mg/kg 4 h before ischemia). *-$p<0.05$, **-$p<0.01$ versus initial level. P – 2 versus 3.

5.2.6 Assessment of LPO in other Ischemic and Reperfused Organs

In recent years, a considerable increase has been observed in the number of publications dealing with the role of LPO processes in pathogenesis of ischemic damage to gastrointestinal organs. This may be accounted for by the fact that these organs are rich in xanthine oxidase, which is involved in the production of ROS (see Chapter 9). LPO intensification proved to play a role in pathogenesis of ischemic and reperfusion damage to mucous membranes of the digestive tract [Desmukh D.R., Mirochnitchenko O., Ghole V.S. et al.; Granger D.N., Rutiti G. and McCord J.M.; Nilsson U.A., Aberg J., Aneman A. and Lundgren O.; Parks D.A., Bulkley G.B., Granger D.N. et al.; Weixiong H., Aneman A., Nilsson U. and Lundgren O.].

Accumulation of LPO products has also been found to occur during hypoxia, ischemia, and reperfusion of lungs [Goroshinskaya I.A. and Mogil'nitskaya L.V.; Kennedy T.P., Rao N.V., Hopkins C. et al.; Török B., Röth E., Zsoldos T. et al.], skin [Im M.J., Shen W.H., Pak C.J. et al.], and other organs.

5.3 Comparative Analysis of Changes in LPO Activity in Organs and Membranes Differing in Sensitivity to Ischemia

The results of our experiments and published data obtained by direct and indirect methods provided conclusive evidence that total or partial ischemia of virtually all the organs examined is accompanied by ROS formation and LPO intensification with the accumulation of its intermediate, primary, secondary, and end products, along with a decrease in ARA and AOA of lipids. Today, LPO activation in ischemic organs is an established fact. However, the researchers working with different organs managed to detect only some of these products, which were not necessarily the same. This fact is explained by the instability of accumulating LPO products, their relatively small amounts, differences in the state of the systems responsible for their metabolism, difficulties in their determination, and a variety of experimental models and methods used for this purpose. Moreover, there also is a widespread (although insufficiently substantiated) opinion that LPO intensification and the involvement of LPO products in organ damage are characteristic mainly (or exclusively) of the reperfusion period.

The available data allow us to compare cases of LPO activation during ischemia and in the postischemic period, the initial periods of this process in organs and membranes with different sensitivity to ischemia, and the relationships between LPO intensification and the severity of ischemic damage in membranes or entire organs.

5.3.1 Comparison of Data on LPO Intensification in Organs during Ischemia and after Reperfusion

In this respect, published data may be divided into two unequal groups that comprise studies demonstrating an increase in the contents of one or several LPO products either *in the course of ischemia* (the first, major group) or *after reperfusion* of organs (the second group).

The first group should apparently include the data on the accumulation of ROS and intermediate free-radical LPO products [Asami O., Ihara I., Shimidzu N. et al.; Rao P.S., Cohen M.V. and Mueller H.S.; Röth E., Török B., Zsoldos T. and Matkovics B.; Van Ye T.M., Roza A.M., Pieper G.M. et al.], and the primary, secondary, and end products of LPO in the kidney [Alessenko A.V., Bilenko M.V., Burlakova E.B. et al.; Bilenko M.V. (1982); Soriano S.G., Coxon A., Wang Y.F. et al.; Green C.J., Healing G., Simpkin S. et al. (1986b); Grune T., Sommerburg O., Petras T. and Siems W.G.], liver [Bilenko N.V., Kagan V.E., Velikhanova D.M. and Komarov P.G.; Velikhanova D.M., Bilenko M.V. and Kagan V.E.; Dudnik L. B., Bilenko M. V. Alessenko A. V. et al. (1981); Zherdev V. P., Sariev A. K., Dvoryaninov A. A. et al.; Kozlov A. V., Shinkarenko L. I., Vladimirov Yu. A. and Azizova O. A.; Konvai V. D., Lukoshkin A. V. and Smirnova V. B.; Lankin V. Z., Tikhaze A. K., Voronina N. V. et al.; Panchenko L.F., Gerasimov A.M., Nozdrachova L.I., Kariakina G.A.; Shilina M.K., Ivashchenko S.P., Shrainberg A.I. et al.], heart [Bilenko M.V., Otverchenko V.M., Astaf'ev V.V. et al.; Bulgakov V.G. and Bilenko M.V.; Zamuruev O. N. (1984); Konvai V. D.; Tavazzi B., Di Pierro D., Bartolini M. et al.; Yakushev V.S. and Lifshits R.I.; del Nido P.J., Nakamura H., Mickle D.A. et al.; Majewska M.D., Strosznajder J. and Lazarewicz J.; Romaschin A.D., Rebeyka I., Wilson G.I. et al.], brain [Bilenko M.V., Tel'pukhov V.I., Churakova T.D. and Komorav P.G.; Churakova T.D., Bilenko M.V., Tel'pukhov V.I. and Komarov P.G.;; Strosznaider J. and Dabrowitecki Z.], limb muscles [Arkhipenko Yu. V., Bilenko M. V., Dobrina S. K. et al.; Bilenko M.V., Churakova T.D., Arkhangelskaya S.A. and Komarov P.G.], and other organs *during exposure to ischemia*. In some works cited above, LPO intensity in the reperfusion period was not analyzed, but most of other studies showed that the moderate or even pronounced accumulation of LPO products continues after reperfusion of organs. An exclusion to this rule concerns several works in which the contents of MDA in the heart during hypoxia [Park Y., Kaneka S. and Kehrer J.P.] or of DC during total ischemia [del Nido P.J., Nakamura H., Mickle D.A. et al.; Hegstad A.C., Strand H. and Ytrehus K.] were found to increase significantly only in the course of exposure to ischemia (hypoxia), remaining unchanged in the early reperfusion period. The data on DC were obtained in experiments in which lipids isolated from the organ were purified chromatographically and separated into triglycerides, nonesterified fatty acids, and phospholipids. Thereafter, only peroxidized PUFA isolated from the phospholipid fraction by HPLC were analyzed for DC. The authors of this work explained this result by a higher specificity of methods used in experiments. It should be noted, however, that the sensitivity of ischemic organs to the repeated initiation of free-radical reactions (the situation observed in the reperfusion period) differs from organ to organ and definitely depends on the severity of ischemia. As noted above, Fe^2–ascorbate- and NADPH-dependent LPO initiation *in vitro* was poorly efficient and produced contradictory results in experiments with the totally ischemic liver [Voronov G.G., Lukienko P.I. and Bushma M.I.; Sharapov V.I., Grek O.R., Dolgov A.V.], whereas partial ischemia proved to result in the increased sensitivity of the rat heart to H_2O_2 [Kogan A. Kh. and Mednykh A. Ya] and of the rabbit spinal cord to Fe^{2+}–ascorbate-induced LPO [Halldorsson A., Kronon M., Allen B.S. et al.]; epithelial cell cultures exposed to anoxia and ischemia became increasingly sensitive to the deleterious effects of H_2O_2 [Bilenko M.V., Ladygina V.G. and Fedosova S.V.(1998a)], TNF-α [Bilenko M.V., Ladygina V.G. and Fedosova S.V.(1998a); Bilenko M.V., Fedosova S.V., and Ladygina V.G.], and oxidized LDL [Bilenko M.V., Ladygina V.G. and Fedosova S.V.(1998b)]. In other words, different organs, depending on the severity, completeness, and duration of ischemia, can demonstrate not only the "depletion" of their potential for the secondary initiation of LPO, but also the "priming" phenomenon increasing the activity of biological systems involved in LPO initiation.

The second group comprises studies in which an increase in the contents of LPO products in the organ or outflowing blood (perfusate) was only observed in the early reperfusion period. The analysis of these works shows that they were mainly performed on models of the least severe ischemia (partial ischemia or hypoxic perfusion) [Gauduel Y. and Duvelleroy M.A.; Guarnieri C., Flamigni F. and Caldarera C.M.; Halat G., Lukacova N. Chavko M. and Marsala J.; Nilsson U.A., Aberg J., Aneman A. and Lundgren O.; Varskeviciene Z.Z., Cerniauskiene R.C. and Grybauskas P.S.] In such cases, either LPO products are formed in amounts insufficient for their reliable detection, or they are diluted and flushed out of the organ with blood or perfusate. Moreover, these authors used as a rule not rather sufficiently specific methods for assessing LPO, such as determination only of MDA or 4-hydroxynonenal, chemiluminescence, and LPO induction *in vitro* [Ballagi-Pordany G., Richter J., Koltai M. et al.; Bilenko N.V., Kagan V.E., Velikhanova D.M. and Komarov P.G.; Velikhanova D.M., Bilenko M.V. and Kagan V.E.; Koller P. and Bergmann S. R.; Ambrosio G., Flaherty J.T., Duilio C. et al.; Grune T., Sommerburg O., Petras T. and Siems W.G.; Grum C.M., Ragsdale R.A., Ketai L.H. et al.; MacMillan V.; Marubayashi S., Dohi K., Ezaki H. et al. (1982); Tsao P.S. and Lefer A.M.; Yamamoto M., Shima T., Uozumi T. et al.; etc].

Primary LPO products are usually detected more frequently than secondary products, as follows from the analysis of works in which the former (HP or DC) were revealed in organs during ischemia, whereas the latter were not (e. g., [Strosznaider J. and Dabrowitecki Z.]). Cases when the methods for assessing primary LPO products failed to reveal their increase during ischemia are relatively rare. They all include also experiments on determining DC in total (crude) lipid fractions of the liver [Hashimoto N., Wu T.W., Au J.X. et al.], brain [Halat G., Lukacova N. Chavko M. and Marsala J.], or heart [Ambrosio G., Flaherty J.T., Duilio C. et al.; Romaschin A.D., Rebeyka I., Wilson G.I. et al.]. As noted above, the analysis of lipids purified by HPLC revealed an obvious increase in DC contents during the ischemic period [del Nido P.J., Nakamura H., Mickle D.A. et al.; Hegstad A.C., Strand H. and Ytrehus K.].

In [Ambrosio G., Flaherty J.T., Duilio C. et al.] study, LPO products were only revealed in the total homogenate of an ischemic organ, rather than in its membrane fraction. This fact may be attributed to active hydrolysis of oxidized PUFA by phospholipases during ischemia and their rapid release from membranes into the cytosol [see Chapter 4, section 5.4). Some authors assessed LPO products only after exposure to both ischemia and reperfusion [Desmukh D.R., Mirochnitchenko O., Ghole V.S. et al.; Lindsay T., Walker P.M., Mickle D.A. and Romaschin A.D.] and, without any plausible reasons, attributed the observed signs of LPO intensification to the effect of reperfusion only, although the corresponding experimental procedure provided no basis for determining the time when LPO was actually activated.

Thus, the concept that free-radical reactions are only activated in the reperfusion period disagrees with abundant experimental data accumulated to date and is apparently erroneous. Although the methods for registering corresponding processes need further improvement, it is obvious that intensification of these reactions begins as early as in the course of exposure to ischemia and continues in the early reperfusion period. Success or failure in their experimental detection depends on many factors, including the severity of ischemia, the type of organ or membrane fraction, and the method used for assessing LPO products.

5.3.2 Comparison of the Time and Degree of Accumulation of LPO Products during Ischemia of Different Organs

The analysis of above presented data (Table 6) [Bilenko M.V. (1988); Bilenko M.V. (1989a)] showed that the increase in the content of LPO products became statistically significant after 5 min of cerebral ischemia (DC, MDA), 15 min of hepatic ischemia (LP, HP, Schiff bases), 30 min of cardiac ischemia (HP, DC, MDA), 60 min of renal ischemia (LP, DC), and 1–3 h of limb muscle ischemia (HP in SR membranes after 1 h; PL and DC in homogenates after 3 h). According to the duration of ischemia resulting in this increase, organs can be arranged in the following ascending series: brain (5 min) < liver (15 min) < heart (30 min) < kidney (60 min) < limb muscles (1–3 h), which corresponds to the arrangement of organs with respect to their sensitivity to ischemia (see Chapter 1).

According to the data shown in Table 6, an increase in the contents of primary, secondary, and end products of LPO occurred virtually simultaneously, except for DC increasing later than HP in SR membranes of limb muscles, as well as MDA and DC increasing earlier than TC in brain lipids. In most cases, however, the accumulated amounts of DC and secondary products were smaller than those of peroxides.

Thus, the exact sequence of accumulation of LPO products in tissues (only primary, then secondary, and, finally, end products), observed in experiments with LPO induction *in vitro*, has not been confirmed in studies on total ischemia of organs *in vivo*. This is probably accounted for by rapid metabolism of primary LPO products, and, hence, experiments on the earlier stages of ischemia and shorter intervals between measurements are required to clarify this situation.

The accumulation of primary LPO products in some organs was characterized by the extreme dynamics in the sense that the initial increase and peak of their contents coincided (Fig. 47). This was observed in the heart, brain and liver (LP in homogenate lipids and mitochondrial and microsomal membranes) and limb muscles (LP in homogenate lipids). In the kidneys, as well as microsomal membrane lipids of the liver and limb muscles, the content of LPO products increased linearly, proportionally to the duration of ischemia. Differences in the type of the curve showing the accumulation of primary LPO products in the homogenate and microsomal membranes of the same organ (liver, heart, or limb muscles) may reflect specific features of the systems responsible for the formation and utilization of these products in microsomal membranes. The highest peak of primary LPO products was observed in the liver (PL and HP increased by a factor of 13 to 6.8), then followed the kidneys (7.4 to 5.5) and the heart (2.9 to 2.8), and the lowest peak was characteristic of limb muscles and in the brain (1.4 to 3), which corresponds to the arrangement of these organs in the series with respect to the initial content of primary LPO products (see Fig. 32). Apparently, the onset of accumulation of LPO products (i.e., the time when the balance between systems responsible for the formation and utilization of LPO products becomes disturbed) is a more informative parameter than the dynamics and degree of this accumulation.

Fig. 47: Primary LPO product accumulation in different completely ischemic organs (% of the initial level)

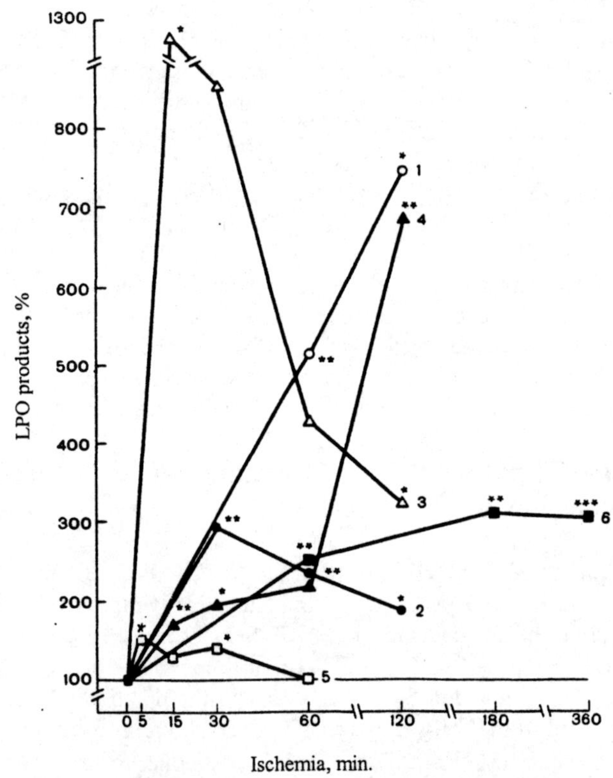

1 - kidney (LP, homogenate); 2 - heart (HP, homogenate); 3 - liver (LP, homogenate); 4 - liver (HP, ER membranes); 5 - brain (DC, homogenate); 6 - limb muscles (HP, SR membranes). *-p<0.05, **-p<0.01 versus initial level. The methods are the same as in Figs. 32 and 33.

In addition, there is published evidence that LPO processes under conditions of E avitaminosis are stronger activated in liver microsomes than in lung microsomes [Shakhman E.V., Donchenko G.V. and Datsenko Z.M.].

5.3.3 Comparison of the Time and Degree of Decrease in AOA during Ischemia of Different Organs

According to experimental data (see Table 6, Fig. 48), the earliest decrease of AOA was revealed in the ischemic liver and heart (after 1 h of ischemia), then followed ischemic kidneys (after 1.5 h) and, finally, ischemic limb muscles (after 6 h); that is, the arrangement of organs by this criterion corresponded to the series reflecting the accumulation of primary LPO products: liver and heart (1 h) < kidneys (1.5 h) < limb muscles (6 h), although the decrease of AOA in any organ began not earlier than 30 min after the onset of increase in LPO products.

Unlike the accumulation of primary LPO products, the decrease of AOA, especially during prolonged ischemia (60–120 min and longer), was proportional to organ sensitivity to ischemia: it was maximal in homogenate lipids of the liver and less pronounced in the heart and kidney homogenates, whereas the AOA of limb muscles retained positive values even after 6 h of ischemia (see Table 6).

As changes in AOA occur during much longer periods than those required for tissue processing, it is apparent that the observed decrease of AOA in organs occurs during their ischemia *in vivo* and is not an artifact resulting from inadequate experimental procedure.

Table 6: LPO Product accumulation and AOA decrease in the lipids of different rat organs during complete ischemia

Organ	Source of lipids	Product	LPO products			AOA		
			Onset of accumulation	MAX accumulation		Onset of decrease	MAX decrease	
				Period	Times °		Period	Amount °°
Kidney	H	LP	1 h	2h	7.4*	1.5h	2h	-1130**
	H	DC	1 h	2h	5.5**			
Heart	H	HP	30 min	30 min	2.9**	1 h	2 h	-1872**
	H	DC	30 min	30 min	2.8**			
	SR	MDA	30 min	1 h	1.3**			
Liver	H	LP	15 min	15 min	13.0*	1 h	2 h	-3710**
	Mt	LP	15 min	15 min	10.0*	15 min	2 h	-625*
	ER	LP	15 min	2 h	7.0*	1 h	2 h	-583**
	N	-	-	-	-	1 h	2 h	-2580**
	ER	HP	15 min	2 h	6.8	-	-	-
	ER	SB	15 min	2 h	2.7**	-	-	-
Skeletal muscles	H	LP	3 h	3-6 h	-	6 h	6 h	+490*
	SR	HP	1 h	3-6 h	3.0**	-	-	-
	SR	DC	3 h	6 h	1.5**	-	-	-
Brain	H	DC	5 min	5 min	1.5*	NOT STUDIED		
	H	TC	15 min	1 h	2.5*			
	H	MDA	5 min	30 min	2.0**			

NOTE: *-P<0.05; **-P<0.01 versus the initial level; °- as compared to the initial value taken as 1; °° - h • ml • mg^{-1} of lipid; + positive, - negative activity (see text); H- homogenate. Mt, ER, SR, N – intracellular membrane. The methods of LPO product determination are the same as in Figs. 32 and 33.

Fig. 48: AOA in homogenate of different ischemic organs (mean ± SEM).

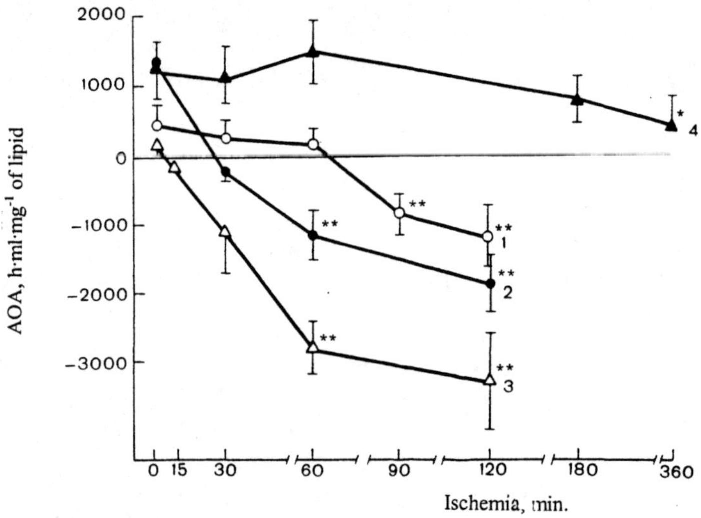

1 - kidney; 2 - heart; 3 - liver; 4 - limb muscles. *-$p<0.05$, **-$p<0.01$ versus initial level. AOA was evaluated on a methyl oleat model [Burlakova E.B., Alessenko A.V., Molochkina E.M. et al.].

The decrease of AOA and the accumulation of LPO products in ischemic organs may be regarded as a case of bilateral cause-and-effect relationships: the decreased AOA may be a prerequisite for LPO intensification and accumulation its products, on the one hand, and the consequence of this intensification and consumption of fat-soluble antioxidants in the course of LPO, on the other (see Chapter 10). In our experiments, the significant AOA decrease always occurred later than an increase in the content of LPO products, which may indicate that this decrease is secondary in relation to LPO processes. In such case, the decrease of AOA serves as additional proof of LPO intensification in ischemic organs; moreover, as this integral indicator characterizes the relationship between all lipid prooxidants and antioxidants, it is more demonstrative that the determination of individual fat-soluble antioxidants (vitamin E, CoQ_{10}, ascorbic acid, etc.), which has been used by some authors for proving the fact of LPO during ischemia [Demopoulos H.B., Flamm E.S., Pietronigro D.D. and Seligman H.L.; Flamm E. S., Demopoulos H. B., Seligman M. L. et al.; Yamamoto M., Shima T., Uozumi T. et al.; Zimmerman B.J., Parks D.A., Grisham M.B et al.].

The decrease of AOA to negative values is also a prognostic sign indicating that ischemic damage to the organ is irreversible, insofar as this decrease occurs during the critical periods of renal, hepatic and cardiac ischemia.

5.3.4 Comparison of the Degree of LPO Intensification during Ischemia in Different Subcellular Membranes of the Same Organ

To this end, the model of hepatic ischemia was used. Taking into account both parameters of LPO intensification (the time and degree of accumulation of LPO products and the decrease of AOA), it becomes apparent that the degree of LPO intensification in subcellular organelles of the ischemic liver decreased in the following series: mitochondria > microsomes > nuclei. The same is true of a decrease in PC, PEA, and AA contents, more pronounced in mitochondrial and microsomal membranes than in nuclei, and that in the cardiolipin level in mitochondrial membranes (see Fig. 13), which provide selective substrates for LPO.

However, the accumulation of LPO products (after the first 15 min) and even the decrease in AOA (after the first 15–60 min) in all membranes occurred much earlier that the decrease in PL and AA and the increase in lysoPL contents (see Figs. 13 and 14); in other words, LPO intensification in membrane lipids of the liver preceded activation of phospholipolysis. These observations are in agreement with current concepts that phospholipases A_2 preferentially hydrolyze oxidized fatty acids and that phospholipolysis is involved in membrane repair after damage (see below).

The earlier and more pronounced intensification of LPO processes in mitochondrial membranes during ischemia was also observed in other [Dzhafarov A. I. (1983); Dzhafarov A. I., Magomedov N. M., Kulieva E. M. et al.], which agrees with the data that these membranes suffer the earliest damage during ischemia and the most severe damage in the postischemic period (see Chapter 2).

At the same time, it was established that mitochondrial membranes of the intact liver have more powerful protective systems (with higher ARA and AOA) than microsomal membranes [51]. Apparently, the prevalent formation of ROS precisely in these membranes (see Chapter 10) makes necessary the presence of a reliable antioxidant protective system, although even such a system is insufficient under conditions of total ischemia (anoxia). On the other hand the study of LPO product content (DC) in subcellular membranes of 45 min ischemic dog heart when tested after 5 min of reperfusion have shown its growth in the following consequence: sarcolemma >> SR>> mitochondria [Romaschin A.D., Wilson G.J., Thomac.U.].

5.3.5 Comparison of the Dynamics of LPO Intensification in Membranes with the Severity of their Functional Damage during Ischemia and Reperfusion

This comparison was performed using membranes of endoplasmic reticulum isolated from three organs differing in sensitivity to ischemia: the liver (ER) and the heart and limb muscles (SR). These membranes were chosen because they are characterized by marked LPO intensification during ischemia, specialized function in each organ, and a high sensitivity to ischemia.

When the patterns of accumulation of LPO products in ER membranes of the liver (Fig. 35), SR membranes of the heart (Fig. 40), and SR membranes of limb muscles (Fig. 44) were compared with their functional parameters, it was found that the increase in the content of LPO products in these membranes during ischemia is in a strong inverse correlation with their specialized function. In SR membranes of limb muscles, the index of correlation between HP accumulation and the parameter characterizing the enzymatic Ca^{2+} transport system (Ca^{2+}/ATP) was –0.73 ($P < 0.01$); in hepatic ER membranes, the index of correlation between HP accumulation and cytochrome P-450 content was –0.94; and the index of correlation between the accumulation of Schiff bases and cytochrome P-450 content was –0.91 ($P < 0.01$ in both cases).

In cardiac SR membranes, the accumulation of MDA coincided in time with the decrease in their capacity for Ca^{2+} uptake. These observations and the improvement of parameters characterizing the functional state of ER membranes in the ischemic liver (increase in the level of aminopyrine N-demethylase) and of SR membranes and lysosomes in ischemic limb muscles (increase in Ca^{2+}/ATP ratio and decrease in nonprecipitable lysosomal enzyme activity) after preventive administration of antioxidants to animals show that the disturbed structural and functional state of the membranes is associated with LPO intensification or, in other words, LPO intensification in an ischemic organ plays a major role in pathogenesis of membrane damage during ischemia. The relationship between the accumulation of ROS and LPO products, on the one hand, and the functional state of membranes and entire organs, on the other, has also been demonstrated in other studies [Salganik R.I., Shabalina I.G., Kolosova N.G. et al.], and especially conclusive data were obtained in experiments with different antioxidants (see Chapter 11).

5.3.6 Comparison of LPO Activity in Different Ischemic Organs in the Early Reperfusion Period

Such a comparison was performed on the basis of data obtained in our experiments (Fig. 49). The results show that the content of LPO products in homogenates of 20-min ischemic brain and 6-h ischemic limb muscles reached peaks as soon as within the first 10–20 min of reperfusion. This mainly concerns primary LPO products (DC, HP), whereas the content of secondary LPO products (MDA, TC) remains at the same high level as before reperfusion (see Figs. 39, 43, 49).

Fig. 49: LPO product content after ischemia (a) and reperfusion (b) of different organs (% of the initial level)

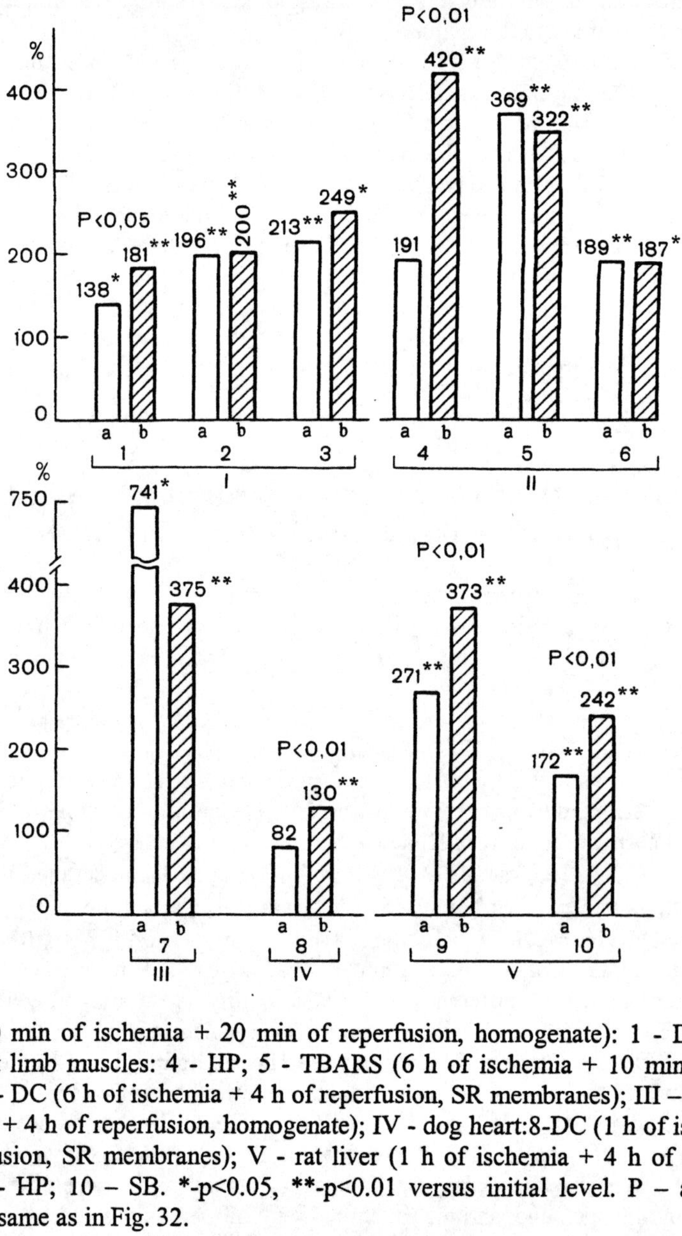

I - rat brain (30 min of ischemia + 20 min of reperfusion, homogenate): 1 - DC; 2 - TC; 3 - TBARS; II - rat limb muscles: 4 - HP; 5 - TBARS (6 h of ischemia + 10 min of reperfusion, homogenate); 6 - DC (6 h of ischemia + 4 h of reperfusion, SR membranes); III – rat kidney:7-LP (2 h of ischemia + 4 h of reperfusion, homogenate); IV - dog heart:8-DC (1 h of ischemia at 20°C + 4 h of reperfusion, SR membranes); V - rat liver (1 h of ischemia + 4 h of reperfusion, ER membranes): 9 - HP; 10 – SB. *-p<0.05, **-p<0.01 versus initial level. P – a versus b. The methods are the same as in Fig. 32.

After a longer period of reperfusion (4 h), the content of LPO products in limb muscles and the kidneys still remained at a level significantly exceeding the norm, and that in the heart and the liver exceeded both the normal level and the level observed before reperfusion (see Fig. 49).

LPO intensification in the kidneys and the heart after reperfusion is also confirmed by the fact that AOA remained at an extremely low, negative level in the former (see Fig. 33) and further decreased down to negative values in the latter (see Fig. 60).

Thus, reperfusion of organs exposed to ischemia for the "critical" (maximum tolerable) period or longer was usually accompanied by further intensification of LPO processes. In some organs, the concentration of LPO products remained high even after 4 h of their reperfusion.

The content of primary LPO products began to increase within the first few minutes of reperfusion, and subsequently both primary and secondary LPO products accumulated. In all the organs, however, the increase in the content of LPO products during the reperfusion period, with the exception of early HP accumulation in limb muscles, never exceeded that during ischemia. This may be due to the fact that the postischemic analysis of the organ was not performed in the "optimal" time or that the potential of the ischemic organ for repeated LPO intensification was temporarily exhausted, which is equivalent to the loss of its capacity for LPO intensification under the effect of various inducers (see above).

Comparing concentrations of LPO products after 4-h reperfusion of the kidney (Fig. 49, III), heart (IV), liver (V), and limb muscles (II), it is seen that these concentrations in the liver and heart increased, compared to those before reperfusion, while in the kidneys and limb muscles they remained unchanged or decreased. This fact shows that in terms of the potential for LPO intensification during reperfusion, these organs can be arranged in the same series as with respect to LPO intensification during ischemia and sensitivity to ischemia.

5.4 Some Mechanisms Responsible for the Damaging Effect of LPO on Membranes

LPO processes are among the factors with the strongest modifying effect on membranes. This effect has been studied in detail on model systems with induction of Fe^{2+}–ascorbate- and NADPH-dependent LPO and, in recent years, LPO induced by lipooxygenase or the Fe^{2+}–xanthine oxidase mixture in lipid (liposomes, BLM) and biological membranes [Vladimirov Yu.A. and Archakov A.I.; Vladimirov Yu.A., Olenev V.I., Suslova T.B. et al.; Vladimirov Yu.A.; Kozlov Yu. P., Danilov V. S. and Kagan V. E.; Kozlov Yu. P., Kagan V. E. and Arkhipenko Yu. V.]. Detailed analysis of these issues is beyond the scope of this book, and we shall limit the discussion to some key aspects of LPO intensification as a factor contributing to membrane damage during ischemia and reperfusion of organs.

The induced LPO processes affect both major biological membrane components—the lipid bilayer and proteins—and the functional properties of glycoproteins.

5.4.1 Effect of LPO on Lipid Membrane Components

This effect is based on two principal mechanisms: (1) increase in passive permeability of the lipid bilayer and (2) changes in its physicochemical properties.

The *increase in passive permeability* for K^+, H^+ and Ca^{2+} ions has been observed in BLM and fragments of biological membranes of different organs both during induction of LPO processes and upon adding exogenous LPO products to the incubation medium [Vladimirov Yu.A. and Archakov A.I.; Vladimirov Yu.A.; Kagan V. E., Azizova O. A., Arkhipenko Yu. V. et al.; Kagan V. E., Churakova T. D., Karagodin V. P. et al.; Kozlov Yu. P., Kagan V. E. and Arkhipenko Yu. V.; Lebedev A. V., Levitsky D. O. and Loginov V. A.; Racay P., Bezakova G., Kaplan P. et al.; Sokolov V.S., Churakova T.D., Bulgakov V.G. et al.]. At present, this increase is attributed to the formation of "passive" polar permeability channels (peroxide clusters) in the lipid bilayer of the membrane [Kagan V. E., Azizova O. A., Arkhipenko Yu. V. et al.; Kagan V. E., Churakova T. D., Karagodin V. P. et al.; Lebedev A. V., Levitsky D. O. and Loginov V. A.; Sokolov V.S., Churakova T.D., Bulgakov V.G. et al.] (Fig. 50). The importance of phospholipid HP incorporation into both BLM layers was demonstrated experimentally, and there is an opinion that LPO products, as well as products of phospholipolysis, can play the role of carrier molecules [Lebedev A. V., Levitsky D. O. and Loginov V. A.].

The damaging action of LPO may also involve the reduction of electrical stability of the membrane lipid bilayer, which leads to its breakdown and either an increase in membrane permeability or rupture and lysis of the cell [Vladimirov Yu.A.]. According to the author of the cited work, such an outcome is typical for the effect of all membrane-damaging mechanisms and may be responsible for so-called LPO-dependent membrane "dismantling" *in vivo* demonstrated for mitochondrial membranes of reticulocytes [Sheve T., Khalanchik V., Khibish C. and Rapaport S.M.], ER membranes of the liver [Archakov A. I. (1975); Bilenko N.V., Kagan V.E., Velikhanova D.M. and Komarov P.G.; Velikhanova D.M., Bilenko M.V. and Kagan V.E.; Vladimirov Yu.A. and Archakov A.I.; Kagan V. E., Kotelevtsev S. V. and Kozlov Yu. P.; Wills E.D.], and SR membranes of skeletal muscles [Kagan V. E., Churakova T. D., Karagodin V. P. et al.] and the heart [Arkhipenko Yu. V. and Shimkovich M. V.].

Fig. 50: Possible mechanisms of membrane damage by LPO products (on a limb muscle SR membrane model) [Kagan V. E., Churakova T. D., Karagodin V. P. et al.]

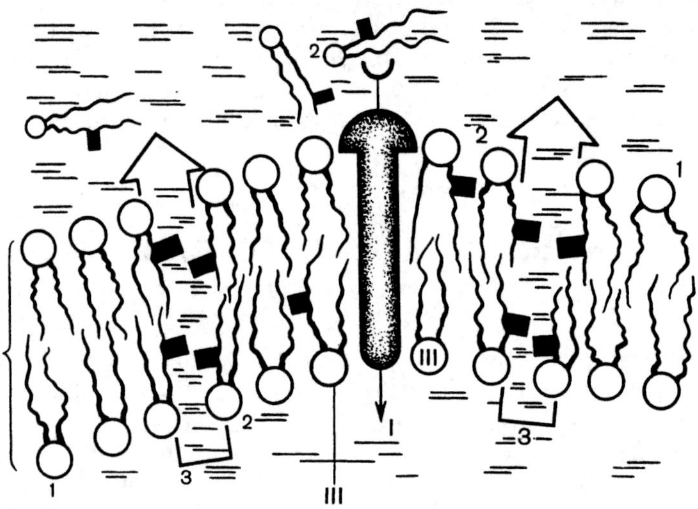

I - integral protein (Ca-ATPase); II - lipid bilayer; III - annular lipids; 1 - PL; 2 - PL HP; 3 - permeability channels (peroxide clusters).

Changes in physicochemical properties of the lipid bilayer (fluidity and electric charge) under the effect of LPO are due to the modified quantitative and qualitative composition of membrane lipids. The induction of Fe^{2+}–ascorbate-, NADPH- and Fe^{2+}–xanthine oxidase-dependent LPO in SR membranes of limb muscles, microsomal membranes of the liver, and microsomal, mitochondrial, and plasma membranes of the brain *in vitro* was accompanied by a decrease in the contents of PUFA (especially 20:4, 22:5 and 22:6) [Kozlov Yu. P., Kagan V. E. and Arkhipenko Yu. V.; Rehncrona S., Smith D.S., Akesson B. et al.; Stirpe F. and Della Corte E.], PEA [Kagan V. E., Azizova O. A., Arkhipenko Yu. V. et al.; Kozlov Yu. P., Kagan V. E. and Arkhipenko Yu. V.; Chan P.H. and Fishman R.A.; Majewska M.D., Strosznajder J. and Lazarewicz J.; Rehncrona S., Smith D.S., Akesson B. et al.], and total PL (membrane depletion of lipids) [Rehncrona S., Smith D.S., Akesson B. et al.], with the decrease of PEA being twice as great as that of PC [Majewska M.D., Strosznajder J. and Lazarewicz J.]. These changes lead to lower fluidity of membranes, which has a strong effect on the function of lipid-dependent enzymes (see below), and the decrease in the content of negatively charged PL affects the charge of the membrane surface and its ability to bind Ca^{2+} ions, as well as the polar and hydrophobic lipid–protein interactions (see Chapter 2).

Changes in lipid composition under the effect of LPO are similar to those caused by phospholipolysis. However, LPO processes are more typically associated with the decrease in the levels of PUFA and PUFA-rich PL (PEA), whereas phospholipolysis usually leads to the reduction of total PL content [Corr P.B., Gross R.W. and Sobel B.E.; Rehncrona S., Smith D.S., Akesson B. et al.].

The incorporation of LPO products into the lipid bilayer also significantly affects its physicochemical properties. As hydroperoxides of FA are more polar than FA themselves, it has been suggested that they come out of the hydrophobic region of the lipid bilayer, thus "loosening" this region and making lipids more accessible to phospholipases and proteins, to proteases [Kagan V.E., Savov V.M. and Serbinova E.A.].

The mechanisms underlying the effect of LPO are similar to mechanisms of phospholipolysis in some respects and differ from them in others. Similarity is manifested in the facts that both these processes: (1) change the composition of membrane lipids by depleting membranes of lipids, primarily of PUFA and PUFA-containing PL, which has a significant effect on physicochemical properties of membranes, protein–lipid interactions, and the functioning of membrane proteins; (2) produce the effect similar to that of detergents, increasing ionic membrane permeability and solubilizing lipids and proteins until the membrane ruptures (effects of FFA, their CoA and carnitine derivatives, lysoPL, FA HP, PL HP); (3)

transfer ions across membranes (similarly to ionophores); and (4) lower the membrane potential, thereby promoting electrical breakdown of the membrane. Differences are as follows: PL of both outer and inner layers of the lipid bilayer are accessible to phospholipases, and their hydrolysis by these enzymes leads to a simultaneous decrease in the PC and PEA levels [Vasdev S. C., Biro G. P., Narbaiz R. and Kako K. J.], whereas the action of LPO processes is mainly manifested in their effect on PEA and cardiolipin. Moreover, LPO products, unlike those of phospholipolysis, are also capable of forming polar ionic channels (peroxide clusters), oxidizing SH groups of proteins, and exerting a cross-linking effect (see below).

Each of these processes is accompanied by the accumulation of specific metabolites, promoted in the presence of various metals [Vladimirov Yu.A. and Archakov A.I.], and suppressed by various inhibitors.

On the other hand, it was shown that phospholipase A_2 hydrolyzes FFA from oxidized membranes in much greater amount than those from unoxidized membranes [Kagan V. E., Shvedova A. A. and Novikov K. N.; Sevanian A., Muakkassah-Kelly S. and Montestruque S.; Sevanian A. and Kim E.]. Moreover, phospholipase C and diacylglycerol lipase also proved to have preferential activity toward oxidized substrates [Gad S.C., Leslie S.W. and Acosta D.]. Therefore, phospholipolysis not only results in membrane damage, but also has a protective effect and plays a role in membrane repair [Davies K.J.A., Wiese A.G., Sevanian A. and Kim E.H.; Sevanian A. and Kim E.].

5.4.2 Effect of LPO on Protein Membrane Components

This effect is more pronounced in the case of integral membrane proteins. In experiments on inducing LPO processes in model membrane systems, their modifying action was demonstrated with respect to Ca-ATPase [Azizova O.A., Maksina A.G., Klaan N.K. et al.; Kagan V. E., Azizova O. A., Arkhipenko Yu. V. et al.; Kaneko M., Hayashi H., Kobayashi A. et al.; Hess M.L., Manson N.H. and Okabe E.; Morgan J., Smith J.A., Wilkins G.M. and Leake D.S.; Xu K.Y., Zweier J.L. and Becker L.C.], Na, K-ATPase [Boldyrev A.A. and Bulygina E.R.; Sazontova T.G., Arkhipenko Yu.V. and Meerson F.Z.; Kramer J.H., Mak I.T. and Weglicki W.B.], cytochrome P-450 [Voronov G.G. and Lukienko P.I. (1982); Gamrekeli D. V., Savov V. M., Stepanova L. I., et al.; Kagan V. E., Kotelevtsev S. V. and Kozlov Yu. P.], Mg-ATPase [Lunets E.F., Kuchuro S.V. and Kostyuk V.A.], adenylate cyclase [Baba A., Lee F., Ohta W. et al.], components of the mitochondrial electron transport chain [Vladimirov Yu.A. and Archakov A.I.; Marshansky V.N., Novgorodov S.A. and Yaguzhinsky L.S.; Salganik R.I., Shabalina I.G., Kolosova N.G. et al.], proteinase C [Mal'tseva E.L., Kurnakova N.V., Burlakova E.B. and Pal'mina N.P.], and creatine kinase (CK) and phosphokinase (CPK) [Spiricheva N.V., Aksenova M.V., Orlov O.N. et al.; Mekhfi H., Veksler V., Mateo P., et al.].

Functions of the above transport protein and enzymes change in a stepwise mode: during the initial step, the function is enhanced; the second step is reversible inhibition; and the third is irreversible inactivation. The transition from one step to another depends on the concentration of accumulating LPO products, i.e., these proteins are lipid peroxide-dependent [Burlakova E.B., Dzhailbova M.I. and Gvakhariya V.O.; Burlakova E.B. and Khrapova N.G.; Kozlov Yu. P., Kagan V. E. and Arkhipenko Yu. V.; Coleman R.; Spector A.A. and Yorek M.A.].

Induction of Fe^{2+}–ascorbate-dependent LPO in SR membranes of limb muscles or the addition of PEA HP to their incubation medium led to activation of Ca-ATPase when HP content increased to 10–14 nmol per mg of lipid; however, further increase to 20 and 26–28 nmol per mg of lipid resulted in reversible inhibition and irreversible inactivation of the enzyme, respectively [Kagan V. E., Azizova O. A., Arkhipenko Yu. V. et al.; Kagan V. E., Churakova T. D., Karagodin V. P. et al.; Kozlov Yu. P., Kagan V. E. and Arkhipenko Yu. V.]. Cytochrome P-450 inactivation in hepatic ER membranes after the onset of NADPH-dependent LPO occurred upon approximately the same increase in PL HP (15-20 molecules per cytochrome P-450 molecule) [Kagan V. E., Kotelevtsev S. V. and Kozlov Yu. P.]. This increase was similar to that in the content of LPO products during hepatic ischemia *in vivo* [Bilenko N.V., Kagan V.E., Velikhanova D.M. and Komarov P.G.], which provides evidence for similarity of corresponding processes. The effect of LPO intensification in membranes on the function of membrane proteins is mediated by three basic mechanisms.

Reversible or irreversible oxidation of protein SH groups does not necessarily affect all lipid-dependent enzymes (e. g., it has virtually no effect on Ca-ATPase in SR of limb muscles) [Kozlov Yu. P.,

Kagan V. E. and Arkhipenko Yu. V.] but plays a key role in disturbing cytochrome P-450 functioning in hepatic microsomes and impairing activities of mitochondrial and other enzymes.

Oxidation of protein SH groups significantly increases ionic permeability of the membrane [Vladimirov Yu.A.].

Alteration of protein–lipid interactions appears to be responsible for the earliest changes in the function of lipid-dependent proteins. In this case, the functions of proteins are disturbed due to changes in physicochemical properties of the lipid membrane component (primarily in the composition, fluidity, and charge of annular lipids), which affect polar and hydrophobic protein–lipid bonds, protein mobility in the membrane, its conformation, substrate accessibility, and other properties (see Chapter 2). The functional alteration of some lipid-dependent proteins (e. g., Ca-ATPase in SR membranes of limb muscles [Kagan V. E., Azizova O. A., Arkhipenko Yu. V. et al.; Kozlov Yu. P., Kagan V. E. and Arkhipenko Yu. V.]) is initially associated with the appearance of polar lipids (PL HP) in the annular layer, which, acting as detergents, "loosen" the lipid layer and facilitate protein functioning. At later stages, PL HP leave the annular layer and are clustered in the bilayer, which results in lower fluidity of annular lipids and worse conditions for Ca-ATPase functioning.

Because of impaired lipid-protein interactions, membrane proteins not only become inactivated, but also come out of membranes (see Chapter 2). Altered composition (primarily in terms of FA) and, consequently, physicochemical properties of the lipid membrane layer lead to significant changes in the functional behavior of glycoproteins, including rhodopsin, cAMP-dependent protein kinase, and transferrin receptors. This mechanism may operate during ischemia and reperfusion of organs.

Protein cross-linking or the formation of covalently bound protein polymers always results in irreversible protein inactivation. Cross-linking is caused by secondary (carbonyl) LPO products, including MDA and hydroxyalkenals, especially 4-hydroxynonenal. This probably accounts for a high toxicity of the latter with respect to cytochrome P-450, aminopyridine methylase, and Ca-ATPase of the liver microsomal membranes [Benedetti A., Comporti M. and Esterbauer M.; Benedetti A., Fulceri R. and Comparti M.; Cadenas E., Müller A., Brigelius R. et al.; Esterbauer H., Cheeseman K.H., Dianzani M.U. et al.].

Such cross-linking (binding with the amino groups of two proteins) results in the formation of stable end products of LPO (Schiff bases).

Thus, LPO intensification in membranes exerts a damaging effect on all their components. A major role in membrane damage belongs to both primary and secondary LPO products, and changes caused by the latter are usually irreversible. Membrane and cytosol proteins modified by oxidation are the first to be digested by proteolytic enzymes (proteases, peptidases, etc.), and this digestion represents the first step in membrane repair [Archakov A. I. and Mokhosoev I. M.; Davies K.J.A., Wiese A.G., Sevanian A. and Kim E.H.; Davies M.G., Fulton G.J. and Hagen P.O.; Stamler J.S., Singel D.J. and Loscalzo J.].

5.5 Conclusions

The data discussed in this chapter clearly show that, notwithstanding intensive worldwide research on LPO processes in ischemic and reperfused organs, there are quite a few controversial and yet unresolved issues that deserve special attention.

First, can LPO processes actually occur in ischemic organs, especially in cases of severe partial or total ischemia, or the ischemic period is merely a prerequisite for subsequent LPO intensification in the course of reperfusion, and the conclusion that LPO processes are intensified during ischemia is based on the results of methodologically imperfect experiments?

Evidence for the existence of conditions necessary for LPO in ischemic tissues (see Chapter 10) and the data reviewed in this chapter show that the accumulation of LPO products during ischemia is not a methodological artifact. Arguments for this conclusion are as follows: (1) the accumulation of substances identified as primary, secondary and end products of LPO (in the case of oxidative lipid degradation in the course of tissue processing, only primary products would have been detected); (2) the decrease of AOA, ARA and the content of natural antioxidants registered along with the accumulation of LPO products (the period of tissue processing is simply insufficient for such a decrease); (3) during ischemia, the membrane is selectively depleted of PUFA (AA) and PL (PEA, PC, cardiolipin), which is more typical for LPO rather than for phospholipolysis; and (4) attempts to induce LPO *in vitro* in tissues already exposed to ischemia are usually unsuccessful, which does not mean that these tissues are not merely predisposed to LPO processes but rather indicates the depletion of substrate, enzymatic and other systems of LPO induction arising during ischemia. The continuous improvement of experimental methods for assessing LPO products will help to find a more definite answer.

Second, even if the accumulation of LPO products does take place during ischemia, the question arises whether they actually produce a damaging effect on membranes or, in other words, whether they are involved in pathogenesis of ischemic membrane damage. The contribution of LPO products to this complex and multistage process was demonstrated by evaluating morphological and biochemical parameters of membranes in experiments with LPO inhibitors (see Chapters 6 and 10) and by analyzing correlations between the content of LPO products and structural or functional parameters of membranes (see above). It should be taken into account, however, that the lack of such correlations does not imply that LPO products have no damaging effect: today, the researchers can only determine a few of the numerous peroxides formed in biological tissues, and these are not necessarily the most toxic products.

Third, the mechanisms of action of LPO products on biological membranes is not yet fully understood, although they have been studied fairly thoroughly under conditions of LPO induction in model systems. Some of possible mechanisms, including changes in physicochemical properties of annular lipids, the formation of permeability channels ("clusters"), cross-linking of lipids and proteins, destruction of certain PL, membrane depletion of lipids, and damage to membrane proteins were described above.

It is apparent that LPO at the early stage plays an adaptive role, as it moderately increases membrane permeability and facilitates the functioning of membrane proteins, whereas subsequent uncontrollable intensification of this process makes it pathological.

Chapter 6

EFFECT OF DIFFERENT BLOOD OXYGEN CONTENT IN EARLY POSTISCHEMIC PERIOD ON THE SEVERITY OF REPERFUSION INJURY IN ORGANS

Experimental studies unequivocally proved that the main factor responsible for both reoxygenation and reperfusion injury is an excessive oxygen influx into the ischemic organ. Reperfusion of ischemic organs with arterial blood (with a normal oxygen level) leads to absolute of relative local hyperoxia because of excessive (for the damaged organ) oxygen concentration, the development of reactive hyperemia, nonuniform blood distribution in functionally different compartments of the organ, a decrease in the oxygen-binding capacity of Hb resulting from acidification of ischemic tissues, and many other reasons (see Chapter 1). Taking into account these facts, we used a very slow, stepwise declamping while transplanting long-term ischemic or preserved kidneys to recipient dogs in the early 1970s [see Bilenko M. V. (1978a)] and performed experiments on reperfusing long-term ischemic kidneys with blood containing only 10% of oxygen, instead of usual 21% [Bilenko M. V. (1978a); Kovanov V. V., Bilenko M. V. and Shelenkova L. N.; Bilenko M.V., Alessenko A.V. and Shelenkova L.N.].

To date, the protective effect of temporary hypoxemia, compared to normoxemia or hyperoxemia, in the early reperfusion period has been studied in our laboratory on three experimental models: ischemia of the kidneys, the heart, and the liver. The results of these studies will be discussed in this chapter.

The experiments on reperfusion of ischemic organs under hypoxic or hyperoxic conditions were performed on dogs (the model of extracorporeal connection of the isolated kidney or the heart to femoral or cervical vessels of the recipient) and rats (reperfusion of the ischemic kidney or the left lobe of the liver *in vivo* by removing microclamps from the vascular peduncle).

To change O_2 level in the arterial blood, experimental animals were made to inhale either hypoxic or enriched gas mixtures (HGM and EGM, containing 10–12% and 80–100% O_2, respectively). In animals of the control group, normal respiration (atmospheric air) was maintained. The partial O_2 pressure (pO_2) in HGM-10 was about 77 mm Hg, which corresponds to that at an altitude of 5800 m above sea level. It was demonstrated that inhalation of HGM-10, as ascending to an altitude of 6000 m, creates moderate hypoxemia fully compensated by the organism [Kolchinskaya A. Z.]. In EGM-80 and EGM-100, pO_2 values were 600 and 760 mm Hg, approximately corresponding to an air pressure of 3.5–4 atm.

Gas mixtures were prepared from O_2 and chemically pure N_2, and their composition was monitored using an oximeter manufactured by Beckman (USA). The recipient dogs received the inhalation mixture through a respirator or an intubation tube; the rats initially breathed through a respirator and then, when the blood flow through the ischemic organ was restored and the abdominal cavity closed, were placed in a chamber constantly supplied with the test gas mixture at a rate of 2 l/min. Relative humidity and temperature in the chamber were maintained at 85–90% and 23–24°C. Inhalation began 7–10 min before reperfusion and continued for 4 h in dogs and 4 or 24 h in rats.

Inhalation of gas mixtures with different O_2 concentrations led to changes in O_2–dependent parameters of the blood. As can be seen from Table 7, inhalation of EGM led to the complete saturation of Hb and a 2.5- to 6.5-fold increase in blood pO_2. Inhalation of HGM was accompanied by a moderate (by 13–16%) decrease in HbO_2 and a more pronounced decrease in pO_2, especially in rats. These changes occurred as early as 5–7 min after the onset of gas mixture inhalation and persisted throughout the inhalation period.

Table 7: Oxygen and pH values in the arterial blood of dogs and rats after inhalation of gas mixtures with different content of oxygen (mean ± SEM).

Inhalated gas mixture	Dogs			Rat		
	HbO_2, %	PO_2, mm Hg	PH	HbO_2, %	pO_2, mm Hg	PH
Air (21% O_2)	94.9 ± 0.6	79.0 ± 3	7.28 ± 0.017	98.0 ± 0.02	102.0 ± 1.8	7.43 ± 0.02
EGM (100% O_2)	100 $P_1 < 0.01$	532.0 ± 10 $P_1 < 0.01$	——	100 $P_1 < 0.01$	258.0 ± 75 $P_1 < 0.01$	7.42 ± 0.01
EGM (80% O_2)	100 $P_1 < 0.01$	281.0±21 $P_1 < 0.01$	7.23±0.04	——	——	——
HGM (10% O_2)	82.2 + 2.1 $P_1, P_2 < 0.01$	49.0 ± 2 $P_1, P_2 < 0.01$	7.31 ± 0.03	82.0 ± 1.7 $P_1, P_2 < 0.01$	47.0 ± 1.7 $P_1, P_2 < 0.01$	7.52 ± 0.02

P_1-EGM or HGM versus air

P_2- HGM versus EGM_{80} or EGM_{100}

Respiration rate in dogs increased upon inhaling HGM (35 ± 5/min) and decreased upon inhaling EGM (15 ± 2/min), compared to that in control air-breathing dogs (21 ± 5/min). However, blood pH in rats and dogs (see Table 7) and arterial pressure in dogs did not change significantly, which allowed comparison between hemodynamic and functional parameters of organs reperfused under different oxygenation conditions.

6.1 Effects of Different Blood Oxygen Content in the Early Postischemic Period on Circulatory, Structural, and Functional Disorders in Reperfused Organs

Experiments with *reperfusion of complete ischemic kidneys* were performed on dogs and rats [Bilenko M. V., Shlifer A. Y. and Epstein I. M.; Bilenko M.V., Shlifer A.Ya. and Ugryumov A.I.; Bilenko M. V. (1978a); Bilenko M.V., Dmitrieva I.S. and Kotlyarov P.M.; Ugryumov A.]. Dog kidneys were reperfused after short (5–10 min) or long (1 h) periods of ischemia at room temperature (20°C). The renal perfusion rate (RPR) in air-breathing animals did not change significantly after brief ischemia (data are not presented) but sharply decreased (by a factor of 2) after 2-h ischemia (Table 8). This no-reflow phenomenon in the kidney ischemic for 2 h was also accompanied by a more marked weakening of cortical circulation (pO_2 in the cortex decreased to 32% of the initial level), blood shunting via the juxtamedullary pathway (a sharp decrease in the cortical : medullar pO_2 ratio), and a significant impairment of diuresis and the concentrating function of the kidneys. Such circulatory and functional changes, typical for the long-term ischemic kidneys in the early postischemic period, were described in detail in our previous publications (see Bilenko M. V. (1978a)).

Table 8: Hemodynamic, oxygen supply and function of 2h ischemic dog kidneys after reperfusion with blood containing different content of oxygen (mean ± SEM).

Parameters	Intial level	After 2h of ischemia		
		Air	EGM	HGM
Renal perfusion rate, ml • min^{-1} per 100g of wet tissue	354±36 (100)	175±30 (49)	146±24 (41)	305±44°* (86)
pO_2 in cortex mm Hg	53±1.5 (100)	17±2.2 (32)	49±14** (92)	27±1.5° (51)
pO_2 in medulla mm Hg	30±2.9 (100)	35±5.5 (117)	124±26** (413)	24±4.2° (80)
Cortical pO_2 medullary pO_2, ratio, rel. units	1.84±0.14 (100)	0.58±0.1 (32)	0.6±0.04 (33)	1.22±0.06° ° ** (66)
Blood pO_2: cortical pO_2 ratio, rel. units	1.7±0.2 (100)	6.9±1.0 (406)	14±3.0 (824)	2.5±1.2 (147)
Diuresis, ml per h	1.4±0.1 (100)	0.65±0.2 (46)	0.9±0.4 (64)	1.7±0.2** ° (121)
Concentration index, rel. units	20±6	2.3±0.45 (12)	5.5±1.7 (28)	13.8±2.0** ° ° (69)

Notes: The initial level is represented by examining kidneys *in situ* [Bilenko M. V., Shlifer A. Y. and Epstein I. M.]; the values in parentheses are % of initial level; *-P<0.05; **-P<0.01 EGM or HGM versus air; o-P<0.05; oo-P<0.01 HGM versus EGM. The methods for pO2 determining in the renal cortex and renal medulla are the same as in Fig. 1 pO2 in the blood was determined using a micro-Astrup apparatus; concentration index was evaluated by urine/blood endogenous creatinin ratio; HGM contained 10% O_2; EGM contained 100 % O2

Pure oxygen inhalation (EGM containing 100% O_2) resulted in a insignificant decrease of RPR in dog kidneys after 5–10 min and 2 h of ischemia as compared to air breathing and a more drastic reduction of the cortical blood flow, which was indicated by a more marked increase of pO_2 in blood (6–7 times), compared to the cortex (approximately 3 times); the blood pO_2 : cortical pO_2 ratio after 5–10 min and 2 h of ischemia increased by factors of 2.8 and 2, respectively, compared to that in air-breathing dogs.

When kidneys were ischemic for 5–10 min, not only cortical, but also medullar vessels displayed a strong vasoconstrictive response to oxygen (a slight increase in medullar pO_2). In 2-h ischemic kidneys, by contrast, pO_2 in medulla increased sharply; i.e., this protective mechanism also failed, which led to more intense blood shunting by the juxtamedullary pathway.

The inhalation of HGM containing 10% O_2 slightly affected the parameters of renal and regional circulation in the kidneys after 5-10 min of ischemia (data are not presented) but caused a significant increase in RPR, enhanced the cortical blood flow (pO_2 decreased in the arterial blood but increased in the cortical layer), and reduced the shunted flow along the juxtamedullary pathway (the cortical pO_2 : medullary pO_2 ratio increased to 1.22 vs. 0.58-0.6 in air- and EGM-breathing dogs).

HGM inhalation, preventing the no-reflow phenomenon and improving the ratio between cortical and juxtamedullary circulation, also significantly stimulated diuresis and the concentrating renal function, especially in 2-h ischemic kidneys.

In the above findings, two observations deserve special attention: first, both the favorable hemodynamic effect of HGM and the adcerse effect of EGM were more pronounced in 2-h ischemic kidneys, compared to briefly ischemic kidneys; second, the favorable effect of HGM inhalation on the hemodynamics and functioning of kidneys reperfused after 2 h of ischemia was manifested more clearly than the adverse effect of EGM inhalation. Both observations confirm the previously proposed hypothesis that the usual O_2 content in the arterial blood is excessive and, therefore, toxic for the ischemic organ [Bilenko M. V. (1978a)]: it creates the state of "relative hyperoxia, " which obviously triggers the development of circulatory disorders after reperfusion. On the other hand, temporary hypoxemia (without any other protective action) abolishes the no-reflow phenomenon, improves blood distribution between the cortex and medulla, and enhances the renal function. The EGM only slightly aggravates hemodynamic disorders that already are strongly manifested in long-term ischemic organs under normoxemic conditions.

Histological and histoenzymochemical studies on reperfused kidneys [Bilenko M.V., Perekrest P.V. and Ugryumov A.I.; Ugryumov A.] showed that EGM inhalation by recipient dogs with extracorporeally connected briefly (5–10 min) ischemic kidneys caused succinate dehydrogenase (SDH) diffusion into the cytoplasm and lactate dehydrogenase (LDH) diffusion into the lumen of renal tubules, which did not occur under air-breathing conditions. In experiments with 2-h ischemic kidney reperfusion, EGM inhalation resulted in more marked suppression of SDH, LDH, and glucose-6-phosphate dehydrogenase (G-6-PDH) activities in cells of the proximal tubules; in addition, it induced multiple hemorrhage into the Bowman capsule. By contrast, HGM inhalation caused a moderate increase in LDH activity in 5–10-min ischemic kidneys (due to intensified glycolysis) but reduced the diffusion of enzymes after 2 h of ischemia.

Gas mixtures with different O_2 concentrations also produced opposite effects on the functioning of rat kidneys reperfused for 4 and 24 h after their 2-h total ischemia at 37°C *in vivo* [Bilenko M.V., Volkova I.A. and Danilov S.P.; Volkova I.A.]. After 4-h reperfusion of the kidneys in rats inhaling EGM (100% O_2), the survival rate was lower than in the control; in rats inhaling HGM for 4 h, it increased slightly (up to 13%); and upon 24-h HGM inhalation, it increased markedly, almost to the same extent as in experiments with preventive administration of BHT at high doses (Fig. 51). EGM or HGM inhalation by rats also had a significant effect on adenine nucleotide synthesis in the kidneys during the early reperfusion period (Fig. 52).

Fig. 51: Rat survival rate after 2 h of complete two kidneys ischemia in vivo an reperfusion with blood containing different content of oxygen (% of the initial level).

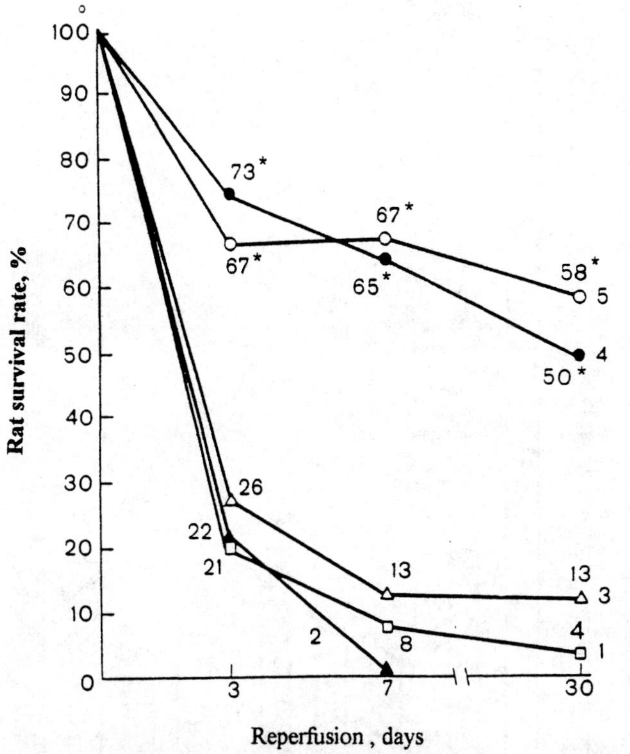

1 - air breathing (control); 2 - EGM (100% O_2) inhalation for 4 h; 3 – HGM (10% O_2) inhalation for 4 h; 4 - HGM inhalation for 24 h; 5 - administration of BHT (240 mg/kg 24 h before the ischemia). *-P< 0.05 control versus.

As shown above (see Chapter 3) and confirmed here, renal ischemia in rats was accompanied by a sharp decrease in ADN and ATP contents. In the postischemic period, ATP content began to increase only after reperfusion for 24 h, whereas ADN content continued to decrease during the first 24 h, and a tendency toward its restoration appeared only after 48–72 h. During 4-h reperfusion of the kidneys in rats inhaling EGM (100% O_2; see Fig. 52), corresponding parameters did not differ significantly from those under air-breathing conditions, but after 24-h reperfusion (4 h of EGM inhalation followed by 20 h of air breathing), ATP and ADP contents were significantly and slightly lower, respectively, than in experiments with air breathing. These results indicate that reperfusion of long-term ischemic kidneys in conditions of both "relative O_2 excess" (air breathing) and "absolute O_2 excess" in blood (EGM inhalation) leads to further degradation of ADN, and the toxic effect of EGM inhalation is much more pronounced after 24 h than after 4 h. Renal reperfusion in rats inhaling HGM for 4 h (see Fig. 52) prevented the postischemic degradation of ADN normally occurring in reperfused organs and caused a significant increase of ATP level in the kidneys. Prolonged (24-h) inhalation of HGM (column 6) intensified ATP resynthesis and prevented the reperfusion degradation of ADN more effectively than 4-h inhalation.

Fig. 52: ATP (I) and ADN (II) contents in the cortex of 2h ischemic rat kidney and after 4 and 24 h of reperfusion with blood containing different content of oxygen (mean ±SEM) (in μmol/g of wet tissue).

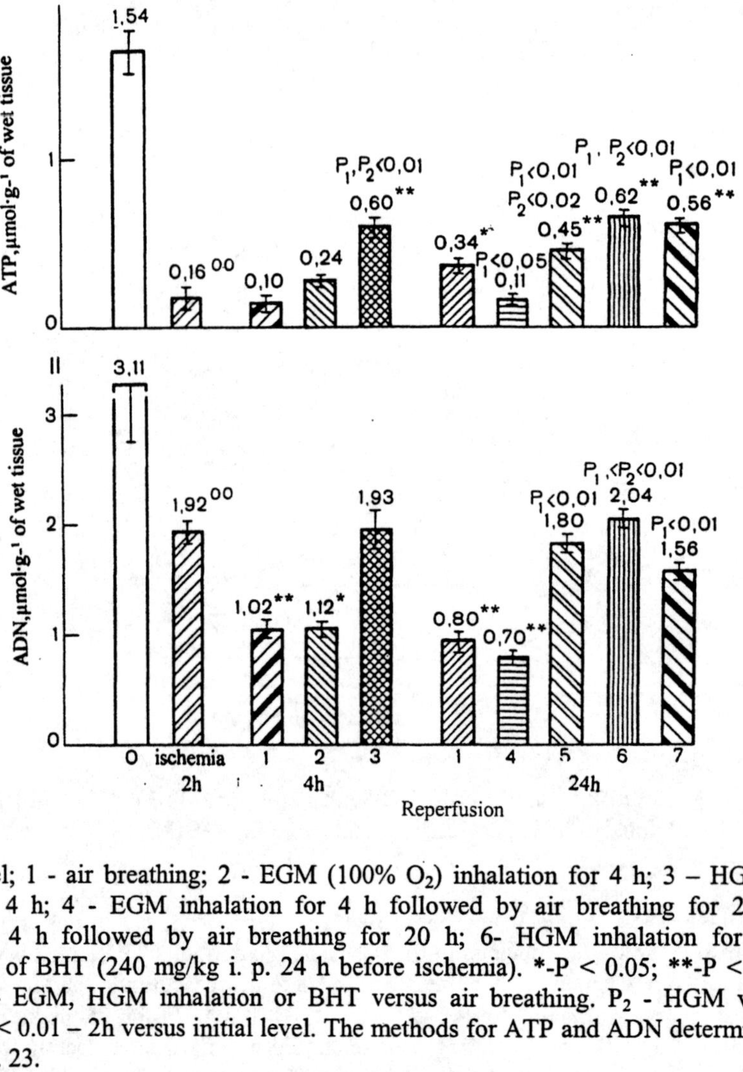

0 - initial level; 1 - air breathing; 2 - EGM (100% O_2) inhalation for 4 h; 3 – HGM (10%O_2) inhalation for 4 h; 4 - EGM inhalation for 4 h followed by air breathing for 20 h;5-HGM inhalation for 4 h followed by air breathing for 20 h; 6- HGM inhalation for 24 h; 7 – administration of BHT (240 mg/kg i. p. 24 h before ischemia). *-P < 0.05; **-P < 0.01 versus ischemia. P_1 - EGM, HGM inhalation or BHT versus air breathing. P_2 - HGM versus EGM inhalation. oo < 0.01 – 2h versus initial level. The methods for ATP and ADN determining are the same as in Fig. 23.

The dependence of the capacity of mitochondria for ATP resynthesis during the early reperfusion period on oxygen content in the inflowing blood is in agreement with the data on animal survival rate (Fig. 53) and the aforementioned results of histoenzymochemical studies on dog kidneys.

Experiments with *reperfusion of the totally ischemic heart* were performed on dogs [Bilenko M.V., Otverchenko V.M., Astaf'ev V.V. et al.; Otverchenko V.N.]. As shown in Fig. 54, cardiac reperfusion under air-breathing conditions after 1 h of total ischemia at 20°C led to incomplete restoration of local blood flow through the left ventricle wall, while pO_2 in the myocardium was significantly higher than the initial level (the state of local hyperoxia). Inhalation of EGM with 80% O_2 caused a significant decrease in CPR and local blood flow but increased pO_2 in the myocardium. HGM inhalation (10% O_2) resulted in a significant increase in both hemodynamic parameters and lower myocardial pO_2, which approached its initial level and became similar to pO_2 in the arterial blood (see Table 6). However, the total O_2 supply to the organ under different oxygenation conditions remained approximately the same: 9.2%/min per 100 g, v/w (data not shown).

Fig. 53: Comparison of ATP contents in kidneys with rat survival rate after 2 h of complete two kidney ischemia and 24 h of reperfusion with blood containing different content of oxygen (% of the initial level).

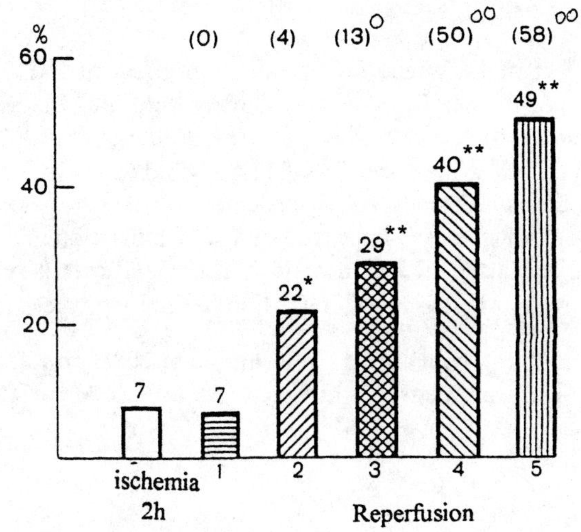

1 - EGM (100% O_2) inhalation for 4 h followed by air breathing for 20 h; 2 - air breathing; 3 - HGM (10% O_2) inhalation for 4 h followed by air breathing for 20 h; 4 – HGM (10% O_2) inhalation for 24 h; 5 - administration of BHT (240 mg/kg i. p. 24 h before ischemia). In parentheses- rat survival rate(%) during 30 days; *-p<0.05; **-p<0.01 versus ischemia. o-P<0.05; o-o-P<0.01-versus air breathing (2).

Fig. 54: Hemodynamic and oxygen supply indexes in canine heart after 1h complete ischemia (at 20°C) and 4 h of reperfusion with blood containing different content of oxygen (mean ± SEM for 4 h of reperfusion).

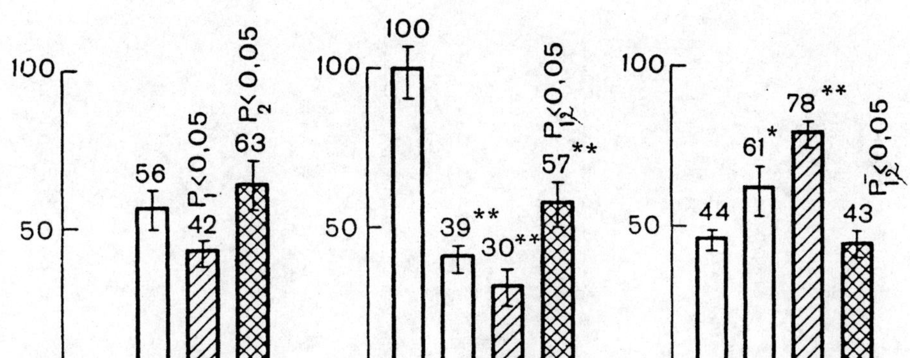

a- coronary perfusion rate(CPR) (ml• min^{-1} per 100 g wet tissue); b – local blood flow (LBF) (ml•min^{-1} per 100 g wet tissue); c – pO_2 in left ventricle muscle (mm Hg) 0- initial level; 1- air breathing; 2 – EGM (80% O_2) inhalation; 3 – HGM(10%) inhalation. *-p<0.05; **-p, 0.01 versus initial level. P_1 – 2 or 3 versus 1. P_2 – 3 versus 2. Reperfusion was performed by extracorporeal connection of ischemic heart to cervical vessels of recipient dog. CPR efflux was measured by efflux from the the pulmonary artery; LBT was measured by the hydrogen polarography method; pO_2 – was determined using LP-7E polarograph and platinum electrodes, implanted to a depth of 0.5 cm. The initial values of LBT and pO_2 were measured in an intact dog heart in situ under artificial air respiration.

The vasoconstrictive effect of EGM and the vasodilating effect of HGM are also apparent from the comparison of pO_2 values in the arterial blood and the myocardium. In the case of EGM, blood pO_2 exceeded that in the myocardium by a factor of 3.6, whereas in the case of HGM, by a factor of only 1.1; that is, this parameter was even below the normal level (1.8–2-fold excess).

Changes of hemodynamic parameters in the heart corresponded to those in the cardiac function. As seen from Fig. 55, the cardiac function was improved by HGM inhalation and adversely affected by EGM inhalation, compared to air breathing. EGM inhalation also significantly reduced the maximal rate of myocardial relaxation ($-dp/dt_{min}$), whereas HGM inhalation had an opposite effect.

Fig. 55: The function of dog heart after 1 h ischemia at 20°C and 4h of reperfusion with blood containing different content of oxygen (mean ± SEM for 4h of reperfusion)

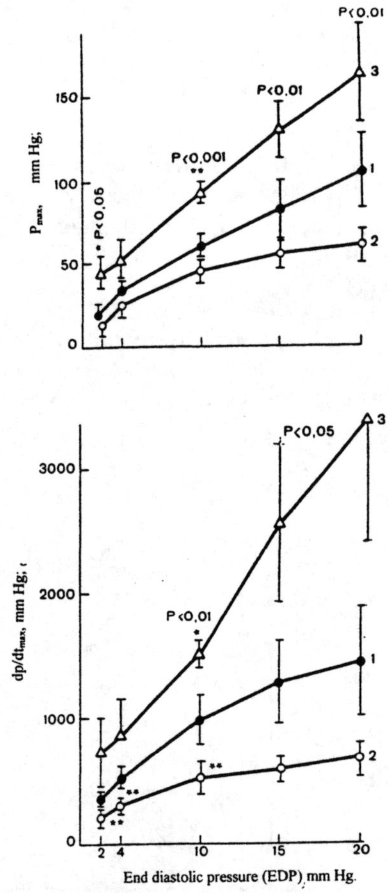

1-air breathing; 2- EGM(80% O_2) inhalation, 3- HGM (10% O_2) inhalation. *-p<0.05;**-p<0.01 versus 1; P - significant differences 2 versus 3.
Heart function was estimated by the isovolume balloon technique under certain EDP; P_{max} and dp/dt_{max} were measured using a Statham electromanometer.

A favorable effect of the ischemic heart reperfusion with low-oxygen blood (pO_2 5–7 mm Hg), compared to that with normal or higher O_2 contents (pO_2 400 mm Hg), was confirmed in experiments on rabbits [Cope J.T., Mauney M.C., Banks D. et al.] and during reperfusion of the isolated rat heart with hypoxic solutions (pO_2 9.2–12.2 mm Hg) [Yamada M., Hearse D.J. and Curtis M.J.]. In the latter case, however, the repeated use of a hypoxic solution after 5-min of reperfusion with a hyperoxic solution (pO_2 > 600 mm Hg) led to a high percentage of reperfusion-induced arrhythmias in the rat hearts [Yamada M., Hearse D.J. and Curtis M.J.]. Earlier, the same group of authors [Hearse D. J., Humphrey S. M. and Bullock G. R] did not observe any decrease of reperfusion-induced damage in the heart (estimated from creatine phosphokinase release to the perfusate) upon a gradual (every 2 min) increase of O_2 content in the perfusing saline [Hearse D. J., Humphrey S. M. and Bullock G. R].

Histological and electron-microscopic studies of the reperfused heart showed that EGM inhalation by recipient dogs aggravated structural damage in the heart: it enhanced pericapillary and interstitial edemas, increased the number and size of local hemorrhages, promoted vacuolization of the cytoplasm in cardiac myocytes and, to a much greater extent, vacuolization of mitochondria, displacement of their cristae, clarification of the matrix, an increase in the number of electron-dense granules, and aggregation and peripheral condensation of chromatin. On the other hand, HGM inhalation alleviated reperfusion-induced structural damage and, in some experiments, even prevented it.

Serious changes of ADN and ATP synthesis in the ischemic heart were also observed when the recipient dogs inhaled gas mixtures during the first 4 h of cardiac reperfusion. ATP and ADN levels dropped sharply during ischemia, showed no significant changes after 4 h of reperfusion under conditions of air breathing and EGM inhalation, and increased significantly when the animals inhaled HGM (Fig. 56) [Volkova I.A.]. The effect of different degrees of blood oxygenation on ATP synthesis is consistent with the above data that gas mixtures with different O_2 contents have opposite effects on the cardiac ultrastructure.

Fig. 56: ATP (I) and ADN (II) contents in isolated canine heart after 1 h of ischemia at 20°C and 4 h of reperfusion with blood containing different content of oxygen (mean ± SEM).

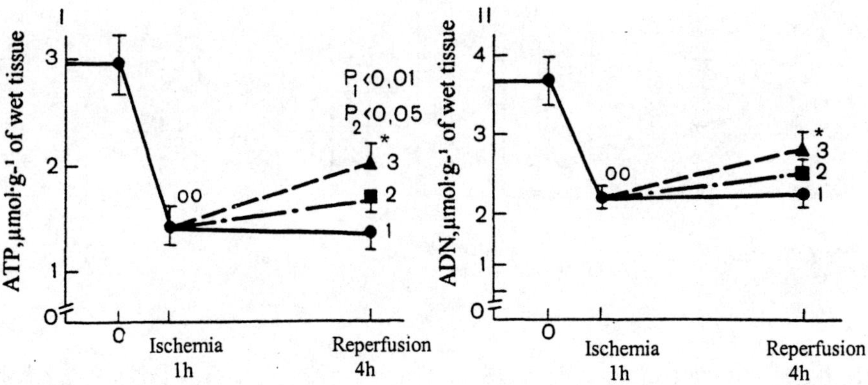

0 - initial level; 1-air breathing; 2-EGM (80% O_2) inhalation, 3-HGM(10% O_2) inhalation; *$p<0.05$ versus 1h ischemia; oo-$p<0.01$ versus initial level; P_1- 3 versus 1; P_2-3 versus 2.

The activity of the enzymatic Ca^{2+} transport system in cardiac SR membranes was also studied [Bilenko M.V., Otverchenko V.M., Astaf'ev V.V. et al.]. As shown in Fig. 57, reperfusion of the heart under conditions of air breathing or EGM inhalation was accompanied by a further significant decrease in the Ca^{2+} uptake capacity and Ca-ATPase activity of SR membranes. Reperfusion of the heart in the case of HGM inhalation led to a significant increase in the Ca^{2+} uptake capacity of SR membranes, with Ca-ATPase activity remaining low; that is, both O_2 excess in the blood, adversely affecting Ca^{2+} transport, and temporary hypoxemia improving the latter resulted from changes (impairment or improvement) in the barrier properties of the lipid bilayer in SR membranes.

Fig. 57: Ca^{2+} - uptake and Ca- ATPase activity in SR membranes of isolated canine heart after 1 h of ischemia at 20°C and 4 h of reperfusion with blood containing different content of oxygen (mean ± SEM).

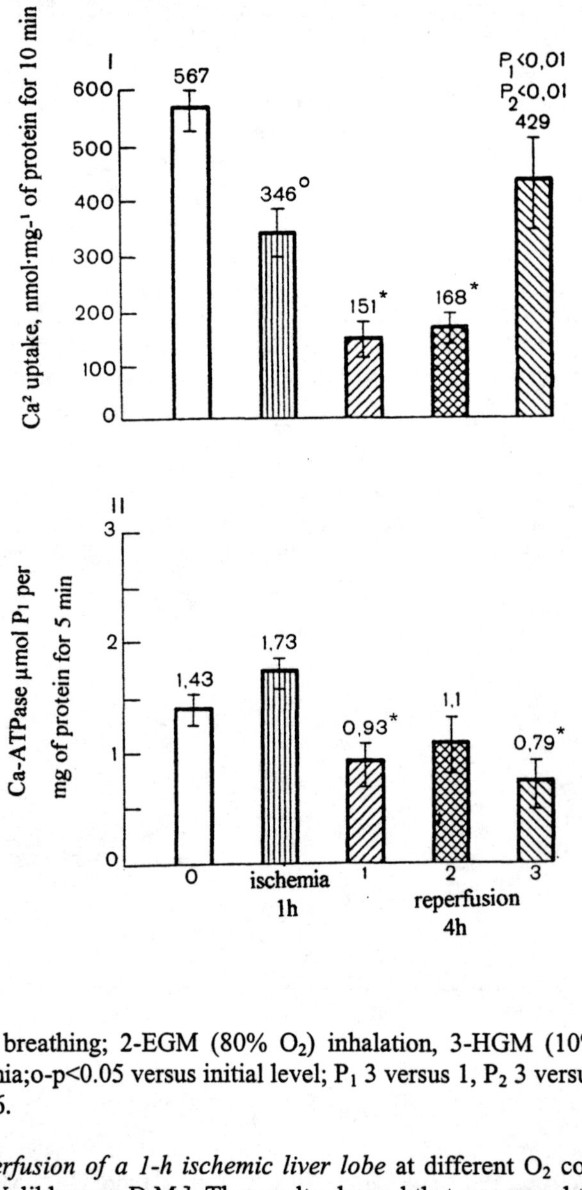

0-initial level; 1-air breathing; 2-EGM (80% O_2) inhalation, 3-HGM (10% O_2) inhalation; *-$p<0.05$ versus ischemia;o-$p<0.05$ versus initial level; P_1 3 versus 1, P_2 3 versus 2. The methods are the same as in Fig. 16.

Experiments on *reperfusion of a 1-h ischemic liver lobe* at different O_2 concentrations in the blood were performed on rats [Velikhanova D.M.]. The results showed that, compared to air breathing, inhalation of EGM and HGM with constant or increasing (stepwise) oxygen concentrations has no significant effect on cytochrome P-450 content during the first 4 h of the reperfusion period (Fig. 58). However, there was a tendency toward an increase in this parameter. It is possible that this increase would have been significant if the time of observation or gas mixture inhalation had been longer. Effects of EGM and HGM were less pronounced than that of a high dose of BHT administered preventively. In experiments with a shorter period of ischemia (15 min), reperfusion of the liver with less oxygenated blood from the portal vein was accompanied by a more intense ATP synthesis than reperfusion with the hepatic artery blood [Wong J.K., Pruett T.L. and Jones R.S.].

Fig. 58: Cytochrome P-450 content in ER membranes of a rat liver after 1 h of complete in vivo ischemia and 4 h of reperfusion with blood containing different content of oxygen (mean ± SEM).

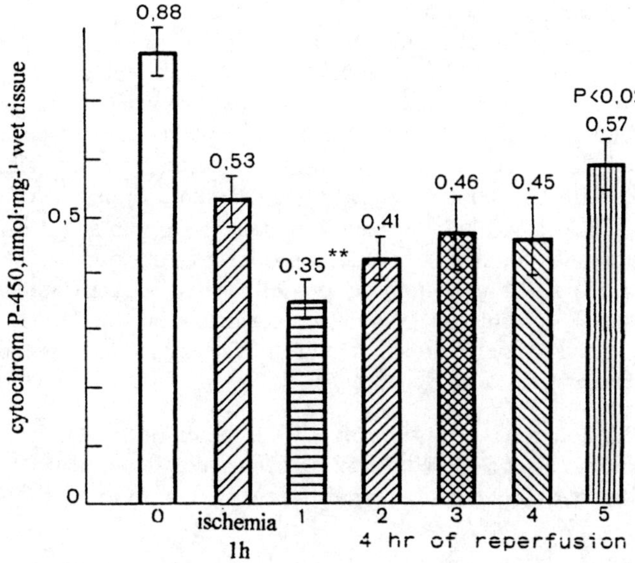

0 - initial level; 1 – air – breathing; 2 – EGM(100%O_2)inhalation for 4h; 3 – HGM(10%O_2) inhalation for 4h; 4 - step-by-step HGM inhalation (increase in pO_2 from 10% to 21% within 4 h); 5 - administration BHT (240 mg/kg i. p. 24 h before ischemia); ** -p<0.01 versus ischemia, P - significant differences 5 versus 1. The determination methods are the same as in Fig. 19.

The results described above show that O_2 content in blood during the early reperfusion period has a significant effect on hemodynamics and functioning of ischemic kidneys and the heart, which are improved upon moderate hypoxemia and impaired upon hyperoxemia. Temporary moderate hypoxemia in the early reperfusion period also caused a decrease in the frequency of morphologically detectable damage to the barrier function of renal and cardiac membranes, intensified ATP resynthesis in ischemic kidneys and the heart, improved parameters of the enzymatic Ca^{2+} transport system in cardiac SR membranes, and produced a favorable (although less pronounced) effect on the cytochrome P-450-dependent monooxygenase system in the liver. By contrast, hyperoxemia aggravated reperfusion damage to mitochondrial membranes, whereas damage to cardiac SR membranes and hepatic ER membranes remained similar to that in air-breathing animals. These findings suggest that (1) in the early reperfusion period, O_2 content in normally oxygenated blood is excessive and toxic for these membranes, and (2) EGM inhalation, reducing blood flow in functionally important compartments of organs, can produce an effect similar to that of HGM inhalation (see sections 6.3 and 6.4).

The favorable effect of low O_2 content in blood during the early reperfusion period was also demonstrated in experiments on other ischemic organs, including rabbit dorsal root ganglion neurons [Fercakova A., Marsala M. and Marsala J.], dog skeletal muscles [Korthuis R.J., Smith J.K. and Carden D.L.], cat stomach [Perry M.A. and Wadhwa S.S.], cultured hippocampal neurons [Khaspekov L.G., Lyzhyn A.A. and Viktorov I.V.]

6.2 Effects of Different Blood Oxygen Content in the Early Postischemic Period on the Intensity of Lipid Peroxidation in Reperfused Organs

The intensity of LPO during reperfusion of ischemic organs at different levels of blood oxygenation was studied in our laboratory on models of renal ischemia in dogs (5–10 min and 2 h of ischemia at 20°C), cardiac ischemia in dogs (1 h at 20°C) and hepatic ischemia in rats (1 h at 37°C), with subsequent reperfusion of the respective organs for 4 h.

Studies on the intensity of LPO during *reperfusion of ischemic kidneys* in dogs [Bilenko M. V. (1978a); Shelenkova L.N.; Bilenko M.V., Alessenko A.V. and Shelenkova L.N.; Bilenko M.V. (1980); Kovanov V. V., Bilenko M. V. and Shelenkova L. N.] showed that when the recipient animal breathed air, the content of primary LPO products (LP) in the kidney exposed to 2-h ischemia and 4-h reperfusion remained at a high level, while AOA had a tendency to decrease further. These results were already presented in Chapter 5. EGM inhalation by dogs (Fig. 59) led to a similar increase in LP contents in the intact kidneys of recipient dogs and kidneys reperfused after 5–10 min and 2 h of ischemia. In all these cases, LP contents were similar to those in the ischemic kidneys or the kidneys reperfused under conditions of air breathing (see Fig. 55). EGM inhalation was also followed by a significant increase of AOA in all the kidneys examined. High LP contents in the kidneys of animals inhaling the EGM, notwithstanding an increased overall AOA in these organs, provided indirect evidence for more pronounced intensification of LPO, compared to that under normoxemic conditions.

Fig. 59: LP content(I) and AOA level(II) in intact kidneys of recipient dogs and in kidneys after 5-10 min and 2 h of complete ischemia at 20°C and 4 h of reperfusion with blood containing different content of oxygen (mean ± SEM).

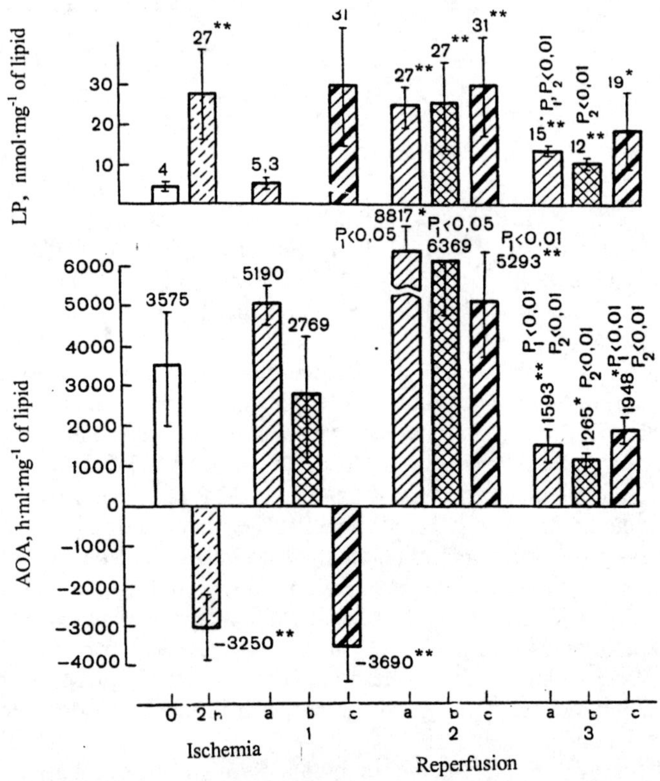

0 - initial level; a - intact kidney of recipient dog; b - kidney after 5-10 min of ischemia; c - kidney after 2 h of ischemia. *-p< 0.05; **-p < 0.01 versus initial level. 1-air breathing; 2- EGN (100% O_2) inhalation;3-HGM(10% O_2) inhalation. Methods are the same as in Figs. 32 and 33. P1 – 2 or 3 versus 1; P2-3 versus 2.

HGM inhalation in dogs led to a moderate (significantly less pronounced than upon EGM inhalation) increase in LP contents in intact kidneys and kidneys ischemic for 5–10 min, accompanied by a simultaneous decrease in AOA. By contrast, LP contents in the kidneys ischemic for 2 h under conditions of HGM inhalation tended to decrease, as compared to 2-h ischemic kidneys reperfused under normoxemic and hyperoxemic conditions, whereas AOA acquired positive values, instead of negative, and significantly (P_1 and $P_2 < 0.01$) exceeded AOA in the kidneys reperfused under air conditions.

Studies on the intensity of LPO during *reperfusion of the ischemic heart* in dogs [Bilenko M.V., Otverchenko V.M., Astaf'ev V.V. et al.; Bulgakov V.G.], showed, that in air-breathing dogs was accompanied by a significant accumulation of DC in lipids of SR membranes and of HP in homogenate lipids and also by a significant decrease of AOA in homogenate lipids (Fig. 60).

Fig. 60: DC (I) and HP (II) contents in homogenate (a) and SR membranes (b) and AOA level in homogenate (III) of canine heart after 1 h of in vitro ischemia at 20°C and 4 h of reperfusion with blood containing different content of oxygen (mean ± SEM).

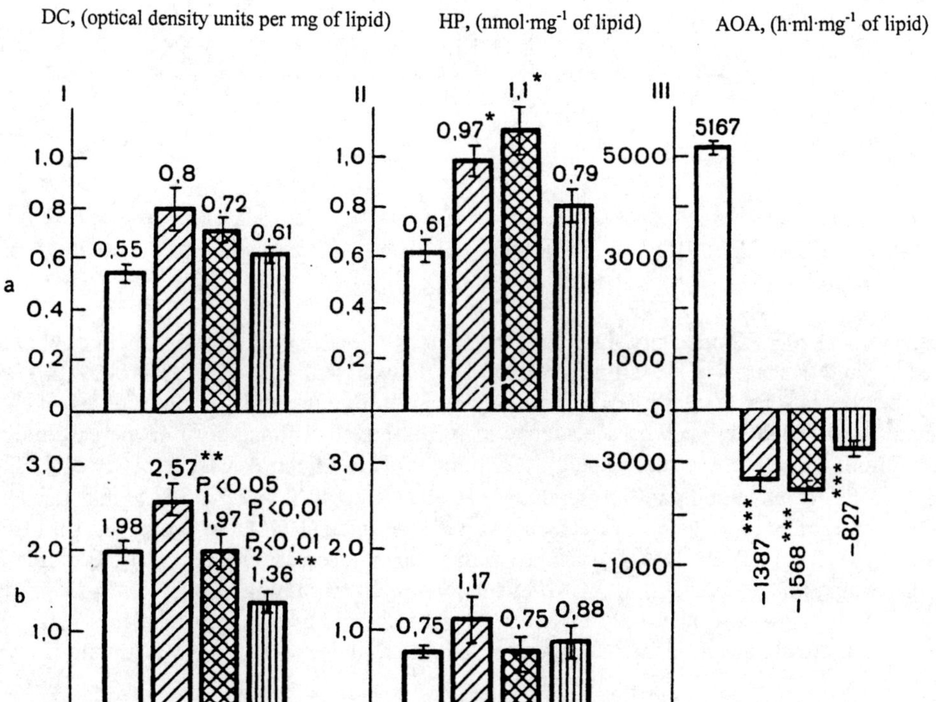

0 - initial level; 1-air breathing; 2-EGM(80%O_2) inhalation; 3-HGM (10%O_2) inhalation. *-p<0.05 versus initial level; P_1 – 2 or 3 versus 1; P_2-3 versus 2.

In dogs inhaling EGM, DC and HP contents and AOA in homogenate lipids were similar to, whereas DC content in SR membranes was lower than those in experiments with air breathing. HGM inhalation led to a slight decrease in HP and DC contents in homogenate lipids and a significant decrease in DC content in SR membrane lipids; AOA also showed a tendency to increase but its values remained negative.

DC contents in lipids of cardiac SR membranes under different conditions of blood oxygenation in the early postischemic period inversely depended on parameters characterizing the functional state of these membranes (Fig. 61).

Fig. 61: Comparison of the contents of DK(a) with Ca^{2+} uptake (b) in SR membranes of canine heart ischemic for 1 h at 20°C after 4 h of reperfusion with blood containing different content of oxygen (% of the initial level)

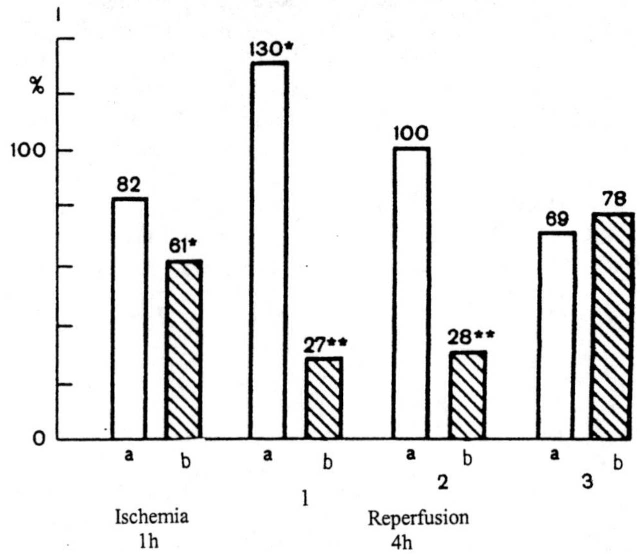

1-air breathing, 2-EGM (80% O_2) inhalation, 3-HGH (10%O_2) inhalation. *-p<0.05;**-p<0.01 versus intial level.

The assessment of LPO intensity during *reperfusion of the ischemic liver* in rats [Velikhanova D.M., Bilenko M.V. and Kagan V.E.; Velikhanova D.M.] showed that upon EGM (100% O_2) and HGM inhalation, the content of secondary LPO products (SB) in lipids extracted from hepatic ER membranes of sham-operated animals either showed a tendency to increase (HGM inhalation) or increased significantly (EGM inhalation) (data are not represented). After reperfusion of the 2-h ischemic liver LPO intensity (Table 9) upon EGM inhalation and air breathing was equally high (2.4–2.7 times higher than initial, $P < 0.01$). At the same time, reperfusion of the ischemic rat liver under HGM inhalation for 4 h (10% O_2 or a stepwise increase from 10 to 21% O_2) led to a significant decrease in HP and SB contents of ER membranes, compared to those upon EGM inhalation or air breathing. Level of LPO products (and hepatocellular injuries) was significantly higher in partial ischemic liver of rats inhalated with 100% O_2 in compare to rats breathing usual air; allopurinol inhibited those changes [Matsumoto F., Sakai H., Yamaguchi M. et al.].

Thus, changes in blood oxygen during the first 4 h of reperfusion affected the intensity of LPO in three different ischemic organs in basically the same manner.

EGM inhalation markedly intensified LPO in intact and briefly ischemic organs (intact or 5–10-min ischemic kidneys and the liver of sham-operated animals) and after reperfusion of long-term ischemic organs. However, LPO intensification under EGM inhalation was more or less similar to that observed under air breathing. Therefore, hyperoxemia led to LPO enhancement in all organs, irrespective of the duration of their ischemia. Equally high contents of LPO products in lipids of long-term ischemic organs during the early period of their reperfusion under normoxemic and hyperoxemic conditions indicate that O_2 level in the arterial blood is excessive for the ischemic organ even under physiological conditions; in other words, this is additional evidence of "relative hyperoxia" ("hyperoxemia").

The reason why EGM inhalation, causing a 3.5- to 6.5-fold increase of pO_2 in the inflowing blood, does not lead to further accumulation of LPO products in long-term ischemic organs remains unclear. This may be due to activation of adaptive reactions in the organism in response to generalized hyperoxemia, one of them being higher AOA in the kidneys (see Fig. 59), although this is only one of the possible adaptive mechanisms. Assuming that the limiting factor here is depletion (even under normal oxygenation) of substrates easily accessible to LPO, it becomes clear why there is no perceptible accumulation of its products even after additional LPO induction in ischemic and reperfused organs *in vitro* (see Table 9). It is also known that a significant increase in pO_2 (100% O_2 under hyperbaric oxygenation) does not affect the radiosensitivity of cells highly susceptible to pO_2 variations only from 0 to 30–40 mm Hg (see Chapter 10).

Table 9. LPO product content in hepatic er membranes of rats after 1h of complete liver ischemia and 4 h of reperfusion with blood containing different content of oxygen.

LPO products	Initial level	Rats after 1 h of ischemia and 4 h of reperfusion			
		Air	EGM	HGM	HGM[1]
HP, nmol•mg-[1] of lipid	3.7±0.7 (100)	9.9±0.3** (267)	9.6±0.6** (259)	7.4±0.2** (197) $P_1<0.01$ $P_2<0.05$	7.42±0.9** (200) $P_1<0.05$
SB, rel. units•mg-[1] of lipid	18.6±0.6 (100)	46.8±4.3* * (251)	44.7±3.4* * (240)	33.4±5.2** (179) $P_1<0.01$ $P_2<0.05$	32.9±5.1** (177)

Notes: In parentheses-percentage of the initial level; HGM[1]-step-by-step increase in pO_2 from 10 to 21%; *-$p<0.05$;**-$p<0.01$ versus intial level; P_1 –HGM versus air; P_2-HGM versus EGM. The methods of isolation hepatic microsomes, HP and SB evalution see[Velikhanova D.M., Bilenko M.V. and Kagan V.E.].

Unlike EGM, HGM produced opposite effects on the intensity of LPO: it intensified corresponding processes in nonischemic or briefly ischemic organs but tended to inhibit or significantly reduced their activity during reperfusion of long-term ischemic organs. In the former case, HGM inhalation created in the organ hypoxemic conditions favorable for LPO intensification, as reported elsewhere [Verbolovich V.P., Teplova L.L., Petrenko E.P. et al.; Maksimova I.A., Maksimov V.M. and Piruzian L.A.; Yoshikawa T., Furukawa Y., Wakamatsu Y. et al.]; in the latter case, it minimized or abolished the state of "relative hyperoxia, " with respective LPO inhibition.

The question why a relatively small decrease in blood O_2 is sufficient for reducing LPO intensity also remains unclear because, as shown for cells *in vitro*, pO_2 of 4–8 mm Hg can provide for the maximal LPO intensification (see Chapter 10). A similar situation (reduction of radiation damage to the surrounding tissues) was observed in patients treated with HGM-10, and it should be noted that the free-radical mechanism of radiation damage is a proven fact, and the sensitivity of tissues to O_2-induced LPO is nearly the same as that to LPO induced by radiation [Demopoulos H.B., Flamm E.S., Pietronigro D.D. and Seligman H.L.].

6.3 Possible Mechanisms Responsible for Effects of Different Blood Oxygen Content in the Early Postischemic Period on the Severity of Reperfusion Injuries in Organs

The data discussed above suggest that inhalation of gas mixtures with different O_2 contents, which create moderate hypoxemia and hyperoxemia in experimental animals has a strong effect on the hemodynamics and functioning of organs, the barrier and structural functions of their membranes, the intensity of LPO, and the severity of reperfusion injuries to organs. The damage to long-term ischemic organs upon EGM inhalation was more severe than under conditions of air breathing, but HGM inhalation by animals with long-term ischemic organs has a protective effect of varying strength (Table 10).

Revealing possible mechanisms responsible for the damaging effect of EGM inhalation and the protective effect of HGM inhalation is important both for gaining a deeper insight into pathogenesis of ischemic and reperfusion injuries and for finding practical applications of these methods. Our own experience and published data confirm the importance of the following two factors: (1) opposite effects of hypoxemia and hyperoxemia on the intensity of LPO in organs and (2) opposite effects of hypoxemia and hyperoxemia on the vascular tone and hemodynamics in organs.

The involvement of LPO intensification in effects of different blood oxygen levels on membrane structures and whole organs has been studied quite extensively but mainly on intact (nonischemic) organs. The results showed that the accumulation of LPO products occurs in virtually all organs of animals breathing 100% O_2 under normal pressure, increased pressure (hyperoxic hyperbaria), and decreased pressure (hyperoxic hypobaria); that is, it appears to be proven that LPO intensification provides the basis for O_2 intoxication, or so-called "oxidative stress" injuries [Lan'shina O. E. . Loginov V. A. and Kovalenko E. A.; Fredlund P.E., Öckerman P.E. and Vang J.O.; Milyutina N.P., Ananyan A.A., Sapozhnikov V.P. et al.; Orrenius S.; Sies H.].

Table 10: Compurative effect of HGM and EGM inhalation in the early reperfusion period on the hemodynamics, function and structure of long term ischemic organs.

Parameter	Organ	Ischemic conditions	Differences from air breathing	
			HGM inhalation	EGM inhalation
Hemodynamics	Dog kidneys	2 h at 20°C	Increase in renal and cortical blood flow, decrease blood shunting through medulla.	Decrease in cortical blood flow, increase blood shunting through medulla.
	Dog heart	1 h at 20°C	Increase in coronary and local blood flows, prevention the growth of pO2 in myocardium.	Decrease in coronary blood flow, higher rate of pO2 increase in myocardium.
Function	Dog kidneys	2 h at 20°C	Enhancement of diuresis and concentrating function.	No differences.
	Rat kidneys	2 h at 37°C	Higher survival rate after 4 and 24 h of inhalation.	No differences.
	Dog heart	1 h at 20°C	Improved cardiac function	Impaired cardiac function.
Morphology	Dog kidneys	2 h at 20°C	Enzymes retain their activity, decreased diffusion of SDH and LDH in epithelium of tubules	Increased diffusion of SDH and LDH, lower activity of SDH, LDH and G-6-PDH, increased hemorrhage.
	Dog heart	1 h at 20°C	Decreased edema in mitochondria improved microstructure of the tissue	Aggravation of edema and ultrastructural changes
State of mitochondrial membranes	Rat kidneys	2 h at 37°C	Increase in ATP content and prevention of further degradation of ADN, after 24 h more pronounced then after 4h	No changes after 4 h; more pronounced decrease in ATP and ADN contents after 24 h
	Dog heart	1 h at 20°C	Increase in ATP content	No differences
State of ER membranes	Dog heart	1 h at 20°C	Increase in Ca^{2+} uptake	No differences
	Rat liver	1 h at 37°C	Tendency to increase in cyt. P-450	Weaker tendency to increase in cyt. P-450.
Intensity of LPO processes	Dog kidneys	2 h at 20°C	Tendency to decrease in LP, increse in AOA	No differences
	Dog heart	1 h at 20°C	Tendency to decrease in HP and DC, increase in AOA in the homogenate; decrease in DC in SR membranes	No differences in DC, HP and AOA in homogenate, decrease DC in SR membran.
	Rat liver	1 h at 37°C	Tendency to decrease in DC.	No differences

Notes: The EGM used contained 80% O_2 in cardiac experiments and 100% O_2 in renal and hepatic experiments. The reperfusion period was 4 h in all experiments, except for those with reperfusion of rat kidneys (4 and 24 h).

The results of experiments described previously, in which intact or sham-operated animals inhaled EGM, are fully consistent with the data cited above, and the increase of AOA observed in the kidneys suggests that these organs are apparently capable of rapidly accumulating antioxidants or enhancing the activity of antioxidant enzyme systems, as demonstrated for hyperoxic lungs [Frank L.; Moak S.A. and Greenwald R.A.] or blood [Lankin V. Z., Tikhaze A. K., Voronina N. V. et al.].

It is obvious that LPO initiation in long-term ischemic organs requires a much smaller amount of O_2 than that in intact organs because (1) the capacity of energy-producing systems (mitochondria) for consuming O_2 becomes reduced during ischemia, and this effect *per se* leads to O_2 excess in the cell, and (2) ischemia creates conditions facilitating LPO initiation: activities of enzymatic and nonenzymatic antioxidant systems decrease, the amount of metabolites initiating LPO increases, systems producing active forms of O_2 become activated, etc. (see Chapter 10).

Indeed, as shown in Chapter 5, the residual amount of O_2 remaining in the ischemic organ (to say nothing of O_2 supplied with the arterial blood during reperfusion) is quite sufficient to initiate LPO, .

An increased sensitivity of organs exposed to ischemia and epinephrine-induced or traumatic damage to physiological or increased O_2 concentrations, which manifests itself in LPO intensification, and the protective effect of antioxidants under these conditions have been reported by many workers in recent years [Demurov E. A., Mil'chakov V. I. and Gerasimov A. M.; Kulitskaya V. I. and Murzin G. V].

In our experiments, however, inhalation treatment with a gas mixture containing 80-100% O_2 during reperfusion of ischemic organs did not result in any additional accumulation of LPO products in the latter, compared to that during air breathing. On the other hand, the barrier function of membrane structures in the heart and kidneys and the cardiac function were impaired to a greater extent and the survival rate of the animals with ischemic kidneys was lower than in air-breathing animals. In other words, no additional LPO intensification under the effect of EGM inhalation took place but the damaging effect of EGM on membranes and organ functioning persisted. LPO inhibition by antioxidants also did not reduce the damaging effect of 100% O_2 during reperfusion of slices of ischemic rat renal cortex or rabbit heart [Asakura K., Ikeda H. and Gemba M.; Cope J.T., Mauney M.C., Banks D. et al.]. These facts may be regarded as an argument against the involvement of LPO in the damaging effect of hyperoxemia on ischemic organs. Nevertheless, it was shown (see above) that hyperoxemia aggravates circulatory disorders and, consequently, makes the distribution of O_2 in the organ more irregular. This may be accompanied by the spatial nonuniformity of LPO intensification, which often remains undiscovered when total LPO products accumulated in membrane or homogenate lipids are determined. However, it is also possible that, upon marked or even maximal LPO intensification in an ischemic organ under normoxemic (more precisely, relatively hyperoxemic) conditions, the damaging effect of excessive blood O_2 is based on mechanisms other than LPO. At the same time, there is evidence for an additional increase in the content of LPO products in blood or different organs reoxygenated or reperfused under hyperoxic conditions [Kontorshchikova K. N. and Solov'eva T. I.; Cope J.T., Mauney M.C., Banks D. et al.; Raguzin A.V.; Salaris S.C. and Babbs C.F.].

The role of less pronounced LPO intensification during the early postischemic period in mechanisms responsible for the protective effect of HGM inhalation on animals is more apparent. This especially concerns cardiac SR membranes in which the significant decrease of DC under the effect of HGM inhalation coincided with a significant increase in Ca^{2+} uptake by these membranes (see Fig. 57 and Table 10).

The inhibitory effect of HGM inhalation on LPO intensity in the early postischemic period is probably associated with the prevention of "relative hyperoxemia" in the organ and the establishment of a certain (at least partial) balance between O_2 supply to the organ and the organ's capacity for its utilization. The effect of moderate hypoxemia preventing LPO in long-term ischemic organs is opposite to its effect on intact organs in which, according to our results and published data, LPO processes are intensified, rather than inhibited [Maksimova I.A., Maksimov V.M. and Piruzian L.A.; Freeman B.A., Topolsky M.K. and Crapo J.D.; Milyutina N.P., Ananyan A.A., Sapozhnikov V.P. et al.; Mogil'nitskaya L.V., Prokofiev V.N. and Morgulis G.L.].

(2) The vasoconstrictive action of O_2 excess and the vasodilating action of O_2 deficiency in the blood also play a major role in the mechanism of the effect produced by inhalation of gas mixtures with different O_2 contents on the severity of reperfusion injuries in organs.

The vasoconstrictive effect of gas mixtures with 50–100% O_2 under normal (normobaric hyperoxia) and elevated (hyperbaric hyperoxia) pressure has already been described in detail [see Bilenko M. V. (1978a)].

Under conditions of "relative hyperoxia" in the early reperfusion period, the vasoconstrictive effect (as well as the prooxidant effect) may also be exerted by much lower, physiological O_2 concentrations in the blood flowing toward the organ; in other words, the postischemic no-reflow phenomenon is determined to some extent by the vasoconstrictive effect of O_2 whose physiological concentrations are excessive for a severely damaged organ. During EGM inhalation in our experiments, blood vessels demonstrated the ability for more pronounced constriction than under air-breathing conditions, which was clearly manifested both in coronary and renal cortical vessels upon long-term ischemia.

The marked vasoconstrictive effect of O_2 excess after ischemia was also observed by other researchers [see Bilenko M. V. (1978a)].

Unlike absolute or relative O_2 excess in blood, hypoxemia exerts a distinct vasodilating effect [see Bershtein S.A., Gurevich M.I. and Soloviev A.N.]. Note that in our experiments with reperfusion of the kidneys, this effect was more pronounced in long-term ischemic (2 h) than in briefly ischemic (5-10 min) organs (see Tables 9, 10). This was apparently due to the fact that hypoxemia, reducing O_2 excess in blood, brought the vessel into the state approaching the physiological norm. The restoration of regular blood flow in the organ, pO_2 reduction, and prevention of abrupt changes in this parameter were the most significant consequences of HGM inhalation that ensured postischemic rehabilitation of the organ because the total O_2 supply to the organ during EGM and HGM inhalation remained approximately the same (see section 6.1).

The free-radical and vasoactive mechanisms underlying the effect of blood O_2 excess or deficiency on ischemic organs in the early reperfusion period are interrelated because, according to the currently accepted concepts, the vasoconstrictive effect of high O_2 concentrations is explained by an increase in Ca^{2+} permeability of the sarcolemma in smooth muscle cells of the vessel, which, in turn, results from intensification of LPO in the cell membrane [see Bershtein S.A., Gurevich M.I. and Soloviev A.N.; Soloviev A.I.; Ebeigde A.B.]. Hypoxemia, precluding O_2 excess (absolute or relative) and LPO reactivation in membranes, protects cells against excessive Ca^{2+} influx and vascular hypertonia.

The data concerning the effect of LPO on vascular tone will be presented in Chapter 8.

Apparently, there are other mechanisms providing for the vasodilating effect of hypoxemia and its protective action during the postischemic period, and some of them may be unknown at present.

6.4 The Use of Methods for Raising and Lowering Blood Oxygen Content in Experiments and Clinical Practice

Basically, three approaches have been used to produce changes of O_2 level in the blood flowing to an ischemic organ: (1) hyperoxygenation, (2) temporary hypooxygenation, and (3) temporary reduction of blood flow rate (so-called controlled reperfusion).

Blood hyperoxygenation by means of O_2 inhalation under normal or increased pressure (hyperbaric oxygenation) is still rather common method of intensive therapy of acute hypoxic conditions, ischemia, and some other pathologies. At the same time, the expediency of using hyperoxemia for antiischemic treatment has not been sufficiently proved by experimental and more or less fundamental evidence; moreover, O_2 inhalation or the exposure of patients in a hyperbaric chamber with air or pure O_2 produced not only favorable, but also adverse results [see Bilenko M. V. (1978a); Korochkin I. M. and Keiyan A. V.; Pakhomov V.J. and Kostyushin V.N.].

The contradictory results of antiischemic treatment by means of blood hyperoxygenation are apparently explained by the lack of definite indications for its use. In this respect, three variants of such treatment can be distinguished, depending on the period when hyperoxemia is induced: (1) before ischemia, to build up a reserve of O_2 (preoxygenation); (2) during ischemia, to achieve a substitutive effect (oxygenation); and (3) during reperfusion, as a therapeutic procedure (postoxygenation).

The first variant is apparently expedient when a moderate increase of pO_2 in inhaled air and blood is produced to prepare the patient for operation on a "dry" organ. It is used in clinical practice, although not very widely. It should be also noted that hyperoxybaric pretreatment (100% O_2, 2.5 atm. absolute, 90 min) produced an ameliorative effect in experiments on 1-h total ischemia of the rat liver [Chen M.F., Chen H.M., Ueng S.W. and Shyr M.H.].

The second variant of using O_2 inhalation or hyperbaric oxygenation is more common, especially in cases of chronic partial ischemia of organs. However, opinions concerning indications for its application, duration, and admissible pO_2 in the gas mixture are contradictory. There is experimental evidence that "substitutive therapy" by means of air inhalation produces a favorable effect during total or partial brain ischemia in rats and gerbils [Grigor'eva A. V., Kazantseva N. V. and Bul'chuk O. V.; Konda A., Baba S.,

Iwaki T. et al.], and the procedure performed at an air pressure of 1.2 in the hyperbaric chamber is more effective than at 2 atm. [Grigor'eva A. V., Kazantseva N. V. and Bul'chuk O. V.].

The published data on the third variant, i.e., O_2 inhalation or hyperbaric oxygenation during the reperfusion (postischemic) period, are scarce. There were reports about favorable effects of O_2 inhalation at a pressure of 2.9–3 atm on the outcome of microsurgical operations on human fingers and toes [Tkachenko S.S., Tikhilov R.M., Belousov A.E. et al.]. Encouraging results were also obtained in recent experiments on the use of ozone or ozone–oxygen mixture for blood or perfusate oxygenation in the reperfusion period [Kontorshchikova K. N. and Solov'eva T. I.].

When patients are operated in a hyperbaric chamber, hyperoxia (hyperoxemia) affects the organ at all three stages of ischemia, and this interferes with the analysis of results and makes them dependent on the extent and duration of ischemia, the type of organ, and other factors.

The above experimental data provide evidence that 100% O_2 inhalation during the early reperfusion period either has no beneficial effect or produces an adverse effect on the structure and function of ischemic organs. Hence, the use of hyperoxemia as a method of treatment is inexpedient.

We repeatedly expressed this opinion in publications [Bilenko M.V. (1983); Bilenko M. V. (1978a); Bilenko M.V., Otverchenko V.M., Astaf'ev V.V. et al.], and its validity was confirmed in recent works of other researchers. For example, they showed that EGM inhalation should be performed at O_2 content of no more than 35–80% and for only a short time, and hyperoxia should only be applied in combination with antioxidant therapy [Gerasimov A.M., Mil'chakov V.I., Demurov E.A. et al.; Demurov E. A., Mil'chakov V. I. and Gerasimov A. M.; Kulitskaya V. I. and Murzin G. V; Meerson F.Z., Saltykova Y.A., Didenko V.V. et al.; Meerson F.Z., Didenko V.V., Savov V.M. et al.].

Blood hypooxygenation, compared to the method described above, is based on an opposite approach and involves inhalation of hypoxic gas mixtures, air inhalation in low-pressure chambers, administration of drugs creating O_2 deficiency, etc. It has been used in clinical practice for a long time but only for protecting tissues surrounding the tumor against radiation damage, rather than for antiischemic treatment.

The radioprotective effect of hypoxemia has been confirmed in experiments on virtually all animal species. In the course of these studies, it was demonstrated that pigs tolerate inhalation of an HGM containing no less than 10% O_2 (HGM-10), dogs are tolerant to HGM-7, mice and rats are more resistant to hypoxemia and tolerate ventilation with HGM-7, HGM-6.3 and, in combination with antihypoxic drugs, even HGM-5 and HGM-4 [see Yarmonenko S.P., Veinson A.A. and Magdon O.].

To date, considerable experience in using HGM-10 inhalations in oncological hospitals has been accumulated. Patients demonstrated good tolerance for repeated procedures lasting for up to 120 min [Strelkov R.B., Kucherenko N.G. and Kozlov V.M.], and, as HGM-8 is twice as effective as HGM-10 [Yarmonenko S.P., Veinson A.A. and Magdon O.], attempts have been made to introduce HGM-9 and even HGM-8 into clinical practice. Extensive experimental studies provided the basis for the successful clinical use of repeated HGM inhalation as a method of preconditioning to subsequent more severe ischemia or hypoxia [see Hypoxia in Medicine].

It should also be noted that repeated hyperbaric oxygenation (100% O_2, 2 atm, five procedures) in experiments with gerbils proved to be successful as a preconditioning method of protection against severe ischemic brain injury [Wada K., Ito M., Miyazawa T. et al.].

Temporary reduction of blood flow rate (controlled reperfusion) is principally similar to the method of blood hypooxygenation because it is also aimed at decreasing oxygen supply to ischemic organs. It was showen that controlled reperfusion performed by temporarily (for the initial 5–30 min) decreasing perfusate volume and pressure with the aid of a pump helped to reduce the severity of reperfusion injuries in the heart, skeletal muscles, and lungs of dogs, pigs, rabbits, and rats [Mitrev Z., Beyersdorf F., Hallmann R. et al.; Anderson R.J., Cambria R., Kerr J. and Hobson R.W.; Bhabra M.S., Hopkinson D.N., Shaw T.E. et al.; Clark S.C., Sudarshan C., Khanna R. et al.; Cope J.T., Mauney M.C., Banks D. et al.]. Gradual reperfusion with stepwise (at 4-min intervals) increase in perfusate flow rate from 13 to 100% of the initial value over 30 min produced a similar effect on the ischemic guinea pig heart [Shul'zhenko V.S., Kuz'min A.I. and Kapel'ko V.I.]. Blood flow reduction by means of temporary mild (34°C) or deep (27°C) hypothermia, on skeletal muscles, liver, spinal cord and the brain [Zar H. A., Tanigawa K., Kim Y. M. and Lancaster J. R. Jr.; Wright J.G., Araki C.T., Belkin M. and Hobson R.W.2nd.; Kakinohana M., Taira Y. and Marsala M.; Boersma E. and Simoons M.L.] or low-pressure reperfusion combined with hemodilution and changes in the composition of the initial perfusate, on the heart, limb muscles, and lungs [Defraigne J.O., Pincemail J., Laroche C. et al.; Halldorsson A., Kronon M., Allen B.S. et al.; Beyersdorf F., Sarai K., Mitrev Z. et al.; Walker P.M., Lindsay T.F., Labbe R. et al.] also showed significant protection.

Intensive research in this field progresses, and some of the methods described above have already found clinical application [e. g., see Beyersdorf F., Sarai K., Mitrev Z. et al.; Defraigne J.O., Pincemail J., Laroche C. et al.].

6.5 Conclusions

The results of our experiments in which animals with long-term ischemic organs (kidneys, liver, heart) received inhalations of oxygen-enriched gas mixtures (EGM) or hypoxic gas mixtures (HGM) and the analysis of published data on the application of methods for raising or lowering blood oxygen level to clinical treatment of different forms of ischemia lead to several conclusions.

(1) During the early reperfusion period, organs exposed to prolonged ischemia come into the state of relative (in many cases, absolute) local hyperoxia, which has a deleterious effect on the structure and functioning of cell membranes and organ as a whole.

(2) Inhalation treatment of animals with gas mixtures containing 80–100% O_2 in this period adversely affects the structure and functioning of long-term ischemic organs, aggravates the reperfusion damage of their membrane structures, impairs local hemodynamic parameters, and makes the no-reflow phenomenon more pronounced. The deleterious effect of hyperoxemia in the early reperfusion period is especially manifested in the ischemic heart.

(3) Inhalation treatment of animals with long-term ischemic organs using the gas mixture containing 10% O_2 (HGM-10) acts beneficially on organ structure and functioning, minimizes reperfusion damage to their membranes, inhibits LPO, improves hemodynamic parameters, and alleviates the no-reflow phenomenon.

(4) Oxygen therapy of patients suffering from ischemic injuries in different organs must be carried out very cautiously and according to definite indications.

As opposed to oxygen inhalation or hyperbaric oxygenation performed for preventive or substitutive purposes, hyperoxygenation in the early reperfusion period is not advisable, especially after restoring circulation in long-term ischemic organs or organs highly sensitive to ischemia.

(5) Inhalation of gas mixtures with a low O_2 concentration (HGM-10) has already been introduced in oncological practice to improve the efficiency of radiation treatment and to minimize damage of tissues surrounding the tumor.

The experimental data presented in this chapter demonstrate the expediency of developing optimal clinical procedures for HGM-10 inhalation during the early period after restoring blood flow in organs exposed to long-term ischemia or highly sensitive to O_2 deficiency (after operations on "dry" organs or organ transplantation).

The use of hypoxemia is also expedient because, as will be shown below (see Chapter 11), antioxidant therapy is mainly effective as a preventive method, and if it is impossible or too late to perform, the possibilities of treating the ischemic organ are rather limited. In this context, the problem of finding new approaches and broadening the spectrum of means for preventing reperfusion damage becomes urgent, and the use of hypoxemia contributes to its solution.

However, hypoxemia as a way to prevent excessive O_2 influx into an ischemic organ should preferably be maintained for a relatively long period of time (at least 4 h). Therefore, it is necessary to develop and improve methods for creating local hypoxemia. Currently, such methods are being developed on the basis of different approaches, including stepwise blood flow restoration in the organ, blood flow restoration under local hypothermia, and regional perfusion with diluted and mildly oxygenated blood, and some of them have already found clinical application.

(6) The data confirming the possibility of preventing or reducing reperfusion injuries in long-term ischemic organs by means of temporary moderate hypoxemia provide additional evidence for the role of O_2 excess in pathogenesis of such injuries and emphasize the expediency of HGM inhalation *per se* or combined with antioxidant therapy and the need for the development and improvement of methods for creating temporary local hypoxemia.

Chapter 7

ROLE OF LIPID PEROXIDATION IN THE PATHOGENESIS OF ISCHEMIC SHOCK

The term "shock" refers to a number of pathological states arising under the effect of different extreme factors (more than 100) and manifested mainly in central and peripheral hemodynamic disorders, progressive generalized circulatory hypoxia, and metabolic, structural and functional disorders in vitally important organs.

Virtually any type of shock comprises three basic components: stressor (pain, fear, neurohumoral disorders), ischemic (bleeding, hypotension, circulatory disorders in organs), and toxic (exogenous, endogenous, or other factors of unidentified nature), although their expression and significance differ depending on the type of shock.

Ischemic shock includes states arising after reperfusion of a large organ ischemic for a long period of time, a group of such organs, or the entire body. These may be the limb tourniquet syndrome (tourniquet shock), the limb or body part compression syndrome (prolonged compression syndrome, positional compression syndrome, immobilization syndrome, crush syndrome), the state following limb revascularization after thrombosis or embolism of a large artery, and the limb replantation syndrome.

The ischemic component also prevails in cold shock, hemorrhagic (hypovolemic) shock, postresuscitation shock, shock caused by a partial block of blood flow in the intestines (intestinal or mesenteric shock), the heart (cardiogenic shock), the brain (cerebrovascular shock), etc.

The course of hemodynamic disorders leading to generalized hypoxia during ischemic shock is closely similar to that during other types of shock and includes: (1) central hemodynamic disturbances (decreasing cardiac output or arterial pressure), (2) redistribution of blood flow between different organs (blood flow centralization), and (3) increasing peripheral resistance and impaired microcirculation. The severity of ischemic shock and mortality associated with it depend on the volume of the ischemic organ as well as on the duration and extent of ischemia [Kulagin B. R.; Sekamova S.M.; Shal'nev A.N., Levandovsky I.V. and Oksman T.M.; Levandovsky I.V. and Epishin Y.N.]. The severe progression of the disease and high mortality during reperfusion of ischemic limbs are also observed in clinical practice [Komarov B. D. and Shimanko I. I.; Nigulyanu V.I., Elsky V.N., Krivoruchko B.I. et al.].

The most common models for inducing ischemic shock under experimental conditions are those of tourniquet shock and crush syndrome, which involve ischemia of a large but not vitally important organ (limb) and, hence, allow researchers to analyze in more detail contributions to the pathogenesis of shock made by the ischemic component and especially by metabolites entering the body from the ischemic organ after its reperfusion (the toxic component). With the aid of these models, the toxic effect of disturbed metabolism products released into blood from the ischemic organ after its reperfusion and their contribution to mortality were clearly demonstrated in experiments with intact animals receiving injections of plasma, perfusate, or supernatant obtained after centrifuging the homogenate of ischemic muscles [Zaks I. O. and Gabrielyan N. I.; Kuzin M. I., Vornovitsky E. G., Zenkova N. A. et al.; Oksman T.H., Dalin M.V. and Kovanov V.V.]. However, the question as to which particular substances flushed out of the organ after prolonged ischemia produce the toxic effect still remains unclear. Among toxic products, there usually are low- and medium-molecular products of protein hydrolysis (peptides and polypeptides), including so-called ischemic toxin [Oksman T.H., Dalin M.V. and Kovanov V.V.], myocardiodepressor [Treshchinsky A.I., Levenets V.N. and Shlapak I.P.; Lefer A.M. (1987)], and renal, hepatic, small-intestinal, necrotic, and other depressors.

At the same time, some data suggest that toxic properties and the ability to induce, aggravate, or complicate the ischemic shock may also be exhibited by products of disturbed lipid metabolism, i.e., those resulting from phospholipid hydrolysis, cyclooxygenase and lipooxygenase pathways of arachidonic acid metabolism [Lefer A.M. (1985); Lefer A.M. (1988); Lefer A.M. and Lefer D.J. (1993)], and substances produced by endothelium and neutrophils, including PAF, TNF–α, IL-1, proteases, etc. In addition, an important role in the pathogenesis of etiologically different types of shock belongs to LPO products.

The initiation of LPO processes and accumulation of toxic LPO products may result from the following factors accompanying the ischemic shock: (1) the virtually obligatory stressor component, (2) generalized hypoxia resulting from central and peripheral hemodynamic disorders, and (3) the release of LPO products from the ischemic organ during its reperfusion. The possibility of LPO intensification in the presence of each factor was confirmed experimentally.

Intensification of LPO processes (increasing content of their products in blood, lipids of the heart and other organs, and the presence of pentane in the exhaled air) was observed on different stress models: cold stress [Kulikov V. Yu., Semenyuk A. V. and Kolesnikova L. I.], emotion stress [Prilipko L.L., Orlov O.N., Ivanova S.M. et al.], emotion–pain stress [Gulyaeva N. V., Dupin A. M., Levshina I. P. et al.; Meerson F.Z.], immobilization stress [Kovacheva-Ivanova S., Bakalova R., Kagan V. and Georgiev G.], traumatic shock [Elsky V.N., Manankov A.K., Mareyeva T.E. and Yakubenko Y.D.; Yakubenko E.D., Mareeva T.E. and Manankov A.K.; Novelli G.P., Martini G., Bordi G. et al.; Schlag G. and Redl H.], thermal burn shock [Aristarkhova S.A., Burlakova E.B. and Zaets T.L.; Badzhinyan S. A. and Aghajanov M. I.; Gerd O.T., Hatherill J.R. and Ward P.A.; Riabinin V.E. and Lifshits R.I.], and circulatory or hemorrhagic shock [Levin G. S. and Kamenetskaya Z. L.; Biasi F., Chiarpotto E., Lanfranco G. et al.; Gasparetto A., Corbucci G.G., Grimi G. et al.].

LPO products can flow out of ischemic organs, subsequently accumulating in other organs, and initiate LPO processes in organism as a whole (generalization of LPO processes), which is based on their better solubility in aqueous solutions, compared to that of their unoxidized precursors [Kagan V. E., Churakova T. D., Karagodin V. P. et al.], and is confirmed by the presence of LPO products in the blood flowing from partially ischemic organs or in the blood and perfusate after reperfusion of totally ischemic organs (see Chapter 5).

At the same time, the degree of LPO intensification during shock in which the ischemic and toxic components prevail over the stressor component has remained unknown for a long time. There were also no studies on individual roles of stress, hypoxia, and the release of LPO products as factors responsible for LPO intensification and on the contribution of LPO processes to the pathogenesis of ischemic shock and the resulting mortality.

These problems were studied in our laboratory on the model of ischemic shock induced by tourniquetting both hind limbs of Wistar rats for 6 h with their subsequent reperfusion [Bilenko M.V. (1986); Bilenko M., Bulgakov V., Komarov P. and Morgunov A.; Bilenko M., Morgunov A., Bulgakov V. and Komarov P.; Bilenko M.V., Morgunov A.A., Churakova T.D. et. al.; Bilenko M.V., Churakova T.D. and Arkhangelskaya S.L.; Bilenko M.V., Churakova T.D., Arkhangelskaya S.A. and Komarov P.G.; Zhdanov R. I. and Komarov P. G.].

We chose the tourniquet shock model because it is simple, reproducible, provides the possibility of inducing relatively severe shock after 6 h of ischemia (most rats died after reperfusion for 6 to 18 h, and all rats died after reperfusion for 24 h), and simulates the pathological state occurring fairly frequently. An additional advantage is that the relative role of pain as a stressor component in this model is less significant than in other types of shock, which makes it possible to perform experiments under mild anesthesia or even without it and to separately evaluate the contributions of the stressor and ischemic components to the pathological process.

7.1 Accumulation of Lipid Peroxidation Products in Blood and Different Organs upon Ischemic Shock

Ischemic shock is known to be accompanied by depression of the central nervous system, respiratory arrest [Myslovatyi B.S., Zairov D.K., Chebotarev A.A. et al.], and drastic changes in the hemodynamics, structure, and function of the liver [Sekamova S.M.; Takacs L. and Scolnik J.H.; Ward P.H., Maldonado M. and Vivaldi E.], kidneys [Istomin N. P., Mishnev O. D., Sergeeva N. A. et al.; Myslovatyi B.S., Zairov

D.K., Chebotarev A.A. et al.; Sekamova S.M.; Shleikin A.G., Shulutko B.I., Oksman T.M. and Orlov A.V.; Chekareva G.A., Mishnev O.D., Serdobintseva T.S. and Istomin N.P.], and myocardium [Dolgikh V. T.; Treshchinsky A.I., Levenets V.N. and Shlapak I.P.; Lefer A.M. (1987)]. On this basis, our experiments with the model of ischemic shock were aimed at analyzing the accumulation of primary (DC, HP) and secondary (MDA) LPO products in the blood, brain, liver, heart, and kidneys of animals (Fig. 62) [Bilenko M.V., Churakova T.D. and Arkhangelskaya S.L.].

After a fixed period of reperfusion of limbs totally ischemic for 6 h, rats were decapitated, and their blood was collected in heparinized tubes. Thereafter, rats were rapidly dissected in the cold to remove the brain (within the first 30 s), other organs, and limbs (within no more than 3 min after decapitation).

The data on accumulation of LPO products in lipids extracted from the homogenate of limbs after their total ischemia for 6 h and in the early postischemic period were described in Chapter 5 (see Figs. 42–44). The contents of LPO products in blood and examined organs increased as well, and two phases of this process were distinguished in all organs: concentrations of LPO products reached a peak by the end of the tourniquet period (the first, ischemic phase) and increased again after reperfusion of the limbs (the second, reperfusion phase).

The amount and spectrum of LPO products and the time of their accumulation differed from organ to organ. *In the blood*, the content of DC became 1.4 times higher by the end of the tourniquet (ischemic) period and increased again after reperfusion for 1 h, whereas HP content increased 1.5-fold only after 12 h of reperfusion. Of the organs examined (Fig. 58), the accumulation of LPO products was maximal in the brain and liver. *In the brain*, a tendency toward an increase in DC content (by a factor of 1.8) and a significant increase in HP content (by a factor of 2.5) were observed by the end of the ischemic period, which was followed by another significant increase after reperfusion for 1 and 6 h (by factors of 2.5 and 3 for DC and HP, respectively). A tendency toward a slight increase in MDA level appeared only after 1 h of reperfusion.

Fig. 62: Content of primary (HP, DC) and secondary (TBARS) LPO products in rat organs of rats after 6 h of complete ischemia and different time of reperfusion two hind limbs (% of the initial level). [Bilenko M.V., Churakova T.D., Arkhangelskaya S.A. and Komarov P.G.]

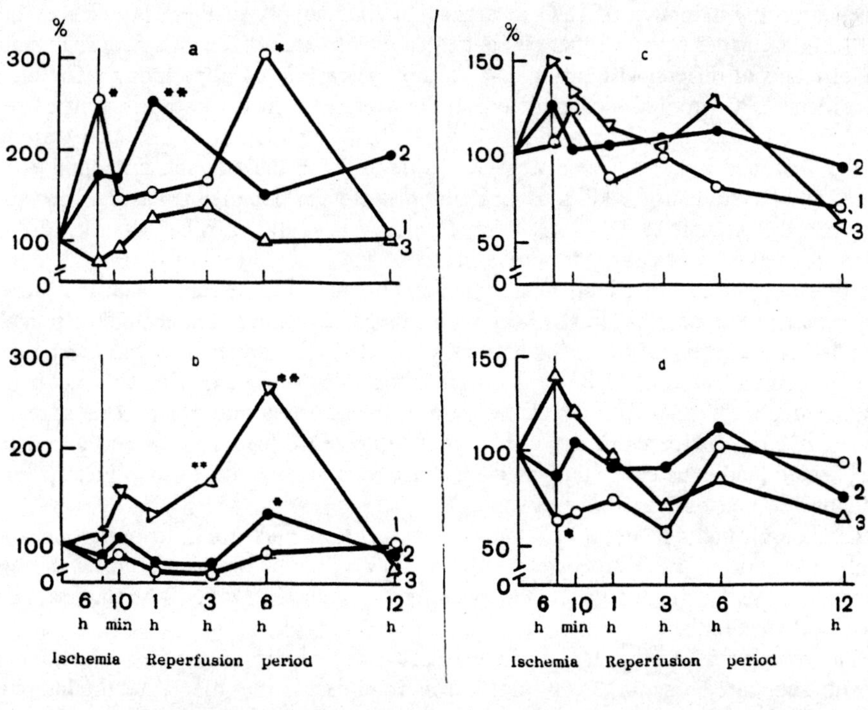

a - brain; b - liver; c - heart; d - kidneys; 1 - HP; 2 - DC; 3 – TBARS;*-p < 0.05; **-p < 0.01 versus initial level. The methods are the same as in Fig. 32.

In the liver, the contents of LPO products remained virtually unchanged when the tourniquet was applied, whereas after reperfusion of the limbs DC and MDA levels increased significantly (bly factors of 1.3 and 2.6 after 6 h).

In the heart, all the LPO products analyzed in experiments, especially MDA, demonstrated no more than a tendency to increase by the end of the ischemic period and in the reperfusion period (after 6 h); *in the kidneys,* this tendency was observed for MDA only during the ischemic period.

Thus, at *the first (ischemic) phase*, LPO processes in the tourniquetted hind limbs were markedly intensified (see Fig. 42) but the limbs retained only neuroreflectory connection with the rest of the body, and stress was the only factor possibly affecting the animal organism. Under these conditions, a moderate accumulation of primary LPO products was observed in the blood and brain, whereas other organs studied either displayed no more than a tendency to accumulate LPO products (heart, kidneys) or remained unchanged in this respect (liver).

At the *second (reperfusion) phase*, the recovery of blood flow in the tourniquetted limbs and their humoral connection with the body was accompanied both by further accumulation of LPO products in the ischemic muscles proper (see Fig. 43) and by repeated and significant increase in the contents of LPO products in the blood, brain, and liver.

Concerning changes in the contents of LPO products in organs during the reperfusion phase, the following observations deserve attention.

(1) After reperfusion of limbs for 1 h, DC content in blood plasma was twice that in blood cells (optical density 3.21 ± 0.57 and 1.5 ± 0.40 units/mg lipid, respectively; $P < 0.05$). This is evidence that LPO products accumulate in blood owing mainly to their release from ischemic limbs, rather than to activation of LPO processes in membranes of blood cells passing through the zone exposed to ischemia. Apparently, the release of LPO products into blood is the main factor responsible for the decrease of their level in ischemic muscles.

(2) The content of LPO products in the brain and liver reached the maximal level as early as during the first 1–6 h of reperfusion, i.e., in the period before the animals began to die of shock. This is consistent with the above evidence that the brain plays a key role in the pathogenesis of mortality resulting from shock and with the data on the severity of shock-dependent changes in the liver. The mere fact that the accumulation of LPO products in different organs during shock occurs at different rates suggests that this process depends on the induction of LPO in organs, and the supply of these products with blood or the induction of LPO occurring when the material is treated *in vitro* are not the only factors responsible for it.

(3) The amounts of different LPO products did not increase in parallel during ischemic shock. In the brain, only primary LPO products accumulated; in the liver, an increase in the content of secondary LPO products was more pronounced. The tendency toward the accumulation of secondary products was also revealed in the heart and kidneys. There were also differences in the amount, time, and even direction of changes in DC and HP contents (see Fig. 62). Similar results were obtained in cases of generalized hypoxia developing when the ACA is switched off [Verbolovich V.P., Teplova L.L., Petrenko E.P. et al.] and in experiments with total ischemia of organs (see Chapter 5).

(4) The dynamics of LPO products in all examined organs had a specific feature: by the 12th hour of reperfusion, their contents decreased to or even below the initial values. The factors responsible for such a decrease are as yet obscure, but two explanations seem plausible: either only the most resistant animals remained in the experiment after 12 h (by that time, about 40% of all experimental rats had survived), or organs became "depleted" of LPO products because of their release into blood. This agrees with the fact that HP content in blood increased by the 12th hour and may result from aggravation of membrane damage by that time, which facilitates either the passive extracellular release of LPO products from cells or their active elimination by phospholipases (see Chapter 4).

Thus, the experimental data described above show that ischemic shock (its tourniquet model) leads to moderate intensification of LPO processes in the blood and brain of animals under the effect of stress factors; however, this intensification becomes much more pronounced (at least in the brain and liver) after reperfusion of ischemic limbs.

The enhancement of LPO in the organism during the reperfusion phase of shock is apparently associated with both the release of LPO products from limb tissues into blood and the induction of LPO in other, nonischemic but vitally important organs by LPO products, which are strong inducers (see Chapter 10), and under the effect of developing generalized hypoxia. There are other data on generalization of LPO processes in the organism after reperfusion of limbs exposed to long-term ischemia [Aoki Y., Nata M., Odaira T. and Sagisaka K.; Sacks T., Moldow C.F., Craddock P.R. et al.] and of ischemic intestines

[Shneivas V.B., Amilov K.S. and Levin G.S.; Zeeb I., Pfenninger E. and Grunert A.]; for reviews, see [Crinnion J.N., Homer-Vanniasinkam S. and Gough M.J.; Redl H., Gasser H., Schlag G. and Marzi I.]. This phenomenon was also revealed in our observations [Churakova T.D., Bilenko M.V., Pokazeeva Z.T. et al.] and studies of many other specialists on patients with acute myocardial infarction (see Chapter 5).

7.2 Role of Lipid Peroxidation in Animal Mortality upon Ischemic Shock

As noted above, the period of accumulation of LPO products in blood and vitally important organs preceded the period of mass animal mortality, which occurred between the 6th and 18th hours of limb reperfusion. To reveal possible relationships between the intensification of LPO processes upon ischemic shock and animal mortality, variations in survival rate were monitored against the background of treatment with the following preparations: (1) prooxidants, (2) antioxidants, and (3) liposomes prepared from exogenous PL.

Prooxidants used in experiments included dimedrol (30 mg/kg), papaverine (50 mg/kg), and ephedrine (50 mg/kg) in the hydrochloride form. These drugs can catalyze LP degradation *in vitro* with a high yield of free radicals and accelerated lipid self-oxidation [Goldenberg V.I., Shmulovich V.G., Yurchenko N.I. et al.]. They were administered to rats 30 min before removing tourniquets to exert their effects on the reperfusion phase of the process. The results showed that all these drugs accelerated death of rats whose hind legs were reperfused after 6 h of ischemia, compared to the data on control rats [Bilenko M.V., Churakova T.D., Arkhangelskaya S.A. and Komarov P.G.; Bilenko M.V., Churakova T.D. and Arkhangelskaya S.L.]. The strongest prooxidant effect and the earliest death of experimental animals were observed after treatment with ephedrine.

Experiments with antioxidants [Bilenko M.V., Komarov P.G., Morgunov A.A. and Zhdanov R.I.; Bilenko M.V., Churakova T.D., Arkhangelskaya S.A. and Komarov P.G.; Bilenko M.V., Churakova T.D. and Arkhangelskaya S.L.] were performed using the synthetic compound BHT (ionol) and a sulfur-containing derivative of 2,2,6,6-tetramethyl piperidine (1-hydroxy-TMP), whose antioxidant effect *in vitro* and *in vivo* was previously confirmed in our laboratory [Komarov P. G., Taskaeva O. N. and Zhdanov R. I.].

Figure 63 shows that preventive BHT administration significantly increased the rate of animal survival during ischemic shock. The strongest effect was observed after administering the drug (at all doses used) 4 h before applying the tourniquet, whereas its administration 24 h or 1 h before tourniquetting (at doses of 30 and 120 mg/kg, respectively) and 2 h after this procedure was less effective. BHT administered 5 h after tourniquetting was ineffective (data not shown).

BHT administered at approximately the same doses and times also had a protective effect during local ischemia of the kidneys (see Chapter 10).

As preventive administration of BHT inhibits the accumulation of LPO products in membrane structures of ischemic limb muscles proper (see Fig. 45), its more pronounced protective effect during ischemic shock, when the drug is administered before tourniquetting, suggests that LPO products released into blood after reperfusion of limbs play the most important role in the generalization of LPO processes and the mortality caused by ischemic shock.

When 1-hydroxy-TMP was administered 1 h before the removal of tourniquets, it also increased the survival rate, and its effect was similar to that of a high dose (240 mg/kg) of BHT administered 4 h before removal of the tourniquets (see Fig. 63). This antioxidant was also very effective in preventing local ischemic injuries in the liver and heart [Komarov P. G., Morgunov A. A., Zhdanov R. I. and Bilenko M. V.; Komarov P., Morgunov A., Shdanov R. and Bilenko M.; Morgunov A.A., Komarov P.G., Bulgakov V.G. and Bilenko M.V.] (see Chapter 11). Note that the protective effect of antioxidants and scavengers of free radicals during the ischemic and other types of shock was confirmed in numerous studies [Redl H., Gasser H., Schlag G. and Marzi I.; Crinnion J.N., Homer-Vanniasinkam S. and Gough M.J.; Zeeb I., Pfenninger E. and Grunert A.; Saez J.C., Cifuentes F., Ward P.H., et al.; Aoki Y., Nata M., Odaira T. and Sagisaka K.; Mitrev Z., Beyersdorf F., Hallmann R. et al.].

Fig. 63: Rat survival rate after 6 h of ischemia and 1-7 days reperfusion of two hind limbs in experiments with BHT (ionol) and 1-hydroxy-TMP administration(%)

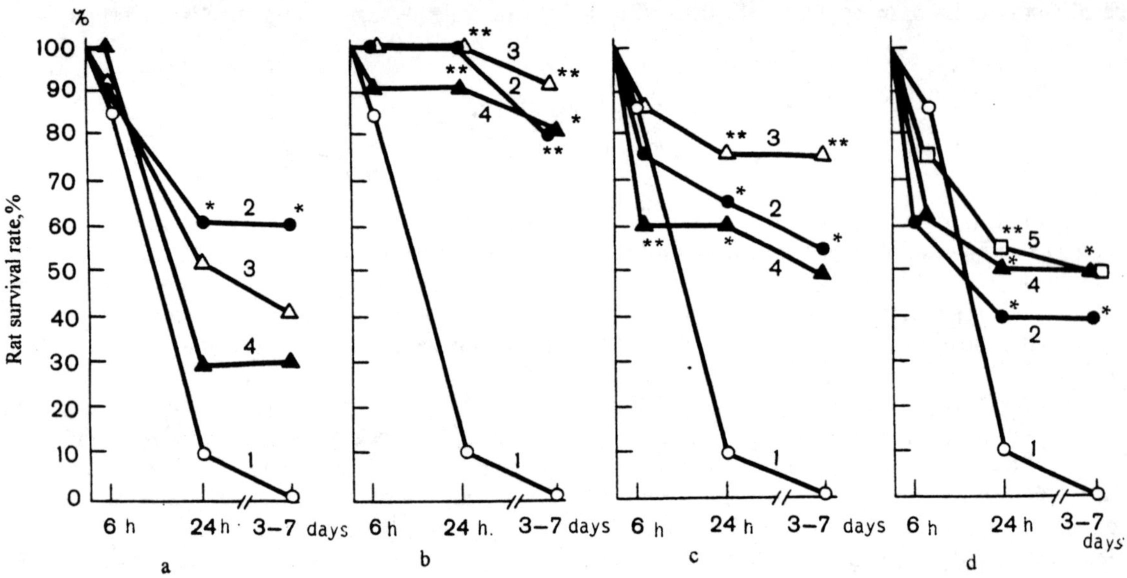

BHT administration: (a) 24 h, (b) 4 h (c) 1 h before tourniqueting the limbs; and d - 2 h after tourniqueting the limbs. 1 - without drug (control); 2 - BHT 30 mg/kg; 3 - BHT 120 mg/kg; 4 – BHT 240 mg/kg; 5 - 1-hydroxy-TMP 36 mg/kg 1 h before removing the tourniquets. *-$p < 0.05$; **-$p < 0.01$ versus control. The drugs were administered i. p.; BHT was administered in the form of an aqueous suspension in Twin 80.

In experiments with administration of liposomes prepared from exogenous PL, unoxidized PC from egg yolk was used along with a PC + PEA mixture (5 : 1 by weight) and oxidized PC [Bulgakov V.G.].

Lipids were injected to rats intraperitoneally, in the form of a suspension of single- or multilayer liposomes in 3–5 ml of isotonic saline, 30 min before the removal of tourniquets. As shown in Fig. 64, single-layer liposomes, irrespective of their composition and doses of constituent lipids, had no effect on the severity of ischemic shock, whereas multiple-layer liposomes (PC, PC + PEA, oxidized PC), being ineffective at dose of 50 mg/kg, significantly increased the rat survival rate when administered at a dose of 100 mg/kg (rats were monitored for 3–5 days).

The ability of liposomes to relieve ischemic shock agrees with other data [Stefanov A.V., Kreines V.M., Dmitrieva A.V. et al.] showing that liposomes from yolk PC are capable of improving the cardiovascular parameters in dogs with the crush syndrome. The authors of this work suggested that liposomes produce the protective effect by adsorbing toxic protein metabolites.

It is even more likely that liposomes absorb substances of lipid nature, i.e., LPO products [Trinus F.P., Pisarev A.A., Chubenko A.V. et al.], with their subsequent transport to the cytochrome P-450-dependent system of monooxygenases in the liver, in which these products are metabolized along the nonradical peroxidase pathway (see below and in Chapter 10). This is consistent with our discovery of the protective effect resulting from intraperitoneal administration of multilayer liposomes, which are more rapidly displaced from blood plasma and accumulate in the liver [Scherphof G., Roerdink F., Hoekstra D. et al.].

Fig. 64: Rat survival rate after 6 h of ischemia and 3-5 days of reperfusion of two hind limbs after liposomes from yolk PL (PC, PCox, PEA) (%) administration.

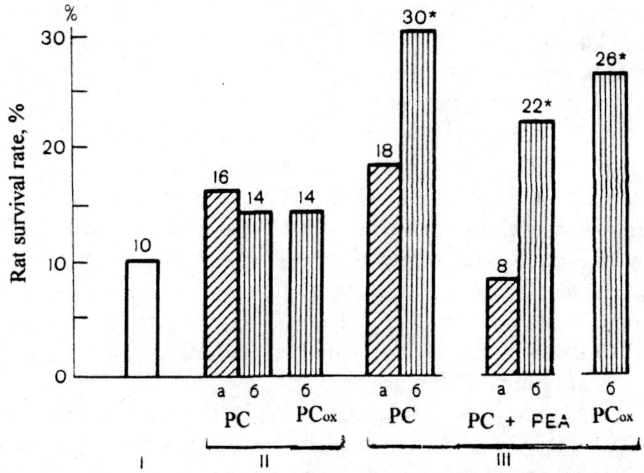

I - without drugs (control); II - administration of single-layer liposomes; III - administration of multiple-layer liposomes; a - 50 mg/kg; b - 100 mg/kg. The PC + PEA mixtures were taken in a ratio of 1:5 parts by weight. *-p < 0.05 versus control. Single-layer liposomes (40-50 nm in diameter) were obtained by ultrasonic treatment of lipids on a UZDN-1 disintegrator at 4°C. Multiple-layer liposomes (0.1-20 μ in diameter) were obtained with the ultrasonic treatment time being limited to 15-20 s. Liposoms were administered i. p.

The protective (rather than damaging) effect of exogenous oxidized PL at doses used in these experiments does not disprove the role of the toxic action of endogenous LPO products for two reasons: (1) they exist in the circulatory system in different forms (the former as liposomes, whereas the latter primarily as high-density lipoproteins), and (2) in addition to peroxides, membrane lipid peroxidation yields other metabolites with even more pronounced membrane-damaging effect (see Chapter 5).

All these data suggest that ischemic shock is accompanied by generalization of LPO processes in some vitally important organs, and, moreover, the accumulating LPO products exert a definite, rather pronounced effect on its severity. Today, it is accepted that LPO inhibitors should be used in programs for treating different forms of shock [Dolinin V. A. and Vinogradov V. M.].

7.3 Role of Lipid Peroxidation-Induced Disturbances of the Hepatic Cytochrome P-450-dependent Monooxygenase System in Animal Mortality upon Ischemic Shock

It is known that endogenous hydrophobic substrates of lipid nature are oxidized in the microsomal cytochrome P-450-dependent monooxygenase system in the liver [Archakov A. I. (1975); Lyakhovich V.V. and Tsyrlov I.B.]. This system also performs the peroxidase function [Hrycay E.G. and O'Brien P.J. (1971); Islam S.J.] and is capable of reducing lipid peroxides via the nonradical pathway, being involved in the interruption of reaction chains and inhibition of peroxidation processes (see Chapter 10).

It has also been established that the induction of cytochrome P-450 by phenobarbital strongly enhances the peroxidase activity of cytochrome P-450 [Kulikov V. Yu., Semenyuk A. V. and Kolesnikova L. I.; Torres M., Jarvisalo J., Hakim J.], whereas the inhibition of the cytochrome P-450-dependent microsomal system suppresses its peroxidase function [Hrycay E.G. and O'Brien P.J. (1972)].

Cytochrome P-450 is highly sensitive to ischemia (see Chapter 2), and ischemic or any other shock accompanied by centralization of the circulation leads to serious hemodynamic disorders in the liver [Kulagin B. R.; Myslovatyi B.S., Zairov D.K., Chebotarev A.A. et al.; Mitrev Z., Beyersdorf F., Hallmann

R. et al.; Shneivas V.B., Amilov K.S. and Levin G.S.]. This may give rise to ischemic injuries in the latter, including damage to the cytochrome P-450-dependent monooxygenase system and the impairment of its capacity for lipid oxidation and peroxidase reactions. Cytochrome P-450 is lipid-dependent, and, hence, the accumulation of LPO products in ER of liver cells, observed upon tourniquet shock (see section 7.1) and other types of shock and stress [Paramonova G.I., Gubsky Yu.I., Goryushko A.G. et al.; Shneivas V.B., Amilov K.S. and Levin G.S.; Mitrev Z., Beyersdorf F., Hallmann R. et al.; Novelli G.P., Martini G., Bordi G. et al.], should lead to a decrease in its activity.

Indeed, it was shown that thermal burn shock [Ciaccio E.I. and Fruncillo R.J.] traumatic shock [Rauckman E.J., Rosen G.M., Post S.E. et al.], emotion–pain stress [Paramonova G.I., Gubsky Yu.I., Goryushko A.G. et al.], and ligation of the aorta above its bifurcation in rats [Griffeth L.K., Rosen G.M. and Rauckman E.J.] result in a decrease in the content of cytochrome P-450 and its capacity for oxidizing xenobiotics both *in vivo* and *in vitro*. Experiments revealed that the contents of different cytochrome P-450 subpopulations (isozymes) selectively decrease during shock, with a simultaneous increase in the content and activity of cytochrome P-448 [Griffeth L.K., Rosen G.M. and Rauckman E.J.]. If such a decrease does occur, the functional disorders of the liver should play a certain role in the peroxidation mechanism responsible for animal mortality upon ischemic shock.

At the same time, some data show that hypoxia can induce cytochrome P-450 rather than inhibit it [Longmuir I.S. and Pashko L.]. The exact degree of ischemic damage to the liver and the prevailing mechanism (disturbing cytochrome P-450 activity or inducing it) are as yet unknown.

Experiments with tourniquet shock in rats, aimed at elucidating the role of the liver in detoxifying endogenous LP metabolites, were performed in our laboratory [Bilenko M.V., Churakova T.D., Arkhangelskaya S.A. and Komarov P.G.; Komarov P., Morgunov A., Shdanov R. and Bilenko M.; Komarov P. G., Morgunov A. A., Zhdanov R. I. and Bilenko M. V.]. They showed that the level of microsomal cytochrome P-450 in the liver increased as soon as by the end of the 6th hour of hind limb ischemia and remained at a significantly higher level throughout the 12-h period of limb reperfusion. The increase in the content of cytochrome P-450 and, possibly, in its peroxidase activity owing to the accumulation in blood of lipid or LP substances (oxidation substrates of the cytochrome P-450-dependent monooxygenase system) appears to be responsible for a slight increase in the contents of primary LPO products (HP, DC) in the liver, because during more severe (total) ischemia of the liver, when the cytochrome P-450 level dropped, the accumulation of both primary and end products of LPO was observed (see Chapter 5, Fig. 35).

The effect of the cytochrome P-450-dependent monooxygenase system on the severity of ischemic shock was also studied in experiments with the use of both inducers (phenobarbital and zixorin) and inhibitors of cytochrome P-450, such as imidazole and $Cu(Lys)_2$. The results showed that both inducers significantly increase animal survival rate, whereas both inhibitors significantly reduce it (Fig. 65).

The data on the protective effect of cytochrome P-450 inducers during ischemic shock correlate with the results obtained by other researchers [Kovalev I.E., Piruzian L. A., Shaternikov V. A. et al.], according to which phenobarbital prevented a decrease in the content and activity of cytochrome P-450 during anaphylactic shock and rendered the animals more shock-tolerant.

Direct evidence for the increase of cytochrome P-450 content in hepatic microsomes, indirect evidence for possible enhancement of its peroxidase function, and the polar effects of cytochrome P-450 inducers and inhibitors on the survival rate of animals suffering from ischemic shock suggest that the cytochrome P-450-dependent monooxygenase system in the liver is involved in the detoxification of lipid metabolites released from ischemic limbs and, consequently, has effect on the severity of ischemic shock. This provides the possibility of a radically new approach to the prevention and treatment of ischemic shock by using pharmacological inducers of the cytochrome P-450- dependent monooxygenase system in the liver.

Note that the protective effect of antioxidants at high doses may also be mediated by this system to some extent because they also cause cytochrome P-450 induction, and, conversely, the protective effect of substances activating the cytochrome P-450-dependent monooxygenase system may produce an increase in the activity of the antioxidative protection system, particularly the accumulation of ascorbic acid [Hahn H.R., Barak A.J., Tuma D.J. et al.] and activation of GSH peroxidase, GSH transferase, and GSH reductase [Kulikov V. Yu., Semenyuk A. V. and Kolesnikova L. I.; Torres M., Jarvisalo J., Hakim J.].

Fig. 65: Rat survival rate after 6 h of ischemia and 1-7 days of reperfusion of two limbs in experiment with administration of inducers and inhibitors of cyt. P-450-dependent monoxygenase system of the liver(%).

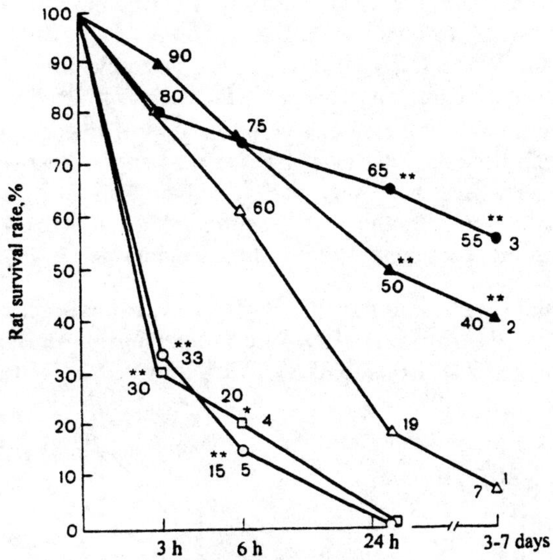

1 - without drugs (control); 2 - administration of PB; 3 - administration of zixorin; 4 - administration of imidazole; 5 - administration of $Cu(Lys)_2$; *-p<0.05; **-p<0.01 versus control. PB was administered once, 70 mg/kg, 40 h before tourniqueting; zixorin in the form of an aqueous suspension in Twin 80, 50 mg/kg, for 3 days before tourniqueting; imidazole 100 mg/kg 30 min before removing the tourniquets; $Cu(Lys)_2$, 30 mg/kg 3 h or 30 min before removing the tourniquets. All drugs were administered i. p.

7.4 Role of Lipid Peroxidation-Induced Disturbances of Coronary Circulation and Myocardial Contractile Function in Animal Mortality upon Ischemic Shock

As shown above, central hemodynamic disorders are necessary, if not principal, signs of all types of shock. Their manifestations include the reduction of minute circulatory volume and cardiac output, accelerated heart rate, and the drop of arterial pressure notwithstanding an increase in overall peripheral resistance, i.e., the signs of cardiac insufficiency [Vornovitsky E.G., Len'kova N.A. and Vasilets L.A.; Zairov D. K., Bukov V. A., Sorokina M. I. et al.; Nazarova T.A. and Zhyetnisbaev B.A.; Nigulyanu V.I., Elsky V.N., Krivoruchko B.I. et al.; Sagach V.F. and Dmitrieva A.V.]. Disturbances of cardiac contraction were observed upon thermal burn shock [Vasilets L.A., Vornovitsky E.G. and Chodorov B.I.;. Vornovitsky E.G., Len'kova N.A. and Vasilets L.A.], hemorrhagic shock [Levin G. S. and Kamenetskaya Z. L.], and resuscitation after clinical death in rats [Dolgikh V. T.].

According to experimental data, plasma or serum of animals suffering from shock produce a toxic effect on the papillary muscle isolated from the rabbit heart [Vornovitsky E.G., Len'kova N.A. and Vasilets L.A.; Kuzin M. I., Vornovitsky E. G., Zenkova N. A. et al.], the "ischemic toxin" alters the amplitude of cardiac contractions in frogs, rats, or dogs [Oksman T.H., Dalin M.V. and Kovanov V.V.], and the "myocardial depression factor" (MDF) exerts a negative inotropic effect during perfusion of an isolated cat heart and papillary muscle [see Lefer A.M. (1987)].

At the same time, the question whether or not the disturbed coronary circulation and impaired cardiac function are in any way related to intensification of LPO processes during ischemic shock remains to be answered.

In our laboratory, a series of experiments was performed to analyze the coronary perfusion rate (CPR) and cardiac function (P_{max}) of the heart taken from rats exposed to 6 h of limb ischemia and different periods of reperfusion [Bilenko M.V. (1992b); Bilenko M.V., Morgunov A.A., Churakova T.D. et. al.; Bilenko M., Morgunov A., Bulgakov V. and Komarov P.; Bilenko M.V. (1992a); Morgunov A.A., Bulgakov V.G. and Bilenko M.V.; Morgunov A.A.; Morgunov A.A. and Bulgakov V.G.]. As follows from Fig. 66, CPR and P_{max} in the hearts removed after 1 h of ischemic limb reperfusion demonstrated no significant differences from corresponding parameters in the hearts of intact animals. In the hearts removed after 6 and 12 h of ischemic limb reperfusion and perfused *in vitro* (the Langendorf–Fallen model), CPR values within the first 15 min became significantly lower than in the hearts of intact rats (62 and 73%, respectively; $P < 0.01$). These values were only slightly higher than CPR revealed in animals suffering from ischemic shock with the aid of radioactive labeling (58%) [Takacs L. and Scolnik J.H.], which indicates that the results obtained *in vitro* closely reflect the situation *in vivo*.

Fig. 66: Coronary perfusion rate (CPR)(I) and cardiac function (P_{max})(II) in the hearts excised from rats after different periods of 6h ischemic limbs reperfusion (mean ± SEM) [Bilenko M.V., Morgunov A.A., Churakova T.D. et. al.].

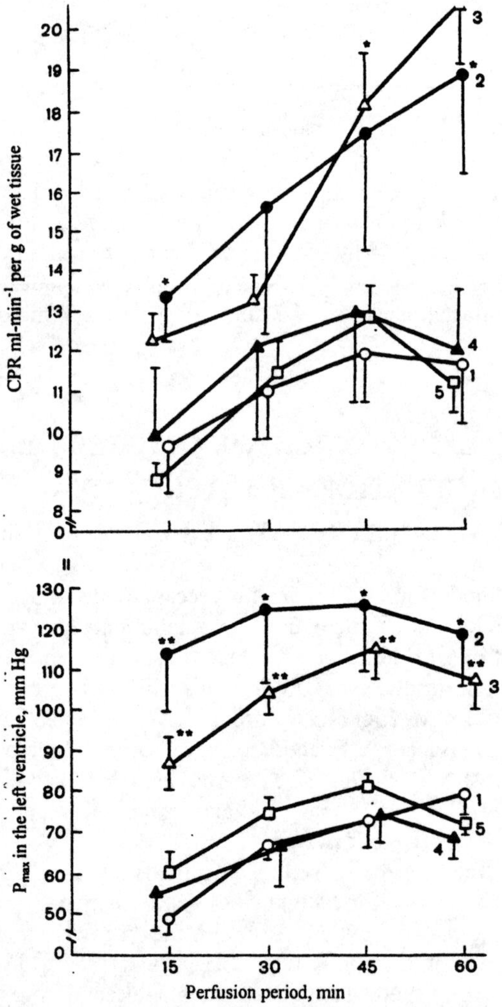

1 - hearts of intact rats; 2 - hearts from rats after 1 h of ischemic limbs reperfusion; 3 - the same after 6 h of reperfusion; 4 – the same after 12 h of reperfusion. *-p < 0.05; **-p < 0.01 versus intact rats. Methods of isolated heart perfusion and determining the CPR and P_{max} are the same as in Fig. 8.

In the hearts removed after 6 and 12 h of ischemic limb reperfusion and perfused *in vitro,* P_{max} was sharply reduced during the first 15 min (39 and 48%, respectively, of that in the heart of intact rats) and demonstrated a weaker tendency toward normalization than CPR (68–63% of the norm after 60 min of reperfusion), notwithstanding the good recovery of the coronary blood flow by that time.

The above data show that ischemic shock developing after 6 and 12 h of reperfusion of limbs exposed to 6-h ischemia is accompanied by a reversible increase in the tone of coronary vessels and a reversible decrease in the myocardial contractile function, changes in the latter being more pronounced and less reversible than those in coronary perfusion. They seem to result not only from disturbances of peripheral hemodynamic or coronary perfusion but also from the damaging effect of circulating toxic metabolites on myocardiocytes proper. As disturbances of both coronary circulation and cardiac function manifested themselves especially strongly after 6 h of limb reperfusion (see Fig. 66), and a tendency toward repeated accumulation of MDA in the myocardium was also observed by that time (see Fig. 62), this particular reperfusion period, critical for the heart, was chosen to study the protective effect of antioxidants and liposomes on this organ.

As follows from Fig. 67, BHT administration at the optimal dose (120 mg/kg 4 h before tourniquetting) and 1-hydroxy-TMP administration 1 h before tourniquet removal resulted in the significant increase of CPR and P_{max} in the hearts taken from rats after 6 h of limb reperfusion. As shown above (see Fig. 63), these drugs significantly increased the survival rate as well.

Phospholipids (unoxydized and oxidized PC) administered to rats intraperitoneally in the form of multilayer liposomes at a dose of 100 mg/kg did not change CPR and P_{max} in the heart removed 6 h after ischemic limb reperfusion, in spite of the simultaneous slight increase in the survival rate (to 30 and 26%, see Fig. 64). This discrepancy is conditioned either by the abovementioned ability of multilayer liposomes to accumulate predominantly in the liver [Scherphof G., Roerdink F., Hoekstra D. et al.] or by the deleterious effect of exogenous PL on the heart, which was observed upon adding PL to the cardiac perfusion medium *in vitro* (see Chapter 8). Another possible explanation is that the relieving effect of liposomes prepared from normal and oxidized PC on the course of ischemic shock was insufficiently strong (by the integral survival rate criterion) and, hence, was not revealed on the *in vitro* cardiac perfusion model used in our experiments.

Comparing the animal survival rate within the first 3 days after reperfusion to the cardiac function (P_{max}) determined after 15 and 45 min of the heart perfusion *in vitro,* a strong positive correlation between these two parameters was revealed ($r_1 = 0.65$, $r_2 = 0.68$; $P < 0.05$), providing evidence that circulatory disorders and disturbed cardiac function make a significant contribution to animal mortality upon ischemic shock.

The improvement of coronary circulation and cardiac function in animals that have received antioxidants also shows that LPO processes are involved in the pathogenesis of cardiac insufficiency during ischemic shock.

Thus, ischemic shock is accompanied by the increased tone of coronary vessels, reduced coronary perfusion rate, and marked deterioration of myocardial function. All these parameters are at the lowest level during the "critical" period of shock, i.e., after 6 h of ischemic limb reperfusion. Treatment with antioxidants improves blood flow and parameters of cardiac function parameters, whereas PL in liposomal form produce no effect. A strong correlation proved to exist between the rate of animal survival *in vivo* and parameters of cardiac function determined *in vitro.*

Fig. 67: Coronary perfusion rate (CPR) and cardiac function (P_{max}) in the hearts excised from rats after 6 h two hind limbs and 6h ischemia reperfusion in experiments with administration of antioxidants and liposomes from yolk PL (mean ± SEM). [Bilenko M.V., Morgunov A.A., Churakova T.D. et. al.].

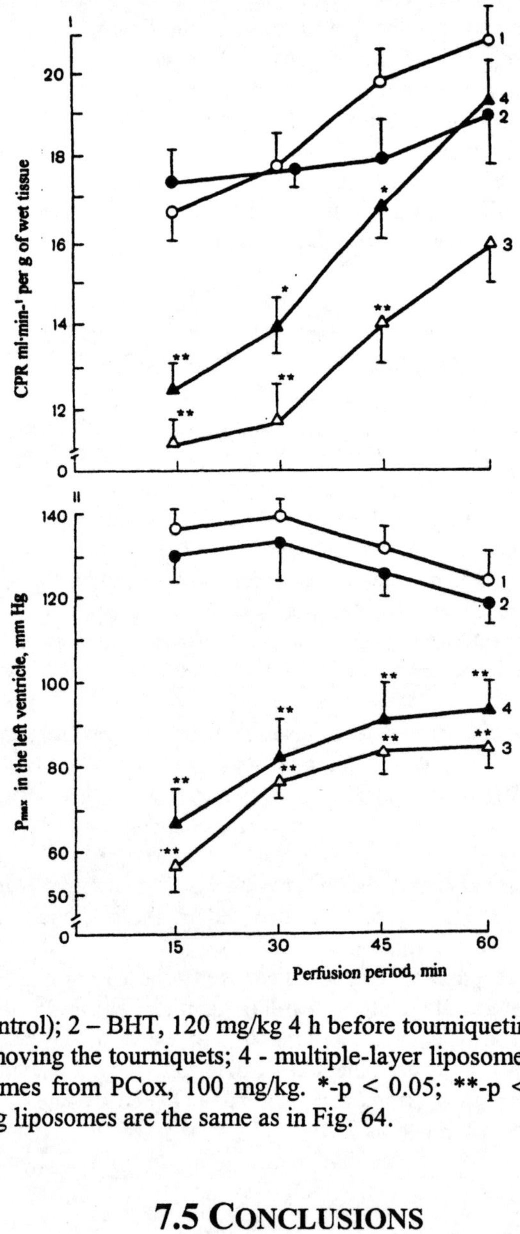

1 - without drugs (control); 2 – BHT, 120 mg/kg 4 h before tourniqueting; 3 - 1-hydroxy-TMP, 36 mg/kg 1 h before removing the tourniquets; 4 - multiple-layer liposomes from PC, 100 mg/kg; 5 - multiple-layer liposomes from PCox, 100 mg/kg. *-p < 0.05; **-p < 0.01 versus control. The methods for obtaining liposomes are the same as in Fig. 64.

7.5 Conclusions

Intensification of LPO processes in organs during ischemia, their repeated activation in the early reperfusion period (see Chapters 5 and 6), and the ability of lipid oxidation products to leave membranes, resulting from their more pronounced hydrophilic properties compared to normal lipids, provide evidence that LPO processes may be involved in the pathogenesis of generalized hemodynamic disorders during ischemic shock.

The results of studies on the role of LPO processes in the pathogenesis of ischemic (tourniquet) shock in rats, presented in this chapter, can be summarized as follows.

(1) Tourniquetting and reperfusion of rat hind limbs are accompanied by the accumulation of LPO products in the blood, brain, and liver, and by a tendency toward their increase in the heart and kidneys; in other words, generalization of LPO processes occurs in the organism. Moreover, at the second

(reperfusion) phase of shock, activation of LPO is more pronounced that at the first (ischemic) phase, i.e., the release of LPO products from ischemic organs plays a more important role than the neuroreflectory response of the organism to stress and pain.

(2) The accumulation of LPO products in blood and vitally important organs of rats after reperfusion of their limbs exposed to long-term ischemia occurs before the period when most animal die. Rat mortality increases upon treatment with prooxidants and decreases under the effect of antioxidants and liposomes prepared from yolk PL. This is evidence that LPO processes have a significant effect on the severity of ischemic shock and contribute to animal mortality upon shock.

(3) The relief of ischemic shock and the increase of survival rate upon treatment with inducers of the liver cytochrome P-450-dependent monooxygenase system and the opposite effect of its inhibitors suggest that this system is involved in the detoxification of lipid metabolites released from ischemic limbs. This apparently depends on the peroxidase function of this system, which may also be responsible for the protective effect of multilayer (but not single-layer) liposomes prepared from yolk PL because such liposomes (together with lipid metabolites probably adsorbed on them) are more readily transferred from blood to the liver [Scherphof G., Roerdink F., Hoekstra D. et al.]. The protective effect of liposomes during ischemic shock appears to be stronger upon simultaneous induction of the cytochrome P-450-dependent monooxygenase system. The latter, however, needs experimental verification.

(4) An increase in the coronary perfusion rate and the improvement of cardiac function under the effect of antioxidants provide additional evidence for the involvement of LPO processes in the pathogenesis of cardiac insufficiency during ischemic shock, and a strong positive correlation between the cardiac function and animal survival rate during shock suggest that cardiac insufficiency has a role in the pathogenesis of shock-dependent animal mortality.

(5) The toxic effect of LPO products released into blood after reperfusion of limbs exposed to long-term ischemia (6 h) may be associated with both induction of peroxidation reactions in nonischemic organs by these products (for the role of primary LPO products in chain branching, see Chapter 9) and their direct damaging action on membrane structures. Possible mechanisms of such damage are discussed in Chapter 5, and the results of studies on the damaging effect of circulating LPO products on blood vessels and the heart are discussed in Chapter 8.

It should be noted that many other factors are involved in both pathogenesis of ischemic shock and mechanisms of adaptation to this state, such as the synthesis of heat shock proteins, which are actively discussed in modern scientific literature. However, these data are beyond the scope of this book.

Chapter 8

VASOACTIVE AND CARDIOTROPIC EFFECTS OF CIRCULATING UNOXIDIZED AND OXIDIZED FREE FATTY ACIDS AND PHOSPHOLIPIDS AND THEIR ROLE IN THE PATHOGENESIS OF LOCAL AND SYSTEMIC CIRCULATORY DISORDERS DURING ACUTE ISCHEMIA AND AFTER REPERFUSION OF ORGANS

Reperfusion of an organ after prolonged ischemia is accompanied by local circulatory disorders at the organic, regional, or microcirculatory level (i.e. by the no-reflow phenomenon, see Chapter 1), while such reperfusion of a large organ entails the development of systemic circulatory disorders in the body, including deterioration of the central hemodynamics, an increase of peripheral resistance, and lower blood flow through nonischemic organs (see Chapter 7). The decrease of blood flow resulting from reduced collateral circulation (onset of its relative insufficiency) occurs also during partial ischemia of organs (see Chapter 1).

As was shown in the preceding chapters, intensification of LPO processes takes place during ischemia and after reperfusion in the ischemic organ itself as well (see Chapter 5), and after reperfusion; an increase of LPO products is also observed in the blood and other nonischemic organs; in other words, LPO processes in the body become generalized (see Chapter 7), their inhibition by an HGM or AO improves the hemodynamics of the organs, as well as the contractile function of the heart (see Chapters 6, 7 and 11). In this connection, it is of particular interest to estimate the possible contribution of newly formed and circulating LPO products to the pathogenesis of local and systemic hemodynamic disorders during ischemia and in the early reperfusion period.

The circulating products of upset lipid metabolism include, apart from undetectable or so far unknown metabolites, unoxidized and oxidized free fatty acids (FFA), phospholipids (PL), secondary and end products of PL and FFA oxidation (see Chapter 5).

8.1 VASOACTIVE EFFECT OF CIRCULATING UNOXIDIZED AND OXIDIZED FREE FATTY ACIDS

In recent years, many workers have established that FFA produce a vasoactive effect whose exact nature depends on the degree of the FFA oxidation and saturation and the type of the target vessel. It has been shown that *unoxidized* FFA, both saturated and monoenoic, do not affect vessels in any perceptible way *in vitro* [Asano M. and Hidaka H.], while polyenoic acids (18:2, 18:3, 20:4, 22:3) exert a strong vasoconstrictive effect on an isolated rabbit aorta, strips of a dog basilar artery [Asano M. and Hidaka H.; Koide T., Noda Y., Nata S. et al.], vessels of an isolated lung in rabbits, cats, dogs, and other animals [Gurther Q.H., Knoblauch A., Smith P.L. et al.], mesenteric vessels in pigs [Houvenaghel A., Schrauwen E. and Wechsung L.], and cerebral vessels in mice [Rosenblum W.I. (1982)]. The vasoconstrictive effect of AA is more pronounced, than that of other acids [Asano M. and Hidaka H.; Koide T., Noda Y., Nata S. et al.; Rosenblum W.I. (1982)]. AA also twofold enhanced contraction and intracellular Ca^{2+} transients in rat ventricular myocytes [Damron D.S., Summers B.A.].

Coronary arteries do not always respond to FFA in the same way as other vessels. As has been found, some FFA acted as vasodilators on coronary vessels during perfusion of isolated rat and guinea pig hearts, as these acids differ in carbon chain length and degree of saturation [Mentz P., Opitz H. and Forster W.], while AA either produced a vasodilating effect on the same model, or caused temporary constriction (biphase reaction) of coronary vessels, in high doses even acting as a vasoconstrictor [Belo S.E. and Talesnik J.]. Arachidonic acid caused also constriction of an isolated coronary artery in cats [Trachte G.J., Lefer A.M., Aharony D. and Smith J.B.]. Biphase action of FFA was observed in cerebral [Rosenblum W.I. (1982)] and mesenteric [Houvenaghel A., Schrauwen E. and Wechsung L.] vessels as well.

The vasoactive effect of FFA *in vitro* is not always equivalent to that *in vivo*, when it often takes a vasodilating rather than vasoconstrictive course, especially when the FFA (including 20:4) is introduced with the vascular tone already high [Hyman A.L., Spannhake E.W. and Kadowitz P.J.; Salvati P., Whittle B.J.R.]. Intravenous administration of AA to dogs, cats, and rats *in vivo* brought about constriction of visceral vessels (pulmonary, mesenteric, etc.) but exerted a hypotensive effect on systemic arterial pressure [Chapnick B.M., Feigen L.P., Hyman A.L. et al.; Hyman A.L., Spannhake E.W. and Kadowitz P.J.; Rose I.C., Johnson M., Ramwell P.W. et al.; Tokumura A., Kume T., Fukusawa K. et al.] - that is, gave rise to a situation close to that arising in the body during ischemic shock (see Chapter 7).

In contrast to unoxidized FFA, all oxidized ones, with few exceptions, act only as vasoconstrictors, this action being more pronounced than in the case of a corresponding unoxidized acid.

The vasoconstrictive effect of AA HP on a rabbit aorta strip was 5 times stronger than that produced by unoxidized AA [Asano M. and Hidaka H.], and 100 times stronger than on isolated coronary arteries in cats [Trachte G.J., Lefer A.M., Aharony D. and Smith J.B.] and increased further with higher AA doses [Asano M., Matsuda T., Suzuki Y. et al.]. Oxidized FFA exerted a vasoconstrictive effect not only *in vitro* but also *in vivo* [Koide T., Noda Y., Nata S. et al.; Sasaki T., Wakai S. and Asano T.].

The references cited attest to the fact that polyunsaturated FFA (PUFA), especially when oxidized, produce a marked, mostly vasoconstrictive effect, which points to their possible involvement in local and systemic postischemic hemodynamic disorders.

8.2 Vasoactive and Cardiotropic Effects of Circulating Unoxidized and Oxidized Phospholipids

The vasoactive effect of phospholipids, also capable of exiting membranes, circulating in the blood, and being incorporated into membranes of other organs, is much less investigated than that of FFA.

An exceptional case is provided by numerous studies to assess the toxic effect of exogenous lysoPL (lysoPC and lysoPEA) on models of perfusion of isolated conducting cardiomyocytes, the papillary muscle, and an isolated heart (see Chapter 4), and also in experiments that have demonstrated the hypertensive effect of lysophosphatidic acid *in vivo* [Tokumura A., Kume T., Fukusawa K. et al.]. There have also been individual reports according to which liposomes from PL of different composition did not affect the hemodynamics of intact animals *in vivo* [Stefanov A.V., Kreines V.M., Dmitrieva A.V. et al.; Shanskaya A.I., Kochetygov W.I., Pozdnyakov P.K. and Puchkova S.H.] but acted beneficially on the hemodynamics of dogs with the crush syndrome [Stefanov A.V., Kreines V.M., Dmitrieva A.V. et al.]. The moderate protective effect of liposomes from PC, oxidized, and PEA on the severity of ischemic shock was also observed in our laboratory (see Chapter 7). However, liposomes from PL *in vivo* quickly leave the circulatory system and accumulate in the liver [Scherphof G., Roerdink F., Hoekstra D. et al.], which makes it difficult to assess their local action on vessels in other organs, including the heart.

In our laboratory, the vasoactive and cardiotropic effects of unoxidized and oxidized PL of different composition, introduced in the form of single- and multiple-layer liposomes, were studied on 4 experimental models: (1) perfusion of an isolated rabbit ear; (2) coronary perfusion of an isolated rat heart; (3) perfusion of an isolated strip of frog myocardium; and (4) coronary perfusion of an isolated rat heart after 30 min of ischemia.

The experiments were conducted on chinchilla rabbits, Wistar rats and frogs *Rana temporaria*. In all cases, total yolk PL were used, as well as PC and PEA extracted from them, both unoxidized and oxidized, with soybean lipoxygenase.

8.2.1 Perfusion of an Isolated Rabbit Ear

The perfusion of an isolated rabbit ear was performed, using as perfusate Ringer solution at 20°C and 100 mm Hg without recirculation [Bilenko M.V. and Churakova T.D.; Bilenko M..V., Bulgakov V.G., Mukumov M.R. and Churakova T.D.; Churakova T.D.]. Unoxidized or oxidized PEA (1 ml) was injected to the perfusate in the form of single-layer liposomes after 15 min of stabilizing perfusion (its parameters were taken for the initial level). The volume perfusion rate (VPR) before and after lipid injection was recorded by the number of drops, using an electromechanical counter. The procedure for obtaining oxidized PEA and secondary LPO products is shown in Fig. 68.

Fig. 68: Procedure for obtaining exogenous LPO products

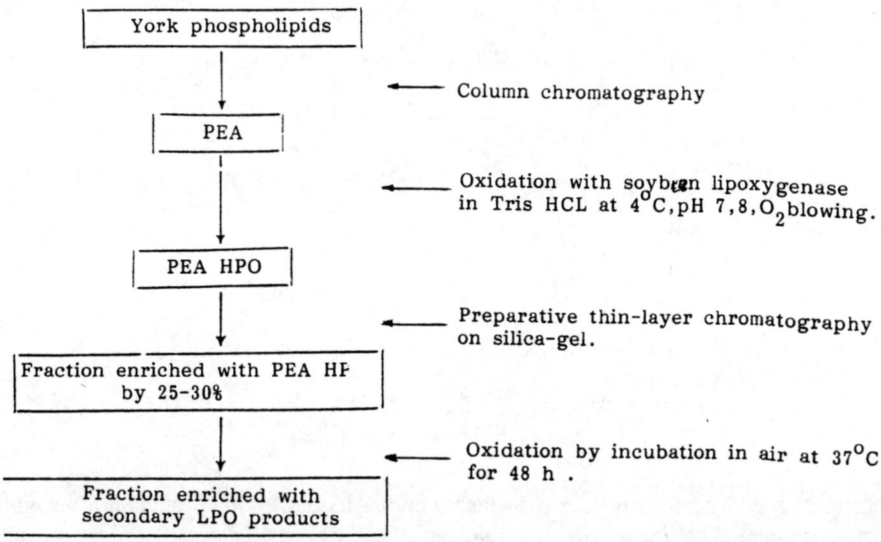

As can be seen in Table 11 and Figure 69, PEA in the doses used increased the VPR in 51% of cases, the vasodilating effect being the most pronounced within the first minutes and persistent throughout the rest of the perfusion period. In contrast, PEA HP in amounts of 0.01 and 0.1 µmol as well as the secondary PEA oxidation products in amounts of 0.02-0.2 µmol caused a significant ($P<0.05$-0.01) decrease in the VPR, their vasoconstrictive effect also being maximal within the 1st min, then to vanish after 3-5 min of perfusion. PEA HP in lower (0.001 µmol) or higher (1.0 µmol) concentrations did not exhibit any clearly defined vasoactive effect in this model, just as AA and AA HP.

Table 11: Effect of oxidized and unoxidized lipids on the volume perfusion rate (VPR) of a rabbit ear.

Lipids	Dose, µmol/ml	Number of experiments	Change, %		
			decrease	increase	no change
PEA	0.01-10.0	37	27	51**	22
PEA HP	0.01-0.1	40	63**	20	17
	0.001, 1.0	13	24	30	46
Secondary LPO products	0.02-0.2	10	60*	20	20
AA	1.0	13	15	31	54
AA HP	0.2-0.5	13	39	39	22

*-$P<0.05$; **-$P<0.01$ % of cases with VPR increase versus % of cases with VPR decrease (X^2 criterium).

Fig. 69: Effect of nonoxidized and oxidized PEA on the volume rate of an isolated rabbit ear (% of the initial level) [Bilenko M.V. and Churakova T.D.]

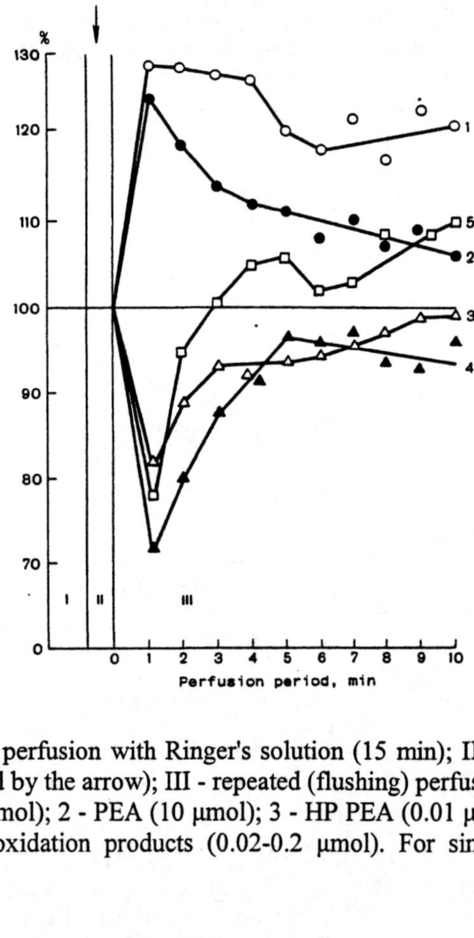

I - initial (stabilizing) perfusion with Ringer's solution (15 min); II – perfusion after addition of lipids (1 min, ingicated by the arrow); III - repeated (flushing) perfusion with Ringer's solution (10 min); 1 - PEA (0.01 μmol); 2 - PEA (10 μmol); 3 - HP PEA (0.01 μmol); 4 - HP PEA (0.1 μmol); 5 - secondary PEA oxidation products (0.02-0.2 μmol). For single-layer liposome obtaining procedure, see Fig. 64.

8.2.2 Coronary Perfusion of an Isolated Rat Heart

The coronary perfusion of an isolated rat heart was carried out in Fallen modification of Langendorf model [Bilenko M.V. (1986); Bilenko M..V., Bulgakov V.G., Mukumov M.R. and Churakova T.D.; Bilenko M.V. (1992b); Bilenko M.V., Bulgakov V.G. and Morgunov A.A. (1987); Bilenko M.V., Bulgakov V.G. and Morgunov A.A. (1989); Bulgakov V.G., Morgunov A.A. and Bilenko M.V.].

Phospholipids of different composition (total PL, PEA + PC, or PC) in the form of single- and multiple-layer liposomes were added to the perfusate (Krebs-Henseleit solution) for 10 min after the stabilizing perfusion was over (its parameters were taken for the initial level).

Figure 70 shows that the addition of unoxidized PL of different composition in the form of single-layer liposomes to the perfusate brought about a marked decrease in both CPR and myocardial contractility (P_{max}). Both indices diminished in parallel, the changes being most pronounced during the 3rd min of perfusion, whereas by the 5th and 10th min the two parameters tended to return to the initial level in spite of the continued addition of PL. The addition of PL (containing about 60% PC) caused the most drastic changes, while the effects produced by the addition of PEA + PC and the same amount of PC alone were more or less the same; that is, the effect of PL on the CPR and P_{max} seemed to stem primarily from PC.

During repeated (flushing) perfusion of the heart, the CPR came back almost to the initial level already after 20 min, and P_{max} in all runs remained 18-24% below the initial level ($P < 0.05$-0.001). Use of multiple-layer liposomes brought about even more drastic and virtually irreversible changes in both parameters. Such changes were apparently due to mechanical obstruction (obturation) of small cardiac vessels.

This model was also used to study the effect of oxidized PEA + PC. The mixture was oxidized with unpurified lipoxygenase in a similar procedure as represented in Scheme 11 but in the presence of sodium deoxycholate (Serva) according to a method described elsewhere [Eskola I. and Laakso S.].

The HP content in oxidized PEA + PC was 42 nmol/mg of lipid. No differences were observed between the effects of oxidized and unoxidized PEA + PC on the CPR and P_{max} in an isolated rat heart (see Fig. 70).

Fig. 70: Effect of nonoxidized and oxidized PL of different composition in the form of single-layer liposomes on the coronary perfusion rate (CPR, a) and cardiac function (P_{max}, b) of an isolated rabbit heart (% of the initial level) [Bilenko M.V., Bulgakov V.G. and Morgunov A.A. (1989)]

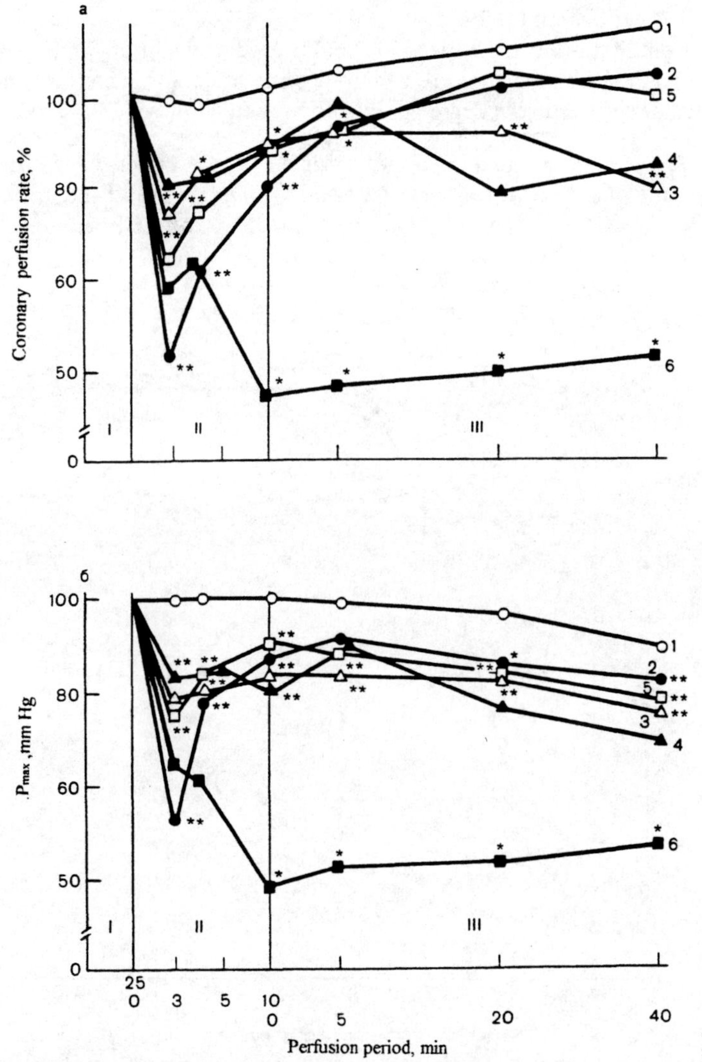

I - initial (stabilizing) perfusion with KH solution (25 min); II - perfusion with PL-containing KH solution (10 min); III - repeated (flushing) perfusion with KH solution (40 min); 1 - without PL (control); 2 - total PL (0.5 mg/ml); 3 - PEA + PC (5:1 parts by weight, 0.5 mg/ml); 4 - oxidized PEA + PC (the same doses); 5 - PC (0.08 mg/ml); 6 - PEA + PL (5:1 parts by weight, 0.5 mg/ml) in the form of multiple-layer liposomes. *-$p < 0.05$; **-$p < 0.01$ versus initial level. For single- and multiple-layer liposome obtaining procedure, see Fig. 64; for cardiac perfusion procedure, see Fig. 8.

8.2.3 Perfusion of an Isolated Strip of Frog Heart

The perfusion of an isolated strip of frog heart [Bilenko M..V., Bulgakov V.G., Mukumov M.R. and Churakova T.D.; Bulgakov V.G., Morgunov A.A. and Bilenko M.V.] was aimed at finding out whether unoxidized and oxidized PL exerted any direct (not through coronary perfusion) negative inotropic effect on myocardial tissue.

The perfusion was performed using Ringer solution at 22°C, pH 7.4, without additional oxygenation with recirculation. The PEA + PC (5:1 parts by weight) mixture was obtained as shown in Fig. 68, and the oxidized PL resulted from self-oxidation of a PL film in air at 37° for 1-2 weeks; different degrees of oxidation were achieved by varying the self-oxidation time. The PL were added to the perfusate in the form of multiple-layer liposomes after 1 h of the preliminary stabilizing perfusion. Perfusion with a PL-containing solution was carried out for 2 h.

As can be seen from Figure 71, unoxidized PEA + PC produced a brief (first 15-30 min) and reversible negative inotropic effect on the myocardium, while the same effect produced by oxidized PEA + PC was monotonic, stable and depended on the degree of oxidation and amount of LPO products in PL.

Fig. 71: Effect of nonoxidized and oxidized PEA + PC on the contraction amplitude of a myocardial strip from the frog heart (% of the initial level) [Bilenko M..V., Bulgakov V.G., Mukumov M.R. and Churakova T.D.].

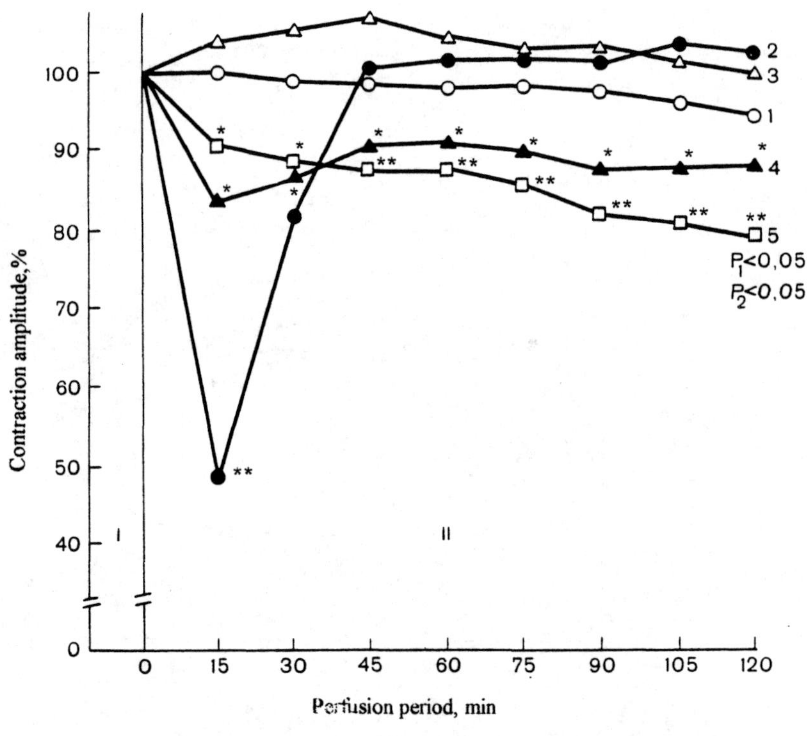

I - initial (stabilizing) perfusion with Ringer's solution (60 min); II - perfusion with PL-containing Ringer's solution; 1 - without PL (control); 2 - nonoxidized PL (PEA + PC); 3-5 – diene conjugate(DC) containing PL (PEA + PC): 3-DC 4 nmol/ml; 4 – DC 8 nmol/ml; 5 – DC 16 nmol/ml; P_1 – 5 versus 1; P_2 – 5 versus 2; *-p<0.05, **-p<0.01 versus initial level. The contraction amplitude of the strip (semi-annular cross section of the heart) was measured using a mechanotron with electrical stimulation (0.5 Hz) under conditions close to isometric.

8.2.4 Coronary Perfusion of an Isolated Ischemic Rat Heart

The coronary perfusion of an isolated ischemic rat heart was used in view of the fact that ischemic organs are more sensitive to LPO inducers [Kogan A. Kh. and Mednykh A. Ya; Goldstein R.A.] (see below, and Chapter 10), and when LPO processes are generalized, precisely such a situation occurs. These experimental results were published in [Bilenko M.V. (1988); Bilenko M.V., Bulgakov V.G. and Morgunov A.A. (1987); Bulgakov V.G., Morgunov A.A. and Bilenko M.V.].

Unoxidized and oxidized PL were added to Krebs-Henseleit solution with restoration of blood flow through the heart, ischemic for 30 min (stop-flow model). PL were oxidized similarly as in the experiments with an ischemic-free heart (see above).

It can be seen from Fig. 72 that reperfusion of the heart after 30 min of ischemia was accompanied by a drop in the CPR and a sharp decrease in myocardial function. Addition of oxidized and unoxidized PL to the perfusate aggravated the no-reflow phenomenon, the effect of oxidized PL being more pronounced, than that of unoxidized PL, especially towards the end of the observation period ($P_1 < 0.01$). In all runs, the myocardial function was restored but to a small extent, and P_{max} by the end of the observation period was 40-49% of the initial level, no differences being seen in the latter parameter between runs with oxidized and unoxidized PL. Importantly, addition of oxidized ΣPL in combination with the AO ionol lowered the CPR to a lesser extent than without ionol and resulted in a significant improvement of myocardial function (P_{max}). Addition of oxidized ΣPL together with a cycloxygenase inhibitor (acetylsalicylic acid) to the perfusate did not change the CPR and P_{max} in the ischemic heart; that is, acetylsalicylic acid did not produce any protective effect.

Fig. 72: Effect of nonoxidized and oxidized PL on the coronary perfusion rate (CPR, a) and P_{max} (b) of a rat heart after 30 min of ischemia (% of the initial level)

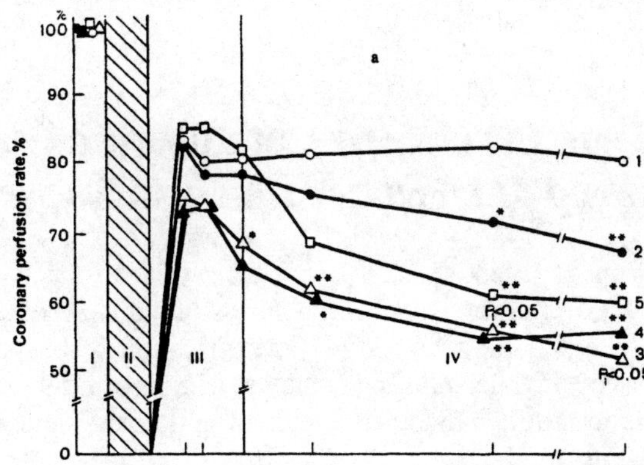

I - initial (stabilizing) perfusion (25 min); II – stop-flow ischemia for 30 min at 37°C; III - reperfusion of the heart with KH solution containing PL of different composition (10 min); IV - flushing reperfusion of the heart with PL-free KH solution (40 min); 1 - without PL (control); 2 - nonoxidized PL; 3 – HP containing PL (410 nmol/mg); 4 – the same + acetylsalicylic acid, 5. 4 10^{-6} mg/ml; 5 – the same + BHT 3 10^{-6} M. *-p < 0.05; **-p < 0.01 versus control. P_1 - 5 versus 2; P_2 - 5 versus 3.

The results of the experiments carried out in four different perfusion models can be summarized as follows.

Firstly, the unoxidized exogenous yolk PL used in the experiments lowered the tone of arteries in the rabbit ear (PEA) but raised the coronary arteries tone in the rat heart (PL, PEA + PC, PC). These differences seem to stem from the use of different experimental models (target vessels) and also different PL, which must be important in view of the most pronounced effect produced by PC.

Secondly, all oxidized PL lowered the volume perfusion rate in the rabbit ear or in the intact and ischemic rat hearts almost to the same degree (by 28-31%, as compared to the control). When acting upon the post-ischemic rat heart, oxidized PL contributed to the failure of coronary circulation to return to the initial level - that is, they aggravated the no-reflow phenomenon.

Thirdly, unoxidized PL produced a temporary negative inotropic effect on the myocardium, which was more stable and dose-dependent in the case of oxidized PL. This effect could be both indirect, partially mediated by the disorders of coronary circulation, and direct. It should be pointed out, however, that the cause-and-effect relationships between the changes in coronary circulation and cardiac function on a model of isolated cardiac perfusion at a constant volume are difficult to assess because of their linear pattern [Neely J. R. and Rovetto M. J.]. In these experiments, the negative inotropic effect of PL and oxidized PL is corroborated by perfusion of the strip of frog cardiac tissue, while the coronaroconstrictive effect is supported by the different patterns (lack of strict parallelism) of changes in the CPR and P_{max} in the experiments with an intact and, especially, ischemic heart. In any case, there is no doubt as to the negative inotropic effect of PL and oxidized PL on the cardiomyocyte, whether it is direct or manifested through the blood flow.

This effect may be the reason why liposomes failed to exert any protective effect on the heart when injected into rats with ischemic shock *in vivo*, but then no negative effect was observed either, possibly because the liposomes quickly moved from the circulating blood into the liver (see Chapter 7).

8.3 Possible Mechanisms of the Vasoactive and Cardiotropic Effects of Circulating Unoxidized and Oxidized Free Fatty Acids and Phospholipids

The effect of circulating in the blood FFA, PL and their LPO products on the vascular wall, may manifest itself in several ways primarily: (1) the triggering of the AA cascade and the formation of cycloxygenase and lipoxygenase AA metabolites; (2) intensification of biosynthesis or release of other vasoactive mediators as well as cyclic nucleotides; (3) intensification of LPO processes and growth of ROS production; (4) direct damage to the membrane structures of endothelial and smooth muscle cells, and a change in the cytoplsms Ca^{2+} content.

8.3.1 The Role of the Increase of Cycloxygenase and Lipoxygenase Pathways of AA Metabolism

The role of the increase of cycloxygenase and lipoxygenase pathways of AA metabolism (the triggering of the arachidonic acid cascade) in the vasoactive effect of lipid metabolites was very clear in the case of unoxidized AA.

Using inhibitors of cycloxygenase (indometacin, acetylsalicylic acid) and thromboxane synthetase (sulfinpyrazon, dopyridamol) reduced the AA vasoactive effect, predominantly its constrictive phase, on the coronary vessels of the isolated rat heart [Belo S.E. and Talesnik J.] and mouse brain [Rosenblum W.I. (1982)], as well as, predominantly in the dilating phase, on the mesenteric vessels in pigs [Houvenaghel A., Schrauwen E. and Wechsung L.].

Addition to the perfusate of a mixture of PUFA (intralipid), and also of tret-butyl HP caused vessel constriction in an isolated rabbit lung and was accompanied by the appearance of TXB_2 and 6-keto-$PgF_{1\alpha}$ in the perfusate; the degree of vasoconstriction correlated with the amount of TXB_2 or with the TXB_2/6-keto-$PgF_{1\alpha}$ ratio, was inhibited by indometacin, and was more pronounced in E-deficient animals [Gurther Q.H., Knoblauch A., Smith P.L. et al.]. At the same time, inhibitors of cycloxygenase and TXA_2 synthetase did not eliminate the constrictive effect of other monoenoic acids and PUFA (18:1, 18:2, 18:3, 22:3, etc.) on coronary arteries [Belo S.E. and Talesnik J.; Rosenblum W.I. (1982)], nor the hypertensive effect of lysophosphatidic acid and the hypotensive effect of lysoPL [Tokumura A., Kume T., Fukusawa K. et al.], and the vasoconstrictive effect of FA HP [Asano M. and Hidaka H.; Koide T., Noda Y., Nata S. et al.; Koide T., Neichi T., Takato M. et al.; Trachte G.J., Lefer A.M., Aharony D. and Smith J.B.]. Acetylsalicylic acid had no influence on the vasoactive effect of unoxidized and oxidized PL in our experiments either (see Fig. 72).

More essential, it would seem, is the relationship between the vasoactive effect of FFA, PL or their oxidation products and the intensification of the lipoxygenase route of AA metabolism.

It has been shown that addition of AA HP to the perfusion medium for a dog isolated basilar artery leads to its constriction and an almost 20-fold increase in the content of hydroxy acids in the medium, with the lipoxygenase inhibitor lowering both the vasoconstrictive effect and accumulation of hydroxy acids [Koide T., Neichi T., Takato M. et al.].

Studying the role of AA lipoxygenase metabolites in the vasoactive effect AA produced is especially important in view of the fact that many experiments have already revealed the coronaroconstrictive and negative inotropic action of leukotrienes (particularly LTD_4 and LTC_4) administered intracoronarily to dogs and pigs *in vivo* [Ezra D., Boyd L.M., Feuerstein G. and Godlstein R.E.; Fiedler V.B., Mardien M. and Abrom T.S.G.] or added to the perfusion medium for an isolated heart or myocardial strip *in vitro* [Burke S.A., Levi R., Guo Z.G. et al.; Letts L.G. and Piper P.J.].

8.3.2 The Role of Intensified Biosynthesis of Vasoactive Mediators and Cyclic Nucleotides

The role of intensified biosynthesis of vasoactive mediators and cyclic nucleotides in evoking the vasoactive action of lipid metabolites is less probable than the AA pathway because the vessel response to FA HP *in vitro* did not change in the presence of α- and β-adrenergic, histamin-, serotonin-, muscarin- and cholinergic inhibitors [Aharony D., Smith S.B., Smith E.F. et al.; Asano M. and Hidaka H.; Koide T., Neichi T., Takato M. et al.], while the action of lysophosphatidic acid on rats *in vivo* was suppressed by administration of inhibitors of the sympathetic nervous system, β-adrenergic inhibitors, and an inhibitor of histamine receptors [Tokumura A., Kume T., Fukusawa K. et al.]. The administration of NE-containing liposomes initiated hypertensive, vasoconstrictive and cardiostimulating effects in spite of the blocking of the α- and β-receptors; that is, PL enhanced the sensitivity of the receptors toward hormones [Dmitrieva A. V., Stefanov A. V., Bojko V. I. et al.; Stefanov A.V. and Soloviev A.I.]. How important is this factor in ischemia and whether it plays a role under *in vitro* conditions is still unclear.

Also still debated in the literature is the possibility of a vasoactive effect being exerted by FFA and FA HP via guanylate cyclase and synthesis of cyclic hexose monophosphate. Additional proof is needed to corroborate this pathway [Asano M. and Hidaka H.].

8.3.3 The Role of Increased Free-Radical Processes and the Production of Reactive Oxygen Species

The role of increased free-radical processes and the production of reactive oxygen species (ROS) in membrane-damaging, vasoconstrictive and permeability-enhancing effects of FFA, PL and, especially, LPO products is very important and is proved by numerous experimental data.

As has already been mentioned, the effect of oxidized FFA and PL on vessels is much more pronounced than that exerted by unoxidized FFA, or radically differs from it.

Clinical experience has also clearly shown the relationship between progressive angiopathy and the accumulation of LPO products (TBARS) in the blood of diabetics [Nishigaki I., Hagihara M., Tsunekawa H. et al.; Sato Y., Hotta N., Sakamoto N. et al.]; between progressing cerebrovascular spasm and the LP content in the cerebrospinal fluid of patients suffering from subarachnoid hemorrhages [Asano M., Matsuda T., Suzuki Y. et al.; Sasaki T., Wakai S. and Asano T.], and between changes in coronary vessels and the level of endogenous LPO products during AMI (see Chapter 5). Circulating lipid metabolites are capable of rapidly incorporating themselves into membranes, unoxidized products – of becoming oxidized, and oxidized products – of inducing LPO processes in membranes.

It has been established that in rats with experimental hypoxia the increase of LPO products in the blood was accompanied by a 2- to 4-fold increase in their content in the arterial wall [Yoshikawa T., Furukawa Y., Wakamatsu Y. et al.]. In our experiments with reperfusion of ischemic limbs, the accumulation of LPO products in the blood was accompanied by intensification of LPO processes in other nonischemic organs (see Chapter 7). In experiments with perfusion of the ischemic rat heart the addition of the AO BHT to the perfusate medium together with PL HP substantially improved the cardiac function (see above). A similar effect was produced by α-tocopherol whose addition to the perfusion medium for an isolated rabbit lung together with AA diminished its vasoconstrictive action and vessel-damaging effects [Seeger W., Wolf H., Stahler G. et al.].

Reactive oxygen species are known to produce vasoconstrictive or vasodilative effects on different kinds of isolated vascular preparations [Ikai I., Chance B. and Kumar C.; Wolin M.S.].

There are data in which both the damaging effect on the vascular structure and the vasoactive effect of FFA and FA HP were precluded through use *in vivo* or addition to the perfusate of such AO enzymes as SOD, catalase, and $^{\cdot}OH$ scavengers [Christman C.W., Wei E.P., Kontos M.A. et al.; Kontos H.A., Wei E.P., Povlishock J.T. et al.; Seeger W., Wolf H., Stahler G. et al.].

Pronounced vasoactive (often biphase, dose-dependent), negative inotropic and vessel-damaging effects of ROS were also demonstrated in experiments using a hypoxanthine/XO mixture during perfusion of an isolated rat heart [Steen E.M., Noronha-Dutra A.M. and Rump S.; Vaage J., Ytrehus K., Myklebust R. et al.; Ytrehus K., Myklebust R. and Mjösi O.D.], a strip of rat myocardium [Burton K.P., McCord J.M. and Ghai G.], a rat thoracic aorta ring [Lawson D.L., Mehta J.L., Nichols W.W. et al.], canine cerebral [Katusic Z.S. and Vanhoutte P.M.], or coronary arteries [Lawson D.L., Mehta J.L. and Nichols W.W.]; during incubation of pialic arteries of mice [Rosenblum W.I. (1983)] and upon injection of this ROS-producing mixture into the pouch of a hamster [Del Maestro R.F., Björk J. and Azfors K.E.]. The administration of SOD and catalase mitigated the damaging and vasoactive effects of ROS.

H_2O_2 and organic peroxides, such as cumene, tert-butyl and others, were also found to exert a constrictive effect on vessels of the human placenta, isolated lung, and coronary vessels [Gurther Q.H., Knoblauch A., Smith P.L. et al.].

The coronaroconstrictive and negative inotropic effects of H_2O_2 were especially manifest under conditions of low cardiac perfusion [Gurther Q.H., Knoblauch A., Smith P.L. et al.]; in other words, a combination of H_2O_2 and ischemia, just as PL HP combined with ischemia in our experiments, produced a marked damaging effect.

8.3.4 Direct Damage to Membrane Structures in Endothelial, Smooth-Muscle or Myocardial Cells

Direct damage to membrane structures in endothelial, smooth-muscle or myocardial cells is the most probable manifestation of the vasoactive and cardiotropic effects of lipid metabolites. Circulating FFA, PL and LPO products can most easily access cell membranes of the vascular endothelium, whose damage (edema, desquamation, necrosis) inevitably results, just as any damaging factor, in constriction of the vessel (see Chapter 1). Indeed, this is precisely what was observed when FFA (18:1, 20:4) were added to the perfusate *in vitro* [Ehrhart I.C., and Hofman W.F.; Seeger W., Wolf H., Stahler G. et al.] and when FFA HP (18:2, 20:4) were administered to animals or applied onto vessels *in vivo* [Christman C.W., Wei E.P., Kontos M.A. et al.; Mizukami M., Aono J., Sakai R. et al.; Nayler W. G., Poole-Wilson P. A. and Williams A.;T., Wakai S. and Asano T.]. The damage effect of lipids and lipid metabolites, as well as other toxic factors, on ischemic endothelial cells are expressed stronger than on nonischemic ones. It is obvious from our experiments with LDL, LDL_{ox}, TNF_{α} and H_2O_2 added to the cultured human umbilical vein endothelial cells during their aerobic or anoxic incubation [Bilenko M.V., Fedosova S.V., and Ladygina V.G.; Bilenko M.V., Ladygina V.G. and Fedosova S.V.(1998a); Bilenko M.V., Ladygina V.G. and Fedosova S.V.(1998b)].

The occurring structural changes also involved vascular smooth muscle cells, with vacuoles and electron-dense corpuscles in cytoplasm, pyknotic changes in nuclei, focal necroses, hyperconstrictions, and ruptured myofibrils [Christman C.W., Wei E.P., Kontos M.A. et al.; Sasaki T., Wakai S. and Asano T.].

The membrane-damaging, vasoconstrictive and permeability-enhancing effects of LPO products may involve active forms of O_2, whose emergence is possible during induction of LPO or reduction of HP in the vascular wall.

One of the mechanisms of the toxic effect produced by PUFA HP on myofibrils (primarily in the heart) is based on their capacity for oxidizing hemoglobin to ferrimyoglobin, which leads to depletion of the short-lived O_2 reserve. Whether this also applies to vascular smooth muscles is not yet clear.

Oxidation of SH groups of membrane proteins was shown to be essential for the contractile response of muscles to LPO products, because pretreatment of rabbit aorta strips with substances rich in SH groups lowered the contractile response of the vessel to FFA (18:2) HP, and treatment with SH inhibitors led to its suppression, whereas the strip ability to contract in response to other vasoconstrictive agents did not change at all [Asano M. and Hidaka H.]. It is obvious now that the changes in the contractile function of any muscle (cross-striated, myocardial, smooth vascular) are related to the cytoplasmic Ca^{2+} content in them [Dhalla N.S., Das P.K. and Sharma G.P.; Morgan K.G.; Weiss S.J.].

The role of changes in the cytoplasm Ca^{2+} content in the realization of the vasoactive and cardiotropic effects of lipid metabolism is obvious from the fact that the vasoconstrictive effect of AA HP in dogs and humans [Asano M., Matsuda T., Suzuki Y. et al.; Van Diest M.I., Verbeuren T.J. and Herman A.G.], as well as the effect of liposomes from PC and PS on the contractile activity of a portal vein strip from a rat [Stefanov A.V. and Soloviev A.I.] were observed only when Ca^{2+} ions were present in the medium, and use of specific inhibitors of slow Ca^{2+} flow channels (verapamil, nifedipin) substantially minimized both the vasoconstrictive effect of FA HP [Asano M., Matsuda T., Suzuki Y. et al.; Koide T., Neichi T., Takato M. et al.] and the effect of liposomes from PC on the contraction of the portal vein strip [Stefanov A.V. and Soloviev A.I.].

Changes in the cell Ca^{2+} content caused by circulating lipid metabolites, and the latter resulting vasoactive effect may be ascribed to the following possible mechanisms.

(1) Superfluous influx of extracellular Ca^{2+} into the cell along the concentration gradient. This, it is assumed, is possible because some FFA and FA HP exhibit properties inherent in Ca^{2+} ionophores, but this contradicts the fact that AA HP with its most pronounced vasoconstrictive effect displays no such properties [Serhan C., Anderson P., Goodman E. et al.].

A more likely mechanism is disruption of the barrier function of the cell membrane with respect to Ca^{2+} by lipid metabolites, which implies incorporation of LPO products into the membrane or at least their retention on the membrane. This assumption is corroborated by the weakening of the vasoconstrictive action of AA HP after the addition of fumaric acid to the perfusate, which, according to [Asano M., Matsuda T., Suzuki Y. et al.], blocks the sites of HP binding on cell membranes.

The accumulation of Ca^{2+} in myofibrils under the effect of exogenous LPO products may also be due to induction of endogenous LPO processes in the vascular wall itself. The permeability of BLM increases under the influence of LPO products, which was demonstrated earlier in our [Sokolov V.S., Churakova T.D., Bulgakov V.G. et al.] and other [Lebedev A. V., Levitsky D. O. and Loginov V. A.] experiments; there is further evidence of BLM permeability toward Ca^{2+} in the presence of $CaCl_2$ [Lebedev A. V., Levitsky D. O. and Loginov V. A.].

(2) Direct Ca^{2+} transport into the cell, mediated by liposomes, just as was demonstrated in an experiment with a myocardiocyte culture [Bkaily G., Sperelakis N., Elishalom Y. et al.], because HP of some FA are ionophores [Serhan C., Anderson P., Goodman E. et al.].

(3) Decreased Ca^{2+} transport into SR vesicles from the cytosol when SR membranes are damaged, as was shown in experiments with induction of LPO processes in SR membranes [Kagan V. E., Azizova O. A., Arkhipenko Yu. V. et al.] or in our experiments with incubation of SR membranes in the presence of PL HP [Kagan V. E., Churakova T. D., Karagodin V. P. et al.].

It is assumed that the increases in the Ca^{2+} permeability of membranes and Ca^{2+} content in the cytoplasm under the effect of unoxidized PL have to do with the specific activation of Ca^{2+} channels, as a result of modification of their lipid environment, or with a nonspecific increase in the Ca^{2+} permeability of membranes due to changes in the PL/cholesterol ratio [Stefanov A.V. and Soloviev A.I.]. Another possibility is oxidation (peroxidation) of unoxidized PL introduced into the blood or perfusate - that is, involvement of endogenous PL HP.

8.4 CONCLUSIONS

Our observations and the references cited in this chapter make it evident that phospholipids, FFA and LPO products, whose content in the circulating blood increases sharply during the reperfusion period, exhibit marked vasoactive and cardiotropic effects and definitely play a certain role in the pathogenesis of local circulatory disorders, such as the no-reflow phenomenon, and also in systemic hemodynamic disorders during ischemic shock. Here, the following key points should be emphasized.

(1) The vasoactive effect of FFA has been extensively studied. The literature contains many reports testifying to its dependence on the degree of FA saturation and oxidation, and the specific sensitivity of the target vessel. The most oxidized FA and particularly AA, both unoxidized and oxidized, are strong vasoconstrictors.

(2) The knowledge about the vasoactive effect of unoxidized and oxidized PL is more limited. As was shown in our experiments, exogenous yolk PL of different composition lower, when added to the perfusate in liposomal form, the volume perfusion rate of an isolated rabbit ear (oxidized PEA), an intact rat heart, and a rat heart ischemic for 30 min at 37°C (unoxidized and oxidized PL, PEA + PC, PC). In experiments with perfusion of intact and ischemic rat hearts and a strip of frog myocardium, unoxidized and, especially, oxidized PL also impaired the contractile function of the myocardium. The vasoconstrictive and negative inotropic effects of oxidized PL were dose-dependent and were not eliminated by administration of acetylsalicylic acid; that is, they did not involve the cyclooxygenase pathway of AA metabolism and were strongly suppressed by the AO BHT.

(3) A leading factor of the vasoconstrictive effect exerted by oxidized FFA and PL seems to be their direct damaging action on endothelial and smooth-muscle cell membranes, namely, impairment of the Ca^{2+}-transporting function of membranes and excessive Ca^{2+} influx into the cell. Apparently, this effect is due to incorporation of oxidized lipids into membranes or induction of endogenous LPO processes in membranes by them. The damaging and vasoconstrictive effects seem to involve active forms of O_2 as well.

(4) The effect produced by unoxidized and oxidized FFA and PL *in vitro* is not always identical with that *in vivo*.

On one hand, these differences may be due to rapid elimination *in vivo* of liposomes from the circulatory system into the liver (see Chapter 7); on the other, they may have to do with the action of lipid metabolites circulating *in vivo* affecting not only the vascular wall but also the formed elements of the blood, which accelerates aggregation of thrombocytes, activates phagocytes, reduces the microcirculation, and reintensifies LPO processes in the vascular wall.

The vasoactive and cardiotropic effects of circulating lipid metabolites are extremely important for understanding their role in the pathogenesis of ischemic and reperfusion injuries in organs and in the body and call for additional in-depth studies.

Chapter 9

ROLE OF CYTOKINES AND GROWTH FACTORS, FREE RADICALS, AND LIPID PEROXIDATION PRODUCTS IN DNA DAMAGE AND REPAIR DURING ACUTE ISCHEMIA AND AFTER REPERFUSION OF ORGANS

DNA, the carrier of genetic information, is very sensitive to damage caused by physical and chemical factors.

DNA damage, depending on the severity of detrimental influence, is manifested in the appearance of individual or multiple single-strand and double-strand breaks. The latter results in the fragmentation of DNA and cell nucleus and in DNA–DNA, DNA–RNA, and DNA–protein cross-linking [Moskaleva E.Yu. and Ilyushina N.O.; Tarasov V.A.; Zhestyannikov V.D.]. The nature and extent of DNA damage determines the mode of cell death (necrosis or apoptosis).

Necrosis is a pathological, uncontrolled form of cell death associated primarily with damage to the plasma membrane and intracellular membranes and the impairment of intracellular homeostasis. Cell organelles and the entire cell swell and undergo lysis (see Chapter 2). In this course, however, DNA suffers little damage or remains unfragmented.

Apoptosis, or programmed cell death, is an active form of cell suicide. Along with subsequent regeneration, it is involved in the mechanism of cell renovation and the maintenance of cell homeostasis. Nuclear DNA damage upon apoptosis is manifested in multiple double-strand breaks leading to internucleosomal DNA cleavage into oligonucleosomal subunits, each 180–200 nucleosomes in length (DNA ladders), which become surrounded by fragments of the cytoplasm and the nuclear membrane. As the result, the nucleus and then the entire cell break up into so-called *apoptotic bodies*, which are easily recognized and eliminated by phagocytes (rather than by lysis). Other membrane structures, including the plasma membrane, are weakly damaged at early stages of apoptosis, and cell homeostasis may remain unchanged. Nuclear chromatin condensation and the formation of apoptotic bodies serve as morphological signs, and extensive *DNA laddering*, as a biochemical sign of cell apoptosis. An increase in the *content of glycoprotein-2 mRNA* is an additional, molecular marker of apoptosis [Linnik M.D., Zobrist R.H. and Hatfield M.D.; Zoli M., Ferraguti F., Zini I. et al.] indicating the expression of certain genes (*ced3, ced 4, ICE, c-fos, c-jun,* etc.) and activation of Ca^{2+}-dependent and Mg^{2+}-dependent endonuclease and protease systems [Hanson K.P.; Zhang Z., Blake D. R., Stevens C. R. et al.]. The involvement of protein-synthesizing systems in the induction of apoptosis is consistent with the protective effect of protein synthesis inhibitors [Pratt R.M.and Greene R.M.]. The role of Ca^{2+} as a secondary messenger in apoptosis is confirmed by the fact that this process accelerates in the presence of Ca^{2+} excess and is delayed when Ca^{2+} concentration in the cytosol is low [Lennon S.V., Martin S.J. and Cotter T.G.; Zhao Z. S., Khan S. and O'Brien P. J.; Thomas C.E. and Jackson R.L.]. There is conclusive evidence that apoptosis is responsible for cell death during physiological and hormonal responses, immune conflicts, cell aging in culture, and the regression of hyperplastic organs. Apoptosis also takes place during many pathological processes and states, such as X-irradiation of tissues, UV-induced tumor pathomorphosis, and stress or shock of different etiology. The extent of apoptosis directly depends on the severity of damage [Zhang W., Naomoto Y., Tanaka N. et al.], Zhizhina G.P., Karobko V.G. and Alessenko A.V.; Harmon B.V., Corder A.M., Collins R.J. et al.]. On the other hand, apoptosis is accompanied or completely substituted by cell necrosis when the intensity of damaging influence exceeds a certain "critical" level, as was observed upon an increase in the duration of UV irradiation (from 10 min to 2 h), in temperature (above 45°C), concentrations of methanol (from 7.5 to 29%) and H_2O_2 (from 15 to 400 μM), and the dose of diethylnitrosamine (from 10 to 100 mg/kg) [Hess M.L., Manson N.H. and Okabe E.; Lennon S.V., Martin S.J. and Cotter T.G.; Daoust R. and Morais R.].

Thus, a weak damage more often leads to dose-dependent apoptosis, whereas a severe injury results in necrosis.

DNA repair depends on the extent of damage and can be used as a parameter characterizing the severity of damaging influence [Binkova B., Topinka G., Sram R. et al.; Yavich M.P., Rozhitskaya I.I., Golubeva L.Yu. et al.]. Apoptosis is a powerful stimulus for DNA repair, and its foci are often the sites of enhanced DNA synthesis and cell proliferation [Fukuda K., Kojiro M. and Chiu J.F.; Kondo S.].

The rate of DNA synthesis *in vivo* in rat organs increases in the following order (in relative units): brain (0.21) < heart (0.41) < liver (1.0) < renal epithelium (1.28) < spleen (5.5) < intestinal epithelium (11.6) [Baril E. and Laszlo J.]. The same authors showed that the activity of DNA polymerase beta, the main enzyme responsible for DNA repair, increases in approximately the same order: brain (0.19) < heart (0.30) < renal epithelium (0.73) < liver (1.0) < spleen (4.3) < intestinal epithelium (9.3).

DNA synthesis can be replicative (prerequisite to cell division) or reparative. The latter is a semiconservative process of repairing one or both DNA strands, with or without the subsequent cell division [Tarasov V.A.; Zhestyannikov V.D.]. In most organs, replicative synthesis normally prevails over DNA repair, and the latter accounts for only 9–12% of total DNA synthesis [Castellot J.J., Miller M.R., Lehtomaki D.M. et al.; Meerson F.Z., Vavich M.P. and Rozhitskaya I.I.; Yavich M.P., Rozhitskaya I.I., Golubeva L.Yu. et al.].

Enzymes of replicative DNA synthesis (DNA polymerases alpha) are significantly more sensitive to adverse environmental effects than DNA polymerases beta, enzymes responsible for DNA repair. This may partially explain the fact that virtually any chemical or physical damage to the cell inhibits DNA replication but leads to the relative or absolute enhancement of reparative DNA synthesis [Castellot J.J., Miller M.R., Lehtomaki D.M. et al.; Meerson F.Z., Vasil'ev K.V. and Dosmagambetova R.S.; Tarasov V.A.; Yavich M.P., Rozhitskaya I.I., Golubeva L.Yu. et al.].

DNA synthesis (the S phase) is a component of the cell cycle (Fig. 73). In adult eukaryotes, cells of most organs are in the resting, or quiescent (nonproliferative) G_0 state characterized by a specifically low metabolic rate. To regenerate, cells require a higher level of metabolic activity; simultaneously, they lose (at least partially) their functional specialization and enter the cell cycle. When progression through the cell (division) cycle ceases, metabolic rate decreases, and cells recover their specialization [Epifanova O.I., Terskikh V.V. and Polunosvky V.A.; Fabrikant J.I. (1968)].

Fig. 73: Cell proliferactive cycle [Fabrikant J.I. (1968)]

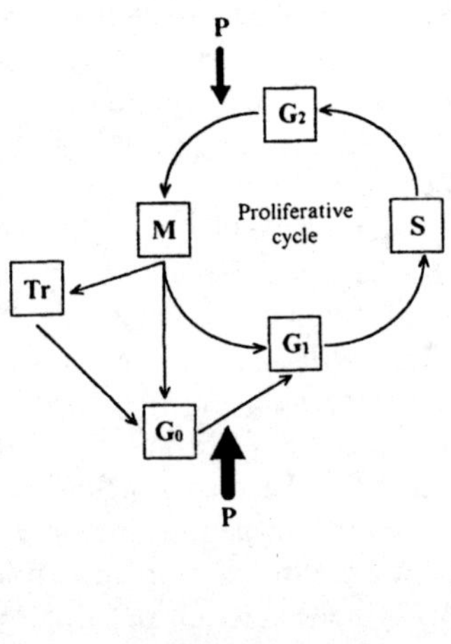

M-phase of mitosis; S- phase of synthesis DNA; G1- phase of presynthesis DNA; G2- phase of premitosis; Tr- Transitional and mature cells; G0 – Potentially proliferative cells, phase of proliferative rest; P- Stimulation

Thus, the degree and nature of DNA damage are important factors determining both the mode of cell death and the ability of cells to regenerate.

Detailed studies on DNA damage upon acute ischemia, the potential for DNA synthesis activation during reperfusion, and the involvement in these processes of major factors stimulating or inhibiting regeneration, such as free radicals, cytokines and growth factors began only a short time ago. Their results will be briefely reviewed in this chapter. Note, however, that the problems discussed here need further investigation.

9.1 DNA Damage and Repair during Acute Ischemia and after Reperfusion of Organs

9.1.1 DNA Damage during Ischemia and after Reperfusion

DNA damage during ischemia and after reperfusion of organs has been investigated in sufficient detail over the last years. When the duration and severity of ischemia exceed a certain limit (specific for each organ) and when the organ or individual populations of its cells are highly sensitive to oxygen deficiency, DNA damage may consist in the appearance of double-strand breaks. This leads to death of a certain cell populations or the entire ischemic organ via apoptosis.

Thus, it was shown that focal ischemia of *the heart* (AMI) is accompanied by single- and double-strand DNA breaks appearing mainly in the nonischemic, periinfarct area. An increase in blood contents of high- and low-molecular products of DNA hydrolysis was observed in dogs after LAD occlusion [Ferreira R., Burgos M., Milei J. et al.] and in patients with AMI [Mozhaitseva A.G.].

In the rat *liver,* the arrest of portal blood flow [Kerr J.F.R.] or occlusion of the hepatic artery and portal vein for 2 or 24 h without reperfusion [Fukuda K., Kojiro M. and Chiu J.F.] were accompanied by morphological and biochemical manifestations of apoptosis. In rats exposed to CCl_4 poisoning, both necrotic and apoptotic changes occurred in the liver. The former were approximately similar to those in the ischemic liver, whereas apoptosis was less pronounced. After aschemia and reperfusion of rat liver lobes sinusoidal endothelial cells showed evidence of apoptosis earlier than hepatocytes [Kohli V., Selzner M., Madden J.F. et al.].

During ischemia of the *kidney,* the signs of apoptosis of renal tubular epithelial cells in rats appeared at the periphery of the infarct zone after 3–24 h [Correa-Rotter R., Hostetter T.H., Manivel J.C., et al.]. Apoptosis was also observed after ligation of the ureter or renal artery stenosis [Sawczuk I.S., Hoke G., Olsson C.A. et al.; Gobe G.C., Axelsen R.A. and Searle J.W.] and after 5–45 min of total kidney ischemia and subsequent reperfusion in mice and rats [Liao H.J., Chen X.M. and Cui S.W.; Schumer M., Colombel M.C., Sawczuk I.S. et al.]. In experiments with total renal ischemia in rats, apoptotic symptoms (the classic triad evaluated after reperfusion for 12, 24, and 48 h) became more severe as the duration of ischemia increased from 5 to 30–45 min, being especially pronounced after 30 min of ischemia and 24 h of reperfusion [Schumer M., Colombel M.C., Sawczuk I.S. et al.]. In mouse kidneys, manifestations of apoptosis after 12 h of reperfusion became significantly more apparent than after 6 h [Liao H.J., Chen X.M. and Cui S.W.].

In rat kidneys reperfused after 60 min of total ischemia, two phases of apoptosis were revealed: a slight increase in the proportion of damaged cells (up to 3%) on the third day of reperfusion, and a considerable increase in this proportion (to 10%) on the eighth day [Schimizu A. and Yamanaka N.]. According to the authors of this study, the first phase directly resulted from ischemia and reperfusion, leading to cell death that stimulated postischemic regeneration, whereas the second phase reflected the involvement of apoptosis in programmed elimination of the excessive regenerated cell mass.

The model of ischemia and reperfusion of *the brain* was extensively used to study the role of apoptosis in organ damage and death under corresponding conditions, probably because apoptosis proved to be involved in the development of the central nervous system [Raff M.C., Barres B.A., Burne J.F. et al.]. There is ample evidence that apoptosis occurs in the ischemic brain of rats [Goto K., Ishige A., Sekigushi K. et al.; MacManus J.P., Hill I.E., Huang Z.-Gao et al.; MacManus J.P., Buchan A.M., Hill I.E. et al.; Shigeno T., Yamasaki Y., Kato G. et al.] and Mongolian gerbils [Okamoto H., Matsumoto M., Ohtsuki T. et al.; Yoshitatsu S., Von Lubitz D.K.J.E., Basile A.S. et al.]. Distinct signs of apoptosis in the rat brain were revealed after 16 min of total ischemia and 48 h of reperfusion [MacManus J.P., Buchan A.M., Hill

I.E. et al.] or 3 h of local ischemia and 24 h of reperfusion [MacManus J.P., Hill I.E., Huang Z.-Gao et al.]. In both cases, an increase in the duration of ischemia from 8 to 12 min [MacManus J.P., Buchan A.M., Hill I.E. et al.] or from 0.5 to 3 h [MacManus J.P., Hill I.E., Huang Z.-Gao et al.] led to more severe manifestations of apoptosis [xx]. In the gerbil brain exposed to total ischemia for 5–10 min, these manifestations reached a peak after 48 h of reperfusion [Yoshitatsu S., Von Lubitz D.K.J.E., Basile A.S. et al.] or in the period from 54 to 72 h of reperfusion [Okamoto H., Matsumoto M., Ohtsuki T. et al.].

The signs of DNA damage and apoptosis in the brain regions most sensitive to ischemia, such as the striatum, thalamus, hypothalamus, and CA1 cells of the hippocampus, were usually more distinct than in other areas [Yoshitatsu S., Von Lubitz D.K.J.E., Basile A.S. et al.; Soriano M.A., Ferre I., Rodriguez-Farre E. and Planas A.M.; Okamoto H., Matsumoto M., Ohtsuki T. et al.; Goto K., Ishige A., Sekigushi K. et al.], and this was often accompanied by an increase in endonuclease activities [Tominaga T., Kure S., Narisawa K. and Yoshimoto T.]. Treatment with inhibitors of protein synthesis (e. g., cycloheximide) decreased the size of the ischemic focus from 15.5 to 11.9% ($p < 0.05$) [Linnik M.D., Zobrist R.H. and Hatfield M.D.] and alleviated symptoms of apoptosis [Goto K., Ishige A., Sekigushi K. et al.; Shigeno T., Yamasaki Y., Kato G. et al.; Thilman P., Xie Y., Kleihues P. and Kiessling M.]. Apparently, only certain types of proteins (killer, stress, or some abnormal proteins) are involved in the initiation of apoptosis because the rate of total protein synthesis in ischemic and reperfused organs is reduced [Fornander J., Hellman A. and Hasselgren P. O.; Kiessling M., Dienel G.A., Jacewicz M. and Pulsunelli W.A.M.; Nowak T.S., Fried R.L., Lust W.D. and Passonneau J.V.]. DNA fragmentation was revealed also in tissues under conditions similar to ischemia, e. g., in Ehrlich ascites carcinoma cells cultivated anaerobically in the glucose-free medium [Proskuryakov S.Ya., Gabai V.L., Mosin A.F. et al.] or mouse cortical neurons cultivated under oxygen deprivation [Copin J.C., Reola L.F., Chan T.Y. et al.].

These data show that ischemia and reperfusion of various organs result in DNA damage with morphological and biochemical signs characteristic of apoptosis. This damage occurs under conditions of incomplete ischemia or after reperfusion of totally ischemic organs, is associated with the active synthesis of certain protein types, and depends on the duration and severity of ischemia.

9.1.2 DNA Repair in Ischemic and Reperfused Organs

DNA repair in ischemic and reperfused organs received less attention. Total ischemia (50 min) of a single kidney in rats resulted in the enhancement of reparative processes estimated from an increase in *c-fos* mRNA level, which reached a peak 1 h after the removal of the clamp and gradually decreased to the baseline level within 4 h [Safirstein R., Price P.M., Saggi S.J. and Harris R.C.]. Temporary cardiac arrest and the resulting hypoxia stimulated DNA synthesis in the brain nuclei [White B.C., De Gracia D.J., Krause G.S. et al.]. The effects of prolonged (75 min) ischemia on regeneration of the dog liver after 70% hepatectomy was estimated indirectly, from the survival rate and the results of biochemical tests for liver functioning [Levandovsky I.V. and Epishin Y.N.]. Ligation of several branches of the portal vein to reduce the portal blood flow in the liver did not affect the normal capacity of the liver for regeneration after subsequent 70% hepatectomy [Kahn D., Kajani M., Zeng Q. et al.]. It was shown that after 15 min of rabbit spinal cord ischemia DNA repair protein (DNA-dependent protein kinase) was activated after 60 min of ischemia cell death programs were initiated [Shackelford D.A., Tobaru T., Zhang S. and Zivin J.A.].

Compared to other organs, the liver has a relatively high regeneration capacity, is highly sensitive to ischemia, and its multilobate structure in rodents allows the analysis of individual and combined effects of ischemia and resection on its regeneration. Hence, this model was chosen in our laboratory for studying the influence of ischemia on DNA repair [Bilenko M.V., Seregina L.A., Komarov P.G. and Abakumova O.Yu.]. Liver ischemia in Wistar rats was produced by clamping the hilar peduncle of the left and median liver lobes (accounting for 70% of the organ weight) under intraperitoneal hexenal anesthesia. Control rats (group 1) were subjected to 30% hepatectomy by removing the left liver lobe. Experimental rats were exposed to 2-h ischemia either without hepatectomy (group 2) or with the simultaneous removal of the right and caudal lobes accounting for a total of 30% of liver weight (group 3). The rate of DNA synthesis was estimated from [^{3}H]thymidine and [^{14}C]thymidine incorporation into the nuclear fraction of ischemic and intact liver lobes; samples were taken 24, 32, 48, and 72 h after surgery.

The results showed (Fig. 74) that after 30% hepatectomy without ischemia, the rate of nuclear DNA synthesis increased two- to fourfold ($p < 0.01$) and remained at a higher level throughout the period of observations. The peak of DNA synthesis was observed 32 h after operation, i.e., approximately 8 h later than in experiments with 70% hepatectomy (20-24 h) [Cheeseman K.H., Collins M., Maddix S. et al.]. After 2-h ischemia without hepatectomy, DNA synthesis rate in ischemic lobes was similar to that observed after 30% hepatectomy, whereas in nonischemic lobes it was significantly higher. In addition, the peak of DNA synthesis in nonischemic lobes occurred after 24 h, compared to 32 h in the control group. In other words, 2-h ischemia stimulated DNA synthesis in both ischemic and nonischemic lobes of the liver.

Fig. 74: DNA synthesis in rat liver nuclei after 30% resection and 2 h ischemia of liver (mean ± SEM). [Bilenko M.V., Seregina L.A., Komarov P.G. and Abakumova O.Yu.].

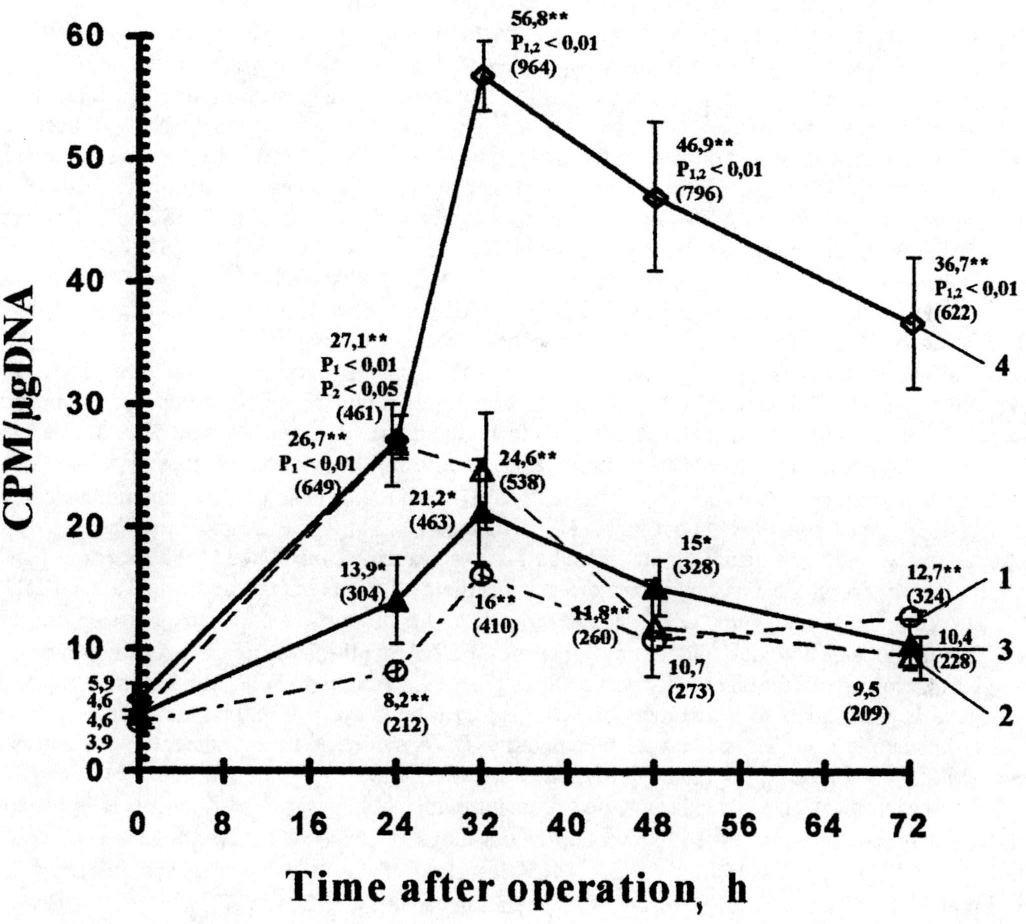

1 — intact lobes in rats with 30% resection of liver, (control, group 1);
2 — intact lobes in rats with 2 h ischemia of two liver lobes, (group 2);
3 — 2 h ischemic lobes of the same liver, (group 2);
4 — 2 h ischemic lobes in rats with ischemia + 30% resection of liver, (group 3).
In parenthesis — % to initial level.
*-p<0.05; **-p<0.01 versus initial level.
P_1 — significant differences with control;
P_2 — significant differences between 4 and 3.
The rate of DNA synthesis was estimated by [^{3}H] thymidine (100 µCu/rat) and [^{14}C] thymidine (50 µCu/rat) incorporation into nuclei 2 h after i. p. injection using scintilar accounter Mark-II (Nuclear Chicago, USA).

In the liver subjected to 2 h of ischemia with simultaneous 30% resection (group 3), the rate of DNA synthesis was higher than in ischemic lobes of the liver in experiments without hepatectomy (group 2) and nonischemic liver after 30% resection (group 1) ($p < 0.05$–0.01) throughout the early reperfusion period (24–72 h). DNA synthesis reached a peak after reperfusion for 32 h and remained at a high level after 72 h.

Thus, the rate of DNA synthesis in the liver regenerating after different types of experimental damage increased in the following series (figures in parentheses show the percentage of normal rate and the period between hepatectomy and the onset of reperfusion and the peak of DNA synthesis): liver after 30% hepatectomy (410%, 32 h) ≤ liver after 2-h ischemia and reperfusion without hepatectomy (463%, 32 h) < nonischemic lobes after 2-h ischemia of other lobes of the same liver (649%, 24 h) << liver after 2-h ischemia and reperfusion with simultaneous 30% hepatectomy (964%, 32 h).

Analyzing these results, the following points should be considered.

First, the nature of DNA synthesis in the lobes remaining after 30% hepatectomy (group 1) and ischemic liver lobes (group 2 and 3) may be different. Upon partial hepatectomy, up to 60–85% of hepatocytes are known to enter the cell cycle and to form the potentially proliferating cell pool [Fabrikant J.I. (1969); Zorin A.V., Isanin N.A. and Yakovlev A.Yu.]; hence, it is apparent that DNA synthesis in them is largely associated with replication, as in the intact liver. This is confirmed by a considerable increase in the mitotic index observed in the liver after partial hepatectomy [Goldenberg. H., Hüttinger, Böck P. et al.; Burke-Gaffney A. and Keenan A.K.]. In the liver lobes ischemic for 2 h (groups 2 and 3), the proportion of hepatocytes entering the cell cycle is much lower because it is known that at least 25% of the liver tissue undergo necrosis after such an exposure [Sokirchenko I.A. and Shkurupii V.A.], and DNA synthesis is mostly reparative. The situation with the nonischemic lobes preserved in experiments on liver ischemia is unclear: DNA synthesis in them may by aimed at both replication and repair.

Second, these data strongly suggest that the functional load is important for stimulating DNA synthesis in the ischemic liver. The rate of DNA synthesis was significantly higher in group 3 than in group 2, although the severity of damage was similar (2-h total ischemia). This is consistent with ample evidence that regeneration becomes more active upon an increase in the volume of tissue removed during hepatectomy (from one to two lobes, i.e., from 30 to 67–70%) and, hence, in the functional load on the liver [Zieve L., Anderson W.R. and Lindblad S.; Anderson W.R., Zieve L. and Lindblad S.]. A higher functional load, as well as treatment with phenobarbital or zyxorine contributed to the severity of ischemic damage to the liver, on the one hand, and the enhancement of its regeneration, on the other [Zakharova M.V., Shkurupii V.A. and Grek O.R.]. However, when the proportion of resected tissue increased to 80–90% (three lobes, subtotal hepatectomy), the rate of DNA synthesis decreased, reaching a peak more slowly, and the onset of mitotic activity was delayed [Zieve L., Anderson W.R. and Lindblad S.; Anderson W.R., Zieve L. and Lindblad S.; Romanova L.K. and Grushevskaya O.O.]. This functional overload can partially explain the decrease in the rate of reparative DNA synthesis in experiments with aortostenosis [Meerson F.Z., Vavich M.P., Rozhitskaya I.I. and Golubeva L.Yu.].

Third, the "critical" mass rendering the organ incapable of regeneration response is approximately equal to that remaining after 70–85% resection of the lungs [Romanova L.K.], 80–83% resection of the kidneys [see Bilenko M. V. (1978a)], and 85–90% resection of the liver [Gaub J. and Iversen J.]. These critical masses determined by the criterion of regeneration capacity are 20–25% lower than those providing for animal survival after resection. Thus, the survival rate in experimental group 3 was very low (only 19%), whereas DNA synthesis in the liver, characterizing its regeneration, has considerably increased (see Fig. 74).

In organs exposed to any deleterious influence, including ischemia, it is difficult to determine the actual cell mortality and, hence, the time of exposure leading to the critical reduction of cell mass. The regeneration capacity of ischemic liver lobes in experiments with simultaneous hepatectomy depends on the severity of ischemia and the proportion of resected tissue. According to our data, 2-h ischemia combined with 30% hepatectomy produced an additive effect on regeneration, stimulating DNA synthesis. In the case of 70% hepatectomy, the additive effect was only achieved after no more than 30–40 min of ischemia [Bolitho G., Engelbrecht G., Lotzz et al.; Maruyama H., Harada A., Kurokawa T. et al.]: 60-min ischemia produced no such effect, and 90–min ischemia inhibited regeneration [Maruyama H., Harada A., Kurokawa T. et al.]. After 80% hepatectomy, 10-min ischemia was sufficient for inhibiting regeneration [Foschi D., Castoldi L., Lesma A. et al.].

Fourth, the relationship between the rates of DNA and adenine nucleotide synthesis deserves special attention. As noted above, regenerating cells demonstrate higher metabolic activity. After partial hepatectomy, the rates of ATP and DNA synthesis in the liver strongly correlated with one another [Karasawa A., Guo J.P., Ma X.L. and Lefer A.M.]. In experiments with the models of 30-min ischemia and reperfusion of left and median liver lobes without (model B) or with simultaneous occlusion or removal of the right lobe (models C and D), the recovery of blood flow and ATP synthesis in the ischemic lobes in model B proceeded significantly slower and less successful than in models C and D, i.e., the "stealing phenomenon" was observed [Canada A.T., Stein K., Martel D. and Watkins W.D.]. This is consistent with our data concerning more active DNA synthesis in experiments on ischemia combined with 30% hepatectomy, compared to experiments without hepatectomy (see Fig. 74). Apparently, not only the functional load on ischemic lobes, but also more complete recovery of blood flow and ATP synthesis in them provided for more active DNA synthesis in this case. After total ischemia of liver lobes for 30, 60, and 90 min with simultaneous 70% hepatectomy, the direction of changes in regeneration activity (increase, no change, or decrease) correlated with the dynamics of ATP synthesis [Maruyama H., Harada A., Kurokawa T. et al.].

Thus, ischemia of organs leads to DNA damage, which is followed by DNA repair at an organ-specific rate. The magnitude of these responses depends on the severity of ischemia, cell sensitivity to oxygen deficiency, regeneration capacity, and critical mass of the organ.

9.2 Roles of Cytokines and Growth Factors in DNA Damage and Repair during Acute Ischemia and after Reperfusion of Organs

Cell reentry into the cycle after the G_0 (nonproliferative) state is now attributed to the effect of soluble mitogens (growth stimulators), and the return to the G_0 state, to the effect of growth inhibitors.

Two groups of substances regulating DNA synthesis, cell proliferation, and organ regeneration after injury deserve special attention: cytokines and growth factors. We shall consider each group with regard to its role in ischemic damage and postischemic reparative DNA synthesis.

9.2.1 Cytokines

Tumor necrosis factor-alpha (TNF-α, or cachectin) and interleukins 1–8 (IL 1–8) are of special importance. These polypeptides are mainly secreted by activated macrophages and, in addition, by cardiac myocytes, endothelial cells, and fibroblasts, in contrast to TNF-β (lymphotoxin) secreted by lymphocytes. At low and moderate concentrations, they produce an immunomodulatory effect, stimulate gene expression, and have an antitumor activity (TNF-α) [Dinarello C.A.]. Low concentrations of TNF-α also stimulate the growth of normal and malignant cells *in vitro*. This effect depends on the state of target cells and interactions between TNF-α and other growth factors [Schepetkin I.A.]. High concentrations of cytokines always have a strong cytotoxic effect, which is manifested in the appearance of single- or double-strand DNA breaks and dose-dependent cell death by apoptosis or necrosis [Higuchi H. and Aggarwal B.B.; Laster S.M., Wood J.G and Gooding L.R.; Wong G.H.W. and Goeddel D.V.; Zhang W., Naomoto Y., Tanaka N. et al.]. TNF-α is especially active in inducing apoptosis of cardiac myocytes [Meldrum D.R.; Squadrito F., Altavilla D., Zingarelli B. et al.] and endothelial cells [Spyridopoulos I., Brogi E., Kearney M. et al.; Polunovsky V.A., Wendt C.H., Ingbar D.H. et al.].

Mechanisms responsible for toxic effects of cytokines on DNA and cells are being extensively studied. The toxicity of TNF-α is mediated, in particular, by an increase in the rate of free-radical reactions and the release of products of lipid metabolism. It was shown that stimulation of TNF-α receptors or exogenous TNF-α administration result in the production of ROS and accumulation of LPO products [Yamauchi N., Kuriyama H., Watanabe N. et al.; Zhizhina G.P., Karobko V.G. and Alessenko A.V.; Schulze-Osthoff K., Bakker A.C., Vanhaesbroek B. et al.], early activation of acid and neutral sphingomyelinase in the nucleus, and the accumulation of sphingomyelin degradation products (ceramide and sphingosine), which stimulate apoptosis [Jarvis W.D., Kolesnick R.N., Fornari F.A. et al.; Kim M.Y., Linardic C.M., Obeid L.M. and Hannun Y.A.; Rusakov S.A., Filippova G.N., Pushkareva M.Yu. et al.; Obeid L.H., Linardic C.M., Karolak L.A. et al.; Zhizhina G.P., Karobko V.G. and Alessenko A.V.]. Apoptosis induced by TNF-α could be prevented by treatment with antioxidants, compounds with thiol groups, and GSH [Satomi N., Sakurai A.,

Haranaka R. et al.; Watanabe N., Kuriyama H., Sone et al.]. The sensitivity of tissues to TNF-α inversely correlated with the content of Mn-SOD in them [Wong G.H.W., Elwell J.H., Oberley L.W. and Goeddel D.V.]. The activation of AA metabolism, accompanied by an increase in the contents of AA and its lipooxygenase products, such as HPETE, 5-HPETE, and 15-HPETE, also belongs to mechanisms mediating the effects of TNF-α [Chang D.J., Ringold G.M. and Heller R.A.; Hayakawa M., Ishida N., Takeuchi K. et al.; Jayadev S., Linardic C.M. and Hannun Y.A.; Matthews N., Neale M.L., Jackson S.K. and Stark J.M.]. Inhibitors of phospholipase A_2 [Suffys P., Beyoerk R., Van Roy F. and Fiers W.; Neale M.L., Fiera R.A. and Matthews N.] and cyclooxygenase [Reid T.R., Torti F.M. and Ringold G.M.; Kettelhut I.C., Fiers W and Goldberg A.L.] decrease the cytotoxicity of TNF-α.

The release of IL-8 proved to be inhibited by oxygen radical scavengers and stimulated by H_2O_2 in a dose-dependent mode [De Forge L.E., Fantone J.C., Kenney J.S. and Remick D.G.], which suggests that an important role in cytokine production and release belongs to free radicals.

There is ample evidence that *ischemia and subsequent reperfusion lead to an increase in the production of cytokines* that cause organ damage. The level of TNF-α in the peripheral blood increased after 90–120-min ischemia and reperfusion of the liver [Colletti L.M., Remick D.G., Burtch C.D. et al.; Scales W.E., Campbell D.A. Jr., Green M.E. and Remick D.G.], 45-60 min ischemia of the liver or kidneys [Daemen M. A., van de Ven M. W., Heineman E. and Buurman W. A.; Sakr M.F., McClain C.J., Gavaler J.S. et al.; Sakr M., Zetti G., McClaio C. et al.; Van-Thiel D.H., Sakr M., Zetti G. and McClain C.], 120-min ischemia of the intestine [Caty M.G., Guice K.S., Oldham K.T. et al.], cardiopulmonary bypass in dogs [Dauber I.M., Parsons P.E., Welsh C.M. et al.], and 24-h occlusion of pulmonary arteries followed by lung reperfusion for 45–60 min [Palace G.P., Del-Vecchio P.J., Horgan M.J. et al.].

The increase of TNF-α, IL-1 and IL-6 was found in the blood flowing out of ischemic rat limbs after 2 h of reperfusion [Strenbergh W.C. 3rd, Tuttle T.M., Makhoul R.G. et al.; Seekamp A., Warren J.S., Remick D.G. et al.]; at the sites of coronary heart microembolia in pigs and rats [Zimmermann R., Schaper J., Mџnkel B. et al.]; in the area of supplied with blood by an occluded coronary artery, before and after the removal of occlusion and in the effluent blood [Arras M., Mohri M., Pich S. et al.; Squadrito F., Altavilla D., Zingarelli B. et al.]; in the endothelial cells of the ischemic heart proper [Frolkis J., Gurevitch J., Yuhas Y.; Kupatt C., Habazettl H., Goedecke A. et al.; Murota S., Nishida M. and Morita J.] and in Kupffer cells of ischemic rat liver [Wanner G.A., Muller P., Ertel W. et al.]. An increase in the level of IL-1 but not TNF-α was found in the effluent blood after reperfusion of ischemic dog muscles [Ascer E., Gennaro M., Cupo S. and Mohan C.]; IL-6, IL-1, and IL-10 (but not TNF-α) in circulating blood significantly increased in surgical patients with tourniquets applied for an average of 105 min [Seekamp A., Jochum M., Ziegler M. et al.]. Conversely, an increase in production of TNF-α, rather than IL-2, was observed after ischemia and reperfusion of the rat intestine [Cohen S., Lahat N., Kinarty A. and Bitterman H.]. An increase in the levels of cytokines (TNF-α, IL-1, and IL-6) occurred prior to the infiltration of ischemic and reperfused organs (lungs or limbs) by neutrophils and the increase of tissue myeloperoxidase activity [Palace G.P., Del-Vecchio P.J., Horgan M.J. et al.; Seekamp A., Warren J.S., Remick D.G. et al.]. The release of IL-6 after 75-min intestinal ischemia and 6-h reperfusion was TNF-dependent [Yao Y.M., Bahrami S., Redl H. et al.].

Treatment with anti-TNF-α immune serum or an inhibitor of TNF-α synthesis considerably increased the survival rate of rats after 60-min LAD occlusion, reduced the infarct area, and suppressed myeloperoxidase activity in the heart [Squadrito F., Altavilla D., Zingarelli B. et al.]; improved the contractile function of the rat heart after 1-h total ischemia and 30-min reperfusion [Gurevitch J., Frolkis I., Yuhas Y. et al.]; and protected rats against muscle and lung injuries after ischemia/reperfusion of their hind limbs [Schulze-Osthoff K., Bakker A.C., Vanhaesbroek B. et al.]. An additional factor contributing to the role of TNF-α in ischemic-reperfusion damage is a sharp decrease of protein synthesis during ischemia [Fornander J., Hellman A. and Hasselgren P. O.; Kiessling M., Dienel G.A., Jacewicz M. and Pulsunelli W.A.M.; Nowak T.S., Fried R.L., Lust W.D. and Passonneau J.V.] because TNF-α-induced apoptosis develops more intensely when protein synthesis is inhibited [Pohlman T.H. and Harlan J.M.; Polunovsky V.A., Wendt C.H., Ingbar D.H. et al.].

Evidence for the *involvement of cytokines in organ regeneration* after damage is scarce and indirect. It was shown that TNF-α production in the culture of rat liver Kupffer cells isolated 24 h after 70% hepatectomy is obviously decreased but increases again after adding indomethacin, a cyclooxygenase inhibitor, to the culture medium. This fact suggested that cyclooxygenase metabolites of AA may regulate the production of TNF-α after partial hepatectomy [Callery M.P., Kamei T. and Flye M.W.].

Recently, it was shown that IL-6 prevents ischemia/reperfusion-induced injuries in the liver and promotes its regeneration [Camargo C.A. Jr., Madden J.F., Gao W. et al.]. This is consistent with the data on a considerable decrease in the levels of IL-1, IL-6, and TNF mRNAs in the liver of patients with primary biliary cirrhosis and other forms of the autoimmune disease that require intensification of liver regeneration [Tovey M.G., Gugenheim J., Guymarho J. and Blanchard B.].

9.2.2 Growth Factors

To date, numerous growth factors stimulating regeneration have been described. They include hepatopoietin B (HPTB); hepatopoietin A (HPTA), (referred to as hepatic growth factor, HGF); scatter factor (SF); basic fibroblast growth factor (bFGF) (alternatively designated as heparin-binding growth factor, HBGF); epidermal growth factor (EGF); platelet-derived endothelial growth factor (PD-EGF); vascular endothelial growth factor (VEGF); transforming growth factors alpha and beta (TGF-α and TGF-β); and many other factors.

These factors do not induce an increase in the rate of DNA synthesis in resting cells *in vivo* but act as active mitogens on cell populations prepared for proliferation [Michalopoulos G., Houck K.A., Dolan M.L. and Loutteke N.C.]. Growth factors are polyfunctional, and their organ specificity is conditional. Nevertheless, PD-EGF, EGF, TGF-α, and bFGF are the most effective mitogens for muscle cells; PD-EGF, bFGF, and VEGF, for endothelial cells; and HGF, EGF and TGF-α, for hepatocytes.

The involvement of growth factors in organ regeneration after damage has been most extensively studied with respect to HGF. This factor was isolated for the first time from blood serum of hepatectomized mice, and its mitogenic properties were discovered in experiments with rat hepatocyte cultures [Michalopoulos G., Houck K.A., Dolan M.L. and Loutteke N.C.]. Thereafter, HGF has been isolated from other sources, purified, and partially characterized. This factor is a heparin-binding glycoprotein consisting of a heavy (70 kDa) and a light (35 kDa) polypeptide chains linked by disulfide bonds [Zarnegar R., Muga S., Rahija R. and Michalopoulos G.; Matsumoto K. and Nakamura T. (1992)]. Its biological activity, mechanism of action, and specific receptors are identical to those of SF isolated from fibroblasts [Naldini L., Weidner K.M., Vigna E. et al.], and, hence, these factors are often referred to as HGF/SF.

Immunohistochemical determinations of HGF in humans, rats, and rabbits revealed its presence in organs differing in regeneration activity, and the patterns of HGF distribution in rat and human organs were nearly identical [Wolf H.K., Zarnegar R. and Michalopulos G.K.]. Specific [^{125}I]HGF binding to corresponding plasma membrane receptors was found in the liver, spleen, kidneys, lungs, adrenals, salivary glands, and the thyroid gland. The density of binding sites varied from 400 to 3000 per nanogram plasma membrane protein, being maximal in the liver.

Many studies demonstrated that HGF is involved in liver regeneration after resection and CCl_4 poisoning. In the case of 70% hepatectomy, HGF proved to be of extrahepatic origin, being synthesized in other organs, most probably in the lungs and kidneys [Matsumoto K. and Nakamura T. (1991)]. After CCl_4 poisoning, both the liver and other organs produced HGF, and the rate of HGF synthesis in the liver directly correlated with the degree of organ damage [Lindroos P.H., Zarnegar R. and Michalopoulos G.K.]. Likewise, no increase of HGF content in partially resected liver but a considerable (4.5-fold) increase in the content of HGF mRNA in the liver of rats poisoned with CCl_4 were revealed [Okajima A., Hiyazawa K. and Kitamura N.]. The HGF gene transcripts were detected in Kupffer and endothelial cells of the normal and CCl_4-treated liver but not in the resected liver [Noji S., Tashiro K., Koyama E. et al.].

There are ample data on the activity of the so-called hepatic stimulator substance, alternatively designated proliferation-stimulating extract (PSE), which is prepared from homogenates of newborn animal liver, liver tissue remaining after 70% hepatectomy, or liver explants from adult animals: it stimulated regeneration in the liver after its partial (30-70%) or even subtotal (80%) resection [Abakumova O.Yu., Kotaev A.Yu., Karagyulyan S.R. et al.; Abakumova O.Yu., Kutsenko M.G., Karagyulyan S.P. et al.; La Brecque D.R. and Pesch L.A.; Starzl T.E., Terblanche J., Porter K.A. et al.; Galperin E.I., Karagyulyan S.R. and. Abakumova O.Yu. et al.], toxic injury [Abakumova O.Yu., Kotaev A.Yu., Karagyulyan S.R. et al.], traumatic crushing [Galperin E.I., Karagyulyan S.R. and. Abakumova O.Yu. et al.], and portacaval shunting [Starzl T.E., Terblanche J., Porter K.A. et al.]. However, the effects of PSE on DNA synthesis and repair in the ischemic and reperfused liver remained unstudied.

In our laboratory, PSE was prepared from the rat liver 48 h after 70% resection (PSE_{res}) and from 2-h ischemic left and median lobes of rat liver after 30% resection nonischemic lobes and 48-h reperfusion of remaining lobes (PSE_{isch}) [Bilenko M.V., Seregina L.A., Komarov P.G. and Abakumova O.Yu.], according to the modified La Bracque method [La Brecque D.R. and Pesch L.A.].

The effect of PSE_{res} on DNA synthesis in the ischemic liver was studied in experiments with male Wistar rats subjected to 2-h ischemia and reperfusion of the left and median liver lobes (70% ischemia) without (group 2) or with (group 3) simultaneous resection of the right and caudal lobes (30% hepatectomy). Control rats (group 1) were subjected to 30% hepatectomy [Bilenko M.V., Seregina L.A., Komarov P.G. and Abakumova O.Yu.].

Figure 75a shows that PSE_{res} stimulated DNA synthesis (apparently, replicative) in the liver of control rats (group 1, 30% hepatectomy): its rate increased two- to fourfold after 24–48 h of reperfusion. After 2-h ischemia without 30% hepatectomy (group 2), PSE_{res} stimulated DNA synthesis in both intact (Fig. 75b) and ischemic (Fig. 75c) lobes. These effects were expressed more strongly but developed later than in group 1 (a five- to sevenfold increase after 48 h). In rats subjected to 2-h ischemia of two liver lobes with simultaneous 30% hepatectomy (group 3), PSE_{res} did not induce any additional increase in DNA synthesis, compared to the control without PSE_{res} (Fig. 75d). Note that the rate of DNA synthesis in the liver of these rats was higher than in any other groups owing to a strong stimulating effect of ischemia combined with 30% hepatectomy, irrespective of PSE_{res} treatment, which is consistent with the results of our previous experiments (see Fig. 74).

Comparing the effects of PSE_{res} and PSE_{isch} obtained by the same methods and tested under similar conditions on the model of 30% hepatectomy, it was found that the stimulating effect of PSE_{isch} is significantly stronger than that of PSE_{res} (Fig. 76a). Neither of the two PSE preparations caused any apparent effect on the ischemic and resected liver (Fig. 76b), probably because liver lobes exposed to long-term ischemia (2 h) combined with a high functional load after 30% hepatectomy reached their regeneration limit. This issue will be discussed in more detail in the following sections.

Taken together, these data demonstrated that the liver exposed to long-term ischemia produces factors stimulating DNA synthesis. Their chemical composition is as yet unknown, and it remains to be elucidated whether or not they belong to HGF or other growth factors.

According to experimental data, the accumulation of monocytes and their adhesion to endothelium followed by transformation into Kupffer cells are strongly increased in reperfused liver [Sokirchenko I.A.], and Kupffer cells produce HGF in the liver upon CCl_4 poisoning [Noji S., Tashiro K., Koyama E. et al.] or ischemia [Hamanoue M., Kawaida K., Takao S. et al.] (see above). Taking this into account, we analyzed the role of activated liver macrophages in the formation of PSE.

In this study [Fedorova L.M., Arkhangelskaya S.L. and Bilenko M.V.], liver macrophages were stimulated with prodigiosan, a lipopolysaccharide promoting the growth of Kupffer cell population and accelerating the peak of proliferation in the rat liver after 70% hepatectomy [Mayansky D.N.]. Polysaccharides can also prevent DNA damage and stimulate DNA synthesis in cultured lymphocytes [Frolov A.F., Varbanets L.D., Antonenko S.V. et al.]. Prodigiosan (0.25 mg/kg) was injected intraperitoneally to unoperated rats or rats subjected to 70% hepatectomy, but after 24 and 48 h we did not observe any increase of DNA synthesis in liver nuclei (data not shown). However, the liver of these rats proved to be a source of highly active PSE. Figure 77 shows that the effects of PSE isolated from livers of nonoperated rats treated only with prodigiosan (PSE_{prod}) and especially from rats subjected to 70% hepatectomy combined with prodigiosan treatment (PSE_{comb}) were significantly stronger than the effect of PSE_{res} ($P_2 < 0.01$). Compared to PSE_{isch}, the effect of PSE_{prod} was similar, and that of PSE_{comb} was stronger (compare Figs. 77 and 76). PSE_{prod} and PSE_{comb} obtained 24 h after corresponding treatment were more active than those obtained after 48 h (data not shown).

Fig. 75: Effect of PSE_{res} on the DNA synthesis in rat liver nuclei after 30% resection and 2 h ischemia of liver (% of the initial level).

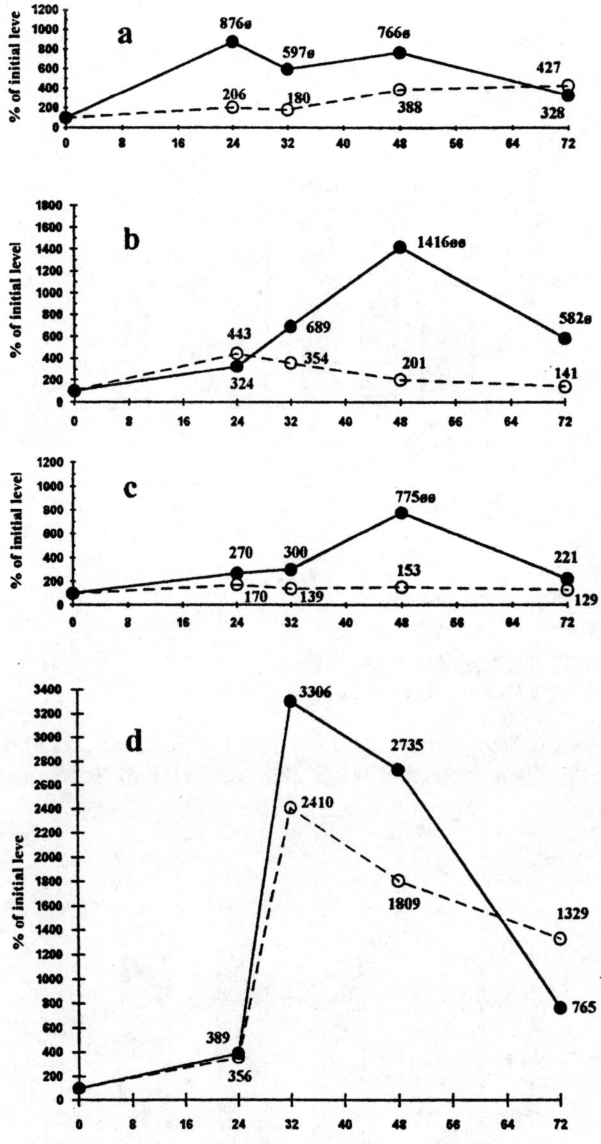

a — intact lobes in rats with 30% resection of liver, (control, group 1);
b — intact lobes in rats with 2 h ishemia of two liver lobes, (group 2);
c — 2 h ischemic lobes of the same liver, (group 2);
d — 2 h ischemic lobes in rats with ischemia + 30% resection of liver, (group 3).
1 — without PSE_{res}: 2 — with PSE_{res}.
ø — significant differences between 1 and 2: ø < 0.05; øø <0.01.
PSE_{res} was obtained from rat liver lobes 48 h after 70% resection by method [La Brecque D.R. and Pesch L.A.] in modification [Abakumova O.Yu., Kotaev A.Yu., Karagyulyan S.R. et al.].

Brifly, liver homogenates were centrifuged in 4 ^{0}C 15 min under 23000 g. Supernatants were warmed 15 min to 100 ^{0}C, centrifuged again, sterilized by 0.45 μ filter and preserved under –20 ^{0}C.

PSE_{res} were injected i. p. 4 ml/rat just after operation. The rate of DNA synthesis estimation is the same as in Fig. 74.

Fig. 76: Comparison of PSE_{res} and PSE_{isch} effects on the DNA synthesis in rat liver nuclei 24 h after 30% resection (a) and in rat liver nuclei 48 h after 2 h ischemia plus 30% resection of liver (b) (% of the initial level).

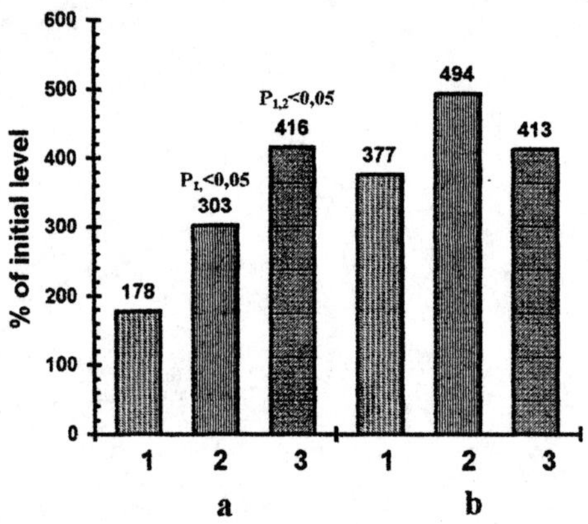

1 — without PSE (control); 2 — PSE_{res};3 — PSE_{isch}.
P_1 — significant differences with control;
P_2 — significant differences between 3 and 2.
PSE obtaining and injection are the same as in fig. 75
The rate of DNA synthesis estimation is the same as in fig. 74.

Fig. 77: Comparison of PSE_{res}, PSE_{prod} and PSE_{comb} effects on the DNA synthesis in rat liver nuclei 24 h after 30% resection of liver(mean ± SEM).

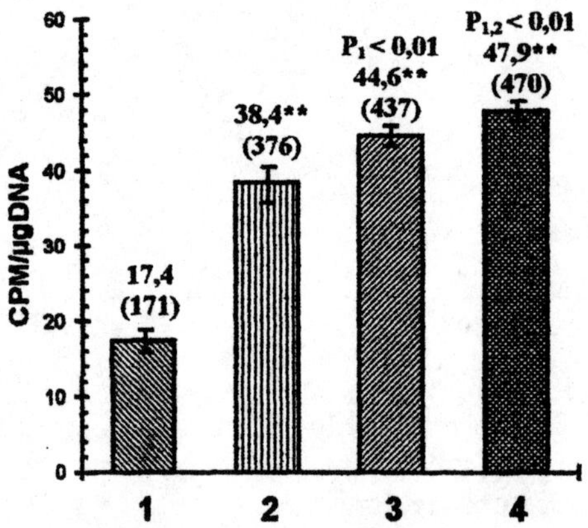

1 — without PSE (control); 2 — PSE_{res}; 3 — PSE_{prod}; 4 — PSE_{comb}.
In parenthesis — % of the initial level.
** — $p < 0.01$, versus control;
P_1 — versus 2; P_2 — versus 3.
PSE_{prod} was obtained from intact rat liver 24 h after prodigiosan (0.25 mg/kg) i. p. injection to rats; PSE_{comb} was obtained from rat liver 24 h after 70% hepatectomy and prodigiosan (0.25 mg/kg) i. p. injection. Methods of PSE preparation and rated of DNA synthesis evaluation are the same as in fig. 74 and fig. 75

These results suggest that the increase in the number of macrophages and their activation in ischemic and reperfused organs (see Chapter 10) may be one of the factors providing for the production of effective PSE_{isch} and the increased rate of DNA synthesis in both ischemic and intact liver lobes (see Fig. 75).

Extrapolating these conclusions to situations *in vivo,* one should take into account that the organ produces factors both stimulating regeneration (HGF) and inhibiting it (TNF-α, etc.), and that the ratio between these factors depends on the degree of damage to the organ. PSE preparations used in this study were heated at 100°C, and, hence, protein factors responsible for their stimulating effect on regeneration should be thermostable. Their exact composition remains unknown, although polyacrylamide gel electrophoresis showed that PSE_{res} contained thermostable proteins with molecular weights of 10 to 70 kDa [Bilenko M.V., Seregina L.A., Komarov P.G. and Abakumova O.Yu.], which is similar to HGF [Zarnegar R., Muga S., Rahija R. and Michalopoulos G.; Asami O., Ihara I., Shimidzu N. et al.]. PSE_{prod} prepared from the liver of nonoperated rats treated with prodigiosan 24 h before sacrifice was separated by ion exchange chromatography, and the resulting fractions were tested on rats subjected to 30% hepatectomy. The results showed that virtually entire initial PSE_{prod} activity was eluted with fraction 1 (Fig. 78). Electrophoresis of this fraction in 7.5% polyacrylamide gel revealed a set of protein bands with molecular weights of 17–100 kDa (unpublished data), but their similarity to or difference from HGF remains to be elucidated.

Fig. 78: Effect of PSE_{prod} and its fractions on the DNA synthesis in rat liver nuclei 24 h after 30% resection of liver (mean ± SEM). [Fedorova L.M., Arkhangelskaya S.L. and Bilenko M.V.].

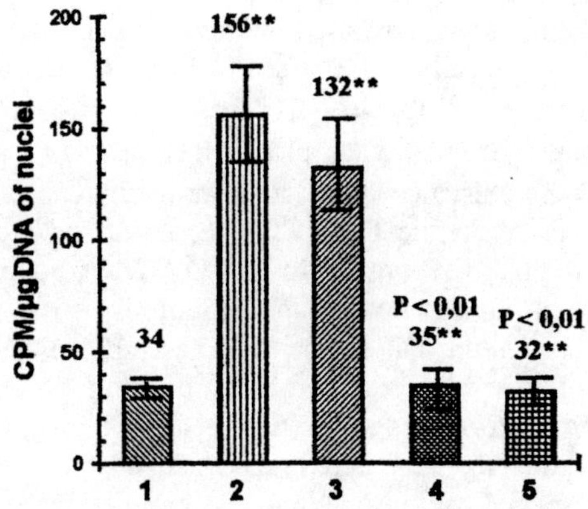

1 — without PSE (control); 2 — PSE_{prod}; 3 — first fraction of PSE_{prod}; 4 — second fraction of PSE_{prod}; 5 — third fraction of PSE_{prod}.
* — $p< 0.05$; **p — < 0.01 versus control P — significant differences versus 2.
The fractions of PSE_{prod} were separated by ion-exchange chromatography using DEAE-tris acril. Elution was performed with NaCl gradient (0 — 0.05 M) at pH 7.5.
Methods of PSE_{prod} preparation, the rate of DNA synthesis estimation are the same as in fig. 77 and fig. 74.

In addition to stimulating DNA synthesis, growth factors produced a strong protective effect on ischemic organs. Thus, bFGF was shown to protect endothelial cells from radiation-induced apoptosis [Fuks Z., Persaud R.S., Alfieri A. et al.] and to alleviate ischemic/reperfusion injury in the rat intestine and liver [Fu X., Sheng Z., Wang Y. et al.]; EGF protected gastric mucosa from such injury [Ishikawa T., Sarfeh J., Tarnawski A. et al.]; addition VEGF to cardioplegic solution protected rat heart against ischemia-reperfusion injury [Luo Z., Diaco M., Murohara T. et al.]; TGF-β treatment led to a decrease in the size of ischemic zone in rabbit and mouse brain [Gross C.E., Bednar M.M., Howard D.B. and Sporn M.B.; Prehn J.H., Backhauss C. and Krieglstein J.], inhibited myeloperoxidase activity in the ischemic myocardium and

$O_2^{\cdot-}$ production in this tissue [Lefer A.M., Ma X.L.; Lefer A.M., Tsao P., Aoki N. et al.], and increased the survival rate of cats subjected to intestinal ischemia, relieving the no-reflow phenomenon [Karasawa A., Guo J.P., Ma X.L. and Lefer A.M.; Thomas G.R. and Thibodaux H.]. TGF-β also protected the heart from damage caused by exogenous or endogenous TNF-α and IL-1 [Lefer A.M., Ma X.L., Weyrich A.S. et al.; Nose P.S.; Roberts A.B., Roche N.S., Winokur T.S. et al.]. Myocardial and cerebral ischemia stimulated the expression of TGF-β mRNA [Klempt N.D., Sirimanne E., Gunn A.J. et al.; Sharma H.S., Wuensch M., Brand T. et al.].

Therefore cytokines and growth factors are actively involved in processes of damage and repair of ischemic organs. Cytokines mainly cause DNA damage that results, depending on its severity, in either cell necrosis or apoptosis followed by regeneration. The effects of cytokines on regeneration are mainly mediated by activation of macrophages and stimulation of free-radical processes. Growth factors, by contrast, apparently have direct and positive effects on DNA synthesis and cell proliferation.

9.3 Role of Free Radicals and Lipid Peroxidation in DNA Damage and Repair during Acute Ischemia and after Reperfusion of Organs

As shown in Chapter 5, acute ischemia and reperfusion of organs lead to activation of free-radical processes. Hence, the involvement of LPO and its products in DNA damage and subsequent repair in ischemic organs deserves special attention.

The hypothesis concerning the role of free radicals and LPO products in DNA damage and the development of apoptosis in ischemic organs is based on the data that ischemia and reperfusion of organs are accompanied by an increase in intracellular Ca^{2+} (see Chapter 2), low-molecular Fe^{2+}, or “”free” Fe^{2+} (see Chapter 10), and in production of TNF-α—the factor implementing its effect via free-radical reactions—by the ischemic organ proper or by macrophages it contains (see section 9.2).

The main clue, however, is the existence in the nuclear membrane of a DNA–lipoprotein complex sensitive to LPO [Comings D.E. and Kakefula T.; Jacksov V., Earnhardt J. and Chalkley R.; Osipov A.N., Yakutova E.Sh. and Vladimirov Yu.A.]. From 75 to 85% of DNA-bound lipids are easily oxidizable (cardiolipin and PEA); their amount increases when the genome is active, especially in the S phase, and decreases when the genome is repressed and in the G_2 phase of the cell cycle [Struchkov V.A. and Strazhevskaya N.B.]. Because of this, changes in the composition of DNA-bound lipids and nuclear membranes can have a strong effect on DNA damage and repair [Cole A., Cooper W.G., Shonka F. et al.; Comings D.E. and Kakefula T.; Struchkov V.A. and Strazhevskaya N.B.].

Among membrane lipids, a special role in regulation of cell division belongs to sphingomyelin. The contents of sphingomyelin and sphingosine, its metabolite, were shown to increase simultaneously with the peak of DNA synthesis in liver cells after hepatectomy [Alessenko A.V., Burlakova E.B. and Pantaz E.A.; Alessenko A.V., Pantaz E.A., Pushkareva M.Ya. et al.]. Low concentrations of exogenous sphingomyelin stimulate cell proliferation, potentiate effects of growth factors, and regulate activities of enzymes responsible for DNA replication and transformation [Zhestyannikov V.D.].

Notwithstanding the above reasons, the involvement of free-radical processes in DNA damage and repair was mainly studied on isolated cells and intact or regenerating organs, whereas the data on ischemic/reperfused organs are scarce.

Experimental research in this field mainly concentrated on (1) effects of free radicals and agents inducing their formation *in vivo* and *in vitro* on DNA synthesis and reparative processes in the cell, (2) parameters characterizing the activity of free-radical processes at different phases of the cell cycle (ROS, AOA, antioxidant enzymes, LPO products), and (3) effects of antioxidants on the rates of DNA damage and repair *in vivo*.

In studies attributed to *the first group,* it was shown that oxygen at increased pressure, other oxidants (NO, alkylating agents), and ionizing radiation reduce the rate of cell metabolism and inhibit the growth of fibroblasts and hemopoietic cells [Albina J.E., Cui S, Mateo R.B. et al.; Bradley T.R., Hodgson G.S. and Rosendaal M.; Buttke T.M. and Sandstrom P.A.]; cumene hydroperoxide reversibly inhibits DNA and protein synthesis [Poot M.]; paraquat and other prooxidants have effect on DNA and induce mutations, especially in SOD-deficient cells and bacteria [Beyer W., Imlay J. and Fridovich I.]. Incubation of various cells with high concentrations of H_2O_2 resulted in their necrosis [Lennon S.V., Martin S.J. and Cotter T.G.; Burdon R.H.], whereas low H_2O_2 concentrations [Murrell G.A.C., Francis M.J.O. and Bromley L.; Burdon R.H., Gill V. and Rice-Evans C.] xanthine oxidase [Spragg R.G.], or moderate Fe^{2+}-ascorbate-induced LPO [Binkova B., Topinka G., Sram R. et al.] stimulated apoptosis and subsequent reparative DNA synthesis in cell cultures, and α-tocopherol partially inhibited this stimulation [Binkova B., Topinka G., Sram R. et al.]. In endothelial cells of the bovine pulmonary artery (or the aorta), the degree of DNA damage (strand breakage) remained approximately the same after their exposure to 50 μM H_2O_2, 25 mU glucose oxidase (H_2O_2 generator), 10 mU xanthine oxidase ($O_2^{\bullet-}$ generator), or X-radiation at a dose of 10 Gy [Spragg R.G.].

The effect of radiation on DNA synthesis proved to be dose-dependent. Irradiation at medium doses stimulated DNA synthesis in HeLa cells, lymphoblasts [Smirnova S.G., Synzynys B.I., Zamulaeva I.A. et al.], and the intact rat liver [Sukolinsky V.N., Konoplya E.F., Morozkina T.S. et al.]. Low and medium doses of radiation, in contrast to high doses, enhanced liver regeneration after 30% hepatectomy [Bossola M., Merrick M.W., Eltaki A. et al.].

The end products of LPO also had a strong damaging effect on DNA: 4-hydroxypentanal and 4-hydroxynonenal inhibited the growth of Ehrlich ascites carcinoma cells, Yoshida ascites hepatoma cells, human bronchial fibroblasts, and hepatocytes or even killed them [Hauptlorenz S., Esterbauer H., Moll W. et al.; Krokan H., Grafstrom R.C., Sundqvist K. et al.; Wawra E., Zollner H., Schaur R.J. et al.]; under the effect of 4-hydroxynonenal, the duration of G_1 phase increased from 25 to 35 h, and the proportion of dividing cells sharply decreased [Poot M.].

Apoptosis can also be induced by oxidized fatty acids, oxidized cholesterol derivatives, and 15-HPETE [Christ M., Luu B., Mejia J.E. et al.; Sandstrom P.A. and Buttke T.M.; Sandstrom P.A., Tebbey P.W., Cleave S.V. et al.]. TNF-α, whose effects are mediated by free-radical processes (see above), is a potent inducer of apoptosis. DNA damage by free-radical species most probably occurs via fatty acid oxidation and modification of the nuclear membrane; however, apoptosis may also be genetically programmed.

Tissues and organs with low levels of GSH, catalase, and SOD are more susceptible to apoptosis than tissues rich in antioxidant enzymes [Zhang Y.H., Takahashi K., Zjang G. et al.; Sandstrom P.A. and Buttke T.M.; Wong G.H.W., Elwell J.H., Oberley L.W. and Goeddel D.V.]. Injection of exogenous antioxidants, such as DMSO, Fe^{2+} chelators, and vitamin E, prevent apoptosis in cells treated with prooxidants [Forrest V.J., Kang. Y.-H., McClain D.E. et al.; Abello P.A., Fidler S.A., Bulkley G.B., et al.].

Emotion–pain stress and surgical stress, which are known to promote the production of free radicals (see Chapter 7), led to the appearance of DNA breaks in cardiac myocytes and to an early increase in the rates of replicative and reparative DNA synthesis [Meerson F.Z. and Vasil'ev V.K.; Meerson F.Z., Vavich M.P. and Rozhitskaya I.I.; Meerson F.Z., Vavich M.P., Rozhitskaya I.I. and Golubeva L.Yu.]. In the rat liver, by contrast, the same types of stress stimulated DNA repair but suppressed replication [Meerson F.Z., Vavich M.P. and Rozhitskaya I.I.]. Similar but weaker effects were observed under surgical stress in the rat brain [Yavich M.P., Rozhitskaya I.I., Golubeva L.Yu. et al.]. The data concerning DNA damage under oxidative stress (single- and double-strand breaks, chromosome aberrations) in different mammalian cell types were reviewed in [Halliwell B. and Aruoma O.J.].

The role of transition metals — iron (Fe^{2+}/Fe^{3+}), copper (Cu^{+}/Cu^{2+}) and chromium (Cr^{2+}/Cr^{3+} and Cr^{5+}/Cr^{+6}) — in the formation of DNA damage has also been shown [Stohes S.J. and Bagchi D.; Wolfe J.T., Ross D., Cohen G.M.].

Experiments of *the second group* were mainly performed on the liver resected by 70%, and free-radical processes were monitored at different stages of the mitotic cycle. Experiments on mice revealed the bimodal dynamics of AOA, with the first peak preceding DNA synthesis, and the second peak preceding mitosis. At the S phase (DNA synthesis), AOA decreased to negative values, which was regarded as evidence for a temporal increase in the rate of free-radical processes [Alessenko A.V. and Burlakova E.B.].

However, in the regenerating rat liver (24 h after a 30 or 70% hepatectomy), the rates of superoxide radical formation in microsomes (Ms), submitochondrial particles (SMP), and cell nuclei decreased, whereas the activities of Mn-SOD and Cu, Zn-SOD remained unchanged [Vartanyan L.S., Sadovnikova I.P., Gurevich S.M. and Sokolova I.S.]. The rates of superoxide radical generation ($V_{O_2^{\bullet-}}$) in Ms and SMP remained low for 4–4.5 days after 70% hepatectomy, whereas in the nuclei this index sharply increased. The activity of cytosolic Cu, Zn-SOD decreased considerably on the second and third days, that is, during the phase of mitosis [Oberley L.W., Bize J.B., Sahus K. et al.; Vartanyan L.S., Sadovnikova I.P., Gurevich

S.M. and Sokolova I.S.]. This dynamics was regarded as evidence that superoxide radicals are not involved in DNA synthesis but rather participate in the decomposition and restoration of the nuclear membrane during karyokinesis [Vartanyan L.S., Sadovnikova I.P., Gurevich S.M. and Sokolova I.S.]. This is consistent with the data discussed above (see section 5.5) on the role of LPO processes in degradation of nonfunctional ER membranes in the liver [Kagan V. E., Kotelevtsev S. V. and Kozlov Yu. P.; Wildhirt S.M., Suzuki H., Horstman D. et al.] and mitochondrial membranes in reticulocytes [Sheve T., Khalanchik V., Khibish C. and Rapaport S.M.].

A decrease (rather than increase) in the rates of free-radical processes in the rat liver 24 h after 70% hepatectomy was determined by several other authors [Cockerill M.J., Player T.J. and Horton A.A.; Dwivedi R.S., Kaur G., Srivastava R.C. and Murti C.R.K.; Thomas C.E. and Jackson R.L.]. Treatment of such rats with antioxidants (α-tocopherol and ascorbic acid), improved the ATP-producing function of mitochondria, and stimulated DNA synthesis in the liver [Obata M.]. In aging mice, the extent of DNA damage inversely correlated with GSH content in the liver [Chaudhuri A., Medhekar T.P.; Stohes S.J., Laxson T.A., Anderson L. and Bulding E.]. Experiments with cell cultures revealed an increase in Cu, Zn-SOD or Mn-SOD activities at the S phase of the cell cycle [Oberley L.W., Bize J.B., Sahus K. et al.].

According to the recent hypothesis [Dmitriev L.F.], a major role in regulating DNA synthesis belongs to malonic dialdehyde, which can establish links between complementary DNA strands, thereby stabilizing DNA and inhibiting cell division [Dmitriev L.F.].

In *the third group* of studies, the effects of antioxidants on DNA synthesis in intact organs and in the liver subjected to 70% resection were examined. Treatment with low doses (0.8–2.8 mg/kg) of BHT led to an increase in liver weight and strongly stimulated DNA synthesis and mitotic activity of hepatocytes. Higher doses of BHT (30 mg/kg) accelerated hepatocytes' entry into the S phase of the cell cycle [Alessenko A.V., Sokolova I.S., Kukushkina G.V. et al.], and 100 mg/kg BHT increased the rate of DNA synthesis in the mouse liver subjected to 70% resection [Pulatova M.K., Todokov I.N., Kosaganova N.Yu. et al.]. BHT stimulated DNA synthesis in intact mouse lungs [Witschi H. and Saheb W.] but inhibited reparative DNA synthesis in irradiated human peripheral blood lymphocytes [Daugherty J.P., Davis S. and Yielding K.L.] and in monkey kidney cell cultures [Milner S.M.].

Keeping hepatectomized rats on the diet enriched with α-tocopherol and santoquin, a synthetic antioxidant, led to a decrease in the level of TBA-reactive substances and increased the rate of DNA synthesis in the liver three and six days after surgery [Gavino V.C., Dillard C.J. and Tappel A.L.]. Holotransferrin, a Fe^{2+} chelator, stimulated DNA synthesis in rat placenta cell culture [De Forge L.E., Fantone J.C., Kenney J.S. and Remick D.G.].

Certain lipid-soluble antioxidants also reduced the extent of natural DNA damage in the liver of aging mice [Lawson T. and Stohes S.], i.e., they directly stimulated DNA repair rather than protected DNA from damage upon aging.

Thus, the data discussed above provide evidence for the inverse relation between the intensity of free-radical reactions and DNA synthesis, i.e., suppression of the latter at a high free radical level and acceleration at a low level. The validity of this conclusion, at least for undamaged growing tissues, is confirmed by the fact that rapidly proliferating tissues and tumors are more resistant to LPO and have lower contents of lipid peroxides and higher contents of antioxidants than slowly growing tissues [Borrello S., Minotti G., Palombini G. et al.; Pal'mina N.P., Mal'tseva E.L., Kuznakova N.V. and Burla-kova E.B.]. However, there are exceptions to this rule: in hepatomas, for example, growth rate is inversely proportional to SOD activity [Bartoli G.M., Bartoli S., Galeotti T. and Bertoli E.].

The role of free-radical processes in DNA damage and repair in ischemic and reperfused organs is poorly understood, although there are data that DNA fragmentation (similar to that during apoptosis) and subsequent repair occur upon partial, subtotal, and total ischemia and reperfusion of most organs studied (see section 9.1). Studies on the effects of antioxidants on free-radical reactions and DNA synthesis in ischemic and reperfused organs could help in establishing the relationship between these processes, but such data are scarce. It was shown that BHT (50 mg/kg) injected 96, 48, and 24 h before or immediately after LAD occlusion in rats reduced the intensity of DNA fragmentation and, hence, the rate of reparative DNA synthesis in nonischemic (periinfarct) zone 24 h after acute myocardial infarction [Meerson F.Z., Vasil'ev K.V. and Dosmagambetova R.S.]. Thus, a decrease in the rate of LPO (and resulting DNA damage) reduced the rate of DNA repair.

To analyze in more detail the influence of free-radical processes on the rate of postischemic DNA synthesis, special series of experiments on the effects of different antioxidants on the ischemic and reperfused liver were performed in our laboratory [Seregina L.A.; Bilenko M., Seregina L., Komarov P. et al.; Liashenko S.M.; Liashenko S.M., Bilenko M.V. and Gefter Ju.V.].

Experimental Wistar rats were divided into three groups and subjected to (1) 30% hepatectomy, (2) 2-h ischemia of the left and median liver lobes (70% of the liver) without resection of nonischemic lobes, and (3) similar ischemia with simultaneous resection of the right and caudal lobes (30% of the liver). Antioxidants used in experiments had different mechanisms of action and produced a strong antiischemic effect (Fig. 79), as estimated from the rat survival rate for seven days after 2-h ischemia combined with 30% hepatectomy (percentage of survived rats is indicated in parentheses). By this criterion, antioxidants formed the following descending series: Se-methionine (93%) > diludin (88%) > 1-HTMP (50%) > K-phenosan and SOD (48 and 47%, respectively) > desferal (42%). In experiments without antioxidants (control), only 19% of rats survived for this period. Differences between the results of control and all experimental series were significant beginning from the first day ($p < 0.05–0.001$).

Fig. 79: Effect of the antioxidants on the viability of rats with two 2 h ischemic lobes of liver and 30% resection of intact liver lobes (%).

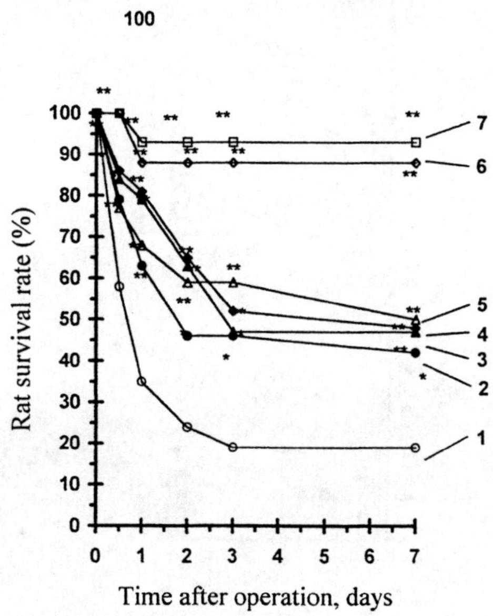

1 — nontreated animals (control);
2 — Desferal (100 mg/kg) 15 min before reperfusion;
3 — SOD (15000 ED/kg×2) 15 min before and 6 h after reperfusion;
4 — K-phenosan (50 mg/kg) 1 h before ischemia;
5 — 1 HTMP (12 mg/kg) 1 h before ischemia;
6 — Diludin (120 mg/kg) 4 h before ischemia;
7 — Se-methionine (1.2 μg/kg×3) 72, 48 and 24 h before ischemia.
* -p < 0.05; ** -p < 0.01 versus control.
All antioxidants, exept Diludin, were injected i. p., Diludin was introduced per os.

In group 1, all antioxidants except 1-HTMP and Se-methionine strongly stimulated DNA synthesis 24 h after 30% hepatectomy (Fig. 80a). Their effects decreased in the series: desferal (304%, $p < 0.01$) > SOD (194%, $p < 0.01$) > K-phenosan (162%, $p < 0.05$) > diludin (138%, $p < 0.05$). Here, the data in parentheses show the proportional increase in DNA synthesis and significance of difference from control (30% hepatectomy without antioxidant treatment). Note that DNA synthesis in the regenerating liver did not differ from control in 1-HTMP-treated rats (164%, $p > 0.05$) and was significantly lower than control (negative effect) in Se-methionine-treated rats (15%, $p < 0.01$).

Fig. 80: Effect of the antioxidants on the DNA synthesis in rat liver nuclei 24 h after 30% resection and 2 h ischemia plus 30% resection of liver (mean + SEM) [Liashenko S.M.; Liashenko S.M., Bilenko M.V. and Gefter Ju.V.].

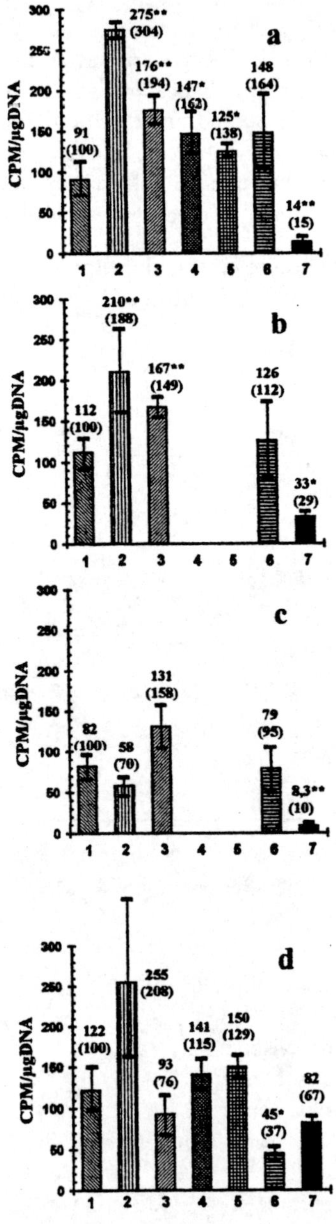

a — intact liver lobes in rats with 30% resection of liver (group 1);
b — intact liver lobes in rats with 2 h ischemia of liver lobes, (group 2);
c — 2 h ischemic lobes of the same liver, (group 2);
d — 2 h ischemic lobes in rats with ischemia + 30% resection of liver, (group 3).
1 — without antioxidants (control); 2 — Desferal; 3 — SOD;
4 — K-phenosan; 5 — diludin; 6 — 1 HTMP; 7 — Se-methionine.
In parentheses — % of the initial level.
* — $p < 0.05$; ** — $p < 0.01$ versus control.
Doses and injection time of antioxidants are the same as in fig. 79. The estimation of rate of DNA synthesis is the same as in fig. 74 and fig. 75.

In experiments with 2-h ischemia of two liver lobes (group 2), the ability of antioxidants to stimulate DNA synthesis in the nonischemic lobes decreased in the same order (see Fig. 80b). The rate of DNA synthesis was the highest in desferal-treated rats (188%, $p < 0.05$), followed by that in SOD-treated rats (149%, $p < 0.01$); 1-HTMP treatment did not result in any significant differences from control, and Se-methionine inhibited DNA synthesis, as in the first group (29%, $p < 0.05$). K-phenosan and diludin were not tested in this experiment. Thus, only antioxidants with moderate or weak antiischemic effects stimulated DNA synthesis in nonischemic liver lobes regenerating after 30% hepatectomy and 2-h ischemia. Compounds with a strong antiischemic activity, i.e., those efficiently promoting animal survival (1-HTMP and Se-methionine) either had no significant effect on DNA synthesis (1-HTMP) or inhibited it (Se-methionine). If we assume that the antiischemic effect of the compound actually depends on its antioxidant properties, then the most active antioxidants proved to be the weakest stimulators or even inhibitors of DNA synthesis under these experimental conditions. Later other authors [Zager R. A., Fuerstenberg S. M., Baehr P. H. et al.] using oxypurinol and OH scavengers (Na benzoate, or dimethylthiourea) found antiproliferative effect of these antioxidants on (^{3}H)thymidine incorporation in nuclei of human tubular cells after renal ischemia and reperfusion; conversely "catalytic" iron had a growth stimulatory effect.

The analysis of DNA synthesis in rat liver lobes subjected to 2-h ischemia either without (group 2) or with (group 3) resection of nonischemic lobes revealed no significant differences from control (see Figs. 80c, 80d), except for the inhibitory effects of 1-HTMP and Se-methionine on DNA synthesis.

Taken together, the data allow several conclusions.

First, most antioxidants were administered before ischemia because this ensured their optimal antiischemic effect (Fig. 80; and Chapter 11). Antioxidants inhibited free-radical processes, thereby decreasing the severity of damage to the liver itself and to its DNA. These effects promoted animal survival but decreased the rate of postischemic regeneration. This is consistent with the experimental data that antioxidants (BHT and inderal) reduced the rate of DNA synthesis in the periinfarct zone of the rat heart [Meerson F.Z., Vasil'ev K.V. and Dosmagambetova R.S.; Meerson F.Z., Vavich M.P., Rozhitskaya I.I. and Golubeva L.Yu.] and that mexamine and especially a vitamin complex inhibited [^{3}H]thymidine incorporation into DNA of irradiated hepatocytes, thus manifesting their hepatoprotective effects [Sukolinsky V.N., Konoplya E.F., Morozkina T.S. et al.].

Second, antioxidants produced no stimulating effect on DNA synthesis in the liver exposed to ischemia combined with simultaneous resection of nonischemic lobes, probably because the liver reached the limit of its regeneration capacity after such a treatment. The latter is confirmed by the results of experiments with PSE_{res} in which it also had no apparent effect on the ischemic liver lobes remaining after 30% hepatectomy, whereas under lower functional load (without resection), both nonishemic and ischemic lobes responded to PSE_{res} by a further increase in DNA synthesis (see Figs. 75 and 76). The ability of some antioxidants (SOD and allopurinol) to stimulate DNA synthesis in the liver was also revealed in experiments with 70% hepatectomy but only after a short-term (15-min) ischemia [Portugal V., Alonso I., Bilbao J. et al.]; after 80% hepatectomy, this period was even shorter (10 min) [Foschi D., Castoldi L., Lesma A. et al.].

Third, we cannot exclude the possibility that antioxidants stimulate only replicative DNA synthesis normally occurring in the liver regenerating after partial hepatectomy, rather than reparative DNA synthesis observed upon exposure to emotion–pain or surgical stress [Meerson F.Z., Vavich M.P. and Rozhitskaya I.I.; Yavich M.P., Rozhitskaya I.I., Golubeva L.Yu. et al.], and ischemia.

Fourth, the inhibitory effect of Se-methionine, which had the strongest antiischemic effect, on DNA synthesis was relative, i.e., it was only significant when comparing the results of similar experiments performed with and without this agent. At the same time, the absolute rate of [^{14}C] thymidine incorporation (cpm/μg DNA) in these experiments was higher than in the nonoperated control, increasing in the series: liver after 30% resection (14 ± 3.9) < intact lobes after 2–h ischemia (32.5 ± 14) < lobes subjected to 2–h ischemia with resection of intact lobes (81.6 ± 7.5). These data indicate that the rate of DNA synthesis in the liver of Se-methionine-treated animals increased proportionally to the severity of damage but this increase was less pronounced than in similar experiments without Se-methionine.

Fifth, there is an inverse relationship between the specialized functioning of the organ and its regeneration activity. After long-term ischemia, the requirements of the whole organism for the specific liver functioning necessary for survival (at least during 24-h period of our observations) may prevail over the need for liver regeneration. These complex relationships between the needs of the whole organism (organ functioning) and the organ itself (regeneration potential) are probably regulated by cytokines and growth factors. To gain better understanding of these processes, it is necessary to analyze in detail the contents of growth factors in ischemic liver lobes and "cell sources" of newly synthesized DNA (in particular, relative contributions of hepatocytes and tissue and blood macrophages to DNA synthesis).

Whether or not these considerations are valid, it is clear that the antiischemic effect of antioxidants is not mediated by stimulation of DNA synthesis and regeneration in the liver. The postischemic DNA synthesis apparently depends on the extent of DNA damage during ischemia or the early reperfusion period. Antioxidants, protecting DNA from ischemic damage, obviate the need for DNA resynthesis, which requires a high metabolic rate, and help the organ to perform its vitally important functions.

Sixth, DNA damage due to free radical reactions or LPO intensification apparently occurs by different mechanisms, including OH generation on DNA molecules by the reaction of H_2O_2 (the most effective ROS in terms of DNA damage) with Cu^{2+} or Fe^{2+}; activation of Ca^{2+}-dependent endonucleases; peroxidation of DNA–lipoprotein complexes; and nuclear membrane injury [Halliwell B. and Aruoma O.J.]. This problem is currently being studied.

9.4 CONCLUSIONS

Published data and the results of our experiments reviewed in this chapter suggest that acute ischemia and reperfusion of organs cause damage of individual cells, cell layers, or the entire organ, and one of mechanisms responsible for this damage is the induction of apoptosis (programmed cell death). The latter involves double-strand DNA breaks resulting in internucleosomal DNA fragmentation and is accompanied (or followed) by less pronounced damage to cell membranes and intracellular membrane structures. DNA damage and fragmentation stimulate postischemic reparative DNA synthesis whose intensity depends on the severity of DNA damage during the ischemic and early reperfusion periods.

The features most characteristic of DNA damage and repair in ischemic and reperfused organs are as follows.

(1) The nature of damage caused by ischemia and reperfusion, especially the development of apoptosis or necrosis, depends on the severity of ischemia and the sensitivity of the organ to oxygen deficiency. Upon more severe damage, cell apoptosis is replaced by necrosis.

(2) Total 2-h ischemia of two liver lobes leads to an increase in DNA synthesis during the early reperfusion period. The stimulating effect of experimental damage on the activity of DNA synthesis in the liver increases in the series: 30% hepatectomy < 2–h lobar ischemia < 2 h of lobar ischemia combined with 30% hepatectomy.

(3) Upon ischemic damage, activation of DNA synthesis in organs under effects of ischemia proper, functional load, or exogenous agents stimulating regeneration is only possible until the mass of cells still capable of synthesizing DNA remains above a certain threshold level, which is specific for each organ.

(4) Cytokines and growth factors play a major role in the regulation of DNA synthesis and tissue regeneration in ischemic and reperfused organs. Some of these substances produce dose-dependent bifunctional effects. Our experiments showed that ischemic liver lobes produce (or accumulate) a thermostable protein stimulating cell proliferation in the liver (PSE_{isch}). Its activity is higher than that of the factor found in the liver after 70% resection (PSE_{res}).

The chemical structure of these factors, their similarity to HGF, and their cell origin are poorly known; however, tissue macrophages are apparently involved in their synthesis. This follows from the fact that injections of prodigiosan, a polysaccharide stimulating Kupffer cells, or such injections combined with 70% hepatectomy resulted in the formation (or accumulation) of proliferation-stimulating factors (PSE_{prod} and PSE_{comb}) in the liver. According to their activities, thermostable proliferation-stimulating factors isolated from liver in our studies formed the following ascending series:

$$PSE_{res} < PSE_{prod} = PSE_{isch} < PSE_{comb.}$$

(5) Free-radical reactions and lipid peroxidation play an important role in DNA damage and repair in ischemic and reperfused organs, affecting the composition of lipids in DNA–lipid complexes and nuclear membranes, damaging DNA directly and activating endonucleases. Our data showed that exogenous

antioxidants stimulate DNA synthesis in intact liver lobes remaining after 30% hepatectomy and in nonischemic lobes of the liver exposed to 2-h partial ischemia. These effects of antioxidants are inversely proportional to their antiischemic (antioxidant) activities. However, no stimulation of DNA synthesis in long-term (2-h) ischemic liver was observed under the effect of antioxidants. This may be attributed to their strong antiischemic action, which prevented DNA damage during ischemia and, hence, obviated the necessity of reparative DNA synthesis upon reperfusion.

Chapter 10

POSSIBLE MECHANISMS OF INTENSIFICATION OF LIPID PEROXIDATION DURING ACUTE ISCHEMIA AND AFTER REPERFUSION OF ORGANS

This chapter provides the theoretical basis for the wealth of accumulated experimental and clinical data attesting to intensification of free-radical peroxidation processes in ischemic organs and their role in the pathogenesis of ischemic and reperfusion injuries (see Chapters 5–8).

The key issue here is the presence of O_2 in tissues in amounts sufficient for these processes to occur in the course of ischemia, and also possible changes in systems responsible for LPO initiation and inhibition, which would promote free-radical reactions despite O_2 deficiency. To clarify it, one must find the answer to the question whether O_2 toxicity is possible not only during absolute, but also relative hyperoxia (reperfusion period), and especially during pronounced hypoxia or anoxia (ischemic period), when speaking about O_2 toxicity or intensification of LPO processes appears paradoxical. In this context, it is especially important to provide theoretical explanation for the existence of conditions and systems promoting LPO initiation and progression during ischemia.

10.1 CONDITIONS PREREQUISITE TO LIPID PEROXIDATION IN MEMBRANES

Conditions prerequisite to free-radical reactions in model systems *in vitro* and biological membranes *in vivo* have been studied in detail and described in numerous reviews and monographs [Burlakova E.B., Alessenko A.V., Molochkina E.M. et al.; Vladimirov Yu.A. and Archakov A.I.; Vladimirov Yu.A., Olenev V.I., Suslova T.B. et al.; Kozlov Yu. P., Danilov V. S. and Kagan V. E.; Lukyanova L.D., Balmukhanov B.S. and Ugolev A.T.; Emanuel N.M. and Lyashovskaya Yu. N.; Cadenas E.; Sevanian A. and Hochstein P.]. Current views on the sequence of corresponding processes in lipids, including those of biological membranes, are summarized in Fig. 81.

As follows from this scheme, the central event in reactions is the formation of lipid radicals—alkyl ($R^{\bullet}$), alkoxyl ($RO^{\bullet}$) and peroxyl radical ($RO_2^{\bullet}$) radicals—at three stages of the process: (I) initiation, (II) propagation, and (III) branching of reaction chains. The scheme also shows that proper conditions for free-radical processes are as follows: (1) the presence of oxidation substrates (RH); (2) the presence of molecular oxygen (O_2) in the medium; (3) the production of reactive oxygen species (ROS) in tissues ($O_2^{\bullet}$, $^{\bullet}OH$, $^1O_2$, H_2O_2), which are stronger oxidizing agents than O_2; (4) the presence of transition metals (Me^n, primarily Fe^{2+}), which catalyze O_2 reduction to its reactive forms and the reactions of chain initiation and branching, and the presence of reducing agents for these metals (NADPH, ascorbic acid, etc.); (5) a steady course of reactions, rather than their progressive self-acceleration, is conditioned by enzymatic (COD, GSH peroxidase, catalase) and nonenzymatic inhibitors of LPO (In), which provide for the reduction of ROS to less active forms ($O_2^{\bullet}$ to H_2O_2) or to H_2O and terminate reaction chains at the level of radicals and peroxides.

The first and second conditions necessary for LPO during acute ischemia and after reperfusion of organs will be discussed here, and others will be considered in the following sections.

The **presence of substrates** for LPO in organs during ischemia and reperfusion is an obvious prerequisite for this process.

Fig. 81: Schematic represention of sequence of LPO reactions in biological membranes.

0. 1. $O_2 + Fe^{2+} \xrightarrow{K_0} Fe^{3+} + \mathbf{O_2^{\bullet}}$

2. $O_2^{\bullet} + O_2^{\bullet} + 2H^+ \rightarrow O_2 + \mathbf{H_2O_2}$ (SOD)

3. $O_2^{\bullet} + O_2^{\bullet} + 2H^+ \rightarrow {}^1\mathbf{O_2} + H_2O_2$ (SD)

4. $H_2O_2 + Fe^{2+} \rightarrow OH^- + Fe^{3+} + \mathbf{{}^{\bullet}OH}$ (Fenton reaction)

5. $H_2O_2 + O_2^{\bullet} \rightarrow OH^- + O_2 + \mathbf{{}^{\bullet}OH}$ (Haber-Weiss reaction)

I. 6. $O_2^{\bullet} + RH \xrightarrow{Fe^{2+} \; Fe^{3+}} \mathbf{R^{\bullet}} + HO_2^-$

7. ${}^{\bullet}OH + RH \xrightarrow{K_1} \mathbf{R^{\bullet}} + H_2O$

II. 8. $R^{\bullet} + O_2 \xrightarrow{K_1} \mathbf{RO_2^{\bullet}}$

9. $\mathbf{RO_2^{\bullet}} + RH \xrightarrow{K_3} \mathbf{ROOH} + R^{\bullet}$

III. 10. $ROOH + Me^n \xrightarrow{K_p} \mathbf{RO^{\bullet}} + Me^{n+1} + OH^-$

11. $\mathbf{RO^{\bullet}} + RH \rightarrow ROH + \mathbf{R^{\bullet}}$; $R^{\bullet} + O_2 \rightarrow \mathbf{RO_2^{\bullet}}$

IV. 12. $R^{\bullet} + R^{\bullet} \xrightarrow{K_4} MP$; $\mathbf{RO_2^{\bullet}} + R^{\bullet} \xrightarrow{K_5} MP$; $\mathbf{RO_2^{\bullet}} + \mathbf{RO_2^{\bullet}} \xrightarrow{K_6} MP$

13. $\mathbf{RO_2^{\bullet}}(R^{\bullet}) + InH \xrightarrow{K_7} \mathbf{ROOH(RH)} + In^{\bullet}$

14. $\mathbf{RO_2^{\bullet}}(R^{\bullet}) + Fe^{2+} \rightarrow Fe^{3+} + RH$

15. $ROOH + GSH \rightarrow ROH + GSSG + H_2O$ (GSH peroxidases I, II)

V. 16. $In^{\bullet} + RO_2^{\bullet} \xrightarrow{K_8} MP$

17. $In^{\bullet} + In^{\bullet} \xrightarrow{K_9} MP$

18. $In^{\bullet} + RH \xrightarrow{K_{10}} InH + \mathbf{R^{\bullet}}$

19. $In + ROOH \rightarrow InH + \mathbf{RO_2^{\bullet}}$

Note: O - production of reactive forms of O_2 (ROS); I - oxidative chain initiation; II - oxidative chain propagation; III - oxidative chain branching; IV - oxidative chain termination; V - reactions of antioxidant radicals. The main reaction products are shown in bold. K_0-K_{10} - reaction rate constants; $O_2^{\bullet}$ - superoxide anion radical H_20_2 - hydrogen peroxide; ${}^{\bullet}OH$ - hydroxyl radical; $R^{\bullet}$ - lipid alkyl radical; $RO^{\bullet}$- lipid alkoxyl radical; $RO_2^{\bullet}$ ($ROO^{\bullet}$) – lipid peroxyl radical; InH - antioxidant; $In^{\bullet}$ - radical of the antioxidant; $GSH^{\bullet}$ and $GSSG^{\bullet}$ -reduced and oxidized glutathione; $SOD^{\bullet}$ – superoxide dismutase; SD - spontaneous dismutation; Me^n and Me^{n+1}. - reduced and oxidized ions of transient metals, respectively; MP -molecular products.

LPO substrates in membranes are PUFA bound to carbon (mostly in position 2), and the detachment of a hydrogen atom and conversion of the lipid (R) into a lipid radical ($R^{\bullet}$) thermodynamically preferable for carbon in α position with respect to the double bond. As a consequence, the rate of LPO increases proportionally to the number of unsaturated bonds in fatty acid molecules, and the main LPO substrates in membranes are PUFA, namely, 22:6, 20:4, and 18:2. As AA content in membranes is much higher, compared to that other PUFA, this acid is the basic LPO substrate.

During ischemia, the content of PUFA, especially AA, in membranes decreases as a result of their oxidation and subsequent hydrolysis by phospholipases (see Chapter 2). It is known, however, that LPO in isolated microsomal membranes may proceed until virtually complete PUFA oxidation; that is, the reduction of PUFA level in membranes, even during prolonged ischemia, can hardly be regarded as a limiting factor for LPO.

The **requirement for oxygen** in amounts sufficient for LPO is obviously fulfilled during reperfusion of ischemic organs; moreover, oxygen is often present in excess (absolute of relative hyperoxia, see Chapter 1). However, the presence of sufficient oxygen during partial and, especially, total ischemia is less obvious and still open to debate.

Molecular oxygen is necessary not only for ROS formation and initiation of oxidation chains, but also at stages of their propagation and branching (see Fig. 81). Hence, it is apparent that the long-lived concept of the complete O_2 absence or insufficiency for LPO during ischemia has been the factor responsible for a serious delay in studies on the role of LPO in ischemic organ injuries.

As follows from observations described in Chapter 1, O_2 concentration during partial ischemia was 50–20% of the initial (pO_2 no lower than 20–10 mm Hg), while pO_2 during total ischemia remained at 2–5 mm Hg for a long period of time. On this basis, the only debatable question is whether the corresponding amount of oxygen is sufficient to initiate and maintain LPO in ischemic tissues.

At present, there is abundant evidence that free-radical reactions may proceed at very low pO_2 values in tissues. It was shown that blowing the chamber with insufficiently purified nitrogen does not terminate cysteine-induced LPO in rat liver mitochondria [Fortney S.R. and Lynn W.S.], and bubbling insufficiently purified argon or nitrogen through the solution does not inhibited UV-induced LPO [Vladimirov Yu.A., Olenev V.I., Suslova T.B. et al.]. At O_2 concentration of 10^{-6} M, all lipid radicals proved to be converted into peroxide ones ($RO_2^{\bullet}$), and free-radical reaction rate (estimated from chemiluminescence intensity) was independent of O_2 concentration and remained virtually unchanged until its almost complete disappearance from the system [Vasiliev R.F. and Vishutinskii A.A.].

Chemiluminescence of oleic acid in a vacuum at 60°C remained fairly intense for four to six days, which was attributed to O_2 release resulting from secondary regrouping reactions (including reduction) with repeated formation of $ROO^{\bullet}$ radicals [Asanov M.A.].

In experiments on Fe^{2+}–ascorbate-dependent LPO in rat brain homogenate, blowing the cell with a gas mixture containing traces of O_2 led to a slight increase in DC content; at O_2 concentration of 1–5%, MDA content in the homogenate increased significantly; at 10% of O_2 in the gas mixture, MDA content continued to increase but at a lower rate [Rehncrona S., Smith D.S., Akesson B. et al.; Smith D.S., Rehncrona S. and Siesjö. Bo.K.].

The effect of O_2 concentration in the gas mixture on the rate of LPO processes was also studied upon LPO induction by oxidation promoters mixed with PUFA (18:2, 18:3, and 20:4) [Dirks D.R., Faiman M.D. and Huyser E.S.]. In these experiments, LPO proceeded at a slow rate even when O_2 was absent (more likely, present in trace amounts), its rate increased tenfold at 10% of O_2 in the gas mixture, but further increase at 21 and 40% O_2 was much less pronounced (by factors of 1.3 and 1.6, respectively, compared to LPO rate at 10% O_2). Even upon exposure of lipids in a pressure chamber with 100% O_2 at 17 ata, LPO rate became only 2.7 times as high as at 10% O_2. If these data are recalculated assuming that LPO rate in the air (21% O_2) is 100%, the observed LPO rates will be as follows: about 8% in the absence of O_2 (or its presence in traces) in the gas mixture, 78% at 10% O_2, 125% at 40% O_2, and 211% at 100% O_2 under a pressure of 7 ATA.

Hence, LPO can actually occur at a pO_2 value much lower than normal, and its rate increases relatively slightly (no more than twofold) upon a marked increase in O_2 concentration in the medium.

As radiation damage is known to be based on a free-radical mechanism, attempts have been made to use its severity as a criterion for determining the amount of O_2 necessary to initiate and maintain LPO.

Experiments showed that radiosensitivity increases at pO_2 ranging from 0 to 30–40 mm Hg, and any further increase in pO_2—aeration with atmospheric air, 100% O_2, or even O_2 at high pressure—produces virtually no effect on cell radiosensitivity. Moreover, radiosensitivity of different cells (bacteria, mammalian cell culture, etc.) reached 50% of the maximal value at pO_2 = 3 mm Hg, that is, at less than 0.5% O_2 in the gas mixture [Hall E.J.] (Fig. 82).

Thus, much more pronounced radiation damage, indirectly confirming intensification of free-radical reactions, was observed after a very slight increase in O_2 concentration. Assuming that cell sensitivity to free-radical reactions during ischemia is equivalent to that under irradiation, the authors [Demopoulos H.B., Flamm E.S., Pietronigro D.D. and Seligman H.L.] calculated that during partial cerebral ischemia (ligation of the middle cerebral artery in cats), when blood flow was reduced to 20–30% of the initial (i.e., at pO_2 = 4–8 mm Hg, vs. 20-40 mm Hg in the norm), LPO rate approached the maximum possible value, and at pO_2 = 1–2 mm Hg (in the center of the ischemic area), this rate was 1.5 times as high as that during anoxia. These data show that there is virtually no O_2 deficiency for LPO during partial ischemia; in cases of total ischemia, residual amounts of O_2 are still retained, being unevenly distributed over hydrophilic and hydrophobic components of the membrane (see Chapter 1), and O_2 deficiency does not seem to be a limiting factor, especially in the early ischemic period.

Fig. 82: Radiosensitivity of cells versus oxygen concentration [Hall E.J.]. For other explanations, see text.

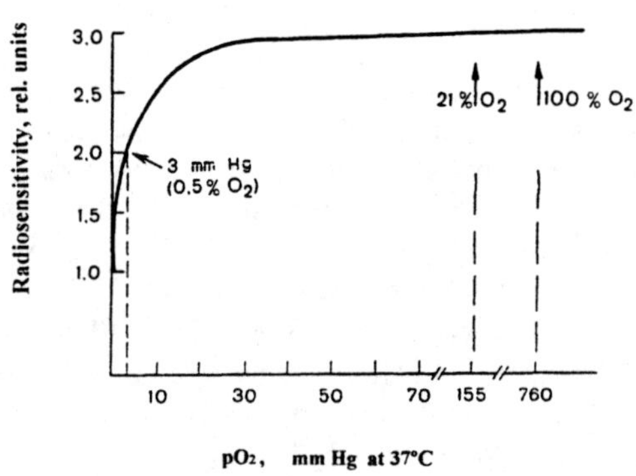

The above findings provide convincing evidence that there is sufficient oxygen for LPO processes in organs during their partial and even total ischemia and suggest that reperfusion with oxygenated blood creates conditions for repeated, although less pronounced, intensification of LPO, which is consistent with our experimental results and the data obtained by other researchers (see Chapter 5).

10.2 Sources of Reactive Oxygen Species

The production of ROS, which are more powerful oxidizing agents than molecular oxygen, is one of prerequisites for LPO processes (see Scheme 12). ROS include the superoxide anion radical ($O_2^{\bullet-}$), which often takes the form of the perhydroxyl radical ($HO_2^{\bullet}$) in solutions; the hydroxyl radical ($^{\bullet}OH$); singlet oxygen ($^1O_2$); and hydrogen peroxide (H_2O_2), which is not a radical but easily forms them (see Fig. 81).

ROS are produced because of incomplete—one-electron (formation of $O_2^{\bullet-}$), two-electron (formation of H_2O_2) or three- electron (formation of $^{\bullet}OH$)—reduction of O_2, instead of its complete four-electron reduction to water by cytochrome oxidase (CCO), the terminal enzyme of the mitochondrial respiratory chain (Fig. 83).

Fig. 83: Production of reactive oxygen species (ROS) in the course of one-, two-, three- and four-electron reduction of the O_2 molecule.

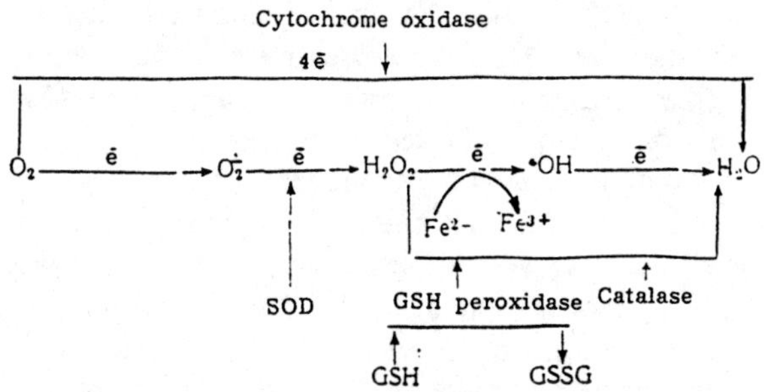

The most reactive O_2 species is $^1O_2$, which can be formed during spontaneous $O_2^{\bullet}$ dismutation in the absence of SOD, especially in the three-component myeloperoxidase + H_2O_2 + Cl^- system of phagocytes, and, in small amounts, in all other ROS-producing systems. However, the role of $^1O_2$ in LPO initiation in biological systems is not yet fully understood.

ROS are unstable compounds. It is known that the lifetime of $O_2^{\bullet}$ is measured in milliseconds; that of $^1O_2$, in microseconds; and that of $^{\bullet}OH$, in fractions of a microsecond [see Gus'kova R. A., Ivanov I. I., Kol'toner V. K. et al.]. Their ability to induce LPO is inversely proportional to their lifetime ($^{\bullet}OH < O_2^{\bullet} < H_2O_2$) and directly proportional to their diffusion capacity. According to [Gus'kova R. A., Ivanov I. I., Kol'toner V. K. et al.], the coefficient of lipid bilayer permeability is 7.6 10^{-8} cm/s for $O_2^{\bullet}$ and 4.9 10^{-4} cm/s for $^{\bullet}OH$.

The short lifetime of $^{\bullet}OH$ or, in other words, its extremely high reaction capacity, is responsible for heavy damage it inflicts on membranes, but this occurs only near the site of radical formation [Halliwell B. and Gutteridge J.M.C. (1984)]. By contrast, $O_2^{\bullet}$ can migrate for some distance through "anionic" channels, and H_2O_2 migrates through virtually all cell membranes, except for mitochondrial ones [Nohl H. and Hegner D.]. The ability of ROS to migrate and, hence, the spread of their damaging and LPO- inducing effects also depend on the state of membranes. Modification of the lipid bilayer with cholesterol, α-tocopherol, or Ca^{2+} reduced the coefficient of lipid membrane permeability for $O_2^{\bullet}$ by a factor of 2 to 3 [Gus'kova R. A., Ivanov I. I., Kol'toner V. K. et al.].

In virtually all ROS-generating systems, $O_2^{\bullet}$ is a primary product, whereas more reactive forms ($^{\bullet}OH$, $^1O_2$, OCl^-) are secondary. Hence, the major role in initiating or suppressing the subsequent chain reactions belongs to catalysts and inhibitors of corresponding processes, to enzymatic and nonenzymatic ROS scavengers. Balance between $O_2^{\bullet}$ production and utilization (reduction) is the reason why the content of ROS in the cell under physiological conditions is extremely low: $O_2^{\bullet}$ concentration in mitochondria is 8 10^{-12} M, and H_2O_2 concentration in erythrocytes is 10^{-9}–10^{-7} M [see Lukyanova L.D., Balmukhanov B.S. and Ugolev A.T.].

Under pathological conditions, especially during ischemia, the disproportionate increase in the content of prooxidants and decrease in the content of antioxidants take place, and this apparently allows the emerging O_2 radicals to initiate LPO.

To day, many systems producing ROS under both physiological and pathological conditions have been revealed. Let us consider some of them in terms of their possible involvement in LPO initiation during acute ischemia and after reperfusion of organs.

10.2.1 Formation of Reactive Oxygen Species in Mitochondrial and Microsomal Electron Transport Chains

In **the mitochondrial respiratory chain**, small amounts of ROS are formed even under physiological conditions due to "leakage" of 1–5% of electrons transported along the chain [McCord J.M.; Weisiger R.A.]. $O_2^{\bullet}$ is involved in the coupling of respiration and oxidative phosphorylation, providing the basis for oxidative processes in mitochondria insensitive to cyanide. The rate of $O_2^{\bullet}$ formation in mitochondria sharply increases when the respiratory chain is blocked, which leads to the reduction of carriers at links located upstream from the block and to more intense electron leakage.

The $O_2^{\bullet}$-producing enzyme complexes of the mitochondrial respiratory chain include NADH-dependent dehydrogenase, alone or combined with ferrosulfate complexes (so-called complex I or NADH-dependent ubiquinone reductase); succinate–FAD; or the site at the level of the second phosphorylation point, i.e., within the b=c segments of the respiratory chain. The production of $O_2^{\bullet}$ in the respiratory chain blocked by rotenone increases moderately; this increase is more pronounced when the blocking agent is cyanide and reaches a peak when this agent is antimycin A. Hence, it appears that $O_2^{\bullet}$ formation proceeds most actively in the segment between the points of respiratory chain inhibition by rotenone and antimycin A (see Fig. 20). These issues were discussed in detail in a number studies [Afanasyev I. B.; Brawn K. and Fridovich I.; Cadenas E., Varsavsky A.I., Boveris A. et al.; Cadenas E.; Chance B., Sies H. and Boveris A.; Nohl H. and Hegner D.; Sevanian A. and Hochstein P.; van den Hoek T.L., Shao Z., Li C. et al.].

In **the microsomal electron-transport chain**, ROS are produced primarily in the NADPH-cytochrome P-450-dependent system responsible for hydroxylation of xenobiotics. The main role in this process belongs to cytochrome P-450, which consecutively forms a series of enzyme-substrate complexes with the hydrophobic substrate. Dissociation of the last, ternary complex ensures oxidation of the substrate by a reactive O_2 species (the formation of a hydrophilic compound), oxidation of the enzyme proper (cytochrome P-450 • Fe^{3+}), and may be accompanied by the release of $O_2^{\bullet}$ with its subsequent reduction to H_2O_2 or $^{\bullet}OH$ [Archakov A. I. (1975); Lyakhovich V.V. and Tsyrlov I.B.; Metelitsa D.I.; Karuzina L.I. and Archakov A.I.; Kuthan H. and Ulrich V.; Sevanian A., Nordenbrand K., Kim E. et al.]. The induction of NADPH–cytochrome P-450-dependent monooxygenase system promotes $O_2^{\bullet}$ production, and the inhibition of this system reduces the latter.

The NADPH–cytochrome P-450-dependent system providing for hydroxylation of xenobiotics is found mainly in liver microsomes; however, similar systems are also found in microsomes of many other organs, except for the myocardium and skeletal muscles. This is why microsomes of these organs do not produce ROS [Koshkin V. V.].

As the total capacity of microsomal electron transport chains is comparable to that of the mitochondrial respiratory chain, and cytochrome P-450 has a very high affinity for O_2 and effectively competes for it with mitochondrial CCO [Archakov A. I. (1975)], the rate of ROS production in microsomes of most organs is almost the same as in mitochondria or even higher [Boveris A.].

Similar mechanisms of $O_2^{\bullet}$-induction were also revealed in the nuclei of liver cells, which suggests their role in LPO initiation in these organelles [Peskin A.V., Zbarsky I.B. and Konstantinov A.A.; Baird M.B., Birnbaum L.S. and Sfeir G.T.] and their involvement in D NA damage (see Chapter 9).

The problem concerning the contribution of mitochondria and microsomes to **ROS production in organs during acute partial or total ischemia and after reperfusion** has been studied insufficiently. Intensified $O_2^{\bullet}$ production in mitochondria of the isolated rat heart was revealed after its storage for 60 min at 37°C [Ledenev A. N. and Runge E. K.] and in the course of postanoxic reoxygenation [Du G., Mouithys-Mickalad A. and Sluse F.E.]; this was accompanied by increasing functional damage inflicted by $O_2^{\bullet}$.

In our laboratory, a model of ischemia of the median and left liver lobes in Wistar rats was used to investigate $O_2^{\bullet}$ production by the microsomal NADPH-dependent electron transport chain. The results were compared with the data on Cu, Zn-SOD activity in the cytosol and MDA accumulation in the liver homogenate and microsomes [Rashba Yu.E., Vartanyan L.S., Seregina L.A. et al. (1986a); Rashba Yu.E., Vartanyan L.S., Seregina L.A. et al. (1986b)].

The results showed (Fig. 84) that the rate of $O_2^{\bullet}$ production ($V_{O_2^{\bullet}}$) decreased to 79 and 65% of the initial level after 30 and 120 min of ischemia, respectively, and continued to decrease in the course of 24-h reperfusion of the liver to 41%; providing evidence that the NADPH–cytochrome P-450-dependent system generating $O_2^{\bullet}$ had not recovered during the reperfusion period. This fact agrees with the aforementioned observations that the content and functions of cytochrome P-450 are not restored for a long period of time in the course of reperfusion (see Fig. 37).

Fig. 84: Cu, Zn-SOD activity in the cytosol (1), and $O_2^{\bullet}$ regeneration rate in microsomes ($V_{O_2^{\bullet}}$) (2), after 2 h of ischemia and 2-24h reperfusion of the rat liver (% of the initial level) [Rashba Yu.E., Vartanyan L.S., Seregina L.A. et al. (1986a)]

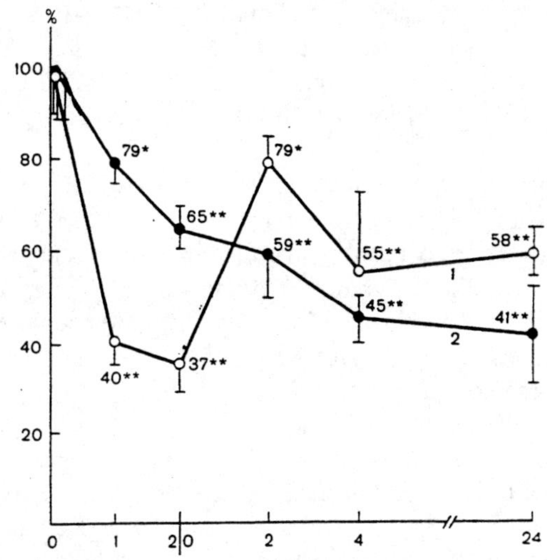

*-P < 0.05; **-P < 0.01 versus the initial level. Microsomes were separated by differential centrifugation. The methods of determination the activity of Cu, Zn-SOD and EPR techniques of determination $V_{O_2^{\bullet}}$ in microsomes based on oxidation of 2,2,6,6-tetramethylpiperidine to a stable nitroxyl radical, see [Rashba Yu.E., Chernikov V.A., Baider L.M. and Vartanyan L.S.].

Along with the decrease of $V_{O_2^{\bullet}}$, Cu, Zn-SOD activity in the cytosol fraction of the liver also decreased to 40 and 37% of the initial after 30 and 120 min of ischemia. This activity was partially recovered during reperfusion but remained significantly below the initial level even after 24 h. The decrease of Cu, Zn-SOD activity during liver ischemia was more pronounced than that of $O_2^{\bullet}$ production rate in microsomes, as followed from a twofold increase in the $V_{O_2^{\bullet}}$ / Cu, Zn-SOD ratio after 30 and 120 min of ischemia, compared to the initial ratio (Table 12). This was also confirmed by the data on TBARS accumulation in liver microsomes and homogenate: after 24-h reperfusion, the $V_{O_2^{\bullet}}$ / Cu, Zn-SOD ratio returned to the initial, whereas TBARS content either decreased (microsomes) or approached the initial level (homogenate).

Table 12: $VO_2^{\bullet}$ /SOD ratio and MDA content after ischemia and reperfusion of rat liver.

Indexes	Initial level	Ischemia		Reperfusion		
		30 min	2h	2h	4h	24h
$V_{O_2^{\bullet}}$ /SOD, rel. Units	1.0	2.0	1.97	0.80	0.90	0.74
MDA in microsomes, nmol•mg-¹ of protein	0.97±0.28 (100)	—	10.71±1.25**	—	—	2.36±0.32
TBARS in homogenate, nmol•g of wet tissue	18.4±1.9 (100)	—	31.1±2.2*	—	—	20.4±1.5

Notes: *-P<0.05, **-P<0.01 versus initial level. TBARS were determinated according to [Mihara M., Uchiyama M. and Fukuzawa K.].

Subsequent experiments on the same model dealt with the relationships between the rate of $O_2^{\bullet}$ production and the activity of Cu, Zn-SOD + Mn-SOD in mitochondrial and nuclear membranes [Vartanyan L.S., Rashba Yu.E., Nayler L.G. et al. (1990); Vartanyan L.S., Rashba Yu.E., Nagler L.G. et al.]. It was found that the rate of $O_2^{\bullet}$ production in microsomes of the intact rat liver (2.0 ± 0.14 nmole/min per mg protein) was 10 and 4 times as high as in mitochondrial and nuclear membranes (0.20 ± 0.02 and 0.50 ± 0.09 nmole/min, respectively). After 2-h ischemia and 24-h reperfusion, however, $V_{O_2^{\bullet}}$ in mitochondrial membranes (more precisely, in submitochondrial particles), in contrast to microsomal membranes, did not decrease but rather increased to 145 and 166% of the initial (Table 13); the ratio of $V_{O_2^{\bullet}}$ to Cu, Zn-SOD + Mn-SOD activity in mitochondria increased 1.9-fold after 2-h ischemia and returned to the initial value (0.97) after 24-h reperfusion, which was due to the 1.8-fold increase of Mn-SOD activity by that time.

In the nuclear fraction, the rate of $O_2^{\bullet}$ production remained virtually unchanged, and the ratio between $V_{O_2^{\bullet}}$ and Cu, Zn-SOD activity in the cytosol (protecting the nuclei against $O_2^{\bullet}$ radicals) increased after 2-h of ischemia (to 3.3) and showed a tendency toward normalization (1.42) after 24-h reperfusion.

These findings show that the electron transport systems of $O_2^{\bullet}$ production in liver membranes remain intact (microsomes and nuclei) or are enhanced (mitochondria) after long-term ischemia and reperfusion; they also indicate the development of relative SOD deficiency, which allows the accumulation of $O_2^{\bullet}$ radicals and initiation of free-radical processes.

The fact that tissues retain their capacity for ROS production during hypoxia, which may even increase in the microsomal NADPH–cytochrome P-450-dependent system, was confirmed by registering the formation of H_2O_2 and by EPR techniques. For instance, the 13-fold decrease of pO_2 in the liver, compared to the initial, reduced the rate of H_2O_2 formation by only 10% [Kuthan H. and Ulrich V.]; in other experiments, an increase in H_2O_2 content was revealed in mitochondria of the ischemic heart [Vandeplassche G., Hermans C., Thone F. and Borgers M.].

In addition to these facts, a number of conditions arising during ischemia also provide evidence for the probability of hydroxy radical production in mitochondria and microsomes. These conditions are as follows.

First, the content of reduced carriers in the mitochondrial respiratory chain sharply increases during ischemia (see Chapter 3). This situation, similar to that created by drug-induced block of respiratory chain, should facilitate electron leakage and promote incomplete (one-electron) O_2 reduction. It was also found that mitochondrial and microsomal radical-forming systems are pH-dependent and become activated during acidosis [Hess M.L., Okabe E. and Kontos H.A.], the activity of mitochondrial NADH-oxidase (producing O_2 and H_2O_2) increases during ischemia [Nohl H.], and that of cytochrome oxidase decreases. During the reperfusion period, damage and impaired functioning of the respiratory chain, the resulting long-term presence of reduced carriers, and the developing relative hyperoxia (see Chapter 6) create even more favorable conditions for $O_2^{\bullet}$ production.

Second, the activity of SOD, which utilizes $O_2^{\bullet}$, decreases, as it occurs in experiments on SOD inhibition or elimination [Nohl H. and Hegner D.]) or under natural conditions, e. g., in SOD-deficient tumors. Such a situation always leads to a relative excess of $O_2^{\bullet}$ radicals, which are involved in LPO induction.

Third, both $O_2^{\bullet}$ production rate and SOD activity decrease, but the decrease in the latter is more pronounced; thus, not only absolute, but also relative SOD deficiency is observed.

This resembles situations arising, for example, in cases of a very rapid (linear) increase in ROS content under hyperbaric oxygenation conditions or a more rapid $O_2^{\bullet}$ increase in cardiac mitochondria of old rats, compared to young, upon blocking the respiratory chain by antimycin A [Nohl H. and Hegner D.]. Such a situation was described above in the context of our experiments with ischemic liver. A factor contributing to the excess of emerging ROS over the capacity of systems utilizing them is the decreased activity of both SOD and other scavengers of O_2 radicals (see below).

Table 13. $VO_2^{\bullet -}$, SOD activity and $VO_2^{\bullet -}$ / SOD ratio in various subcellular organelles of rat liver after 2 h of ischemia and 24 h of reperfusion (mean ± SEM). [Vartanyan L.S., Rashba Yu.E., Nayler L.G. et al. (1990)].

Experimental condition	Microsomes		Submitochondrial particles		nuclei		Cu, Zn-SOD activity	Mn-SOD activity
	$VO_2^{\bullet -}$	$VO_2^{\bullet -}$ / Cu, Zn-SOD	$VO_2^{\bullet -}$	$VO_2^{\bullet -}$ / Cu, Zn-SOD+ Mn-SOD	$VO_2^{\bullet -}$	$VO_2^{\bullet -}$ / Cu, Zn-SOD		
Intact live control	1.0	1.0	1.0	1.0	1.0	1.0	1.0	1.0
2h of ischemia	0.64±0.03**	1.88	1.45±0.13**	1.88	1.13±0.26	3.3	0.34±0.05	1.02±0.19
2h of ischemia plus 24 h of reperfusion	0.41±0.10**	0.72	1.66±0.21**	0.97	0.81±0.25	1.42	0.57±0.07	1.80±0.07

Notes: $VO_2^{\bullet -}$ - the rate of $O_2^{\bullet -}$ generation, nmoles • min^{-1} • mg^{-1} proteins, here is represented in relative units.
*-P<0.05; **-P<0.01 versus control.

It should be noted that the direct relationship between the rate of $O_2^{\bullet}$ production in mitochondria and microsomes, on the one hand, and the intensity of LPO in them, on the other, is not necessarily observed in every organ. It is pronounced in the liver [Rashba Yu.E., Vartanyan L.S., Seregina L.A. et al. (1986a); Nohl H. and Hegner D.] because the NADPH–cytochrome P-450-dependent $O_2^{\bullet}$-producing system and the NADPH-dependent LPO system are spatially close to one another, and the sites of LPO initiation are near those occupied by cytochrome P-450 [Archakov A. I. (1975); Vladimirov Yu.A. and Archakov A.I.]. At the same time, the sites of $O_2^{\bullet}$ production and LPO initiation in mitochondria and microsomes of the heart [Takeshiga H., Takayanagi R. and Manakami S.] or skeletal muscles [Koshkin V. V.] are separated, which determined the lack of direct relationship between the rate of $O_2^{\bullet}$ production and the intensity of LPO. In microsomes of skeletal muscles, no ROS are produced, although SR membranes readily undergo peroxidation attributed to intermembrane $O_2^{\bullet}$ diffusion [Koshkin V. V.]. In mitochondria of skeletal muscles, $O_2^{\bullet}$ production reaches its peak when the respiratory chain is blocked by antimycin A, and the accumulation of MDA is maximal when the blocking agent is rotenone, which the author [Koshkin V. V.] explains by the fact that LPO in mitochondria of skeletal muscles is induced owing to direct reduction of the NADP–Fe^{3+} complex of the respiratory chain, rather than by $O_2^{\bullet}$–Note also that the relationships between the rate of hydroxylation (and, consequently, $O_2^{\bullet}$ production) and LPO in microsomes are not only direct but also reciprocal [Kagan V. E., Kozlov Yu. P., Bilenko M. V. et al.], and the excess of LPO products has a damaging effect on cytochrome P-450 or even leads to its degradation [Archakov A. I. (1975); Kagan V. E., Kotelevtsev S. V. and Kozlov Yu. P.]. The rate of accumulation of LPO products (estimated by MDA) and the extent of cytochrome P-450 degradation during different types of LPO in microsomes can be described by the following descending series: NADPH-dependent LPO >> NADH-dependent LPO > ascorbate-dependent LPO [Kagan V. E., Savov V. M., Didenko V. V. et al. (1984); Kagan V.E., Savov V.M. and Serbinova E.A.]. Susceptibility to the toxic effect of ROS produced by these systems or products of intensified LPO is characteristic of not only cytochrome P-450, but also of other $O_2^{\bullet}$-producing enzymes.

Thus, the involvement of mitochondrial and microsomal electron transport chains in the induction of ROS production and LPO initiation is highly probable during acute ischemia of organs and almost certain during the reperfusion period, although studies on this problem continue.

10.2.2 Formation of Reactive Oxygen Species in the Xanthine-Xanthine Oxidase Reaction

The term "xanthine oxidase" actually refers to two structurally similar molybdenum- and iron-containing unstable cytosolic enzymes: xanthine oxidase (xanthine : oxygen oxidoreductase) proper (XO) and xanthine dehydrogenase (XDH). They occur in most organs, are characterized by broad substrate specificity, and participate in many oxidation reactions. The latter include metabolism of purines (hypoxanthine and xanthine) with the formation of uric acid; oxidation of pyrimidines, fatty acids, glutathione and epinephrine; dehydrogenation of NADH and NADPH to NAD and NADP; etc. In the rat intestine, kidneys, lungs, spleen, and liver *in vivo,* most of the enzyme (up to 85–95%) proved to be in the XDH form, and only 5–15%, in the XO form; in the rat heart, however, XDH and XO contents were found to be almost equal (55 and 45%); i.e., the initial XO level in the heart was higher than in other organs [Roy R.S. and McCord J.M.].

According to other data [Kehrer J.P., Piper H.M. and Sies H.], XO content in the rat heart is 34.8% of the total amount, and that in the rat liver is 12.3%; in the heart, however, XDH and XO are only found in endothelial cells, being absent from cardiomyocytes. The finding that XO in different organs is present mainly in the endothelium of capillaries is consistent with those of other workers [Betz A.L.; Beetsch J.W., Park T.S., Dugan L.L. et al.; Granger D.N., McCord J.M., Parks D.A. and Hollwarth M.E.; Jarasch E.D., Bruder G. and Heid H.W.]. In the small intestine, XDH + XO content in the duodenum is three times as high as in the ileum, but the proportion of XO in both divisions is the same: 19% of the total enzyme activity [Parks D.A., Williams T.K. and Beckman J.S.].

The total enzyme content demonstrates species- and organ-specific variation, being high in intestinal mucous membranes and the liver and much lower in the brain and skeletal muscles [Stirpe F. and Della Corte E.]. Species-specific differences of the enzyme content in the heart are especially pronounced. By this criterion, species can be arranged in the following descending series: mouse > rat > guinea pig > cow >

rabbit > pig > man [De Jong J.W., Vander Meer P., Huizer T. et al.]. Some authors failed to reveal any XO activity in the human and rat hearts [Grum C.M., Ragsdale R.A., Ketai L.H. et al.; Downey J.M., Hearse D.J. and Yellon D.M.; Eddy L., Steward J., Jones H. et al.], whereas others managed to detect a weak activity [De Jong J.W., Vander Meer P., Huizer T. et al.; Terada L.S.]. This activity in rabbit kidneys proved to be lower than that in rat kidneys [see Doctor R.B. and Mandel L.J.].

Hypoxanthine oxidation to uric acid by XDH is NAD-dependent, whereas that by XO occurs without NAD but yields $O_2^{\bullet}$ (see Fig. 85 and reactions *1* and *2* in Fig. 86). The latter provided the basis for the use of xanthine–xanthine oxidase reaction in the assessment of vasoactive and damaging effects of ROS.

Fig. 85: Involvement of xanthine dehydrogenase and xanthine oxidase in reactions yielding uric acid and some pathways for their intensification and inhibition

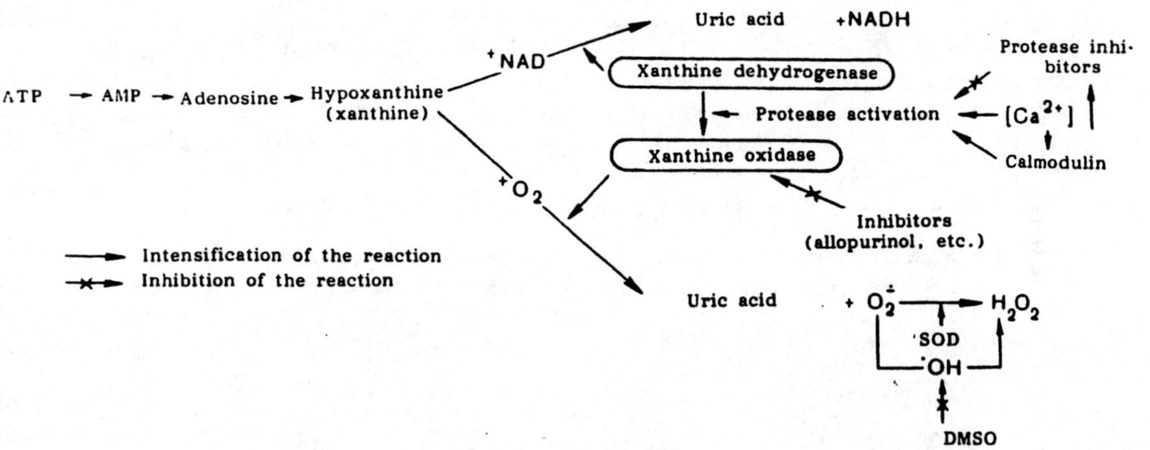

Fig. 86: Main reactions producing reactive oxygen species (ROS) in the hypoxanthine xanthine oxidase (XO) system, according to [Free radicals in biology; Fridovich I. (1983); Roy R.S. and McCord J.M.]

I

1. $EH_2 + 2O_2 \longrightarrow E + 2O_2^{\dot{-}} + 2H^+$

2. $EH_2 + O_2 \longrightarrow E + H_2O_2$

II

3. $Me^{n+1} + O_2^{\dot{-}} \longrightarrow Me^{n+} + {}^1O_2$

4. $Me^{n+} + H_2O_2 \longrightarrow Me^{n+1} + {}^{\cdot}OH + OH^-$

III

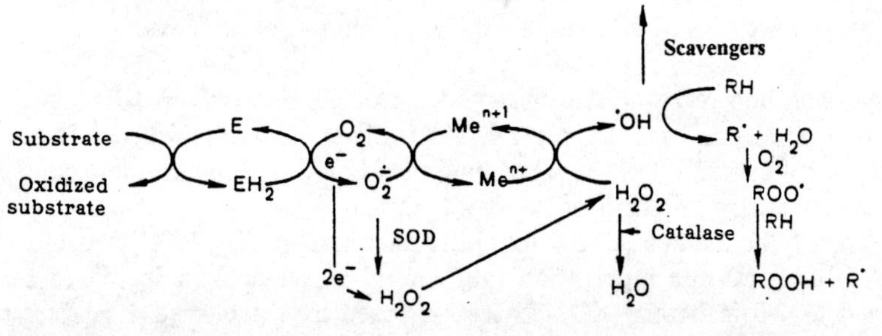

I: 1, 2 - production of $O_2^{\bullet}$ and H_2O_2 from O_2; II - 3, 4 - production of ${}^1O_2$ and $\bullet OH$ in the presence of transient metals; III - overall scheme and initiation of LPO processes.

When transition metals are present in the medium, the XO reaction leads to the formation of •OH or even $^1O_2$ as secondary products and to LPO induction [Del Maestro R.F., Björk J. and Azfors K.E.]. The role of XO in the production of ROS may also be conditioned by its ability to raise free iron content in blood plasma by recruiting iron from hepatic ferritin [Topham R.W., Walker M.C. and Calisch M.P.].

The concept that the xanthine–xanthine oxidase reaction is involved in ROS production during ischemia is based on at least three groups of data confirming (1) XDH conversion into XO in the ischemic organ, (2) increase in the content of reaction substrates in the ischemic organ, and (3) antiischemic and antiradical effects of XO inhibitors.

The *possibility of XDH conversion into XO at low tissue* pO_2 was described for the first time in [Granger D.N., Rutiti G. and McCord J.M.; McCord J.M. and Roy R.S.].

The ability of the enzyme to catalyze the formation of uric acid under different experimental conditions (with and without NAD) and after different periods of total ischemia was studied on the model of isolated and stored organs. The results showed that almost complete XDH conversion into XO in the small intestine occurs within the first 10 s, complete conversion takes 1 min, the process does not lead to any detectable loss of enzyme activity and is irreversible [Roy R.S. and McCord J.M.]. In other organs, this process occurred at a lower rate (Fig. 87) and was partially reversible. The half-time of XDH conversion into XO was 4 s in the small intestine mucosa, 10 min in the brain, 5 min in the heart, and 30 min in the liver, lungs, and kidneys; no conversion was observed in skeletal muscles [McCord J.M.; Roy R.S. and McCord J.M.]. The above data suggest that the contribution of such conversion to ROS production during ischemia differs depending on the organ, being most significant during ischemia of the intestine.

Fig. 87: Conversion of xanthine dehydrogenase (XD) into xanthine oxidase (XO) in different organs during total ischemia *in vitro* [Roy R.S. and McCord J.M.]

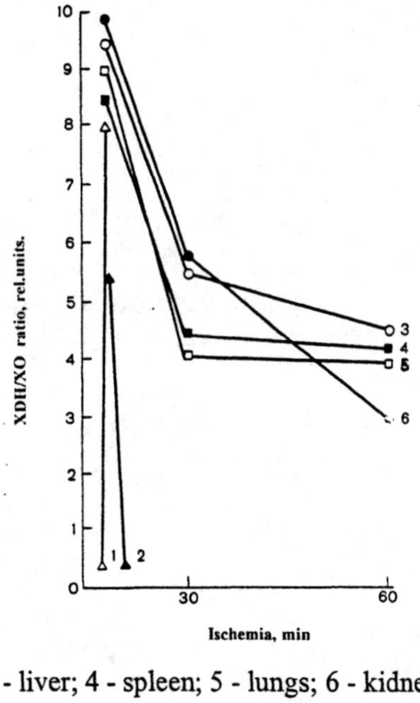

1 - small intestine; 2 - heart; 3 - liver; 4 - spleen; 5 - lungs; 6 - kidneys.

XDH conversion into XO was also observed during cardiac ischemia in dogs [Chambers D.S., Parks D.A., Patterson G. et al.] and rats [Granger D.N., Hollwarth M.E. and Parks D.A.] and in the frozen rat liver stored for 24 h [Kehrer J.P., Piper H.M. and Sies H.]. Subsequently, this conversion in the course of ischemia was confirmed by a number of authors.

In the rat liver, XO content after 180-min ischemia increased by 26% [Gonzalez-Flecha B., Cutrin J.C. and Boveris A.]; after 60-min reperfusion following 1-h ischemia, it increased by 18%. In rat skeletal muscles exposed to 4-h ischemia, XO activity increased by a factor of 2.5 [Kawasaki S., Sugiyama S., Ishiguro N. et al.].

More precise data were obtained on the timing of XDH conversion into XO. As shown in [Parks D.A., Williams T.K. and Beckman J.S.], this process in the rat intestine takes much more time than it was assumed previously [see McCord J.M. and Roy R.S.; Roy R.S. and McCord J.M.]. Irrespective of the part exposed to ischemia, the proportion of XO after 1, 2, and 3 h reached 34, 46, and 61% of the total enzyme activity, respectively, increasing by approximately 13% per hour. In other words, 50% conversion took about two hours. According to other authors [Engerson T.D., McKelvey T.G., Rhyne D.B. et al.; Yokoyama Y., Beckman J.S., Beckman T.K. et al.], 50% irreversible conversion occurs only after 3–4-h ischemia at 37°C, and this process *in vivo* is more rapid than *in vitro,* at least as concerns the liver and kidneys [McKelvey T.G., Hollwarth M.E., Granger D.N. et al.].

XDH conversion into XO is explained by either oxidation of SH groups (reversible conversion) [Kehrer J.P., Piper H.M. and Sies H.] or activation of Ca^{2+}-calmodulin-dependent proteases and limited proteolysis of the enzyme (irreversible conversion) [Stirpe F. and Della Corte E.] (see Fig. 85). The role of Ca^{2+}-dependent proteolysis in the XDH-to-XO conversion in the rat intestine is confirmed by the data on protective effects of calmodulin and trypsin inhibitors [Roy R.S. and McCord J.M.] and the antiischemic effect of trypsin inhibitor revealed in experiments on ischemia of the intestines and other organs [Dadoukis I., Angouridakis K. and Aletras M.].

According to our data, the total XDH + XO activity in the rat liver decreased after 2-h total ischemia, irreversible XDH conversion into XO was weak (XO content increased from 8.7 to 15.3% of the total enzyme activity), and reversible conversion was absent [Rashba-Step J., Tatoyan A., Duncan R. et al.]. Additional evidence for the absence of XDH conversion into XO or its low extent has been recently obtained in experiments with the totally ischemic rat kidneys [Okabe H.], rat heart [Xia Y. and Zweier J. L.; Coudray C., Boucher F., Pucheu S. et al.], and dog brain (the latter with respect to irreversible conversion) [Mink R.B., Dutka A.J., Kumaroo K.K. and Hallenbeck J.M.].

The timing and extent of conversion largely depend on the severity (degree) of ischemia. Thus, the increase of XO activity in the rat liver *in vitro* after 3-h hypoxia and 2-h ischemia was similar (by 45%) [Brass C.A., Narciso J. and Gollan J.L.]; in the hypoxic rat liver *in vitro,* XDH conversion into XO was weak (≈ 8%) [Siems W., Schmidt H., Müller M. et al.]; and in the rat heart exposed to 60-min hypoxic perfusion and 30-min oxygenation, no such conversion was revealed [Bindoli A., Cavallini L., Rigobello M.P. et al.; Kehrer J.P., Piper H.M. and Sies H.].

Damage to the barrier function of membranes in ischemic organs leads to the increase of XO content in the outflowing blood (perfusate) during reperfusion. This phenomenon was observed in patients after reconstructive surgery on limbs [Friedl H.P., Smith D.J., Till G.O. et al.; Tan S., Gelman S., Wheat J.K. et al.] and in rats exposed to hemorrhagic shock [Tan S., Yokoyama Y., Dickens E. et al.] and hypoxia or ischemia of the liver [Frederiks W.M. and Bosch K.S.; Tan S., Yokoyama Y., Wang Z. et al.; Yokoyama Y., Beckman J.S., Beckman T.K. et al.] and intestine [Caty M.G., Schmeling D.J., Friedl H.P. et al.]. In rabbits with the descending aorta occlusion accompanied by ischemia of the liver and intestine, the increase of XO activity in the heart was observed [Nielsen V.G., Tan S., Baird M.S. et al.]. This was attributed to the binding of circulating XO by the endothelium of heart vessels. The possibility of XO binding to endothelial cells in culture, demonstrated previously [Tan S., Yokoyama Y., Dickens E. et al.], makes this explanation quite logical.

The second important factor conditioning ROS production in the xanthine–xanthine oxidase system during ischemia is a *marked increase in the amount of reaction substrates, xanthine and hypoxanthine,* due to increasing degradation of adenine nucleotides during ischemia and accumulation of nucleotides and purine bases in the tissue (see Chapter 3, Fig. 27). The accumulation of adenine nucleotide degradation products, especially hypoxanthine, is particularly intensive during total ischemia, when nucleotide synthesis is inhibited, hypoxanthine oxidation by XO is limited by NAD^+ deficiency, and no substances are flushed out of the organ [Jennings R.B., Reimer K.A., Hill M.L. and Mayer S.E.; Van Belle H., Goossens F. and Wynantsi J.; Warnick C.T. and Lazarus H.M.].

On this basis, many specialists assumed that the xanthine–xanthine oxidase reaction provides the basic source of hydroxy radicals and, consequently, is the main factor responsible for free-radical processes in ischemic organs.

It was also found that changes produced by the xanthine–xanthine oxidase reaction in isolated membranes, cells, or organs are similar or even identical to those observed during ischemia [Hess M.L., Okabe E. and Kontos H.A.; Parks D.A., Bulkley G.B., Granger D.N. et al.].

It should be noted that both factors promoting the xanthine–xanthine oxidase reaction (XDH conversion into XO and substrate accumulation) operate during ischemia (hypoxia) of organs, although the prevailing view is that substrate oxidation by xanthine oxidase and ROS production occur only in the reperfusion period. This view is based on the data that the xanthine–xanthine oxidase reaction is strongly oxygen-dependent ($pO_2 \geq 2$ mm Hg) [Hille R. and Massey V.].

The third factor, *antiischemic and antiradical effects of XO inhibitors,* will be discussed in detail below (see Chapter 11). Here, it should be noted that XO inhibitors—allopurinol, oxipurinol, etc. —not necessarily produced a desirable effect, and their use often led to adverse consequences [Kehrer J.P., Piper H.M. and Sies H.; Nakashima M., Niwa M., Iwai T. and Uematsu T.; Zager R. A. and Gmur D. J.]. In many cases, XO inhibitors did produce the protective antiischemic effect but had no influence on ROS production and accumulation of LPO products during ischemia and reperfusion [Ferrari R.P., Battiston B., Brunelli G. et al.; Pisarenko O.I., Lakomkin V.L., Studneva I.M. et al.]. This effect of XO inhibitors was observed in patients after aortocoronary shunting [Arduini A., Eddy L. and Hochstein P.; , Canada A.T., Stein K., Martel D. and Watkins W.D.; Johnson W.D., Kayser K.L., Brenowitz J.B. and Saedi S.F.] and in pigs with the transplanted heat–lung complex [Qayumi A.K., Jamieson W.R., Godin D.V. et al.], although XO activity in the human and porcine heart is very low or absent (see above).

Thus, the data on the existence and extent of XDH into XO are as yet inconclusive; XO at the probable peak of its activity in the reperfusion period can be washed out of the organ with blood or perfusate; the data on the ability of XO inhibitors to reduce ROS production and accumulation of LPO products are contradictory; and XO inhibitors (allopurinol) proved to produce some effects unrelated to their basic function, i.e., XO inhibition (see Chapter 11). All these circumstances prevent us from any definite conclusions concerning the active role of the xanthine–xanthine oxidase reaction in initiating free-radical processes in ischemic organs.

10.2.3 Formation of Reactive Oxygen Species during Arachidonic Acid Metabolism along the Cyclooxygenase and Lipoxygenase Pathways

As follows from Fig. 88, ROS and LPO products are formed at the initial steps of both cyclooxygenase and lipoxygenase pathways of AA metabolism.

Fig. 88: Production of reactive oxygen species (ROS) and LPO products along AA metabolism pathways, according to [Varfolomeev S.D. and Mezch A.T.; Demopoulos H.B., Flamm E.S., Pietronigro D.D. and Seligman H.L.] with modifications.

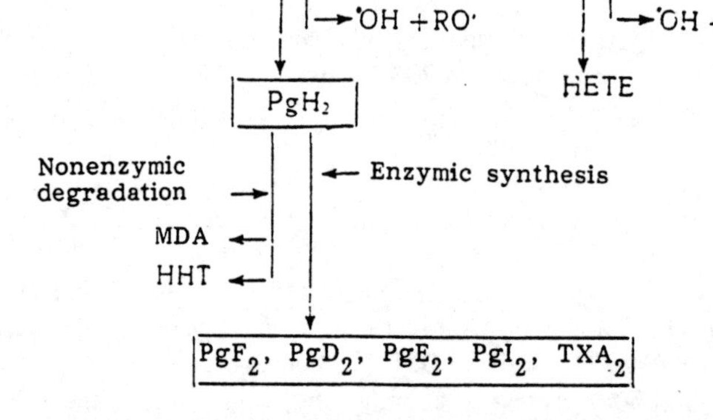

(1) In the cyclooxygenase pathway, this occurs in the course of PgG_2 conversion into PgH_2 (the peroxidase function of PgH synthetase).

(2) In the lipoxygenase pathway, this occurs during the conversion of AA hydroperoxide (HPETE) into a hydroxy acid (HETE) (the peroxidase function of GSH peroxidase).

(3) This also occurs because PgH synthetase is nonspecific and can metabolize not only endoperoxides (PgG_2 into PgH_2), but also hydroperoxides of AA and other acids. Under anaerobic conditions, such a peroxidase function may also be performed by lipoxygenase.

Thus, the initial stages of AA metabolism are accompanied by the production of ROS ($^{\bullet}OH$), which can initiate LPO in tissues.

In addition, the scheme shows that cyclooxygenase and lipoxygenase produce hydroperoxides and MDA and, hence, can induce enzymatic LPO themselves

The role of these enzyme systems in ROS induction and their involvement in organ damage during ischemia and reperfusion have been confirmed by abundant data on the protective effect of agents inhibiting phospholipase A_2, cyclooxygenase, and lipoxygenase (see Chapter 11). However, the expression of AA metabolism via different pathways varies depending on the organ (cell type). The cyclooxygenase pathway is especially active in endothelial cells of small vessels and in blood platelets, which account for the synthesis of basic cyclooxygenase products, prostacyclin and thromoxane. The exposure of these cells to ischemia and reperfusion sharply intensifies AA metabolism and ROS production ($O_2^{\bullet}$, $^{\bullet}OH$) and induces platelet activation and spontaneous aggregation [Leo R., Pratico D., Iuliano. L. et al.], whereas cyclooxygenase blockers inhibit these processes in ischemic and reperfused organs (see Chapter 4 and 11).

The lipoxygenase pathway of AA metabolism is very active in polymorphonuclear leukocytes (PMN): their functional activity, oxidative burst, and chemotaxis are regulated by lipoxygenase products (LTA_4, LTA_3, etc.).

On this basis, inhibitors of different enzyme systems involved in AA metabolism are not equally effective during ischemia of different localization.

10.2.4 Formation of NO$^{\bullet}$ and Reactive Oxygen Species by Nitric Oxide Synthases

Nitric acid synthase (NOS) is a cytochrome P-450-type enzyme catalyzing oxidation of L-arginine nitrogen with the formation of L-citrulline and nitric oxide ($NO^{\bullet}$), a primary radical species.

Three isoforms of NOS were described: one inducible and Ca^{2+}-independent (iNOS) and two constitutive calmodulin/Ca^{2+}-dependent (cNOS) localized either in the vascular endothelium (eNOS) or in nerve endings and brain cells (ncNOS). The inducible iNOS form was found in virtually all cell types, being especially abundant in macrophages, blood platelets, mast cells, Kupffer cells, smooth muscle cells, and kidney tubules. It is inactive under physiological conditions but is readily inducible under the effect of various factors, including IL-1, TNF-α, etc. Upon one-time induction, iNOS generates $NO^{\bullet}$ 1000 times as actively as eNOS and retains its activity for up to five days [Alexander B.; Davies M.G., Fulton G.J. and Hagen P.O.] mains much first form

Nitrogen monoxide (NO) is water- and lipid-soluble, diffusible, and short-lived molecule (half-life $\approx$ 5 s) with messenger, cytotoxic, and vasodilative properties. It exists in the forms of $NO^{\bullet}$(nitric oxide), NO^+ (nitrosonium, an oxidized form), and NO^- (nitroxyl anion, a reduced form) [Stamler J.S., Singel D.J. and Loscalzo J.]. $NO^{\bullet}$ can react with O_2, producing NO_2^-, NO_3^-, N_2O_3, and N_2O_4; with thiols and transition metals, forming NO–thiol and dinitrosyl iron–cysteine (DNIC) complexes; etc. However, the most harmful effect on tissues arises from the ability of $NO^{\bullet}$ to react with $O_2^{\bullet}$ [Huie R.E. and Padmaja S.] (reaction rate $3.4 \cdot 10^7\ M^{-1} \cdot s^{-1}$ [Radi R., Beckman J.S., Bush K.M. and Freeman B.A.] or even $0.7 \cdot 10^9\ M^{-1} \cdot s^{-1}$), producing peroxynitrite anion ($ONOO^-$) and peroxynitrous acid (ONOOH):

$^{\bullet}NO + O_2^{\bullet} \rightarrow ONOO^- + H^+ \leftrightarrow ONOOH \rightarrow {}^{\bullet}NO_2 + {}^{\bullet}OH$ [Radi R., Beckman J.S., Bush K.M. and Freeman B.A.].

$NO^{\bullet}$ and $ONOO^-$ are one-electron oxidants actively initiating LPO [Alexander B.; Radi R., Beckman J.S., Bush K.M. and Freeman B.A.]. Sulfhydryl groups react with $ONOO^-$ 1000 times as actively as with H_2O_2 [Beckman J.S. and Crow J.P.]. $ONOO^-$ and especially ONOOH (after decomposition to $^{\bullet}NO_2$+ $^{\bullet}OH$)

are stronger oxidants and LPO inducers, compared to $O_2^{\bullet-}$, and their oxidant effect is comparable to that of $^{\bullet}OH$ [Beckman J.S. and Crow J.P.; Beckman J.S.]. On this basis, excess $NO^{\bullet}$ synthesis in cells, especially combined with excess $O_2^{\bullet-}$ formation, results in the substitution of more reactive oxidants ($ONOO^-$ and $^{\bullet}OH$) for less reactive $O_2^{\bullet-}$. A direct correlation was revealed between NOS activity and generation of free radicals [Onufriev M.V., Stepanichev M.Yu., Lazareva N.A. et al.].

Like other free radical species, $NO^{\bullet}$, $ONOO^-$, and ONOOH at high doses are toxic for cells: they induce LPO (with all its consequences), cause nitrosation of iron-containing enzymes, inhibit mitochondrial respiration and aconitase activity, induce DNA damage and mutations, etc. [see Alexander B.]. On the other hand, their small doses can have a favorable cytoprotective effect. Thus, $NO^{\bullet}$ can serve as $O_2^{\bullet-}$ scavenger, enhance DNA and protein synthesis, and regulates vascular tension, ensuring endothelium-dependent relaxation of small arteries and arterioles (a very important property) (see Chapter 1). Peroxynitrite at concentrations of 100 to 1000 nM inhibits thrombin activity and adhesion of polymorphonuclear leukocyte–endothelial cell adhesion induced by H_2O_2or ischemia and reperfusion, has a vasodilative effect, and improves the functioning of ischemic organs, e. g., the rat and cat heart [Lefer D. J., Scalia R., Campbell B. et al.; Liu S., Beckman J.S. and Ku O.D.; Nossuli T.O., Hayward R., Scalia R. et al.]. However, the boundary between high (toxic) and low (protective) concentrations of endogenous NOS products shifts within a broad range under pathological conditions. Apparently, this is the factor explaining controversial results obtained in experiments on using NOS inhibitors or, $NO^{\bullet}$ donors as protective drugs (see Chapter 11).

Effects of ischemia and reperfusion on NOS activity and $NO^{\bullet}$ production have been actively studied during the past three to five years. On the one hand, the results showed that the production of endothelium-dependent vasodilators ($NO^{\bullet}$ or $NO^{\bullet}$ + S-nitrosocysteine) in the vascular endothelium of ischemic organs is reduced, which may be a factor responsible for the postischemic no-reflow phenomenon (see Chapter 1). On the other, the data on the role of eNOS in ischemic/reperfusion damage are somewhat contradictory. In experiments with eNOS-knockout transgenic mice, damage to the heart after 16-min total ischemia and 60-min reperfusion was less pronounced than in wild mice [Flogel U., Decking U.K., Godecke A. and Schrader J.]; however, in other experiments on a similar model (20–30-min ischemia and 120-min reperfusion), this damage in transgenic mice was relatively stronger [Jones S.P., Girod W.G., Palazzo A.J.et al.]. Most of other studies, however, provide evidence for the increase of NOS activity during ischemia. Thus, total NOS activity in the rabbit heart increased as soon as after 5 min of incomplete (low-flow) ischemia [Depre C., Fierain L. and Hue L.]; total NOS and iNOS activities in the rat brain increased after 40-min ischemia and 4-h reperfusion [Forman L.J., Liu P., Nagele R.G. et al.]; iNOS activity increased in neutrophils infiltrating the ischemic human brain [Fortster C., Clark H.B., Ross M.E. and Iadecola C.] and in glial cells and macrophages of the ischemic neurohypophysis, reaching a peak after 24 h; eNOS activity in rat kidneys increased after 1-h ischemia and 2–24-h reperfusion [Shoskes D.A., Xie Y. and Gonzalez-Cadavid N.F.]; iNOS (but not eNOS) activity in the rat heart increased 72 h after coronary artery occlusion [Wildhirt S.M., Suzuki H., Horstman D. et al.]. The increase of $NO^{\bullet}$ content was observed in the ischemic and reperfused rat heart, liver, and brain [Ma X.L., Lopez B.L., Liu G.L. et al.; Islam S.J.; Lu Y.T., Liu S.F., Mitchell J.A. et al.] and in the brain of piglets exposed to 90-min ischemia and 120-min reperfusion [Segawa D., Hatori N., Yoshizu H. et al.]. The content of $NO^{\bullet}$ and peroxynitrite increased in blood plasma and tissues of the ischemic rat brain [Forman L.J., Liu P., Nagele R.G. et al.] and rat intestine 60 min after its reperfusion [Cuzzocrea S., Zingarelli B. and Caputi A.P.]. NOS inhibition by means of L-NAME reduced the accumulation of LPO products in the rat brain after 10–60-min occlusion of the middle cerebral artery [Gumuslu S., Serteser M., Ozben T., et al.] and, on the same model, reduced by 46.4% the formation of $^{\bullet}OH$ after 2-h ischemia and 2-h reperfusion [Wei J.and Quast M.J.]. NOS inhibitors also reduced the accumulation of TBARS and prevented the decrease of SOD activity in the ischemic rat heart [Yang B.C. and Mehta J.L.]. Recent studies showed that $NO^{\bullet}$ formation in acidic and highly reduced conditions (characteristic of ischemia) can also occur nonenzymatically, by direct disproportionation or reduction of nitrite to $NO^{\bullet}$. Such NOS-independent $NO^{\bullet}$ generation at pH dropping from 7.4 to 5.5 increased 100-fold, and in the ischemic cardiac tissue it increased more than 40-fold, i.e., exceeded the rate of $NO^{\bullet}$ generation by NOS [Zweier J.L., Samouilov. And Kuppusamy P.].

These data show that ischemia, damaging endothelial cells, inhibits eNOS activity in them but activates iNOS, which leads to the increase of the total NOS activity and $NO^{\bullet}$ and peroxynitrite contents in all the cells. Combined with enhanced $O_2^{\bullet-}$ production and possible NOS-independent $NO^{\bullet}$ generation, this creates additional sources of highly active LPO inducers. It may well be that NO generated by different NOS isoforms have different or opposing effects in physiological or pathological states. This question, however, remains open and is currently being studied with the use of new approaches, such as selective inhibition of different NOS isoforms and experiments with transgenic NOS-deficient animals [see Huang P.L. and Lo E.H.].

10.2.5 Other Intracellular Sources of Reactive Oxygen Species

ROS formation during biosynthesis and oxidation of catecholamines. As shown above (see Chapter 5), large amounts of exogenous CA or increase in the content of endogenous CA (especially epinephrine of adrenal origin) in the blood and organs during stress and shock (see Chapter 7) cause LPO intensification. The LPO-promoting action of CA may be a consequence of two processes: (1) generalized or local hemodynamic disorders in individual organs under the effect of CA, which lead to ischemia, and the latter, in turn, causes generalized or local intensification of LPO processes; and (2) direct LPO induction by CA.

The first pathway of LPO initiation by CA is most probable in target organs, the heart being one of them.

The second pathway may be accounted for by ROS ($O_2^{\bullet}$) production both during biosynthesis of epinephrine (with thyramine being converted into octopamine) and during oxidation or autoxidation of epinephrine to adrenochrome (with epinephrine being converted into semiquinone and semiquinone into quinone). It is also known that stimulation of CA-ergic neurons leads to accumulation of CA analogues, such as hydroxydopamine (6-OHDA) and 6-aminodopamine (6ADA) which also produce $O_2^{\bullet}$, H_2O_2 and $^{\bullet}OH$ when undergoing autoxidation. Of particular importance in this case is the quantitative ratio between ROS producers and the content of norepinephrine, which may act as a ROS scavenger, as well as the activities of SOD and catalase, which inhibit 6-OHDA autoxidation under certain conditions. CA oxidation may be catalyzed by transition metals (Cu^{2+}, Fe^{2+}), iron proteins, and iron–heme proteins combined with H_2O_2.

The participation of CA in ROS production may also be mediated by intensification of HMPS and generation of $O_2^{\bullet}$ in neutrophils. In this case, adrenochrome intensifies HMPS 10 times as actively as epinephrine, and its formation may be promoted by epinephrine oxidation by myeloperoxidase.

The question concerning the probability of $O_2^{\bullet}$ production via this pathway *during ischemia of organs* remains open, but there are conditions obviously favorable for this process. Thus, after LAD occlusion (partial cardiac ischemia) in rats, epinephrine content in the left ventricle reached 160–240% of the initial after 10-120 min and continued to grow 40 min after reperfusion. By contrast, norepinephrine content during ischemia increased for only a short period of time (to 126% within the first 10 min), falling below the initial level after 20–120 min and during the postischemic period [Olbinskaya L.I. and Litvitsky P.F.]. A similar pattern was observed in other experiments [Popov V.G., Lazutin V.K., Khitrov N.K. et al.]. However, in experiments on a similar model of cardiac ischemia in dogs [Rao P.S., Cohen M.V. and Mueller H.S.], the contents of both epinephrine and norepinephrine increased after 30 and 45 min to 132–135 and 170–185%, respectively.

After complete ischemia or hypoxia of the isolated guinea pig or rat heart, conversely, the content of norepinephrine in the perfusate increased much stronger than those of epinephrine and dopamine [Kuzmin A. I., Shulzhenko V. S. and Medvedev O. S.], and the accumulation of endogenous CA during ischemia and their release into the perfusate within the first few minutes of reperfusion were proportional to the duration of ischemia and Ca^{2+}-independent.

In the ischemic rat brain, CA accumulation and increasing ROS production were observed [Oh S.M. and Betz A.L.; Simonsen S.G., Zhang J., Canada A.T. et al.].

CA oxidation during ischemia may be promoted by a decrease in pH factor and the release of Fe^{2+} in the cytosol (see below), and the involvement of CA (both epinephrine and norepinephrine) in ROS production is conditioned by their ability (especially manifested under anoxic and ischemic conditions) to release iron from ferritin and to react with iron and xanthine oxidase, producing $^{\bullet}OH$ [Allen D.R., Wallis G.L. and McCay P.B.].

Although mechanisms of this process have been studied insufficiently, the role of CA in the initiation of free-radical reactions is obvious, especially taking into account that β-adrenergic blockers (β-antagonists), preventing CA binding to receptors, exert a protective effect during ischemia and reperfusion of organs, the heart in particular (see Chapter 11).

ROS production can be regulated by peroxidases, and a major role in this case belongs to three enzymes: GSH peroxidase, myeloperoxidase of phagocytes, and cytochrome P-450 peroxidase.

GSH peroxidase breaks down peroxides by a nonradical mechanism and, hence, is an antioxidant enzyme.

As myeloperoxidase metabolizes H_2O_2, it produces highly reactive compounds, which stimulate LPO (see below).

The peroxidase function of cytochrome P-450 is only manifested with respect to organic peroxides and is combined with its ability to produce ROS (see above).

Some other enzymes also have peroxidase activity. Follicular cells of the thyroid contain thyroperoxidase, salivary and mammary glands contain lactoperoxidase; peroxidases in small amounts are also found in the uterus, intestinal mucosa, and nerve tissue. Their presence in all these organs makes their secretions bactericidal by enriching them with ROS or other highly reactive compounds. The ability of lactoperoxidase to induce LPO was demonstrated in experiments with microsomes and artificial lipid membranes [Buege J.A. and Aust S.D.].

Peroxidase or pseudoperoxidase activity is also characteristic of many other heme-containing compounds, such as hemes and metals, Hb, Mb, cytochromes c and b_5, cytochrome oxidase, hemocyanine, and Pg synthetase (cyclooxygenase). In most cases, this activity is supplementary rather than primary.

Enzymes located within peroxisomes can also participate in LPO induction, particularly during ischemia. Experiments performed in our laboratory showed that incubation of the dog renal cortex with peroxisomal D-amino acid oxidase for 3 h at 25°C is accompanied by a fourfold increase in MDA content, and such incubation in the presence of D-alanine as a substrate, by a fivefold increase in MDA, compared to its initial content [Komarov O. S.].

An additional source of hydroxy radicals is provided by autoxidation of Hb (to MetHb), glutathione, thiols, flavins, cytochrome c, ascorbic acid, and other compounds, but its role is insignificant.

10.2.6 Production of Reactive Oxygen Species by Blood and Tissue Phagocytes

The functional activity of blood and tissue phagocytes (neutrophils, monocytes, and macrophages) is known to largely depend on their ability to generate ROS. The bactericidal effect of phagocytes is accounted for by the action of ROS within phagolysosomes, and their cytotoxic effect, by extracellular ROS release promoted by a rise in the intracellular Ca^{2+} level [Campbell A.K. and Hallett M.]. According to current views, other forms of phagocyte activity, including proliferation, blast transformation, and chemotaxis, are also based on the free-radical mechanism [Beklemishev N. D.; Klebanoff S.J. and Seymour J.].

ROS production in phagocytes, accompanied by an increase in oxygen consumption (respiratory burst), is initiated by activation of the NADPH oxidases located on the outer plasma membrane, which catalyze electron transfer from NADPH to molecular oxygen. This results in the decrease of NADPH level and the formation of $O_2^{\bullet}$ in the reaction:

$$NADPH + 2O_2 \rightarrow NADP^+ + 2H^+ + 2O_2^{\bullet},$$

as well as stimulates glucose metabolism in the HMPS (Fig. 89). Thereafter, $O_2^{\bullet}$ react with HO, producing $^{\bullet}OH$ (see figure 81).

Another source of ROS in phagocytes is the myeloperoxidase (MPO)–H_2O_2–halogen (Cl^-, Br^-, I^-) or thiocyanate (SCN^-) system. This system is also triggered by phagocyte activation and gives rise to the most reactive forms, namely, $^1O_2$, OCl^- and HOCl. The latter, reacting with amines, produces chloramines, which are long-lived oxidants [see Arndt H., Kubes P. and Granger D. N.; Rossi F.].

MPO activity is especially high in neutrophils, slightly lower in macrophages (decreasing in the course of cell maturation: bone marrow promonocyte > bone marrow monocyte > blood monocyte > macrophage), and still lower in eosinophils and blood platelets [Mayansky A.N. and Mayansky D.N.].

Fig. 89: Production of reactive oxygen species (ROS) in the NADPH- dependent electron-transport chain of blood and tissue phagocytes.

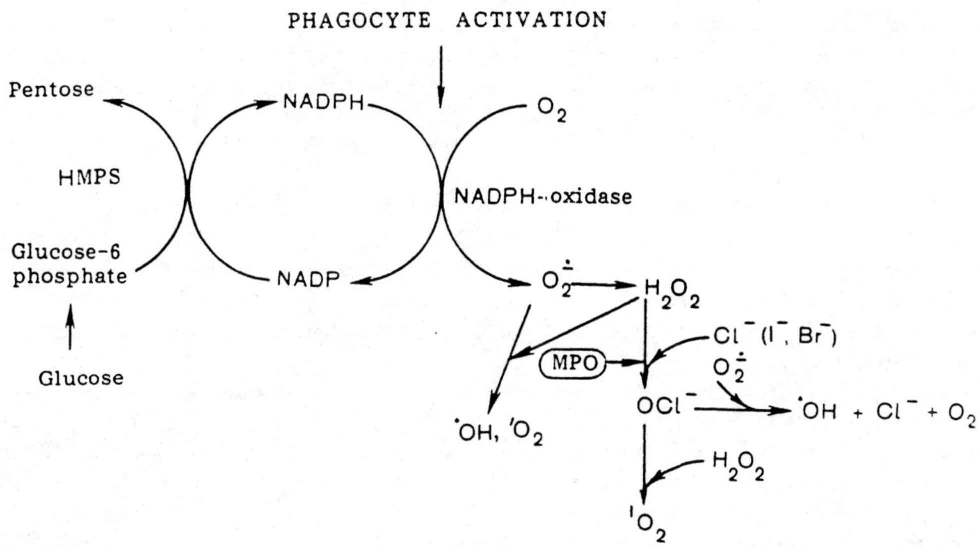

ROS production in phagocytes (especially in neutrophils) may also be promoted by other auxiliary systems of HMPS stimulation, including NADPH and NADH oxidation catalyzed by Mn^{2+}, MPO, NAD kinase, ascorbic acid, or H_2O_2 combined with MPO, catalase, and ascorbic acid.

The more powerful MPO system, the presence of other $O_2^{\bullet}$- producing auxiliary systems, and SOD deficiency in neutrophils apparently account for the fact that oxygen-dependent mechanism play a greater role in the functional activity of these cells, as compared to macrophages. It was shown that up to 90% of O_2 consumed by neutrophils is expended for the formation of $O_2^{\bullet}$ and H_2O_2, compared to 30% in monocytes and less than 10% in alveolar macrophages [see Beklemishev N. D.; Lukyanova L.O., Kudrin A.N. and Kogan A.Kh.; Mayansky A.N. and Mayansky D.N.].

To trigger the system of ROS production in neutrophils, monocytes, and macrophages, these cells must be activated. The functional activation of phagocytes (to chemotaxis, adhesion, phagocytosis, etc.) and ensuing ROS production by them are achieved upon cell contact with foreign bodies or pathologically transformed proteins; interaction with FFA and diacylglycerols formed due to activation of phospholipases; chemotactic stimulation; and interaction with membrane-active substances altering the molecular topography of the plasma membrane, such as LPO products, cytokines (TNF-α, IL-1), or oxidized LDL [Beklemishev N. D.; Bilenko M.V., Klebanov G.I., Dolgina E.N., and Nikankina L.V.; Mayansky A.N. and Mayansky D.N.; Klebanoff S.J. and Seymour J.].

Recent studies revealed a number of new stimulants, which include vascular adhesion molecules expressed on endothelial cells: vascular cell adhesion molecule-1 (VCAM-1), intercellular adhesion molecule-1 (ICAM-1), endothelin-1 (ET-1), P-selectin, and others. In response to stimulation, phagocytes form specific receptors. Among the latter, CD11b/CD18 adhesion glycoproteins have been studied in more detail.

Phagocyte activation is a rapid process. It was observed within few minutes after protein damage [Shipp J.C., Opie L.H. and Challoner D.], and $O_2^{\bullet}$-producing systems were triggered 20–60 s after stimulation [Beklemishev N. D.]. Upon activation, the rates of $O_2^{\bullet}$ and H_2O_2 production increase by a factor of 2 to more than 12 [Klebanoff S.J. and Seymour J.].

Upon *ischemia*, phagocytes confined to the organ become activated, and *reperfusion* leads to the additional accumulation and activation of these cells in microvessels. Phagocytes remaining in the organ during ischemia (neutrophils and tissue macrophages) are more tolerant to this condition than organ-specific cells (hepatocytes, cardiomyocytes, etc.) because both neutrophils (receiving energy supply mainly through glycolysis) and macrophages (partly facultative anaerobes) are highly resistant to hypoxia. Neutrophils retain some of their functions (e. g., phagocytosis) even under anaerobic conditions [Mayansky

A.N. and Mayansky D.N.]; their cytotoxic effect during hypoxia is reduced by 29–73% [Hafeman D.G. and Lucas Z.J.] but does not disappear completely even in totally anoxic conditions.

Reperfusion is accompanied by changes of hemodynamics and microcirculation in the organ, its edema, and endothelial cell damage (see Chapter 1). Under these conditions, endothelium–blood cell interactions change significantly, which leads to the accumulation of blood cells (including neutrophils), the enhancement of their chemotactic and adhesive properties, and, eventually, occlusion of capillaries (more frequently, of postcapillary venules) and extravasation of leukocytes [Arndt H., Kubes P. and Granger D. N.; Gute D. C., Ishida T., Yarimizu K. and Korthuis R. J.; Menger M.D., Lehr H.A. and Messmer K.; Schmid-Shonbein G.W.].

The accumulation of neutrophils increases with the time of ischemia and reperfusion, whereas the use of antibodies to cell adhesion molecules, CD11b/CD18 glycoprotein blockers, and other methods of leukocyte depletion prevent changes in the chemotactic properties of neutrophils that are responsible for their accumulation [Arndt H., Kubes P. and Granger D. N.; Harris A.G., Leiderer R., Peer F. and Messmer K.; Korthuis R.J. and Gute D.C.; Mullane K.M., Read N., Salmon J.A. and Moncada S.; Romson J., Hook B., Kunkel S. et al.].

Accumulated leukocytes contribute to reperfusion injury in several ways, aggravating the no-reflow phenomenon (see Chapter 1), producing proteolytic enzymes (catalase, collagenase, gelatinase) and inflammatory mediators, and, finally, promoting respiratory burst with ROS formation and LPO initiation. The degree and mechanisms of this contribution differ from organ to organ. It was shown that the deleterious effect of phagocytes is strongly manifested during reperfusion of the ischemic intestine [Arndt H., Kubes P. and Granger D. N.; Hernandez L.A., Grisham M.B., Twohig B. et al.; Yoshikawa T., Ueda S., Naito Y. et al.], heart [Campo G.M., Squadrito F., Campo S. et al.; Romson J., Hook B., Kunkel S. et al.; Röth E., Lantos J., Temes G. et al.], skeletal muscles [Belkin M., LaMorte W.L., Wright J.G. and Hobson R.W.; Ciuffetti G., Mercuri M., Mannarino E. et al.; Ferrante R.J., Hobson R.W. 2nd, Miyasaka M. et al.; Korthuis R.J. and Gute D.C.], and liver [Jaeschke H., Farhood A. and Smith C. W.; Chavez-Cartaya R.E., Metcalfe S., Ramirez-Romero P. et al.], being far less pronounced in lungs [Deeb G.M., Grum C.M., Lynch M.J. et al.] and kidneys [Thornton M.A., Winn R., Alpers C.E. and Zager R.A.]. In addition, the type of phagocytes responsible for ROS production and organ damage and the degree of their involvement in these processes depend on the duration of ischemia and reperfusion.

During partial ischemia of the rat brain (middle cerebral artery occlusion), $O_2^{\bullet}$ production by neutrophils was more pronounced after two hours than after one hour [Fabian R.H. and Kent T.A.]. During early reperfusion of the totally ischemic liver (after 20–60 min), the main role in ROS production and reperfusion injury belonged to tissue macrophages (Kupffer cells) [Jaeschke H., Bautista A. P., Spolarics Z. and Spitzer J. J.; Rymsa B., Wang I.F. and de Groot H.; Lindert K.A., Caldwell-Kenkel J.C., Nukina S.et al.; Shiratori Y., Kiriyama H., Fukushi Y. et al.; Walsh T.R., Rao P.N., Makowka L. and Starzl T.E.]; after 6–24 h of reperfusion, this role was assumed by PMN accumulated in the liver during this period [Jaeschke H., Farhood A. and Smith C. W.; Jaeschke H., Farhood A., Bautista A. P. et al.]. Antibodies to ICAM-1 + LFA-1 or ICAM-1 + CD18 significantly inhibited PMN accumulation in the rat liver exposed to 90-min ischemia and 6-h reperfusion, but the decrease of MDA content in tissues (compared to the control) was revealed not earlier than after 12 h of reperfusion [Oshiro Y., Marubayashi S., Maeda T. et al.].

In the heart and skeletal muscles, in contrast to the liver, PMN were the main cell type responsible for organ damage and ROS production, although this was manifested after long-term reperfusion, as in the previous case. PMN accumulation in the infarct area of the dog heart reached a peak after 3–6-h reperfusion [de Lorgeril M., Rousseau G., Basmadjian A. et al.], whereas chemiluminescence of blood PMN in dogs with AMI was activated only after reperfusion for 12–14 h [Klebanov G.I., Kreinina M.V., Pozin V.M. et al.]; in hamster striated muscles exposed to 4-h ischemia, a peak of PMN accumulation and a 20-fold increase in their adhesion to the endothelium of postcapillary venules was revealed after 24 h of reperfusion [Menger M.D., Pelikan S., Steiner D. and Messmer K.]. An enhanced ROS production by circulating PMN was also observed in other studies on AMI in dogs [Röth E., Lantos J., Temes G. et al.] and human patients [Korochkin I. M., Klebanov G. I. Chukaeva I. I. et al.; Klebanov G.I., Turkmenov E.M. and Kreinina M.V.]. On the other hand, the intensity of ROS production and the severity of reperfusion injuries were comparatively low in leukocyte-depleted animals with AMI [Röth E., Lantos J., Temes G. et al.; Romson J., Hook B., Kunkel S. et al.], MAC-1 (CD11b/CD18)-deficient mice exposed to partial ischemia and reperfusion of the brain [Sorensen V., Nilsson U., Pettersson S. et al.], and P-selectin-deficient mice with total hepatic ischemia [Yadav S.S., Howell D.N., Steeber D.A. et al.].

In addition to direct activation of ROS production in Kupffer cells and PMN, ischemia caused the priming of these cells, i.e., the state preceding cell activation and characterized by a reduced latent period and a greater strength of cell response for repeated stimuli. Thus, the priming of Kupffer cells for enhanced ROS production was observed *in vitro* [Jaeschke H., Bautista A. P., Spolarics Z. and Spitzer J. J.] and *in vivo* [Liu P., Vonderfecht S.L., Fisher M.A. et al.] after hepatic ischemia, the priming of PMN was revealed in dogs with AMI [Klebanov G.I., Kreinina M.V., Pozin V.M. et al.], and that of human monocytes, after their incubation with endothelial cells under hypoxic conditions [Herbert J.M., Corseaux D., Lale A. and Bernat A.].

As ROS formation and the accumulation of LPO products in organs were mainly observed in the course of ischemia or during the first 5–30 min of reperfusion (see Chapter 5), it appears that phagocytes become involved in ROS production only at the late stage of reperfusion (at least as concerns PMN). It is even more probable that the very processes of PMN accumulation, adhesion to the endothelium, and activation prerequisite to ROS production are conditioned by ROS previously formed in the organ by other systems, e. g., xanthine oxidase of endothelial cells, and, hence, the involvement of blood phagocytes in ROS production is a secondary process. The fact that antioxidants prevent PMN accumulation in reperfused organs confirms the validity of this assumption [Arndt H., Kubes P. and Granger D. N.; Bienvenu K. and Granger D.N.; Jones H.P., Grisham M.B., Bose S.K. et al.; Menger M.D., Pelikan S., Steiner D. and Messmer K.; Nose P.S.], but more data are needed for drawing any definite conclusions.

Relationships between activation of phagocytes to ROS production and the intensity of LPO processes in them are not yet absolutely clear as well.

On the one hand, the bactericidal and cytotoxic activities proved to be conditioned by not only ROS themselves, but also by LPO processes [Mayansky A.N. and Mayansky D.N.]. In particular, a strong bactericidal activity is characteristic of MDA, which plays a major role in macrophages [Gordienko S.M.].

On the other, activation of macrophages in animals treated with zymozan led to the pronounced accumulation of MDA in the liver only during the first 24 h; after 7 days, MDA content decreased below the initial level, although this period was characterized by a sharp increase in phagocyte activity and the number of nonparenchymal cells in the liver [Semenyuk A.V., Voronin A.Yu., Kulikov V.Yu. et al.]. It should be noted that the excess H_2O_2 formed in phagocytes can damage their membranes and, being released from these cells, affect the surrounding tissues [Mayansky A.N. and Mayansky D.N.].

10.2.7 Production of Reactive Oxygen Species by Endothelial Cells

Endothelial cells (EC) account for only 1% of body weight [Davies M.G. and Hagen P.] but have a considerable total surface area of plasmalemma. In the heart, for example, this area is two to four times as great as that of cardiomyocytes [Sasaki R., Watanabe Y., Morishita T. and Yamagata S.].

Endothelial cells are characterized by a low number of mitochondria, low mitochondrial oxygen consumption [Spector W.S.], and a high rate of aerobic glycolysis, which makes their metabolism similar to that of granulocytes and macrophages [Dobrina A. and Rossi F.]. A high activity of hexose monophospate shunt (HMPS) [Dobrina A. and Rossi F.] in endothelial cells maintains the reduced state of GSH peroxidase and GSH reductase, which are important factors protecting EC from H_20_2 [Suttorp N., Toepten W. and Roka L.].

On the other hand, there are prerequisites for the active involvement of endothelium in generation of ROS and initiation of free-radical reactions. Corresponding evidence can be summarized as follows.

(1) EC of virtually all organs contain xanthine dehydrogenase, and many EC are characterized by the maximal content of xanthine oxidase [Phan S.H., Gannon D.E., Varani J. et al.; Kehrer J.P., Piper H.M. and Sies H.]. In several bovine and human organs (liver, heart, lungs, intestine, and kidneys), xanthine dehydrogenase capable of converting into xanthine oxidase is found exclusively in the endothelium of capillaries and small blood vessels, rather than in the cells of epithelial or mesenchymal origin [Jarasch E.D., Grund C., Bruder G. et al.; Jarasch E.D., Bruder G. and Heid H.W.]. Xanthine oxidase content in microvascular endothelium is much higher than in the endothelium of large vessels [Jarasch E.D., Bruder G. and Heid H.W.], and the levels of xanthine dehydrogenase and xanthine oxidase in cultivated endothelial cells of bovine brain capillaries is 20–25 times as high as in the brain homogenate; xanthine dehydrogenase and xanthine oxidase contents in EC of the brain and lungs is more or less similar [Terada L.S., Willinghaw I.R., Rosandich M.E. et al.]. In EC of human lung, heart, and kidney microvessels, xanthine oxidase contents proved to be sufficiently high, although there are data that the total xanthine

dehydrogenase and xanthine oxidase activities in humans are much lower than in animals. In addition to cytosolic xanthine oxidase, this enzyme was also found on the outer surface of EC plasma membranes (porcine aorta), being bound to heparin-like proteoglycans [Adachi T., Fukushima T., Usami Y et al.].

There is a hypothesis that EC, owing to the presence of xanthine oxidase, have an antimicrobial defense mechanism similar to that mediated by NADPH-oxidase system in PMN [Jarasch E.D., Bruder G. and Heid H.W.].

It was also suggested that the presence of xanthine dehydrogenase easily convertible to xanthine oxidase makes EC the source of the earliest generation of ROS under unfavorable conditions; these ROS can then initiate chemotaxis of PMN to EC, activate PMN, followed by respiratory burst [Jones H.P., Grisham M.B., Bose S.K. et al.]. Other data, by contrast, indicate that it is exactly activated PMN that cause the conversion of xanthine dehydrogenase to xanthine oxidase in EC [Phan S.H., Gannon D.E., Varani J. et al.].

(2) EC are characterized by a high rate of ADN metabolism and high contents of ADN degradation products (adenosine, inosine, hypoxanthine, and adenine contents are 40–60 times as high as in cardiomyocytes) [Nees S. and Gerlach E.]. Compared to other cell cultures, cultivated EC release considerably more adenine nucleosides into the medium [Pearson J., Carleton S. and Gordon J.L.]. The activities of enzymes involved in nucleotide degradation (ATPase and 5'-nucleotidase) and nucleotide synthesis (adenylate kinase, adenosine deaminase, and purine nucleoside phosphorylase) in cardiac EC are two to nine times as high as in cardiomyocytes [Nees S. and Gerlach E.]. High rates of formation and accumulation of ADN catabolism products (upon ischemia, mainly of hypoxanthine and xanthine) were demonstrated in a series of studies [Jennings R.B., Reimer K.A., Hill M.L. and Mayer S.E.; Van Belle H., Goossens F. and Wynantsi J.]; against the background of a high xanthine oxidase activity (xanthine dehydrogenase conversion into xanthine oxidase), this is an important factor predisposing cells to ROS generation.

Recent studies showed that xanthine oxidase of EC differs from the enzyme of other origin by its ability to use NADH (rather than hypoxanthine) as an oxidation substrate, and this process is suppressed by FAD site inhibitors (e. g., diphenyleneiodonium) [Zhang Z., Blake D. R., Stevens C. R. et al.]; hence, the conversion of xanthine dehydrogenase into xanthine oxidase is not necessary for ROS production in EC.

(3) EC contain iron, have transferrin receptors, and are capable of receptor-specific binding of transferrin to the cell surface, with subsequent endocytosis, acidification, and reduction (by ferrireductase) of transferrin-bound Fe^{3+} to Fe^{2+}. Inhibition of intracellular SOD potentiates this process [Brieland J.K., Clarke S.J., Karmiol S. et al.]. The accumulation of Fe^{2+} can facilitate ROS production by the endothelium.

(4) EC, which produce many biologically active substances regulating the vascular tone (see Chapter I), have active cyclooxygenase and lipoxygenase systems of arachidonic acid metabolism, in the course of which active radicals and hydroperoxides are formed.

(5) Like macrophages and nerve cells, EC contain NO synthases responsible for NO production. The latter is a short-lived radical actively interacting with $O_2^{\bullet}$, which results in the formation of an peroxynitrite that is more reactive than $O_2^{\bullet}$ and induces LPO more efficiently (see section 10.2.4).

(6) Due to their borderline location, EC are in permanent contact with blood phagocytes. Any damage to EC, including ischemic, leads to phagocyte activation accompanied by ROS production (see section 10.2.6).

Ample experimental evidence accumulated to date confirmed that *various adverse physical, chemical, or mechanical influences on EC sharply enhance ROS production by these cells* [Babbs C.F., Cregora M.D. and Badylak S.F.; Kinnula V.L., Everitt J.I., Whorton A.R. and Crapo J.D.; Kinnula V.L., Whorton A.R., Chang L.Y. and Crapo J.D.; Kramer J.H., Dickens B.F., Misik V. and Weglicki W.B.; Palermartinez A., Panus P.C., Chumley P.H. et al.].

The ability of EC to produce ROS also drastically increases under the effect of exogenous substances initiating free radical reactions, which makes these cells very sensitive targets for H_2O_2, the xanthine–xanthine oxidase system, paraquat, TNF-α, arachidonic acid, low density lipoproteins, and many other compounds.

The cytotoxic effect of H_2O_2 on cultivated EC of the rat pulmonary artery was manifested in rapid ATP depletion and cell death [Hyslop P.A., Hinshaw D.B., Halsey W.A. et al.; Markey B.A., Phan S.H., Varani J. et al.; Varani J., Phan S.H., Gibbs D.F. et al.; Ward P.A.], and that on EC of the bovine pulmonary artery and aorta, in DNA strand breaks [Spragg R.G.]. At low concentrations (5 μM), H_2O_2 proved to stimulate PgI_2 synthesis in EC of the porcine aorta, whereas high concentrations (25 and 50 μM) inhibited this synthesis [Buckley B.J., Kent R.S. and Whorton A.R.; Whorton A.R., Montgomery M.E. and Kent R.S.]. The addition of xanthine oxidase alone or together with xanthine led to a rapid decrease in the content of ATP, the accumulation of intracellular Ca^{2+}, and functional disorders in EC of the human umbilical vein (HUVEC) [Aalto T.K. and Raivio K.O.; Shatos M.A., Doherty J.M. and Hoak J.C.; Andreoli S.P. and McAteer J.A.; Geeraerts M.D., Ronveaux-Dupal M.F., Lemasters J.J. and Herman B.] and the bovine pulmonary artery or aorta [Franceschi D., Graham D., Sarasua M. and Zollinger R.M.; Spragg R.G.]. The damaging effect of the xanthine–xanthine oxidase system on HUVEC was manifested stronger than that on human renal epithelial cells [Andreoli S.P. and McAteer J.A.], and the effect of this system on human saphenous vein was stronger than that on human ventricular myocytes or fibroblasts [Mickle D.A., Li R.K., Weisel R.D. et al.]. Specific heparin-reversible sites of exogenous xanthine oxidase binding to the vascular endothelium were revealed [Freeman B.A., Whitfield K., Gelman D. and Parks D.A.].

Extracellular initiators of free-radical processes in EC also include cytokines, especially TNF-α produced by activated macrophages and monocytes in response to various stimuli, including stress, ionizing radiation, and infection. TNF-α selectively damaged the microvascular endothelium of various animal organs when administered *in vivo* (Persidskii Yu.V., Kudriavets Yu.I. and Barshtein Yu.A.;); *in vitro,* this factor altered cell membrane permeability in microvascular EC of the bovine and rat pulmonary arteries [Goldblum S.E., Ding X. and Campbell-Washington J.; Partridge C.A., Horvath C.J., Delvecchio P.J. et al.] and caused dose-dependent death of EC from the human umbilical vein and bovine aorta [Von Asmuth E.J., Leeuwenberg J.F., Vander-Linden C.J. and Buurman W.A.; Estrada C., Gomez C., Martin C. et al.; Bilenko M.V., Ladygina V.G. and Fedosova S.V.(1998a)].

Substances of lipid nature can also initiate ROS production in EC and have a cytotoxic effect on them. Incubation of EC from cerebral microvessels with arachidonic acid was accompanied by pronounced edema, which could not be prevented by treatment with cyclooxygenase inhibitors but was strongly aggravated upon adding H_2O_2 [Villicara A., Kempski O. and Spatz M.].

Another class of compounds cytotoxic to EC are LDL, which proved to undergo oxidation by EC in the course of long-term coincubation and, thereafter, to acquire cytotoxicity with respect to these cells and fibroblasts [Chisolm G.M. and Morel D.W.; Henriksen T., Mahoney E.M. and Steinberg D.; Morgan J., Smith J.A., Wilkins G.M. and Leake D.S.; Steinbrecher U.P.].

The cytotoxic effect of the aforementioned initiators of free-radical reactions on EC is explained by the formation of ROS (H_2O_2, $O_2^{\bullet}$, and $^{\bullet}OH$) in these cells, as SOD, catalase, iron chelators, free radical scavengers, or combinations of these agents proved to have dose-dependent protective effects [Shatos M.A., Doherty J.M. and Hoak J.C.; Markey B.A., Phan S.H., Varani J. et al.; Mickle D.A., Li R.K., Weisel R.D. et al.; Spragg R.G.; Varani J., Phan S.H., Gibbs D.F. et al.].

The response of EC to different exogenous oxidation promoters proved to be nonuniform. It was found that H_2O_2 mainly induces the formation of $O_2^{\bullet}$ [Markey B.A., Phan S.H., Varani J. et al.] or $^{\bullet}OH$ in EC [Franzini E., Sellak H., Marguetty C. et al.]; xanthine–xanthine oxidase, as well as activated phagocytes, rapidly (within approximately 1 min) generate H_2O_2, $O_2^{\bullet}$ and $^{\bullet}OH$ in reactions involving intracellular iron [Andreoli S.P. and McAteer J.A.; Kuppusamy P. and Zweier J.L.]; TNF-α induces $O_2^{\bullet}$ [Galley H.F., Walker B.E., Howdle P.D. and Webster N.R.] and $NO^{\bullet}$ generation in EC [Burke-Gaffney A. and Keenan A.K.; Estrada C., Gomez C., Martin C. et al.; Galley H.F., Walker B.E., Howdle P.D. and Webster N.R.]. Lipoxygenase-catalyzed reactions are the most probable source of free radicals during LDL oxidation because several inhibitors of the lipoxygenase system were very effective in blocking cell-induced oxidation of LDL by EC [Parthasarathy S., Wieland E. and Steinberg D.]. Although EC contain few mitochondria, these organelles play an important role in exogenously induced ROS production by these cells. Indeed, inhibition of mitochondria by oligomycin and rotenone or of EC metabolism by reducing glucose content in the incubation medium increased H_2O_2 cytotoxicity for EC of the bovine pulmonary artery [Delius R.E. and Hinshaw D.B.] and promoted ROS generation in EC of sheep pulmonary microvessels in response to hyperoxia [Sanders S.P., Zweier J.L., Kuppusa M.Y. and Harrison S.J.].

Ischemia and reperfusion of organs or local ischemia and reperfusion of the vascular wall enhance the ability of EC to generate ROS, whereas treatment with antioxidants and enzyme system inhibitors suppress ROS production and has a protective effect.

ROS formation on the luminal surface of EC in arteries, capillaries, and venules (but not in other cells of the organ) was demonstrated histochemically as soon as after 2–3-min reperfusion following 1-h ischemia of the rat heart, lungs, and intestine [Goldhaber I.J. and Weiss I.N.], brain [Mori T., Asano T., Matsui T. et al.], and liver [Rao P.S., Walsh T.R., Makowka L. et al.].

During reperfusion, the formation of free radicals in conserved and transplanted kidneys apparently occurred only near the vessels, as SOD prevented kidney injury only after being injected intravascularly [Koyama I., Bulkley G. B., Williams G. H. and Im M. J.].

ROS production (H_2O_2, $O_2^{\bullet}$, and $^{\bullet}OH$) by ischemic EC was also demonstrated in the absence of other sources of ROS, namely, in the cultivated endothelium of rat pulmonary artery [Ratych R.E., Chuknyiska R.S. and Bulkly G.B.], bovine aorta [Zweier J.L., Kuppusamy P. and Lutty G.A.], rabbit aorta [Mackenzie R.J., Furnival C.M., Wood C.B. et al.], and human umbilical vein [Michiels C., Arnould T., Houbion A. and Remacle J.; Nishida K.; Schinetti M.L., Sbarbati R. and Scarlattini M.].

In cultivated EC of the rat pulmonary artery, xanthine dehydrogenase was found to be almost fully converted to xanthine oxidase after 45-min anoxia followed by 30-min reoxygenation [Ratych R.E., Chuknyiska R.S. and Bulkly G.B.], whereas 24-h of anoxia EC from the bovine pulmonary microvessels led to twofold increase in the content of xanthine dehydrogenase (but not of xanthine oxidase). Reoxygenation of cultivated EC, especially under hyperoxic conditions (95% O_2), sharply increased the release of xanthine oxidase into the incubation medium [Partridge C.A., Horvath C.J., Delvecchio P.J. et al.].

In EC of the bovine pulmonary artery incubated for 48 h under hyperoxic (95% O_2), normoxic (21% O_2), hypoxic (3-10% O_2), or anoxic conditions (0% O_2), the activities of xanthine dehydrogenase and xanthine oxidase, as well as their total activity, increased along with a decrease in O_2 content in the gas mixture and incubation medium, although the xanthine dehydrogenase/xanthine oxidase ratio remained unchanged [Terada L.S., Guidot D.M., Leff J.A. et al.]. After anoxic incubation of EC for 48 h, the level of $O_2^{\bullet}$ in the incubation medium was significantly higher than after incubation under normoxic conditions. Xanthine oxidase inhibitors (tungsten and allopurinol) reduced the increase of $O_2^{\bullet}$ content after anoxic incubation [Terada L.S., Guidot D.M., Leff J.A. et al.].

The leading role of the xanthine–xanthine oxidase system in ROS production by ischemic and reperfused EC was demonstrated by other authors [Beetsch J.W., Park T.S., Dugan L.L. et al.; Inauen W., Payne D.K., Kvietys P.R. and Granger D.N.], although this finding was not confirmed in several other studies [Palermartinez A., Panus P.C., Chumley P.H. et al.; Phan S.H., Gannon D.E., Varani J. et al.]. In addition to xanthine oxidase, ROS generation by EC proved to depend on cyclooxygenase activity [Wu S., Nagashima T., Ikeda K. et al.]. The role of TNF-α in this process during EC ischemia and reperfusion apparently also increases significantly, as follows from an increase in the content of this factor in the blood flowing out of the organ and the production of TNF-α by ischemic EC themselves (see Chapter 9).

According to our data, H_2O_2 and rhTNF-α added to an HUVEC culture incubated under ischemic conditions had a more pronounced cytotoxic effect on these cells, compared to EC cultivated under normal conditions, i.e., the effects of H_2O_2 and ischemia or rhTNF-α and ischemia proved to be synergistic [Bilenko M.V., Ladygina V.G. and Fedosova S.V.(1998a); Bilenko M.V., Fedosova S.V., and Ladygina V.G.]. We also observed synergism in the effect of combinations LDL + ischemia and LDL_{ox} + ischemia on HUVEC; under ischemia, LDL oxidation by EC, estimated from the accumulation of LPO products (TBARS and SB) in LDL, was more intense than upon incubation under aerobic conditions [Bilenko M.V., Ladygina V.G. and Fedosova S.V.(1998b); Bilenko M.V., Vakhrusheva T.V. and Fedosova S.V.] and affected more intensely macrophage chemiluminescence [Bilenko M.V., Klebanov G.I., Dolgina E.N., and Nikankina L.V.]. Xanthine + xanthine oxidase added to the medium damaged EC of the human saphenous vein incubated for 14 days at pO_2 = 40 mm Hg, and this damage was more severe than upon incubation at pO_2 = 150 mm Hg [Li R.K., Shaikh N., Weisel R. et al.].

Thus, ample experimental data accumulated today provide evidence that EC are sensitive target cells for various agents and oxidation promoters; moreover, under aerobic conditions and, especially, upon ischemia and reperfusion, these cells can provide an active source of ROS.

Therefore, EC can be the initial site of ROS production, free-radical initiation, and tissue injury in ischemic and reperfused organs. That is, precisely endothelial cells appear to play the primary role in ROS production and initiation of free-radical processes, whereas the roles of phagocytes and parenchymal cells are apparently secondary.

It should be noted that the problem discussed in this section has many unresolved aspects. Thus, it remains unknown whether or not EC of different organs are heterogeneous enough to account for the existing organ-specific differences in sensitivity to ischemic and reperfusion injuries; what are the main sources of ROS formation in EC; etc.

10.3 CHANGES IN THE ACTIVITY OF ANTIOXIDANT ENZYME SYSTEMS

Antioxidant enzyme systems are formed by superoxide dismutase (SOD), catalase, and glutathione peroxidase (GSH peroxidase). The systems of GSH S-transferases and GSH reductases are supplementary, and the secondary protective role belongs to phospholipases A_2 and C. These enzymes prevent overproduction of ROS and participate in nonradical degradation of lipid peroxides.

The main sites where these enzymes are active and the reactions catalyzed by them are shown in Fig. 81 and 90, and the corresponding data were summarized in several reviews [Petrovich Yu.A. and Gutkin D.V.; Beyer W., Imlay J. and Fridovich I.; Downey J.M.; Marklund S.L.; Poberezkina N.B. and Osinskaya L.F.].

Fig. 90: Schematic representation of principal reactions with reactive oxygen species catalysed by antioxidant enzymes.

I. $ECu^{2+} + O_2^{\bullet-} \rightarrow ECu^{+} + O_2$; $ECu^{+} + O_2^{\bullet-} + 2H^+ \rightarrow ECu^{2+} + H_2O_2$

$EMn^{3+} + O_2^{\bullet-} \rightarrow EMn^{2+} + O_2$; $Emn^{2+} + O_2^{\bullet-} + 2H^+ \rightarrow EMn^{3+} + H_2O_2$

$EFe^{3+} + O_2^{\bullet-} \rightarrow EFe^{2+} + O_2$; $EFe^{2+} + O_2^{\bullet-} + 2H^+ \rightarrow EFe^{3+} + H_2O_2$

Overall reaction: $O_2^{\bullet-} + O_2^{\bullet-} + 2H^+ \rightarrow O_2 + H_2O_2$

II. $EFe^{3+} + H_2O_2 \rightarrow$ complex 1

Complex 1 + $H_2O_2 \rightarrow EFe^{3+} + 2H_2O + O_2$ (1)

Complex 1 + $AH_2 \rightarrow EFe^{3+} + 2H_2O + A$ (2)

III. $EcysSe + H^+ + H_2O \rightarrow ECysSeOH + H_2O$

$$ECysSeOH + GSH \xrightarrow{H_2O_2} ECysSeSG + H_2O$$

$ECysSeSG + GSH \rightarrow ECysSe^- + GSSG + H^+$

Overall reaction: $2GSH + H_2O_2 \rightarrow GSSG + 2H_2O$

IV. $ROOH + 2GSH \rightarrow ROH + GSSG + H_2O$

Note: I - dismutation of $O_2^{\bullet-}$ - to H_2O_2, catalysed by SOD, principle of interaction between different Me^{n+1} forming part of the active site of SOD isoenzymes; II - reduction of H_2O_2 to water with catalase: catalase (1) and peroxidase (2) reactions types [Chance B., Sies H. and Boveris A.]; E - enzyme; AH2 - electron donor; III - reduction of H_2O_2 to water with Se-containing GSH peroxidase [Free radicals in biology]; IV - reduction of lipid HP with GSH peroxidase.

Superoxide dismutase is the most powerful natural suppressor of $O_2^{\bullet-}$, which is univalently reduced (dismutated) by this enzyme to H_2O_2 [Fridovich I. (1981)].

Spontaneous $O_2^{\bullet-}$ dismutation (in the absence of SOD) proceeds much slower and is qualitatively different because it yields $^1O_2$ [Free radicals in biology; Brawn K. and Fridovich I.; Fridovich I. (1983)]. Acidosis accelerates this process [Hess M.L., Okabe E. and Kontos H.A.].

Depending on enzyme structure and the metal occupying its active site, distinction is made between Cu, Zn-SOD (found in the cytoplasm and, in smaller amounts, mitochondria and nuclei; inhibited by cyanide), Mn-SOD (found in the mitochondrial matrix, cyanide-independent), and Fe-SOD (in eukaryotes, found only in small amounts in mitochondria). With respect to Cu, Zn-SOD content, human organs (tissues) form the following descending series: liver > lungs > renal cortex > erythrocytes > uterus > gray matter of the brain > heart > skeletal muscles > thyroid > testes > spleen > lymph nodes > fat. The content of Mn-SOD is lower than that of Cu, Zn-SOD, and, since Cu^{2+} is a stronger reducing agent than other metals, Cu, Zn-SOD is 1.5–2 times as active as Mn-SOD.

When $O_2^{\bullet}$ production in the rat heart mitochondria was increased by blocking the respiratory chain with antimycin A, SOD transformed into H_2O_2 about 80% of the emerging $O_2^{\bullet}$, and the remaining 20% migrated into the cytosol [Nohl H. and Hegner D.]. Copper and zinc added in the form of fine powder changed the activity of SOD, zinc exhibiting more pronounced antioxidant properties.

SOD-like properties are also characteristic of peroxidase, cytochrome oxidase, ascorbic acid, hemoglobin, ceruloplasmin, transferrin, and other molecules containing ions of transition metals [Babizhaev M. A., Savov V. M. and Kagan V. E.; Goldstein I.M., Kaplan H.B., Edeeson M.S.and Weissman G.; Poberezkina N.B. and Osinskaya L.F.].

Catalase reduces H_2O_2 to water (see Fig. 90, *II*). Active sites of the enzyme accommodate Fe^{3+}-protoporphyrin, which interacts with H_2O_2 by either catalase or peroxidase mechanism at high and low H_2O_2 concentrations, respectively [Panchenko L.F., Gerasimov A.M. and Antonenkov V.D.]. The enzyme, is found in virtually all tissues. It is mainly located in peroxisomes, where its concentration may reach 10^{-6} M, and catalase reactions probably proceed in these organelles.

Glutathione peroxidase, like catalase, reduces H_2O_2 to H_2O and breaks down lipid hydroperoxides by a nonradical mechanism [Crake T. and Poole-Wilson P.A.] (see Fig. 90, *III* and *IV*).

More than 70% of GSH peroxidase are in the cytosol, and 25–30% are in the mitochondrial matrix [Panchenko L.F., Gerasimov A.M. and Antonenkov V.D.; De Marchena O., Guarnieri M. and McKhann G.]. Activation of the enzyme located in different cell compartments may be sequential, rather than simultaneous: first in the cytosol and then, upon increase in peroxide concentration, in the mitochondria [Chance B., Sies H. and Boveris A.]. With respect to GSH peroxidase content, organs may be arranged in the order: liver > kidneys > lungs > heart > testes > adrenals > spleen. In the brain of different animals, the enzyme is present in very small amounts, 50–200 times as small as in the liver; its activity in the liver and brain depends on animal species and decreases in the following series: mice > rabbits > fishes > guinea pigs > dogs, goats, chickens [De Marchena O., Guarnieri M. and McKhann G.]. GSH peroxidase has two isoforms: Se-dependent (GP-I) and Se-independent (GP-II). Se-dependent form is involved primarily in H_2O_2 reduction and, to a lesser extent, reduces hydroperoxides of PUFA derived from membrane phospholipids after their hydrolysis by phospholipase A_2 (PLA_2). There also is another form of Se-dependent GP, which reduces hydroperoxides of membrane phospholipids, cholesterol, and cholesterol esters [Karli J.N., Karikas G.A., Hatzipavlou P.K. et al.; Maiorino M., Thomas J.P., Girotti A.W.and Ursini F.].

GP-II is involved exclusively in the reduction of organic peroxides (those of FFA, steroids, and cholesterol) and, hence, is referred to as lipoperoxidase. There is evidence that GP-II is similar or identical to GSH S-transferase, and this enzyme (or its isoform isolated from the rat and bovine lungs) can utilize hydroperoxides of membrane phospholipids as well as those of FFA [Fisher A.B., Dobia Ch., Monovichy Y. et al.]. GP-II content is especially high in the liver, adrenals, testes, kidneys, and brain, whereas some other organs (tissues) are virtually devoid of this enzyme (this concerns the heart, lungs, intestinal mucosa, spleen, and skeletal muscles). The activity of GP-II depends on animal species: this form accounts for 33–35% of the total GSH peroxidase activity in the rat and rabbit liver and for 67–84% in the pig and human liver [Irvin and O'Brien, Morrissey P. and O'Brien P.J.]. As GSH peroxidase isozymes are involved in LPO at different stages (GP-I at the stage of ROS initiation, and GP-II at the stage of lipid peroxide decomposition), their ratio is an important factor. It varies depending on animal age [Lankin V. Z., Tikhaze A. K., Voronina N. V. et al.].

GP-I activity depends on the content of selenium in tissues: the greater this content, the higher the activity, and vice versa [Sunde R.A.].

Glutathione S-transferase reduces organic peroxides (hydroperoxides, endoperoxides) to alcohols in the reaction similar to that catalyzed by GSH peroxidase (Fig. 90, *IV*). This enzyme is highly GSH-specific and occurs in virtually all cell types of all animal species. It is located mostly in the cytosol but can also be detected in mitochondria, microsomes, and cell nuclei [Wahlländer A., Soboll S. and Sies H.]. GSH S-transferase, especially in liver mitochondria, is easily activated under the effect of limited proteolysis, thiol–disulfide exchange, ROS, and other stimuli [Aniya Y. and Naito A.].

Glutathione reductase is located near GSH peroxidases and catalyses the reduction of oxidized glutathione (GSSG) in a NADPH-dependent reaction (Fig. 91; the names of enzymes catalyzing the process are in frames). Reduced glutathione (GSH) is an electron donor and an active site for both GSH peroxidases, and its presence in the cell is a prerequisite for reactivation of the enzyme. Therefore, insufficient NADPH in the cell may be a factor limiting the reduction of H_2O_2 and lipid hydroperoxides and, hence, responsible for LPO initiation.

Our studies on the distribution of enzymes involved in H_2O_2 detoxification in the renal cortex of dogs (GSH peroxidase, GSH reductase, catalase; Table 14) showed that most of them (37–62%) are concentrated in the cytosol. In the mitochondrial–lysosomal fraction, the content of GSH peroxidase was extremely low, and the same concerns all the enzymes in the microsomal fraction. The analysis of competition between catalase and GSH peroxidase for H_2O_2 endogenously induced by D-alanine in the canine renal cortex showed that a major role belongs to GSH peroxidase. Direct correlations were revealed between SOD and catalase activities [Ahnja B.S. and Kiran U.] and SOD and GSH peroxidases [Lankin V. Z., Tikhaze A. K., Lemeschko V. V. et al.]. Inverse correlations proved to exist between antioxidant enzyme activities (SOD and GSH peroxidase) and the contents of hydroperoxides, diene conjigates, and Schiff bases in the blood of different animals. A similar correlation was also observed between SOD, GP-I, GP-II and catalase activities, on the one hand, and the content of hydroperoxides and Schiff bases, on the other, in the left and right ventricles of the rat heart [Kagan V. E., Savov V. M., Didenko V. V. et al. (1984)]. Thus, it is quite obvious that the multienzyme antioxidant system plays a key role at both early and late stages of LPO initiation in the cell.

Fig. 91: Oxidation-reduction cycle of glutathione.

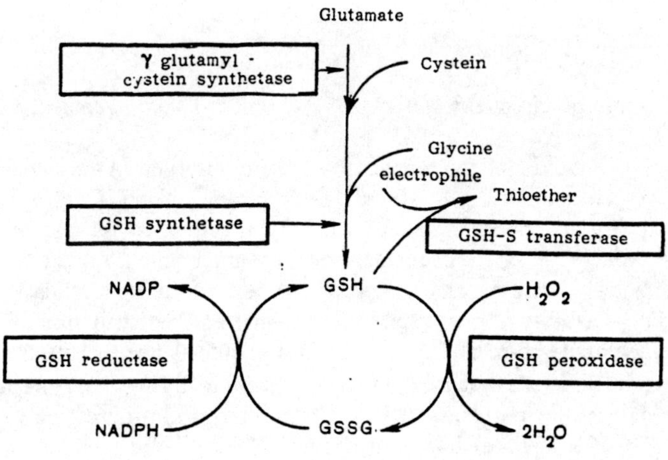

Table 14: Distributionof AO enzymes in subcellular fractions of the canine renal cortex (% of activity in the homogenate)

Fraction	GSH peroxidase	GSH reductase	Catalase
Homogenate	100	100	100
Mitochondria + lysosomes	4.8	37.5	25
Microsomes	7.7	0.16	5
Cytosol	50.5	62	37

Ischemia and reperfusion lead to a significant reduction of antioxidant enzyme activities. To date, this concept has been repeatedly confirmed by a number of experimental and clinical observations.

Studies performed in our laboratory as early as in as 1975 showed that the total free and, especially, nonprecipitable catalase activity in the renal cortex drastically decreases when an isolated dog kidney is stored at 25°C for a long period of time, whereas the activity of D-amino acid oxidase, the $O_2^{\bullet}$-producing enzyme, increases [Komarov O. S.] (Fig. 92).

Fig. 92: Activity of D-amino acids oxidase (a) and catalase (b) in the complete ischemic dog kidney cortex (isolated storage at 25°C) (% of the initial level)

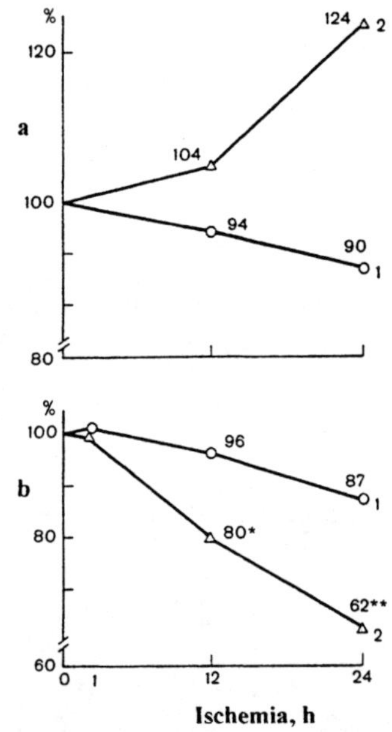

1 - total activity; 2 - non-precipitatable activity; *-P<0.05; **-P<0.01 versus the initial level.

Significant changes in antioxidant enzyme activities were observed in our experiments with long-term conservation of kidneys [Gerasimov A.M., Bilenko M.V., Panchenko L.F. et al.]. Table 15 shows that conservation of dog kidneys by perfusing them with cryoprecipitated plasma or Collins' solution was accompanied by a sharp decrease in GSH reductase activity, a tendency toward decrease (or a significant decrease) in GSH peroxidase activity, and a significant increase in free catalase activity. The latter is indicative of disturbed permeability of peroxisomal membranes and redistribution of the enzyme in the cell. The disturbed plasma membrane permeability for enzymes manifests itself in the fact that large amounts of GSH peroxidase and reductase are released into the perfusate during a 12–24-h perfusion period. In experiments with plasma used for perfusion, MDA content increased progressively in the renal cortex (by factors of 3.5, 6.7, and 7.9 after 2, 24, and 48 h, respectively).

Changes in the activity of antioxidant enzyme systems during total ischemia of organs *in situ* were also studied for the first time in joint experiments performed in our laboratory [Dudnik L.; Dudnik L. B., Tikhaze A. K., Alessenko A. V. et al.; Alessenko A.V., Bilenko M.V., Dudnik L.B. et al.].

As shown on the model of 15–120-min occlusion of the hilar pedicle of the median liver lobe in August rats (Fig. 93), Cu, Zn-SOD activity decreased at the highest rate during the first 15 min of ischemia, continued to decrease after 30 and 45 min, and became as low as 50 and 48% of the initial level after 60 and 120 min, respectively. GSH peroxidase activity began to decrease after 30 min of ischemia; after 60 and 120 min, it was 64 and 50% of the initial. Antioxidant activity decreased at approximately the same rate as GSH peroxidase activity, which was lower than the highest rate of accumulation of lipid hydroperoxides (within the first 15 min of ischemia), whereas the decrease in Cu, Zn-SOD activity coincided with the peak of the latter. Thus, Cu, Zn-SOD activity decreased and the content of lipid hydroperoxides increased as soon as within the first 15 min of ischemia.

Table 15: Variations in the activity of AO enzymes in dog kidneys preserved by perfusion methods (mean ± SEM)

Perfusion	Perfusion period, h	GSH peroxidase μmol•min-¹•g-¹	GSH reductase	Catalase, total activity, c-¹per•mg of protein	Catalase free activity (% of total activity).
Initial level	—	3.4±1.2	3.6±0.4*	0.150±0.006	41±3
Cryopreci-pitated plasma	3	—	2.0±0.4*	0.144±0.012	58±3*
	24	2.9±0.7	2.0±0.4*	0.156±0.007	61±3.8*
	48	2.9±0.8	1.8±0.5*	0.138	52
Solution Collins	24	1.8±0.2*	1.9±0.2*	0.117±0.010	63±4*

Note: *-P<0.05 versus initial level, the methods for determination of the enzymes are similar to [Gerasimov A.M., Bilenko M.V., Panchenko L.F. et al.], the renal perfusion procedure is similar to [Bilenko M.V., Perekrest P.V. and Ugryumov A.I.].

Fig. 93: The rate of decrease in the activity of antioxidant enzymes and AOA and accumulation of LP content in 15-120 min complete ischemic rat livers [Dudnik L.]

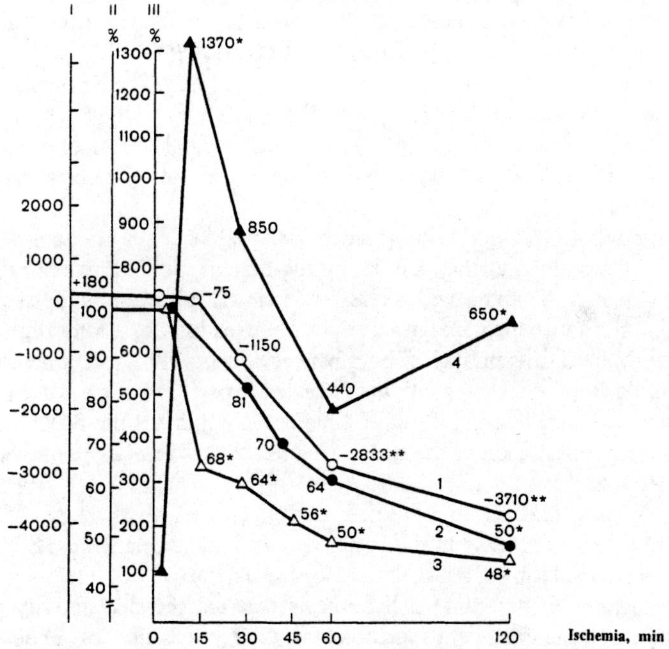

I - AOA (h • ml • mg^{-1} of lipids); II - activity of Cu, Zn-SOD and GSH peroxidase (% of the initial level); III - LP content (% of the initial level); 1 - AOA; 2 - GSH peroxidase; 3 - Cu, Zn-SOD; 4 - LP *-P<0.05; **-P<0.01 versus the initial level. The methods of determination Cu, Zn-SOD and GSH peroxidase activities, see [Dudnik L. B., Tikhaze A. K., Alessenko A. V. et al.]; the AOA determination method is the same as in Fig. 32; LP is the same as in Fig. 33.

The early and sharp decreases in Cu, Zn-SOD activity during total ischemia of the median and left liver lobes of Wistar rats and its incomplete recovery during the first 24 h of reperfusion were observed in our subsequent [Bilenko M. V., Alessenko A. V., Burlakova E. B. et al. (1991); Rashba Yu.E., Vartanyan L.S., Seregina L.A. et al. (1986a)]. In this case, Cu, Zn-SOD that had retained its activity after 2-h ischemia had the same electrophoretic properties and thermostability (65°C) as the enzyme isolated from an intact liver. In the same experiments, Mn-SOD activity in mitochondria remained more or less stable during 2-h ischemia but increased significantly (up to 180% of the initial, $P < 0.05$) after 24-h reperfusion of the liver. At the same time, such an increase in Mn-SOD activity was apparently insufficient for inhibiting $O_2^{\bullet}$

production in mitochondria because $V_{O_2^{\bullet}}$ in these organelles and the $V_{O_2^{\bullet}}$/Mn-SOD ratio increased (see above), and LPO in the mitochondria of reperfused organs was more intense than in other membranes.

To date, decrease in the activity of antioxidant enzyme systems during ischemia of different organs has been confirmed by the data obtained in many studies. It was shown that a simulated climb to a high altitude in a pressure chamber brings about a decrease in the activity of SOD in the liver of experimental animals [Chumakov V.N. and Osinskaya L.F.]. In studies on the model of 6.5-min asphyxia caused by clamping the intratracheal tube, the activity of SOD in the liver of experimental rats decreased to 65% of the initial; that of GSH peroxidase, to 75%; and that of catalase, to 43%. This was accompanied by a decrease in antioxidant activity (to 64%) and an increase in LPO intensity (by a factor of 1.8) [Konvai V. D., Lukoshkin A. V. and Smirnova V. B.].

In experiments with hypoxic perfusion of the isolated rat heart, cytoplasmic SOD and GSH peroxidase activities remained unchanged for 20 min and then decreased to 70 and 89% after 40 min, respectively, and to 58–55 and 70% after 60–80 min [Guarnieri C., Flamigni F. and Caldarera C.M.]. LAD occlusion in rats resulted in the reduction of cytoplasmic SOD activity in the ischemic tissue after 2 min and of GSH peroxidase activity after 15 min [Lankin V. Z., Kogan A. Kh., Kovalevskaya A. L. et al.]. On the same model of ischemia, SOD activity after 1 and 3 h decreased to 51 and 38% of the initial, and that of GSH peroxidase decreased to 12.5 and 17%, respectively [Gutkin D. V. and Petrovich Yu. A.]; after 24 h, the activities of cytosolic SOD, GP-I, and GP-II were 82, 43, and 34%, respectively [Meerson F.Z., Kagan V.E., Azkhipenko Yu.V. et al.], or, according to other data, SOD and GSH peroxidase activities became three and two times as low as the initial level [Röth E., Török B., Zsoldos T. and Matkovics B.]. Catalase activity during cardiac ischemia decreased at a slower rate and not earlier than 3 h after coronary occlusion [Gutkin D. V. and Petrovich Yu. A.]; after 24 h, it was 64–75% of the initial [Gutkin D. V. and Petrovich Yu. A.; Meerson F.Z., Kagan V.E., Azkhipenko Yu.V. et al.]. The activity of GSH S-transferase began to decrease only on day 3 after occlusion.

Clinical observations on patients with AMI revealed the decrease of SOD activity in the blood after 48 h, whereas GSH peroxidase activity began to decrease only 12 days after AMI [Loeper J., Goy J., Rozensztajn L. et al.].

When the model of LAD occlusion (moderate partial ischemia) in dogs was used (see Chapter 1), GSH peroxidase activity remained unchanged during the first 30 min but decreased to 25% of the initial after 45 min [Rao P.S., Cohen M.V. and Mueller H.S.], which was followed by the accumulation of LPO products.

Our data that the antioxidant enzyme activity remains low for a long time during the reperfusion period are consistent with the results published by other researchers. This phenomenon was demonstrated for SOD and GSH peroxidase after reoxygenation of the heart perfused for a long period of time under hypoxic conditions [Guarnieri C., Flamigni F. and Caldarera C.M.], cytosolic SOD, and GSH peroxidase and GSH S-transferase during reperfusion of the tissue affected by AMI in rats [Konovalova G. G., Cherpachenko N. M., Lankin V. Z. et al.].

Thus, our results and those obtained by other specialists suggest that ischemia of organs is accompanied by a sharp decrease in the activity of antioxidant enzymes; SOD is the first to decrease, and then follow GSH peroxidase, GSH S-transferase, and catalase.

These observations have led to a hypothesis that the reduced activity of these enzymes, especially SOD, is one of the basic factors responsible for LPO intensification under conditions of O_2 deficiency [Dudnik L. B., Tikhaze A. K., Alessenko A. V. et al.; Alessenko A.V., Bilenko M.V., Dudnik L.B. et al.]. Today, many specialists share this opinion.

At the same time, an increasing amount of data provides evidence that LPO intensification during ischemia is not necessarily preceded by a decrease in the activity of antioxidant enzyme systems and that both phenomena are not always in a one-way cause-and-effect relationship. Thus, the marked accumulation of primary and secondary LPO products against the background of unchanged activities of antioxidant enzymes was observed in the following models: the rat brain during asphyxia and after resuscitation [Konvai V. D.], the rabbit liver and aorta after storing the dead animal at 4–8°C [Lankin V. Z., Tikhaze A. K., Voronina N. V. et al.], and the cat intestine during 2-h of the mesenteric artery occlusion and after reperfusion [Schoenberg M.H., Yones M., Muhl E. et al.].

It cannot be excluded that the sequence of events during ischemia may be inverse, i.e., early LPO intensification (within the first few seconds or minutes) caused by other known or still unknown factors leads to a decrease in the activity of antioxidant enzyme systems, and this, in turn, additionally intensifies LPO, thus closing the vicious circle.

Assumptions concerning the possible existence of such a free-radical mechanism of damage to antioxidant enzymes are based on indirect and direct published evidence. The most convincing arguments in its favor are as follows:

(1) The inhibitory and dose-dependent effect of ionizing radiation [Lipin'ski S., Lipetska K., Donets Y. et al.] and γ-radiation [Bartosz G., Soszynski M. and Retelewska W.] on the activity of SOD.

(2) Long-term decrease in the activity of SOD in the liver and other organs of rats after single administration of linoleic and, especially, linolenic acids or their hydroperoxides [Mkhitaryan V.G. and Badalyan G.E.].

(3) A decrease in SOD activity (by 22%) and, simultaneously, in the antioxidant activity of lipids (by 82%) on day 3 after single administration of epinephrine to rabbits [Demurov E. A., Mil'chakov V. I. and Gerasimov A. M.], and a decrease in SOD and GSH peroxidase activities (to 54 and 77% of the initial) in the heart of sham-operated rats [Gutkin D. V. and Petrovich Yu. A.].

(4) The inhibitory effect of H_2O_2 on Cu, Zn-SOD, Fe-SOD, and, to a lesser extent, Mn-SOD [Beyer W., Imlay J. and Fridovich I.; Bartosz G., Soszynski M. and Retelewska W.; Hodgson E.K. and Fridovich I.]. As shown in [Sinet R.M. and Garber P.], the damaging effect of H_2O_2 on Cu, Zn-SOD is enhanced when H_2O_2 is combined with $O_2^{\bullet}$; moreover, SOD inactivated by H_2O_2 becomes more vulnerable to proteolytic enzymes, which cause its fragmentation and denaturation.

(5) Other evidence confirming the free-radical mechanism of antioxidant enzyme damage includes the less pronounced decrease in SOD and GP activities observed during the storage at 37°C of the rat liver isolated from animals injected with antioxidants [Dzhafarov A. I. (1983)]; according to other data, however, administration of α-tocopherol and sodium selenite over a long period of time reduced chemiluminescence intensity but did not offer any protection against the decrease of SOD and GP activities in the infarct area [Gutkin D. V. and Petrovich Yu. A.].

Another important factor probably responsible for the decrease of SOD activity is acidosis developing in the ischemic organ. As demonstrated experimentally [Gorbatenkova E.A., Azizova O.A., Paramonov N.V. and Vladimirov Yu.A.], pH decrease to 6.2 led to complete SOD inhibition, whereas laser irradiation of samples, causing deprotonation of the SOD active site, restored the activity of this enzyme. The fact that the activity of antioxidant enzymes during the postischemic period remains low for a long time can be accounted for by their removal with blood or perfusate [Petrovich Yu.A. and Gutkin D.V.], as it occurs with other enzymes (see Chapter 2).

The reduction of antioxidant enzyme activities can be preceded by their short-term increase. Thus, the activities of mitochondrial and microsomal Mn-SOD and cytosolic Cu, Zn-SOD in the rat liver increased after 30 min but decreased after 60 min of total ischemia [Gupta M., Dobashi K., Greene E.L., et al.]; GSH S-transferase activity increased under moderate oxidative stress but decreased under more severe stress (aggravated by chemical inactivation of catalase, GSH content reduced by more than 50%, etc.) [Aniya Y. and Naito A.]. In rat kidneys, GSH peroxidase activity increased after 4-h ischemia but returned to the initial level after 8-h ischemia [Barnard M.L., Snyder S.J., EngersonT.D., and Turrens J.F].

The role of antioxidant enzyme activities in pathogenesis of LPO activation and in the free-radical mechanism of ischemic and reperfusion organ damage was convincingly demonstrated in recent experiments on transgenic mice with constitutively increased or decreased expression of antioxidant enzymes.

In mice overexpressing GSH peroxidase-1, pathological changes in the brain region exposed to 1-h ischemia and 24-h reperfusion were expressed to a lesser extent than in nontransgenic mice [Weisbrot-Lefkowitz M., Reuhl K., Perry B., et al.]. The same effect was observed in mice overexpressing Cu, Zn-SOD in experiments on 10-min global ischemia and 1–3-h reperfusion of the brain [Chan P.H., KawaseM., Murakami K.et al.] and on 30-min global ischemia and reperfusion of the heart [Wang P., Chen H., Qin H., et al.]. Conversely, in GSH peroxidase-1 knockout mice, ROS formation and pathological changes in the heart after 30-min global ischemia and reperfusion were more pronounced than in wild-type animals [Yoshida T., Maulik N., Engelman R.M. et al.]. This also applies to animal mortality and pathological changes in a 1-h ischemic and reperfused brain region in experiments with Cu, Zn-SOD-depleted (100%) or deficient (50%) mice [Kondo T., Reaume A.G., Huang T.T., et al.], the release of mitochondrial cytochrome c into the cytosol and DNA fragmentation (laddering) after 1–2-h and 24-h focal cerebral ischemia in Mn-SOD knockout mice (deficient in mitochondrial Mn-SOD) [Fujimura M., Morita-Fujimura Y., Kawase M. et al.], and pathological changes in kidneys after their 40-min ischemia and 60-min reperfusion in catalase-inhibited rats [Paller M.S.]. The toxic effect of H_2O_2, LDL or LDLox was expressed much more in SOD-deficient strain of Escherichia coli than in that of wild strain [Bilenko M.V., Ladygina V.G. and Tarakanova I.A.].

Thus, it has been conclusively demonstrated that the activity of antioxidant enzyme systems protecting the cell against ROS and lipid peroxides changes (decreases) during ischemia and is not restored in the early reperfusion period. The extent and time of this decrease vary and depend both on the type of the enzyme and on the severity, extent and duration of ischemia. This process obviously plays a major role either in the initiation or further intensification of LPO in the organ. A probable factor responsible for decrease in the activity of antioxidant enzymes is their high sensitivity to ROS or lipid peroxides; it should be noted, however, that interactions between enzymes and these agents are reciprocal. The role of suppression of antioxidant enzyme systems in the induction of ROS and LPO is additionally confirmed by wide experience in the successful use of antioxidant enzymes as antiischemic drugs (see Chapter 11).

10.4 Changes in the Contents of Water- and Fat-soluble Nonenzymatic Prooxidants and Antioxidants

In addition to antioxidant enzymes, there are many nonenzymatic compounds with prooxidant, antioxidant, or both properties, which are involved in the regulation of LPO in cells. When a compound displays both properties simultaneously, the final effect mainly depends on its absolute or relative tissue content, and increase or decrease in the latter may have a strong effect on LPO intensity.

During local and, especially, complete ischemia, antioxidant compounds are no more supplied to the organ with blood, whereas available antioxidants continue to be metabolized (catabolized) or irreversibly expended in the course of LPO. All this creates a local deficiency of antioxidants, whereas the accumulation of prooxidants due to the lack of outflow gives rise to their local excess. Such a disturbance of balance between prooxidant and antioxidant agents in the organ may be a major cause of LPO intensification during acute ischemia and after reperfusion.

The interpretation of factors leading to antioxidant deficiency and effects of this deficiency on LPO is not always unambiguous.

On the one hand, if a compound is involved in LPO reactions (e. g., GSH, ascorbic acid, or α-tocopherol), its deficiency may serve as indirect evidence that LPO proceeds in the ischemic organ (see Chapter 5). However, an antioxidant may become deficient because of its catabolism; in this case, a decrease in its content creates condition for LPO initiation in the tissue. Both situations often occur simultaneously, making it difficult to interpret the cause-and-effect relationships between the level of antioxidants and the intensity of LPO. This situation is similar to that discussed above in the context of SOD and LPO.

Let us consider the main groups of compounds that modulate free-radical reactions and whose contents were found to change during ischemia.

10.4.1 Changes in the Contents of Transition Metals, their Reductants, and Complexing Agents

Transition metals, especially those capable of monovalent oxidation (conversion of Me^n into Me^{n+1}), was shown to exert their prooxidant effect during ROS production and at the first stage of LPO chain branching, whereas their antioxidant effect manifests itself at the chain termination stage (see Fig. 81).

Transition metal redox pairs are formed by iron (Fe^{2+}/Fe^{3+}), copper (Cu^{+}/Cu^{2+}), chromium (Cr^{2+}/Cr^{3+} and Cr^{5+}/Cr^{6+}), manganese (Mn^{2+}/Mn^{3+}), cobalt (Co^{2+}/Co^{3+}), vanadium (V^{4+}/V^{5+}) and other metals [Stohes S.J. and Bagchi D.; Halliwell B.; Liochev S.I. and Fridovich I.]. They undergo redox cycling, resulting in the production of ROS. LPO-catalyzing effects of these metals in chemical model systems (PUFA) are more pronounced than in biological model systems (microsomes, mitochondria, etc.) and, especially, *in vivo* conditions. The reasons why Me^n lose their LPO-catalyzing ability in biological membranes and some of them are converted from prooxidants to antioxidants (Co^{2+}, Cu^{2+}, and Fe^{2+}) include the complexing of the metal with chelators, change in its concentration, a lack or decrease in the level of reducing agents, etc.

Some other metals, such as cadmium, mercury, and nickel, do not undergo redox cycling but deplete GSH and protein-bound SH groups, thereby influencing ROS production and stimulating LPO [Stohes S.J.

and Bagchi D.]. Zinc-induced cortical neurotoxicity associated with DNA fragmentation proved to result from the increasing free radical generation and membrane LPO, as Trolox, a vitamin E analogue, blocked it almost completely [Kim Y.H., Kim E.Y., Gwag B.J, et al.].

Metal ions may also enhance the production of TNFα (see Chapter 9).

The crucial role in the regulation of LPO processes in both chemical and biological systems belongs to the redox pairs Cu^{+}/Cu^{2+} and, especially, Fe^{2+} (ferrous iron) / Fe^{3+} (ferric iron), because these metals are essential elements and widely occur in nature.

Fe^{2+} is an indispensable component of all the systems producing $O_2^{\bullet}$ from O_2 (mitochondria, microsomes, xanthine–xanthine oxidase, etc.); it is especially important for $^{\bullet}OH$ production in the Fenton and Haber-Weiss reactions (see Fig. 81).

Fe^{2+} acts as a catalyst in the $^{\bullet}OH$-producing reaction with OCl^{-} [Osipov A.N., Yakutova E.Sh. and Vladimirov Yu.A.; Yakutova E.Sh., Osipov A.N., Kostenko O.V. et al.]: $Fe^{2+} + OCl^{-} + H^{+} \rightarrow {}^{\bullet}OH + Fe^{3+} + Cl^{-}$.

During autoxidation, iron is also capable of initiating lipid radicals without generation of ROS [Vladimirov Yu.A. and Archakov A.I.; Bucher J.R., Tien M. and Steven D.A.]. This is particularly true in the case of enzymatic and other systems incapable of initiating $O_2^{\bullet}$ (e. g., SR membranes in skeletal muscles) [Koshkin V. V.].

Iron plays a key role in LPO intensification during degradation of lipid peroxides (ROOH) yielding a reactive alkoxy radical (RO•), which leads to chain branching (see Fig. 81). This is principally similar to the involvement of iron in the reaction with H_2O_2 (the Fenton reaction).

The effect of Fe^{2+} is dose-dependent; the "threshold" dose $[Fe^{2+}]^{\bullet}$ at which it changes its prooxidant properties for antioxidant depends on the type of membrane and, particularly, on whether enzymatic NADPH-dependent or nonenzymatic Fe^{2+}-ascorbate-dependent LPO processes prevail in it. In a mitochondrial suspension under conditions of Fe^{2+}-induced LPO, Fe^{2+} acted as a prooxidant at concentrations of 10^{-6}–10^{-4} M and as an antioxidant at concentrations higher than 10^{-4} M (10^{-3}–10^{-2} M) [Vladimirov Yu.A. and Archakov A.I.; Brogan W.C., Miles P.R. and Colby H.D.]. $[Fe^{2+}]^{\bullet}$ for nuclear membranes from the rat liver proved to be three times as low as that for mitochondria [Pogosian G.A., Dremina E.S., Sharov V.S. et al.], i.e., Fe-induced LPO in different membranes is expressed to different degrees.

A greater part of endogenous iron in cells is known to exist as Fe^{3+} bound to the iron-porphyrin complex (heme), which is the active site of hemoglobin, myoglobin, cytochromes, cytochrome oxidase, catalase, and peroxidase; iron is also included into the Fe-sulfur complex which is the active site of NADH dehydrogenase, xanthine oxidase, and other enzymes. Soluble iron in the cytosol is found as a component of ferritin, an iron-storage protein; in blood plasma and extracellular fluid, iron is transported by other proteins, transferrin and lactoferrin; in mitochondria, it may accumulate in the insoluble hemosiderin granules [Lehninger A.; Gutteridge J.M.C., Halliwell B., Treffry A. et al.].

Iron incorporated into the heme, Fe–sulfur complexes, ferritin, transferrin (lactoferrin), or hemosiderin is biologically inactive. It becomes catalytically active only after its recruitment from these storage and transport protein complexes and simultaneous reduction from Fe^{3+} to Fe^{2+}.

Agents such as $O_2^{\bullet}$, ascorbic acid, reduced flavins, low-density molecular ligands, and a variety of xenobiotics (paraquat, adriamycin, nitrofurantoin, etc.) facilitate the release of iron from ferritin [Yoshida T., Tanaka M., Sotomatsu A. and Hirai S.; Harris L.R., Cake M.H. and Macey D.J.; Burton K.P., McCord J.M. and Ghai G.; Gutteridge J.M.C., Halliwell B., Treffry A. et al.].

The $O_2^{\bullet}$-dependent release of iron from ferritin in the human, rat, and trout liver accounts for 86, 92 and 84% of the total release, respectively [Harris L.R., Cake M.H. and Macey D.J.]. Slow protonation of the amino acid ligands of two iron-binding sites in acidic media (low pH) facilitates the release of iron from transferrin [El Hage Chahine J.M. and Fain D.; El Hage Chahine J.M. and Pakdaman R.].

It was shown that a small mobile pool of nonheme and protein-unbound iron exists in the cytosol, which represents easily dissociating iron complexes with low-molecular chelators, such as nucleotides, organic acids, glycine, cysteine, dipeptides, glucose, etc. [Bakkeren D.L., De Jen-Jaspars C.H.H., Van Der Heul, C and Van Eijk, H.G.; Bucher J.R., Tien M. and Steven D.A.; Mulligan M., Althaus B. and Linder M.C.; Tien M., Svingen B.A. and Aust S.D.; Weaver, J. and Pollack, S.]. This biologically active and easily soluble iron is referred to as the transit pool of iron [Healing G., Gower J., Fuller B. and Green C.], desferal-available iron [Gower J.D., Healing G. and Green C.J.; Healing G., Gower J., Fuller B. and Green C.], "free" iron [Yakutova E.Sh., Osipov A.N., Kostenko O.V. et al.; Otani H., Prasad M.R., Jones R.M.

and Das D.K.], loosely-bound chelatable iron [Otani H., Prasad M.R., Jones R.M. and Das D.K.; Van Jaarsveld H., Potgieter G.M., Kuyl J.M. et al.], and, most often, low-molecular-weight complexes of iron (LMWI) [Sevanian A. and Hochstein P.; Bralet J., Schreiber L. and Bouvier C.; Otani H., Prasad M.R., Jones R.M. and Das D.K.; Ozaki M., Fuchinoue S., Teraoka S. and Ota K. (1984)]. Iron incorporated in LMWI is considered to be mainly Fe^{2+}. Iron conversion from Fe^{3+} into Fe^{2+} upon its transfer from ferritin into cytosolic LMWI is mediated by reduced flavins, ascorbic acid, and cysteine [Kirshner R.E. and Fantini G.A.]. LMWI content is relatively small: in the rat cerebral cortex, it accounts for approximately 1% of total tissue iron [Bralet J., Schreiber L. and Bouvier C.; Otani H., Prasad M.R., Jones R.M. and Das D.K.]. Depending on physiological or pathological conditions, LMWI can increase owing to iron recruitment from protein complexes (mainly from ferritin). Hb can serve as a source of catalytically active (free) iron in the reaction with OCl^- and H_2O_2 [Yakutova E.Sh., Osipov A.N., Kostenko O.V. et al.].

It was also shown that the major fraction of LMWI is contained in closed membrane compartments, whereas a minor portion is dissolved in the cytoplasm. Heating a tissue to 45–50°C, a temperature too low for protein denaturation but sufficient for causing phase changes in membrane lipids, was shown to displace free iron from organelles and enlarges the cytoplasmic pool [Vanin A.F., Kubrina L.N., Kiladze S.V. et al.; Tarasova N.I., Kovalenko O.A. and Vanin A.F.]. These investigations, however, have not progressed as yet.

Cytoplasmic LMWI is essential for ROS production in the Fenton and Haber–Weiss reactions and during LPO initiation [Halliwell B. and Gutteridge J.M.C. (1984)], whereas iron released from heme-containing compounds (Hb or Mb) is more important for degrading ROOH [Mimnaugh E.G., Trush M.A. and Gram T.E.].

The most active forms of iron are ferryl (FeO^{2+}, $FeOH^{3+}$, $Fe^{2+}O$) and perferryl radicals ($FeO^{3+} \cdot O_2^{\bullet}$, $Fe^{2+}O_2$) formed in the Fenton (peroxidase) reaction and, especially, complexes of ferryl or perferryl radicals with AMP [Tien M., Svingen B.A. and Aust S.D.] or ADP [Aust S.D. and Svingen B.A.].

Recent studies showed that LPO is also controlled by Ca^{2+} ions. Thus, LPO was intensified by Ca^{2+} added to the medium for incubating rat heart mitochondria or microsomes in the presence of LPO stimulators, such as Fe^{2+}-ascorbate, NADPH, or NADH [Kagan V. E., Savov V. M., Didenko V. V. et al. (1983); Kagan V. E., Savov V. M., Didenko V. V. et al. (1984)] and their addition to lipids (yolk phospholipids or linoleic acid) in the presence of Fe^{2+}-ascorbate [Savov V.M., Babijaev M.A. and Kagan V.E.]. Ca^{2+} produced a moderate stimulatory effect (a 30-35% increase of MDA production) at concentrations of 10^{-6}–$5 \cdot 10^{-5}$ M, whereas higher concentrations ($5 \cdot 10^{-3}$–$2 \cdot 10^{-3}$ M) inhibited LPO [Kagan V. E., Savov V. M., Didenko V. V. et al. (1983); Kagan V. E., Savov V. M., Didenko V. V. et al. (1984); Savov V.M., Babijaev M.A. and Kagan V.E.]. The LPO-stimulating effect of Ca^{2+} was attributed to the ability of Ca^{2+} to interact with negatively charged groups of lipids and to displace Fe^{2+}, converting it into a "free" and, consequently, active form; the LPO-inhibiting effect of Ca^{2+} was explained by its ability to interact with $O_2^{\bullet}$ [Savov V.M., Babijaev M.A. and Kagan V.E.]. The LPO-initiating effect of Ca^{2+} also seems to involve intensification of phospholipolysis.

Because it is Fe^{2+} that acts as prooxidant at all stages of LPO, compounds reducing Fe^{3+} to Fe^{2+} are as important in LPO processes as iron itself.

The group of **transition metal reducing agents** includes compounds with pronounced electron-donor properties. The most important of them are (1) thiol compounds (glutathione, cysteine, sulfite, SH groups of proteins, thiourea, etc.) (2) ascorbic acid, and (3) reduced pyridine nucleotides (NADPH and NADH).

Thiol compounds (RSH) of different composition are the strongest agents reducing Fe^{3+} to Fe^{2+}. These compounds participate in LPO by not only reducing iron, but also through other mechanisms; hence, they will be discussed in more detail (see below).

A major role in Fe^{3+} reduction during nonenzymatic LPO is also played by *ascorbic acid*. Its concentration of 10^{-3} M is sufficient for this function under physiological conditions [Vladimirov Yu.A. and Archakov A.I.]. However, the absolute amount of ascorbate is not as important as the Fe^{2+}/ascorbate ratio: in a model system (guinea pig adrenal mitochondria), ascorbate ($5 \cdot 10^{-6}$–$5 \cdot 10^{-4}$ M) interacted synergistically with a small (10^{-6} M), rather than large, amount of Fe^{2+} [Brogan W.C., Miles P.R. and Colby H.D.]. In biological membranes (liver microsomes and mitochondria) and liposomes with a Fe^{2+} concentration of 10^{-5}–10^{-4} M, the rate of LPO reached a peak when ascorbic acid was added at a concentration of (1–7.5) 10^{-4} M, but its concentrations above $7.5 \cdot 10^{-3}$ M arrested the accumulation of LPO products [see Free radicals and biological stabilizators].

Similarly to Fe^{2+}, ascorbic acid at high concentrations can produce an antioxidant effect owing to its ability to reduce oxidized forms of α-tocopherol, GSH, and other inhibitors of phenolic type [Burlakova E.B., Alessenko A.V., Molochkina E.M. et al.]. Ascorbic acids also has chelating, scavenging, and antioxidant properties [Ozaki M., Fuchinoue S., Teraoka S. and Ota K. (1995)].

Other strong Fe^{3+}-reducing agents are *reduced nucleotides NADPH and NADH* providing for enzymatic LPO in microsomes [Vladimirov Yu.A. and Archakov A.I.]. The prooxidant effects of NADPH and NADH are also important for the production of ROS and induction of LPO in mitochondria; LPO processes are mainly NADH-dependent in mitochondria of skeletal muscles [Koshkin V. V.] and NADPH-dependent in myocardial mitochondria [Takeshiga H., Takayanagi R. and Manakami S.] and circulating and tissue phagocytes (see section 10.2). Pyridine nucleotides may also act as antioxidants by reducing GSSG to GSH with the aid of GSH reductase (see below).

The group of **transition metal chelating agents** includes AMP, ADP, ATP, EDTA, ceruloplasmin, and other compounds, both prooxidant and antioxidant.

AMP, ADP, and ATP form active complexes with Fe^{2+}, probably due to their easy solubility [Aust S.D. and Svingen B.A.]. However, accumulation of excess pyrophosphate relative to Fe^{2+} inhibits LPO [Vladimirov Yu.A. and Archakov A.I.].

EDTA complexes exhibit dose-dependent prooxidant or antioxidant properties *in vitro* [Vladimirov Yu.A. and Archakov A.I.; Gutteridge J.M.C., Richmond R. and Halliwell B. (1979)], and the prooxidant effect of EDTA in the Fe^{2+}-dependent $^{\bullet}OH$ production in the Fenton and Haber–Weiss reactions is much more pronounced than those of other chelators.

Among chelating agents of the antioxidant type is *ceruloplasmin*, which is the main antioxidant in blood plasma and extracellular fluid [Govorova M.Yu., Sharanov B.P., Iankovskii O.Yu. and Lyzlova S.N.; Kozlov A. V., Shinkarenko L. I., Vladimirov Yu. A. and Azizova O. A.]. Ceruloplasmin can oxidize Fe^{2+} (to Fe^{3+}), ascorbic acid, and phenols; this is why it is also referred to as ferroxidase, ascorbate oxidase, and aminooxidase [Sanina O.L. and Berdinskykh N.K.]. Ceruloplasmin is the strongest inhibitor of ROS production in the myeloperoxidase–H_2O_2–Cl^- system of neutrophils: 50% decrease in the intensity of chemiluminescence in this system was achieved under the effect of $6 \cdot 10^{-8}$ M ceruloplasmin, compared to $6.1 \cdot 10^{-7}$ M albumin and $1.7 \cdot 10^{-6}$ M SOD [Govorova M.Yu., Sharanov B.P., Iankovskii O.Yu. and Lyzlova S.N.]. Ceruloplasmin is also capable of scavenging $O_2^{\bullet}$, displaying an SOD-like activity [Afanasyev I. B.; Goldstein I.M., Kaplan H.B., Edeeson M.S.and Weissman G.; Gutteridge J.M.C., Richmond R. and Halliwell B. (1980)] and protecting cells from neutrophil-mediated injury. however, the main mechanism of the antioxidant effect produced by ceruloplasmin *in vivo* is as yet unknown.

Some transition metals can also be chelated by natural *histidine-containing dipeptides* (carnosine and anserine), especially in muscles where their concentrations are high. These compounds can remove metals from the hydrophobic phase of membranes, thereby minimizing their effects on LPO [Dupin A. M., Bemanandzara M., Stvolinsky S. L. et al.; Kozlov Yu. P., Kagan V. E. and Arkhipenko Yu. V.]. In addition, they can regenerate vitamin E, allowing it to be repeatedly used as an antioxidant. In model systems, these dipeptides inhibit spontaneous and Fe^{2+}–ascorbate-induced LPO and reduce the contents of already accumulated products. This antioxidant effect is stronger in a complex with Cu and Zn. Urocanic acid, a metabolite of histidine, also has antioxidant properties [see Free radicals and biological stabilizators].

Changes in the contents of transition metals and their reducing and chelating agents during acute ischemia and after reperfusion of organs have been extensively studied over recent years. Of ischemia-induced changes, the most significant is the release of iron bound to proteins (mainly ferritin) into the catalytically active cytosolic iron-chelating pool (LMWI), which leads to the increase of the latter.

It was shown that LMWI increases during long-term storage of isolated organs [Kozlov A. V., Shinkarenko L. I., Vladimirov Yu. A. and Azizova O. A.; Goto K., Ishige A., Sekigushi K. et al.; Healing G., Gower J., Fuller B. and Green C.]. Incubation of the isolated rat liver for 6 h in isotonic NaCl solution under nitrogen at 37°C led to almost twofold increase in the content of total "free" iron [Kozlov A. V., Shinkarenko L. I., Vladimirov Yu. A. and Azizova O. A.]; its accumulation followed the same pattern as that of MDA.

Accumulation of "free" cytoplasmic iron was also observed during a shorter periods of ischemia and after storage of isolated liver for 30 min, whereas total "free" iron increased only after 60 min of ischemia; the latter effect was accompanied by MDA accumulation during the same period of time [Kozlov A. V., Vdovin A. V., Azizova O. A. and Vladimirov Yu. A.].

These results were obtained in experiments with rat liver stored *in vitro,* when the postischemic period was simulated by incubating liver slices first under nitrogen and then in the air. We repeated these experiments but under *in vivo* conditions [Bilenko M. V., Alessenko A. V., Burlakova E. B. et al. (1991)]. The dynamics of LMWI was studied in the median and left lobes of the Wistar rat liver subjected to total ischemia for different periods of time. The results showed (Fig. 94) that ischemia *in vivo* was accompanied by a rapid growth of cytoplasmic "free" iron pool: after 15, 30, and 45 min of ischemia, it exceeded the initial pool by factors of 2.5, 3.9, and 5.7, respectively. Subsequent increase in "free" iron (between 45 and 180 min) was relatively small, whereas the content of lipid peroxides sharply increased after 60 min of ischemia.

Fig. 94: Content of "free" cytoplasmic Fe^{2+} and lipid HP (2) in a complete 15-180 min ischemic rat liver in vivo(mean ± SEM).

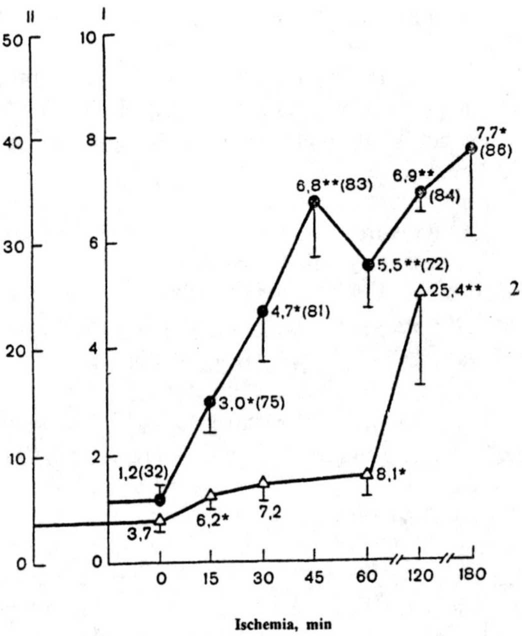

I –"free" Fe^{2+} (LMWI) content (in relative units); II - HP content in microsomal membrane (in nmol • mg-¹ of lipid). *-P < 0.05; **-P < 0.01 versus the initial level. Fe^{2+} determination method, see [Kozlov A. V., Shinkarenko L. I., Vladimirov Yu. A. and Azizova O. A.]; HP determination method is the same as in Fig. 32

In addition to aforementioned triggers ($O_2^{\bullet}$ and others), the release of Fe^{2+} from ferritin can be initiated by acidosis, which develops during ischemia [Vanin A.F., Kubrina L.N., Kiladze S.V. et al.; Bralet J., Schreiber L. and Bouvier C.; Kirshner R.E. and Fantini G.A.; Oubidar M., Boquillon M., Maril Ch. et al.; Voogd A., Sluiter W., Van Eijk H.G. and Koster J.F.]; additional iron can also be released from blood cells retained in microvessels of the organ. Indeed, the accumulation of LPO products (HP, DC, MDA and SB) was more pronounced during incubation of nonperfused rat liver, in comparison to perfused liver [Dzhafarov A. I. (1983)]. The catecholamine-induced release of iron from ferritin also increases under conditions of ischemia and reperfusion [Allen D.R., Wallis G.L. and McCay P.B.].

A similar increase in the level of LMWI in ischemic liver lobes was demonstrated again in our recent study on the antiischemic effect of Desferal (see Chapter 11).

Intracellular redistribution of iron and an increase in the level of LMWI were found in rabbit kidneys upon ischemia and cold storage [Gower J.D., Healing G. and Green C.J.; Gower J.D., Healing G., Fuller B.J. and Green C.J.; Paller M.S.and Hedlund B.E.], in the ischemic heart [Holt S., Gunderson M., Joyce K.

et al.; Voogd A., Sluiter W., Van Eijk H.G. and Koster J.F.], liver [Rothman R.J., Serroni A. and Farber J.L.], brain [Krause G. S., Joyce K. M., Nayini N. R. et al.; Bralet J., Schreiber L. and Bouvier C.; Kondo T., Reaume A.G., Huang T.T., et al.], and skeletal muscles [Chiao J.J., Kirschner R.E., Fantini G.A.].

The content of iron increases in the first portions of solution after reperfusion of ischemic organs, as demonstrated on the heart and intestine [Chevion M. Jiang Y. Har-El R. et al.; Robinson E. and Hedlund B.], but it remains high in the organ proper too, e. g., in the kidney and skeletal muscles [Chiao J.J., Kirschner R.E., Fantini G.A.; Healing G., Gower J., Fuller B. and Green C.]; moreover, the longer the duration of ischemia and the more severe damage to the organ (kidney), the longer the period characterized by a high LMWI content [Healing G., Gower J., Fuller B. and Green C.].

The rat hears isolated from iron-overloaded animals after 15-min ischemia and 15-min reperfusion demonstrated more severe structural disturbances than the hearts of control rats [Pucheu S., Coudray C., Tresallet N. et al.].

Not only LMWI, but also iron contained in myoglobin (Mb) changes considerably during ischemia/reperfusion. Under physiological conditions, Mb is known to exist mainly in the form of oxymyoglobin, oxy-Mb(II) or $Mb^{2+} \cdot O_2$, in which iron is in the ferrous form (Fe^{2+}). The redox properties of Mb are also well-known. In its molecule, the heme group may change its oxidation state in the presence of various oxidizing and reducing agents. Mb undergoes the "myoglobin redox cycle" during ischemia/reperfusion [Hochstein P.] (Fig. 95), which plays a major role in muscle cell injury. The first stage of this cycle is $Mb^{2+} \cdot O_2$ conversion into deoxygenated Mb (Mb^{2+}). As Mb^{2+} is much more susceptible to two-electron oxidation by H_2O_2, compared to $Mb^{2+} \cdot O_2$ (by a factor of up to 100) [Yusa K. and Shikama K.; Galaris D., Eddy L., Arduini A. et al.], during early ischemia or reperfusion it is oxidized into Mb molecules in a hypervalent state: met Mb (Mb^{3+}), ferryl-Mb (Mb^{4+}), or its transient free-radical form ($\cdot Mb^{4+}$). Mb^{4+} and $\cdot Mb^{4+}$ are very strong oxidants with nonspecific peroxidase activity that can have effect on many molecules essential for the cell (PUFA, cholesterol, etc.), oxidizing them in direct interactions. The formation of $\cdot Mb^{4+}$ is accompanied by ROS production and generation of electronically-excited state. Thus, the emergence of Mb^{4+} and $\cdot Mb^{4+}$ is a critical event in pathogenesis of ischemic/reperfusion damage [Galaris D., Eddy L., Arduini A. et al.; Hochstein P.].

Fig. 95: The myoglobin redox cycle [Hochstein P.].

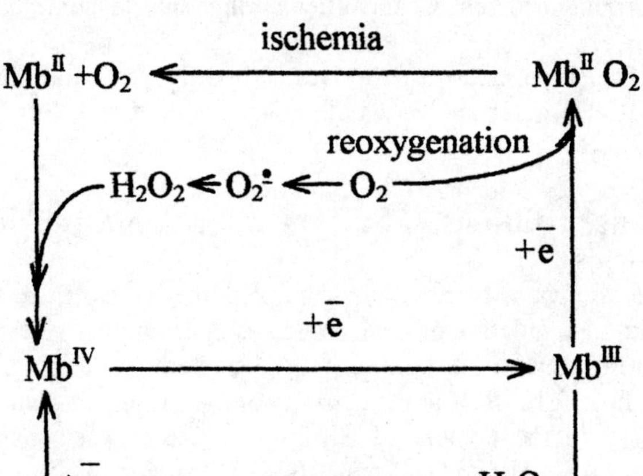

Mb^{II} — deoxy-myoglobin;
$Mb^{II}\,O_2$ — oxymyoglobin;
Mb^{III} — met-myoglobin;
Mb^{IV} — ferrylmyoglobin;
Mb^{IV} — ferrylmyoglobin radical.

At the second stage of the myoglobin redox cycle, hyperoxidized forms of Mb (Mb^{4+} or $\bullet Mb^{4+}$) are rapidly converted into met-Mb or Mb^{2+} in presence of various one-electron reductants, such as ascorbic acid [Galaris D., Cadenas E. and Hochstain P. (1985)], ergothionein (thiolhistidine) [Arduini A., Eddy L. and Hochstein P.] or GSH [Galaris D., Cadenas E. and Hochstain P. (1989)]. Owing to this rapid reduction, these heme proteins serve as "electron sinks" scavenging hydrogen peroxide or organic peroxides, thereby creating a mechanism protecting muscles against oxidative injury. The protective effect of these reducing agents was demonstrated in experiments on the rat heart exposed to a brief period of global ischemia followed by reperfusion [Galaris D., Eddy L., Arduini A. et al.; Arduini A., Eddy L. and Hochstein P.; Hochstein P.] and in cardiological patients operated on cardiopulmonary bypass [Eddy L., Hurvitz R. and Hochstein P.].

Changes in the amount and properties of not only transient metals, but also of their reducing and complexing agents belong to the factors promoting LPO intensification during ischemia and reperfusion.

Ischemia is known to be accompanied by the accumulation of ADP (at least relative) and a sharp increase in the level of AMP (see Chapter 3), which lead to the formation of Fe^{2+} chelates that more readily initiate LPO.

Ischemia is also marked by a sharp increase in the content of reduced pyridine nucleotides (NADPH, NADH) (see Chapter 3), which cause Fe^{3+} reduction to Fe^{2+} and determine its involvement in the reaction. As observed in our experiments with SR fragments of skeletal muscles [Arkhipenko Yu. V., Bilenko M. V., Dobrina S. K. et al.], adding NADPH and especially NADH to the incubation medium leads to a marked increase in the rate of accumulation of LPO products.

It was discovered that the concentration of ascorbic acid, one of the main iron-reducing agents, strongly changes already at the early stages of ischemia. During partial ischemia of the cat brain (ligation of the middle cerebral artery), the content of ascorbate decreased to 60% of the initial as soon as after 1 h and remained at that level after 5 and 24 h of ischemia [Demopoulos H.B., Flamm E.S., Pietronigro D.D. and Seligman H.L.]. In experiments on LAD occlusion in dogs, ascorbate content decreased significantly after 15 min of ischemia, simultaneously with generation of free radicals, whereas lower GSH peroxidase activity and higher MDA content in the ischemic area were observed later, only after 45 min of ischemia [Rao P.S., Cohen M.V. and Mueller H.S.]. As ascorbate in tissues is depleted, it loses its antioxidant properties and becomes a prooxidant.

In muscle tissue, the level of histidine-containing dipeptides proved to decrease during hypoxia, and this decrease was more pronounced during and after strong muscle contractions [see Free radicals and biological stabilizators].

Thus, the contents of transition metals themselves (primarily Fe^{2+}) and their reducing and complexing agents change during acute ischemia in such a way that their prooxidant effects increase.

10.4.2 Changes in Concentrations of Sulfur- and Selenium-Containing Compounds

These compounds are similar in terms of the functions they perform in the organism. They play an especially important role in the regulation of LPO processes in biological systems.

Sulfur-containing (sulfhydryl, thiol) compounds act as both oxidation promoters and antioxidants.

The prooxidant effect manifests itself in the capacity of SH groups to reduce transition metals, which is accompanied by a decrease in the content of SH groups due to their oxidation to disulfides (-SS-), sulfenates (-SOH), sulfinates ($-SO_2H$), and sulfonates ($-SO_3H$, irreversible form).

Oxidation may affect both low-molecular nonprotein thiols (amino acids, glutathione, etc.) and protein SH groups, impairing specific functions of membrane proteins (see Chapter 5). Many researchers observed that the content of SH groups decreases during ischemia [Kulagin B. R.; Torchinsky Yu.M.]. Thus, the involvement of protein SH groups in iron reduction during ischemia is one of the factors contributing to membrane damage.

The prooxidant effect can also be exerted by reduced glutathione (GSH), which participates in free-radical reactions, such as $GSH + O_2 \rightarrow GS^{\bullet} + HO_2^{\bullet}$; $HO_2^{\bullet} + RH \rightarrow R^{\bullet} + H_2O_2$; $R^{\bullet} + O_2 \rightarrow RO_2^{\bullet}$, or promotes oxidation of epinephrine to adrenochrome.

However, the antioxidant effect of sulfur-containing compounds is much stronger than their prooxidant action. It is accounted for by either "reactive" protein SH groups, which undergo reversible S-thiolation in the presence of ROS (followed by dethiolation) and, thus, protect proteins against more drastic transformations, or (predominantly) by GSH, which constitutes a major portion of nonprotein low-molecular thiols and participates in GSH peroxidase and GSH S-transferase reactions (see Fig. 91).

Reduced and oxidized glutathiones represent a conjugate oxidation–reduction pair (GSH ↔ GSSG); hence, a change in their ratio may be critical for regulation of peroxidation processes in tissues. Under normal conditions, GSH content is almost hundred times that of GSSG, to say nothing of the fact that about 30% of GSH are present in a complex with proteins. The highest concentrations of GSH are found in the liver, kidneys, erythrocytes, and eye lens.

Two glutathione pools are distinguished: cytoplasmic and mitochondrial. The latter accounts for no more than 10–13% of the total cell glutathione and consists primarily (by 90%) of reduced GSH [Lefer A.M. (1985); Meredith M.J. and Reed D.J. (1982)]. The mitochondrial GSH pool is characterized by turnover rate 15 times as slow as that of the cytoplasmic pool but is much more difficult to deplete [Meredith M.J. and Reed D.J. (1982)].

The reduction of GSSG and regeneration of GSH are catalyzed by GSH reductase, and GSH synthesis *de novo,* by GSH synthetase (see Fig. 91).

The capacity for glutathione synthesis and reduction differs from one organ to another. The liver is the main source, and the kidneys are primarily consumers of GSH; the lungs both synthesize GSH and receive it from blood plasma [Berggren M., Dawson J. and Moldeus P.].

In addition to being involved in the functioning of GSH peroxidase, GSH is capable of nonenzymatic H_2O_2 inactivation and ROS scavenging [McNeil C.J., Banford J.C., Drown D.M. and Smith W.F.].

The data on the relationship between the contents of GSH and other SH groups, on the one hand, and LPO intensity, on the other, are contradictory [Gulak P. V., Dudchenko A. M., Zaysev V. V. et al.], which is apparently explained by the fact that these compounds can act as antioxidants and oxidation promoters.

The decrease of GSH content in hepatocytes under the effect of different toxic factors proved to be accompanied by the accumulation of LPO products in the tissue. In this case, the depletion of mitochondrial GSH pool was especially important, as only a sharp decrease in both pools, rather than in the cytoplasmic pool alone, could lead to MDA accumulation and membrane damage [Meredith M.J. and Reed D.J. (1983)].

Conversely, LPO induction in the rat brain *in vitro* [Rehncrona S., Smith D.S., Akesson B. et al.] and in isolated hepatocytes [Anundi T., Högberg J. and Stead A.H.] was accompanied by a decrease in GSH content and an increase in that of GSSG, although no change in GSH content was observed in other experiments with hepatocytes.

In our experiments [Gerasimov A.M., Bilenko M.V., Panchenko L.F. et al.], most of total and protein-unbound low-molecular thiols, including GSH, in the renal cortex were found in the mitochondrial fraction and lysosomes (54 and 65%, respectively); smaller amounts (26 and 30%), in the cytosol; and the smallest amounts (7 and 9%), in the microsomal fraction. In the kidneys preserved for 3, 24 and 48 h (perfusion with oxygenated plasma or Collins' solution, storage under hypothermia at 4 ATA O_2 without perfusion), the contents of these thiols sharply decreased, but this was not accompanied by MDA accumulation in the kidney, or this accumulation was delayed.

The above findings suggest that the decrease of GSH and SH groups in tissues always creates conditions for LPO intensification, whereas the mere fact that the content of SH or GSH decreases can hardly be used as a solid argument in favor of LPO intensification, unless it is supplemented by direct measurements of LPO products.

To date, many experiments have been performed to study changes in low-molecular thiols, GSH, and the GSH/GSSG ratio (reduction potential index) during ischemia and reperfusion of organs, as well as the relationship between these changes and LPO processes during ischemia. It was found [Guarnieri C., Flamigni F. and Caldarera C.M.] that hypoxic perfusion of the heart in rats reduces the content of thiols, both protein-bound and free low-molecular, including GSH, without being accompanied by the accumulation of GSSG and MDA in the tissue. After reoxygenation, the level of sulfur-containing compounds continued to decrease, GSSG release into the perfusate increased, and the MDA level went up. Thus, the decrease in the GSH content preceded LPO intensification and appeared to be responsible for peroxidation damage to the ischemic cardiac tissue during reoxygenation.

Both GSH content and GSH/GSSG ratio were found to decrease during ischemia or conservation of rabbit and rat kidneys with their subsequent reperfusion, which was accompanied by the accumulation of LPO products in the kidneys [Green C.J., Healing G., Lunec J. et al.; Green C.J., Healing G., Simpkin S. et al. (1986a)]. The same phenomena were observed in the ischemic tissue 3 and, especially, 24 h after LAD occlusion in dogs; however, the maximum increase in MDA content, to 143.5% of the initial, was observed only 3 h after the occlusion [Röth E., Török B., Zsoldos T. and Matkovics B.].

The level of nonprotein low-molecular thiols was found to drop during focal ischemia and reperfusion of the rat heart [Maezawa H., Manaka K., Yamakawa K., et al.], isolated storage of the rat liver [Panchenko L.F., Gerasimov A.M., Nozdrachova L.I., Kariakina G.A.], as well as in the rat liver and brain during the resuscitation period following 6.5-min asphyxia [Konvai V. D., Lukoshkin A. V. and Smirnova V. B.].

In experiments on a model of total (compression) cerebral ischemia in Wistar rats [Rehncrona S., Smith D.S., Akesson B. et al.], GSH content in the brain tissue ischemic for 30, 60 and 120 min gradually decreased to 87, 77 and 45% of the initial, respectively. Reperfusion of the brain led to further decrease in GSH level. A similar decrease during the reperfusion period was also observed after partial ischemia. However, no increase in the GSSG content was observed in these experiments during either ischemia or reperfusion, whereas such an increase was well-manifested during LPO induction in the brain tissue *in vitro* [Rehncrona S., Smith D.S., Akesson B. et al.]. The authors concluded that the decrease of GSH content during ischemia is due to the primary disturbance of GSH synthesis (because of energy deficiency) or its release into the extracellular space, rather than to GSN consumption in peroxidation reactions. The role of GSH deficiency as a factor of LPO intensification during ischemia, rather than its consequence, is also confirmed by the data on the protective antiischemic effect of GSH (see Chapter 11).

At the same time, many researchers revealed the increase of GSSG or increased GSSG/GSH ratio content in the blood or perfusate flowing out of ischemic organs [Tritto I., Duilio C., Santoro G., et al.; Lesnefsky E.J., Repine I.E. and Horwitz L.D.; Sies H. and Akerboom P.M.T.] in the early reperfusion period and, on this basis, suggested that GSSG formation can serve as a sensitive marker of oxidative stress.

Selenium-containing compounds, in contrast to sulfur- containing ones, have only antioxidant properties. This is because (1) selenium (Se) is incorporated into the active site of Se-dependent GSH peroxidase (GP-I) and (2) selenium (Se-containing amino acids) *per se* can act as an antioxidant by scavenging free radicals or participating in nonradical degradation of lipid peroxides.

Animal organisms receive selenium with food in the form of selenates (Na_2SeO_4), selenites (Na_2SeO_3), and selenides (Na_2Se), as well as Se-containing amino acids (Se-cysteine, Se-methionine) and proteins. It is an essential element, and clinical manifestations of selenium deficiency are close to those of vitamin E deficiency.

In the organism, selenium undergoes complex metabolic transformations (Fig. 96) [Sunde R.A.]. It is incorporated into the active site of GSH peroxidase in the form of Se-cysteine, and this is why the form in which selenium enters the organism is important for its activity. Se-cysteine and, especially, Se-methionine are more susceptible to rapid degradation, which yields selenide suitable for the formation of GSH peroxidase.

In organs rich in GSH peroxidase (rat liver, sheep erythrocytes), most of selenium (about 70-100%) is bound to this enzyme. In contrast, Se in organs containing small amounts of GSH peroxidase is associated with proteins or amino acids [Sunde R.A.]. The data on changes in the content of Se-containing compounds and disturbances of their metabolism during ischemia are scarce, but it is apparent that the deficiency of these compounds (during partial ischemia) or block of their supply (during total ischemia) must affect the content of Se in the ischemic organ. This follows from the aforementioned decrease in the activity of Se-dependent GSH peroxidase during and after ischemia, as well as from the data on the protective antiischemic effect of selenium-based drugs (see Chapter 11). Thus, the involvement of primary or secondary (resulting from LPO intensification) deficiency or excess of sulfur- and selenium-containing compounds in pathogenesis of ischemic and reoxygenation injuries is an established fact, which has already provided the basis for practical measures to control the corresponding processes.

Fig. 96: Possible selenium metabolism pathways [Sunde R.A.].

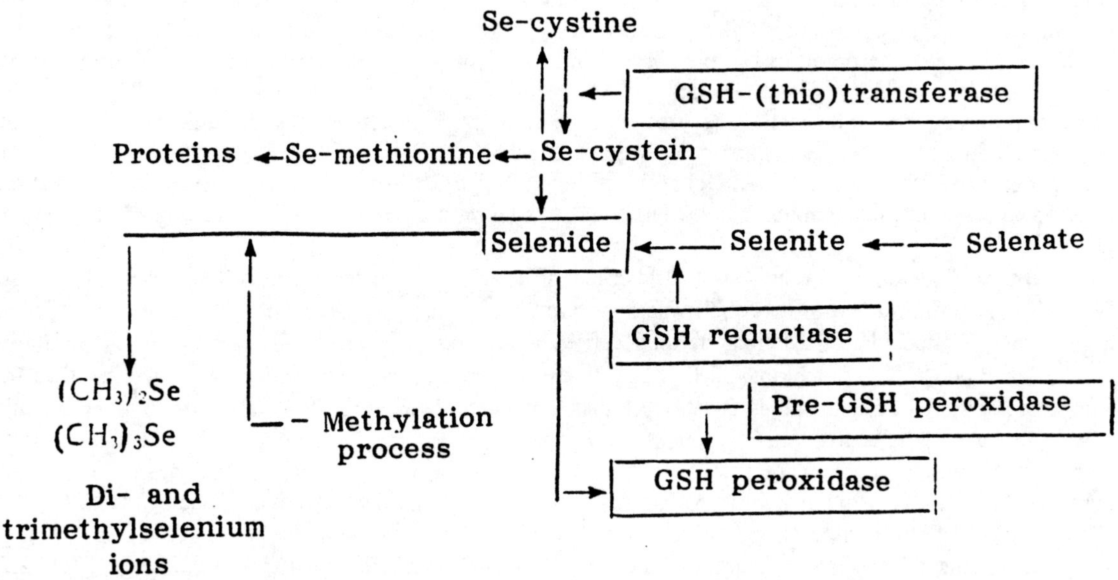

10.4.3 Changes in the Content of Fat-Soluble Compounds

Fat-soluble compounds act either as antioxidants, exclusively (tocopherols, ubiquinones, steroid hormones, etc.) or predominantly, depending on dosage (vitamins A, K, etc.), or as oxidation promoters (vitamin D, etc.).

The mechanism of antioxidant action of these compounds is determined by their pronounced donor properties and the ability to reduce lipid radicals ($RO_2^{\bullet}$), which is why they are classified as antiradical substances, or "true" antioxidants (InH), and their antiradical activity is characterized by the constant K_7 (see Fig. 81). However, their overall antioxidant activity or, in other words, the ability to slow down LPO in general, is determined not as much by their antiradical activity as by the ability of the newly emerging radical ($In^{\bullet}$) to participate not only in recombination reactions yielding inactive molecular products, but also in those with the new molecules of the oxidized compound (the constant K_{10}), i.e., to initiate new oxidation chains. Substances with active radicals (high K_{10}) exhibit lower antioxidant activity, compared to substances with "weak" radicals [Burlakova E.B., Alessenko A.V., Molochkina E.M. et al.].

As follows from the above data, the antioxidant and antiradical activities of these compounds are not always identical, and the antioxidant activity (compared to antiradical) is a more reliable indicator of the involvement of a natural fat-soluble compound in LPO control in the cell.

Active fat-soluble LPO regulators include vitamins, ubiquinones, steroid hormones, complex phospholipids, and many other compounds. Let us discuss only a few of them, those the contents of which change significantly during ischemia and reperfusion.

Tocopherols are a group of exogenous compounds (not synthesized in animal organism) comprising eight fat-soluble vitamins differing in the number of methyl groups (1–3) and their spatial position in the molecule (α, β, γ, etc.). All of them have antioxidant properties accounted for by their high antiradical activity. Radicals of these compounds are "strong, " and this makes their antioxidant activity much lower. The antiradical activity of tocopherols is based on the constituent phenol (chromane) ring capable of reversible oxidation yielding semiquinone (tocopherylquinone) and quinone [Burlakova E.B. and Khrapova

N.G.; Emanuel N.M.]. The reduction of quinones to phenols, i.e., regeneration of tocopherol in tissues, is possibly mediated by ascorbic acid.

A distinguishing feature of tocopherols is their long side (isoprenoid) chain comparable in length with the fatty-acid chain of phospholipids [Ivanov I. I., Merzlyak M. N. and Tarusov B. N.]. This chain apparently accounts for the membrane-stabilizing (nonradical) effect of tocopherols. According to recent evidence, however, the chromane ring is also necessary for this effect [see Free radicals and biological stabilizators]. The incorporation of tocopherols into a membrane is accompanied by the formation of complexes between methyl groups of their side chain and double bonds of PUFA [Lucy A.J.], so that the more unsaturated the fatty acid, the stronger the linkage between the membrane and tocopherol. Notwithstanding numerous mechanisms by which tocopherol can perform its important function in the cell [see Burton G.W., Cheeseman K.H., Doba T. and Ingold K.U.; Willson R.L.], some researchers regard the antioxidant effect of this compound as its most fundamental property [Cadenas E., Ginsberg M., Rabe U. et al.].

The highest antioxidant activity is characteristic of α- and γ-tocopherols [Lindsey J., Zhang H., Kaseki H. et al.]. As demonstrated in model chemical systems, the antiradical activity of α-tocopherol is twice as high as that of γ-tocopherol, whereas its total antioxidant activity, conversely, is twice as low [Burlakova E.B., Alessenko A.V., Molochkina E.M. et al.]. It was shown that the contribution of α-tocopherol to the antiradical activity of total lipids is 60%, while that to the antioxidant activity is only 10–20%. In addition to antiradical activity, some other mechanisms may account for the antioxidant effect of α-tocopherol (especially under pathological conditions). They include its abilities to scavenge active forms of O_2, to form complexes with PUFA (which stabilizes membranes structurally and improves their resistance to LPO), and to enhance the action of other natural antioxidants [Free radicals in biology].

Tocopherols are present in all organs. In rats, α-tocopherol content is maximal in the lungs, heart, and brain; lower in the kidneys and skeletal muscles; and minimal in the liver. Different concentrations of α-tocopherol were also revealed in microsomes isolated from different rat organs: in pulmonary microsomes, its content was six times as high as in hepatic and cerebral microsomes and twice as high as in cardiac microsomes [Kornbrust D.J. and Mavis R.D.]. It was demonstrated that the highest concentrations of α-tocopherol are accumulated in the inner mitochondrial membrane. Membranes have specific sites for binding tocopherol [Wimalasena J., Davis M. and Kitabchi A.E.].

The role of tocopherols in LPO reactions was thoroughly studied in experiments on E-deficient animals *in vivo*, in which an increase in the concentration of LPO products in the blood and different organs was revealed [Donchenko G. V., Kuzmenko I. V., Kovalenko V. N. and Sidorenko L. V.; Free radicals in biology]. When LPO in membranes of E-deficient animals was initiated *in vitro,* researchers observed intensification of this process or five- to sixfold shortening of the induction period [Cadenas E., Ginsberg M., Rabe U. et al.].

The contents of α-tocopherol and MDA in the rat liver and lungs were inversely proportional; upon LPO induction, the content of tocopherol sharply decreased (even to its complete disappearance) [Varskeviciene Z.Z., Cerniauskiene R.C. and Grybauskas P.S.]

Retinols are also phenol-containing vitamins (vitamins of group A). However, they can act as both antioxidants and oxidation promoters. Their antioxidant effect was observed during LPO induction in hepatic microsomes and erythrocytes of vitamin A-deficient rats; in a model chemical system (yolk phospholipids), retinol at small doses (0.15 mole/100 mole PL) accelerated Fe^{2+}- ascorbate-induced LPO, while at large doses (1 mole/100 mol PL) it acted as an inhibitor.

The precursors of retinol are carotenoids (primarily β-carotene), which enter the animal organism with food and have a high antiradical activity. It is known that the amount of retinol accumulating in the liver is 10–20 times that in other organs.

The mechanism of antioxidant activity of retinol is not yet fully understood. It may be based on the ability of vitamin A to preserve thiol compounds or to scavenge ROS. Antioxidant derivatives of retinols have been studied less thoroughly, but many specialists recommend their use as a component of integrated antioxidant therapy.

Ubiquinones are essentially fat-soluble coenzymes differing in the length of the side chain ($CoQ_1 \rightarrow CoQ_{10}$). In mammals, the prevailing form is CoQ_{10}, which exists both in a free state and in a complex with a protein. The specific function of ubiquinones consists in electron transfer in the mitochondrial respiratory chain, during which ubiquinones are reduced to ubiquinols ($CoQH_2$); the reduced form is unstable and is rapidly oxidized again. According to recent data, mitochondrial ubiquinones, in addition to their specific oxidation–reduction function, also act as antioxidants [Burlakova E.B. and Khrapova N.G.; Landi L., Cabrini L., Tadolini B. et al.] and regulate membrane fluidity. The antiradical and antioxidant activities are characteristic of not only reduced, but also of oxidized forms, although to a lesser extent [Landi L., Cabrini L., Tadolini B. et al.]. According to different authors, the antiradical and antioxidant activities of ubiquinones with side chains of different lengths are either more or less the same [Landi L., Cabrini L., Tadolini B. et al.] or slightly decrease in the following sequence: $Q_1 > Q_6 > Q_{10}$ [see Free radicals and biological stabilizators].

In experiments on comparing antioxidant effects of CoQ_{10} and α- tocopherol, CoQ_{10} (2 mg/kg) and α-tocopherol (10 mg/kg) administered to rats over a period of 5 days caused a similar decrease in the content of lipid peroxides in cardiac mitochondria; in hepatic mitochondria, however, this decrease was more pronounced under the effect of α-tocopherol, which suggested that the antioxidant effect of both substances is organ-dependent. It was also found that CoQ_{10} protects mitochondria against phospholipases A_2 and C [Kobayashi M., Shimomura I., Suzuki H et al.; Nagai S., Miazaki Y., Ogawa K. et al.], although the exact mechanism of this effect (stabilization of membranes, antioxidant activity, or direct inhibition of phospholipases) is as yet unknown.

Analysis of radiation-induced LPO in liposomes prepared from PC [Zaslavsky Yu. S.] showed that K_7 of reduced CoQ is $3.2 \cdot 10^5$ l • mole^{-1}• s, and that of oxidized CoQ equals $0.3 \cdot 10^3$ l • mole^{-1}• s, which is much lower than K_7 of α-tocopherol ($4 \cdot 10^5$ l • mole^{-1}• s). The mild antioxidant effect of CoQ_6 and CoQ_9 was also observed on the model of methyl oleate oxidation. The antioxidant activity of ubiquinone was only 4–7% of that of ionol. This provided a basis for the conclusion that the contribution of ubiquinones to the antioxidant activity of liver lipids does not exceed 5–10%.

Steroid hormones (thyroxine, estrogens) exhibit both antioxidant (antiradical) and prooxidant properties [Sergeev P.V.]. In rat liver mitochondria, thyroxine at concentrations exceeding $5 \cdot 10^{-7}$ M inhibited Fe^{2+}-induced chemiluminescence, being in this respect only slightly inferior to α-tocopherol; at lower concentrations (10^{-8} M), however, thyroxine acted as an oxidation promoter [Parnev O.M., Naumov O.G., Ivkov N.N. et al.].

Lipids, cholesterol, and cholesterol derivatives may also act as both oxidation promoters and antioxidants.

As shown above, phospholipids, especially those containing PUFA, are the main substrate for LPO processes.

Cholesterol is also a substrate for enzymatic (NADPH- dependent) and nonenzymatic (Fe^{2+}-ascorbate-dependent) LPO, which was demonstrated in direct experiments on ER membranes of the liver [Vladimirov Y.A.; Lankin V. Z., Zakirova A. N., Akhmetova B. Kh. et al.]. The resulting LPO products (7-ketocholesterol, 7β- and 7α-hydroxycholesterols, etc.) can initiate LPO in phospholipids and alter the properties of biomembranes. Cholesterol peroxidation in the cytochrome P-450-dependent system of liver monooxygenases, just as LPO in phospholipids, inhibits other oxidative functions of this system, particularly the monooxygenase pathway of oxidation of cholesterol itself to steroid hormones and bile acids [Lankin V. Z., Zakirova A. N., Akhmetova B. Kh. et al.].

A marked prooxidant effect is produced by cholesterol derivatives, such as calciferol and cholecalciferol, which are metabolized into vitamins of group D in the cytochrome P-450-dependent system of liver microsomes, accumulate in the liver, and, readily undergoing autocatalytic free-radical oxidation (particularly under conditions of hypervitaminosis), contribute to the accumulation of lipid peroxides [Spirichev V.B. and Kon' I.Ya.].

The prooxidant effect of vitamin D is especially pronounced when other antioxidants are deficient, which may serve as indirect evidence that its prooxidant function is apparently enhanced when other systems of antioxidant protection are impaired, as during ischemia. This is particularly true in the case of ischemia of the liver, in which the concentration of vitamin D is maximal. At the same time, phospholipids (especially PC and PEA) in membranes may act not only as oxidation promoters, but also as antioxidants [Burlakova E.B., Alessenko A.V., Molochkina E.M. et al.; Zhuravlev A. I.].

Cholesterol can also act as an antioxidant by either increasing viscosity of membranes and stabilizing them or exerting an antiradical effect [Vladimirov Yu.A.].

The number of publications dealing with **changes in the content of fat-soluble compounds during acute ischemia and after reperfusion of organs** has been steadily increasing in recent years, although the results described are as yet controversial. For instance, in rats kept under hypobaric conditions (200 mm Hg in air) for 2 h, the content of α-tocopherol remained unchanged in the liver, kidneys, heart, brain, and skeletal muscles, and that in the lungs even increased significantly [Varskeviciene Z.Z., Cerniauskiene R.C. and Grybauskas P.S.]. The content of α-tocopherol also remained unchanged upon another LPO-inducing disturbance of the oxygen regime, namely, hyperbarism in newborn rats [Bucher J.R. and Roberts R.J.].

In contrast to these observations, total cerebral ischemia in rats (occlusion of two vertebral and two carotid arteries) led to a significant decrease in α-tocopherol content, which was observed as soon as after 30-min ischemia and continued after 15–30 min of reperfusion [Yoshida S., Abe K., Busto R. et al.]. The α-tocopherol level dropped in the infarct area after LAD occlusion in rats [Meerson F.Z.]. Five days after AMI, vitamin A (retinol) concentration in the aortic intima and blood of patients proved to decrease, with a corresponding increase in the content of lipid peroxides and a decrease in the antioxidant activity of lipids [Kalmykova V. I. and Burlakova E. B.].

Methods of assessing the total antiradical (ARA) or total antioxidant activity (AOA) of lipids extracted from an ischemic or reperfused organ, rather than the content of individual fat-soluble compounds, are more widespread and, apparently, more informative. Thus, we revealed a sharp decrease in the total antiradical activity (ARA) in dog kidneys after their long-term conservation [Gerasimov A.M., Bilenko M.V., Panchenko L.F. et al.]. Other researchers demonstrated that the total antiradical activity of lipids in the rat brain decreased by 33% 5 min after clinical death and resuscitation [Semchenko V.V., Poluektov L.V. and Konvai V.D.], and that in the rat liver decreased by 36% 6.5 min after asphyxia and by 20% during the first 5–30 min of reperfusion [Konvai V. D., Lukoshkin A. V. and Smirnova V. B.].

The data obtained by the method of thermal oxidation of methyl oleate showed that total ischemia and reperfusion of the kidneys, liver, heart, and skeletal muscles are accompanied by a sharp decrease in the total antioxidant activity (AOA) of lipids, but this decrease, however, occurs 30–60 min after the initial significant accumulation of LPO products (see Chapter 5). Subsequent experiments performed by the same method revealed a decrease in the content of fat-soluble antioxidants during ischemia (isolated storage) and prolonged (more than 4 h) preservation of the liver [Dzhafarov A. I. (1983)].

The total amount of fat- and water-soluble antioxidants decreased by 30% in the brain of rats 5 min after their decapitation [Majewska M.D., Strosznajder J. and Lazarewicz J.].

The fact that a decrease in the content of fat-soluble natural compounds enhances LPO processes during ischemia and after reperfusion has been additionally confirmed by the successful use of such compounds as antiischemic drugs (see Chapter 11).

Thus, natural fat-soluble compounds play an important role in the regulation of LPO in the cell. This fat-soluble antioxidant system comprises some already known and still unknown compounds involved in a complex interrelationship with each other.

The proportion of tocopherols is fairly high but differs from one organ to another; hence, determination of the total antiradical and, especially, antioxidant activity of lipids may yield more informative results than that of each individual component of the system.

The tocopherol content and total antiradical activity and antioxidant activity of lipids decrease during ischemia, which may be an indication of their "consumption" during LPO or, in other words, additional evidence for LPO occurring in the ischemic organ. The low antiradical and antioxidant activities of fat-soluble compounds are also the factor facilitating LPO reintensification during the reperfusion period.

10.5 Conclusions

The data discussed in this chapter have mostly been obtained in recent years. The body of knowledge in this field is rapidly expanding, which is accounted for by great interest and intensive research in corresponding problems, and many of the findings soon become outdated. As a consequence, many aspects are yet to be fully elucidated, although others have been completely clarified. Nevertheless, generalizing the available data can help to explain some phenomena and to outline the most promising areas of research.

The results presented here lead to the following conclusions.

(1) A reduced or even residual O_2 concentration in organs affected by partial and even total ischemia seems to be quite sufficient for initiating LPO processes and maintaining them for a variable period of time. Since the toxic effect of LPO products on membranes is extremely strong (see Chapter 5), LPO initiation and progression for even a short period of time course are sufficient for causing membrane injury and/or triggering other damaging mechanisms.

(2) Ischemia of organs creates conditions stimulating several O_2 radical-generating enzyme systems and enhancing the production of ROS, which then initiate LPO in membranes. These are the systems responsible for electron transport in mitochondria and microsomes, hypoxanthine and xanthine oxidation (xanthine oxidase), arachidonic acid metabolism (cyclooxygenase and lipoxygenase), L-arginine nitrogen oxidation (nitric oxide synthases), epinephrine oxidation, the system of peroxidases, and others.

However, it is still unclear which of these ROS-generating systems operate under oxygen deficiency, i.e., during ischemia proper, and which are activated only during the postischemic period. It appears that ROS production in mitochondria and microsomes, in the course of arachidonic acid metabolism, catecholamine oxidation, $^{\bullet}$NO and peroxynitrite (its derivative) formation, and subsequent ROS metabolism and conversion can occur during ischemia proper; ROS generation in the xanthine oxidase and NADP-oxidase reactions in tissue macrophages is enhanced during the early reperfusion period; and that in the NADP-oxidase reaction in polymorphonuclear leukocytes, during the late reperfusion period. Thus, different systems of ROS generation are switched on in a certain sequence.

The involvement of different system in ROS generation varies depending on the organ, animal species, the type of membranes, and the severity (completeness and duration) of ischemia).

(3) Cells of different origin also differ in the ability to generate ROS and to initiate LPO. To date, it has been demonstrated that endothelial cells are the most active ROS producers; on the other hand, these cells are the most sensitive target to the toxic action of ROS and represent the site of initial LPO intensification during ischemia and reperfusion. The explanation is as follows. Endothelial cells, compared to parenchymal cells, contain greater amounts of xanthine oxidase, and their xanthine dehydrogenase is more easily converted into xanthine oxidase; they accumulate significant amounts of hypoxanthine; are highly capable of producing NO$^{\bullet}$; contain powerful enzyme systems of arachidonic acid metabolism and the synthesis of prostaglandins and leukotrienes, which is accompanied by the formation of ROS and LPO products; are in close contact with transferrin (an iron carrier), leukocytes, and blood platelets, which are the potential sources of ROS; being located at the blood–vessel interface, they interact with the blood characterized by higher residual Hb O_2content and pO_2during ischemia and their sharp increase upon the onset of reperfusion.

There also is ample direct evidence that LPO in endothelial cells is easily initiated under the effect of ROS, iron, neutrophils, and low-density lipids. In this context, temporal differences in LPO intensification during ischemia of different organs can only be explained by the existence of organ-specific differences in the sensitivity of microvascular endothelium to ischemia and in its ability to respond to LPO induction. No proof to this effect has been obtained as yet, although the heterogeneity of different vessels in terms of sensitivity to FFA and LPO products (see Chapter 8) gives every reason to believe that the microvascular endothelium is also heterogeneous in terms of its properties.

(4) Ischemia is accompanied by a decrease in the activity of antioxidant enzyme systems (SOD, GSH peroxidase, catalase) and in the content of water- and fat-soluble antioxidants, whereas the content of prooxidant compounds increases. Such a tipping of the prooxidant/antioxidant balance during ischemia was clearly demonstrated in experiments on assessing the $V_{O_2^{\bullet}}$/SOD, Fe^{2+}/ascorbate and ferritin/ceruloplasmin ratios and on determining the total antioxidant activity. In the latter, the antioxidant activity (AOA) values changed to negative, i.e., fat-soluble oxidation promoters began to prevail over antioxidants. The change (increase) in the prooxidant/antioxidant ratio appears to be the most important factor responsible for LPO initiation during ischemia and the development of oxygen toxicity ("oxidative stress") even under conditions of acute O_2 deficiency in the organ.

Imbalance of oxidation promoters and antioxidants is conditioned not only by accumulation of the former and consumption of the latter in the half-closed or closed system of the ischemic organ, but also by the opposite effects exerted by many LPO-regulating compounds (Fe^{2+}, Ca^{2+}, SH groups, GSH, ascorbic acid, etc.). Most of these compounds intensify LPO processes at small doses and inhibit them at large doses. Therefore, a decrease in the content of some of them not only enhances their effect quantitatively, but also reverses it, which is essentially a qualitative change.

The analysis of changes in the content of some LPO-regulating compounds or in the activity of antioxidant enzyme systems has shown that changes occurring during severe ischemia are directed toward LPO intensification.

Some changes in the relationship of radical-producing and antioxidant systems, which promote LPO intensification during ischemia, are shown in Table 16.

(5) Enzyme systems responsible for ROS production (cytochrome P-450, enzymes of the mitochondrial respiratory chain, xanthine oxidase, enzymes of arachidonic acid metabolism, NADPH-

oxidase of phagocytes, cyclooxygenase, lipoxygenase, peroxidases) and those preventing excessive ROS production and catalyzing nonradical degradation of lipid peroxides (SOD, GSH peroxidase, GSH S-transferase, GSH reductase, catalase) are highly ROS-dependent. They become activated at low concentrations of ROS, whereas high concentrations inhibit them; i.e., these are "suicide" enzymes.

(6) A certain role in LPO initiation belongs to the increase of intracellular Ca^{2+} content, which is a major symptom characterizing both ischemic and, especially, postischemic periods. Ca^{2+}-induced LPO intensification, demonstrated in direct experiments on chemical and biological model systems *in vitro*, is apparently explained by the ability of Ca^{2+} (1) to activate Fe^{2+} by converting it from the protein-bound to the catalytically active "free" form; (2) to promote the conversion of xanthine dehydrogenase into xanthine oxidase; (3) to accelerate the release of ROS from neutrophils; (4) to activate phospholipid hydrolysis by phospholipase A_2, which, in turn, intensifies LPO processes; and (5) to damage mitochondrial membranes mechanically by forming precipitates (calcium phosphate) (see upper and Chapters 2, 4, and 5).

Table 16: Conditions promoting LPO intensification during acute ischemia and reperfusion of organs.

Conditions promoting free-radical reaction	Enhancement of oxidation-promoting factors	Mitigation of AO factors
Activation of ROS-generating system.	Increase in reduced carriers, enhancing electron "leakage" in electron transport chains of mitochondria, microsomes, and nuclei. Express activation of phagocytes and endothelial cells. Conversion of xanthine dehydrogenase into xanthine oxidase, increase in hypoxanthine content. Ca^{2+} accumulation. intensification of PL hydrolysis and cyclooxygenase and lipoxygenase pathways of AA metabolism. Activation of NOS producing $NO^{\bullet}$ and peroxynitrite. Increase in epinephrine content and the epinephrine/norepinephrine ratio. Enhancement of peroxidase mechanisms of ROS production upon destabilization of lysosomes and peroxisomes.	Decrease in SOD, GSH peroxidase and catalase activities. Increase $Vo_2^{\bullet}$ /SOD ratio
Activation of transition metals.	Accumulation of "free" cytoplasmic Fe^{2+}, accumulation of ADP and AMP forming active complexes with iron. Accumulation of NAXDPH and NADH reducing Fe^{3+} to Fe^{2+}. Changes in the Fe^{2+}/ascorbate ratio.	Decrease in the content ascorbic acid, natural dipeptides, and ceruloplasmin.
Degradation of lipid peroxidates and branching of oxidative chains.		Decrease in the contents of GSH, SH groups, ascorbic acid, α-tocopherol and retinol, local SE deficiency. Decrease in GSH peroxidase and GSH-S transferase. Decrease in the total ARA and AOA of fat-soluble AO.

(7) The intensification of LPO processes and their damaging effect on membranes are closely related to other membrane-damaging mechanisms, such as energy deficiency, acidosis, lipolysis and phospholipolysis, the release of lysosomal enzymes (lipolysis, phospholipolysis, and proteolysis), and reactions of the arachidonic acid cascade. One of the factors responsible for synergism of these mechanisms is that accumulating products inflict damage to membrane structure (injuring of the "structural antioxidant") that makes polyene acyls more accessible to phospholipases and LPO inducers (see Chapter 2).

(8) The relationship between LPO and disturbances of high-energy phosphate metabolism during ischemia is based on both energy deficiency and acidosis developing in the ischemic organ. LPO intensification upon energy deficiency is associated with a number of factors, such as disturbed functioning of ATPases and the loss of ionic homeostasis (cell overloading with Ca^{2+}); structural rearrangements in membranes due to their dephosphorylation; disturbed synthesis and resynthesis of membrane phospholipids; degradation of ADN and accumulation of hypoxanthine, substrate for the xanthine oxidase reaction; suppressed β-oxidation of FFA and their detergent-like action on membranes. The damaging effect of LPO on energy synthesis is based primarily on structural and functional damage to mitochondria. In particular, this follows from the fact that preventive administration of antioxidants (α-tocopherol, BHT, CoQ_{10}) does not affect the rate of ATP degradation during ischemia of the kidneys and liver but promotes ATP resynthesis in these organs during the reperfusion period (see Chapters 3 and 11).

Acidosis can also promote LPO initiation. Mechanisms involved in this process are as follows: the reduction of oxygen-binding coefficient for Hb and the growth of residual pO_2 in the blood during ischemia, as well as for the growth of pO_2 during reperfusion period (hyperoxia phenomenon) (see Chapter 1); the enhancement of phospholipolysis (see Chapter 4); the accumulation of reduced carriers in mitochondrial and microsomal electron transport chains, which promotes electron "leakage" and ROS generation; the accumulation of NADPH and NADH that promote Fe^{2+} regeneration; activation of transition metals and iron decompartmentalization and conversion from the bound to the catalytically active "free" state; catecholamine-induced iron release from ferritin and oxidation of catecholamines themselves accompanied by ROS production; the establishment of a new, nonenzymatic pathway of $NO^{\bullet}$ generation and peroxynitrite formation; SOD inhibition and the acceleration of spontaneous $O_2^{\bullet}$ dismutation with the formation of more reactive oxygen species (see above).

(9) LPO intensity and the phospholipase mechanism of membrane damage are in complex relationships. On the one hand, ROS generation and LPO intensification in model systems are known to promote phospholipolysis, which is explained by structural changes in membranes, Ca^{2+} accumulation in the cell, or possible conversion of membrane-bound phospholipases from the inactive into the active form under the effect of LPO products. Activation of phospholipolysis following LPO intensification is also due to the fact that phospholipases hydrolyze fatty acids of phospholipids, which contain peroxide groups, more actively than ordinary fatty-acid acyls], thus repairing biological membranes (see Chapter 4).

The delayed accumulation of lysoPL, relative to LPO products, observed in some experiments, as well as the ability of some antioxidants to prevent activation of phospholipases during ischemia and, conversely, the inability of some phospholipase inhibitors to slow down LPO processes (see Chapter 11) provide evidence that LPO is the primary process during ischemia, whereas phospholipolysis is secondary.

It has also been shown, however, that the accumulation of FFA and lysoPL in the organ can occur during the early ischemic period (see Chapter 4); exogenous phospholipases (e. g., those of scorpion, viper, and other venoms) intensify LPO, while FFA accumulating during activation of phospholipases serve as a good substrate for LPO [Burlakova E.B., Alessenko A.V., Molochkina E.M. et al.; Burlakova E.B. and Khrapova N.G.; Vladimirov Yu.A. and Archakov A.I.; Slater T.F.] (see Chapter 5). Thus, the close relationship of LPO and phospholipolysis and their mutually potentiating effect in the pathogenesis of ischemic and reperfusion injuries are obvious.

(10) The relationships between LPO and solubilization of lysosomal enzymes, which have a toxic effect on both lipid (lipases, phospholipases) and protein (proteases) components of membranes are also of considerable interest.

It is known that lysosomal membranes, due to their specific lipid composition (high contents of sphingomyelin, cholesterol, and saturated fatty acids) [Lizenko E. I.], are more resistant to LPO induction *in vitro* than mitochondrial and microsomal membranes [Demopoulos H.B., Flamm E.S., Pietronigro D.D. and Seligman H.L.; Tappel A.L.] but are sensitive to that *in vivo* and easily break down under the effect of lipid peroxides [Desai I.D., Sawant P.L. and Tappel A.L.]. According to the electron-microscopic, cytochemical, and immunofluorescent data, changes in lysosomes during ischemia or anoxic perfusion occurred later than in mitochondria, nuclei, and ER [Decker R.S. and Wildenthal R.; Hearse D.J., Humphrey S.M. and Chain E.B.; Kloner R.A., Ganote C.E, Whalen D.A. et al.]; the release of lysosomal enzymes into the cytosol occurred either later than or simultaneously with the accumulation of LPO products [Sharapov V.I., Grek O.R., Dolgov A.V.], and LPO inhibition by antioxidants prevented solubilization of lysosomal enzymes (see Chapter 5). It has also been shown that anoxic mitochondria coming in contact with lysosomes can initiate LPO in the latter, whereas the reverse process is manifested very weakly [Dzhafarov A. I., Magomedov N. M., Kulieva E. M. et al.].

These observations suggest that LPO processes are primary and play a major role in lysosomal membrane damage, although the release of phospholipases and, especially, proteases from lysosomes after this damage markedly enhances the effect of LPO products on membranes [Kagan V.E., Savov V.M. and Serbinova E.A.].

(11) Contributions of different mechanisms of membrane damage to the pathogenesis of ischemic injuries is to some extent dependent on the model of ischemia, although their comparison in parallel experiments was made with respect to only a few parameters. As follows from Table 17, the absence of metabolite outflow from the organ during total ischemia leads to more severe acidosis, the pronounced accumulation of products of LPO and phospholipid hydrolysis, and earlier disturbances of mitochondrial function, as compared to other forms of ischemia. At the same time, the maintenance of blood flow or perfusion at a reduced (during incomplete ischemia) or normal rate (during hypoxia) leads to more marked ATP and glycogen depletion and the outflow of ADN degradation products, i.e., to the development in these models of more pronounced energy deficiency, disturbances of electrolyte composition, the prevalence of epinephrine over norepinephrine in the heart, and earlier development of its ischemic contracture.

(12) As noted above, principal targets for the oxidative action of ROS are membrane lipids (phospholipids), proteins, and nuclear DNAs. Hence, the effect of ROS is manifested in LPO intensification and inactivation, denaturation, or fragmentation of proteins and DNAs, which can be responsible for the development not only necrosis, but also apoptosis in ischemic and reperfused organs.

Table 17: Relative activities of Different Mechanisms Responsible for Membrane and Cell Damage during Different Types of.

Mechanism of damage	Parameter studied	Organ	Type of ischemia			Possible factors responsible for difference	Source
			Total	Partial	Hypoxia		
Disturbance of metabolism	ATP depletion	Heart	Later	Earlier	—	Maintenance of cardiac function during PI and H	[Jennings R.B., Reimer K.A., Hill M.L. and Mayer S.E.]*
	Glucose depletion	—	—	Less pronounced	More pronounced	—	[Cobbe S.M. and Poole-Wilson P.A.]*
	Disturbed mitochondrial function	—	More pronounced, earlier	Less pronounced, later 20 min.	—	—	[Toleikis A.I., Vavzhinkova Kh. and Tutterova M.; Rouslin W.]*
	ATP resynthesis during reerfusion	Brain	More pronounced Same after 20 min	Less Pronounced	—	Outflow of ADN degradation products during PI.	[Nordström C.H. and Siesjö.Bo.K.]*
	Acidosis	Heart	Pronounced	pronounced	Still less pronounced	Lack of metabolite outflow and lactate accumulation during TI	[Cobbe S.M. and Poole-Wilson P.A.; Wexler L.F., Weinberg E.O., Ingwall J.S. et al.]*
Intensification of PL hydrolysis	Accumulation of FFA	Heart and Brain	—	—	Leys pronounced	Lacl of metabolite outflow during TI.	[Corr P.B., Gross R.W. and Sobel B.E.; Sutton H.C.; Van der Vusse G.I., Roemen Th.A., Prinzen F.W. et al.]* [Rehncrona S., Folbergrova J., Smith D. and Siesjö B.K.]*
	Accumulation of FFA metabolites	Heart	Less pronounced	More pronounced	More pronounced	Maintenance of FFA metabolism during PI and H.	[Moore K.H., Koen A.E. and Hull F.E.; Neely J.R., Rovetto M.J. and Whitmer J.T.; Shug A.L., Thomsen J.H. and Folts J.D.]**
LPO intensification	Accumulation of LPO products	Heart, brain, etc.	More pronounced	Less pronounced	Less pronounced or absent	Lack of metabolite outflow during TI, acidosis	See Ch. 5

Disturbed ion balance	Changes of membrane potential	Skeletal muscles	Less pronounced	More pronounced	—	—	[Roberts G.P., Perry M.O., Hariri R.G. and Shires G.T.]*
	Ca^{2+} accumulation in the cell	Heart	Less pronounced	More pronounced	—	—	[Sharov V.G., Severin V.V. and Mogilevsky G.M.; Nayler W.G. and Elz J.S.; Poole-Wilson P.A., Harding D.P., Bourdillon P.D.V. et al.]**
	K^{+} efflux from the cell	—	—	Less pronounced	More pronounced, earlier	Plasma membrane depolarization during H	[Shine K.I.]*
	Ischemic contracture	—	Later	—	Earlier	Ca^{2+} accumulation in the cell during H	[Wexler L.F., Weinberg E.O., Ingwall J.S. et al.]*
Changes in CA content	Accumulation of E, increase in E/NE ratio	—	—	Pronounced *in vivo*	—	Epinephrine release from adrenals during PI, NE release from nerve during TI endings	[Olbinskaya L.I. and Litvitsky P.F.; Popov V.G., Lazutin V.K., Khitrov N.K. et al.; Rao P.S., Cohen M.V. and Mueller H.S.]**
	Accumulation of NE, dopamine	—	Pronounced *in vitro*	—	Pronounced *in vitro*	—	[Kuzmin A. I., Shulzhenko V. S. and Medvedev O. S.; Schömig A., Dart A.M., Dietz R. et al.]**
		Brain	Pronounced *in vivo*	—	—	—	[Oh S.M. and Betz A.L.; Simonsen S.G., Zhang J., Canada A.T. et al.]*

Note: (*) results of parallel experiments on different types of ischemia; (**) results of experiments not performed in parallel with one another.

Chapter 11

THE USE OF FREE-RADICAL INHIBITORS FOR PROTECTING ORGANS AGAINST ISCHEMIC AND REPERFUSION INJURIES

The complexity of pathogenesis of ischemic and reperfusion injuries and the fact that it involves not just one but many factors involved in reciprocal cause–effect relationships explain why so many drugs have been used with greater or lesser success to prevent and treat ischemic injuries in organs and why there is no satisfactory classification of such drugs, notwithstanding their abundance. Taking into account only the main, clearly observable action of antiischemic drugs, they can be divided into three groups: (1) drugs improving energy metabolism of cells during ischemia, (2) drugs enhancing the resistance of cell membrane structures toward ischemia, and (3) drugs intensifying reparative processes in the cell.

The first group includes compounds improving the blood flow, microcirculation, and O_2 supply to the cell (vasodilators, adrenoblockers, substances increasing the O_2 capacity of Hb and modifying its dissociation curve, etc.), promoting energy production (activators of phosphorylation in the mitochondrial respiratory chain, activators of SDH activity and electron transport shunting; artificial electron carriers; activators of anaerobic glycolysis and HMPS; ATP precursors; etc.), and reducing energy expenditures (substances facilitating tissue adaptation to hypoxia, reducing the load on enzymatic systems of electrolyte transport, inhibiting energy-consuming functions that are not vitally important, etc.).

The second group is represented by compounds stabilizing membrane structures; inhibiting free-radical processes, activation of phospholipases, and arachidonic acid metabolism; Ca^{2+} channel blockers; and others.

The third group comprises chemical and biological factors stimulating DNA synthesis (hormones, immunostimulants, growth factors, etc.).

Such a division is obviously conventional: drugs of the first group (there is a tendency to distinguish an individual group of "antihypoxic drugs, " although they effectively protect ischemic organs as well) also exhibit pronounced membrane-stabilizing and antioxidant properties [Seifulla R.D., Onishchenko N.A., Krasova E.I. et al.], whereas drugs of the second and, partially, third groups markedly affect the rate of energy consumption during ischemia and resynthesis of energy in the early postischemic period and participate in the regulation of local and systemic hemodynamics (see Chapter 7 and 8). It is timely to accept that most antiischemic drugs have "universal" effect, acting upon different stages of severe pathological processes developing upon acute ischemia of organs.

Emphasis should also be made on the need for a more comprehensive approach to the problem of pharmacological antiischemic protection of organs. Drugs are traditionally used for *prevention ischemic injuries in organs* (methods of long- or short-term adaptation) and *treatment in the course of ischemia* (obviously, this is possible only in cases of hypoxia or partial ischemia). However, it is necessary to perform *postischemic* (*postanoxic*) *treatment of the organ*, aimed at minimizing reoxygenation or reperfusion injuries largely contributing to the total damaging effect of ischemia. As mechanisms of postischemic injuries are not always identical to those operating in the course of ischemia (see Chapter 10), principles of their prevention may also be different, which makes necessary further research on the development of corresponding methods. Some of methods for postischemic treatment have already been represented in Chapter 6.

As publications on methods for correcting hemodynamic and metabolic disorders in ischemic organs are extremely numerous and attention in this book is focused on the role of free radicals in ischemic and reperfusion injuries, only the methods of antiischemic protection based on inhibition primary of free-radical processes in membranes will be considered in this chapter.

Today, *inhibitors of free-radical processes* is the category referring not only to substances directly interacting with active centers of the oxidation chain, i.e., with lipid radicals R·· RO· and ROO· (that is, true antioxidants), but also other compounds inhibiting LPO in model and biological systems by acting on one of the links in this process. In other words, the terms "antioxidant, " and "LPO inhibitor" virtually became synonymous. Indeed, the term "antioxidant" has been widely discussed in the literature. According to Halliwell and Gutteridge, this term can be defined as "any substance that, when present at low concentrations compared to those of an oxydizable substrate, significantly delays or prevents oxidation of that substrate. " The authors also added that "The relative importance of these as protective agents depends on which ROS is generated, how it is generated, where it is generated, and what target of damage is measured" [Halliwell B. and Gutteridge J.M.C. (1995); p. 125]. It is only possible to add to this definition that this is absolutely true not only for oxygen radicals, but also for all other radicals (NO·· R·· RO·· ROO·· etc.) which can initiate, continue and promote the branching of oxidative chains or prevent radical's toxity. Effects of different drugs on ROS and LPO, which will be discussed below, may also be realized via the closely related processes of phospholipolysis, arachidonic acid metabolism, synthesis of prostaglandins and leukotrienes, and enzymatic production of reactive oxygen species. This possibility of indirect action considerably broadens the spectrum of compounds that can inhibit LPO.

The action of an antioxidant strongly depends on its chemical structure determining its affinity to membranes, which is necessary for interaction with oxidation substrates (fatty-acid acyls of phospholipids), and parameters of its molecule: the ratio of hydrophobic and hydrophilic sites determining its incorporation into different regions of the membrane, volume of molecule, and the presence of functional groups such as hydroxyl (OH) or those affecting the mobility of the hydrogen atom [Sherstnev M.P., Klebanov G.I. and Vladimirov Yu.A.].

11.1 Natural and Synthetic Antiradical Drugs

Antiradical drugs break lipid oxidation chains by interacting with: (1) alkyl radicals R· (quinones, iminoquinones, methylene quinones, stable nitroxyl radicals, etc.; efficient in conditions of low oxygen content); (2) peroxyl radicals-ROO· (phenols, naphthols, aromatic amines, aminophenols, etc.); (3) various lipid radicals—alkyl, alkoxyl (RO·), and peroxyl (anthracene, methylene quinone, etc.); or (4) lipid hydroperoxides (ROOH), breaking them without free radical formation (sulfides, phosphides, arsenites, etc.) [Denisov E. T. and Azatjain V. V.]. Note, however, that many compounds of these types are too toxic to be used *in vivo* as antiischemic drugs.

We will concentrate on LPO inhibitors that have actually been used as antiischemic drugs, characterized in detail in our experiments or available publications, or promising in terms of subsequent analysis and probable practical application. Structural forms of some of these drugs are shown at Fig. 97.

11.1.1 Tocopherols

Among antiradical drugs of natural origin the most important are tocopherols. Among them plant origin enter the animal organisms with food and, being hydrophobic, easily incorporate into all kinds of membranes. Out of their eight existing forms, α-tocopherol (vitamin E) is the most active [Burlakova E.B., Alessenko A.V., Molochkina E.M. et al.], its chroman ring apparently being responsible for antiradical activity, and the phytol side chain, for its membrane-stabilizing effect [Kelly M.J.].

Since the main, antiradical activity is usually displayed by reduced form of natural antioxidant (quinol), whereas α-tocopherol is an oxidized form (quinone), it was suggested that the latter undergoes several changes in the organism and, being converted into conjugated hydroquinone, acquires antiradical activity and interacts with R·. This activity is accounted for by the fact that α-tocopherol itself yields active radicals in the course of reaction. that makes it less efficient as antioxidant.

Fig. 97: Chemical structural formulas of some natural or synthetic antiradical drugs.

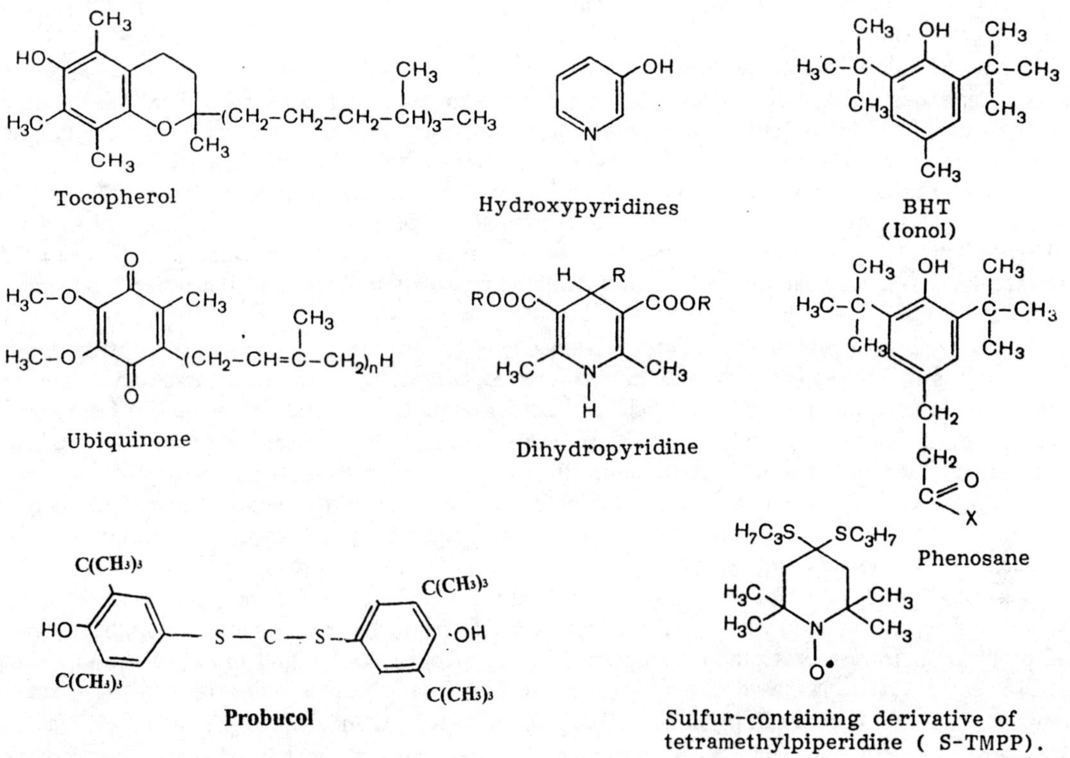

Water-soluble α-tocopherol esters do not display any antioxidant activity *in vitro* [Burlakova E.B. and Khrapova N.G.], and their antioxidant properties *in vivo* appear to be accounted for by their active metabolites. One of the latter is probably α-tocopheryl acetate in which the side chain is reduced to six carbon atoms: in experiments with rats suffering from E avitaminosis, its antioxidant activity were higher than those of the initial compound and α-tocopheryl quinone (without the side chain) [Kunitsa N.I., Kuz'menko I.V., Alekseev S.M. et al.].

In recent years, it has been synthesized a large number of various tocopherols and their derivatives, such as dl-α-tocopheryl phosphates and 6-hydroxychromanols with carbon side chains of different length (C_1–C_{16}) [see Free radicals and biological stabilizators]. Experiments on model chemical and biological systems demonstrated that short-chain tocopherol derivatives are rapidly transformed and released from the organism because they serve as substrates for cytochrome P-450 [Viner R.I., Novikov K.N., Kozlov Yu.P. et al.]. Their LPO-inhibiting effect decreases in the series C_1–$C_3 \geq C_6 > C_{11}$–C_{16}, which is apparently associated with their different mobility in biological membranes [Kagan V. E., Serbinova E. A., Bakalova R. A. et al.]. At the same time, in contrast to α-tocopherol, its derivatives with a short side chain or without it did not prevent the damaging effect of FFA on liver mitochondria [Simidzhev I., Kagan V.E. and Minkov I.] and Ca^{2+} transport in the SR of skeletal and heart muscles but proved to have an uncoupling effect on oxidative phosphorylation [Simidzhev I., Kagan V.E. and Minkov I.].

The second generation of α-tocopherol derivatives includes Lazaroids (U-74500A, U-74006F, U-74389G and U-83836E, Raxofelast) in which the ring portion of α-tocopherol molecule is bonded with amino group, and water-soluble vitamin E analogue (MDL 74, 722, Trolox), which are much more efficient in trapping peroxyl radicals than the initial compound. For example, Lazaroid reduced the content of LPO products and exerted an antiischemic effects on the ischemic and reperfused heart in rabbits [Nishida T., Morita S., Masuda M. et al.], dogs [Takahashi T., Takeyoshi I., Hasegawa Y. et al.], and rats [Campo G.M., Squadrito F., Campo S. et al.]. Experiments with Trolox *in vitro* showed that it prevents DNA fragmentation and apoptosis of mouse thymocytes under the effect of hydrogen peroxide [Forrest V.J.,

Kang. Y.-H., McClain D.E. et al.]. In rats MDL 74, 722 decreased the focus of necrosis by 49% after MCA occlusion [van der Worp H.B., Thomas C.E., Kappelle L.J. et al.] and by 80% after global or partial liver ischemia [Wu T.W., Hasmihoto N., Au J.X. et al.]. Intensive research on new α-tocopherol derivatives is currently progressing.

Both α-tocopherol and its numerous known derivatives have been extensively studied as antiischemic drugs under experimental and clinical conditions. It was shown that α-tocopherol or Raxofelast reduce the degree of functional and ultrastructural damage to the myocardium and the intensity of LPO in cases of AMI [Kudrin A. N., Kogan A. Kh., Korolev V. V. et al.; Tripathy Y. and Hegde B.M.; Campo G.M., Squadrito F., Campo S. et al.], aid in maintaining the level of adenine nucleotides in the ischemic skeletal muscle and heart [Artamonov S. D. and Onishchenko N. A.; Seifulla R.D., Onishchenko N.A., Krasova E.I. et al.; Punz A., Nanobashvili J., Fuegl A., et al.], improve ATP and creatine phosphate resynthesis and decrease membrane permeability for enzymes during reoxygenation of the heart [Guarnieri C., Flamigni F. and Caldarera C.M.].

Preventive parenteral administration of α-tocopherol to rats during partial ischemia of the brain (20 mg/kg 30 min before the onset of ischemia) lowered LPO intensity in the brain tissue and blood serum during the reperfusion period, improved the neurological status, and intensified ATP resynthesis [Yamamoto M., Shima T., Uozumi T. et al.]. A similar treatment by α–tocopherol proved to preclude PL imbalance and accumulation of FFA in the rat liver during acute hypobaric hypoxia (30-50 mg/kg 24 h before experiment) [Yakhnina D.N.] and to aid in maintaining the activity of cytochrome P-450-dependent monooxygenase system in ER membranes during total hepatic ischemia (50 mg/kg 2 days before the onset of ischemia) [Voronov G.G. and Lukienko P.I. (1983); Grek O.R. and Sharapov V.I.].

When used as an antiischemic drug, α–tocopherol was particularly effective in combination with other antioxidants: its combination with sodium selenite reduced chemiluminescence intensity in the miocardium and protected it from necrotization during AMI in rats [Kogan A. Kh. and Mednykh A. Ya], reduced TBARS content and increased the proportion of functionally active strips of cardiac tissue after cryopreservation of the rat heart [Matthes G.H., Hackensellner U.A. and Dahlke A.]. In clinical practice, α-tocopherol, α–tocopheryl acetate, or their combination with other antioxidants or vitamins (ascorbic acid, rutin, glutamic acid, GSH) have been used for treating cardiological patients (with AMI and angina pectoris) [Rysmendiev A.Zh., Tukeshova V., Grabarova I.P. et al.] and in operations on organ transplantation, for preparing donors or as ingredients added to the conservation medium for transplants [Kirpatovskii V.I., Nikiforova N.V., Kudryavtsev Yu.V. and Nadtochii O.N.; Nagel E., Meyer zu Vilsendorf A., Bartels M. and Pichlmayr R.]; combination of α-tocopherol with β-carotene and ascorbic acid protected rat ischemic hearts [Van Waarde A. Avison M.J. Thulin G. et al.].

Since α-tocopherol is a natural antioxidant, its administration is not accompanied with adverse side effects, which makes it especially suitable for clinical use.

11.1.2 Ubiquinones

Ubiquinone (coenzymes Q, CoQ_{10}), which is close to tocopherol in chemical structure (see Fig. 97), also interacts with $R^{\bullet}$.

The antiischemic action of CoQ_{10} has been studied on the models of cardiac, hepatic and renal ischemia. In experiments with temporary (15–min) LAD occlusion in dogs, preventive administration of the ubiquinone CoQ_{10} led to the increase of CoQ_{10} content in the heart and had a dose-dependent antiarrhythmic effect, which the authors attributed to protection of membranes against phospholipase [Nagai S., Miazaki Y., Ogawa K. et al.]. The protective effect of CoQ_{10} was also observed during hypoxic perfusion of an isolated papillary muscle of the guinea pig heart [Aomine M. and Arita N.].

CoQ_{10} exerted a protective effect also on the totally ischemic and reperfused kidneys [Takenaka M., Tatsukawa Y., Dohi K. et al.; Tatsukawa Y., Dohi J., Yamada K. et al.] and liver [Marubayashi S., Dohi K., Ezaki H. et al. (1982)]. In addition to increasing animal survival rate, improving the functioning of mitochondria, and preventing Ca^{2+} accumulation in them, CoQ_{10} administration also prevented accumulation of LPO products in tissues and mitochondria during the postischemic period, i.e., these effects were associated with LPO inhibition by ubiquinone and its $^{\bullet}OH$ radical scavenger action [Marubayashi S., Dohi K., Ezaki H. et al. (1983); Wildhirt S.M., Suzuki H., Horstman D. et al.].

CoQ_{10} has already been used in clinical practice [Greenberg S. and Frishman W.H.], and possible aspects of its application are currently being investigated.

Synthetic antiradical drugs are mostly derived from natural antioxidants (vitamin B, dihydronicotinic acid) by adding or substituting certain functional groups that impart or enhance certain properties (e. g., make initially fat-soluble substances water-soluble). In this category, substances studied in more detail as antiischemic drugs include BHT and its water-soluble derivatives, nitroxyl radicals, and others.

11.1.3 BTH

BTH (butylated hydroxytoluene; 2,6-di-tert-butyl-4-methylphenol, ionol, Dibunol) is a fat-soluble, sterically screened phenol (see Fig. 97) and, in free-radical reactions, yields radicals stabilized by two tert-butyl side groups; that is, in contrast to natural antioxidant (e. g., α-tocopherol), it forms stable, almost inactive radicals, which accounts for its high antioxidant activity. This made it possible to use BHT as a stabilizer in food, cosmetic, and pharmaceutical industries, as well as in experimental and clinical medicine for treating free-radical pathologies such as radiation injury and malignant tumors [Emanuel N.M. and Lyashovskaya Yu. N.].

In our laboratory, effects of BHT were studied on different models of long-term total ischemia of the kidneys, liver, heart, brain, and limb muscles and in experiments with ischemic shock, and antiischemic properties of this drug were demonstrated for the first time [Bilenko M. V. (1978a); Bilenko M.V. (1982); Shelenkova L.N. and Bilenko M.V.].

Analysis of pharmacokinetics showed that BHT administered i. p. or per os at high doses (240 mg/kg) reaches a peak of concentration in blood as soon as after 1–2 h, which is followed by a gradual decrease during 24 h (Fig. 98) [Bilenko M.V., Shelenkova L.N., Barsel V.N. and Matveeva S.A.]. In the brain, kidneys, heart, and liver, the maximal level of BHT was observed 1 h after intraperitoneal and 2h after per os administration, whereas in the adipose tissue this was observed after 12 h. When BHT was administered perorally, its peak concentration decreased in the series: liver > heart > brain > kidneys > blood, while after intraperitoneal administration the order was different: brain > liver > heart > kidneys > blood. The accumulation of BHT in the same organ also depended on the way of administration: in the case of peroral administration, its concentration in the liver, heart, and kidneys was much higher (4.8, 3.3, and 2.5 times, respectively) than after intraperitoneal administration (in blood and brain, however, no such dependence was observed).

As has been recently established, BHT is unevenly distributed not only in different organs, but also in different subcellular fractions of tissue extracts: microsomes > light mitochondria > nuclei> supernatant > heavy mitochondria [Degterev I. A. and Zaikov G. E.].

Studies on the protective effect of BHT during *renal ischemia* [Bilenko M.V., Shelenkova L.N., Dubur G.Ya. and Velena A.H.; Shelenkova L.N. and Bilenko M.V.] showed that a single dose of 120 mg/kg administered 4–24 h before the onset of ischemia provided for a high animal survival rate, and the highest protective effect was achieved when a dose of 240 mg/kg was injected 24 h before ischemia (Table 18). Low doses of BHT were ineffective. Administration perorally also produced a significant protective effect but at higher doses (240–360 mg/kg). BHT proved to also protect mitochondrial membranes of ischemic kidneys: administered per os at the optimal dose of 240 mg/kg one day before the onset of 2-h ischemia, it did not prevent a decrease in the content of adenylate pool components in the course of ischemia but promoted a slight increase in ATP and the significant accumulation of AMP and total ADN (instead of their degradation) within the first 24 h of reperfusion [Bilenko M.V., Volkova I.A. and Danilov S.P.] (see Fig. 52). This is consistent with the findings that other antioxidant, such as α-tocopherol and CoQ_{10}, have no influence on the rate of consumption of high-energy phosphates during ischemia but do have effect on their resynthesis in the reperfusion period [Guarnieri C., Flamigni F. and Caldarera C.M.; Marubayashi S., Dohi K. and Kawasaki T.].

Fig. 98: BHT (ionol) content in the blood and rat organ homogenates after single intraperitoneal (a) or intragastric (b) administration. [Bilenko M.V., Shelenkova L.N., Barsel V.N. and Matveeva S.A.]

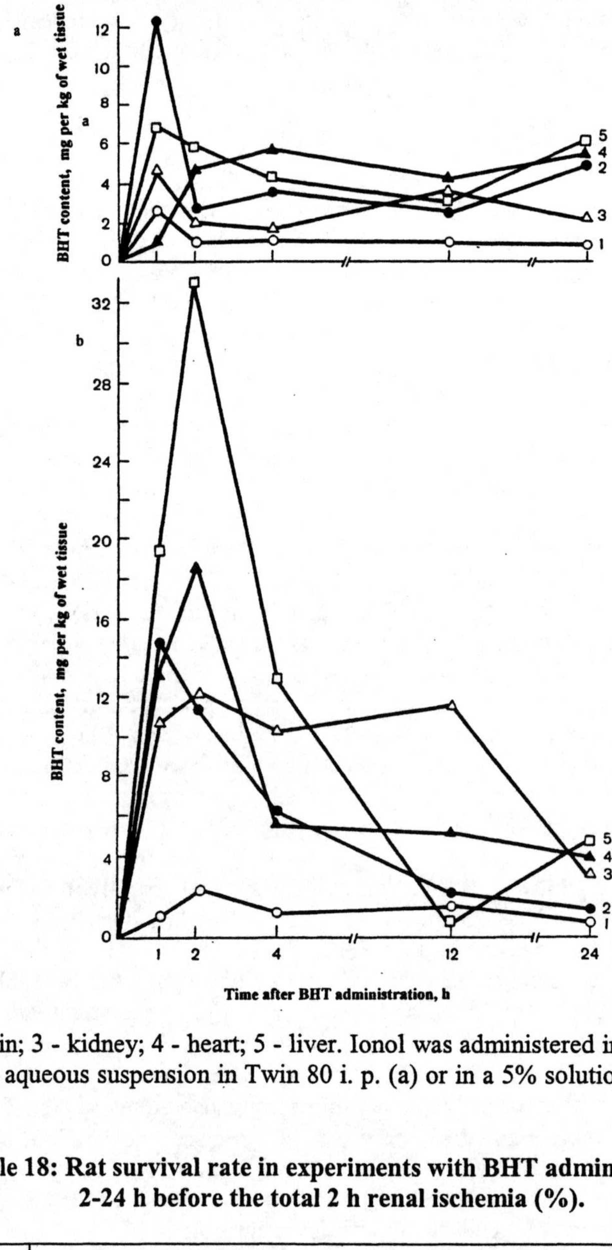

1 - blood; 2 - brain; 3 - kidney; 4 - heart; 5 - liver. Ionol was administered in a dose of 240 mg/kg in the form of an aqueous suspension in Twin 80 i. p. (a) or in a 5% solution in sunflower oil per os (b).

Table 18: Rat survival rate in experiments with BHT administration 2-24 h before the total 2 h renal ischemia (%).

Time of BHT administration before ischemia, h	Rat survival rate, %						
	i. p. mg/kg				per os, mg/kg		
	30	60	120	240	120	240	360
2	0	0	17	33		17	
4	0	0	50**	17	14	28*	25**
12	0	40**	50**	50**		50**	
24	20	50**	50**	60**	0	0	33

*-P<0.05: **-P<0.01 versus control.

A model of renal ischemia was used to study the mechanisms responsible for the antiischemic effect of BHT. Animals received intraperitoneal injections of the drug at ineffective (30 mg/kg) and effective (120 mg/kg) doses 4 h before the onset of ischemia, and the total antioxidant activity (AOA) was analyzed in lipids extracted from kidney homogenates before ischemia, 2 h after its onset, and after 2-h ischemia followed by 4-h reperfusion [Bilenko M.V., Shelenkova L.N., Barsel V.N. and Matveeva S.A.]. The results showed that BHT at a dose of 30 mg/kg did not prevent the decrease of AOA during ischemia, although promoted its recovery in the reperfusion period. At a dose of 120 mg/kg, however, it provided for the maintenance of antioxidant activity during ischemia and markedly raised it (twofold) during the postischemic period (Fig. 99). Apparently it was the ability of BHT to maintain antioxidant activity in the ischemic organ that ensured the antiischemic effect of the drug.

Fig. 99: AOA of lipids in kidney homogenates after 2 h of rat kidney complete ischemia and 4 h of reperfusion in experiments with i. p. BHT administration in different doses 4 h before the ischemia (% of the initial level).

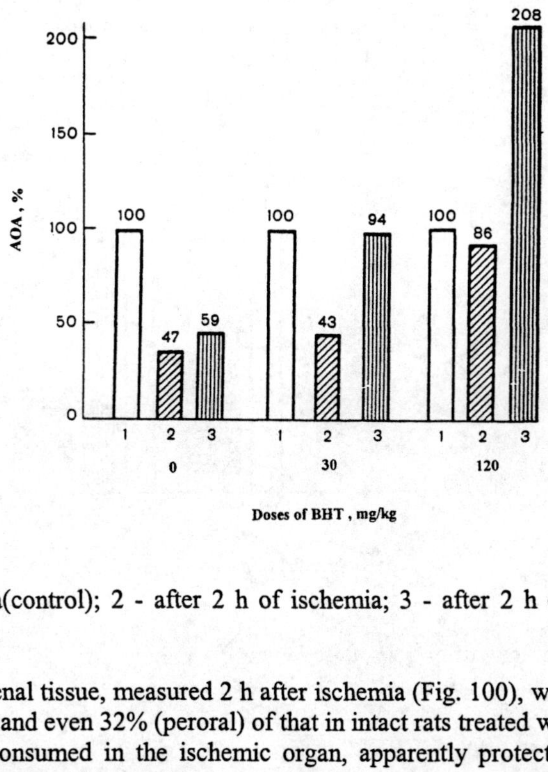

1 – without ischemia(control); 2 - after 2 h of ischemia; 3 - after 2 h of ischemia and 4 h of reperfusion.

BHT content in the renal tissue, measured 2 h after ischemia (Fig. 100), was sharply decreased to 55% (intraperitoneal injection) and even 32% (peroral) of that in intact rats treated with BHT. This was evidence that BHT was actively consumed in the ischemic organ, apparently protecting reserves of natural fat-soluble antioxidants (α-tocopherol, ubiquinones, etc.), which maintained the antioxidant activity at a high level (close to initial). During the reperfusion period, the BHT content in the kidney sharply increased (especially after intraperitoneal administration), possibly due to redistribution from other organs or the adipose tissue, which provided for a high antioxidant activity in the postischemic organ.

The fact that the antiischemic effect of BHT and its ability to inhibit free-radical reactions are determined by its antioxidant activity, rather than its influence on antioxidant enzyme systems or hormonal level, was observed in other studies as well [Burlakova E.B., Alessenko A.V., Molochkina E.M. et al.; Gorban E.N. and Kol'tover V.K.; Konovalova G. G., Cherpachenko N. M., Lankin V. Z. et al.].

BHT metabolites are also highly active [Degterev I. A. and Zaikov G. E.]; this may partially explain a discrepancy between two facts: the decrease of BHT content in the organ after 4–24 h and a high efficiency of the drug upon its administration namely within the same period of time prior to ischemia.

The protective effect of BHT and its combination with phenobarbital during *liver ischemia* was discussed in Chapter 5 (see Fig. 37). Briefly repeate that it was found [Bilenko N.V., Kagan V.E., Velikhanova D.M. and Komarov P.G.] that the drug raised the content and improved the function of components of the cytochrome P-450-dependent monooxygenase system in ER membranes of the ischemic liver and increased the survival of rats with the single ischemic liver lobe (from 15 to 47%). Both effects manifested themselves against the background of a decrease in the level of LPO product, i.e., were apparently determined by antioxidant properties of the drug [Bilenko N.V., Kagan V.E., Velikhanova D.M. and Komarov P.G.].

Fig. 100: BHT content in rat kidney homogenates after 2 h of ischemia and 4 h of reperfusion, in experiments with single administration BHT i. p. (I) or per os (II) in a dose of 240 mg/kg (mean ± SEM)

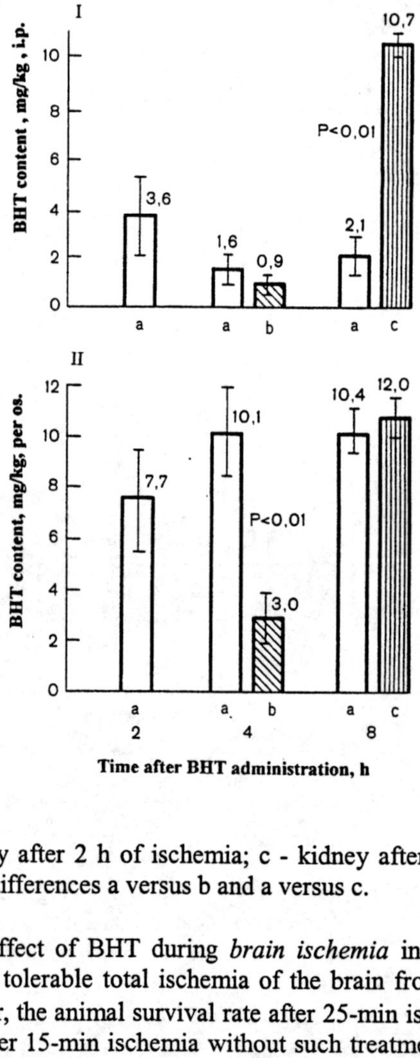

a - intact kidney; b - kidney after 2 h of ischemia; c - kidney after 2 h of ischemia and 4 h of reperfusion. P - significant differences a versus b and a versus c.

Studies on the protective effect of BHT during *brain ischemia* in rats showed that BHT (60–240 mg/kg) prolonged the period of tolerable total ischemia of the brain from 15 min (without treatment) to 25–30 min (Table 19). Moreover, the animal survival rate after 25-min ischemia upon BHT administration was significantly higher than after 15-min ischemia without such treatment. The protective effect of BHT on the brain, as in the case of ischemic kidneys, reached a peak when the drug was administered 4 h before ischemia at doses of 120 and 240 mg/kg. Treatment with BHT according to the optimal scheme (120 mg/kg 4 h before ischemia) also inhibited DC and MDA accumulation in the brain both during its 30-min ischemia and after 20-min reperfusion (Fig. 101).

Table 19: Rat survival rate in experiments with BHT or diludin administration 4 h before the complete brain ischemia (%).

Antioxidant	Ischemia, min	Dose mg/kg	Rat survival rate over than 7 days, %
(Control)	10	-	100
	15	-	30
	25	-	0
BHT (ionol, i. p.)	25	120	82*°
	30	60	30 §
	30	120	44#
	30	240	40
	35	120	9
Diludin	20	120	73*
	25	120	33
	25	240	44
	30	120	7

*-P<0.05 in experiments with BHT or diludin versus control (15 min of ischemia);
o-P< 0.05 in experiments with BHT or diludin versus control (25 min of ischemia);
#-P<0.05 in experiments with BHT versus diludin (120 mg/kg, 25 min of ischemia);
§-P<0.05 in experiments with BHT versus diludin (120 mg/kg, 30 min of ischemia);

Fig. 101: DC (I) and TBARS (II) contents in the rat brain after 30 min of ischemia and 20 min of reperfusion in experiments with BHT administration (% of the initial level)

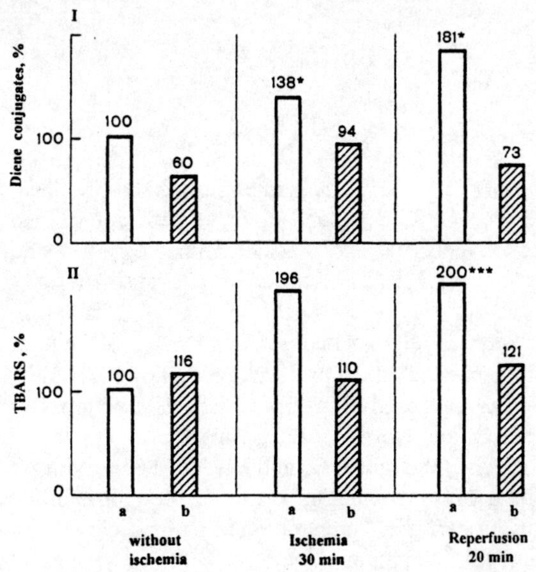

I – without ischemia; II - 30 min of ischemia; III – 30 min of ischemia plus 20 min of reperfusion; a - without drug; b – with BHT administration. BHT was administered in a dose of 120 mg/kg 4 h before ischemia. *-P < 0.05; **-P < 0.01 versus initial level.

The protective effect of BHT during *heart ischemia* was studied on several models. In our experiments on perfusing the isolated rat heart by Langendorf–Fallen's method and the working rat heart by Neely's method, BHT (240 mg/kg) administered 24 h before ischemia improved the cardiac function and diastolic properties of the heart exposed to 30-min ischemia at 20°C (Fig. 102). The same doses of BHT tested on a model of heterotopic heart transplantation (on abdominal vessels) in rats [Bilenko M.V., Bulgakov V. and Otverchenko V.; Bilenko M.V. (1992b); Bilenko M.V. and Otverchenko V.N.; Otverchenko V.N.] improved the cardiac function after 60-min heart ischemia at 20°C and reperfusion (Fig. 103) and also significantly increased Siegel's contractility index (193 ± 4 vs. 167 ± 6 in the control; $P < 0.05$). In the case of more severe ischemic injury (60 min at 37°C), administration of BHT did not led to any increase in the time necessary for the development of ischemic contracture or the number of hearts with restored cardiac function; however, contractions in experiments with BHT began sooner (after 47 s versus to 106 s; $P < 0.05$) and continued for a longer time than in the control. Thus, BHT improved the condition of the heart even after extremely severe total ischemia, although this effect was expressed to a lesser extent.

Fig. 102: Coronary perfusion rate (CPR, I) cardiac function(Pmax, II and dp/dtmin, III) and end diastolic pressure (EDP, IV) of 30 min ischemic rat heart, in experiments with BHT administration (mean±SEM).

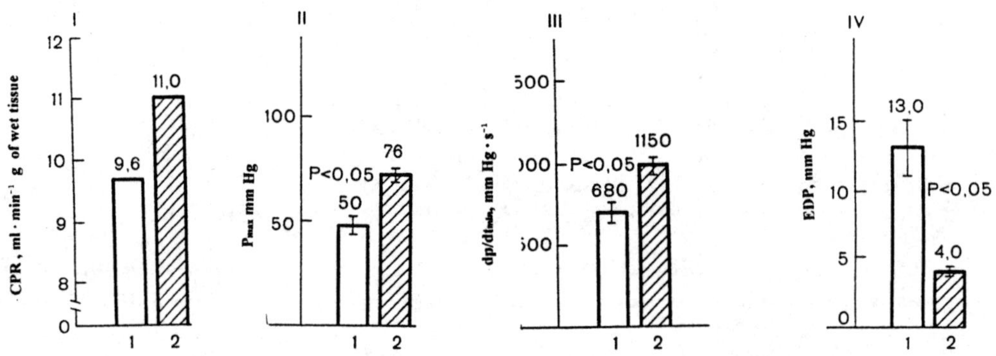

1 - without BHT (control); 2 – BHT, 120 mg/kg 24 h before the ischemia. P - significant differences 1 versus 2. The CPR, P_{max}, and dp/dt_{min} were determined on a model of isolated working heart according to Neely; the EDP was determined on a model of isolated heart according to Langendorf-Fallen [see Neely J. R. and Rovetto M. J.].

Experiments with perfusion of an isolated rat heart after 30-min of ischemia, according to the Langendorf–Fallen's model, demonstrated the protective effect of BHT added to the perfusing saline in the form of single-layer liposomes prepared from the total PL fraction (see Chapter 8) [Bilenko M.V., Tel'pukhov V.I., Churakova T.D. and Komorav P.G.; Bulgakov V.G., Morgunov A.A. and Bilenko M.V.; Bulgakov V.G.; Morgunov A.A., Bulgakov V.G. and Bilenko M.V.; Morgunov A.A. and Bulgakov V.G.; Morgunov A.A.]. This method of administration opens up new possibilities for using fat-soluble drugs during preservation of organs by perfusion, preparation of cardioplegic solutions, and further improvement of methods for their parenteral administration.

We also demonstrated the strong antiischemic effect of BHT during *ischemia of limb muscles*. A single dose of 240 mg/kg injected 12 h before applying a tourniquet provided for the maintenance of more active Ca^{2+} transport in SR membranes (equivalent to that observed upon administering 50 mg/kg of α-tocopherol for three days), lower release of lysosomal enzymes, and slower accumulation of HP and DC after 3- and 6-h total ischemia (tourniqueting) of limb muscles in Wistar rats [Arkhipenko Yu. V., Bilenko M. V., Dobrina S. K. et al.; Bilenko M.V. (1978b)]. It also led to a dose- and time-dependent increase in the rat survival rate and the improvement of coronary perfusion and cardiac function during ischemic shock (see Chapter 7).

Fig. 103: Pmax (I), compliance (1/EDP, II), and relaxation (dp/dt min, III) of a rat heart after 60 min of ischemia at 20°C and reperfusion in experiments with BHT (mean ± SEM).

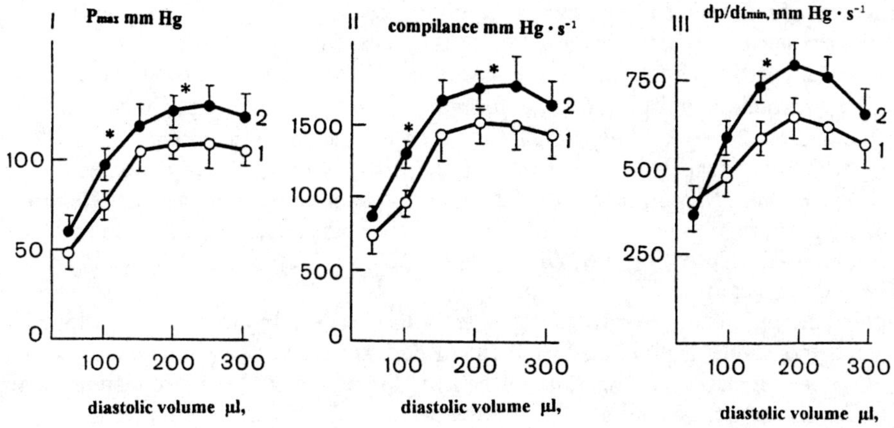

1 - without BHT (control); 2 - BHT administration in a dose of 120 mg/kg 24 h before ischemia. *-P < 0.05 2 versus 1. The heart was reperfused by heterotopic transplantation onto abdominal vessels of recipient rat.

The protective antiischemic effect of BHT used preventively (240 mg/kg 1-2 h before the onset of ischemia) was also observed by other authors in experiments with ischemia of the small intestine in rats [Melikyants A.G. and Kucheryanu V.G.]. The drug enhanced the transport and barrier properties of the intestine damaged by total ischemia (50–75 min) and increased the survival rate of test animals, administration perorally being naturally more effective in this case than intraperitoneal injection.

Our experience in using BHT as an antiischemic drug has led to the following conclusions: (1) the drug exerts antiischemic action with respect to all ischemic organs; (2) it is effective in cases of both brief and prolonged total ischemia, but if the ischemic damage is too severe, the protection is only partial; (3) the drug produces a protective effect if administered at a single dose but only if this dose is high (120-240 mg/kg), and an important role belongs to the time and way of administration. The greatest effect is achieved when the drug is administered preventively 4–24 h before the onset of ischemia, whereas its therapeutic application (after the onset of ischemia and before reperfusion) is less effective, which provides evidence that the presence of the drug in the organ during its ischemia and complete humoral isolation from the organism is important; (4) the protective effect of BHT is detected by monitoring both the state of organism as a whole (animal survival rate) and specialized functions of the organ (heart contraction and relaxation); it is also observed while analyzing the structural and functional state of different subcellular membranes (e. g., liver mitochondria and ER, SR and lysosomes of skeletal muscles); (5) the antiischemic effect of BHT is determined by its antioxidant properties and its capacity for substituting natural antioxidants in free-radical processes and concentrating (via redistribution from depots) in the organ damaged by ischemia during the postischemic period; (6) the way of administering BHT should be chosen depending on the target organ: the parenteral way (in our experiments, intraperitoneal injections) is more effective for the kidneys and, presumably, for the heart, skeletal muscles, and other organs as well; while the intragastric mode is more appropriate for organs of the portal system (in particular, intestine and the liver). A promising approach consists in using the drug in the liposomal form: it offers new perspectives for therapy with fat-soluble antioxidants, which can be added to perfusates, cardioplegic solutions, and even the blood. This approach can provide for more precisely directed transport of antioxidants to the membrane lipid bilayer.

The antiischemic effect of BHT has been confirmed by other specialists in experiments with AMI, myocardial ischemia and hypoxia in rats and rabbits [Golikov A.P., Avilova O.A., Polumiskov V.Yu et al.;

Konovalova G. G., Cherpachenko N. M., Lankin V. Z. et al.; Meerson F.Z.], emotion–pain stress [Meerson F.Z.], traumatic and hemorrhagic shock [Nigulyanu V.I., Elsky V.N., Krivoruchko B.I. et al.; Elsky V.N., Manankov A.K., Mareyeva T.E. and Yakubenko Y.D.; Yakubenko E.D., Mareeva T.E. and Manankov A.K.]. In these studies, BHT was effective at doses ranging from 20 to 120 mg/kg administered before ischemia or both before and after ischemia. The reported results are of theoretical interest, proving the free-radical mechanism of ischemic and reperfusion injuries in organs, and of practical significance, providing evidence for the existence of a fairly strong antioxidant suitable for clinical use.

Indeed, clinical tests in patients with acute vast myocardial infarction showed that the course of BHT (Dibunol) therapy peroral administration in the form of 20% oil solution in gelatin capsules, 20–40 mg/kg upon admission and 10–20 mg/kg daily for 21 days—led to a decrease in the content of LPO products in blood plasma, which was especially pronounced on the first day of the disease, i.e., at the peak of AMI [Churakova T.D., Bilenko M.V., Pokazeeva Z.T. et al.]. Dibunol also improved the electrophysiological parameters of the heart, reduced the size of infarct, and lowered the release of CPK and its MB fraction into the blood [Golikov A.P., Polumiskov V.Yu., Berestov A.A. and Ryabinin V.A.; Golikov A.P., Avilova O.A., Polumiskov V.Yu et al.].

The problem of drug toxicity at doses acceptable for its chronic administration is extremely important.

It has been demonstrated that Dibunol itself and its derivatives have very low toxicity and do not cause any changes in protein metabolism, functions of the liver, kidneys, and blood coagulation system, the level of immunoglobulins, and the general state of healthy persons after administration for 21 days at a dose of 20 mg/kg [Nadjaryan T.L., Pikovskaya G.N., Barsel V.A. et al.]. At the same time, it has been reported that high doses of the drug ($3 \cdot 10^{-5}$–10^{-3} M) added *in vitro* into a cardiomyocyte culture [Leslie S.W., Gad S.C. and Acosta D.] or used for perfusing an isolated rat auricle [Gad S.C., Leslie S.W. and Acosta D.] had an adverse effect on the processes of myocardium contraction and relaxation. High doses of BHT ($3 \cdot 10^{-5}$ M) *in vitro* also disrupted Ca^{2+} transport in SR membranes of skeletal muscles [Mukanov K.O., Buldygerova N.V., Kovaleva E.S. et al.], led to aggregation of thrombocytes [Mukanov K.O., Buldygerova N.V., Kovaleva E.S. et al.] and hemolysis of erythrocytes [Kumar P.S., Georg T., Bai N.J. et al.]. When BHT was administered *in vivo*, its toxic effect on the rat heart was manifested either at an extremely high dosage (500 mg/kg) or at doses of 30 mg/kg but administered over a long period of time [Golikov A.P., Polumiskov V.Yu., Berestov A.A. and Ryabinin V.A.].

Apparently, it is necessary to continue studies on selecting optimal doses of this drug, especially when it is used in the liposomal form *in vitro,* which provides for direct and selective transport of unmetabolized BHT to membranes. Indications for the use of BHT an antiischemic drug should also be defined more explicitly. As BHT is much more efficient as a preventive (rather than curative) agent, its use during operations on dry heart or heart transplantation may prove to be more appropriate than in cases of AMI. However, clinical effects of BHT treatment under such conditions has not yet been studied in detail.

To date, specialists have synthesized a number of BHT derivatives—hydrophobic screened phenols, which interact with $ROO^\bullet$ at approximately the same rate (i.e., are similar in antiradical activity) but strongly differ from one another in the activity of yielded radicals ($In^\bullet$, see Fig. 81) capable of initiating new reaction chains, i.e., in the antioxidant activity [Serbinova E., Kharfus M., Ukhin L.Yu. et al.]. Another interesting BHT derivative is Probucol (see Fig. 97), which demonstrates both the ability to reduce cholesterol content and the antiradical activity comparable to that of α-tocopherol [Weglicki W.B., Mak I.T. and Simic M.G.].

11.1.4 Phenosanes

Phenosanes (K^+, Na^+, or Li^+ salts of 4-hydroxy-3,5-di-tert-butyl-phenyl propionic acid), first synthesized at the Institute of Chemical Physics, Russian Academy of Sciences (Moscow), are water-soluble BHT (ionol) derivatives. As demonstrated in our laboratory [Morgunov A.A. and Bulgakov V.G.], K^+-phenosane (10^{-5} M) added to the solution for perfusing the isolated rat heart before and then after 30-min total ischemia at 37°C (the stop-flow model)—increased the CPR by 30–40% and the parameter of cardiac function (P_{max}) by 23–41% (Fig. 104). At 10^{-6} M the drug was ineffective, while at 10^{-4} M it did cause an increase in CPR and P_{max} but, at the same time, cardiac function was resumed later, i.e., this dose of drug had a certain toxic effect.

Fig. 104: Coronary perfusion rat (CPR, a) and cardiac function (P_{max}, b)of an isolated complete ischemic in experiments with K-phenosane (mean ± SEM).

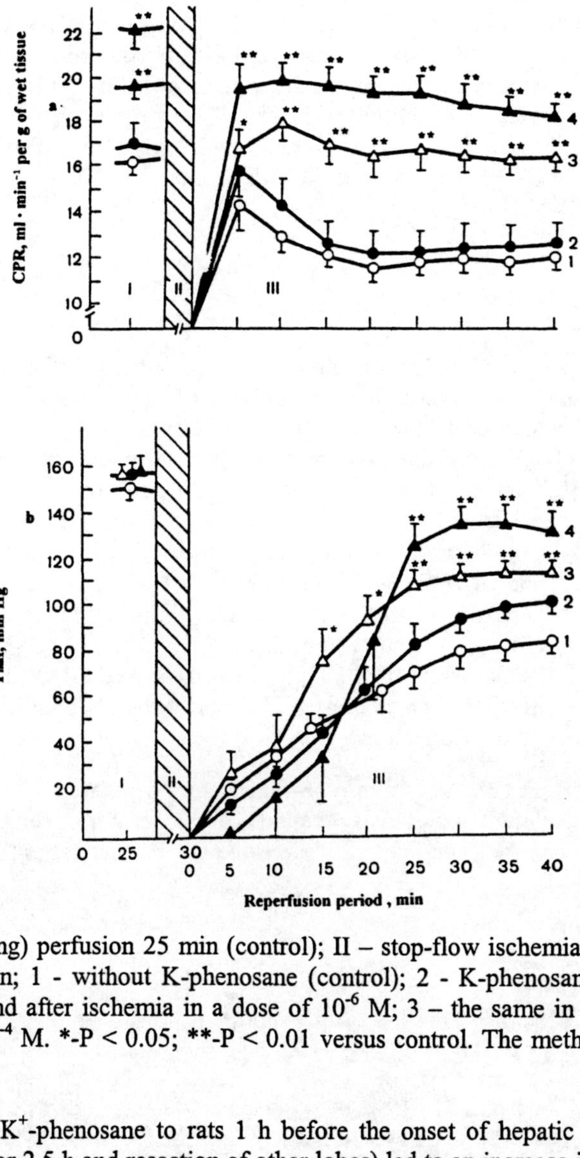

I - initial (stabilizing) perfusion 25 min (control); II – stop-flow ischemia, 30 min at 37°C; III – reperfusion, 40 min; 1 - without K-phenosane (control); 2 - K-phenosane addition to the K H perfusate before and after ischemia in a dose of 10^{-6} M; 3 – the same in a dose 10^{-5} M; 4 – the same, in a dose 10^{-4} M. *-$P < 0.05$; **-$P < 0.01$ versus control. The methods, are the same as at Fig. 8.

Administration of K^{+}-phenosane to rats 1 h before the onset of hepatic ischemia (occlusion of the median and left lobes for 2.5 h and resection of other lobes) led to an increase in animal survival rate to 43 and 50%, compared to 25% in the control. This effect was observed at doses of 50 and 100 mg/kg, whereas 150 mg/kg were toxic [Seregina L.A.].

11.1.5 Hydroxypyridines

Hydroxypyridines are nitrogen-containing heterocyclic phenols, synthetic analogues of vitamin B_5 [Smirnov L.D. and Dyumaev K.N.], which are convenient for use because of their solubility in water. Their antioxidant activity is determined by the polarity of the OH bond; electron-donating substituents in their molecules render the hydroxyl hydrogen atom more mobile and improve their antioxidant properties.

Studies on the pharmacokinetics of 2-ethyl-3-methyl-HP, one of the compounds belonging to this class, showed [Zherdev V. P., Sariev A. K., Dvoryaninov A. A. et al.] that the drug administered to rats intraperitoneally at a dose of 150 mg/kg was rapidly absorbed from the abdominal cavity ($T_{1/2}$ of absorption being 0.94 h), reaching the peak concentration in blood plasma, liver, and brain after 2–3 h. The rate of drug removal was the highest in the brain. The time of its removal from liver ER membranes (72 h) exceeded that from the cytosol (24 h), providing evidence that the drug is membranotropic.

The data on the antiischemic properties of drugs belonging to this group are as yet scarce. It was shown, however, that hydroxypyridine administered for 3 days at a dose of 100 mg/kg prevented hypoxic contracture of the myocardium and CPK release [Meerson F.Z., Kagan V.E., Azkhipenko Yu.V. et al.]; 3-hydroxypyridine injected to rats intraperitoneally at doses ranging from 100 to 200 mg/kg 40 min before the animals were taken to an altitude of 11 000 m increased the time of their survival [Val'dman A.V., Voronina T.A., Smirnov L.D. et al.]; 2-ethyl-6-methyl-3-hydroxypyridine (Emoxipine) administered to dogs with AMI at a low dose (10 mg/kg) reduced the size of infarct [Konorev E. A., Polumiskov V. Yu., Avilova O. A. and Golikov A. P.].

11.1.6 Derivatives of 1,4-Dihydropyridine

Derivatives of 1,4-dihydropyridine (2,6-dimethyl-3,5-dietoxicarbonil, 1,4-DNP, diludin) analogues of 1,4-dihydronicotinamide were first synthesized at the Institute of Organic Synthesis, Latvian Academy of Sciences. They include both water-insoluble (in particular, Diludin) and water-soluble compounds.

In model systems, 1,4-DHP derivatives demonstrated various membranotropic properties, such as antiradical activity and the ability to scavenge ROS, to inhibit phospholipase A_2, and to suppress LPO.

In our laboratory, the antiischemic properties of Diludin were tested on models of 2-h renal ischemia [Bilenko M.V., Shelenkova L.N., Dubur G.Ya. and Velena A.H.], 2-h hepatic ischemia [Seregina L.A.], and 25–30-min cerebral ischemia [Bilenko M.V., Tel'pukhov V.I., Churakova T.D. and Komorav P.G.]. All the experiments were conducted on rats, with Diludin being administered per os 4 h and 24 h before the onset of ischemia at single doses of 60, 120 and 240 mg/kg.

In the experiments with renal ischemia (Fig. 105, I), the drug caused a considerable increase in animal survival rate. All doses were effective when administered 4 h prior to ischemia; when this period was 24 h, only doses of 120 and 240 mg/kg were effective.

Fig. 105: Effect of diludin on the rat survival rate after 2 h of complete two kidney ischemia (I) and after 2 h of complete two rat liver lobe ischemia with resection of two other lobes (II) (%).

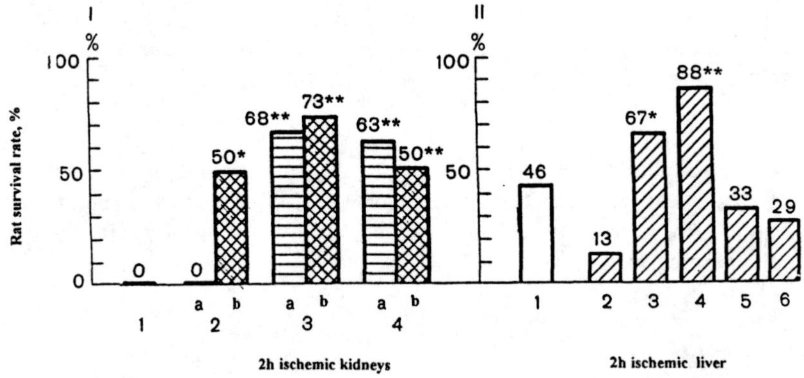

I: 1 - without diludin (control); 2 - diludin in a dose of 60 mg/kg; 3 – the same, in a dose of 120 mg/kg; 4 – the same, in a dose of 240 mg/kg. Diludin was administered per os 24h (a) or 4h (b) before ischemia. II: 1 - without diludin (control); 2 - administration of diludin in a dose of 120 mg/kg 24 h before ischemia; 3 – the same dose- 12 h before ischemia; 4 – the same dose- 4 h before ischemia; 5 – the same dose - 2 h before ischemia; 6 – the same dose 2 h after ischemia. *-$P < 0.05$; **-$P < 0.01$ versus control. The survival rate was determined on the 7th day of ischemia.

In the experiments with hepatic ischemia (see Fig. 105, II), a significant increase in animal survival rate was observed when the drug (120 mg/kg) was administered 12 and 4 h before ischemia, whereas its administration 2 and 24 h before ischemia or before reperfusion was ineffective.

In the experiments with cerebral ischemia (see Table 19), the tolerable period of total ischemia upon Diludin administration increased to 25 min (survival rate 33–44%), but the antiischemic effect of this drug was singnificantly weaker than that of BHT.

The antiischemic effect of water-soluble 1,4-DHP derivatives was also observed on the "vasopressor spasm" model, as followed from the improved parameters of hemodynamics and prevention of DC and MDA accumulation in myocardial membranes [Frolkis R.A., Voronkov G.S., Dubur G.Ya. et al.; Tsyomik V.A.]. One of such derivatives, Cerebrocrast, significantly improved blood flow in the rat brain after its ischemia [Plotnikov M.B., Vaizova O.E. and Suslov N.].

11.1.7 Nitroxyl Radicals formed by Derivatives of 2,2,6,6-Tetramethylpiperidine

Nitroxyl radicals formed by derivatives of 2,2,6,6-tetramethylpiperidine (TMP) have antioxidant properties apparently associated with their ability to react with alkoxyl ($RO^{\bullet}$) and peroxyl ($RO_2^{\bullet}$) radicals, thus breaking the reaction chain. Experiments performed in our laboratory showed that both nitroxyl radicals proper and compounds forming them in aqueous solutions—1-hydroxy-TMP and especially its sulfur-containing derivative (see the formula above)—can inhibit LPO in model systems (spontaneous and NADPH-dependent LPO in rat liver microsomes, ascorbate-dependent LPO in liposomes) [Komarov P. G., Taskaeva O. N. and Zhdanov R. I.]. Since 1-hydroxy-TMP is more readily soluble in water and less toxic than nitroxyls, its use as an antioxidant is more feasible. In studies on the antiischemic effect of sulfur-containing 1-hydroxy-TMP [Bilenko M.V., Komarov P.G., Morgunov A.A. and Zhdanov R.I.; Komarov P., Morgunov A., Shdanov R. and Bilenko M.; Komarov P. G., Morgunov A. A., Zhdanov R. I. and Bilenko M. V.], its addition to the perfusate and the medium for preserving (at 37°C) the isolated rat liver reduced TBARS accumulation, and 1-hydroxy-TMP added to the medium with the liver microsome fraction stored at 4°C reduced TBARS accumulation and slowed down degradation of cytochrome P-450, being more effective in this respect than BHT. Administration of 1-hydroxy-TMP to rats significantly increased the survival rate after 2.5-h total hepatic ischemia *in situ* (see Fig. 79).

In experiments on the model of stop-flow ischemia of the rat heart [Morgunov A.A., Komarov P.G., Bulgakov V.G. and Bilenko M.V.; Morgunov A.A.], 1-hydroxy-TMP administered throughout the perfusion period at concentrations of $5 \cdot 10^{-6}$–$5 \cdot 10^{-5}$ M increased the CPR and cardiac function (P_{max}) both before and after ischemia (Fig. 106) and effectively lowered diastolic pressure in the left ventricle during the postischemic period, which increased as the result of developing ischemic contracture of the myocardium. An analogue of this drug (conventionally designated TMP), which lacked the antioxidant effect, did not exhibit any antiischemic activity. On the model of ischemic shock, 1-hydroxy-TMP also effectively increased the survival rate and improved the cardiac function [Bilenko M.V., Morgunov A.A., Churakova T.D. et. al.] (see Chapter 7). In addition, the protective effect of nitroxil radicals (TEMPO) in regional ischemia of the isolated rat heart, combined with lower postischemic release of $^{\bullet}OH$ radicals, was demonstrated in [Gelvan D., Saltman P. and Powell S.R.].

The data discussed above suggest that hydroxy derivatives of TMP deserve closer attention as antiischemic drugs promising for clinical use.

Fig. 106: Effect of 1-hydroxy-TMP on the coronary perfusion rate (CPR, a) and cardiac function (P_{max}, b) of an isolated ischemic rat heart (mean±SEM) [Bilenko M.V., Komarov P.G., Morgunov A.A. and Zhdanov R.I.]

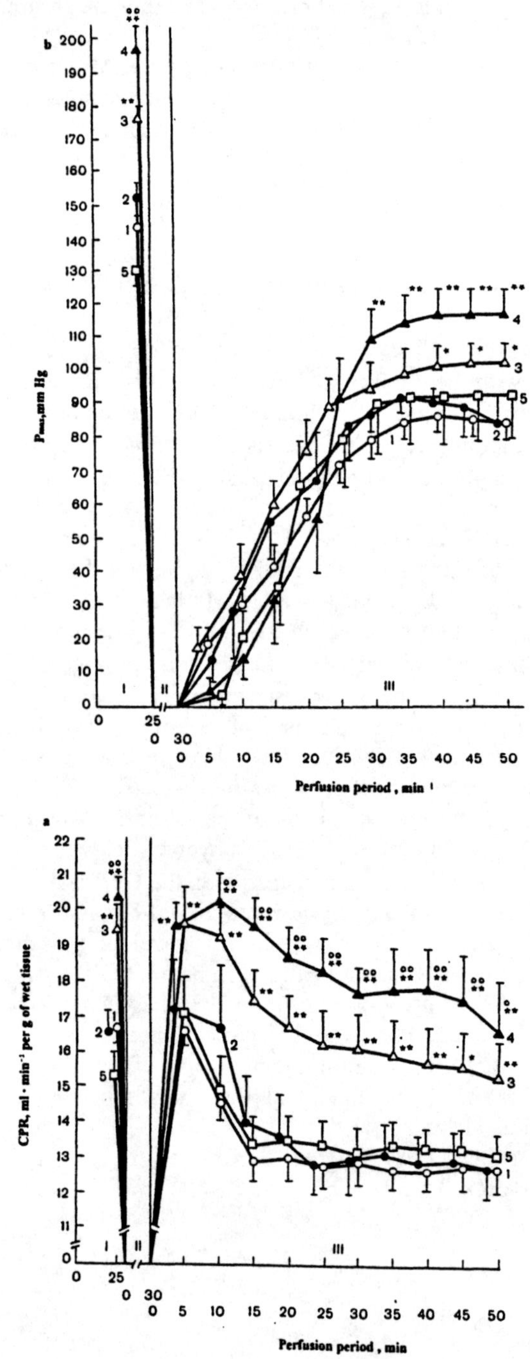

I - initial (stabilizing) perfusion (25 min); II – stop-flow ischemia (30 min at 37°C); III - reperfusion (50 min); 1 - without drug (control); 2 - addition of 1-hydroxy-TMP to the K H perfusate in a dose of 10^{-6} M; 3 –the same in a dose of 5 10^{-6} M; 4 – the same in a dose of 10^{-5} M; 5 - TMP (inactive analogue) in a dose of 10^{-5} M; *-P<0.05; **-P<0.01 versus control; o-P< 0.05; oo-P < 0.01. 4 versus 5.

11.2 ANTIOXIDANT ENZYMES AND SCAVENGERS OF REACTIVE OXYGEN SPECIES

As described in Chapter 9, the enzymatic system of antioxidant cell protection against free-radical reactions includes superoxide dismutase (SOD; $O_2^{\bullet}$ reduction to H_2O_2), catalase (H_2O_2 reduction to H_2O), Se-dependent GSH peroxidase (mainly H_2O_2 reduction to H_2O), and Se-independent GHS peroxidase (nonradical breakdown of lipid peroxides). Their activity decreases during ischemia, which makes logical attempts to compensate deficiency in these enzymes in order to ensure antiradical and antiischemic protection.

11.2.1 SOD and Catalase

SOD and catalase have been extensively tested as potential means of antiischemic protection, especially in recent years. These enzymes, used separately or, in most cases, in combination, exhibited partial protective action during subtotal hepatic ischemia in dogs see [Workshop on: Oxygen free-radicals in shock] and after administration to cats and dogs with 2-h intestinal ischemia [Parks D.A. and Granger D.N.; Schoenberg M.H., Yones M., Muhl E. et al.]. The protective action of antioxidant enzymes has also been demonstrated on different models of myocardial ischemia. In animals with 15–90-min coronary occlusion followed by reperfusion, treatment with SOD or SOD + catalase improved the ECG parameters [Gross G.J., Farber N.E. Hardman H.F. and Warlher D.C.], reduced the size of infarct [Näslung U., HäggmarkS., Johansson G. et al.], enhanced hemodynamics, lowered the content of LPO products, and provided for the maintenance of the total ADN pool [Breyer R.H., Otani H., Engelman R.M. et al.; Gross G.J., Farber N.E. Hardman H.F. and Warlher D.C.].

The protective effect of SOD and catalase on the cardiac function was also observed in experiments with perfusion of isolated rat and rabbit hearts under total ischemia or with their hypoxic perfusion [Ytrehus K., Gunnes S., Myklebust R. et al.]. However, antioxidant enzymes applied after the onset of ischemia or only during the reperfusion period did not produce any antiischemic effect [Przyklenk K. and Kloner R.A.; Uraizee A., Reimer R.A., Murry Ch.E. and Jennings R.B.].

In our experiments, administration of SOD isolated from human erythrocytes increased the survival rate of rats exposed to ischemic shock [Bilenko M.V. (1992a)] and ischemia of liver lobes [Liashenko S.M., Bilenko M.V. and Gefter Ju.V.] and also significantly intensified reparative processes in liver lobes after their 2-h ischemia, as estimated from ^{14}C-thymidine incorporation into nuclear DNA [Liashenko S.M., Bilenko M.V. and Gefter Ju.V.] (see Fig. 79). SOD (30 000 U) added to the perfusate before 30-min total ischemia of the isolated rat heart and after its reperfusion improved the contractile function of the myocardium, increasing P_{max} and dp/dt_{max} by 33 and 26%, respectively ($P < 0.05$) [Morgunov A.A.].

In recent years, SOD and SOD + catalase both have been used in the increasing number of studies aimed at elucidating the contribution of ROS to pathogenesis of ischemic and reperfusion injuries in organs. Success was achieved in using SOD + catalase in ischemic dog heart [Bando K. and Senoo Y.] or SOD + allopurinol [Concannon.M.J., Dooley T.W., Puckett C.L.]; SOD +acidified sodium nitrite ($NaNO_2$) in LAD occlusion of dog and cat hearts [Tripathy Y., Hegde B.M. and Raghuveer C.V.; Johnson G.3d., Tsao P. and Lefer A.M.]. However, not all combinations were effective: thus, combined administration of Cu, Zn–SOD and GSH in experiments on kidney ischemia and reperfusion proved to be inefficient and had a toxic effect [Paller M.S. and Eaton J.W.], combination of Mn–SOD + GSH was more effective and was not toxic. SOD can be toxic not only in "inappropriative" combinations, but also at high doses. The protective effect of Mn–SOD on the isolated ischemic rabbit heart changed to toxic when the dose was increased from 5-20 mg/l to 40-50 mg/l [Omar B.A. and McCord J.M.; Nelson S.K., Bose S.K and McCord I.M.]. Cu, Zn–SOD changed its effect from protective to damaging approximately in the same dose interval [Bernier M., Manning A.S. and Hearse D.I.].

As SOD has a large molecular weight (31,000-33,000) and a short circulatory half-life (1/2 min) [Ratych R.E., Chuknyiska R.S. and Bulkly G.B.], it was successfully used not only in its usual form, but also as conjugates: SOD – albumine complex (SM–SOD, half life 6h) [Kawamoto S., Inoue M., Tashiro S.et al.]; SOD– polyethylene glycol conjugates (PEG–SOD, half-life 30h) [Chi L.G., Tamura Y., Hoff

P.T.et al.; Cronenwett J.L., Lee K.R., Shlafer M. and Zelenock G.B.; Tan S., Yokoyama Y., Dickens E. et al.], SOD conjugates with carboxymethyl (SOD–CMD), diethylaminoethyl (SOD–DEAED), galactosylate (GAL–SOD) and mannosylate (Man–SOD) [Fujita T., Furitsu H., Nishikawa M.et al.], or with succinylgelatin (Sac–Gel–SOD) [Kojima Y., Haruta A., Imai T.et al]. Studies were also performed on salen-manganese complexes, a class of synthetic SOD/catalase mimetics, such as EUK-8, EUK-134 [Baker K., Marcus C. B., Huffman K. et al.; Pucheu S., Boucher F., Malfroy B. and De Leiris J.] or cross-linked hemoglobin-superoxyde dismutase-catalase (PolyHb–SOD–CAT) [D'Agnillo F. and Chang T.M.; Razack S., D'Agnillo F.and Chang T.M.], or as recombinant Cu, Zn–SOD hybrid protein (SOD:Tet 451) [Francis J.W., Ren J., Warren L.et al.]; or encapsulated into rat or rabbit erythrocytes (RBS–SOD, half-life 13.7 day) [Wang X., Yang J. and Liu F.; Wang X., Zhou H. and Yang J.]. Along with usual (intracellular) SOD, it was separated extracellular isozyme (EC–SODc), which *in vivo* has a high affinity to heparin sulfate proteoglycans in glycocalix of various cell types (more often, in endothelial cells) [Ohta H., Adachi T. and Hirano K.]. Half-life of EC–SODc in the circulation is near by 20h [Karlsson K., Sandstrom J., Edlund A.et al.]. The antiischemic effect of EC–SODc is more expressed when it is administered before ischemia [Nilson U.A., Haraldson G., Bratell S.et al.].

11.2.2 Se-Independent and Se-Dependent GSH Peroxidase

The activity of GSH peroxidase depends on the tissue content of reduced glutathione (GSH), and the activity of Se-dependent GSH peroxidase, on both GSH and selenium contents, which decrease during ischemia (see Chapter 10). Hence, instead of adding exogenous GSH peroxidase, it is possible to stimulate the activity of endogenous GSH peroxidase with the aid of GSH and its precursors (e. g., S-adenosyl-L-methionine, N-acetylcystein, etc.), which exert antioxidant and antiischemic effects [Paller M.S. and Eaton J.W.; Evans P.J., Whiteman M., Tredger J.M. and Halliwell B.; Marubayashi S., Dohi K. and Kawasaki T.].

Pronounced antiischemic effect was also demonstrated for Se-containing compounds. Sodium selenite at dose of 30 μg/kg, especially combined with α-tocopherol, administered to rats 2 h before and for 7 days after AMI increased their survival rate, improved ECG parameters and ultrastructure of cardiomyocytes, and lowered chemiluminescence intensity and DC content in myocardial lipids [Kogan A. Kh., Kudrin A. V. and Nikolaev S. M. (1976); Kogan A. Kh., Kudrin A. N., Kakturskii L. V. and Losev N. I.; Kudrin A. N., Kogan A. Kh., Korolev V. V. et al.]. The authors of these studies suggested that introduced selenium can accumulate in the ischemic tissue and exert a stabilizing effect on membranes, although it is more probable that selenium either is incorporated into Se-dependent GHS peroxidase or stimulates the activity of GSH reductase, which restores the GSH pool in the cell (see Fig. 91). The Se-containing preparation Ebselen or selenium enriched diet also prove to have a strong protective effect [Tanguy S., Boucher F., Besse S. et al.; Watanabe T., Nishiyama M., Hori T. et al.].

In our laboratory, experiments on the models of hepatic and cardiac ischemia revealed a similar effect of another Se-containing compound, Se-methionine. Intraperitoneal administration of its aqueous solution at 0.6 μg/kg doses for 3 days before creating ischemia of the left and median liver lobes was not enouph effective, whereas at doses of 1.2 μg/kg (10^{-6} M, 1ml) increased animal survival rate to 93% versus 25% in the control ($P < 0.01$) (See Fig. 79).

In experiments with perfusion of the isolated rat heart, Se-methionine in a concentration of $1 \cdot 10^{-9}$ M did not produce any effect on the CPR and cardiac function of the intact heart (P_{max}) before ischemia but markedly improved both parameters (especially P_{max}) after ischemia [Morgunov A.A., Komarov P.G., Bulgakov V.G. and Bilenko M.V.; Morgunov A.A.]. As Se-methionine is readily soluble in water and is efficient at very low doses, it needs further study as a promising antiischemic agent.

In addition to antioxidant enzymatic systems, detoxification (reduction) of reactive oxygen species ($O_2^{\bullet}$, $^{\bullet}OH$, $^1O_2$) is performed by a number of compounds collectively known as ***ROS scavengers***. As a rule, interaction with ROS is a supplementary activity of these agents, which makes it difficult to evaluate their contribution to the overall antiischemic effect. The antiischemic action was demonstrated for the following ROS scavengers: mannitol and aminazine, which were tested in experiments on ischemia and hypoxic perfusion of the rat skeletal muscle, rabbit heart or kidney and LAD occlusion in dogs, and clinically used as additives to the cardioplegic solution during surgical operations [Afanasyev S. A., Laptev B. I., Prokopyeva V. D. and Larionov N. P.; Barclay J. K. and Hansel M.; Zager R. A., Mahan J. and Merola A.

J.; Schlag M.G., Clarke S., Carson M.W. et al.; Ferreira R., Burgos M., Llesuy S. et al.; Ouriel K., Ginsburg M.E., Patti C.S. et al.]; DMSO in cardiac, cerebral, skeletal muscle and renal ischemia [Feller A.M., Roth A.C., Russell R.C. et al.; Iakovlev G.M., Svistov A.S. and Mironova M.I.; Plotnikov M.B., Avdoshin A.D., Chernysheva G.A. and Saratikov A.S.]; DMPO (5.5-dimethyl-pirroline-N-oxide) in cardiac ischemia [Tosaki A., Blasig I.E., Pali T. and Ebert B.]; uric acid in cerebral ischemia [Yu Z.F., Bruce-Keller A.J., Goodman Y. and Mattson M.P.]; histidine-containing dipeptides (carnosine and others) used in experiments on ischemia and reperfusion of different organs and in clinical cardiology [Boldyrev A.A. (1992); Lee J.W., Miyawaki H., Bobst E.V., et al.; Pavlov A.R., Revina A.A., Dupin A.M. et al.]; and several other agents, such as L-methionine, glucose, dimethylthiourea, caffeine, albumin, pyridoindoll (stobadine), aprotonine, metallothionein, and alcohols. However, as noted above, the contribution of their ROS scavenger activity to the overall antiischemic effect is as yet unknown.

11.3 Chelators and Reducers of Transition Metals

Chelators of transition metals can inhibit free-radical reactions and, hence, have the antiischemic effect. These are natural and synthetic compounds that form dissociation-resistant complexes with transition metals and thereby prevent their involvement in free-radical reactions. Among several known chelators (desferrioxamine, EDTA, bipiridyl, *o*-phenanthroline - PHE, etc.), and especially desferrioxamine (Deferoxamine mesylate 2,3-dihydroxybenzoic acid, Desferal, DFO) and *o*-phenanthroline (PHE) have been studied most extensively.

DFO, naturally produced by *Streptomyces pillosus,* selectively interacts with Fe^{3+} to form a highly stable complex with $K_a = 10^{31}$ [Kozlov A. V., Meschanov A. Yu., Nikolayeva I. G. et al.]. It effectively inhibits $^{\bullet}OH$ generation in the Fenton and Haber–Weiss reactions, as well as Fe-dependent lipid peroxidation *in vitro* [Gutteridge J.M.C., Richmond R. and Halliwell B. (1979); Halliwell B.].

Both DFO and PHE are more active than other chelators in inhibiting luminol-dependent chemiluminescence *in vitro* (ADP < EDTA < PGA < DFO < PHE) and self-oxidation of lipids in bovine brain homogenates [ADP < ACK < EDTA < PGA < DFO < PHE) [Kozlov A. V., Meschanov A. Yu., Nikolayeva I. G. et al.].

The ability of iron chelators (DFO and PHE) to inhibit the formation of LPO products and to produce antiischemic effects was also demonstrated *in vivo*. DFO added to the perfusate inhibited in a dose-dependent manner the formation of primary and secondary LPO products (DC, MDA and SB) in rabbit kidneys after 24 h of cold storage [Green C.J., Healing G., Simpkin S. et al. (1986c)]. DFO i. v. injected to rabbits inhibited the formation of DC and SB in kidneys reperfused after 60–120 min of warm ischemia [Green C.J., Healing G., Simpkin S. et al. (1986a)] or after 24 h of cold preservation [Fuller B.J., Lunec J., Healing G. et al.]. DFO also protected against the breakdown of membrane phospholipids in ischemic–reperfused rat hearts [Lui X.K., Prasad M.R., Engelman R.M. et al.] and inhibited the accumulation of low molecular weight iron (LMWI) and MDA in the brain after cardiac arrest and subsequent resuscitation in dogs [Nayini N.R., White B.C., Aust S.D. et al.]. It reduced the content of free radicals in the effluent blood and had a protective antiischemic effect in experiments on cardiac and intestinal ischemia in dogs and rabbits [Spencer K.T., Lindower P.D., Buettner G.R. and Kerber R.E.; Williams R.E., Zweier J.L. and Flaherty J.T.], ischemia of the liver [Omar R., Nomikos I., Piccorelli G. et al.; Drugas G.T., Paidas C.N., Jahanda A.M. et al.] and skeletal muscles [Fantini G.A. and Yoshioka T.] in rats, and in cases of AMI, cardiopulmonary bypass, and revascularization of the human heart [Beyer W., Imlay J. and Fridovich I.; Ferreira R., Burgos M., Milei J. et al.; Bel A., Martinod E. and Menasche P.].

DFO can enter cell membranes but has a short half-life period in blood plasma and, hence, was conjugated with some carriers (such as hydrohyethyl starch (HES-DFO)) to improve its efficiency [Drugas G.T., Paidas C.N., Jahanda A.M. et al.; Mousa S.A., Ritger R.C. and Smith R.O.; Hedlund Bo.E. and Hallaway P.E.]. Other synthetic and natural Fe^{2+} and Fe^{3+} chelators, such as hexadentate pyridinone (CP 130) [Gower J.D., Ambrose I.J., Manek S. et al.], albumin [Strubelt O., Younes M. and Li Y.] bipyridyl [Oubidar M., Boquillon M., Maril Ch. et al.; Ikeda J., Ikeda K. and Long D.M.], deferiprone [Arthur A. S., Fergus A. H., Lanzino G. et al.], kojic acid, catechol, mimosine etc. were also shown to have a protective antiischemic effects [Vander Kraaij A.M., van Eijk H.G. and Koster J.F.]. In particular, *o*-phenanthroline proved to protect mammalian cells from DNA damage induced by ROS [De Mello Filho A. C. and Menehgini R.].

In the recent study, we analyzed effects of DFO on the survival rate, LMWI, and TBARS levels during complete ischemia and after reperfusion of liver lobes in male Wistar rats [Bilenko M.V., Bulgakov V.G. and Morgunov A.A. (1989); Bilenko M.V., Lyashenko S.M., Kozlov A.V. et al.; Liashenko S.M.]. Complete ischemia of the left and median liver lobes was created by clamping hilar peduncles. After 1- and 2-h exposure to ischemia, clamps were removed, and liver lobes were reperfused for 5–60 min. In experiments performed to determine rat survival, two nonischemic (wright and caudatum) liver lobes (30% of liver) were immediately excised as described previously [Bilenko N.V., Kagan V.E., Velikhanova D.M. and Komarov P.G.] (see Chapter 5).

Among rats not treated with DFO, 53% remained alive 12 h after the operation and only 7% survived on day 7 after operation (Fig. 107). DFO increased significantly the survival of rats with ischemic liver lobes only when administered at a dose of 100 mg/kg 15 min before reperfusion, being ineffective at lower doses and when injected before ischemia. In studies performed by other authors [Drugas G.T., Paidas C.N., Jahanda A.M. et al.], the same dose of 100 mg/kg injected before liver lobe reperfusion also proved to be very effective in preventing hepatic ischemic injury. It was shown that this is the highest nontoxic dose of DFO *in vivo* [Mousa S.A., Ritger R.C. and Smith R.O.].

Fig. 107 Effect of DFO on rat survival rate after 2 h of ischemic two liver lobes and 30% resection of intact liver lobes (%).

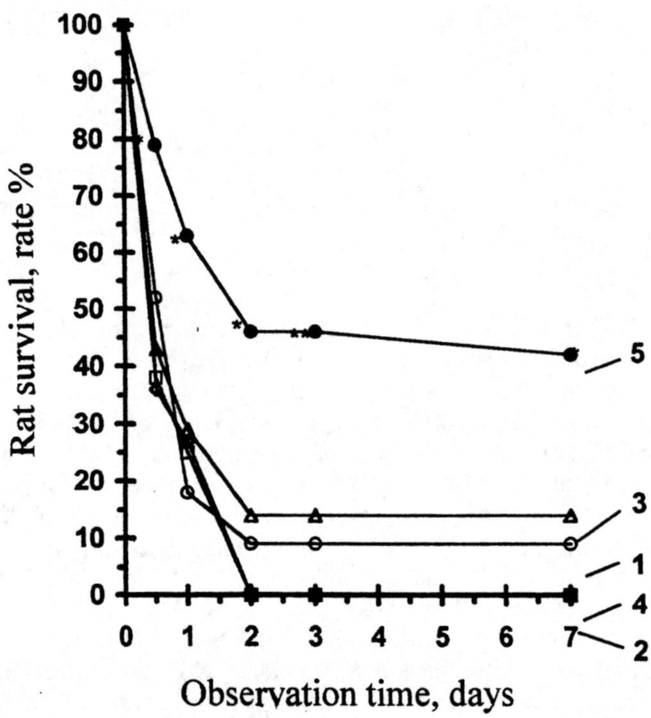

1 — without DFO, (control);
2 — DFO 50 mg/kg 15 min before reperfusion;
3 — DFO 50 mg/kg 15 min before ischemia and 15 min before reperfusion;
4 — DFO 100 mg/kg 15 min before ischemia;
5 — DFO 100 mg/kg 15 min before reperfusion.
* — $P < 0.05$; ** — $P < 0.01$ versus control (Wilcoxon-Mann-Uitny test).
DFO was injected i. p.

Determination of LMWI content in the liver tissue by a new technique (EPR analysis of nonhomogenized liver samples) [Kozlov A. V., Yegorov D. Yu., Vladimirov Yu. A. and Azizova O. A.] showed that this parameter increases by factors of 2.5 and 3.1 after 1-h and 2-h ischemia, respectively (Fig. 108), which was consistent with our previous data (see Chapter 10), and decreases to the control level during reperfusion. This decrease manifested itself after 15 min of reperfusion, and the lowest LMWI level was observed after 60 min. Intraperitoneal injection of DFO to rats at a dose of 100 mg/kg 15 min before ischemia prevented the increase in the level of LMWI in only 1-h ischemic liver lobes. Injections made 15 min before ischemia or 15 min before reperfusion did not lead to any changes in LMWI content in the reperfusion period, compared to that in untreated rats (See Fig. 108, I, II).

Fig. 108: Content of LMWI in rat liver lobes after 1 and 2h of ischemia (I) and after 2h of ischemia and different time of reperfusion (II) (mean ± SEM).

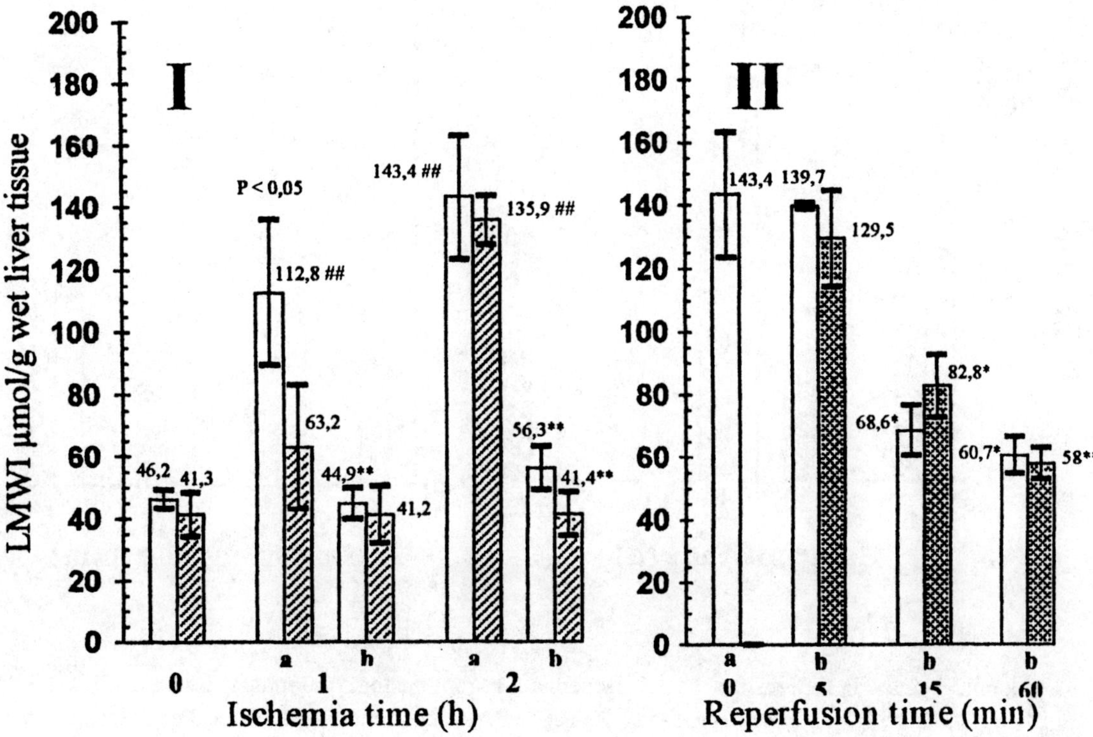

I: Ishemia 1 or 2 h, reperfusion 1 h; II: Ischemia 2 h, reperfusion 5-60 min; a — ischemia; b — reperfusion; 1. □ — nontreated rats, 2. ▨ — DFO — 100 mg/kg 15 min before ischemia (I); 3. ▩ — DFO — 100 mg/kg 15 min before reperfusion (II). # -P < 0.05, ## -P < 0.01;-versus control. * -P< 0.05, ** - P< 0.01 reperfusion versus ischemia; P — DFO — treated rats versus non-treated rats. LMWI were measured as described by [Kozlov A. V., Yegorov D. Yu., Vladimirov Yu. A. and Azizova O. A.]. Briefly, small pieces of excised liver after 20 min incubation at 20 ^{0}C in 8, 7 M sodium nitrite solution were frozen in the liquid nitrogen. EPR spectra were recorded using a Varian E-4 spectrometer (oscillator klystron frequency 9.12 GHz, power 10 mW, time constant 1.0 s). The curve was plotted using EPR signals of intact liver lobe homogenate mixed with known amount of $FeSO_4$.

TBARS levels did not increase in liver lobes during ischemia, but significantly increased (by a factor of 1.6–3.0) within 15 min of reperfusion (Fig. 109, I, II). DFO administration suppressed the increase in TBARS content only when injected 15 min before reperfusion, and this effect was noticed as soon as after 5 min of reperfusion.

Fig. 109: Content of TBARS in rat liver lobes after 1 and 2h of ischemia (I) and after different time of reperfusion (II) (mean ± SEM).

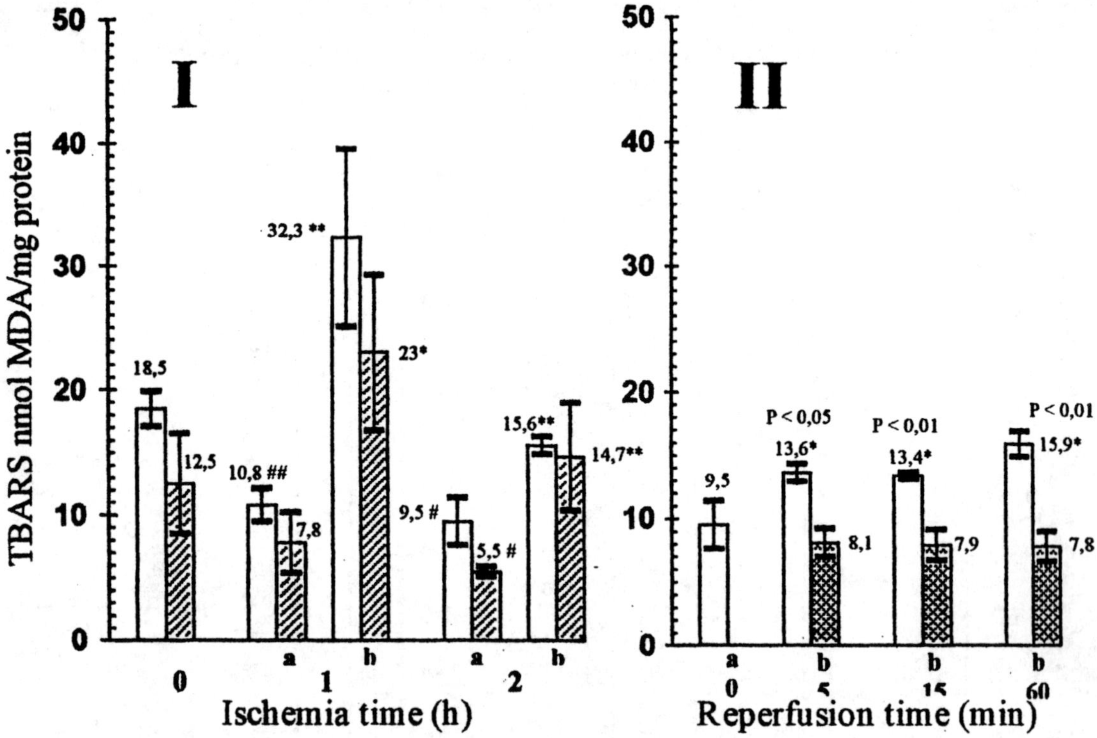

Ischemia 1 and 2 h, reperfusion 1 h; II. Ischemia 2 h, reperfusion 5-60 min. a, b, 1. □, 2. ▨, 3. ▩, #, ##, *, **, P — are the same as in Fig. 108. TBARS content was determined as described by [Mihara M., Uchiyama M. and Fukuzawa K.], protein content was determined by Lowry.

Therefore, our data showed that the content of LMWI reaches the highest level during ischemia, and that of TBARS, in the early reperfusion period. Virtually the same dynamics of free iron and LPO products was described in [Ozaki M., Fuchinoue S., Teraoka S. and Ota K. (1984)]. These data suggest that LMWI accumulating during ischemia activates TBARS production as soon as tissue oxygen level becomes sufficient for metabolizing primary LPO products into secondary. Second, our data provided evidence that DFO is an effective antiischemic agent because it significantly suppressed (practically blocked) the growth of TBARS content within the first 5 min of reperfusion and also promoted the rat survival rate when injected at a dose of 100 mg/kg 15 min before reperfusion. As well as other authors [Van Jaarsveld H., Potgieter G.M., Kuyl J.M. et al.], we did not observe any decrease of LMWI in 2-h ischemic liver lobes in experiments with DFO; therefore, its role as a chelator of iron was not evident during prolong ischemia. The structure of the LMWI pool and its availability for helation are not clear enough as yet; however it may well be that DFO binds only a small but the most active (in LPO initiation) fraction of LMWI. The antiischemic effect of DFO, as well as those of copper chelators (bathocuproin, neocuproin, copper (II) coumarin, and others) can also be conditioned by its ability to scavenge reactive oxygen species, as shown in many publications [Appelbaum Y.J., Kuvin J., Borman J.B. et al.; Vladimirov Yu.A., Parfenov E.A., Epanchintseva O.M. et al.; Spencer K.T., Lindower P.D., Buettner G.R. and Kerber R.E.; Halliwell B. and Chirico S.; Hartley A., Davies M. and Rice-Evans C.]: thus, catecholic iron complexes proved to have a greater scavenging activity than catechol itself [Zhao Z. S., Khan S. and O'Brien P. J.]. The relationship of the antiischemic effect of DFO with its ability to suppress or block the formation of reactive oxygen species and the accumulation of LPO products was also demonstrated by other authors on the models of ischemic newborn lambs myocardium [Shadid M., Van Bel F., Steendijk P. et al.], ischemic rat skeletal muscles [Fantini G.A. and Yoshioka T.], cerebral canine partial ischemia [Hurn P.D., Koehler R.C., Blizzard K.K. and Traystman R.J.], ischemic human myocardium [Drossos G., Lazou A., Panagopoulos P. and Westaby S.; Ferreira R., Burgos M., Milei J. et al.]. See also some review devoted to this problem [Bergeron R.J. and Brittenham G.M.,; Kirshner R.E. and Fantini G.A.].

11.4 Inhibitors of Enzyme Systems Involved in Production of Reactive Oxygen Species

11.4.1 Inhibitors of Phospholipase, Cyclooxygenase, and Lipoxygenase

As noted in Chapter 4, the activities of phospholipase, cyclooxygenase, and lipoxygenase increase during ischemia and in the early postischemic period. This results in the accumulation of arachidonic acid and its metabolic products in the ischemic organ and the effluent blood. These metabolites, such as prostaglandins and thromboxanes (the cyclooxygenase pathway), hydroxy acids (HPETE and HETE), and leukotrienes (the lipoxygenase pathway), produce vasoactive, membranotoxic, and cell-damaging effects.

Intensification of both the cyclooxygenase and lipoxygenase pathways of arachidonic acid metabolism is followed by an increase in ROS production and LPO. The production of ROS increases (see chapters 4.4 and 9.2.4) during conversion of PgG_2 into PgH_2 and HPETE into HETE, due to the peroxidase properties of PgH_2 synthetase and lipoxygenase. Therefore, compounds inhibiting these enzymes can be expected to have antioxidant and antiischemic effects.

Indeed, the most active *phospholipase inhibitors,* especially the inhibitors of phospholipase A_2 (such as chlorpromazine and quanacrine) were shown to have an antiischemic action. Their effects during ischemia and after reperfusion were studied on the models of AMI, middle cerebral artery occlusion, and hepatic ischemia [Chien K.R., Abrams J., Serroni A. et al.; Chien K.R., Pfau R.G. and Farber J.L.; Estevez A.Y. and Phillis J.W.] (see also Chapter 4).

Cyclooxygenase inhibitors include numerous nonsteroid antiinflammatory substances, such as acetylsalicylic acid, indomethacin, flurbiprofen, naprofen, ibuprofen, diclofenac, naproxen, etc. Most of them proved to have protective properties: they enhanced collateral circulation and reduced the zone of AMI in dogs and baboons [Jugdutt B.I., Hutchins G.M., Bulkle B.H. et al. (1981); Ruf W., McNamara J.J., Saehiro A. et al.]; had a protective effect on myocardial function and high-energy phosphate metabolism during partial cardiac ischemia and reperfusion in dogs and rabbits [Kawabata H., Sugiyama K. and Katori R.; Seth S.D., Maulik M., Manchada S.C. and Maulik S.K.], prevented arrhythmia after AMI in pigs [Das D.K., Engelman R.M., Rauson J.A. et al. (1986)]; improved the functional state, enhanced blood reflow, and mitigated edema during brain ischemia and reperfusion in dogs and Mongolian gerbils [Hallenbeck J.M. and Furlaw T.W.; Jannotti F., Crockard A. and Ladds G.]; prolonged the time of survival in mice during generalized acute hypoxia [Nicolov R.]; and abolished the postischemic tail hyperalgesia in rats [Gelgor L., Phillips S., Butkow N. and Mitchell D.]. In some studies, however, such an effect of selective cyclooxygenase inhibitors was not revealed [Hallenbeck J.M. and Furlaw T.W.; Jugdutt B.I., Grover Ch., Hutchins G.M. et al.].

In recent years, acetylsalicylic acid (ASA, aspirin) and especially its water-soluble forms (Bayer aspirin, claridine, and other substances) have been widely used in clinical practice as platelet-inhibiting agents (antiaggregants). These drugs were administered to patients before and after reconstructive vascular surgery on the coronare, carotid, cerebral, and low limb arteries and the aorta, aortocoronary bypass, endarterectomy. They were also used for treating patients with high risk of these diseases and for primary or secondary prevention of cerebral stroke [Boysen G., Sorensen P.S., Juhler M. et al.; Chaudhuri A., Medhekar T.P.; UK TIA Study Group] or myocardial infarction [Garcia-Dorado D., Theroux P., Tornos P. et al.; Mahony Ch.; Mehta J.L. and Conti C.R.]. In some subgroups of patients, this treatment failed, and aspirin was combined with additional antiplatelet agents, such as ticlopidine, dipyridamole, heparin, clopidogrel, triflusat, and sulfinpyrazone

There was no difference between the effects of ASA administered at daily doses of 160 and 325mg [Chaudhuri A., Medhekar T.P.], 300, and 1200 mg [UK TIA Study Group], but very low doses (40–75 mg ASA per day) were reported to be ineffective [Boysen G., Sorensen P.S., Juhler M. et al.]. Thus, the most effective dosage of aspirin remain to be determined.

The anticoagulant effect of aspirin and aspirin-like drugs is attributed to their diverse inhibitory effects on prostaglandin biosynthesis in different tissues, including irreversible inhibition of platelet cyclooxygenase and delayed reversible inhibition of vascular cyclooxygenase. Indeed, administration of aspirin in experiments and clinical trials was accompanied by a decrease in serum TxA_2 and changes in the TxA_2/PgI_2 index [Montalescot G., Maclouf J., Drobinski G. et al.; Vejar M., Hackett D., Brunelli C. et al.]. However, this mechanism is apparently not the only and even the main one responsible for the antiischemic effect of ASA because selective inhibition of TXA_2 synthetase (by OKY-046) during myocardial ischemia in rabbits was ineffective [Evers A.S., Murphree S., Saffitz J.E. et al.; Sheker A. and Singh M.], whereas preventive inhibition of this enzyme by OKY-046 and CV-4151 or introduction of TXA_2 receptor antagonists (KW-3635, BMS-180 291, sulotroban, etc.) produced only a moderate positive effect on the myocardium, as compared to that of aspirin [Grover G.J., Schumacher W.A. and Ogletree M.L.; Kawabata H., Ryomoto T. and Katori R.; Savage M.P., Goldberg S., Bove A.A. et al.; Schrör K., Smith E.F., Bickerton M. et al.]. Aspirin (Aspisol, i. v., 150mg kg^{-1} /day) was shown to decrease PgI_2 and especially TXA_2 production, and at higher doses, also decreased the iNOS activity in the infarcted rabbit heart [Yamamoto T., Cohen A.M., Kakar N.R. et al.]; Salicylates also inhibited mitochondrial damage and preserved ascorbate and α-tocopherol in the ischemic/reperfused rat heart [van Jaarsveld H., Kuyl J.M., van Zyl G.F. and Barnard H.C.], whereas indomethacin activated eNOS in piglet cerebral vessels [Beasley T.C., Bary F., Thore C. et al.].

Most of numerous recently synthesized water-soluble forms of aspirin (aspisol, aspidol, delgesic, dolomega, flectadol, lasdol, solusprin, venopirin, aspegic, acelizine, etc.) are ASA complexes with L-arginine, DL-lysine, histidine, glutamic acid, or ornithine. They all retain specific properties of ASA but are much less harmful to the stomach and, hence, can be used at significantly high doses, compared to water-insoluble drugs. After parenteral (intramuscular or intraperitoneal) administration, they appear in blood within a very short time.

To study the possibility of using cyclooxygenase inhibitors as antiischemic drugs, we examined the protective effect of acelizine, a water-soluble form of aspirin (DL-lysine) synthesized in the Research Institute of Chemical Technology of Drugs (Kharkov, Russia)

Acelizine proved to produce a strong inhibitory effect on ADP-induced platelet aggregation as soon as 15 min after intramuscular injection at a dose of 100 mg ASA/kg [Bilenko M.V., Komarov P.G., Tel'pukhov V.I. et al.] (Fig. 110). This effect reached a peak after 30 min and disappeared after 60 min. The significant effect of usual insoluble aspirin administered orally at the same dose developed much later, only after 120 min.

Fig. 110: The effects of acelizine (1) and unsoluble aspirin (2) on ADP-induced aggregation of blood thrombocytes in rabbits (%)

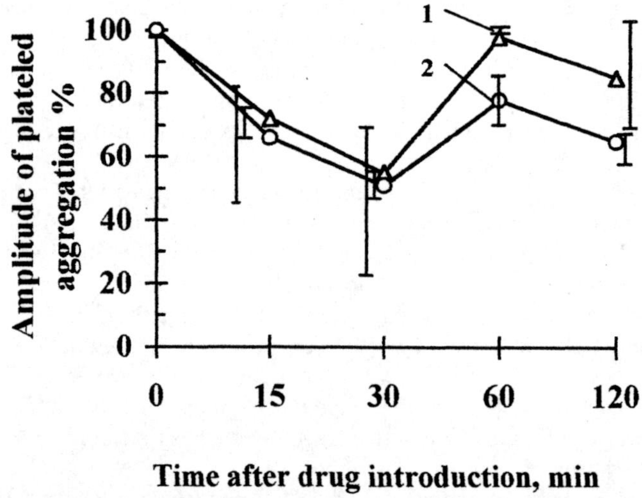

*-P<0.05;**-P<0.01 versus control. Acelizine (100 mg ASA/kg) was injected to rabbits i. m., aspirin (100 mg/kg) was introduced per os. ADP — induced aggregation was measured in chronolog aggregometer (USA).

Acelizine (100 mg ASA/kg, 30 min after injection) was also more effective than insoluble aspirin in suppressing platelet aggregation and MDA production during thrombin-induced platelet aggregation in rats (Fig. 111).

To evaluate the antiischemic effect of acelizine and to determine its optimal dosage and time of administration, experiments with ischemia/reperfusion of limb muscles, heart, liver, and brain in Wistar rats were carried out.

Fig. 111: The effects of acelizine (1) and unsoluble aspirin (2) on thrombin-induced aggregation of thrombocytes (I) and thrombocyte MDA production (II) in rats (% to control).

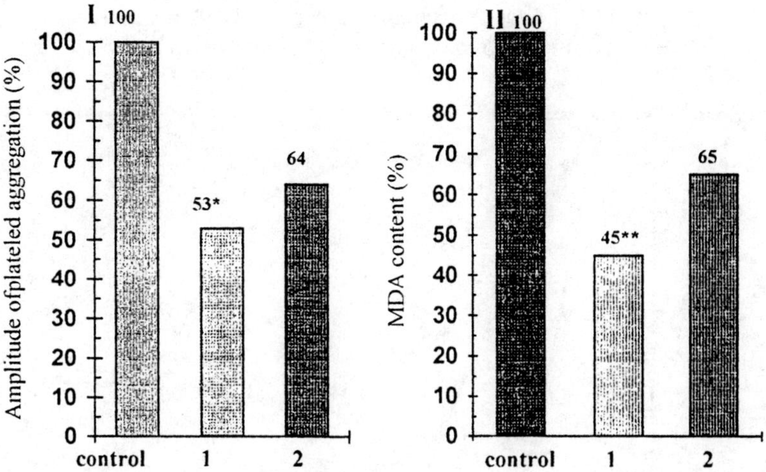

*-P<0.05; **-P<0.01 versus control.
Amplitude of thrombocyte aggregation was measured in Specol-10;
Blood was taken from rat heart under ether anesthesia.

Fig. 112: Effect of acelizine on acute postischemic edema of rat leg muscles (% to preischemic leg volume).

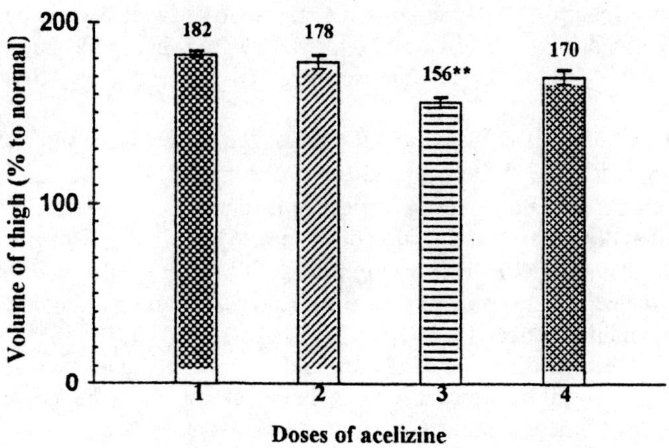

1 — without acelizine (control); 2 — acelizine 100 mg ASA/kg twice 3 h before ischemia and 3 h before reperfusion i. p.; 3 — acelizine 200 mg ASA/kg in the same manner; 4 — acelizine 300 mg ASA/kg thrice-72, 48 and 24 h before ischemia. **-P<0.01 versus control.

In experiment with total ischemia and reperfusion of rat limb muscles [Bilenko M.V., Kovaleva V.L., Morgunov A.A. et al.], the model of 3-h tourniquet ischemia of two legs in the upper one-third, followed by 3-h reperfusion, was used. Figure 112 shows that acelizine significantly reduced acute edema of schemic/reperfused rat thighs only when administered at two doses of 200 mg ASA/kg (30 min before ischemia and 30 min before reperfusion). Lower and higher doses, as well as continuous treatment, were ineffective.

By contrast, in experiments on a model of more severe 6-h tourniquet ischemia of two legs (ischemic shock with lethal outcome within 24-48 h in most rats), acelizine produced a protective antiischemic effect only when administrated preventively (72, 48 and 24 h before ischemia) at high doses of 300 mg ASA/kg (Fig. 113).

Fig. 113: Effect of acelizine on rat survival rate after 6 h tourniquet ischemia and reperfusion of two legs (%).

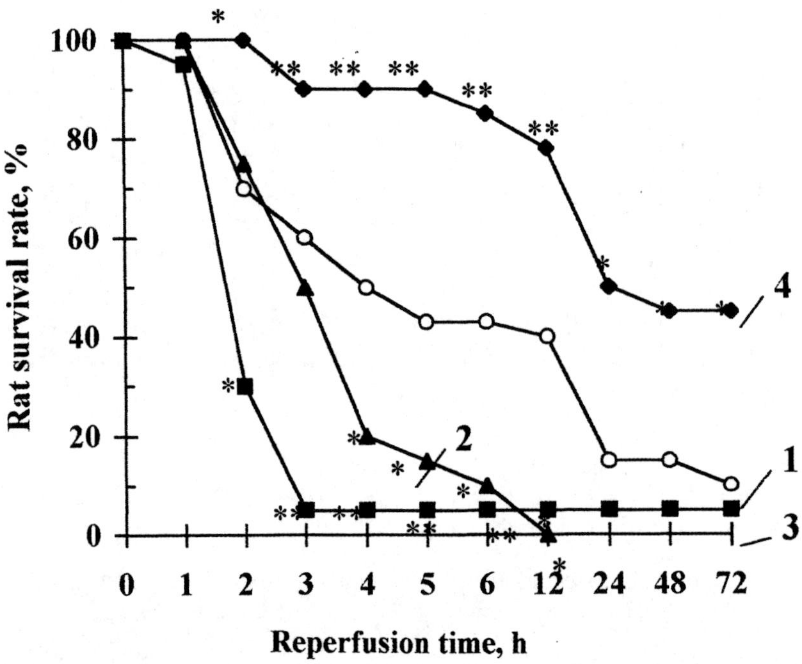

1 — without acelizine (control); 2 — acelizine 100 mg ASA/kg twice 30 min before ischemia and 30 min before reperfusion; 3 — acelizine 200 mg ASA/kg in the same manner; 4 — acelizine 300 mg ASA/kg thrice 72, 48 and 24 h before ischemia:*P - <0.05; **P - < 0.01-versus control.

Studies on the protective effect of acelizine on the isolated totally ischemic and saline-reperfused rat heart (stop-flow for 30 min at 37°C in the Langendorff–Fallen model) showed that high doses of acelizine (200–400 mg ASA/l) added to saline during preischemic perfusion produced a short-term, moderate improvement in the contraction-relaxation function of the heart [Bilenko M., Komarov P., Tel'pukhov V. et al.]. Doses of 50–150 mg/l were ineffective. Approximately the same results were obtained with acelizine injected to rats intraperitoneally at doses of 300 mg ASA/kg 48, 24 h and 30 min before dissecting the heart for isolated perfusion (unpublished data).

In experiments with 2-h total ischemia of the left and central liver lobes and resection of other lobes (30% of liver weight) followed by reperfusion, acelizine at none of the doses (50, 100, 200, and 300 mg ASA/kg one to three times) administered 30 min before ischemia or reperfusion had any effect on the survival of rats [Liashenko S.M.].

The strongest protective effect of acelizine was observed on the model of total cerebral ischemia in rats (bilateral ligation of the carotid and subclavian arteries) [Bilenko M.V., Komarov P.G., Tel'pukhov V.I. et al.; Bilenko M., Komarov P., Tel'pukhov V. et al.; Tel'pukhov V.I., Bilenko M.V., Tikhomirova A.I. et al.; Tel'pukhov V.I., Bilenko M.V., Khokhlov A.V. and Komarov P.G.].

In rats exposed to 17-min total cerebral ischemia, acelizine (25, 50, 100 and 150 mg ASA/kg) injected intramuscularly 30 min before brain reperfusion significantly accelerated the recovery of early neurological responses (nociceptive sensation, spontaneous respiration, seizures) and increased the survival rate. High doses (200 and 250 mg ASA/kg) were ineffective. At optimal doses of 100 and 150 mg ASA/kg, the protective effect of acelizine was observed not only upon 17-min cerebral ischemia, but also upon 25-min and, partially, even 30-min ischemia (Figs. 114, 115).

Fig. 114: Effect of acelizine on recovery time of nociceptive sensation (I), spontaneous respiration (II), and seizure activity appearance (III) in rats after 17, 25 and 30 min of total rat cerebral ischemia and reperfusion (mean ± SEM). [Tel'pukhov V.I., Bilenko M.V., Tikhomirova A.I. et al.]

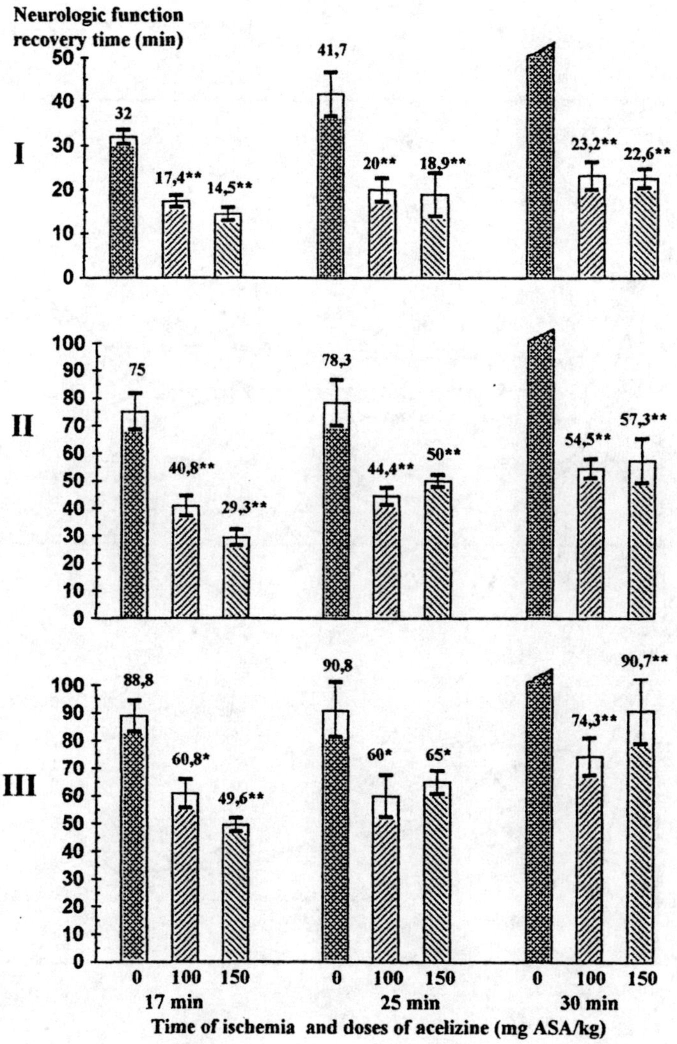

-P <0.05; ** -P < 0.01, versus control. Acelizine was injected i.m. in doses 100 and 150 mg/kg 30 min before reperfusion.

Fig. 115: Effect of acelizine on rat survival rate after 17 (I), 25 (II) and 30 min (III) of total cerebral ischemia and reperfusion (%). [Tel'pukhov V.I., Bilenko M.V., Tikhomirova A.I. et al.]

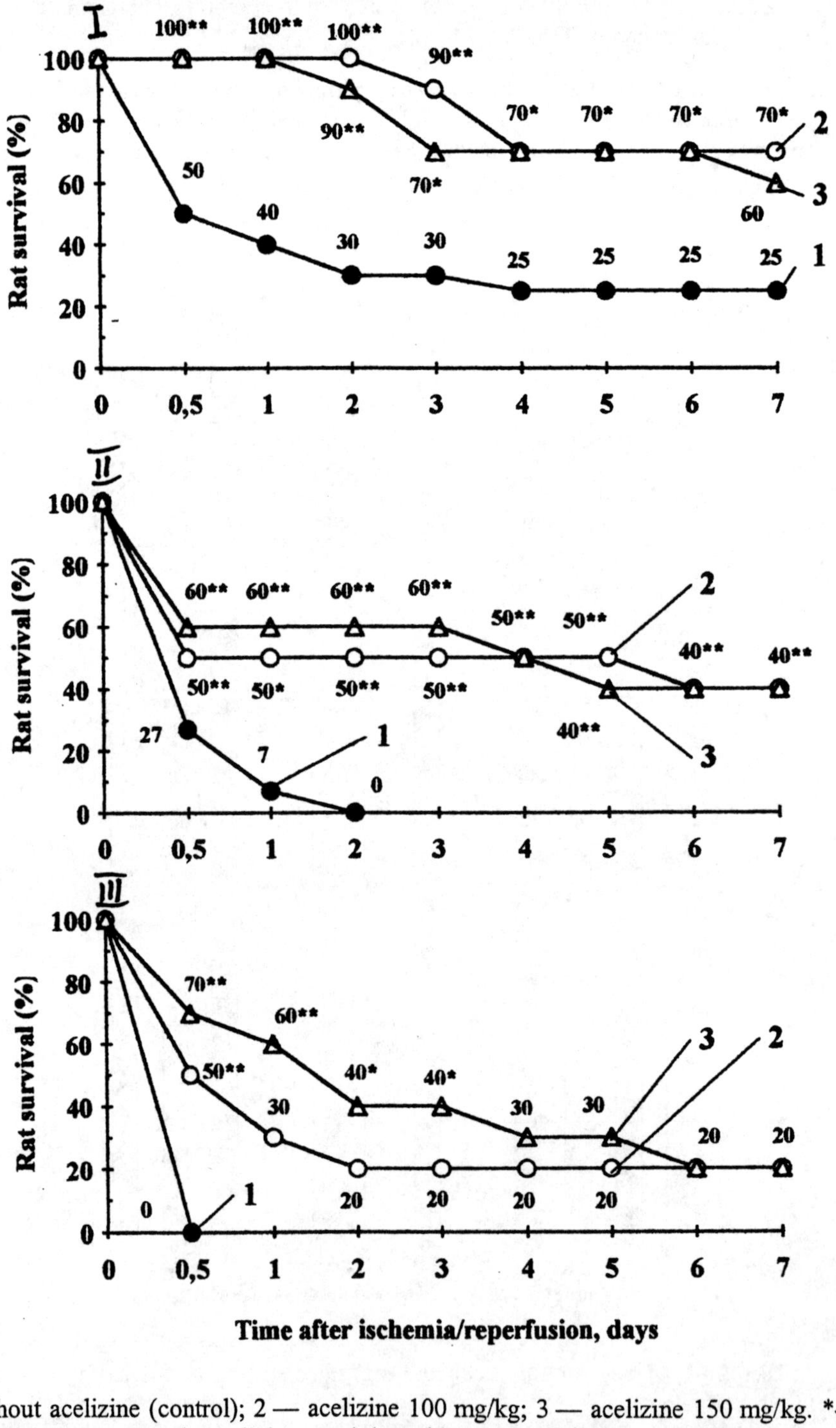

1 — without acelizine (control); 2 — acelizine 100 mg/kg; 3 — acelizine 150 mg/kg. *-P <0.05; **-P < 0.01, versus control. Acelizine was injected i. m. 30 min before reperfusion.

If total cerebral ischemia in rats lasted 35 min, acelizine was ineffective (data was not shown). A study on the optimal time of acelizine injection to rats with 25-min ischemia (Fig. 116) showed that preventive administration produced effect in any period of time, but the best results were obtained when acelizine was injected 30 min before ischemia. Injections made 15 min before reperfusion or immediately after it were less effective or absolutely ineffective (see Fig. 116). Analysis by the criterion of rat survival produced similar results (Fig. 117). The data obtained in these experiments suggest that the antiischemic effect of acelizine (and, possibly, any other form of ASA) is associated not only with inhibition of platelet aggregation in the circulating blood during the postischemic period, but also with the influence of acelizine on arachidonic acid metabolism in the ischemic brain cells themselves.

Fig. 116: Effect of different time of acelizine injection on recovery time of nociceptive sensation (I), spontaneous respiration (II), and seizure activity appearance (III) in rats after 25 min of total cerebral ischemia and reperfusion (mean ± SEM).

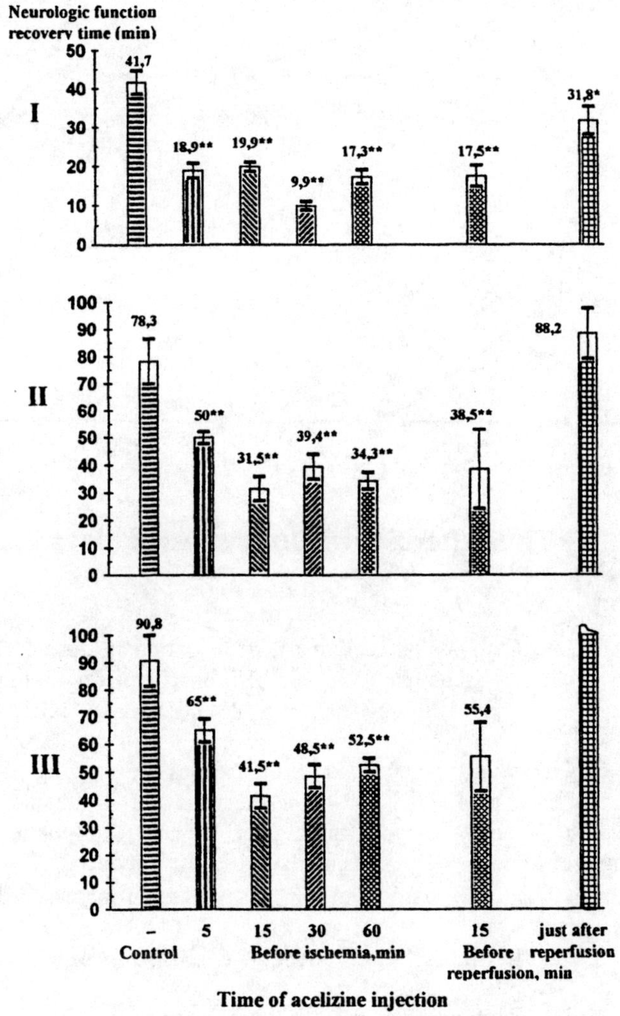

* -P <0.05; ** -P < 0.01, versus control. Acelizine was injected i. m. 150 mg ASA/kg.

Fig. 117: Effect of different time of acelizine injection on rat survival rate after 25 min of total cerebral ischemia and reperfusion (%)

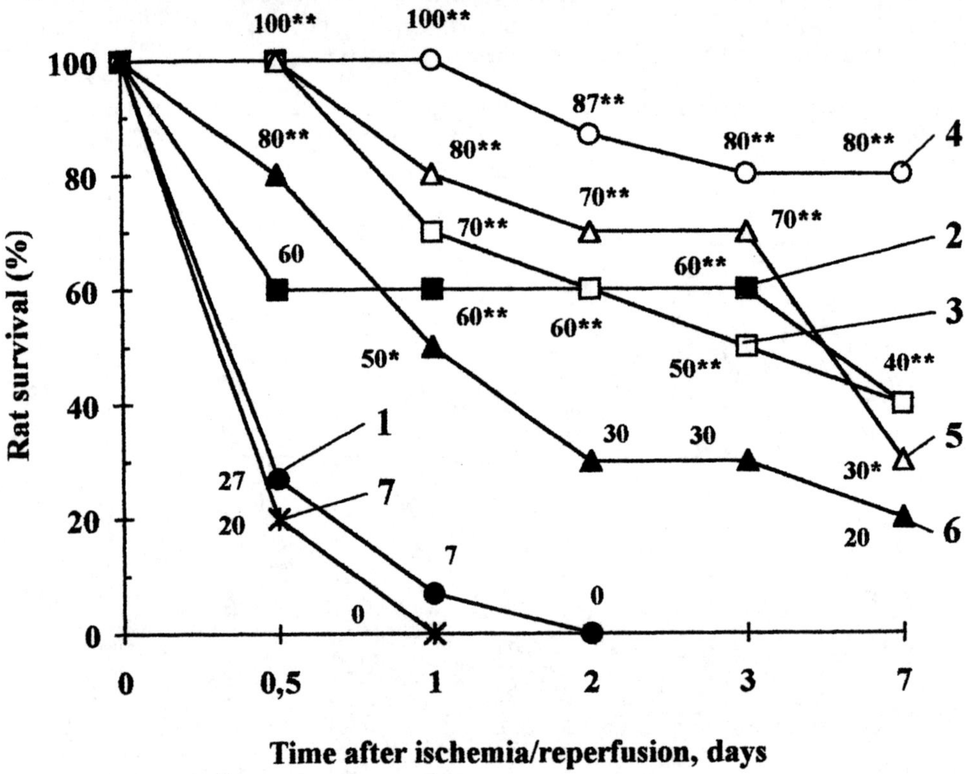

1 — without acelizine (control); 2 – 7 — acelizine i. m. injection in dose 150 mg/kg: 2. 5 min before ischemia; 3. 15 min before ischemia; 4. 30 min before ischemia; 5. 60 min before ischemia; 6. 15 min before reperfusion; 7. just after reperfusion. * -P <0.05; ** -P < 0.01, versus control.

Acelizine injected at experimentally determined optimal doses (150 mg/kg 30 min before ischemia) also promoted the recovery of EEG parameters and higher nervous activity in surviving rats, as estimated from their ability to develop an active escape reflex in a V-maze [Tel'pukhov V.I., Bilenko M.V., Khokhlov A.V. and Komarov P.G.]; in gerbils, it significantly reduced the period of recovering spontaneous respiration (from 37.1 ± 4.0 to 16 ± 2.1 min, $p < 0.01$) and increased the survival rate (up to 50% vs. 0 without treatment).

Thus, as described above, the antiischemic effect of acelizine in our experiments had certain specific features.

First, its magnitude varied depending on the organ studied, reaching a peak in the case of total cerebral ischemia: acelizine was effective in a broad dose range (25 to 150 mg/kg) and upon injection during a fairly long period of time, 5 to 60 min before ischemia and, to a lesser extent, 15 min before reperfusion. The antiischemic effect on the heart and limb muscles was expressed only when acelizine was administered repeatedly (twice) or at high doses (200-300 mg/kg). It should be noted that aspirin administered orally and intravenously was also shown to have no effect of tissue viability and even to increase edema in experiments on the model of 6-h hind limb tourniquet ischemia and 18-h reperfusion [Braithwaite B.D., Petrik P.V., Moore W.S. et al.]. In the case of the liver, acelizine in none of the doses injected before ischemia or reperfusion had any protective effect.

Second, the protective effect developed only if acelizine was administered in the preischemic period. In experiments with cerebral ischemia the drug had the strongest effect when injected 30 min before the induction of ischemia; that is, ischemia was induced at the peak of acelizine concentration in circulating blood (as estimated from the maximal decrease of platelet aggregation) and, hence, at the peak of its activity as a cyclooxygenase inhibitor. When administered after the beginning of reperfusion, acelizine had only a weak effect or was ineffective. These data indicate that, to achieve a protective effect, the cyclooxygenase pathway of arachidonic acid metabolism in blood cells accumulating in an ischemic organ and in cells of the organ proper should be inhibited before or during ischemia; this inhibition in reperfusion period is inefficient.

Third, acelizine had a short-term and moderate effect when added to saline used for preischemic perfusion of the isolated heart. In this case, protective properties of acelizine cannot be attributed by its effect on platelet aggregation. A possible explanation is that acelizine directly inhibits the cyclooxygenase pathway of arachidonic acid metabolism in cardiomyocytes and decreases their production of cyclic endoperoxides (PGG_2 and PGH_2) which initiate free radical reactions (see Chapter 4). Additional evidence for this explanation is that acelizine appears to reduce MDA content in platelets during their thrombin-induced aggregation [Fig. 111].

This mechanism of antiischemic action can explain the fact that the effects of acelizine depends on the type of ischemic organ and is more pronounced during ischemia of the brain: the cyclooxygenase pathway of arachidonic acid metabolism prevails in brain cells, and the rate of phospholipid hydrolysis and the activity of cyclooxygenase in them additionally increase upon transition from aerobic conditions to hypoxia and anoxia (ischemia) (see Chapter 4).

Thus, cyclooxygenase inhibitors, especially water-soluble, can be successfully used for preventing ischemia. Their effect is apparently explained (at least partly) by prevention or inhibition of free-radical processes associated with arachidonic acid metabolism in the organ.

The group of ***lipoxygenase inhibitors*** comprises chemical compounds differing in their activity. Some of them inhibit both lipoxygenase and cyclooxygenase—the dual inhibitor BW 755c, or 3-amino-1[*n*-(trifluoromethyl)]-2-pyrazoline; phenidone; FP-α-755c; AHR-6293; some selectively inhibit lipoxygenase only (iloprost, an PgI_2 analogue; nordihydroquaiaretic acid; NDGA); and some are specific inhibitors of 5-lipoxygenase activity. The latter include Zileuton; MK-886; nafazatrom, or 1-[2-α-(naphthyloxy)ethyl-3-methyl-2-pyrazolin-5-one; REV-5901, or α–pentyl-3-(2-quinolinylmethoxy)benzene-methanol; SK&F 86002, or 5-(4-pyridyl)-6-(4-fluorophenyl)-2,3-dihydroimidazo-(2,1-b) thiazole; A-861; diethylcarbamazin; WY-50295; troinethomine; etc. In addition, there are synthetic selective antagonists of C4/D4 leukotriene receptors, which have been tested experimentally (MK-571, FPL-55712, SC-41930).

Some inhibitors of the lipoxygenase pathway of arachidonic acid metabolism additionally act as antioxidants (e. g., NDGA), whereas others lack such properties (SK&F 86002, REV-5901). The antiischemic effect of lipoxygenase inhibitors is largely due to their opportunity to suppress the oxidative metabolic activity of neutrophils, their adhesiveness and transendothelial migration, and LTB_4 production, thus reducing the role of PMN in ischemic/reperfusion injury; this also applies to inhibitors that lack any antioxidant activity [Shappell S.B., Taylor, A.A., Hughes H. et al.]. The effect of these compounds was tested on the models of cardiac ischemia and reperfusion in dogs (BW 755c, iloprost, REV-590l, nafazatrom) [Bednar M., Smith B., Pinto A. and Mullane K.M.; Maulik M., Seth S.D., Maulik S.K. and Manchanda S.C.; Mullane K.M., Read N., Salmon J.A. and Moncada S.], rats (FPL-55712) [Sheker A. and Singh M.], and rabbits (nafazatrom, FPL-55712) [Fiedler V.B.; Maclouf J., Sala A., Rossoni G. et al.]; under ischemia of canine small intestine (Zileuton, A 64077) [Mangino M.J., Mangino J.E., Kotadia B. and Sielczak M.].

There is still an open question concerning the antiischemic effect of substances inhibiting different pathways of arachidonic acid metabolism: is it due not only to their properties as anticoagulant and antileukocyte agents, but also to their inhibitory influence on ROS production and lipid peroxidation? As follows from the data discussed above, the answer is positive. This is additionally confirmed by the data showing that at least some oxygenase inhibitors, such as BW 755c, AA-861, nafazatrom, dipyridamol, and propyl gallate, exerted a cardioprotective effect by directly preventing phospholipid peroxidation in myocardial membranes in a concentration-dependent manner, and, conversely, cyclooxygenase inhibitors ineffective against myocardial infarction (aspirin, indomethacin, naproxen, and sulfinpyrazone) proved to lack antioxidant properties.

11.4.2 Inhibitors of Xanthine Oxidase

As many authors regard the xanthine/xanthine oxidase reaction as a major source of ROS initiating ischemic organ damage during reperfusion (see Chapter 10), xanthine oxidase (XO) inhibitors—allopurinol, oxipurinol, tungsten, amflutizol, BOF-4272, etc. —have been widely used as antiischemic agents in experimental and clinical practice. This is especially true of allopurinol and its metabolite, oxipurinol.

All the data obtained with the use of XO inhibitors may be conventionally divided into three groups. In the first, the antiischemic effect of these agents was observed against the background with simultaneous reduction (or complete suppression) of XO activity, ROS production, and (or) the accumulation of LPO products. These were experiments on total ischemia and reperfusion of rat small intestine [Siems W., Kowalewski J., Werner A. et al.], liver [Cohen P.J.], brain [O'Regan M.H., Smith-Barbour M., Perkins L.M. et al.], heart [Thompson-Gorman S.L. and Zweier J.L.], and skeletal muscles [Punch J., Rees R., Cashmer B. et al.; McCutchan H.J., Schwappach J.R., Enquist E.G. and Walden D.L.]; rabbit kidneys [Haraldsson G., Nilsson U., Bratell S. et al.]; dog liver [Nakano M., Sugano M., Terasaki M. et al.]; endothelial cell culture derived from the newborn piglet brain [Beetsch J.W., Park T.S., Dugan L.L. et al.]; and on hypoxia and reoxygenation of rat heart [Miwa-Nishimura N., Kanaide H., Abe S. and Nakamura M.].

In studies included in the second group, XO inhibitors exerted an antiischemic effect without preventing ROS production and the accumulation of LPO products. Such observations were made in experiments with the ischemic rat heart [Pisarenko O.I., Lakomkin V.L., Studneva I.M. et al.] and limbs [Ferrari R.P., Battiston B., Brunelli G. et al.]. We can attribute to the same group the data on the positive antiischemic effect of XO inhibitors during transplantations of the heart–lung complex in pigs [Qayumi A.K., Jamieson W.R., Godin D.V. et al.] and surgery on the "dry" heart in human patients [Bochenek A., Religa Z., Spyt T.J. et al.; Canada A.T., Stein K., Martel D. and Watkins W.D.; Johnson W.D., Kayser K.L., Brenowitz J.B. and Saedi S.F.], although XO activity in both humans and pigs was shown to be very weak or absent (see Chapter 10).

The authors of studies included in the third group described the antiischemic effect of XO inhibitors (mainly allopurinol) but did not investigated the processes of ROS formation and LPO product accumulation. The corresponding experiments were performed on the models of cardiac ischemia in dogs [Werns S.W., Grum C.M., Ventura A et al. and Lucochesi B.R.] and rabbits [LoBalsamo L., Bergsland J., Lajos P. and Feldman M.J.], partial cerebral ischemia in rat pups [Palmer C., Vannucci R.C. and Towfighi J.], cerebral ischemia in newborn lambs [Shadid M., Moison R., Steendijk P. et al.], and coronary occlusion in dogs [Werns S.W., Grum C.M., Ventura A et al. and Lucochesi B.R.].

On the other hand, there are data on the absence of any effect of XO inhibitors on ischemic and reperfusion damage to canine heart [Werns S.W., Grum C.M., Ventura A et al. and Lucochesi B.R.], gerbil brain [Arai T., Mori H., Ishii H. et al.], rabbit skeletal muscle [Hickey M.J., Knight K.R., Lepore D.A. et al.], and isolated proximal tubules of rat kidney [Doctor R.B. and Mandel L.J.].

The pronounced antiischemic effect of XO inhibitors was usually accompanied by accelerated recovery of ADN and ATP synthesis in the reperfusion period [Miwa-Nishimura N., Kanaide H., Abe S. and Nakamura M.; O'Regan M.H., Smith-Barbour M., Perkins L.M. et al.; Nakano M., Sugano M., Terasaki M. et al.; Pisarenko O.I., Lakomkin V.L., Studneva I.M. et al.; Siems W., Kowalewski J., Werner A. et al.], although there are exceptions to this rule. Most authors observed a strong antiischemic effect upon administering XO inhibitors prior to the onset of ischemia [Hearse D.J., Manning AS, Downey J.M. et al.; Qayumi A.K., Jamieson W.R., Godin D.V. et al.]; in some cases, however, this effect was stronger when the inhibitor (allopurinol) was administered before or soon after the onset of reperfusion [Haraldsson G., Nilsson U., Bratell S. et al.; LoBalsamo L., Bergsland J., Lajos P. and Feldman M.J.]

In experiments with cerebral ischemia, allopurinol alone or combined with indomethacin and iron chelators produced the same effect [Shadid M., Moison R., Steendijk P. et al.]; however, its combination with SOD and catalase during coronary occlusion or hepatic ischemia in rats [Crome R. and Manning A.S.; Minor T., Isselhard W., Yamamoto Y. et al.; Nathan C.F.] proved to be more effective than allopurinol alone.

It is quite obviouse that the authors of studies included in the first group attribute the protective effect of XO inhibitors to their suppression of XO activity and acknowledge the important role of the xanthine/xanthine oxidase reaction in ROS production and that of ROS themselves in the pathogenesis of ischemic and reperfusion injuries. In the second group, the antiischemic effect of XO inhibitors (in particular, allopurinol) is explained by their well known involvement in the resynthesis of purine bases, restoration of energy supply, electron transport, and, at high doses, by their fairly strong scavenger function [Das D.K., Engelman R.M., Clement R. et al.; Klein A.S., Joh J.W., Rangan U. et al.; Peterson D.A., Kelly B. and Gerrard J.M.]. Recent studies showed that allopurinol and oxipurinol increase sensitivity of the rat myocardium to Ca^{2+}, i.e., have an inotropic effect [Perez N.G., Gao W.D. and Marban E.].

11.4.3 Inhibitors and Activators of Nitric Acid Synthase

Activation of nitric oxide synthase (NOS), especially of its inducible form (iNOS), and increased production of $NO^{\bullet}$ and its toxic derivatives during ischemia (peroxynitrite, peroxynitrous acid, etc.; see Chapter 10) stimulated some authors to test antiischemic properties of iNOS inhibitors, both nonselective—N-omega-nitro-L-arginine methyl ester (L-NAME), N-omega-nitro-L-arginine (L-NA), N(G)-nitro-L-arginine (L-NOARG), N(G)-monomethyl-L-arginine (L-N-MMA), 1-(2-trifluoromethylphenyl) imidazole (TRIM), 3-bromo-7-nitroindazole—and selective, such as 7-nitroindazole (7-NI), aminoquanidine and S-methylisothiourea sulfate (SMT) et al.

These studies produced contradictory results. In one group of experiments, L-NAME, aminoquanidine, 7-NI or SMT, reduced the contents of $NO^{\bullet}$ and peroxynitrite and mitigated injuries of rat heart, intestine and brain after their ischemia and reperfusion [Coert B.A., Anderson R.E. and Meyer F.B.; Cuzzocrea S., Zingarelli B. and Caputi A.P.; Forman L.J., Liu P., Nagele R.G. et al.; Nishikawa T., Kirsch J.R., Koehler R.C. et al.; Wang D., Yang X.P., Liu Y.H. et al], L-NAME decreased TRABS and CD content after MCA occlusion in rat brain [Gumuslu S., Serteser M., Ozben T., et al.] and decreased Ca^{2+} accumulation in anoxic endothelial cells from capillaries of the rat brain [Ikeda K., Nagashima T., Wu S. et al.]. 7-NI administered to rats immediately after the onset of 20-min total cerebral ischemia and one hour later or immediately after the onset of partial ischemia decreased morphologically registered ischemic injuries [Nanri K., Montecot C., Springhetti V. et al.; Escott K.J., Beech J.S., Haga K.K. et al.], and TRIM had a protective effect on the rat brain when administered either 5 or 90 min after the occlusion of brain vessels [Escott K.J., Beech J.S., Haga K.K. et al.]. Combinations of NOS inhibitors with SOD or ROS scavengers proved to be more effective than each of these agents used separately [see Beckman J.S. and Crow J.P.].

According to other group of authors, NOS inhibitors either had no protective effect, or even exerted adverse influence on organs, impairing blood flow, function and metabolism in the reperfused rat heart [Walker P.M., Lindsay T.F., Labbe R. et al.], skeletal muscle [Joneschild E.S., Chen L.E., Seaber A.V. et al.] and liver [Pannen B.H., Al-Adili F., Bauer M., et al.] and reducing survival rate and cerebral blood flow in rats with cerebral ischemia [Beckman J.S. and Crow J.P.; Prado R., Watson B.D. and Wester P.], whereas their protective effect (if any) was manifested only at very low concentrations and in cases of partial ischemia [du Toit E.F., McCarthy J., Miyashiro J. et al.]. In addition to dosage, the effect (or its absence) depended on the time when a NOS inhibitor was administered: this effect could be adverse upon its administration prior to the onset of ischemia [Beckman J.S. and Crow J.P.] and favorable upon administration during ischemia or after reperfusion [Escott K.J., Beech J.S., Haga K.K. et al.; Segawa D., Hatori N., Yoshizu H. et al.].

Finally, a group of authors observed that the protective effect was exerted by NOS stimulators or $NO^{\bullet}$ donors, rather than by NOS inhibitors. These were L-arginine during heart, kidney and liver ischemia in rats [Agullo L., Garcia-Dorado D., Inserte J. et al.; Shoskes D.A., Xie Y. and Gonzalez-Cadavid N.F.; Cottart C.H., Do L., Blanc M.C. et al.; Meldrum D.G., Stephenson L.L. and Zamboni W.A.], and liver ischemia in dogs [Shimamura T., Zhu Y., Zhang S. et al.]; L-arginine+SOD during canine heart ischemia [Mori E., Haramaki N., Ikeda H. and Imaizumi T.]; L-arginine + aspirin during cardiac ischemia in rats [Wanna F.S., Obayashi D.Y., Young J.N.and DeCampil W.M.]; sodium nitroprusside during ischemia and reperfusion of skeletal muscles [Wang W.Z., Anderson G., Fleming J.T. et al.] and the heart [du Toit E.F., McCarthy J., Miyashiro J. et al.] in rats, and SIN-1 after limb ischemia in rabbits [Johnson D., Freischlag J.A., Lesniak R. et al.]. Note, however, that SIN-1 administered immediately after reperfusion of the ischemic rat heart aggravated reperfusion injuries [Ma X.L., Lopez B.L., Liu G.L. et al.]. $NO^{\bullet}$ donors produced the strongest effect when administered at very low doses in the early reperfusion period, i.e., against the background of most pronounced hemodynamic disturbances in the organ, such as the no-reflow phenomenon, angiospasm, blood cell adhesion, etc. (see Chapter 1).

The factors responsible for such a contradiction in experimental results, which was also observed in the practice of cardiac surgery (see Noji S., Tashiro K., Koyama E. et al.;) can be rather explained by cytotoxic effect of high and protective effect of low NO concentration (see Chapter 10) and by difficulties in determining the optimal drug dose (concentration). Improvment in the field of antiischemic therapy will be possibly provided by using selective inhibitors (against $_i$NOS and $_{nc}$NOS), not affecting $_e$NOS activity. There are some experimental data, confirming this hypothesis [Wildhirt S.M., Suzuki H., Horstman D. et al.; Tan S., Yokoyama Y., Dickens E. et al.].

11.5 Other Antiischemic and Antiradical Agents

In addition to the agents described above, there are other substances which, along with their basic specific activity, can inhibit free-radical reactions at different phases owing to their antioxidant (antiradical) or membrane-stabilizing properties.

11.5.1 Membrane-Stabilizing Agents

The protective effect of membrane-stabilizing agents is explained by the fact that any loosening of membranes—because of changes in their composition, balance of charged molecules, the ratio of hydrophilic and hydrophobic components, etc. —is a factor promoting LPO intensification, and these agents counteract this process. The ability to stabilize membranes is characteristic of tocopherols, steroids, histidine-containing peptides, and other compounds that typically have a heterocyclic structure and are membranotropic. Being incorporated into the membrane, they make it more rigid and, hence, less vulnerable to damage inflicted by LPO. Membrane-protecting properties probably provide for the antiischemic effect of glucocorticosteroids [Kanellopoulos G. K., Kato H., Wu Y. et al.; Ekert P., MacLusky N., Luo X.P., et al.], phosphocreatine [Konorev A. E., Saks V. A., Sharov V. G. and Makhotina L. A.; Saks V.A., Dzhaliashvili I.V., Konorev E.A. and Strumia E.; Mogilevskii G.M., Sharov V.G., Severin V.V. and Semenovskii M.L.], liposomes prepared from phospholipids [nov A.V., Khromov A.S., Zhukova A.V. et al.; Downey J.M., Hearse D.J. and Yellon D.M.], and a complex of phospholipids isolated from marine organisms [Datchenko Z.M., Donchenko G.V., Shakhman E.V. et al.].

11.5.2 Barbiturates

Some barbiturates also have antioxidant properties, as demonstrated on the system of Fe^{2+}–ascorbate-induced LPO in brain homogenates. Thiopental proved to be the most active, whereas phenobarbital and pentobarbital had no antioxidant activity, although all these barbiturates had a protective effect during ischemia of the brain [Flamm E. S., Demopoulos H. B., Seligman M. L. et al.; Smith D.S., Rehncrona S. and Siesjö. Bo.K.].

11.5.3 Lidocaine

Lidocaine also has antioxidant activity, as it protected myocardium and brain from ischemic/reperfusion injury and prevented the accumulation of LPO products in them [Lantos J., RÖth E. and Temes G.; Lesnefsky E.J., Van Benthuysen K.M., McMurtry I.F. et al.; Okamura I., Sunamori M.and Suzuki A; Rasool N., Faroqui M. and Rubinstein E.H.].

11.5.4 Cardiovascular Agents

Many cardiovascular agents, in addition to their basic activity, also exert a direct or indirect antiradical effect. Thus, β-adrenoblockers demonstrated the ability to inhibit Fe–ADP-induced LPO in sarcoplasmic membranes *in vitro,* which decreased in the series: propranolol > pindolol > metoprolol > atenolol > sotalol, i.e., proportionally to their hydrophobicity. The antiradical effect of these compounds is apparently explained by their ability to stabilize membranes, especially in the case of propranolol [Weglicki W.B., Mak I.T. and Simic M.G.].

11.5.5 Calcium Channel Blockers

Calcium channel blockers (Ca^{2+} antagonists), such as bepridil, prenylamine, flunarizine, cinnarizine, and others, inhibited *in vitro* ROS-induced LPO in lysosomal membranes [Mak J.T. and Weglicki W.B.]; nifedipine, verapamil, and diltiazem inhibited LPO in sarcolemmal membranes [Mak J.T. and Weglicki W.B.]; cinnarizine, flunarizine, nicardipine, verapamil, and diltiazem acted as cytoprotectors on isolated cardiomyocytes incubated with a ROS-producing system [Ver Donck L., Van Reempts J., Vandeplasche G. and Bogers M.A.]; the antiischemic effect of these agents has been repeatedly demonstrated on the models of ischemia and reperfusion almost all organs. As hydrophobic Ca^{2+} antagonists are aromatic compounds, their antioxidant effect is apparently explained by their concentration in membranes and lipophilic interaction with them [Weglicki W.B., Mak I.T. and Simic M.G.].

11.5.6 Angiotensin-Converting Enzyme Inhibitors

Angiotensin-converting enzyme inhibitors (ACE-I) (captopril, epicaptopril, enalapril, imidaprilat, lozartan, zofenoprilat, cilazaprilat) proved to act as $^{\bullet}OH$ scavengers in experiments on incubating *in vitro* different membrane preparations and cells with a ROS-generating system [Mak J.T. and Weglicki W.B.; Weglicki W.B., Mak I.T. and Simic M.G.] and demonstrated an antioxidant effect when administered to patients with diabetes [Bain S. C., Le Guen C. A., Lunec J. and Barnett A. H.]. These properties are probably determined by the scavenging effect of the sulfhydryl moiety [Weglicki W.B., Mak I.T. and Simic M.G.] present in molecules of these compounds. ACE-I (captopril) also provided for the protection of SH-groups in subcellular sarcoplasmatic and mitochondrial membranes in ischemic canine heart [Yanagishita T., Tomita M., Itoh S.et al.]; another ACE-I (cilazaprilat prevented reperfusion injury in the quinea pig heart due to inhibition of postcapillary leukocyte adhesion [Kupatt C., Habazettl H., Zahler S.et al.]; cilazaprilat or imidaprilat direcrly protected the primary culture of cardiac myocytes, as well as ischemic canine myocardium *in vivo*, against hypoxia [Matoba S., Tatsumi T., Keira N.et al.].

11.5.7 Substances Extracted from Herbs

Substances extracted from herbs demonstrated antioxidant and, in some cases, antiischemic effects. Honokiol (extracted from *Magnolia officinalis*) inhibited MDA formation during Fe-induced LPO in rat liver mitochondria *in vitro,* being 550–680 times as effective as α-tocopherol; *in vivo,* this agent protected the respiratory function of mitochondria in the rat liver exposed to 60-min ischemia and 60-min reperfusion [Chiu J.H., Ho C.T., Wei Y.H. et al.]. Antioxidant properties also have a numerous other plant extracts [Bolshakova I.V., Lozovskaya E.L. and Sapeshinskii I.I.].

11.5.8 Agents Preventing DNA Damage and Apoptosis

Agents preventing DNA damage and apoptosis are largely similar to membrane protectors. Along with substances affecting leukocyte–endothelium interactions, decreasing of PNM adhesion, PAF and cytokin's antagonists P-selective inhibitors, growth factors, and vasodilators, they were briefly considered in corresponding sections of the book and will not be discussed in detail in this chapter.

The described effects are not always associated with basic (specific) functions of corresponding agents; on the contrary, they are mostly independent of them, although Ca^{2+}-blocking and vasodilating activities may also affect certain links of such a complex process as initiation and inhibition of free-radical reactions.

11.6 CONCLUSIONS

Thus, the protective antiischemic effect of various chemical and biological preparations capable of inhibiting LPO processes at different stages has been demonstrated on models of ischemia of virtually all organs.

The agents studied in more detail include primarily natural and synthetic substances with direct antiradical activity, although the antiischemic effect has also been demonstrated for nonenzymatic ROS scavengers, antioxidant enzymes, membrane stabilizers, inhibitors of ROS-generating systems, and chelators of transition metals. In addition, antioxidant properties *in vitro* have been revealed to date in most of phospholipase inhibitors, calcium channel blockers, barbiturates, α– and β-adrenoblockers, other cardiotropic drugs, steroid and nonsteroid antiinflammatory drugs, analgesics, and many other agents. Such a diversity is apparently explained by intricate interrelations of mechanisms responsible for membrane damage *in vitro* and, especially, *in vivo*, and the possibility of inhibiting the corresponding processes by acting on their different links.

Some aspects of using LPO inhibitors during ischemia deserve special attention.

(1) Natural fat-soluble antiradical substances (α–tocopherol, CoQ_{10}, other vitamins) are less toxic but also less efficient than synthetic drugs. Synthesis of their analogues is an important field of modern medical and biological chemistry.

(2) Fat-soluble drugs are more efficient than water-soluble due to their more rapid and directed incorporation into biological membranes. Hence the importance of preparing fat-soluble antioxidants in a stable liposomal form, which offers possibilities for administering antioxidant drugs parenterally and adding them to perfusion or cardioplegic solutions and blood.

(3) The mechanism underlying the antioxidant effect of corresponding drugs is rather intricate and cannot be simply confined to their main—antiradical—activity.

As shown above, many synthetic antioxidants (e. g., BHT, CoQ_{10}), being expended earlier than natural antioxidants present in tissues, help to maintain the level of the latter and, hence, tissue antioxidant activity during ischemia. Another well-known feature of antioxidants (α-tocopherol, sodium selenite, oxypyridine-6, K-phenosane, diludin) is stimulation of regenerative processes, which should promote repair of membranes damaged during ischemia and reperfusion.

(4) Virtually all LPO inhibitors are much more effective when administered before than after ischemia. This is evidence for an urgent need to develop additional methods for controlling LPO during the postischemic period. One of them is inhalation of hypoxic gas mixtures in the early postischemic period or the method of controlled reperfusion (see Chapter 6).

(5) The most effective approach is to administer not one but several LPO inhibitors because the process of LPO involves many stages, and each inhibitor mainly affects only one of them.

Further efforts are necessary to elaborate such combinations with due account for the additive effect of the constituent agents, different times of their administration, and differentiation of their action with respect to LPO stages as well as to avoid possible toxic effect. Unfortunetly, the most of studies with antioxidants were performed to prove the free-radical mechanism of ischemic and reperfusion damage. It is a matter of particular urgency to accelerate the work on preparing clinical forms of antioxidant drugs, as the list of antioxidant agents accepted for clinical use is now rather limited.

It should also be emphasized that theoretical and experimental studies providing evidence for the involvement of LPO in the mechanism of ischemic and reperfusion injuries in organs proved to be rather fruitful and opened up an entirely new field of effective antiischemic therapy based on the use of LPO inhibitors. This field is being actively explored, and the spectrum of antiischemic drugs is expected to expand from one year to another.

The data discussed in this chapter show that the integrated approach to treatment of ischemic organs should apparently involve antioxidant therapy performed against the background of created conditions of hypoxemia, hypocalcemia, leukopenia, and hyperosmia. Further study should help to test its efficiency and to optimize all its components.

REFERENCES*

Aalto T. K. and Raivio K. O. Adenine nucleotide depletion from endothelial cells exposed to xanthine oxidase. *Am. J. Physiol.,* 1990, v. 259, No. 6, Pt. 1, pp. C883-C888.

Abakumova O. Yu., Kotaev A. Yu., Karagyulyan S. R. et al. [Effect of cytoplasmic proliferation stimulating factors and of partial hepatectomy on DNA synthesis in liver tissue of rats poisoned with CCl_4 for a long time]. (Russian). *Vopr. Med. Khim.*, 1985, v. 31, No. 5, pp. 95-97.

Abakumova O. Yu., Kutsenko M. G., Karagyulyan S. P. et al. [Stimulation of DNA and RNA synthesis in partially hepatectomized rats by means of the liver cytosole fractions as well as the preparation of medium, obtained after cultication of rat liver explants]. (Russian). *Vopr. Med. Khim.*, 1989, v. 35, No. 1, pp. 69-74.

Abd-Elfattah A. S., Maddox R. P., Jessen M. E. et al. Role of nucleoside transport and purine release in a rabbit model of miocardial stunning. *Mol. Cell. Biochem.*, 1998, v. 180, No. 1-2, pp. 145-151.

Abello P. A., Fidler S. A., Bulkley G. B., et al. Antioxidants modulate induction of programmed endothelial cell death (apoptosis) by endotoxin. *Arch. Surg.*, 1994, v. 129, No. 2, pp. 134-40.

Adachi T., Fukushima T., Usami Y et al. Binding of human xanthine oxidase to sulphated glycosaminoglycans on the endothelial cell surface. *Biochem. J.,* 1993, v. 289, pp. 523-527.

Adamson G. M. and Billings R. Tumor necrosis factor induced oxidative stress in isolated mouse hepatocytes. *Arch. Biochem. Biophys.*, 1992, v. 294, pp. 223-229.

Afanasyev I. B. [Oxygen radicals in biological processes. Review]. (Russian). *Khim. Farm. Zh.*, 1985, v. 19, No. 1, pp. 11-23.

Afanasyev S. A., Laptev B. I., Prokopyeva V. D. and Larionov N. P. [Study of aminazine protective action at ischemia and reperfusion of isolated heart]. (Russian). *Biol. Nauki*, 1986, No. 3, pp. 25-29.

Ageta M., Yamasaki S., Ueki K., et al. Lipid peroxide in myocardial infarction. *Jpn. Circ. J.*, 1982, v. 46, p. 903.

Agullo L., Garcia-Dorado D., Inserte J. et al. L-arginine limits myocardial cell death secondary to hypoxia-reoxygenation by a cGMP-dependent mechanism. *Am. J. Physiol.* 1999, v. 276, No. 5, Pt 2, pp. H1574-H1580.

Agureev A. P., Altukhov N. D., Mokhova E. N. and Savelyev I. A. [Activation of the external pathway of oxidation of NADH in the mitochondria with decreasing pH]. (Russian). *Biokhimiya*, 1981, v. 46, No. 11, pp. 1945-1956. English transl. see[3*], pp. 1531-1540.

Aharony D., Smith S. B., Smith E. F. et al. Effects of arachidonic acid hydroperoxides on vascular and nonvascular smooth muscle. *Prostaglandins and Medicine*. N. Y., 1981, v. 7, pp. 527-535.

---

* Notice;

Translation from Russian see in:

1 - Bulletin of Experimental Biology and Medicine*, Plenum Publishing Corp., Consultants Bureau, N. Y.

2 - Biophysics*. Pergamon Press Inc. Oxford, England.

3 - Biochemistry*. Plenum Publishing Corp. Consultants Bureau, N. Y.

4 - Pharmaceutical Chemistry* J. Plenum Publishing Corp. Consultants Bureau, N. Y.

5 - Ukranian Biochemistry*. Allerton Press Inc., N. Y.

6 - Doklady Biochemistry*. Plenum Publishing Corp. Consultants Bureau, N. Y.

7 - Doklady Biological Sciences*. Plenum Publishing Corp. Consultants Bureau, N. Y.

8 - Doklady Biophysics*. Plenum Publishing Corp. Consultants Bureau, N. Y.

Everywhere year, volume and number correspond to original. The pages are mentioned in References.

Ahnja B. S. and Kiran U. Correlation between superoxide dismutase, and catalase in red blood cells of different animals. *Curr. Sci.*, 1978, v. 47, No. 15, pp. 544-545.

Alanen K., Lipasti J. and Nevalainen T. ATP-content and ischaemic contracture in isolated rat heart. *Pathol. Res. Pract.* 1982, v. 176, No. 1, pp. 3-4.

Albina J. E., Cui S, Mateo R. B. et al. Nitric oxide-mediated apoptosis in murine peritoneal macrophages. *J. Immunol.,* 1993, v. 150, No. 11, pp. 5080-5085.

Alessenko A. V. and Burlakova E. B. [The connection of lipid antioxidative activity with DNA synthesis in regenerating liver of mice]. (Russian). *Dokl. Akad. Nauk SSSR*, 1972, v. 207, No. 6, pp. 1471-1475.

Alessenko A. V., Bilenko M. V. and Shelenkova L. N. [Use of natural and synthetic antioxidants to prevent ischemic injuries in organs]. (Russian). In: [*Free radical oxidation of lipids in norma and pathology*]. Moscow, Nauka, 1976, pp. 4-5.

Alessenko A. V., Bilenko M. V., Burlakova E. V. and Dudnik L. B. Correlation between lipid peroxidation and content of phospholipids and free fatty acids in subcellular fractions of ischemic liver. *33rd Intern. Conference on the Biochemistry of Lipids*. Abstract Lyon-Villeurbanne, France, 1992, No. 43.

Alessenko A. V., Bilenko M. V., Burlakova E. B. et al. [Study of peroxidase levels and antioxidative activity of lipids in ischaemic and preserved kidneys]. (Russian). *Vest. Akad. Med. Nauk SSSR*, 1976, No. 8, pp. 61-67.

Alessenko A. V., Bilenko M. V., Dudnik L. B. et al. Some biochemical aspects of lipid peroxidation accelerated in acute organ ischemia. *VII Intern. Symp. Med. Chem. Abstracts*. Malaga, 1980, p. 117.

Alessenko A. V., Burlakova E. B. and Pantaz E. A. [Effect of sphingomyelin on the RNA polymerase activity in cell nuclei of normal and regenerating rat liver]. (Russian). *Biokhimiya*, 1984, v. 49, No. 4, pp. 621-628. English trans. see[3*], pp. 529-535.

Alessenko A. V., Pantaz E. A., Pushkareva M. Ya. et al. [Changes in the level of endogenous sphingosine in cell nuclei of regenerating rat liver]. (Russian). *Biokhimiya*, 1993, v. 58, No. 3, pp. 461-470. English trans. see[3*], pp. 298-305.

Alessenko A. V., Sokolova I. S., Kukushkina G. V. et al. [Stimulating effect of dibunol on the biosynthesis of macromolecules]. (Russian). *Dokl. Akad. Nauk SSSR*, 1980, v. 254, No. 6, pp. 1472-1476. English transl. see[6*], pp. 307-310.

Alexander B. The role of nitric oxide in hepatic metabolism (Review). *Nutrition*, 1998, v. 13, No. 4, pp. 376-390.

Alexandrov P. N., Shinkarenko U. S., Khugaeva U. K. and Morozov S. E. [Microvascular responses to short-term ischemia]. (Russian). *Fiziol. Zh. Akad. Nauk SSSR*, 1986, v. 72, No. 9, pp. 1237-1243.

Alferov Yu. M. and Berezovskiy V. A. [Effect of medium ion content on the mass transfer of oxygen across the electrolite-phospholipid film interface]. (Russian). In: Berezovskiy V. A., ed., *Lung surfactants in norma and pathology*. Kiev, Naukova Dumka, 1983, pp. 78-85.

Allen D. R., Wallis G. L. and McCay P. B. Catechol adrenergic agents enhance hydroxyl radical generation in xanthine oxidase systems containing ferrition: implications for ischemia/reperfusion. *Arch. Biochem. Biophys.*, 1994, v. 315, No. 2, pp. 235-243.

Allison T. B. and Holsinger V. W. Myocardial metabolism and reginal myocardial blood flow in the canine left ventricle following twenty minutes of circumflex artery occlusion and reperfusion. *J. Molec. Cell. Cardiol.*, 1983, v. 15, pp. 151-161.

Ambrosio G., Flaherty J. T., Duilio C. et al. Oxygen radicals generated at reflow induce peroxidation of membrane lipids in reperfused hearts. *J. Clin. Invest.*, 1991, v. 87, No. 6, pp. 2056-2066.

Ames A., Wright R. Z., Kowada M. et al. Cerebral Ischemia. II. The no-reflow phenomenon. *Am. J. Pathol*, 1968, v. 52, pp. 437-453.

Anderson R. J., Cambria R., Kerr J. and Hobson R. W. 2nd. Sustained benefit of temporary limited reperfusion in skeletal muscle following ischemia. *J. Surg. Res.*, 1990, v. 49, No. 3, pp. 271-275.

Anderson W. R., Zieve L. and Lindblad S. Ultrastructural study of hepatic regeneration following one lobe, 2 lobe and subtotal hepatectomy in the rat. *Expt. Pathol.*, 1990, v. 38, No. 1, pp. 61-72.

Andreoli S. P. and McAteer J. A. Reactive oxygen molecule-mediated injury in endothelial and renal tubular epithelial cells in vitro. *Kidney Internat.*, 1990, v. 38, No. 5, pp. 785-794.

Andreoni K. A., Kazui M., Cameron D. E. et al. Ethane : a marker of lipid peroxidation during cardiopulmonary bypass in humans. *Free Radic. Biol. Med.,* 1999, v. 26, No. 3-4, pp. 439-445.

Aniya Y. and Naito A. Oxidative stress-induced activation of microsomal glutathione S-transferase in isolated rat liver. *Biochem. Pharm.,* 1993 v. 45, No. 1, pp. 37-42.

Anundi T., Högberg J. and Stead A. H. Glutathione depletion in isolated hepatocytes: its relation to lipid peroxidation and cell damage. *Acta Pharmacol. Toxicol.*, 1979, v. 45, pp. 45-51.

Aoki Y., Nata M., Odaira T. and Sagisaka K. Suppression of ischemia-reperfusion injury by liposomal superoxide dismutase in rats subjected to tourniquet shock. *Intern. J. Leg. Med.*, 1992, v. 105, No. 1, pp. 5-9.

Aomine M. and Arita N. Pretreatment with coenzyme Q10 protects guinea pig ventricular muscle from hypoxia-induced deterioration of action potentials and contraction. *Gen. Pharmacol.* 1985, v. 16, pp. 91-96.

Appelbaum Y. J., Kuvin J., Borman J. B. et al. The protective role of neocuproine against cardiac damage in isolated perfused rat hearts. *Free Rad. Biol. Med.*, 1990, v. 8, pp. 133-143.

Arai T., Mori H., Ishii H. et al. Oxypurinol, a xanthine oxidase inhibitor and a superoxide scavenger, did not attenuate ischemic neuronal damage in gerbils. *Life Sci.*, 1998, v. 63, No. 7, pp. PL 107-PL 112.

Araki R., Tamura M. and Yamazaki J. The effect of intracellular oxygen concentration on lactate release pyridine nucleotide reduction and respiration rate in the rat cardiac tissue. *Circ. Res.* 1983, v. 53, No. 4, pp. 448-456.

Arai A. E., Pantely G. A., Anselone C. G. et al. Active downregulation of myocardial energy requirements during prolonged moderate ischemia in swine. *Circ. Res.*, 1991, v. 69, No. 6, pp. 1458-1469.

Archakov A. I. [*Microsomal oxidation*]. (Russian). Moscow, Nauka, 1975, 327 p.

Archakov A. I. [*Oxygenesis of biological membranes*]. (Russian). Moscow, Nauka, 1983, 55 p.

Archakov A. I. and Mokhosoev I. M. [Modification of proteins by active oxygen and their decomposition]. (Russian). *Biokhimiya*, 1989, v. 54, No. 2, pp. 179-186. English transl. see[3*], pp. 125-130.

Arduini A., Eddy L. and Hochstein P. The reduction of ferrylmyoglobin by ergothioneine: a new function for ergothioneine. *Arch. Brochem. Biophys.*, 1990, v. 281, pp. 41-43.

Aristarkhova S. A., Burlakova E. B. and Zaets T. L. [Lipid peroxidation in liver subcellular organelles in thermal burns]. (Russian). *Vopr. Med. Khim.*, 1983, v. 28, No. 4, pp. 102-106.

Arkhipenko Yu. V. and Shimkovich M. V. [Role of lipid peroxidation in regression of the hypertrophied heart]. (Russian). *Biull. Eksp. Biol. Med.*, 1989, v. 108, No. 11, pp. 556-558. English transl. see[1*], pp. 1566-1568.

Arkhipenko Yu. V., Bilenko M. V., Dobrina S. K. et al. [Damage to the sarcoplasmic reticulum of the skeletal muscles in ischemia: role of lipid peroxidation]. (Russian). *Biul. Eksp. Biol. Med.*, 1977, v. 83, No. 6, pp. 683-686. English transl. see[1*], pp. 810-813.

Arndt H., Kubes P. and Granger D. N. Involvement of neutrophils in ischemia-reperfusion injury in the small intestine. (Review). *Klinische Wochenschrift*, 1991, v. 69, No. 21-23, pp. 1056-1060.

Arnsdorf M. F. and Sawicki G. J. The effects of lysophosphatidylcholine, a toxic metabolite of ischemia, on the components of cardiac excitability in sheep Purkinje fibers. *Circ. Res.*, 1981, v. 49, pp. 16-30.

Arras M., Mohri M., Pich S. et al. Tumor necrosis factor-a (TNF-a) is produced in monocytes/macrophages in ischemic porcine myocardium. *J. Mol. Cell Cardiol.*, 1993, v. 25, Suppl 1., p. S. 86.

Arroyo C. M., Kramer J. H., Leiboff R. H. et al. Spin trapping of oxygen and carbon-centered free radicals in ischemic canine myocardium. *Free Rad. Biol. Med.*, 1987, v. 3, pp. 313-316.

Artamonov S. D. and Onishchenko N. A. [Energy balance of myocardium in norma and ischemia]. (Russian). In: Shumakov V. I. and Onishchenko N. A., eds. *Actual questions of organs conservation.* Moscow, 1982, pp. 4-21.

Arthur A. S., Fergus A. H., Lanzino G. et al. Systemic administration of the iron chelator deferiprone attenuates subarachnoid hemorrhage-induced cerebral vasospasm in the rabbit. *Neurosurg.*, 1997, v. 41, No. 6, pp. 1391-1392.

Asami O., Ihara I., Shimidzu N. et al. Purification and characterization of hepatocyte growth factor from injured liver of carbon tetrachloride-treated rats. *J. Biochem.*, 1991, v. 109, No. 1, pp. 8-13.

Asakura K., Ikeda H. and Gemba M. *In vitro* oxygenation injury to slices prepared from ischemic kidney in rats. *Jpn. J. Pharmacol.*, 1992, v. 60, No. 2, pp. 149-151.

Asano M. and Hidaka H. Contractile responses of isolated rabbit aortic strips to unsaturated fatty acid peroxides. *J. Pharmacol. Exp. Ther.*, 1979, v. 208, No. 3, pp. 347-353.

Asanov M. A. [Dependence of spontaneous and the spectral characteristics of oleic acid on the chemiluminescence partial oxygen pressure and temperature]. (Russian). *Biofizika*, 1993, v. 38, No. 3, pp. 397-405. English transl. see[2*] pp. 397-405.

Asano M., Matsuda T., Suzuki Y. et al. Contractile responses of mammalian cerebral arteries to 15-hydroperoxyarachidonic acid vary in the presence of fusaric acid and verapamil. *Pharmacology*, 1984, v. 28, No. 5, pp. 251-261.

Ascer E., Gennaro M., Cupo S. and Mohan C. Do cytokines play a role in skeletal muscle ischemia and reperfusion? *J. Cardiovasc. Surg.*, 1992, v. 33. No. 5, pp. 588-598.

Asimakis G. K. and Conti V. R. Myocardial ischemia: Correlation of mitochondrial adenine nucleatide and respiratory function. *J. Mol. Cell. Cardiol.*, 1984, v. 16, No. 5, pp. 439-448.

Aust S. D. and Svingen B. A. The role of iron in enzymatic lipid peroxidation. In: Pryor W. A., ed. *Free Radicals in Biology*. N. Y., Acad. Press, 1982, v. 5, pp. 1-28.

Avdonin P. V., Men'shikov M. Yu., Orlov S. N. et al. [Mechanisms of increase in Ca^{2+} consentration platelet cytoplasm induced by aggregation factors]. (Russian). *Biokhimiya*, 1985, v. 50, No. 8, pp. 1241-1248. English transl. see[3*], pp. 1055-1062.

Azin A. L., Mitagvariya N. P., Medvedev V. G. et al. [Effect. of tumor necrosis cerebral arterial tone]. (Russian). *Biul. Exsp. Biol. Med.*, 1990, v. 110, No. 8, pp. 121-124. English transl. see[1*], pp. 1001-1004.

Azizova O. A. and Ritov V. B. [*Methods of of spine marks and zonds*]. (Russian). Moscow, Nauka, 1986, pp. 143-163.

Azizova O. A., Maksina A. G., Klaan N. K. et al. [Modification of the enzyme system on Ca^{2+} transport in the sarcoplasmic reticulum in lipid peroxidation. Change in the molecular organization of the lipoprotein complex of Ca-ATPase]. (Russian). *Biokhimiya*, 1983, v. 48, No. 5, pp. 861-869. English transl. see[3*], pp. 741-748.

Azizova N. N., Muratova U. Z., Gagel'gans A. I. and Saatov T. S. [Influence of calmodulin and chlorpromazine on the activity of phospholipases A_2 from various sources]. (Russian). *Biokhimiya*, 1990, v. 55, No. 1, pp. 23-28. English transl. see [3*], pp. 17-21.

Baba A., Lee F., Ohta W. et al. Activation of adenylate cyclase of rat brain by lipid peroxidation. *J. Biol. Chem.*, 1981. v. 250, pp. 3679-3684.

Babbs C. F., Cregora M. D. and Badylak S. F. Histochemical demonstration of endothelial superoxide and hydrogen peroxide generation in ischaemic and reoxygenated rat tissues. *Cardiovasc. Res.*, 1992, v. 26, No. 6, pp. 593-602.

Babizhaev M. A., Savov V. M. and Kagan V. E. [Formation of complexes of superoxide anion-radicals with Ca^{2+}]. (Russian). *Biul. Eksp. Biol. Med.*, 1983, v. 96, No. 11, pp. 47-48. English transl. see[1*], pp. 1546-1548.

Badzhinyan S. A. and Aghajanov M. I. [Effect of vitamin E on the lipids peroxide oxidation level and permeability of bilayer membrane in the rat heart after burn injuries]. (Russian). *Zh. Eksp. Klin. Med.*, 1983, v. 2, pp. 105-107.

Bain S. C., Le Guen C. A., Lunec J. and Barnett A. H. Comparison of the free radical scavenging activity of captopril versus enalapril: a three month *in vivo* study in hypertensive diabetic patients. *J. Human Hypertension*, 1991, v. 5, No. 6, pp. 511-515.

Baird M. B., Birnbaum L. S. and Sfeir G. T. NADPH-driven lipid peroxidation in rat liver nuclei and nuclear membranes. *Arch. Biochem. Biophys.*, 1980, v. 200, pp. 108-115.

Baker K., Marcus C. B., Huffman K. et al. Synthetic combined superoxide dismutase/catalase mimetics are protective as a delayed treatment in a rat stroke model: a key role for reactive oxygen species in ischemic brain injury. *J. Pharmacol. Exp. Ther.*, 1998, v. 284, No. 1, p. 215-221.

Bakkeren D. L., De Jen-Jaspars C. H. H., Van Der Heul, C and Van Eijk, H. G. Analysis of iron-binding components in the low molecular weight fraction of rat reticulocyte cytosol. *Int. J. Biochem.*, 1985, v. 17, pp. 925-930.

Ballagi-Pordany G., Richter J., Koltai M. et al. Is malondialdehyde a marker of the effect of oxygen free radicals in rat heart tissue? *Basic Res. Cardiol.*, 1991, v. 86, pp. 266-272.

Bando K. and Senoo Y. Oxygen-derived free radical damage in canine heart transplantation. *J. Surg. Res.*, 1989, v. 46, iss. 2, pp. 152-156.

Barclay J. K. and Hansel M. Free radicals may contribute to oxidative skeletal muscle fatigue. *Canadian J. Phys. and Pharm.*, 1991, v. 69, No. 2, pp. 279-284.

Baril E. and Laszlo J. Subcellular localization and characterization of DNA polymerases from rat liver and hepatomas. In: Weber. G., ed. *Adv. Enz. Regul.* 1971, v. 9, pp. 183-206. N. Y., Pergamon Press.

Barnard M. L., Snyder S. J., EngersonT. D., and Turrens J. F. Antioxidant enzyme status of ischemic and postischemic liver and ischemic kidney in rats. *Free Radic. Biol. Med.*, 1993, v. 15, No. 2, pp. 227-232.

Bartoli G. M., Bartoli S., Galeotti T. and Bertoli E. Superoxide dismutase content and microsomal lipid composition of tumours with different growth rates. *Biochim. Biophys. Acta.*, 1980. v. 620, No. 2, pp. 205-211.

Bartosz G., Soszynski M. and Retelewska W. Aging of the erythrocyte. X. Immunoelectrophoretic studies on the denaturation of superoxide dismutase. *Mechanism of Aging and Development*, 1981, v. 17, pp. 237-251.

Baskurt O. K., Temiz A. and Meiselman H. J. Effect of superoxide anions on red blood cell rheologic properties. *Free Radic. Biol. Med.*, 1998, v. 24, No. 1, pp. 102-110.

Bazan N. G. and Rodriquez de Furco E. B. Membrane lipids in the pathogenesis of brain edema: phospholipids and arachidonic acid, the earliest membrane components changed at the onset of ischemia. *Adv. Neurol.*, 1980, v. 28, pp. 197-205.

Beasley T. C., Bary F., Thore C. et al. Cerebral ischemia/reperfusion increases endothelial nitric oxide synthase levels by an indomethacin-sensitive mechanism. *J. Cereb. Blood. Flow. Metab.*, 1998, v. 18, No. 1, pp. 88-96.

Beckman J. R., Owens K., Knauer T. E. et al. Hydrolysis of sarcolemma by lysosomal lipases and inhibition by chlorpromazine. *Am. J. Physiol.*, 1982, v. 242, H652-H656.

Beckman J. S. Peroxynitrite versus hydroxyl radical: the role of nitric oxide in superoxide-dependent cerebral injury. *Ann. N. Y. Acad. Sci.*, 1994, v. 738, pp. 69-75.

Beckman J. S. and Crow J. P. Pathological implications of nitric oxide, superoxide and peroxynitrite formation (Review). *Biochem. Soc. Trans.*, 1993, v. 21, No. 2, pp. 330-334.

Bednar M., Smith B., Pinto A. and Mullane K. M. Nafazatrom-induced salvage of ischemic myocardium in anesthetized dogs is mediated through inhibition of neutrophil function. *Circ. Res.*, 1985, v. 57, pp. 131-141.

Beetsch J. W., Park T. S., Dugan L. L. et al. Xanthine oxidase-derived superoxide causes reoxygenation injury of ischemic cerebral endothelial cells. *Brain Res.*, 1998, v. 786, No. 1-2, pp. 89-95.

Beklemishev N. D. [*Immunology and immunoregulation*]. (Russian). Moscow, Meditsina, 1986, 256 p.

Bekmukhamedova Z. U. [Problems of adaptation to Na-K-ATPase]. (Russian). In: Biological membranes and membrane-active compounds. Tashkent, *FAN*, 1985, pp. 259-266.

Bel A., Martinod E. and Menasche P. Cardioprotective effect of desferrioxamine. *Acta Haematol.*, 1996, v. 95, No. 1, pp. 63-65.

Belkin M., LaMorte W. L., Wright J. G. and Hobson R. W. 2nd. The role of leukocytes in the pathophysiology of skeletal muscle ischemic injury. *J. Vasc. Surg.*, 1989, v. 10, No. 1, pp. 14-18.

Belo S. E. and Talesnik J. Coronary vasoconstrictor and vasodilator actions of arachidonic acid in the isolated perfused heart of the rat. *Br. J. Pharmac.*, 1982, v. 75, No. 2, pp. 269-286.

Belousova V.V., Dudchenko A.M. and Luk'yanova L.D. [Energy consumption and energy production in rat hepatocytes in various oxygen deficiency states]. (Russian). *Byull. Eksp. Biol. Med.*, 1992, v. 114, No. 12, pp. 588-590. English transl. see[1*], pp. 1772-1775.

Benedetti A., Comporti M. and Esterbauer M. Identification of 4-hydroxynonenal as a cytotoxic product originating from the peroxidation of liver microsomal lipids. *Biochim. Biophys. Acta*, 1980, v. 620, pp. 281-296.

Benedetti A., Fulceri R. and Comparti M. Inhibition of calcium sequestration activity of liver microsomes by 4-hydroxyalkenals originating from the peroxidation of liver microsomal lipids. *Biochim. Biophys. Acta*, 1984, v. 793, pp. 489-493.

Benevolensky D. S. and Levitsky D. O. [Changes in calcium transport ability of cardiac sarcoplasmic reticulum induced by acidosis]. (Russian). *Biul. VKNC AMN SSSR*, 1981, v. 4, No. 2, pp. 15-18.

Berezovsky V. A. and Sushko B. S. [Oxygen concentration profile in a cell and certain disputable problems of free oxygen transport in biological objects]. (Russian). *Fiziol. Zh. Acad. Nauk SSSR*, 1984, v. 30, No. 3, pp. 345-355.

Bergeron R. J. and Brittenham G. M., eds. *The development of iron chelators for clinical use.* Crc. Press, Boca Raton, Fl., 1994, 416p.

Berggren M., Dawson J. and Moldeus P. Glutathione biosynthesis in the isolated perfused rat lung: utilization of extracellular glutathione. *FEBS Lett.*, 1984, v. 176, No. 1, pp. 189-182.

Bergmann S. R., Ferguson T. B. and Sobel B. E. Effects of amphiphiles on erythrocytes, coronary arteries and perfused hearts. *Am. J. Physiol.*, 1981, v. 9, H229-H237.

Berne R. M., Belardinelli L., Winn H. R. et al. The adenosine concept in ischemia of heart and brain. In: Wauquioz A. et al., eds. *Protection tissues against hypoxia*, Amsterdam, 1982, pp. 105-114.

Bernier M., Manning A. S. and Hearse D. I. Reperfusion arrhythmias: dose-related protection by anti-free radical interventions. *Am. I. Physiol.*, 1989, v. 256, pp. H1344-H1352.

Bershtein S. A., Gurevich M. I. and Soloviev A. N. [Deficite of oxygen and vascular tonus] (Russian). Kiev, *Naukova Dumka*, 1984, 263 p.

Bershtein S. A., Gurevich M. I. and Soloviev A. N. [Molecular mechanisms for contraction of the vascular smooth muscles and the role of Ca^{2+} translocation in the development of contractile reactions on oxygenation variations]. (Russian). *Fiziol. Zh.*, Kiev, 1985, v. 31, No. 4, pp. 473-487.

Bersohn M. M., Philipson K. D. and Weiss R. S. Lysophosphatidylcholine and sodium-calcium exchange in cardiac sarcolemma: comparison with ischemia. *Am. J. Physiol.*, 1991, v. 260, No. 3, Pt. 1, pp. C433-C438.

Betz A. L. Identification of hypoxanthine transport and xanthine oxidase activity in brain capillaries. *J. Neurochem.*, 1985, v. 44, pp. 574-579.

Beyer W., Imlay J. and Fridovich I. Superoxide dismutases. In: Cohn W. E. and Moldare K., eds. *Progress in nucleic acid research and molecular biology*, Acad. Press Inc., 1991, v. 40, pp. 221-253.

Beyersdorf F., Sarai K., Mitrev Z. et al. Studies of reperfusion injury in skeletal muscle: controlled limb reperfusion to reduce post-ischaemic syndrome. *Cardiovasc. Surg.*, 1993, v. 1, No. 4, pp. 330-336.

Bhabra M. S., Hopkinson D. N., Shaw T. E. et al. Controlled reperfusion protects lung grafts during a transient early increase in permeability. *Ann. Thorac. Surg.*, 1998, v. 65, No. 1, pp. 187-192.

Biasi F., Chiarpotto E., Lanfranco G. et al. Oxidative stress in the development of human ischemic hepatitis during circulatory shock. *Free Rad. Biol. Med.*, 1994, v. 17, No. 3, pp. 225-233.

Bienvenu K. and Granger D. N. Leukocyte adhesion in ischemia/reperfusion. [Review]. *Blood Cells*, 1993, v. 19, No. 2, pp. 279-288.

Bilenko M. V. [Ischemic shock is a free-radical pathology; methods of prevention and treatment]. (Russian), *Bioantioxidant, IV Conference.* Abstracts. v. 2, Moscow. 1992, pp. 2-3.

Bilenko M. V. Lipid peroxidation (LPO) and myocardial ischemia/reperfusion injury. Abstract., *J. Mol. Cell. Cardiol.*, 1992, v. 24, Suppl. 1, P01-29, p. S. 146.

Bilenko M. V. [Experience in studying of lipid peroxidation processes during ischemia and reperfusion of different organs and comparative analysis of methods for their antiradical protection]. (Russian). *Antioxidants. 3rd All-Union Conference.* Abstracts. Moscow. 1989, v. 2, pp. 100-101.

Bilenko M. V. [*Ischemic and reperfusion injury to organs*]. (Russian). Moscow, Meditsina, 1989, 368 pp.

Bilenko M. V. A comparative study of the role of free-radical lipid peroxidation (LPO) in ischemic and reperfusion injury to various organs. Abstract. *J. Mol. Cell. Cardiol.*, 1988, v. 20, suppl. V, No. 50, p. S. 35.

Bilenko M. V. [Role of circulating endogenous and exogenous lipid peroxidation products in the pathogenesis of cardiovascular disorders and protective effect of antioxidants]. (Russian). In: *Bioantioxidant. 2nd All – Union Conference.* Abstracts. Chernogolovka, 1986, v. 2, pp. 25-26.

Bilenko M. V. [Theoretical and experimental substantiation of antioxidant therapy application for acute ischemic injuries in organs]. (Russian). In: Zhuravlev A. I., ed. *Bioantioxidants in metabolism regulation in norma and pathology*. Moscow, Nauka, 1982, pp. 195-213.

Bilenko M. V. Aspects of the hyperbaric oxygenation procedure. (Discussion.) *7th Intern. Congress "Hyperbaric Medicine"*, Moscow, 1981. Moscow, Nauka, 1983, v. 1, p. 72.

Bilenko M. V. *Prophylactics and therapy of damage in long-term ischaemized or preserved organs*. Intern. Meeting of the Soc. for Low Temperature Biology. Prague, 1980, pp. 33-38.

Bilenko M. V. [*Biological aspects of renal allotransplantation*]. (Russian). Moscow, Meditsina, 1978, 384 pp.

Bilenko M. V. [Role of lipid peroxidation in pathogenesis of ischemic and postischemic disorders in organs and prospects of using antioxidant therapy]. (Russian). In: *Acute Ischemia in Organs and Early Postischemic Disorders*. 2nd All-Union Symposium. Abstracts. Moscow, 1978, pp. 51-52.

Bilenko M. V. and Churakova T. D. [Effect of lipid peroxidation products on the vascular tone]. (Russian). *Biul. Eksp. Biol. Med.*, 1982, v. 94, No. 7, pp. 22-25. English transl. see[1*], pp. 876-879.

Bilenko M. V. and Otverchenko V. N. [Protective effect of antioxidant ionol in total cardiac ischemia]. (Russian). *Biul. Eksp. Biol. Med.*, 1988, v. 106, No. 10, pp. 412-415. English transl. see[1*], pp. 1395-1398.

Bilenko M. V., Alessenko A. V. and Shelenkova L. N. Role of lipid peroxidation in pathogenesis of changes in ischemic and preserved organs: methods of antioxidant therapy. Abstracts. Warsaw. Eur. *Surg. Res.*, 1977, v. 9, No. 1, p. 82.

Bilenko M. V., Alessenko A. V., Burlakova E. B. et al. [The phenomenon of intensification of lipid peroxidation in ischemic (myocardial and renal) tissues]. (Russian). Diploma on Discovery, No. 393, December 13, 1990. State Committee for Inventions and Discoveries. *Biul. Discovery and Invention*, 1991, No. 30, p. 3.

Bilenko M. V., Alessenko A. V., Burlakova E. B. et al. In: [Role of lipid peroxidation processes during ischemia and preservation of kidneys]. (Russian). Structure, Biosynthesis and Transformation of lipids in animal and human organism. *2nd All-Union Symposium. Abstracts.* Moskow, 1975, p. 29.

Bilenko M. V., Bulgakov V. and Otverchenko V. Role of lipid peroxidation in ischemic improvement of myocardial contractility. *9th World Congress of Cardiology. Abstracts.* Moscow, 1982, Vol. 2, No. 129, p. 36.

Bilenko M. V., Bulgakov V. G. and Morgunov A. A. [Negative inotropic and vasoconstrictive effects of oxidized phospholipids]. (Russian). *Kardiologiya*, 1989, v. 29, No. 6, pp. 88-94.

Bilenko M. V., Bulgakov V. G. and Morgunov A. A. Vasoconstriction and negative inotropic effects of exogenous unperoxidized phospholipids on the heart. Abstract. *J. Mol. Cell. Cardiol.*, 1987, v. 19, Suppl. III, No. 17, p. S. 6.

Bilenko M. V., Bulgakov V. G. and Velikhanova D. M. Membrane injures in ischemic tissues and role of lipid peroxidation. IV Symposium uber Probleme der Tieftemperaturkonservierung von Zellen, Geweben und Organen. Proceeding Berlin, May 1981, *Probleme der Hematologie, Transfusion und Transplantation*, 1981, v. 8, No. 2, pp. 53-62.

Bilenko M. V., Bulgakov V. G., Mukumov M. R. and Churakova T. D. [Effect of lipid peroxidation products on vascular tone and myocardial contractility]. (Russian). *1st All-Union Biophysical Congress. Abstracts.* Moscow, 1982, p. 82.

Bilenko M. V., Bulgakov V., Komarov P. and Morgunov A. Lipid peroxidation-induced myocardial disorder in ischemic shock. Abstract. *J. Mol. Cell. Cardiol.*, 1989, v. 21, suppl. IV, pp. S. 37, p. 84.

Bilenko M. V., Burlakova E. B., Alessenko A. V. and Shelenkova L. N. [Lipid peroxidation and use of antioxidants during ischemia and preservation of organs]. (Russian). *Organ and Tissue Transplantation. 7th All-Union Conference.* Abstracts. (Russian). Rostov-on-Don, 1976, p. 117.

Bilenko M. V., Churakova T. D. and Arkhangelskaya S. L. The role of processes of free-radical lipid peroxidation during ischemic shock. Workshop on: *Oxygen Free Radicals in Shock. Abstracts.* Florence, 1985, p. 42.

Bilenko M. V., Churakova T. D., Arkhangelskaya S. A. and Komarov P. G. [The pathogenesis and approaches to prevention of ischemic shock during reperfusion of long ischemized limbs]. (Russian). *Vestn. Akad. Med. Nauk SSSR*, 1985, No. 4, pp. 24-33.

Bilenko M. V., Dmitrieva I. S. and Kotlyarov P. M. [Classification of vascular change severity in ischemic renal graft at early reperfusion period]. (Russian). In: *Acute ischemia in organs and prevention of early postischemic disorders*. All-Union Symposium. Abstracts. Moscow, 1973, pp. 96-98.

Bilenko M. V., Fedosova S. V., and Ladygina V. G. Comparative study of cytotoxic effect of TNF-α and TNF-α + LDL on nonischemic and ischemic human umbilical vein endothelial cells (HUVEC). Abstract. *J. Mol. Cell. Cardiol.*, 1998, v. 30, Abstract book. No. 769, A195.

Bilenko M. V., Gushcha I. A., Managadze L. G. et al. [Use of diuretics in transplantation of long-term ischemic kidneys]. (Russian). *Khirurgiya*, 1982, No. 7, pp. 71-76.

Bilenko M. V., Kagan V. E., Velikhanova D. M. and Komarov P. G. [Protective effect of antioxidants and microsomal monooxygenase inducers in ischemic and reoxygenation liver damage]. (Russian). *Biul. Eksp. Biol. Med.*, 1983, v. 95, No. 4, pp. 30-32. English transl. see[1*], pp. 438-441.

Bilenko M. V., Klebanov G. I., Dolgina E. N., and Nikankina L. V. [The effect of native and oxidized LDL on macrophage chemiluminescence]. (Russian). *Biul. Eksp. Biol. Med.*, 1999, vol. 128, N11, pp. 514-517

Bilenko M. V., Komarov P. G., Morgunov A. A. and Zhdanov R. I. [Anti-ischemic action of the 1-hydroxy-derivative of 2, 2, 6, 6, -tetramethylpiperidine, a nitroxyl bioantioxidants]. (Russian). *Biul. Eksp. Biol. Med.*, 1991, v. 111, No. 5, pp. 500-503. Engl. transl. see[1*], pp. 648-652.

Bilenko M. V., Komarov P. G., Tel'pukhov V. I. et al. [Effect of water-soluble forms of acetylsalicylic acid on the aggregation of thrombocytes and severity of ischemic injuries]. (Russian). In: *Current aspects of clinical use of acetylsalicylic acid.* All-Union Symposium. Abstracts. 1990, Kharkov. pp. 9-8.

Bilenko M. V., Komarov P., Tel'pukhov V. et al. Evaluation of the protective effect of water-soluble aspirins in total ischemia and reperfusion of organs. Abstract. *J. Mol. Cell. Cardiol.*, 1990, v. 22, suppl. III, PT. 47, p. S. 43.

Bilenko M. V., Kovaleva V. L., Morgunov A. A. et al. [Use of SOD and new water-soluble forms of acetylsalicylic acid (ASA) for prevention of acute ischemic oedema in limbs]. (Russian). *Bioantioxidant. IVth All-Union Symposium. Abstracts*. Moscow, 1992, p. 234-235.

Bilenko M. V., Ladygina V. G. and Fedosova S. V. [Comparative evaluation of cytotoxic effect of hydrogen peroxide and tumor necrosis factor alpha on nonischemic and ischemic endothelial cells]. (Russian). *Vopr. Med. Khim.*, 1998, v. 44, No. 3, pp. 229-240.

Bilenko M. V., Ladygina V. G. and Fedosova S. V. [Cytotoxic effects of low density lipoproteins on nonischemic, ischemic and reperfused endothelial cells.] (Russian). *Biul. Eksp. Biol. Med.*, 1998, v. 126, No. 9, pp. 302-306. English transl. see[1*], pp. 911-914.

Bilenko M. V., Ladygina V. G. and Tarakanova I. A. Assessment of the toxic and protective effect of initiators and inhibitors of free radical reactions using a wild type strain of Escherichia coli and a strain deficient for superoxide dismutase. *Biul. Eksp. Biol. Med.*, 1996, v. 121, No. 1, pp. 74-79. English transl. see[1*], pp. 67-71.

Bilenko M. V., Lyashenko S. M., Kozlov A. V. et al. [Protective effect of desferal in liver ischemia]. (Russian). *Bioantioxidant. IVth All-Union Symposium*. Abstracts. Moscow, 1992, p. 155-156.

Bilenko M. V., Morgunov A. A., Churakova T. D. et. al. [Disorders of cardiac contractile function in ischemic shock; the protective effect of antioxidants and liposomes made from egg phospholipids]. (Russian). *Biul. Eksp. Biol. Med.*, 1989, v. 108, No. 12, pp. 660-663. English transl. see[1*], pp. 1700-1704.

Bilenko M. V., Otverchenko V. M., Astaf'ev V. V. et al. [The pathogenesis of reoxygenation injuries in the ischaemic myocardium: the influence of the blood oxygenation level and the role of lipid peroxidation.]. (Russian). *Grudnaya Khir.*, 1983, No. 6, pp. 20-26.

Bilenko M. V., Perekrest P. V. and Ugryumov A. I. [Comparative evaluation of a number of perfusion methods for preservation of the kidney]. (Russian). *Vest. Akad. Med. Nauk SSSR*, 1975, No. 7, pp. 61-70.

Bilenko M. V., Seregina L. A., Komarov P. G. and Abakumova O. Yu. [Regenerating ability of ischemic liver tissue and effect of resection and of proliferation-stimulating factors on liver tissue regeneration]. (Russian). *Vopr. Med. Khim.*, 1989, No. 2, pp. 93-97.

Bilenko M. V., Seregina L., Komarov P. et al. Effects of antioxidants and h-SOD on the reparative processes in ischemic and resected liver. Regulation of free-radical reactions (Biomedical aspects). *Intern. Conference. Abstracts*. Varna. 1989, No. 13.

Bilenko M. V., Shelenkova L. N., Barsel V. N. and Matveeva S. A. In: [Pharmacokinetics and antiischaemic effect of antioxidant dibunol]. (Russian). *Methods of individualization and optimization in medicine application based on the study of their pharmacokinetics.* All-Union Conference. Proceedings. Moscow, 1982, Part 2, pp. 47-51.

Bilenko M. V., Shelenkova L. N., Dubur G. Ya. and Velena A. H. [Application of antioxidants in the prevention of renal injury during acute ischemia and reperfusion]. (Russian). *Biul. Eksp. Biol. Med.*, 1983, No. 9, pp. 8-11. English transl. see[1*], pp. 1195-1198.

Bilenko M. V., Shlifer A. Y. and Epstein I. M. [Investigation of oxygen regime in the intact, ischaemic and preserved kidneys]. (Russian). *Eksp. Khir. Anestez.*, 1976, v. 2, pp. 46-52.

Bilenko M. V., Shlifer A. Ya. and Ugryumov A. I. [Effect of partial atmospheric oxygen pressure on the hemodynamics, function and structure of ischemic kidney]. (Russian.) In: *Oxygen supply of the body and mechanisms for it maintaining.* All-Union Conference. Abstracts. Barnaul, 1978, Part 2, pp. 48-49.

Bilenko M. V., Tel'pukhov V. I., Churakova T. D. and Komorav P. G. [Effect of total ischemia and reperfusion on lipid peroxidation of the rat brain and the protective effect of antioxidants]. (Russian). *Biul. Eksp. Biol. Med.*, 1988, v. 105, No. 4, pp. 394-396. English trans. see[1*], pp. 466-469.

Bilenko M. V., Vakhrusheva T. V. and Fedosova S. V. [Endothelial cells induce oxidation of low density lipoproteins]. (Russian). *Biul. Eksp. Biol. Med.*, 1998, v. 126, No. 9, pp. 314-317. English trans. see[1*], pp. 921-924.

Bilenko M. V., Vartanyan L., Rashba Y. U. et al. Some mechanisms of lipid peroxidation (LPO) in organ ischemia and reperfusion. Abstracts. *J. Mol. Cell. Cardiol.*, 1989, v. 21, Suppl. IV, P83, p. S. 37.

Bilenko M. V., Volkova I. A. and Danilov S. P. [Protection of kidneys after prolonged ischemia against reoxygenation injury by temporary inhalation of a hypoxic gas mixture]. (Russian). *Biul. Eksp. Biol. Med.*, 1985, v. 100, No. 8, pp. 147-151. English transl. see[1*], pp. 1014-1018.

Bilenko M. V., Volkova I. A. and Gushcha I. A. [Protective effect of diuretics early in the post-transplantation period during long-term renal ischemia under experimental conditions]. (Russian). *Khirurgiya*, 1982, No. 8, p. 122.

Bilenko M., Morgunov A., Bulgakov V. and Komarov P. [Lipid-peroxidation induced myocardial disorder in ischemic shock]. Abstract. *J. Mol. Cell. Cardiol.*, 1989, v. 21, suppl. IV, P84, p. S. 37.

Bindoli A., Cavallini L., Rigobello M. P. et al. Modification of the xanthine-converting enzyme of perfused rat heart during ischemia and oxidative stress. *Free Rad. Biol. Med.*, 1988, v. 4, No. 3, pp. 163-167.

Binkova B., Topinka G., Sram R. et al. [Damage and repare of human lymphocyte DNA under activation of molecular oxygen]. (Russian). *Biul. Eksp. Biol. Med.*, 1988, v. 106, No. 12, pp. 676-678. Russian transl. see[1*], pp. 1705-1707.

Bird R. P. and Draper H. H. Comparative studies on different methods of malonaldehyde determination. Oxygen Radicals in Biological Systems In: *Methods in Enzymology*, 1984, v. 105. Oxygen Radicals in Biological systems. Packer L., ed. N. Y. Acad. Press. Inc. pp. 229-305.

Biro G. P. and Beresford-Kroeger D. The effect of hemodilution with stroma-free hemoglobin and dextrin on collateral perfusion of ischemic myocardium in the dog. *Am. Heart. J.*, 1980, v. 99, No. 1, pp. 64-75.

Bkaily G., Sperelakis N., Elishalom Y. et al. Effect of Na^+-or Ca^{2+}-filled liposomes on electrical activity of cultured heart cells. *Am. J. Physiol.*, 1983, v. 245, No. 5, H756-H761.

Bligh E. G. and Dyer W. J. A rapid method of total lipids extraction and purification. *Can. J. Biochem. Physiol.*, 1959, v. 37, No. 8, pp. 991-917.

Bochenek A., Religa Z., Spyt T. J. et al. Protective influence of pretreatment with allopurinol on myocardial function in patients undergoing coronary artery surgery. *Eur. J. Cardiothorac. Surg.*, 1990, v. 4, No. 10, pp. 538-542.

Bodr K. F., Baylis C., Pfeffer S. M. et al. Renal and systemic hemodynamic responses to intravenous infusion of leukotriene C_4 in the rat. *Circ., Res.*, 1984, v. 54, No. 5, pp. 492-499.

Boersma E. and Simoons M. L. Reperfusion strategies in acute myocardial infarction. *Eur. Heart J.*, 1997, v. 18, No. 11, pp. 1703-1711.

Boime I., Smith E. S. and Hunter F. E. The role of fatty acids in mitochondrial changes during liver ischemia. *Arch. Biochem. Biophys.* 1970, v. 139, No. 2, pp. 425-443.

Boldyrev A. A. [*Biological membranes and transport of ions*]. (Russian). Moscow, MGU, 1985, 208 p.

Boldyrev A. A. [*Introduction in biochemistry of membranes*]. (Russian). Moscow, Vysshaya Shkola, 1986, 112 p.

Boldyrev A. A. [Carnosine: biological role and potential applications in medicine]. (Russian). *Biokhimiya*, 1992 v. 57, No. 9, pp. 1302-1310. English transl. see[3*], pp. 898-904.

Boldyrev A. A. and Bulygina E. R. Na/K-ATPase and oxidative stress. *Ann. N. Y. Acad. Sci.*, 1997, v. 834, pp. 666-668.

Boldyrev A. A. and Mel'gunov V. J. [Transport ATPases]. (Russian). *Itogi Nauki i Tekhniki. Ser. Biophys.* Moscow, VNIIMI, 1985, v. 17, 245 p.

Boldyrev A. A. and Tverdislov V. A. [Molecular structure and mechanism of functioning of Na-pump]. (Russian). *Itogi Nauki i Tekhniki. Ser. Biophys.*, Moscow, VNIIMI, 1978, v. 10, 148 p.

Bolitho G., Engelbrecht G., Lotzz et al. Liver regeneration after hepatic ischemia and reduced liver autotransplantion in the rat. *Hepatology*, 1993, v. 17, No. 2, pp. 273-279.

Bolland J. L. and Koch H. J. The course of auto-oxidation reactions in polyisoprenes and allied compounds. *J. Chem. Soc.*, 1945, v. 7-12, pp. 445-447.

Bolshakova I. V., Lozovskaya E. L. and Sapeshinskii I. I. Antioxidant properties of a number of plant extracts. *Biofizika*, 1998, v. 43, No. 2, pp. 186-188.

Bore P. J., Paratheofanis I. and Sells R. A. Adenosine triphosphate regeneration and function in the rat kidney following warm ischaemia. *Transplantation*, 1979, v. 27, No. 4, pp. 235-237.

Borgeat P. and Sirois P. Leukotrienes: a mayor step in the understanding of immediate hypersensitivity reactions. *J. Med. Chem.*, 1981, v. 24, No. 2, pp. 121-126.

Borovskikh N. A., Filev L. V. and Luk'yanov Yu. V. [Changes in blood oxygen transport in different methods of blood oxygenation and nociceptive pulsation blockade]. (Russian). *Fiziologiya Cheloveka,* 1993, v. 19, No. 2, pp. 76-83.

Borrello S., Minotti G., Palombini G. et al. Superoxide-dependent lipid peroxidation and vitamin E content of microsomes from hepatomas with different growth rates. *Arch. Biochem. Biophys*, 1985, v. 238, No. 2, pp. 588-595.

Bossola M., Merrick M. W., Eltaki A. et al. Rat-liver tolerance for partial resection and intraoperative radiation-therapy regeneration is radiation dose dependent. *J. Surg. Oncol.,* 1990, v. 45, No. 3, pp. 196-200.

Boveris A. Mitochondrial production of superoxide radical and hydrogen peroxide. *Adv. Exp. Med. Biol,* 1977, v. 78, pp. 67-82.

Boysen G., Sorensen P.S., Juhler M. et al. Danish very-low-dose asperin after carotid endarterectomy trial. *Stroke*, 1988, v. 19, pp. 1211-1215.

Bradford B. U., Marotto M., Lemasters J. J. et al. New simple models to evaluate zone-specific damage due to hypoxia in the perfused rat liver: time course and effect of nutritional state. *J. Pharmacol. Exp. Ther.*, 1986, v. 236, No. 1, pp. 263-265.

Bradley T. R., Hodgson G. S. and Rosendaal M. The effect of oxygen tension on haemapoietic and fibroblast cell proliferation in vitro. *J. Cell. Physiol.*, 1988, v. 97, No. 3, Pt 2, pp. 517-522.

Braithwaite B. D., Petrik P. V., Moore W. S. et al. Aspirin increases tissue oedema after skeletal muscle ischaemia and reperfusion. *Eur. J. Vasc. Endovasc. Surg.*, 1996, v. 12, No. 1, pp. 76-80.

Bralet J., Schreiber L. and Bouvier C. Effect of acidosis and anoxia on iron delocalization from brain homogenates. *Biochem. Pharmacol.* 1992, v. 43, pp. 979-983.

Brass C. A., Narciso J. and Gollan J. L. Enhanced activity of the free radical producing enzyme xanthine oxidase in hypoxic rat liver. Regulation and pathophysiologic significance. *J. Clin. Invest.*, 1991, v. 87, No. 2, pp. 424-431.

Braunwald E. and Kloner R. A. Myocardial reperfusion: a double-edged sword? *J. Clin. Invest.*, 1985, v. 76, pp. 1713-1719.

Brawn K. and Fridovich I. Superoxide radical and superoxide dismutases: threat and defense. *Acta Physiol. Scand.,* 1980, Suppl. 492, pp. 9-18.

Brecher P. The interaction of long-chain acyl CoA with membranes. *Molec. Cell. Biochem.* 1989, v. 5, pp. 3-15.

Bretschneider H. J., Hübner G., Knoll D. et al. Myocardial resistance and tolerance to ischemia: physiological and biochemical basis. *J. Cardiovas. Surg.*, 1975, v. 16, pp. 241-260.

Breyer R. H., Otani H., Engelman R. M. et al. Superoxide dismutase and catalase pretreatment prior to cardioplegic arrest in the face of regional ischemia. *Surg. Forum,* 1985, v. 36, pp. 269-271.

Brezis M., Rosen S., Silva P. et al. Renal ischemia: a new perspective. *Kidney Int.,* 1984, v. 26, No. 4, pp. 375-783.

Brieland J. K., Clarke S. J., Karmiol S. et al. Transferrin: a potential source of iron for oxygen free radical-mediated endothelial cell injury. *Arch. Biochem. Biophys.*, 1992, v. 294, No. 1, pp. 262-270.

Brogan W. C., Miles P. R. and Colby H. D. Factors affecting lipid peroxidation in guinea-pig adrenal microsomes. *Biochim. Biophys. Acta*, 1981, v. 663, pp. 230-238.

Bucher J. R. and Roberts R. J. Alpha-tocopherol (vitamin E) content of lung, liver and blood in the newborn rat and human infant: influence of hyperoxia. *J. Pediatr.,* 1981, v. 98, pp. 806-811.

Bucher J. R., Tien M. and Steven D. A. The requirement for ferric in the initiation of lipid peroxidation by chelated ferrous iron. *Biochem. Biophys. Res. Comm.*, 1983, v. 111, No. 3, pp. 777-784.

Buckley B. J., Kent R. S. and Whorton A. R. Regulation of endothelial cell prostaglandin synthesis by glutathione. *J. Biol. Chem.*, 1991, v. 266, No. 25, pp. 16659-16666.

Buege J. A. and Aust S. D. Lactoperoxidase-catalyzed lipid peroxidation of microsomal and artificial membrans. *Biochim. Biophys. Acta*, 1976, v. 44, No. 1, pp. 192-201.

Buhl M. R. and Jörgensen S. Breakdown of 5'-adenin nucleotides in ischaemic renal cortex estimated by oxypurine excretion during perfusion. *Scand. Clin. Lab. Invest.,* 1975, v. 35, pp. 211-217.

Bulgakov V. G. *Role of lipid peroxidation in disorder of membranes and the contractile function of an ischemic myocardium.* [Dissertation]. (Russian). Moscow, 1985.

Bulgakov V. G. and Bilenko M. V. [Lipid peroxidation and the calcium transport enzyme system in the sarcoplasmic reticulum of ischemic myocardium]. (Russian). *Biul. Eksp. Biol. Med.*, 1988, v. 106, No. 9, pp. 272-274. English transl. see[1*], pp. 1224-1227.

Bulgakov V. G., Morgunov A. A. and Bilenko M. V. [Effect of phosholipids of varied composition on coronary vascular tone and myocardial contractility]. (Russian). *Biul. Eksp. Biol. Med.*, 1987, v. 104, No. 12, pp. 643-646. English transl. see[1*], pp. 1637-1640.

Bulgakov V. G., Volkova I. A., Astafyev V. V. and Bilenko M. V. [Effect of different ischemic periods on lipid peroxidation processes: calcium transport and metabolism of the heart transplant]. (Russian). In: Lopukhin Yu. M., ed. Current Problems of Hemosorption and Transplantation. *All-Russian Meeting. Proceedings.* Moscow, 1980, p. 77.

Burdon R. H. Superoxide and hydrogen in relation to mammalian cell proliferation. (Review). *Free Radic. Biol. Med.*, 1995, v. 18, No. 4, pp. 775-794.

Burdon R. H., Gill V. and Rice-Evans C. Cell proliferation and oxidative stress. *Free Rad. Res. Commun.*, 1989, v. 7, pp. 149-159.

Burke S. A., Levi R., Guo Z. G. et al. Leukotrienes C_4, D_4 and E_4: effects on human and guinea pig cardiac preparations *in vitro*. *J. Pharmacol. Exp. Ther.*, 1982, v. 221, No. 1, pp. 235-241.

Burke-Gaffney A. and Keenan A. K. Does TNF-alpha directly increase endothelial cell monolayer permeability? *Agents and Actions.* 1993, v. 38, pp. C83-C85.

Burlakova E. B. and Khrapova N. G. [Lipid peroxidation of membranes and natural antioxidants]. (Russian). *Usp. Khim.*, 1985, v. 54, No. 9, pp. 1540-1558.

Burlakova E. B., Alessenko A. V., Molochkina E. M. et al. [*Bioantioxidants in radiation injury and tumor growth*]. (Russian). Moscow, Nauka, 1975, 211 p.

Burlakova E. B., Dzhailbova M. I. and Gvakhariya V. O. [The influence of membrane lipids on an enzymatic activity]. (Russian). In: Zhuravlev A. I., ed. *Bioantioxidants in regulation of metabolism in norma and pathology.* Moscow, Nauka, 1982, pp. 113-140.

Burlakova E. B., Kruglyakova K. E. and Shishkina L. N., eds. [*Investigation of synthetic and natural antioxidants in vitro and in vivo*]. (Russian). Moscow, Nauka, 1992, 110 p.

Burlakova E. B., Zaslavsky Yu. A. and Shishkina L. N. [Influence of changes in membrane lipids on Mg^{2+}-dependent ATPase activity]. (Russian). *Radiobiologiya*, 1984, v. 24, No. 4, pp. 505-508.

Burton G. W., Cheeseman K. H., Doba T. and Ingold K. U. Vitamin E as an antioxidant *in vitro* and *in vivo*. *Ciba Found. Symp.*, 1983, v. 101, pp. 4-18.

Burton K. P., McCord J. M. and Ghai G. Myocardial alterations due to free-radical generation. *Am. J. Physiol.*, 1984, v. 246, No. 6, Pt. 2, H776-H783.

Buttke T. M. and Sandstrom P. A. Oxidative stress as a mediator of apoptosis. *Immunology Today*, 1994, v. 15, No. 1, pp. 7-10.

Cadenas E. Biochemistry of oxygen toxicity. *Ann. Rev. Biochem.*, 1989, v. 58, pp. 79-110.

Cadenas E., Ginsberg M., Rabe U. et al. Evaluation of α-tocopherol antioxidant activity in microsomal lipid peroxidation as detected by low-level chemiluminescence. *Biochem. J.*, 1984, v. 223, pp. 755-759.

Cadenas E., Müller A., Brigelius R. et al. Effects of 4-hydroxynonenal on isolated hepatocytes. *Biochem. J.*, 1983, v. 214, pp. 479-487.

Cadenas E., Varsavsky A. I., Boveris A. et al. Oxygen- or organic hydroperoxide-induced chemiluminescence of brain and liver homogenates. *Biochem. J.*, 1981, v. 198. pp. 645-654.

Callery M. P., Kamei T. and Flye M. W. Kupffer cell tumor necrosis factor-alpha production is suppressed during liver regeneration. *J. Surg. Res.*, 1991, v. 50, No. 5, pp. 515-519.

Camargo C. A. Jr., Madden J. F., Gao W. et al. Interleukin-6 protect liver against warm ischemia/reperfusion injury and promotes hepatocyte proliferation in the rodent. *Hepatology*, 1997, v. 26, No. 6, pp. 1513-1520.

Camilleri J. P., Joseph D., Fabiani J. N. et al. Microcirculatory changes following early reperfusion in experimental myocardial infarction. *Virchows Arch. Path. Anat. Histol.*, 1976, v. 369, No. 4, pp. 315-333.

Campbell A. K. and Hallett M. Measurement of intracellular calcium ions and oxygen radicals in polymorphonuclear leucocyte - erythrocyte "grost" hybrids. *J. Physiol.* (Lond.), 1983, v. 338, pp. 537-550.

Campo G. M., Squadrito F., Campo S. et al. Beneficial effect of raxofelast, an hydrophilic vitamin E analogue, in the rat heart after ischemia and reperfusion injury. *J. Mol. Cell Cardiol.*, 1998, v. 30, No. 8, pp. 1493-1503.

Canada A. T., Coleman L. R. Jr., Fabian M. A. and Bollinger R. R. Adenine nucleotides of ischemic intestine do not reflect injury. *J. Surg. Res.*, 1993, v. 55, No. 4, pp. 416-421.

Canada A. T., Stein K., Martel D. and Watkins W. D. Biochemical appraisal of models for hepatic ischemia-reperfusion injury. *Circ. Shock.,* 1992, v. 36, No. 3, pp. 163-168.

Cascarano J. and Seidman I. Anaerobic cell function: latent capacity for Na^+ and K^+ transport in rat liver slices. *Biochim. Biophys. Acta,* 1965, v. 100, No. 1, pp. 301-303.

Cascarano J., Chick W. L. and Seidmann I. Anaerobic rat heart: effect of glucose and Krebs cycle metabolites on rate of beating. *Proc. Soc. Exp. Biol. Med.,* 1968, v. 127, No. 1, pp. 25-30.

Castellot J. J., Miller M. R., Lehtomaki D. M. et al. Comparison of DNA replication and repair enzymology using permeabilized baby hamster kidney cells. *J. Biol. Chem.*, 1979, v. 254, pp. 6904-6908.

Caty M. G., Guice K. S., Oldham K. T. et al. Evidence for tumor necrosis factor - induced pulmonary microvascular injury after intestinal ischemia-reperfusion injury. *Ann. Surg.,* 1990, v. 212, No. 6, pp. 694-700.

Caty M. G., Schmeling D. J., Friedl H. P. et al. Histamine: a promoter of xanthine oxidase activity in intestinal ischemia/reperfusion. *J. Pediatr. Surg.,* 1990, v. 25, No. 2, pp. 218-222.

Ceconi C., Cargnoni A., Pasini E. et al. Evaluation of phospholipid peroxidation as malondialdehyde during myocardial ischemia and reperfusion injury. *Am. J. Physiol.*, 1991, v. 260, No. 4, Pt. 2, pp. H1057-H1061.

Chambers D. S., Parks D. A., Patterson G. et al. Xanthine oxidase as a source of free radical damage in myocardial ischemia. *J. Mol. Cell. Cardiol.*, 1985, v. 17, pp. 145-152.

Chan P. H. and Fishman R. A. Alterations of membrane integrity and cellular constituent by arachidonic acid in neuroblastoma and glioma cells. *Brain Res.,* 1982, v. 248, pp. 151-157.

Chan P. H., Yurko M. and Fishman R. A. Phospholipid degradation and cellular edema induced by free radicals in brain cortical slices. *J. Neurochem.*, 1982, v. 38, pp. 528-531.

Chan P. H., KawaseM., Murakami K. et al. Overexpression of SOD1 in transgenic rats protects vulnerable neurons against ischemic damage after reperfusion. *J. Neurosci,* 1998, v. 18, No. 20, pp. 8292-8299.

Chance B. Reaction of oxygen with the respiratory chain in cells and tissues. *J. Gen. Physiol.,* 1965, v. 49, No. 1-2, pp. 163-188.

Chance B., Sies H. and Boveris A. Hydroperoxide metabolism in mammalian organs. *Physiol. Rev.*, 1979, v. 59, No. 3, pp. 527-605.

Chang D. J., Ringold G. M. and Heller R. A. Cell killing and induction of manganous superoxide dismutase by tumor necrosis factor-α is mediated by lipoxygenase metabolites of arachidonic acid. *Biochem. Biophys. Res. Commun.,* 1982, v. 188, No. 2, pp. 598-546.

Chapnick B. M., Feigen L. P., Hyman A. L. et al. Differential effects of prostaglandins in the mesenteric vascular bed. *Am. J. Physiol.*, 1978, v. 235, No. 3, H326-332.

Chappell S. P., Lewis M. J. and Henderson A. H. Myocardial reoxygenation damage: can it be circumvented. *Cardiovasc. Res.*, 1985, v. 19, No. 5, pp. 299-303.

Chavez-Cartaya R. E., Metcalfe S., Ramirez-Romero P. et al. Rat liver blood flow after ischemia and reperfusion. The effects of the platelet-activating factor antagonist WEB-2170 and of removing circulating leukocytes. *Transplantation*, 1994, v. 57, No. 10, pp. 1440-1444.

Chaudhuri A., Medhekar T. P. The effect of aspirin on neurodeficit and platelet aggregability in acute ischemic cerebral strokes. *J. Assoc. Physic. India*, 1994, v. 42, No. 6, pp. 458-460.

Cheeseman K. H. Tissue injury by free radicals. [Review] *Toxicol. Indust. Health*, 1993, v. 9, No. 1-2, pp. 39-51.

Cheeseman K. H., Collins M., Maddix S. et al. Lipid peroxidation in regenerating rat liver. *FEBS Lett.,* 1986, v. 209, No. 2, pp. 191-196.

Chekareva G. A., Mishnev O. D., Serdobintseva T. S. and Istomin N. P. [Changes in renal enzymatic activity in temperary ischemia of the extremities]. (Russian). *Arkh. Patol.,* 1981, v. 43, No. 8, pp. 27-34.

Chen M. F., Chen H. M., Ueng S. W. and Shyr M. H. Hyperbaric oxygenation pretreatment attenuateas hepatic reperfusion injury. *Liver*, 1998, v. 18, No. 2, pp. 110-116.

Chernysheva G. V., Stoida L. V., Amakantova G. G. et al. [Effect of disseminated necrosis of the myocardium on ATPase activity of Ca^{2+} transport and lipid peroxidation of mitochondrial and microsomal membranes of the heart]. (Russian). *Biul. Eksp. Biol. Med.*, 1980, v. 89, No. 5, pp. 563-565. English transl. see[1*], pp. 615-618.

Chevion M. Jiang Y. Har-El R. et al. Copper and iron are mobilized following myocardial ishemia: possible predictive criteria for tissue injury. *Proceed. Nat. Acad. Sci. USA*, 1993, v. 90, No. 3, pp. 1102-1110.

Chi L. G., Tamura Y., Hoff P. T. et al. Effect of superoxide dismutase on myocardial infarct size in the canine heart after 6 hours of regional ischemia and reperfusion: a demonstration of myocardial salvage. *Circ. Res.*, 1989, v. 64, No. 4, pp. 665-675.

Chiao J. J., Kirschner R. E., Fantini G. A. Iron delocalization occurs during ischemia and persists on reoxygenation of skeletal muscle. *J. Lab. Clin. Med.*, 1994, v. 124, No. 3, pp. 432-438.

Chien K. R., Abrams J., Serroni A. et al. Accelerated phospholipid degradation and associated membrane dysfunction in irreversible, ischemic liver cell injury. *J. Biol. Chem.*, 1978, v. 253, pp. 4809-4817.

Chien K. R., Cherman S. C., Mittnacht S. et al. Microsomal membrane structure and function subsequent to calcium activation of an endogenous phospholipase. *Arch. Biochem. Biophys.*, 1980, v. 205, No. 2, p. 614-622.

Chien K. R., Pfau R. G. and Farber J. L. Ischemic myocardial cell injury. Prevention by chlorpromazine of an accelerated phospholipid degradation and associated membrane dysfunction. *Am. J. Pathol.*, 1979, v. 97, No. 3, pp. 505-529.

Chien K. R., Reeves J. P., Buja L. M. et al. Phospholipid alterations in canine ischemic myocardium. Temporal and topographical correlations with Tc-99m-PPi accumulations and an *in vitro* sarcolemmal Ca^{2+} permeability defect. *Circ. Res.*, 1981, v. 48, No. 5, pp. 711-719.

Chimowitz M. I., Furlan A. J., Nayak S. and Sila C. A. Mechanism of stroke in patients taking aspirin. *Neurology*, 1990, v. 40, pp. 1682-1685.

Chisolm G. M. and Morel D. W. Lipoprotein oxidation and cytotoxicity: effect of probucol on streptozotocin-treated rats. *Am. J. Cardiol.*, 1988, v. 62, pp. 20B-26B.

Chiu J. H., Ho C. T., Wei Y. H. et al. *In vitro* and *in vivo* protective effect of honokiol on rat liver from peroxidative injury. *Life Sci.*, 1997, v. 61, No. 19, pp. 1961-1971.

Christ M., Luu B., Mejia J. E. et al. Apoptosis induced by oxysterols in murine lymphoma cells and in normal thymocytes. *Immunol.*, 1993, v. 78, No. 3, pp. 455-460.

Christman C. W., Wei E. P., Kontos M. A. et al. Effects of 15-hydroperoxy-eicosatetraenoic acid (15 HPETE) on cerebral arterioles on cats. *Am. J. Physiol.*, 1984, v. 247, No. 4, Pt. 2, pp. H631-H637.

Chumakov V. N. and Osinskaya L. F. [Activity of zinc-, copper-containing superoxide dismutase from rat tissues in normal state and under hypoxia]. (Russian). *Vopr. Med. Khim.*, 1979, v. 25, No. 3, pp. 261-266.

Churakova T. D. [*Influence of lipid peroxidation products on membrane permeability contractile function*]. Dissertation. (Russian). Moscow, 1980.

Churakova T. D., Bilenko M. V., Pokazeeva Z. T. et al. [Blood content of lipid peroxidation products in acute myocardial infarct patients treated and untreated with the antioxidant dibunol]. (Russian). *Vestn. Akad. Med. Nauk SSSR*, 1985, No. 4, pp. 71-74.

Churakova T. D., Bilenko M. V., Tel'pukhov V. I. and Komarov P. G. [Lipid peroxidation processes and protective effect of antioxidants during ischemia and reperfusion of the rat brain]. (Russian). *Pharmacological Correction of Hypoxic States. I All-Union Conference.* Abstracts. Moscow, 1988, pp. 144-145.

Ciaccio E. I. and Fruncillo R. J. Decreased aryl hydrocarbon hydroxylase after a 15% burn. *Biochem. Pharmacol.*, 1979, v. 28, pp. 3151-3152.

Ciuffetti G., Mercuri M., Mannarino E. et al. Are leucocyte-derived free radicals involved in ischaemia in human legs? Eur. *J. Clin. Invest.*, 1991, v. 21, No. 1, pp. 111-117.

Clark S. C., Sudarshan C., Khanna R. et al. Controlled reperfusion and pentoxifylline modulate reperfusion injury after single lung transplantation. *J. Thorac. Cardiovasc. Surg.*, 1998, v. 115, No. 6, pp. 1335-1341.

Clarkson C. W. and Ten Eick R. E. On the mechanism of lysophosphatidylcholine-induced depolarization of cat ventricular myocardium. *Circ. Res.*, 1983, v. 52, No. 5, pp. 543-556.

Cobbe S. M. and Poole-Wilson P. A. The time of onset and severity of acidosis in myocardial ischemia. *J. Mol. Cell. Cardiol.*, 1980, v. 12, No. 8, pp. 745-760.

Coceani F., Bishai I., White E. et al. Action of prostaglandins endoperoxides and thromboxanes on the lamb ductus artheriosus. *Am. J. Physiol.*, 1978, v. 234, No. 2, pp. H117-H122.

Cockerill M. J., Player T. J. and Horton A. A. Studies on lipid peroxidation in regenerating rat liver. *Biochim. Biophys. Acta*, 1983, v. 750, No. 1, pp. 208-213.

Coert B. A., Anderson R. E. and Meyer F. B. A comparative study of the effects of two nitric oxide synthase inhibitors and two nitric oxide donors on temporary focal cerebral ischemia in the Wistar rat. *J. Neurosurg*, 1999, v. 90, No. 2, pp. 332-338.

Cohen P. J. Allopurinol administered prior to hepatic ischaemia in rat prevents chemiluminescence following restoration of circulation. *Can. J. Anaesthesia,* 1992, v. 39, No. 10, pp. 1090-1093.

Cohen S., Lahat N., Kinarty A. and Bitterman H. *Ex vivo* secretion of tumor necrosis factor and interleukin-2 by rat splenocytes after intestinal ischemia and shock. *Lymphokine Cytokine Res.,* 1992, v. 11, No. 5, pp. 215-220.

Cole A., Cooper W. G., Shonka F. et al. DNA scission in hamster cells and isolated nuclei studied by low-voltage electron beam irradiation. *Radiation. Res.,* 1974, v. 60, pp. 1-33.

Coleman R. Membrane-bound enzymes and membrane ultrastructure. *Biochem. Biophys. Acta,* 1973, v. 300, No. 1, pp. 1-30.

Colletti L. M., Remick D. G., Burtch C. D. et al. Role of tumor necrosis factor-alpha in the pathophysiologic alterations after hepatic ischemia/reperfusion injury in the rat. *J. Clin. Invest.*, 1990, v. 85, No. 6, pp. 1936-1943.

Comings D. E. and Kakefula T. Initiation of deoxyribo-nucleic acid replication at the nuclear membrane in human cells. *J. Mol. Biol.,* 1968, v. 33, pp. 225-229.

Concannon. M. J., Dooley T. W., Puckett C. L. Improved survival in a replantation model containing ischemic muscle. *Microsurgery,* 1991, v. 12, iss. 1, pp. 18-22.

Cope J. T., Mauney M. C., Banks D. et al. Controlled reperfusion of cardiac grafts from non-heart-beating donors. *Ann. Thorac. Surg.*, 1996, v. 62, No. 5, pp. 1418-1423.

Copin J. C., Reola L. F., Chan T. Y. et al. Oxygen deprivation but not a combination of oxygen, glucose, and serum deprivation induces DNA degradation in mouse cortical neurons *in vitro*: attenuation by transgenic overexpression of CuZn-superoxide dismutase. *J. Neurotrauma,* 1996, v. 13, No. 5, pp. 233-244.

Corr P. B., Gross R. W. and Sobel B. E. Amphipathic metabolites and membrane dysfunction in ischemic myocardium. *Circ. Res.*, 1984, v. 55, No. 2, pp. 135-154.

Corr P. B., Lee B. I. and Sobel B. E. Electrophysiological and biochemical derangements in ischemic myocardium: interactions involving the cell membrane. *Acta Med. Scand.,* 1981, v. 210, Suppl. 651, pp. 59-68.

Correa-Rotter R., Hostetter T. H., Manivel J. C., et al. Intrarenal distribution of clusterin following reduction of renal mass. *Kidney Intern.*, 1992, v. 41, pp. 938-950.

Cottart C. H., Do L., Blanc M. C. et al. Hepatoprotective effect of endogenous nitric oxide during ischemia-reperfusion in the rat. *Hepatology* 1999., v. 29, No. 3, pp. 809-813.

Coudray C., Boucher F., Pucheu S. et al. Xanthine oxidase activity and lipid peroxide content following different types of ischemia in the isolated rat heart. *Agents Action,* 1994, v. 41, No. 3-4, pp. 144-150.

Cowan J. C. and Vaughan Williams E. M. The effects of various fatty acids on action potential shortening during sequential periods of ischaemia and reperfusion. *J. Mol. Cell. Cardiol.*, 1980, v. 12, No. 3, pp. 347-369.

Crake T. and Poole-Wilson P. A. Evidence that calcium influx on reoxygenation is not due to cell membrane disruption in the isolated rabbit heart. *J. Mol. Cell Cardiol.,* 1986, Suppl. 4, pp. 31-35.

Crinnion J. N., Homer-Vanniasinkam S. and Gough M. J. Skeletal muscle reperfusion injury: pathophysiology and clinical considerations. (Review.) *Cardiovasc. Surg.*, 1993, v. 1, No. 4, pp. 317-324.

Cristol J. P., Thiemermann C., Mitchell J. A. et al. Support of renal blood flow after ischemic-reperfusion injury by endogenous formation of nitric oxide and of cyclo-oxygenase vasodilator metabolits. *Brit. J. Pharmacol.,* 1993, v. 109, No. 1, pp. 188-194.

Crome R. and Manning A. S. Allopurinol and reperfusion-induced arrhythmias: increased protection by simultaneous administration of anti-oxidant enzymes. *Cardiovasc. Drugs Ther.*, 1988, v. 2, No. 3, pp. 295-304.

Cronenwett J. L., Lee K. R., Shlafer M. and Zelenock G. B. The effect of ischemia-reperfusion derived oxygen free radicals on skeletal muscle calcium metabolism. *Microcirculation. Endothelium. Lymphatics.,* 1989, v. 5, No. 3-5, pp. 171-187.

Cunarro J. A., Schults S. E., Johnson W. A. and Weiner M. W. Effects of ischemia on metabolite concentrations in dog renal cortex. *Renal Physiol.,* 1982, v. 15, No. 3, pp. 143-155.

Cuzzocrea S., Zingarelli B. and Caputi A. P. Role of constitutive nitric oxide synthase and peroxynitrite production in a rat model of splanchnic artery occlusion shock. *Life Sci.,* 1998, v. 63, No. 9, pp. 789-799.

Dadoukis I., Angouridakis K. and Aletras M. The action of the basic trypsin inhibitor, trasylol, on shock resulting from occulusion of the superior mesenteric artery: experimental study. *J. Int. Med. Res.,* 1981, v. 9, pp. 31-39.

D'Agnillo F. and Chang T. M. Polyhemoglobin-superoxide dismutase-catalase as a blood substitute with antioxidant properties. *Nat. Biotechnol.,* 1998, v. 16, No. 7, pp. 667-671.

Dagys A. I., Toleykis A. I. and Praskevičius A. K. [Fatty acid composition of phospholipids in ischemia-damaged cardiac mitochondria]. (Russian). *Biul. Eksp., Biol. Med.,* 1985, v. 99, No. 1, pp. 43-45. English transl. see[1*], pp. 53-55.

Daemen M. A., van de Ven M. W., Heineman E. and Buurman W. A. Involvement of endogenous interleukin-10 and tumor necrosis factor-alpha in renal ischemia-reperfusion injury. *Transplantation,* 1999, v. 67, No. 6, pp. 792-800.

Damron D. S., Summers B. A. Arachidonic acid enhances contraction and intracellular Ca^{2+} transients in individual rat ventricular myocytes. *Am. J. Physiol.,* 1997, v. 272, No. 1, Pt. 2, pp. H350 – H359.

Danilov V. S., Sitkovsky M. V., Kagan V. E. and Kozlov Yu. P. [Investigation of the lipid peroxidation under normal and pathological conditions using the polarographic method]. (Russian). *Izv. Akad. Nauk SSSR,* Ser. Biol., 1972, No. 4, pp. 574-579.

Daoust R. and Morais R. Degenerative changes, DNA synthesis and mitotic activity in rat liver following single exposure in diethylnitrosamine. *Chem. Biol. Interact.,* 1986, v. 57, pp. 55-64.

Darius H., Osborne J. A., Reibel D. K. et al. Protective actions of a stable prostacyclin analog in ischemia induced membrane damage in rat myocardium. *J. Mol. Cell. Cardiol,* 1987, v. 19, No. 3, pp. 243-250.

Darsee J. R., Kloner R. A. and Braunwald E. Time course of regional function after coronary occlusions of 1 to 120 min duration. *Am. J. Physiol.,* 1981, v. 240, No. 3, pp. 399-407.

Das D. K., Engelman R. M., Clement R. et al. Role of xanthine oxidase inhibitor as free radical scavenger: a novel mechanism of action of allopurinol and oxypurinol in myocardial salvage. *Biochem. Biophys. Res. Commun.,* 1987, v. 148, pp. 314-319.

Das D. K., Engelman R. M., Rouson J. A. et al. Pathophysiology of superoxide radical as potential mediator of ischemic and reperfusion injury in pig heart. *Basic. Res. Cardiol.,* 1986, v. 81, No. 2, pp. 155-166.

Das D. K., Engelman R. M., Rauson J. A. et al. Aerobic vs. anaerobic metabolism during ischemia in heart muscle. *Ann. Chir. Gynaecol.,* 1987, v. 76, No. 1, pp. 68-76.

Datchenko Z. M., Donchenko G. V., Shakhman E. V. et al. [The role of phospholipids in functionally different cell membranes under the conditions of antioxidants system disturbance]. (Russian). *Ukr. Biokhim. Zh.,* 1996, v. 68, No. 1, pp. 49-54.

Dauber I. M., Parsons P. E., Welsh C. M. et al. Peripheral bypass-induced pulmonary and coronary vascular injury. Association with increased levels of tumor necrosis factor. *Circulation,* 1993, v. 88, No. 2, pp. 726-735.

Daugherty J. P., Davis S. and Yielding K. L. Inhibition by butilated hydroxytoluene of excision repair synthesis and semiconsertative DNA synthesis. *Biochem. Biophys. Res. Commun.,* 1978, v. 80, pp. 963-969.

Davies M. G. and Hagen P. The vascular endothelium: a new horizon. (Review). *Ann. Surg.,* 1993, v. 218, No. 5, pp. 593-609.

Davies K. J. A., Wiese A. G., Sevanian A. and Kim E. H. Repair systems in oxidative stress. In: Finch C. E. and Johnson T. E., eds. *Molecular biology of aging.* N. Y., A. R. Liss Inc., 1990, pp. 123-141.

Davies M. G., Fulton G. J. and Hagen P. O. Clinical biology of nitric oxide (Review). *Br. J. Surg.,* 1995, v. 82, N12 pp. 1598-1610.

Davydov V. V., Tverdokhleb V. P. and Yakushev V. S. [Creatine phosphate content and creatine phosphokinase activity in cardiomyocyte mitochondria in myocardial infarction in rats exposed to emotional and pain stress]. (Russian). *Patol. Fiziol. Eksp. Ter.,* 1983, No. 1. pp. 33-36.

De M., Hunt J. S. and Soares M. J. Stimulation of rat placental cell DNA synthesis by transferrin. *Biol. Reprod.,* 1988, v. 38, No. 5, pp. 1123-1128.

Decker R. S. and Wildenthal R. Lysosomal alterations in hypoxic and reoxygenated hearts. I. Ultrastructural and cytochemical changes. *Am. J. Pathol.,* 1980. v. 98, pp. 425-444.

Decker R. S., Poole A. R., Crie J. S. et al. Lysosomal alterations in hypoxic and reoxygenated hearts. II. Immunohistochemical and biochemical changes in cathepsin D. *Am. J. Pathol.,* 1980, v. 98, No. 2, pp. 445-456.

Deeb G. M., Grum C. M., Lynch M. J. et al. Neutrophils are not necessary for induction of ischemia/reperfusion lung injury. *J. Appl. Physiol.,* 1990, v. 68, pp. 374-381.

De Forge L. E., Fantone J. C., Kenney J. S. and Remick D. G. Oxygen radical scavengers selectively inhibit interleukin 8 production in human whole blood. *J. Clin. Invest.*, 1992, v. 90, No. 5, pp. 2123-2129.

Defraigne J. O., Detry O., Pincemail J. et al. Direct evidence of free radical production after ischaemia and reperfusion and protective effect of desferrioxamine: ESR and vitamin E studies. Eur. *J. Vasc. Surg.*, 1994, v. 8, No. 5, pp. 537-543.

Defraigne J. O., Pincemail J., Laroche C. et al. Successful controlled limb reperfusion after severe prolonged ischemia. *J. Vasc. Surg.*, 1997, v. 26, No. 2, pp. 346-350.

Degterev I. A. and Zaikov G. E. [Ionol: distribution in the body, metabolism and biological action. I. Distribution in the body and metabolism. Review]. (Russian). *Khim. -Farm. Zh.,* 1985, v. 19, No. 8, pp. 910-919.

De Jong J. W., Vander Meer P., Huizer T. et al. Does xanthine oxidase cause damage during myocardial ischemia? *Bratisl. Lek. Listy,* 1991, v. 92, No. 1, pp. 41-47.

De La Lastra C. A. Melatonin protects against gastric ischemia-reperfusion injury in rats. *J. Pineal. Res*., 1997, v. 23, No. 2, pp. 47-52.

De Leiris J. and Opie L. H. Effect of substrates and of coronary artery ligation on mechanical performance and on release of lactate dehydrogenase and creatine phosphokinase in isolated working rat hearts. *Cardiovasc. Res.*, 190, No. 12, p. 585-596.

De Leyn P. R., Lerut T. E., Schreinmakers H. H., et al. Effect of inflation on adenosine truiphosphate catabolism and lactate production during normothermic lung ischemia. *Ann. Thor. Surg.,* 1993, v. 55, No. 5, pp. 1074-1078.

del Nido P. J., Nakamura H., Mickle D. A. et al. Maturational difference in functional/metabolic sequelae of free radical formation on reperfusion. *J. Surg. Res*., 1989, v. 46, No. 6, pp. 532-536.

Delius R. E. and Hinshaw D. B. Metabolic inhibition potentiates oxidant injury. *J. Surg. Res*., 1991, v. 50, No. 4, pp. 314-322.

Del Maestro R. F., Björk J. and Azfors K. E. Increase on microvascular permeability induced by enzymatically generated free radicals. I. *In vivo* study. *Microvasc. Res.,* 1981, v. 22, No. 3, pp. 239-270.

de Lorgeril M., Rousseau G., Basmadjian A. et al. Spacial and temporal profiles of neutrophil accumulation in the reperfused ischemic myocardium. *Am. J. Cardiovasc. Pathol.,* 1990, v. 3, No. 2, pp. 143-154.

Del Zoppo G. J. The role of platelets in ischemic stroke. *Neurology*, 1998, v. 51, No. 3 (Suppl. 3), pp. S9-S14.

Del Zoppo G. J., Schmid-Schonbein G. W., Mori E. et al. Polymorphonuclear leukocytes occlude capillaries following middle cerebral artery occlusion and reperfusion in baboons. *Stroke*, 1991, v. 22, No. 10, pp. 1276-1283.

De Marchena O., Guarnieri M. and McKhann G. Glutathione peroxidase levels in brain. *J. Neurochem*., 1974, v. 22, pp. 773-776.

De Medio G. E., Goracci G. and Horrocks L. A. et al. The effect of transient ischemia on fatty acid and lipid metabolism in the gerbil brain. *Ital. J. Biochem.,* 1980, v. 29, pp. 412-432.

De Mello Filho A. C. and Menehgini R. Protection of mammalian cells by o-phenanthroline from lethal and DNA-damaging effects produced by active oxygen species. *Biochim. Biophys. Acta*, 1985, v. 847, pp. 82-88.

Demopoulos H. B., Flamm E. S., Pietronigro D. D. and Seligman H. L. The free radical pathology and the microcirculation in the major central nervous system disorders. *Acta Physiol. Scand.,* 1980, v. 492, Suppl., pp. 91-119.

Demurov E. A. and Ignatova V. A. [Metabolic and neurohumoral mechanisms of miocardium ischemic injury]. (Russian). *Itogy Nauki i techniky. Ser. Fiziol.* Moscow, VINITI, 1985, v. 30, 159 p.

Demurov E. A., Mil'chakov V. I. and Gerasimov A. M. [The effect of hyperbaric oxygenation combined with α-tocopherol on the contractile function and some components of the heart affected by adrenalin]. (Russian). *Kardiologiya*, 1984, v. 24, No. 1, pp. 67-70.

Denisov E. T. and Azatjain V. V. [The inhibition of chain reactions]. (Russian). *Chernogolovka, RAN,* 1997, 268 p.

Depre C., Fierain L. and Hue L. Activation of nitric oxide synthase by ischaemia in the perfused heart. *Cardiovasc. Res.*, 1997, v. 33, No. 1, pp. 82-87

Desai I. D., Sawant P. L. and Tappel A. L. Peroxidative and radiation damage to isolated lysosomes. *Biochem. Biophys. Acta*, 1964, v. 86, No. 2, pp. 277-285.

Desmukh D. R., Mirochnitchenko O., Ghole V. S. et al. Intestinal ischemia and reperfusion injury in transgenic mice overexpressing copper-zinc superoxide dismutase. *Am. J. Physiol.*, 1997, v. 273, No. 4, Pt. 1, pp. C1130-C1135.

De Windt L. J., Reneman R. S., Van der Vusse G. J. and Van Bilsen M. Phospholipase A2-mediated hydrolysis of cardiac phospholipids: the use of molecular and transgenic techniques. *Mol. Cell Biochem.,* 1998, v. 180, No. 1-2, pp. 65-73.

Dhalla N. S., Das P. K. and Sharma G. P. Subcellular basis of cardiac contractile failure. Concise review. *J. Mol. Cell. Cardiol.*, 1978, v. 10, pp. 363-385.

Dinarello C. A. Inflammatory cytokines: interleukin-1 and tumor necrosis factor as effector molecules in autoimmune diseases. *Current Opinion in Immunology*, 1991, v. 3, pp. 941-948.

Dirks D. R., Faiman M. D. and Huyser E. S. The role of lipid, free redical initiator, and oxygen on the kinetics of lipid peroxidation. *Toxicol. Appl. Pharmac.,* 1982, v. 63, pp. 21-28.

Dmitriev L. F. [Malondyaldehyde can control cell division on the DNA replication stage (hypothesis)]. (Russian). *Zh. Evol. Biokhim. Fiziol.,* 1992, v. 28, No. 6, pp. 720-730.

Dmitrieva A. V., Stefanov A. V., Bojko V. I. et al. [Cardiostimulating effect of the norepinephrine contained in liposomes in condition of adrenoreceptor blockade]. (Russian). *Fiziol. Zh. Akad. Nauk SSSR,* 1983, v. 69, No. 8, pp. 1023-1030.

Dobrina A. and Rossi F. Metabolic properties of freshly isolated bovine endothelial cells. *Biochim. Biophys. Acta*, 1983, v. 762, pp. 295-301.

Doctor R. B. and Mandel L. J. Minimal role of xanthine oxidase and oxygen free radicals in rat tubular reoxygenation injury. *J. Am. Soc. Nephrol.,* 1991, v. 1, pp. 959-969.

Dolgikh V. T. [The antioxidant ionol as an agent preventing pustresuscitation cardiac insufficiency]. (Russian). *Patol. Fiziol. Eksp. Ter.,* 1984, No. 1, pp. 45-50.

Dolgova N. P. [Changes in glycogenolysis in the zone of ischemia in experimental myocardial infarction]. (Russian). *Biul. Eksp. Biol. Med.*, 1980, v. 89. No. 3, pp. 304-307. English transl. see[1*], pp. 300-303.

Dolinin V. A. and Vinogradov V. M. [Principal trends in research in the area of prevention of trautamic shock by means of pharmocologic agents]. (Russian) *Voen. Med. Zh.,* 1983, No. 4, pp. 26-30.

Donchenko G. V., Kuzmenko I. V., Kovalenko V. N. and Sidorenko L. V. [Determination of adenine nucleotide and inorganic phosphate contents in the rat liver with E-hypovitaminosis and administration of α-tocopherol and ubiquinone-g]. (Russian). *Ukr. Biokhim. Zh.,* 1981, v. 53, No. 3, pp. 94-98.

Dougherty V. H., Levy D. E. and Weksler B. B. Experimental cerebral ischemia produces platelet aggregates. *Neurology*, 1979, v. 29, pp. 1460-1465.

Downey J. M. Free radicals and their involvement during long-term myocardial ischemia and reperfusion. *Annu. Rev. Physiol.* 1990, v. 52 HoffmanJ. F., De Weer P., eds. pp. 487-504.

Downey J. M., Hearse D. J. and Yellon D. M. The role of xanthine oxidase during myocardial ischemia in several species including man. *J. Mol. Cell Cardiol.*, 1988, v. 20 (Suppl. II), pp. 55-63.

Drossos G., Lazou A., Panagopoulos P. and Westaby S. Deferoxamine cardioplegia reduces superoxide radical production in human mycardium. *Ann. Thor. Surg.,* 1995, v. 59, No. 1, pp. 169-172.

Drugas G. T., Paidas C. N., Jahanda A. M. et al. Conjugated desferoxamine attenuates hepatic microvascular injury following ischemia/reperfusion. *Circulat. Shock,* 1991, v. 34, No. 2, pp. 278-283.

Du G., Mouithys-Mickalad A. and Sluse F. E. Generation of superoxide anion by mitochondria and impairment of their functions during anoxia and reoxygenation in vitro. *Free Radic. Biol. Med.* 1998, v. 25, No. 9, pp. 1066-1074.

Dudchenko A. M., Chernobaeva G. N., Belousova V. V. et al. [Bioenergetic parameters of the brain in rats with different resistance to hypoxia]. (Russian). *Biul. Eksp. Biol. Med.*, 1993, v. 115, No. 3, pp. 251-254. English transl. see[1*], pp. 263-267.

Dudnik L. *Intensification of lipid peroxidation processes and their influence on the physico-chemical properties and structural organization of ischemic liver membranes.* [Dissertation]. (Russian). Moscow. 1982.

Dudnik L. B., Bilenko M. V. Alessenko A. V. et al. [Intensification of peroxidation and changes in lipid composition in the homogenates and subcellular fractions of ischemic liver]. (Russian). *Biul. Eksp. Biol. Med.*, 1980, v. 89, No. 5, pp. 556-558. English transl. see[1*], pp. 605-607.

Dudnik L. B., Bilenko M. V., Alessenko A. V. et al. [Role of peroxidation in destruction of membrane lipids under conditions of liver tissue ischemia]. (Russian). *Vopr. Med. Khim.*, 1981, v. 22, No. 3, pp. 380-383.

Dudnik L. B., Tikhaze A. K., Alessenko A. V. et al. [Changes in superoxide dismutase and glutathione peroxidase activity during intensification of lipid peroxidation in ischemic liver]. (Russian). *Biul. Eksp. Biol. Med.*, 1981, v. 91, No. 4, pp. 451-453. English transl. see[1*], pp. 488-489.

Dupin A. M., Bemanandzara M., Stvolinsky S. L. et al. [Muscle dipeptides: natural inhibitors of lipid peroxidation]. (Russian). *Biokhimiya*, 1987, v. 52, No. 5, pp. 782-787. English transl. see[3*], pp. 672-676.

Dusting G. J. and Vane J. R. Some cardiovascular properties of prostacyclin (PgI_2) which are not shared by PGE_2. *Circ. Res.*, 1980, v. 46, Suppl. 1, pp. 1-183.

du Toit E. F., McCarthy J., Miyashiro J. et al. Effect of nitrovasodilators and inhibitors of nitric oxide synthase on ischemic and reperfusion function of rat isolated hearts. *Br. J. Pharmacol.,* 1998, v. 123, No. 6, pp. 1159-1167.

Dwivedi R. S., Kaur G., Srivastava R. C. and Murti C. R. K. Lipid peroxidation in tin intoxicated partially hepatechtomized rats. *Bull. Inviron. Contam. Toxicol,* 1984, v. 33, No. 2, pp. 200-209.

Dzhafarov A. I. [Effect of anoxia under different conditions on kinetics of lipid peroxidation in cell organelles]. (Russian). *Biul. Eksp. Biol. Med.*, 1981, v. 92, No. 10, pp. 425-427. English transl. see[1*], pp. 1347-1351.

Dzhafarov A. I. [*Lipid peroxidation in isolated storage tissues*]. (Russian). Dissertation. Minsk, 1983.

Dzhafarov A. I., Magomedov N. M., Kulieva E. M. et al. [Lipid peroxidation in external and internal mitochondrial membranes during anoxia]. *Biul. Eksp. Biol. Med.*, 1985, v. 100, No. 10, pp. 433-435. English Transl. see [*], pp. 1359-1361.

Ebeigde A. B. Influence of hypoxia on contractility and calcium uptake in rabbit aorta. *Experientia*, 1982, v. 38, No. 8, pp. 935-987.

Eddy L., Hurvitz R. and Hochstein P. A protective role for ascorbate in induced ischemic arrest associated with cardiopulmonary bypass. *J. Appl. Cardiol.,* 1990, v. 5, pp. 409-414.

Eddy L., Steward J., Jones H. et al. Xanthine oxidase is detected in ischemic rat heart but not human hearts. *Physiologist*, 1986, v. 29, p. 166.

Edgar A. D., Strosznajder J., and Horrocks L. A. Activiation of ethanolamine phospholipase A_2 in brain during ischemia. *J. Neurochem.*, 1982, v. 39, pp. 1111-1116.

Edlund A., Sahlin K. and Wennmalm A. Effect of prostacyclin on the severity of ishemic injury in rabbit hearts subjected to coronary ligation. *J. Mol. Cell. Cardiol.*, 1986, v, 18. No. 10, pp. 1067-1076.

Ehrhart I. C., and Hofman W. F. Oleic acid dose-related edema in isolated canine lung perfused at constant pressure. *J. Appl. Physiol.,* 1981, v. 50, No. 6, pp. 115-120.

Ekert P., MacLusky N., Luo X. P., et al. Dexamethasone prevents apoptosis in a neonatal rat model of hypoxic-ischemic encephalopathy (HIE) by a reactive oxygen species-independent mechanism. *Brain Res.* 1997, v. 747, No. 1, pp. 9-17.

Ekholm R., Kerstell I., Olson R. E. et al. Morphologic and biochemical studies of dog heart mitochondria after short period of ischemia. *Am. J. Cardiol.,* 1986, v, 22, No. 3, pp. 312-318.

El Hage Chahine J. M. and Fain D. The mechanism of iron release from transferrin slow-proton-transfer-induced loss of nitrilotriacetatoiron (III) complex in acidic media. *Eur. J. Biochem.,* 1994, v. 223, No. 2, pp. 581-587.

El Hage Chahine J. M. and Pakdaman R. Transferrin, a mechanism for iron release. *Eur. J. Biochem.* 1995, v. 230, No. 3, pp. 1102-1110.

Elsky V. N., Manankov A. K., Mareyeva T. E. and Yakubenko Y. D. The role of lipid peroxidation in shock pathogenesis. In: Workshop on: *Oxygen free-radicals in shock.* Florence, 1985, p. 34.

Emanuel N.M. [Physicochemical basis of phenolic compounds application in chemistry and biology]. (Russian). *Phenolic compounds and their biological functions. 1st All-Union Symposium. Proceedings.* Moscow, Nauka, 1968, pp. 311-331.

Emanuel N. M. and Lyashovskaya Yu. N. [*Inhibition of lipid. peroxidation*]. (Russian). Moscow, Pishchepromizdat, 1961, 359 p.

Engelman R. M., Adler S., Gouge T. H. et al. The effect of normothermic anoxic arrest and venticular fibrillation on the coronary blood flow distribution of the pig. *Thorac. Cardiovasc. Surg.*, 1975, v. 69, No. 6, pp. 658-869.

Engerson T. D., McKelvey T. G., Rhyne D. B. et al. Conversion of xanthine dehydrogenase to oxidase in ischemic rat tissues. *J. Clin. Invest.*, 1987, v. 79, pp. 1564-1570.

Enseleit W. H., Domer F. K., Jarott D. M. et al. Cerebral phospholipid content and Na^+, K^+ATPase activity during ischemia and postischemic reperfusion in the mongolian gerbil. *J. Neurochem.*, 1984, v. 43, No. 2, pp. 320-327.

Epifanova O. I., Terskikh V. V. and Polunosvky V. A. [Rugulational mechanisms of cell proliferation]. (Russian). *Itogi Nauki i Techniki. Ser. Gen. Probl. Phys. Chem. Biol.*, Moscow, 1988, v. 10, 163 p.

Epshtein I. M. [Role of barrier function of plasmatic membrane in mechanism of regulation of tissues respiration]. (Russian). In: *Biological membranes in norma and pathology*. Moscow, 1972, pp. 69-70.

Eriksson E, Replagle R. L. and Geagov S. Reperfusion of skeletal muscle after warm ischemia. *Ann. Plast. Surg.*, 1977, v. 18, No. 3, pp. 224-229.

Escott K. J., Beech J. S., Haga K. K. et al. Cerebroprotective effect of the nitric oxide synthase inhibitors, 1-(2-trifluoromethylphenyl) imidazole and 7-nitro indazole, after transient focal cerebral ischemia in the rat. *J. Cereb. Blood Flow Metab.*, 1998, v. 18, No. 3, pp. 281-287.

Eskola I. and Laakso S. Bile salt-dependent oxygenation of polyansaturated phosphatidylcholines by soybean lypoxigenase-1. *Biochim. Biophys. Acta*, 1983, v. 751, pp. 305-311.

Esterbauer H., Cheeseman K. H., Dianzani M. U. et al. Separation and characterization of the aldehydic products of lipid peroxidation stimulated by ADP-Fe^{2+} in rat liver microsomes. *Biochem. J.*, 1982, v. 208, pp. 129-140.

Estevez A. Y. and Phillis J. W. The phospholipase A2 inhibitor, quinacrine, reduces infarct size in rats after transient middle cerebral artery occlusion. *Brain Res.*, 1997, v. 752, No. 1-2, pp. 203-208.

Estrada C., Gomez C., Martin C. et al. Nitric oxide mediates tumor necrosis factor-alpha cytotoxicity in endothelial cells. *Biochem. Biophys. Res. Commun.*, 1992, v. 186, No. 1, pp. 475-482.

Evans N. T. S., Naylor PFO, and Rowlinson G. Oxygen diffusion transients in the mouse ear. *Respir. Physiol.*, 1982, v. 48, pp. 387-394.

Evans P. J., Whiteman M., Tredger J. M. and Halliwell B. Antioxidant properties of S-adenosyl-L-methionine: a proposed addition to organ storage fluids. *Free Radic. Biol. Med.*, 1997, v. 23, No. 7, 1002-1008.

Evers A. S., Murphree S., Saffitz J. E. et al. Effects of endogenously produced leukotrienes, thromboxane, and prostaglandins on coronary vascular resistance in rabbit myocardial infarction. *J. Clin. Invest.*, 1985, v. 75, No. 3, pp. 992-999.

Ezra D., Boyd L. M., Feuerstein G. and Godlstein R. E. Coronary construction by leukotriene C_4, D_4 and E_4 in the intact pig heart. *Am. J. Cardiol.*, 1983, v. 51, No. 8, pp. 1451-1454.

Fabian R. H. and Kent T. A. Superoxide anion production during reperfusion is reduced by an antineutrophil antibody after prolonged cerebral ischemia. *Free Radic. Biol. Med.*, 1999, v. 26, No. 3-4, pp. 355-361.

Fabrikant J. I. The kinetics of cellular proliferation in regenerating liver. *J. Cell. Biol.*, 1968, v. 36, No. 3, pp. 551-565.

Fabrikant J. I. Size of proliferating pools in regenerating liver. *Exp. Cell. Res.*, 1969, v. 55, No. 2, pp. 277-279.

Fantini G. A. and Yoshioka T. Deferoxamine prevents lipid peroxidation and attenuates reoxygenation injury in postischemic skeletal muscle. *Am. J. Physiol.*, 1993, v. 264, No. 6, Pt. 2, pp. H1953-H1959.

Farber J. L., Chien K. R and Mittnacht S. The pathogenesis of irreversible cell injury in ischemia. *Am. J. Pathol.*, 1981, v. 102, pp. 271-281.

Fedorova L. M., Arkhangelskaya S. L. and Bilenko M. V. [Evaluation of possibility to use prodigiosan for obtaining extracts stimulating proliferation in the resected liver]. (Russian). *Biul. Eksp. Biol. Med.*, 1992, v. 64, No. 10, pp. 405-406. English transl. see[1*], pp. 1503-1505.

Feller A. M., Roth A. C., Russell R. C. et al. Experimental evaluation of oxygen free radical scavengers in the prevention of reperfusion injury to skeletal muscle. *Ann. Plastic Surg.* 1989, v. 22, No. 4, pp. 321-331.

Fercakova A., Marsala M. and Marsala J. Influence of graded postischemic reoxygenation on reperfusion alterations in rabbit dorsal root ganglion neurons. *J. Hirnforschung.*, 1994, v. 35, No. 2, pp. 295-302.

Ferrante R. J., Hobson R. W. 2nd, Miyasaka M. et al. Inhibition of white blood cell adhesion at reperfusion decreases tissue damage in postichemic striated muscle. *J. Vasc. Surg.*, 1996, v. 24, No. 2, pp. 187-193.

Ferrari R., Albertini A., Curello S. et al. Myocardial recovery during postischaemic reperfusion: effects of nifedipine, calcium and magnesium. *J. Mol. Cell. Cardiol.*, 1986, v. 18, pp, 487-497.

Ferrari R. P., Battiston B., Brunelli G. et al. The role of allopurinol in preventing oxygen free radical injury to skeletal muscle and endothelial cells after ischemia-reperfusion. *J. Reconstr. Microsurg.*, 1996, v. 12, No. 7, pp. 447-450.

Ferreira R., Burgos M., Llesuy S. et al. Reduction of reperfusion injury with mannitol cardioplegia. *Ann. Thorac. Surg.*, 1989, v. 48, No. 1, pp. 77-83.

Ferreira R., Burgos M., Milei J. et al. Effect of supplementing cardioplegic solution with deferoxamine on reperfused human myocardium. *V. Thorac. Cardiovasc. Surg.*, 1990, v. 100, No. 5, pp. 708-714.

Ferrero M. E., Orsi R., Bernelli-Zazzera A. Effect of ischemia on drug-metabolizing microsomal enzymes in rat liver. *Exp. Molec. Path.*, 1978, v. 28, pp. 256-266.

Fiedler V. B. Reduction of acute myocardial ischemia in rabbit hearts by nafazatrom. *J. Cardiovasc. Pharmacol.,* 1984, v. 6, pp. 318-324.

Fiedler V. B., Mardien M. and Abrom T. S. G. Leukotriene D_4-induced vasoconstriction of coronary arteries in anesthetized dogs. *Eur. Heart. J.,* 1984, v. 5, No. 3, pp. 253-260.

Fiehn W. and Hasselbach W. The effect of phospholipase A on the calcium transport and the role of unsaturated fatty acids in ATPase activity of sarcoplasmic vesicles. *Eur. J. Biochem.,* 1970, v. 13, pp. 510-518.

Finkelstein S. D., Gilfor D. and Farber J. L. Alterations in the metabolism of lipids in ischemia of the liver and kidney. *J. Lipid Res.,* 1985, v. 26, No. 6, pp. 726-734.

Fiolet J. W., Baartscheer A., Schumacher C. A. et al. Transmural inhomogeneity of energy metabolism during acute global ischemia in the isolated rat heart: dependence of environmental conditions. *J. Mol. Cell. Cardiol.*, 1985, v. 17, No. 1, pp. 87-92.

Fisher A. B., Dobia Ch., Monovichy Y. et al. Non-selenium phospholid hydroperoxide glutathione peroxidase (PHGPx): a novel anti-oxydant enzyme. *Free Rad. Biol. Med.* 1998, v. 25, suppl. 7, p. S29, No. 60

Fisher D. J. Comparative effects of metabolism acidemia and hypoxemia on cardiac output and regional blood flow in unanesthetized newborn lambs. *Pediatr. Res.,* 1986, v. 20, No. 8, pp. 756-760.

Flamm E. S., Demopoulos H. B., Seligman M. L. et al. Free radical in cerebral ischemia. *Stroke*, 1978, v. 9, No. 5, pp. 445-447.

Flamm E. S., Seligman M. L. and Demopoulos H. B. Barbiturate protection of the ischemic brain. In: Cottrell J. E. and Turndorf M., eds. *Anesthesia and neurosurgery.* 1980, pp. 248-266.

Flogel U., Decking U. K., Godecke A. and Schrader J. Contribution of NO to ischemia-reperfusion injury in the saline-perfused heart: a study in endothelial NO synthase knockout mice. *J. Mol. Cell. Cardiol.*, 1999, v. 31, No. 4, pp. 827-836

Flohe L. Glutation peroxidase brought into focus. In: Pryor W. A., ed. *Free Radicals in Biology*. N. Y., Acad. Press., 1982, v. 5, pp. 223-254.

Flyorov M. A., Borovitskaya A. E. and Tolstukhina T. I. [Phosplolipase A action on ATPase systems of subcellular fractions from cranial brain]. (Russian). In: *Biochemical and physiological basis of training processes in nervous system.* L., 1982, No. 23, pp. 91-95.

Folch B. J., Lees M. and Stanley H. S. A simple method for the isolation and purification of total lipids from animal tissues. *Biol. Chem.*, 1957, v. 226, pp. 497-509.

Forman L. J., Liu P., Nagele R. G. et al. Augmentation of nitric oxide, superoxide, and peroxynitrite production during cerebral ischemia and reperfusion in the rat. *Neurochem. Res.*, 1998, v. 23, No. 2, pp. 141-148.

Fornander J., Hellman A. and Hasselgren P. O. Effects of methylprednisolone on protein synthesis and blood flow in the postischemic liver. *Circulat. Shock*, 1984, v. 12, No. 4, pp. 289-295.

Forrest V. J., Kang. Y. -H., McClain D. E. et al. Oxidative stress-induced apoptosis prevented by trolox. *Free Rad. Biol. Med.*, 1994. v. 16, No. 6, pp. 675-684.

Fortster C., Clark H. B., Ross M. E. and Iadecola C. Inducible nitric oxide synthase expression in human cerbral infarcts. *Acta. Neuropathol.* (Berl)1999, v. 97, No. 3, pp. 215-220.

Fortney S. R. and Lynn W. S. Role of ascorbate and cysteine on swelling and lipid peroxidation in rat liver mitochondria. *Arch. Biochem. Biophys.*, 1964. v. 104, No. 2, pp. 224-247.

Foschi D., Castoldi L., Lesma A. et al. Effects of ischaemia and reperfusion on liver regeneration in rats. *Eur. J. Surg.*, 1993, v. 159, No. 8, pp. 393-398.

Fox A. C., Reed G. E., Meilman H. et al. Release of nucleosides from canine and human hearts as an index of prior ischemia. *Am. J. Cardiol.*, 1979, v. 43, No. 1, pp. 52-58.

Fox K. A. A., Saffitz J. E. and Corr P. B. Pathophysiology of myocardial reperfusion. *Cardiol. Clin.*, 1987, v. 5, No. 1, pp. 31-48.

Franceschi D., Graham D., Sarasua M. and Zollinger R. M. Mechanisms of oxygen free radical-induced calcium overload in endothelial cells. *Surgery*. 1990, v. 108, No. 2, pp. 292-297.

Francis J. W., Ren J., Warren L. et al. Postischemic infusion of Cu/Zn superoxide dismutase or SOD: Tet451 reduces cerebral infartion following focal ischemia/reperfusion in rats. *Exp. Neurol.*, 1997, v. 146, No. 2, pp. 435-443.

Frank L. Prolonged survival after paraquate. Role of the lung antioxidant enzyme systems. *Biochem. Pharmacol.*, 1981, v. 30, No. 16, pp. 2319-2324.

Franzini E., Sellak H., Marguetty C. et al. Inhibition of human neutrophil binding to hydrogen peroxide-treated endothelial cells by cAMP and hydroxyl radical scavengers. *Free Rad. Biol. Med.*, 1996, v. 21, No. 1, pp. 15-23.

Frederiks W. M. and Bosch K. S. The role of xanthine oxidase in ischemia reperfusion damage of rat liver. *Histol. Histopathol.*, 1995, v. 10, pp. 111-116.

Fredlund P. E., Öckerman P. E. and Vang J. O. Acidosis and increased plasma levels of b-D-galactosidase after hepatic inflow occlusion in the pig. *Acta Chir. Scand.*, 1974, v. 140, No. 3, pp. 234-241.

Freeman B. A., Topolsky M. K. and Crapo J. D. Hyperoxia increases oxygen radical production in rat lung homogenates. *Arch. Biochem. Biophys.*, 1982, v. 216, pp. 477-489.

Freeman B. A., Whitfield K., Gelman D. and Parks D. A. Heparin-reversible binding of xanthine oxidase to vascular endothelium. *Free Rad. Biol. Med.*, 1990, v. 9, Suppl. 1, p. 94, 10. 15.

[Free radicals and biological stabilizators]. (Russian). *The 1st Bulgarian-Soviet Symposium*. Abstracts. Sophia, 1987.

[*Free radicals in biology*]. Pryor W., ed. (Russian transl. from English). Moscow, Mir, 1979, v. 1, 308 p., v. 2, 322 p.

Fridovich I. Superoxide radical and superoxide dismutase. In: *Oxygen Living process: Interdiscip. Approach*. NY., 1981, pp. 250-272.

Fridovich I. Superoxide radical: an endogenous toxicant. *Ann. Rev. Pharmacol. Toxicol.*, 1983, v. 23, pp. 239-257.

Friedl H. P., Smith D. J., Till G. O. et al. Ischemia-reperfusion in humans. Appearance of xanthine oxidase activity. *Am. J. Pathol.*, 1990, v. 136, No. 3, pp. 491-495.

Frolkis R. A. [Some indices of myocardial energy processes in dosed restriction of coronary blood flow]. (Russian). *Kardiologiya*, 1979, v. 19, No. 9, pp. 109-107.

Frolkis J., Gurevitch J., Yuhas Y. Cellular origin for *de novo* synthesis of TNF-α in the ischemic isolated rat heart. Abstract. *J. Mol. Cell. Cardiol.*, 1997, v. 29, No. 5, p. A107.

Frolkis R. A., Voronkov G. S., Dubur G. Ya. et al. [Creatin phosphokinase activity in membranes of the myocardial sarcoplasmic reticulum in experimental coronary insufficiency]. (Russian) *Biul. Eksp. Biol. Med.*, 1983, v. 95, No. 5, pp. 53-55. English transl. see[1*], pp. 628-630.

Frolov A. F., Varbanets L. D., Antonenko S. V. et al. [Polysaccharide — a stimulator of repair synthesis of DNA]. (Russian). *Microbiol. Zh.* (Kiev), 1989, v. 51, No. 4, pp. 98-99.

Fu X., Sheng Z., Wang Y. et al. Basic fibroblast growth factor reduces the gut and liver morphologic and functional injuries after ischemia and reperfusion. *J. Trauma*, 1997, v. 42, No. 6, pp. 1080-1085.

Fujimura M., Morita-Fujimura Y., Kawase M. et al. Manganese superoxide dismutase mediates the early release of mitochondrial cytochrome C and subsequent DNA fragmentation after permanent focal cerebral ischemia in mice. *J. Neurosci.*, 1999, v. 19, No. 9, pp. 3414-3422.

Fujikawa M., Kamiike W., Hatanaka N., et al. Changes in biliary glutathione level during ischemia reperfusion of rat liver. *J. Surg. Res.*, 1994, v. 57, No. 5, pp. 569-573.

Fujita T., Furitsu H., Nishikawa M. et al. Therapeutic effects of superoxide dismutase derivatives modified with mono-or polysaccharides on hepatic injury induced by ischemia/ reperfusion. *Biochem. Biophys. Res. Commun.*, 1992, v. 189, No. 1, pp. 191-196.

Fujiwara H., Ashraf M., Sato S. and Millard R. W. Transmural cellular damage and blood flow distribution in early ischemia in pig hearts. *Circ. Res.*, 1982, v. 51, No. 6, pp, 683-693.

Fuks Z., Persaud R. S., Alfieri A. et al. Basic fibroblast growth factor protect endothelial cells against radiation-induced programmed cell death *in vitro* and *in vivo*. *Cancer Res.*, 1994, v. 54, No. 10, pp. 2582-2590.

Fukuda K., Kojiro M. and Chiu J. F. Demonstration of extensive chromatin cleavage in transplanted morris hepatoma 7777 tissue: apoptosis or necrosis? *Am. J. of Pathol.*, 1993, v. 142, No. 3, pp. 935-946.

Fuller B. J., Lunec J., Healing G. et al. Reduction of susceptibility to lipid peroxidation by desferioxamine in rabbit kidneys subjected to 24-hour cold ischemia and reperfusion. *Transplantation*, 1987, v. 43, No. 4, pp 604-606.

Fyodorova Z. P., Fyodorov V. I. and Shmyreva T. I. [The influence of sodium succinate on the kidney activity after acute massive bleeding]. (Russian). *Fiziol. Zh.* (Kiev), 1980, v. 26, No. 5, pp. 607-611.

Gabrielyan E. S. and Amroyan E. A. [Prostaglandins and the regulation of cerebral circulation under conditions of altered gasious composition of the blood]. (Russian). *Vestn. Akad. Med. Nauk SSSR*, 1982, No. 5, pp. 38-45.

Gad S. C., Leslie S. W. and Acosta D. Inhibitory actions of butylated hydroxytoluene on isolated ileac, atrial, and perfused heart preparations. *Toxicol. Appl. Pharmacol.*, 1979, v. 49, No. 1, pp. 45-52.

Gajkowska B., Viron A. and Cholewinski M. Immunocytochemical localization of endothelial nitric oxide synthase (eNOS) and inducible nitric oxide synthase (iNOS) in rat neurohypophysis after transient cerebral ischemia. *Folia Neuropathol* 1999, v. 37 No. 1., pp. 10-19.

Galaris D., Cadenas E. and Hochstain P. Redox cycling of myoglobin and ascorbate: a potential protective mechanism against oxidative reperfusion injury in muscle. *Arch. Biochem. Biophys.*, 1985, vol. 273, No. 2, pp. 497-504.

Galaris D., Cadenas E. and Hochstein P. Glutathione-dependent reduction of peroxides during ferryl- and met-myoglobin interconversion: a potential protective mechanism in muscle. *Free Rad. Biol. Med.*, 1989, v. 6, No. 2, pp. 473-478.

Galaris D., Eddy L., Arduini A. et al. Mechanisms of reoxygenation injury in myocardial infarction: implications of a myoglobin redox cycle. *Biochem. Biophys. Res. Commun.*, 1989, v. 160, No. 3, pp. 1162-1168.

Galley H. F., Walker B. E., Howdle P. D. and Webster N. R. Regulation of nitric oxide synthase activity in cultured human endothelial cells: effect of antioxidants. *Free Rad. Biol. Med.*, 1996, v. 21, No. 1, pp. 97-101.

Galperin E. I., Karagyulyan S. R. and. Abakumova O. Yu. et al. [Stimulation of restorative processes in the liver in its massive affections and resections]. (Russian). *Khirurgiya*, 1985, No. 4, pp. 82-87.

Gamrekeli D. V., Savov V. M., Stepanova L. I., et al. [The correlation of enzymatic hydroxylation reactions of benzo-(a) pyrene and lipid peroxidation in liver microsomes]. (Russian). *Biol. Nauki,* 1980, No. 8, pp. 21-24. Moscow, Vysshaya Shkola.

Ganote C. F. Cell to cell interactions. Contributing to the "Oxygen paradox". *Basic Res. Cardiol.*, 1983, v. 80. Suppl 2, pp. 141-146.

Garcia J. H., Conger K. A., Morawetz R. et al. Postischemic brain edema: quanxitation and evolution. *Adv. Neurol.*, 1980, v. 28, pp. 147-169.

Garcia-Dorado D., Theroux P., Tornos P. et al. Previous aspirin use may attenuate the severity of the manifestation of acute ischemic syndromes. *Circulation*, 1996, v. 92, No. 7, pp. 1743-1748

Gasparetto A., Corbucci G. G., Grimi G. et al. Aspects of mitochondrial oxidative damage and of oxygen free radical mechanisms in the pathogenesis of circulatory shock in humans. In: Workshop on: *Oxygen free-radicals in shock*. Florence, 1985, pp. 19-20.

Gasteva S. V., Raize T. E., and Sharagina L. M. [Phospholipid metabolism in rat brain subcellular structures during ischemia and in different periods after postischemic reperfusion]. (Russian). *Patol. Fiziol. Eksp. Ter.*, 1986, No. 6, pp. 9-11.

Gatsura V. V. and Pichugin V. V. [Intermediates of the Krebs cycle and electronacceptor systems as novel cardiopharmacologic agents]. (Russian). *Vestn. Akad. Med. Nauk SSSR*, 1982, No. 5, pp. 34-38.

Gatsura V. V. and Sidorenko A. F. [Coronarogenic metabolical acidosis and the problem of its pharmacological correction. Review]. (Russian). *Usp. Fiziol. Nauk*, 1986, v. 17, No. 4, pp. 38-56.

Gaub J. and Iversen J. 90% partial hepatectomy (PHx) in rats: A model of reversible liver insufficiency. *Hepatology*, 1982, v. 2, No. 1. p. 178.

Gaudio K. M., Thulin G., Ardito T. et al. Redistribution of cellular energy following renal ischemia. *Pediatric. Nephrol.*, 1991, v. 5, No. 5, pp. 591-596.

Gauduel Y. and Duvelleroy M. A. Role of oxygen radicals in cardiac injury due to reoxygenation. *J. Mol. Cell. Cardiol.*, 1984, v. 16, No. 5, pp. 459-470.

Gavin J. B., Thomson R. W., Humphrey S. M. et al. Changes in vascular morphology associated with the no-reflow phenomenon in ischaemic myocardium. *Virchows Arch. Path. Anat. Histol.,* 1983, v. 399, p. 325.

Gavino V. C., Dillard C. J. and Tappel A. L. Effect of dietary vitamin E and santoquin on regenerating rat liver. *Life Sci.,* 1985, v. 36, pp. 1771-1777.

Gazitt Y., Ohad I. and Logter A. Phosphorylation and dephosphorylation of membrane protein as a possible mechanism for structural rearragement of membrane componets. *Biochim. Biophys. Acta*, 1976, v. 436, pp. 1-14.

Geeraerts M. D., Ronveaux-Dupal M. F., Lemasters J. J. and Herman B. Cytosolic free Ca^{2+} and proteolysis in lethal oxidative injury in endothelial cells. *Am. J. Physiol.*, 1991, v. 261, No. 5, Pt. 1, pp. C889-C896.

Geimonen E. R., Rubtsov A. M. and Boldyrev A. A. Mechanism of thermal uncoupling of the sarcoplasmic reticulum Ca^{2+} pump: change in functional properties and structural state of the enzyme molecule. (Russian). *Biokhimiya*, v. 58, No. 8, pp. 1296-1306. English transl. see[3*], pp. 934-942.

Geisbuchler T., Altschuld R. A., Trewyn R. W. et al. Adenine nucleotide metabolism and compartmentalization in isolated adult rat heart cells. *Circ. Res.*, 1984, v. 54, No. 5, pp. 536-546.

Gelgor L., Phillips S., Butkow N. and Mitchell D. Injectable aspirin and mepyramine abolish post-ischemic hyperalgesia in rats. *Pain*, 1986, v. 26, pp. 353-359.

Gelvan D., Saltman P. and Powell S. R. Cardiac reperfusion damage prevented by a nitroxide free radical. Proc. Natl. Acad. Sci. USA, 1991, v. 88, No. 11, pp. 4680-4684.

Gerasimov A. M., Bilenko M. V., Panchenko L. F. et al. [The state of glutathione-depending way of destruction of peroxides of the preserved kidney]. (Russian). *Vestn. Akad. Med. Nauk SSSR*, 1975, No. 7, pp. 70-74.

Gerasimov A. M., Mil'chakov V. I., Demurov E. A. et al. [Inhibition of superoxide dismitase as a factor promoting myocardial dysfunction in oxygen loading]. (Russian). *Biul. Eksp. Biol. Med.*, 1984, v. 98, No. 10, pp. 400-403. English transl see[1*], pp. 1318-1321.

Gerd O. T., Hatherill J. R. and Ward P. A. Evidence for role of hydroxyl radical in tissue damage and lipid peroxidation following thermal injury. In: Workshop on: *Oxygen free-radicals in shock.* Florence, 1985, pp. 28.

Gerlach E., Nees S. and Becker B. F. The vascular endothelium: a survey of some newly evolving biochemical and physiological features. *Basic Res. Cardiol.*, 1985, v. 80, pp. 459-474.

Gevers W. Generation of protone by metabolic process in heart cells. *J. Mol. Cell. Cardiol.*, 1977, v. 9, No. 11, pp. 867-873.

Giacomini K. M., Swezey S. E. and Giacomini J. C. et al. Administration of heparin causes *in vitro* release of non-esterified fatty acids in human plasma. *Life Sci.*, 1980, v. 27, No. 9, pp. 771-780.

Gobe G. C., Axelsen R. A. and Searle J. W. Cellular events in experimental unilateral ischemic renal atrophy and in regeneration after contralateral nephrectomy. *Lab. Invest.,* 1990. v. 63. No. 6, pp. 770-779.

Gogvadze V. G., Bruetovetskii N. N. and Zhukova A. A. [Participation of phospholipase A2 in uncoupling of rat liver mitochondria inducible by lipid peroxidation products.] (Russian). *Biokhimiya*, 1990, v. 55, No. 12, Pt. 1, pp. 2195-2199. English transl. see [3*], pp. 1634-1637.

Goldblum S. E., Ding X. and Campbell-Washington J. TNF-alpha induces endothelial cell F-actin depolymerization, new actin synthesis, and barrier dysfunction. *Am. J. Physiol.*, 1993, v. 264, No. 4, Pt. 1, pp. C894-C905.

Goldenberg. H., Hüttinger, Böck P. et al. Inflence of subtotal hepatectomy on peroxisomes and peroxisomal enzymes of rat liver and isolated liver cell fractions. *Histochemistry*, 1975, v. 44, pp. 47-56.

Goldenberg V. I., Shmulovich V. G., Yurchenko N. I. et al. [Homolytic reactions of drugs. III. Biologically active salts of amines as catalysts of the homolytic decomposition of hydroperoxides]. (Russian). *Khim. Farm. Zh.*, 1983, v. 17, No. 7, pp 786-791. English transl. see[4*], pp. 464-468.

Goldhaber I. J. and Weiss I. N. Oxygen free radicals and cardiac reperfusion abnormalities. Hypertension, 1992, v. 20, No. 1, pp. 118-127.

Goldstein I. M., Kaplan H. B., Edeeson M. S. and Weissman G. Ceruloplasmin. A scavenger of superoxide anion radicals. *J. Biol. Chem.*, 1979, v. 254, No. 10, pp. 4040-4045.

Goldstein R.A. Kinetics of rubidium-82 after coronary occlusion and reperfusion: assessment of potency and viability in open-chested dogs. *J. Clin. Invest.*, 1985, v. 75, No. 4, pp. 1131-1137.

Golikov A. P., Avilova O. A., Polumiskov V. Yu et al. [Effect of dibunol and verapamil on serum creatine kinase and myoglobin levels in dogs during postischemic coronary reperfusion]. (Russian). *Biul. Eksp. Biol. Med.*, 1987, v. 104, No. 10, pp. 413-416. English transl. see[1*]. pp. 1361-1364.

Golikov A. P., Polumiskov V. Yu., Berestov A. A. and Ryabinin V. A. [The first clinical experience with the antioxidant dibunol used in the acute period of myocardial infarction]. (Russian). *Kardiologiya*, 1984, v. 24, No. 1, pp. 15-18.

Gonzalez-Flecha B., Cutrin J. C. and Boveris A. Time course and mechanism of oxidative stress and tissue damage in rat liver subjected to *in vivo* ischemia reperfusion. *J. Clin. Invest.*, 1993, v. 91, No. 2, pp. 456-464.

Gonzalez-Flecha B., Evelson P., Sterin-Speziale N. and Boveris A. Hydrogen peroxide metabolism and oxidative stress in cortical, medullary and papillary zones of rat kidney. *Biochim. Biophys. Acta.*, 1993, v. 1157, No. 2, pp. 155-161.

Gorban E. N. and Kol'tover V. K. [Effect of the antioxidant butylated hydroxytoluene (dibunol) on EPR signals in rat tissues at different age]. (Russian). *Biul. Eksp. Biol. Med.*, 1985, v. 99, No. 6, pp. 694-696. English transl. see[1*], pp. 745-747.

Gorbataya O. N., Ziyatdinova R. Kh., Akhmedjanov R. et al. [Exposition two kinds of phospholipase A_2 in rat liver mitochondrias]. (Russian). *Uzb. Biol. Zh.*, 1994, No. 4, pp. 7-9.

Gorbatenkova E. A., Azizova O. A., Paramonov N. V. and Vladimirov Yu. A. [The mechanism of superoxide dismutase photoreactivation by helium-neon laser light]. (Russian). *Dokl. Acad. Nauk SSSR*, 1988, v. 299, No. 4, pp. 995-1000.

Gordienko S. M. [One hundred years of the development of the theory of phagocytosis. Current concepts of the role phagocytes in nonspecific cellular immunity]. (Russian). *Biul. VKNC AMN SSSR*. 1983, v. 8, pp. 144-150.

Gorgoshidze M. L. and Megreladze I. I. [Dynamics of blood antioxidation activity depending on EC 6 changes and character of acute myocardial infarction]. (Russian). *Kardiologiya*, 1981, v. 21, No. 6, pp. 48-51.

Gorman M. W. and Sparks H. V. Progressive coronary vasoconstriction during relative ischemia in canine myocardium. *Circ. Res.*, 1982, v. 51, No. 4, pp. 411-420.

Goroshinskaya I. A. and Mogil'nitskaya L. V. [Lipid peroxidation induced by intermittent hypoxia and the protective effect of Pyrazidole]. (Russian). *Biokhimiya*, 1994, v. 59, No. 7, pp. 974-982. English transl. see[3*], pp. 727-732.

Goto K., Ishige A., Sekigushi K. et al. Effects of cycloheximide on delayed neuronal death in rat hippocampus. *Brain Res.*, 1990, v. 534, pp. 299-302.

Govorova L. V., Kucherenko A. G., Tashchaev Sh. S. et al. [Dynamics of prostaglandin level when blood circulation is disturbed and sympatic innervation of cerebrum is ceased]. (Russian). In: *Prostaglandins and circulation of blood.* Erevan, 1980, pp. 29-30.

Govorova M. Yu., Sharanov B. P., Iankovskii O. Yu. and Lyzlova S. N. [Inactivation of myeloperoxidase-generated active oxygen forms by serum proteins]. (Russian). *Dokl. Acad. Nauk SSSR*, 1986, v. 290, No. 2, pp. 480-483. English transl. see[6*], pp. 308-310.

Gower J. D., Ambrose I. J., Manek S. et al. The effect of a synthetic hexadentate iron chelator (CP130) and desferrioxamine on rabbit kidneys exposed to cold and warm ischemia. *Agents & Actions*, 1993, v. 40, No. 1-2, pp. 96-105.

Gower J. D., Healing G., Fuller B. J. and Green C. J. Iron redistribution and lipid peroxidation in the ischaemic kidney. *Biochem. Soc. Trans.*, 1989, v. 17, pp. 487-488.

Gower J. D., Healing G. and Green C. J. Determination of desferrioxamine-available iron in biological tissues by high-pressure liquid chromatography. *Analyt. Biochem.*, 1989, v. 180, No. 1, pp. 126-130.

Granger D. N., Hollwarth M. E. and Parks D. A. Ischemia/reperfusion injury: role of oxygen-derived free radicals. *Acta Physiol. Scand.*, 1986, v. 548, Suppl., pp. 47-63.

Granger D. N., McCord J. M., Parks D. A. and Hollwarth M. E. Xanthine oxidase inhibitors attenuate ischemia-induced vascular permeability changes in the cat intestine. *Gastroenterology*, 1986, v. 90, pp. 80-84.

Granger D. N., Rutiti G. and McCord J. M. Superoxide radicals in feline intestinal ischemia. *Gastroenterology*, 1981, v. 81, pp. 22-29.

Grayson J., Irvine M., Parratt J. R. et al. Vasospasmic elements in myocardial infarction following coronary occlusion in the dog. *Cardiovasc. Res.* 1968, v. 2, pp. 54-62.

Green C. J., Healing G., Lunec J. et al. Evidence of free-radical-induced damage in rabbit kidneys after simple hypothermic preservation and autotransplantation. *Transplantation*, 1986, v. 41, No. 2, pp. 161-165.

Green C. J., Healing G., Simpkin S. et al. Desferrioxamine reduces susceptibility to lipid peroxidation in rabbit kidneys subjected to warm ischaemia and reperfusion. *Comp. Biochem. Physiol.* [B], 1986, v. 85, No. 1, pp. 113-117.

Green C. J., Healing G., Simpkin S. et al. Increased susceptibility to lipid peroxidation in rabbit kidneys: a consequence of warm ischaemia and subsequent reperfusion. *Comp. Biochem. Physiol.* [B], 1986, v. 83, No. 3, pp. 603-606.

Green C. J., Healing G., Simpkin S. et al. Reduced susceptibility to lipid peroxidarion in cold ischemic rabbit kidneys after addition of desferrioxamine, mannitol, or uric acid to the flush solution. *Cryobiology*, 1986, v. 23, No. 4, pp. 358-365.

Greenberg S. and Frishman W. H. Co-enzyme Q10: a new drug for cardiovascular disease. *J. Clin. Pharmacol.*, 1990, v. 30, No. 7, pp. 596-608.

Grek O. R. and Sharapov V. I. [Prevention of ischemic lesions of the liver monooxygenase system by lidocaine and α-tocopherol]. (Russian). Farmakol. Toksikol., 1987, v. 50, No. 5, pp. 60-64.

Grek O. R., Sharapov V. I., Litvinov V. S. et al. [Ischemic disorders of xenobiotic metabolism in rat liver in early and late periods of rehabilitation]. (Russian). *Farmakol. Toksikol.,* 1984, v. 47, No. 3, pp. 114-118.

Griffeth L. K., Rosen G. M. and Rauckman E. J. Effects of model traumatic injury on hepatic drug metabolism in the rat. *Drug Metab. Disposit.,* 1984, v. 12, No. 5, pp. 588-595.

Grigor'eva A. V., Kazantseva N. V. and Bul'chuk O. V. [Effect of different hyperbaric oxygenation schedules on transcription activity and morphology of frontal cortical neurons in rats with occlusion of the common carotid artery]. (Russian.) *Biul. Eksp. Biol. Med.*, 1992, v. 114, No. 7, pp. 88-90. English transl. see[1*], pp. 1031-1034.

Grigoryants R. A., Meitina R. A. and Gasilin V. S. [Blood oxygen transport function in patients with ischemic heart disease]. (Russian). *Kardioligiya*, 1980, v. 20, No. 1, pp. 50-53.

Grinde B. Autophagy and lysosomal proteolysis in the liver. *Experientia*, 1985, v. 41, No. 9, pp. 1089-1095.

Grinwald P. M, and Nayler W. G. Calcium entry in the calcium paradox. *J. Mol. Cell. Cardiol.*, 1981, v. 13, pp. 867-880.

Gronow G. H. J. and Cohen J. J. Substrate support for renal functions during hypoxia in the perfused rat kidney. *Am. J. Physiol.*, 1984, v. 247, No. 41-42, pp. F618-F631.

Gross C. E., Bednar M. M., Howard D. B. and Sporn M. B. Transforming growth factor-beta 1 reduces infarct size after experimental cerebral ischemia in a rabbit model. *Stroke*, 1993, v. 24, No. 4, pp. 558-562.

Gross G. J., Farber N. E. Hardman H. F. and Warlher D. C. Benefical action of SOD and catalase in stunned myocardium of dogs. *Am. J. Physiol.*, 1986, v. 250, pp. H372-H377.

Gross R. W., Corr P. B., Lee B. I. et al. Incorporation of radiolabelled lysophosphatidyl choline into canine Purkinje fibers and ventricular muscle: electrophysiological, biochemical and autoradiographic correlations. *Circ. Res.*, 1982, v. 51, No. 1, pp. 26-36.

Grover G. J., Schumacher W. A. and Ogletree M. L. Thromboxane receptor antagonist BMS-180291, but not aspirin, reduces the severity of pacing-induced ischemia in dogs. *J. Cardiovasc. Pharmacol.*, 1994, v. 24, No. 3, pp. 493-499.

Grum C. M., Ragsdale R. A., Ketai L. H. et al. Absence of xanthine oxidase or xanthine dehydrogenase in the rabbit myocardium. *Biochem. Biophys. Res. Commun.,* 1986, v. 141, pp. 1104-1108.

Grune T., Sommerburg O., Petras T. and Siems W. G. Postanoxic formation of aldehydic lipid peroxidation products in human renal tubular cells. *Free Rad. Biol. Med.*, 1995, v. 18, No. 1, pp. 21-27.

Guarnieri C., Flamigni F. and Caldarera C. M. Role of oxygen in the cellular damage induced by re-oxygenatian of hypoxic heart. *J. Mol. Cell. Cardiol.*, 1980, v. 12, No. 8, pp. 797-808.

Gudbjarnason S., Cowan Ch., Braasch V. and Bing R. J. Changes in enzyme pattern of infarcted heart muscle during tissue repair. *Cardiologia*, 1967, v. 51, No. 3, pp. 148-159.

Gugushvili L. L. [*Retrograde circulation of liver and portal hypertension*]. (Russian). Moscow, Meditsina, 1972, 196 p.

Gulak P. V., Dudchenko A. M., Zaysev V. V. et al. [*Hepatocyte: functional and metabolic properties*]. (Russian). Moscow, Nauka, 1985, 270 p.

Gulyaeva N. V., Dupin A. M., Levshina I. P. et al. [Carnosin prevents activation of free-radical lipid oxidation during stress]. (Russian). *Biul. Eksp. Biol. Med.*, 1989, v. 107, No. 2, pp. 144-147. English transl. see[1*], pp. 148-152.

Gumuslu S., Serteser M., Ozben T., et al. Inhibitory role of N omega-nitro-L-arginine methyl ester (L-NAME), a potent nitric oxide synthase inhibitor, on brain malondialdehyde and conjugated diene levels during focal cerebral ischemia in rats. *Clin. Chim. Acta.*, 1997, v. 267, No. 2, pp. 213-223.

Gupta M., Dobashi K., Greene E. L., et al. Studies on hepatic injury and antioxidant enzyme activities in rat subcellular organelles following in vivo ischemia and reperfusion. *Mol. Cell. Biochem.*, 1997, v. 176, No. 1-2, pp. 337-347.

Gurevitch J., Frolkis I., Yuhas Y. et al. Anti-tumor necrosis factor-alpha improves myocardial recovery after ischemia and reperfusion. *J. Am. Cell. Cardiol.*, 1997, v. 30, No. 6, pp. 1554-1561.

Gurther Q. H., Knoblauch A., Smith P. L. et al. Oxidant- and lipid-induced pulmonary vasoconstriction mediated by arachidonic acid metabolites. *J. Appl. Physiol.*, 1983, v. 55, No. 3, pp. 919-951.

Gushcha I. A. The antiischemic protective effect of saluretics in kidney transplantation. [Dissertation]. (Russian). Moscow, 1984.

Gushcha I. A., Borzov D. B., Volkova I. A. and Bilenko M. V. [Therapeutic effect of succinic acid during prolong warm ischemia]. (Russian). In: *Transplantation of organs and tissues. IX All-Union Conference.* Abstracts. Tbilisi, 1982, p. 197.

Gus'kova R. A., Ivanov I. I., Kol'toner V. K. et al. [Permeability of bilayer phospholipid membranes for superoxide radicals of oxygen]. (Russian). *Biokhimiya*, 1984, v. 49, No. 5, pp. 758-766. English transl. see[3*], pp. 650-657.

Gute D. C., Ishida T., Yarimizu K. and Korthuis R. J. Inflammatory responses to ischemia and reperfusion in skeletal muscle. *Mol. Cell. Biochem.*, 1998, v. 179, No. 1-2, pp. 169-187.

Gutkin D. V. and Petrovich Yu. A. [Activity of antioxidant enzymes of the myocardium during ischemia]. (Russian). *Biul. Eksp. Biol. Med.* 1982, v. 93, No. 1, pp. 33-35. English transl. see[1*], pp. 38-40.

Gutteridge J. M. C., Halliwell B., Treffry A. et al. Effect of ferritin-containing fractions with different iron loading on lipid peroxidation. *Biochem. J.*, 1983, v. 209, No. 2, pp. 557-560.

Gutteridge J. M. C., Richmond R. and Halliwell B. Inhibition of the iron-catalysed formation of hydroxyl radicals from superoxide and of lipid peroxidation by Desferrioxamine. *Biochem. J.*, 1979, v. 184, No. 2, pp. 469-472.

Gutteridge J. M. C., Richmond R. and Halliwell B. Oxygen free-radicals and lipid peroxidation: inhibition by the protein caeruloplasmin. *FEBS Lett.*, 1980, v. 112., No. 2, pp. 269-272.

Hafeman D. G. and Lucas Z. J. Polymorphonuclear leukocyte-mediated, antibody-dependent, cellular cytotoxicity against tumor cells: dependence on oxygen and the respiratory burst. *J. Immunol.*, 1979, No. 123. pp. 55-62.

Hahn H. R., Barak A. J., Tuma D. J. et al. Effects of phenobarbital administration on levels of physiological antioxidants in rat liver. *Pharmacology*, 1978, v. 17, pp. 341-344.

Halldorsson A., Kronon M., Allen B. S. et al. Controlled reperfusion after lung ischemia: implications for improved function after lung transplantation. *J. Thorac. Cardiovasc. Surg.*, 1998, v. 115, No. 2, pp. 415-424.

Halat G., Lukacova N. Chavko M. and Marsala J. Effect of severe incomplete ischemia and postischemic reperfusion on phospholipids and unesterified cholesterol in rabbit spinal cord. *Physiologia Bohemoslovaka,* 1990, v. 39, No. 4, pp. 351-360.

Haljamäe H. and Enger E. Human skeletal muscle energy metabolism during and after complete tourniquet ischemia. *Ann. Surg.*, 1975, v. 182, pp. 9-14.

Hall E. J. *Radiology for the radilogist.* 1978, N. Y., Harper & Row Publisher Inc., pp. 81-92.

Hallenbeck J. M. and Furlaw T. W. Prostaglandin I_2 and indomethacin prevent impairment of postischemic brain reperfusion in the dog. *Stroke*, 1979, v. 10, No. 6, pp. 629-637.

Halliwell B. Reactive oxygen species in living systems: source, biochemistry, and role in human disease. [Review]. *Am. J. Med.*, 1991, v. 91, pp. 14S-22S.

Halliwell B. and Aruoma O. J. DNA damage by oxygen-derived species: its mechanism and measurements in mammalian systems. *FEBS Lett.*, 1991, v. 281, No. 1, 2, pp. 9-19.

Halliwell B. and Chirico S. Lipid peroxidation: its mechanism, measurement, and significance. [Review]. *Am. J. Clin. Nutr.*, 1993, v. 57, No. 5, Suppl., pp. 715S-725S.

Halliwell B. and Gutteridge J. M. C. Oxygen toxicity, oxygen radicals, transition metals and disease. Review. *Biochem. J.*, 1984, v. 219, pp. 1-14.

Halliwell B. and Gutteridge J. M. C. Oxygen radicals and the nervous system. *Trends Neurosci.*, 1985, v. 8, No. 1. pp. 22-26.

Halliwell B. and Gutteridge J. M. C. The definition and measurement of antioxidants in bilogical systems. Letter to the Editors. *Free Rad. Biol. Med.*, 1995, v. 18, No. 1, pp. 125-126.

Hamanoue M., Kawaida K., Takao S. et al. Rapid and marked induction of hepatocyte growth factor during liver regeneration after ischemic or crush injury. *Hepatology*, 1992, v. 16, No. 6, pp. 1485-1492.

Hamberg M., Hedqvist P. and Radergranb K. Identification of 15-hydroxy-5, 8, 11, 13-eicosatetraenoic acid (15-HETE) as a major metabolite of arachidonic acid in human lung. *Acta Physiol. Scand.*, 1980, v. 110, No. 2. pp. 219-221.

Hanson K. P. [Programmed cell death (apoptosis): molecular mechanisms and the role in biology and medicine]. (Russian.) *Vopr. Med. Chem.*, 1997, v. 43, No. 5, pp. 402-415.

Harada H., Azuma J., Hasegawa H. et al. Enhanced suppression of miocardial slow action potentials during hypoxia by free fatty acids. *J. Mol. Cell. Cardiol.*, 1984, v. 16, pp. 261-276.

Haraldsson G., Nilsson U., Bratell S. et al. ESR-measurement of production of oxygen radicals *in vivo* before and after renal ischaemia in the rabbit. *Acta Physiol. Scand.*, 1992, v. 146, No. 1, pp. 99-105.

Harmon B. V., Corder A. M., Collins R. J. et al. Cell death induced in a murine mastocytoma by 42-47° heating *in vitro*: evidence that the form of death changes from apoptosis to necrosis above a critical heat load. *Int. J. Radiat. Biol.*, 1990, v. 58, No. 5, pp. 845-858.

Harris A. G., Leiderer R., Peer F. and Messmer K. Skeletal muscle microvascular and tissue injury after varying durations of ischemia. *Am. J. Physiol.*, 1996, v. 271, No. 6, Pt. 2, pp. H2388-H2398.

Harris L. R., Cake M. H. and Macey D. J. Iron release from ferritin and its sensitivity to superoxide ions differs among vertebrates. *Biochem. J.*, 1994, v. 301, Pt. 2, pp. 385-389.

Harrison-Shostak D. C., Lemasters J. J., Edgell C. J. and Herman B. Role of ICE-like proteases in endotelial cell hypoxic and reperfusion injury. *Biochem. Biophys. Res. Commun.*, 1997, v. 231, No. 3, pp. 844-847.

Hartley A., Davies M. and Rice-Evans C. Desferrioxamine as a lipid chain-breaking antioxidant in sickle erythrocyte membranes. *FEBS Lett.*, 1990, v. 264, No. 1, pp. 145-148.

Hashimoto N., Wu T. W., Au J. X. et al. Radiochemical quantitation of conjugated diens during ischemia and reperfusion in the rat liver. *Clin. Biochem.*, 1991, v. 24, No. 2, pp. 153-158.

Hashizume H., Hoque A. N., Magishi K. et al. A new approach to the development of anti-ischemic drugs. Substances that counteract the deleterious effect of lysophosphatidylcholine on the heart. *Jpn. Heart J.*, 1997, v. 38, No. 1, pp. 11-25.

Hasselgren P., Jehnische E., Fornander J. and Hellman A. No beneficial effect of ATP-MgC_{l2} on impaired transmembrane potential and protein synthesis in liver ischemia. *Acta Chir. Scand.*, 1982, v. 148, pp. 601-607.

Hauptlorenz S., Esterbauer H., Moll W. et al. Effects of the lipid peroxidation product 4-hydroxynonenal and related aldehydes on proliferation and viability of cultured Ehrlich ascites tumor cells. *Biochem. Pharmacol.*, 1985, v. 34, No. 21, pp. 3803-3809.

Hayakawa M., Ishida N., Takeuchi K. et al. Arachidonic acid-selective cytosolic phospholipase A_2 is crucial in the cytotoxic action of TNF. *J. Biol. Chem.*, 1993, v. 208, pp. 11290-11295.

Healing G., Gower J., Fuller B. and Green C. Intracellular iron redistribution. An important determinant of reperfusion damage to rabbit kidneys. *Biochem. Pharmacol.*, 1990, v. 39, No. 7, pp. 1239-1245.

Hearse D. J. Reperfusion of the ischemic myocardium. *J. Molec. Cell. Cardiol.*, 1977, v. 9, pp. 605-616.

Hearse D. J., Humphrey S. M. and Chain E. B. Abrupt reoxygenation of the anoxic potassium arrested perfused rat heart: a study of myocardial enzyme release. *J. Mol. Cell. Cardiol.*, 1973, v. 5, No. 4, pp. 395-407.

Hearse D. J., Humphrey S. M. and Bullock G. R. The oxygen paradox and the calcium paradox. Two furets of the same problem? *J. Mol. Cell. Cardiol.*, 1978, v. 10, pp. 641-668.

Hearse D. J., Manning AS, Downey J. M. et al. Xanthine oxidase: a critical mediator of myocardial injury during ischemia and reperfusion? *Acta Physiol. Scand.*, 1986, v. 548, Suppl., pp. 65-78.

Hedlund Bo. E. and Hallaway P. E. High-dose systemic iron chelation attenuates reperfusion injury. Biochem. Soc. *Transactions*, 1999, v. 21, No. 2, pp. 340-343.

Hegstad A. C., Strand H. and Ytrehus K. Phospholipid peroxidation after 60 min of global ischaemia and 10 min of reperfusion. A study in the isolated rat heart. *J. Mol. Cell. Cardiol.*, 1994, v. 26, No. 5, pp. 569-578.

Hellstrom H. R. The injury-spasm (ischemia-induced hemostatic vasoconstrictive) and vascular autoregulatory hypothesis of ischemic disease. *Am. J. Cardiol.*, 1982, v. 49, pp. 802-810.

Henriksen T., Mahoney E. M. and Steinberg D. Enhanced macrophage degradation of biologically modified low density lipoprotein. *Arteriosclerosis*, 1983, v. 3, pp. 149-159.

Herbert J. M., Corseaux D., Lale A. and Bernat A. Hypoxia primes endotoxin-induced tissue factor expression in human monocytes and endothelial cells by a PAF-dependent mechanism. *J. Cell. Physiol.*, 1996, v. 169, No. 2, pp. 290-299.

Hernandez L. A., Grisham M. B., Twohig B. et al. Role of neutrophils in ischemia-reperfusion-induced microvascular injury. *Am. J. Physiol.*, 1987, v. 253, pp. H699-H703.

Hess M. L. and Manson N. H. Molecular oxygen: friend and foe. *J. Mol. Cell Cardiol.*, 1984, v. 16, pp. 969-985.

Hess M. L., Krause S. M. and Greenfild L. J. Assessment of hypothermic, cardioplegic protection of the global ischemic canine myocardium. *J. Thorac. Cardiovasc. Surg.*, 1980, v. 80, pp. 293-301.

Hess M. L., Manson N. H. and Okabe E. Involvement of free radicals in the pathophysilogy of ischemic heart disease. *Can. J. Physiol. Pharmacol.*, 1982, v. 60, No. 11. pp. 1382-1389.

Hess M. L., Okabe E. and Kontos H. A. Proton and free oxygen radical interaction with the calcium transport system of cardiac sarcoplasmic reticulum. *J. Mol. Cell Cardiol.*, 1981, v. 13, pp. 767-772.

Heyndrickx G. R, Amano J., Patrick Th. A. et al. Effects of coronary artery reperfusion on regional myocardial blood flow and function in conscious baboons. *Circulation*, 1985, v. 71, No. 5, pp. 1029-1031.

Hickey M. J., Knight K. R., Lepore D. A. et al. Influence of postischemic administration of oxyradical antagonists on ischemic injury to rabbit skeletal muscle. *Microsurgery*, 1996, v. 17, No. 9, pp. 517-523.

Higgins A. J. and Blackburn K. J. Prevention of reperfusion damage in working rat hearts by calcium antagonists and calmodulin antagonists. *J. Mol. Cell. Cardiol.*, 1984., v. 16, pp. 427-438.

Higginson L. A. J., Beanlands D. S., Nair R. C. et al. The time course and characterization of myocardial hemorrhage after coronary reperfusion in the anesthetized dog. *Circulation*, 1983, v. 67, p. 1024.

Higuchi H. and Aggarwal B. B. P-80 form of the human tumor necrosis factor receptor is involved in DNA fragmentation. *FEBS Lett.*, 1993, v. 331, No. 3, pp. 252-255.

Hill J. L. and Gettes L. S. Effect of acute coronary artery occlusion on local myocardial extracellular K^+ activity in swine. *Circulation*, 1980, v. 61, pp. 768-778.

Hille R. and Massey V. Studies on the oxidative half-reaction of xanthine oxidase. *J. Biol. Chem.*, 1981, v. 256, pp. 9090-9095.

Hirashima Y., Koshu K., Kamiyama K et al. The activities of phospholipase A_1, A_2, lysophospholipase and acyl CoA; lysophospholipid acyltransferase in ischemic dog brain. *Recent Progr. Study and Ther. Brain Edema. Proc. 5th Int. Symp. Proceedings*. 1982, N. Y., Lond. 1984, pp. 213-221.

Hirata Y., Taguchi T., Nakao M. et al., The relationship between the adenine nucleotide metabolism and the conversion of the xanthine oxidase enzyme system in ischemia-reperfusion of the rat small intestine. *J. Pediatr. Surg.*, 1996, v. 31, No. 9, pp. 1199-1204.

Hirche H., Bissig R., Friedrich R. et al. Measurement of myocardial extracellular Na^+, K^+, Ca^{2+} and H^+ using ion-selective electrodes during ischemia. In: *Progress in Enzyme and Ion-Selective Electrodes*. Berlin, 1981, pp. 164-170.

Hochachka P.W. Metabolic arrest as a mechanism of protection against hypoxia. In: Wanquier A. et al., eds. *Protection of tissues against hypoxia*. Amsterdam, 1982, pp. 1-12.

Hochstein P. Protective mechanisms in reperfusion injury: one-electron reductants and the myoglobin redox cycle. In: Rice-Evans C., ed. *Free Radicals, Diseased States and Anti-Radical Interventions*. London, Richelieu Press. 1989, pp. 413-430.

Hodgson E. K. and Fridovich I. The interaction of bovine erythrocyte superoxide dismutase with hydrogen peroxide: inactivation of the inzyme. *Biochemistry*, 1975, v. 14, No. 24, pp. 5294-5299.

Holt S., Gunderson M., Joyce K. et al. Myocardial tissue iron delocalization and evidence for lipid peroxidation after two hours of ischemia. Ann. Emerg. Med., 1986, v. 15, pp. 1155-1159.

Horrocks L. A., Spanner S., Mozzi R. et al. *Myelination and demyelination: recent chemical advances*. N. Y., Plenum Press. 1978, pp. 342-347.

Houston R. J., Heerschap A., Skotnicki S. H. et al. ATP content measured by 31P NMR correlates with post-ischaemic recovery of external work output in an isolated rat heart perfused with eruthrocyte suspension. *J. Mol. Cell. Cardiol.*, 1997, v. 29, No. 6, pp. 1763-1766.

Houvenaghel A., Schrauwen E. and Wechsung L. Influence of primary prostaglandins prostacyclin and arachidonic acid on mesenteric hemodynamics in the pig. *Prostaglandins Med.*, 1979, v. 2, pp. 83-95.

Hrycay E. G. and O'Brien P. J. Cytochrome P-450 as a microsomal peroxidase utilizing a lipid peroxide substrate. *Arch. Biochem. Biophys.*, 1971. v. 147, No. 1, pp. 14-27.

Hrycay E. G. and O'Brien P. J. Cytochrome P-450 as a microsomal peroxidase in steroid hydroperoxide reduction. *Arch. Biochem. Biophys.*, 1972, v. 153, No. 2, pp. 480-494.

Huang P. L. and Lo E. H. Genetic analysis of NOS isoforms using nNos and eNOS knockout animals. *Prog. Brain. Res.*, 1998, v. 118., pp. 13-25

Huie R. E. and Padmaja S. The reaction of NO with superoxide. *Free Rad. Res. Commun.*, 1993, v. 18, pp. 195-199.

Humphrey S. M., Gavin I. B. and Herdson P. B. The relationship of ischaemic contracture to vascular reperfusion in the isolated rat heart. *J. Mol. Cell. Cardiol.*, 1980, v. 12, pp. 1397-1406.

Humphrey S. M., Gavin J. B. and Herdson P. B. Catecholamine-depletion and the no-reflow phenomenon in anoxic and ischemic rat hearts. *J. Mol. Cell. Cardiol.*, 1982, v. 14, pp. 151-161.

Hurn P. D., Koehler R. C., Blizzard K. K. and Traystman R. J. Deferoxamine reduces early metabolic failure associated with severe cerebral ischemic acidosis in dogs. *Stroke.*, 1995, v. 26, No. 4, pp. 588-695.

Hyman A. L., Spannhake E. W. and Kadowitz P. J. Divergent responses to arachidonic acid in the feline pulmonary vascular bed. *Am. J. Physiol.*, 1980, v. 239, pp. H40-H46.

Hypoxia in Medicine, Abstracts of 3rd Intern. Conference, Moscow, 1998. In: *Hypoxia medical J.*, 1998, v. 6, No. 2, pp. 74-95.

Hyslop P. A., Hinshaw D. B., Halsey W. A. et al. Mechanisms of oxidant-mediated cell killing: the glycolytic and mitochondrial pathways of ADP phosphorylation are major targets of H_2O_2-mediated injury. *J. Biol. Chem.*, 1988, v. 263, pp. 1665-1675.

Iakovlev G. M., Svistov A. S. and Mironova M. I. [Dimethyl sulfoxide in experimental and clinical medicine. The achievements and prospects. Review]. (Russian). *Eksper. Klin. Farmakol.*, 1992, v. 55, No. 4, pp. 74-76.

Ichigara K., Abiko Y. and Neely J. R. Effect of ischemia on fatty acid metabolism in isolated perfused rat hearts. *Jpn. J. Pharmacol.*, 1982, v. 32, pp. 1-143.

Idström J. P., Soussi B., Elander A. and Bylund-Fellenius A. C. Purine metabolism after *in vivo* ischemia and reperfusion in rat skeletal muscle. *Am. J. Physiol.*, 1990, v. 258, pp. H1668-H1673.

Ikai I., Chance B. and Kumar C. Reactive oxygen inducing vasoconstriction in the isolated perfused rat liver. *Free Rad. Biol. Med.*, 1994, v. 16, No. 5, pp. 539-545.

Ikeda K., Nagashima T., Wu S. et al. The role of calcium ion in anoxia/reoxygenation damage of cultured brain capillary endothelial cells. *Acta Neurochir.*, 1997, v. 70 (Suppl.), pp. 4-7.

Ikeda J., Ikeda K. and Long D. M. Comparative study of different iron-chelating agents in cold-induced brain edema. *Neurosurgery*, 1989, v. 24, pp. 820-824.

Im M. J., Shen W. H., Pak C. J. et al. Effect of allopurinol on the survival of hyperemic island skin flaps. *Plast. Reconstr. Surg.*, 1984, v. 73, pp. 276-278.

Inauen W., Payne D. K., Kvietys P. R. and Granger D. N. Hypoxia/reoxygenation increases the permeability of endothelial cell monolayers: role of oxygen radicals. *Free Rad. Biol. Med.*, 1990, v. 9, No. 3, pp. 219-223.

Irvin C. and O'Brien P. J. Glutathione peroxidase II of guinea pig liver cytosol: relationship to GSH-transferases. *Arch. Biochem. Biophys.*, 1980, v. 205, No. 1, pp. 122-131.

Ishiharajima S., Aida T., Nakagawa R. et al. Early membrane damage during ischemia in rat heart. *Exp. Mol. Pathol.*, 1984, v. 44, No. 1, pp. 1-6.

Ishikawa T., Sarfeh J., Tarnawski A. et al. Epidermal growth factor protects portal hypertensive gastric mucosa in ischemia/reperfusion: the role of capillary endothelia and prostaglandins. *Surgery*, 1992, v. 112, No. 2, pp. 341-346.

Islam S. J. The effect of various modifiers on rat microsomal peroxidase kinetics. *Biochem. Pharmacol.*, 1982, v. 31, pp. 1823-1826.

Istomin N. P., Mishnev O. D., Sergeeva N. A. et al. [Functional and morphological changes in the kidneys after acute embologenic occlusion of the limb arteries]. (Russian). *Biul. Eksp. Biol. Med.*, 1987, v. 103, No. 2, pp. 168-171. English transl. see[1*], pp. 189-192.

Ivanov I. I., Merzlyak M. N. and Tarusov B. N. [Vitamin E: biotechnology, the function concerning antioxidant properties]. (Russian). In: *Bioantioxidants*. Moscow, Nauka, 1975, pp. 30-52.

Iwakura K., Ito H. and Takiuchi S. et al. Alternation in the coronary blood flow velocity pattern in patients with no-reflow and reperfused acute myocardial infarction. *Circulation* 1996, v. 94, No. 6, pp. 1269-1275.

Jacksov V., Earnhardt J. and Chalkley R. A DNA-lipid protein containing material isolated from calf thymus nuclear chromatin. *Biochem. Biopys. Res. Commun.*, 1968, v. 33, No. 2, pp. 253-259.

Jaeschke H., Farhood A. and Smith C. W. Neutrophils contribute to ischemia/reperfusion injury in rat liver *in vivo*. *FASEB J.*, 1990, v. 4, pp. 3355-3359.

Jaeschke H., Bautista A. P., Spolarics Z. and Spitzer J. J. Superoxide generation by Kupffer cells and priming of neutrophils during reperfusion after hepatic ischemia. *Free Rad. Res. Commun.*, 1991, v. 15, No. 5, pp. 277-284.

Jaeschke H., Farhood A., Bautista A. P. et al. Functional inactivation of neutrophils with a Mac-1 (CD11b/CD18) monoclonal antibody protect against ischemia/reperfusion injury in rat liver. *Hepatology*, 1993, v. 17, pp. 915-923.

Jaeschke H., Schini V. B. and Farhood A. Role of nitric oxide in the oxidant stress during ischemia/reperfusion injury of the liver. *Life Sci.*, 1992, V. 50, pp. 1797-1804.

Janero D. R., Burghardt B. and Lopez R. Protection of cardiac membrane phospholipid against oxidative injury by calcium antagonists. *Biochem. Pharmacol.*, 1988, v. 37, pp. 4197-4203.

Jannotti F., Crockard A. and Ladds G. Are prostaglandins involved in experimental ischemic edema in gerbils? *Stroke*, 1981, v. 12, pp. 301-306.

Jarasch E. D., Bruder G. and Heid H. W. Significance of xanthine oxidase in capillary endothelial cells. *Acta Physiol. Scand.*, Suppl. 1 1986, v. 548, pp. 39-46.

Jarasch E. D., Grund C., Bruder G. et al. Localization of xanthine oxidase in mammary gland epithelium and capillary endothelium. *Cell*, 1981, v. 25, pp. 67-82.

Jarvis W. D., Kolesnick R. N., Fornari F. A. et al. Induction of apoptotic DNA damage and cell death by activation of the splingomyelin pathway. *Proc. Natl. Acad. Sci. USA*, 1994, v. 91, p. 77.

Jayadev S., Linardic C. M. and Hannun Y. A. Identification of arachidonic acid as a mediator of splingomyelin hydrolysis in response to tumor necrosis factor-alpha. *J. Biol. Chem.*, 1994, v. 269, pp. 5757-5763.

Jayakumari N., Ambikakumari V., Balakrishnan K. G. and Iyer K. S. Antioxidant status in relation to free radical production during stable and unstable anginal syndromes. *Atherosclerosis*, 1992, v. 94, No. 2-3, pp. 183-190.

Jennings R. B. and Reimer K. A. Lethal myocardial ischemic injury. *Am. J. Path.*, 1981, v. 102, pp. 241-255.

Jennings R. B. and Reimer K. A. Lethal reperfusion injury: fact or fancy? In: J. R. Parratt, ed. *Myocardial Response to Acute Injury*. London, MacMillan Press, 1992, pp. 17-34.

Jennings R. B., Reimer K. A. and Steenbergen C. Effect of inhibition of the mitochondrial ATPase on net myocardial ATP in total ischemia. *J. Mol. Cell. Cardiol.*, 1991, v. 23, pp. 1383-1395.

Jennings R. B., Reimer K. A., Hill M. L. and Mayer S. E. Total ischemia in dog hearts *in vitro*. 1. Comparison of high energy phosphate production, utilization and depletion and of adenine nucleotide catabolism in total ischemia *in vitro* vs. Severe ischemia *in vivo*. *Circ. Res.*, 1981, v. 49, pp. 892-900.

Jennings R. B., Reimer K. A. and Steenbergen C. Jr. Myocardial ischemia and reperfusion: role of calcium. In: Parratt J. R., ed. *Control and manipulation of calcium movement*. N. Y., Raven Press, 1985, pp. 273-302.

Johnston D. G. and Alberti K. G. Acid-base balance in metabolic acidosis. *Clin. Endocr. Metabol.*, 1983, v. 12, No. 2, pp. 267-285.

Johnson D., Freischlag J. A., Lesniak R. et al. Endothelial damage due to ischemia and reperfusion is prevented with SIN-1. *Cardiovasc. Surg.*, 1998, v. 6, No. 5, pp. 527-532.

Johnson G. 3d., Tsao P. and Lefer A. M. Synergism between superoxide dismutase and sodium nitrite in cardioprotection following ischemia and reperfusion. *Am. Heart J.*, 1990, v. 119, No. 3, Pt 1, pp. 530-537.

Johnson W. D., Kayser K. L., Brenowitz J. B. and Saedi S. F. A randomized controlled trial of allopurinol in coronary bypass surgery. *Am. Heart J.*, 1991, v. 121, No. 1, Pt. 1, pp. 20-24.

Jolly S. R. and Lucchesi B. R. Effect of BW-755C in an occlusion-reperfusion model of ischemic myocardial injury. *Am. Heart. J.*, 1983, v. 106, pp. 8-13.

Jones D. P. The role of oxygen concentration in oxidative stress: hypoxic and hyperoxic models. In: H. Sies, ed. *Oxidative stress.* London, Academic Press., 1985, pp. 152-195.

Jones D. P. and Mason H. S. Gradients of O_2 concentration in hepatocytes. *J. Biol. Chem.*, 1978, v. 258, No. 14, pp. 4874-4880.

Jones H. P., Grisham M. B., Bose S. K. et al. Effect of allopurinol on neutrophil superoxide production, chemotaxis or degranulation. *Biochem. Pharm.*, 1985, v. 34, No. 20, pp. 3673-3676.

Jones S. P., Girod W. G., Palazzo A. J. et al. Myocardial ischemia-reperfusion injury is exacerbated in absence of endothelial cell nitric oxide synthase. *Am. J. Physiol.* 1999, v. 276, No. 5, Pt 2, pp. H1567-H1573.

Joneschild E. S., Chen L. E., Seaber A. V. et al. Effect of a NOS inhibitor, L-NAME, on the contractile function of reperfused skeletal muscle. *J. Reconstr. Microsurg.* 1999, v. 15, No. 1, pp. 55-60.

Jugdutt B. I., Grover Ch., Hutchins G. M. et al. Myocardial infarction in the conscious dog: three-dimensional mapping of infarct, collateral flow and region at risk. *Circulation*, 1979, v. 60, No. 5, pp. 1141-1150.

Jugdutt B. I., Hutchins G. M., Bulkle B. H. et al. Effect of indomethacin on collateral blood flow and infarct size in the conscious dog. *Circulation*, 1979, v. 59, pp. 734-743.

Jugdutt B. I., Hutchins G. M., Bulkle B. H. et al. Dissimilar effects of prostacyclin, prostaglandin E_1, and prostaglandin E_2 on myocardial infarct size after coronary occlusion in conscious dogs. *Circ. Res.*, 1981, v. 49, pp. 685-700.

Kagan V. E., Azizova O. A., Arkhipenko Yu. V. et al. [Interrelations between structural and functional modifications in the membranes of the sarcoplasmic reticulum following lipid peroxidation]. (Russian). *Biofizika*, 1977, v. 22, No. 4, pp. 625-630. English transl. see[2*], pp. 648-654.

Kagan V. E., Churakova T. D., Karagodin V. P. et al. [Disruption of the Ca^{2+} transport enzyme system in the membranes of the sarcoplasmic reticulum upon exposure to phospholipid hydroperoxides and fatty acids hydroperoxide]. (Russian). *Biul. Eksp. Biol. Med.*, 1979, Vol. 87, No. 2, pp. 145-149. English transl. see[1*], pp. 124-128.

Kagan V. E., Kotelevtsev S. V. and Kozlov Yu. P. [Role of enzymatic peroxidation of phospholipids in the breakdown of endoplasmic reticulum membranes *in vivo*]. (Russian). *Dokl. Acad. Nauk SSSR*, 1974, v. 217, No. 1, pp. 213-216. English transl. see[8*], pp. 45-47.

Kagan V. E., Kozlov Yu. P., Bilenko M. V. et al. [On the involve of the lipid peroxidation process in mechanisms of the rhythmic daily changes of cytochrome P-450 content in liver endoplasmic reticulum membranes]. (Russian). *Biol. Nauki*, 1986, No. 3, pp. 26-29.

Kagan V. E., Savov V. M., Didenko V. V. et al. [Calcium and lipid peroxidation in the heart mitochondrial and microsomal membranes]. (Russian). *Biul. Eksp. Biol. Med.*, 1983, Vol. 95, No. 4, pp. 46-48. English transl. see[1*], pp. 458-461.

Kagan V. E., Savov V. M., Didenko V. V. et al. [Correlation between activities of antioxidant systems and endogenous lipid peroxidation in the left and right ventricular myocardium]. (Russia). *Biul. Eksp. Biol. Med.*, 1984, v. 97, No. 6, pp. 664-666. English transl. see[1*], pp. 726-727.

Kagan V. E., Savov V. M. and Serbinova E. A. Mechanisms of disassembly of a mixed function oxygenase system in liver endoplasmic reticulum. *Acta Physiol. Pharmacol. Bulg.*, 1984, v. 10, No. 2, pp. 7, 73-89.

Kagan V. E., Serbinova E. A., Bakalova R. A. et al. [Mechanisms of the stabilization of biomembrane by α-tocopherol. Role of the isoprenoid chain in the inhibition of lipid peroxidation]. (Russian). *Dokl. Acad. Nauk SSSR*, 1987, v. 295, No. 3, pp. 728-731. English transl. see[8*], pp. 135-138.

Kagan V. E., Shvedova A. A. and Novikov K. N. [Participation of the phospholipases in the "repair" of photoreceptor membranes exposed to lipid peroxidation]. (Russian). *Biofizika*, 1978, v. 23, No. 2. pp. 279-284. English transl. see[2*], pp. 281-287.

Kagava Y. [*Biological membranes*]. (Russian transl. from English). Moscow, Vysshaya Shkola, 1985, p. 302.

Kahn D., Kajani M., Zeng Q. et al. Effect of partial portal vein ligation on hepatic regeneration. *J. Invest. Surg.*, 1988, v. 1, No. 4, pp. 267-276.

Kakinohana M., Taira Y. and Marsala M. The effect of graded postischemic spinal cord hypothermia on neurological outcome and histopathology after transient spinal ischemia in rat. *Anesthesiology*, 1999, v. 90, No. 3, pp. 789-798.

Kalmykova V. I. and Burlakova E. B. [Disturbances in the interrelationship between the lipid content and their peroxidation products and approaches to its correction in atherosclecosis of the coronary arteries and myocardial infarction]. (Russian). *Sov. Med.*, 1982, No. 4, pp. 3-7.

Kamiike W., Fujikawa M., Koseki M. et al. Different patterns of leakage of cytosolic and mitochondrial enzymes. *Clin. Chim. Acta*, 1989, v. 185, No. 3, pp. 265-270.

Kaminski P. M. and Proctor K. G. Attenuation of no-reflow phenomenon, neutrophil activation, and reperfusion in intestinal microcirculation by topical adenosine. *Circ. Res.*, 1989, v. 65, No. 2, pp. 426-435.

Kaneko M., Hayashi H., Kobayashi A. et al. Inhibition of heart sarcolemmal Ca^{2+}-pump activity by oxygen free radicals. *Bratisl. Lek. Listy*, 1991, v. 92, No. 1, pp. 48-56.

Kanellopoulos G. K., Kato H., Wu Y. et al. Neuronal cell death in the ischemic spinal cord: the effect of methylprednisolone. *Ann. Thorac. Surg.* 1997, v. 64, No. 5, 1279-1285.

Kapelko V. I., Gorina M. S. and Novikova N. A. [Comparative estimation of isolated heart muscle contraction and establishment when Ca^{++} concentration in perfusion solution decreases at acidosis and metabolic blocade]. (Russian). In: Chazov E. I. and Morgan H. E., eds. *Myocardium metabolism.* Moscow, Meditsina, 1983, pp. 45-59.

Karasawa A., Guo J. P., Ma X. L. and Lefer A. M. Beneficial effects of transforming, growth factor-beta and tissue plasminogen activator in splanchnic artery occlusion and reperfusion in cats. *J. Cardiovasc. Pharmacol.*, 1991, v. 18, No. 1, pp. 95-105.

Karli J. N., Karikas G. A., Hatzipavlou P. K. et al. The inhibition of Na^+ and K^+ stimulated ATPase activity of rabbit and dog heart sarcolemma by lysophosphatidyl choline. *Life Sci.*, 1979, v. 24, pp. 1869-1876.

Karlsson K., Sandstrom J., Edlund A. et al. Pharmacokinetics of extracellular-superoxide dismutase in the vascular system. *Free Radic. Biol. Med.*, 1993, v. 14, No. 2, pp. 185-190.

Karnaukhov V. N. [*Functions of carotinoides in animal cells*]. (Russian). Moscow, Nauka, 1973. 104 p.

Karuzina L. I. and Archakov A. I. Hydrogen peroxide- mediated inactivation of microsomal cytochrome P-450 during monooxygenase reactions. *Free Rad. Biol. Med.*, 1994, v. 17, pp. 557-567.

Karwinski W., Bolann B., Ulvik R. et al. Normothermic liver ischemia in rats: xanthine oxidase is not the main source of oxygen free radicals. *Res. Exper. Med.*, 1993, v. 193, No. 5, pp. 275-283.

Kashkarov K. P., Vasil'eva E. V. and Ruuge E. K. [Generation of superoxide radicals by the mitochondrial respiratory chain of isolated cardiomyocytes]. (Russian). *Biokhimiia*, 1994, v. 59, No. 6, pp. 813-818. Russian transl. see[3]* pp. 601-605.

Katagiri Y., Sasai N., Nakamura M. et al. Acid hydrolases in the initiation of ischemic myocardial necrosis. *J. Mol. Cell. Cardiol.*, 1980, v. 12, Suppl. 7, p. 69.

Katoh S., Toyama J., Kodama I. et al. Protective action of iron-chelating agents (catechol, mimosine, deferoxamine, and kojic acid)against ischemia-reperfusion injury of isolated neonatal rabbit hearts. *Eur. Surg. Res.*, 1992, v. 24, No. 6, pp. 349-355.

Katusic Z. S. and Vanhoutte P. H. Anoxic contractions in isolated canine cerebral arteries: contribution of endothelium-derived factors, metabolites of arachidonic acid, and calcium entry. *J. Cardiovasc. Pharmacol.*, 1986, v. 8, Suppl., pp. S97-S101.

Katusic Z. S. and Vanhoutte P. M. Superoxide anion is an endothelium derived contracting factor. *Am. J. Physiol.*, 1989, v. 257, No. 1, Pt. 2, pp. H33-H37.

Katz A. M. and Messineo F. C. Lipid-membrane interactions and the pathogenesis of ischemic damage in the myocardium. *Circ. Res.*, 1981, v. 48, No. 1, pp. 1-16.

Katz L. M., Callaway C. W., Kagan V. E. and Kochanek P. M. Electron spin resonance measure of brain antioxidant activity during ischemia/reperfusion. *NeuroReport*, 1998, v. 9, No. 7, pp. 1587-1593.

Kawabata H., Ryomoto T. and Katori R. Effect of thromboxane A2 synthetase inhibitor on metabolism and conractility in ischemic reperfused rabbit heart. *Angiology*, 1997, v. 48, No. 8, pp. 689-697.

Kawabata H., Sugiyama K. and Katori R. Effect of acetylsalicylic acid on metabolism and contractility in the ischemic reperfused heart. *Jpn. Circ. J.*, 1996, v. 60, No. 12, pp. 961-971.

Kawamoto S., Inoue M., Tashiro S. et al. Inhibition of ischemia and reflow-induced liver injury by an SOD derivative that circulates bound to albumin. *Arch. Biochem. Byophys.*, 1990, v. 277, No. 1, pp. 160-165.

Kawasaki S., Sugiyama S., Ishiguro N. et al. Implication of superoxide radicals on ischemia-reperfusion-induced skeletal muscle injury in rats. *Eur. Surg. Res.*, 1993, v. 25, No. 3, pp. 129-136.

Kazui M., Andreoni K. A., Norris E. J. et al. Breath ethane: a specific indicator of free-radical-mediated lipid peroxidation following reperfusion of the ischemic liver. *Free Rad. Biol. Med.*, 1992, v. 13, No. 5, pp. 509-515.

Kehrer J. P., Piper H. M. and Sies H. Xanthine oxidase is not responsible for reoxygenation injury in isolated-perfused rat heart. *Free Rad. Res. Commun.*, 1987, v. 3, No. 1-5, pp. 69-78.

Kelly M. J. α-Tocopherol as agonist in hypoxia. *J. Pharm. Pharmacol.*, 1986, v. 38, pp. 66-68.

Kelly R. F., Hursey T. L., Schaer G. L. et al. Cardiac endothelin release and infarct size, myocardial blood flow, and ventricular function in canine infarction and reperfusion. *J. Investig. Med.*, 1996, v. 44, No. 9, pp. 575-582.

Kempski O. and Behmanesh S. Endothelial cell swelling and brain perfusion. *J. Trauma.*, 1997, v. 42, No. 5, Suppl., pp. S38-S40.

Kennedy T. P., Rao N. V., Hopkins C. et al. Role of reactive oxygen species in reperfusion injury of the rabbit lung. *J. Clin. Invest.*, 1989, v. 83, No. 4, pp. 1326-1335.

Kennedy Ch. H., Maples K. R. and Mason R. P. *In vivo* detection of free radical metabolites. *Pure Appl. Chem.*, 1990, v. 62, No. 2, pp. 295-299.

Kerr J. F. R. Shrinkage necrosis: a distinct mode of cellar death. *J. Pathol.*, 1971, v. 105, No. 1, pp. 13-20.

Kettelhut I. C., Fiers W and Goldberg A. L. The toxic effects of tumor necrosis factor *in vivo* and their prevention by cyclooxygenase inhibitors. *Proc. Natl. Acad. Sci. USA, 1987,* v. 84, pp. 4273-4277.

Khachatur'yan M. L., Gukasov V. M., Komarov P. G. et al. [Comparison of lipid peroxidation parameters in the heart, liver and brain of rats with different degrees of resistance to hypoxia]. (Russian). *Biul. Eksper. Biol. Med.*, 1996, v. 121, No. 2, pp. 138-143. English transl. see[1*], pp. 126-131.

Khaspekov L. G., Lyzhyn A. A. and Viktorov I. V. [The role of posthypoxic reoxygenation in destruction of cultured hippocampal neurons]. (Russian). *Biull. Eksp. Biol. Med.*, 1993, v. 115, No. 2, pp. 130-132. English transl. see:[1*], pp. 120-122.

Kiessling M., Dienel G. A., Jacewicz M. and Pulsunelli W. A. M. Protein synthesis in postischemic rat brain: a two-dimensional electrophoretic analysis. *J. Cerebral Blood Flow Metabolism*, 1986, v. 6, pp. 642-649.

Kikugawa K., Kojima T., Yamaki S. and Kosugi H. Interpretation of the thiobarbituric acid reactivity of rat liver and brain homogenates in the presence of ferric ion and ethylendiaminetetraacetic acid. *Anal. Biochem.*, 1992, v. 202, pp. 249-255.

Kim E. A., Danilenko M. P., Murzakhmetova M. K. et al. [Na, K-ATPase activity in myocardium sarcolemma during ischemia]. (Russian). *Ukr. Biokhim. Zh.*, 1989, v. 61, No. 4, pp. 60-65. English trans. See[5*], pp. 69-75.

Kim M. Y., Linardic C. M., Obeid L. M. and Hannun Y. A. Identification of sphingomyplin turnover as an effector mechanism for the action of tumor necrosis factor alpha and gamma-interferon. *J. Biol. Chem.*, 1991, v. 266, No. 1, pp. 484-489.

Kim Y. H., Kim E. Y., Gwag B. J, et al. Zinc-induced cortical neuronal death with features of apoptosis and necrosis: mediation by free radicals. *Neurosci*, 1999, v. 89, No. 1, pp. 175-182.

Kinnula V. L., Everitt J. I., Whorton A. R. and Crapo J. D. Hydrogen peroxide production by alveolar type II cells, alveolar macrophages and endothelial cells. *Am. J. Physiol.*, 1991, v. 261, No. 2, Pt. 1, pp. L84-L91.

Kinnula V. L., Whorton A. R., Chang L. Y. and Crapo J. D. Regulation of hydrogen peroxide generation in cultured endothelial cells. *Am. J. Respir. Cell. Mol. Biol.*, 1992, v. 6, No. 2, pp. 175-182.

Kinoshita T., Mira O. S., Matsumoto K. and Nakamura T. Possible endocrine control by hepatocyte growth factor of liver regeneration after partial hepatectomy. *Biochem. Biophys. Res. Commun.*, 1991, v. 177, No. 1, pp. 330-335.

Kintner D., Fitzpatrick J. H., Louie J. A et al. Cerebral oxygen and energy metabolism during and after 30 minutes of moderate hypoxia. *Am. J. Physiol.*, 1984, v. 247, No. 4, Pt. 1, pp. E475-E482.

Kirpatovskii V. I., Nikiforova N. V., Kudryavtsev Yu. V. and Nadtochii O. N. [Use of α-tocopherol emulsion for antioxidant protection of ischemic or stored kidneys]. (Russian). *Biul. Eksp. Biol. Med.*, 1996, v. 121, No. 5, pp. 499-503. English transl. see[1*], pp. 452-455.

Kirshner R. E. and Fantini G. A. Role of iron and oxygen-derived free radicals in ischemia-reperfusion injury. *J. Am. Coll. Surg.* 1994, v. 179, No. 1, pp. 103-117.

Kislev G. V., Pavlinova Z. I. and Chetverikov D. A. [Properties of high-energy bond of triphosphoinoside]. (Russian). *Biokhimiya*, 1978, v. 43, No. 4, pp. 614-621. English transl. see[3*], pp. 489-495.

Kiselev V. D. [Correlation of the properties of actomyosin and myofibril ATPases of dog, rat and frog hearts with their physiological peculiarities]. (Russian). *Kardiologiya*, 1979, v. 19, No. 1, pp. 107-112.

Kitakaze M., Node K., Minamino T. et al. Inhibition of angiotensin-converting enzyme increases the nitric oxide levels in canine ischemic myocardium. *J. Mol. Cell. Cardiol.*, 1998, v. 30, No. 11, pp. 2461-2466. Biochem., 1992, v. 202, pp. 249-255.

Klebanov G. I., Kreinina M. V., Pozin V. M. et al. [Time course of blood polymorph function in dogs with reversible myocardial ischemia]. (Russian). *Biul. Eksp. Biol. Med.*, 1988, v. 106, No. 9, pp. 297-299. English transl. see[1*], pp. 1260-1263.

Klebanov G. I., Turkmenov E. M. and Kreinina M. V. Hetorogeneity of receptors on polymorphonuclear blood leikocytes. *Biol. Membrane,* 1987, v. 4, No. 10, pp. 1084-1092.

Klebanoff S. J. and Seymour J. Oxygen metabolism and the toxic properties of phagocytes. *Ann. Intern. Med.*, 1980, v. 93, pp. 480-489.

Klein A. S., Joh J. W., Rangan U. et al. Allopurinol: discrimination of antioxidant from enzyme inhibitory activities. *Free Radic. Biol. Med.*, 1996, v. 21, pp. 713-717.

Klein M. H., Nebendahl K., Lindert S. et al. A modified regionally ischemic porcine heart preparation with eligible residual blood-flows. *Basic Res. Cardiol.*, 1986, v. 81, No. 4, pp. 384-393.

Kleinman J. G., Brown W. W., Ware R. A. et al. Cell pH and acid transport in renal cortical tissue. *Am. J. Physiol.*, 1980, v. 239, pp. 440-444.

Klempt N. D., Sirimanne E., Gunn A. J. et al. Hypoxia-ischemia induces transforming growth factor beta 1 mRNA in the infant rat brain. *Brain Res.*, 1992, v. 13, No. 1-2, pp. 93-101.

Klimovitskaya P. L. [Distribution of metabolic ways in the dog kidney cortical layer in acute ischemia]. (Russian). *Vestn. Akad. Med. Nauk SSSR*, 1975, No. 7., pp. 52-56.

Kloner R. A., De Boeh L. W. V, Darsel J, R. et al. Recovery from prolonged abnormalities of canine myocardium salvaged from ischemic necrosis by coronary reperfusion. *Proc. Nat. Acad. USA Biol. Sci.,* 1981, v. 78, No. 11, pp. 7152-7156.

Kloner R. A., Fishtein M. C., Hare C. M. et al. Early ischemic ultrastructural and histochemical alterations in the myocardium of the rat following coronary artery occlussion. *Exp. Mol. Pathol.,* 1979, v. 30, pp. 129-143.

Kloner R. A., Ganote C. E, Whalen D. A. et al. Effect of a transient period of ischemia on myocardial cells: II. Fine structure during the first few minutes of reflow. *Am. Pathol.*, 1974, v. 74, pp. 399-422.

Klyachkin M. L. and Zakharova N. B. [Reversibility of structural-functional changes in experimental acute ischemia of the limb tissues]. (Russian). *Patol. Fiziol. Eksp. Ter.*, 1984, No. 6, pp. 34-37.

Kobayashi M., Shimomura I., Suzuki H et al. Structural effect of ubiquinones on the mitochondrial inner membrane. *J. Appl. Biochem.*, 1980, v. 2., pp. 270-279.

Kogan A. Kh. and Mednykh A. Ya. [Research methods of resistance of tissues (normal and pathological) to initiation free-radical oxidation by H_2O_2]. (Russian). *Free radical lipid oxidation in norma and pathology*. Moscow, Nauka, 1976, pp. 74-76.

Kogan A. Kh., Kudrin A. V. and Nikolaev S. M. [Lipid free radical peroxidation and the pathogenesis of coronarocclusive myocardial infarction]. (Russian). *Free radical lipid oxidation in norma and pathology*. Moscow, Nauka, 1976, pp. 68-71.

Kogan A. Kh., Kudrin A. V. and Nikolaev S. M. [On the role played by free radical lipid peroxidation on the pathogenesis of adrenaline cardionecroses]. (Russia). *Patol. Fiziol. Eksp. Ter.,* 1979, No. 1, pp. 66-67.

Kogan A. Kh., Kudrin A. N., Kakturskii L. V. and Losev N. I. [Free radical peroxidation mechanisms of the pathogenesis of ischemia and myocardial infarction and their pharmacologic regulation. Review]. (Russian). *Patol. Fiziol. Eksper. Ter.*, 1992, No. 2, pp. 5-15.

Kogan A. Kh., Losev N. I., Kudrin A. N. et al. [Changes in the size of the myocardial injury area during, postischemic reperfusion]. (Russian). *Biul. Eksp. Biol. Med.*, 1984, v. 97, No. 3, pp. 270-272. English transl. see[1*], pp. 240-242.

Kogan A. Kh., Miyeegombyn A., Losev N. I. and Kudrun A. N. [Dynamics of "no-reflow" phenomenon in reperfusion of the myocardium, depending on the duration of myocardial ischemia]. (Russian). *Biul. Eksp. Biol. Med.* 1986, v. 101, No. 1, pp. 16-18. English transl. see[1*], pp. 19-21.

Kohli V., Selzner M., Madden J. F. et al. Endothelial cell and hepatocyte deaths occur by apoptosis after ischemia-reperfusion injury in the rat liver. *Transplantation*, 1999, v. 67, No. 8, pp. 1099-1105.

Koide T., Neichi T., Takato M. et al. Possible mechanisms of 15-hydroperoxy-arachidonic acid-induced contraction of the canine basilar artery in vitro. *J. Pharm. Exp. Ther.*, 1982, v. 221, No. 2, pp. 481-488.

Koide T., Noda Y., Nata S. et al. Contraction of the canine basilar artery following linoleic, arachidonic, 13-hydroperoxylinoleic, or 15-hydroperoxy-arachidonic acid (41294). *Proc. Soc. Exp. Biol. Med.*, 1981, v. 168, pp. 399-402.

Kojima Y., Haruta A., Imai T. et al. Conjugation of Cu, Zn-superoxide dismutase with succinylated gelatin: pharmacological activity and cell-lubricating function. *Bioconjugate Chemistry,* 1993, v. 4, No. 6, pp. 490-498.

Kolchinskaya A. Z. [Classification of hypoxic states]. (Russian). *Patol. Fiziol. Eksp. Ter.,* 1981, No. 4, pp. 3-10.

Koller P. and Bergmann S. R. Reduction of lipid peroxidation in reperfused isolated rabbit heart by dietiazem. *Circ. Res.*, 1989, v. 65, pp. 838-846.

Komara J. S., Nayini N. R., Bialick H. A. et al. Brain iron delocalization and lipid peroxidation following cardiac arrest. *Ann. Emerg. Med.,* 1986, v. 15, pp. 384-389.

Komarov B. D. and Shimanko I. I. [*Trench compression of tissues*]. (Russian). Moscow, Meditsina, 1984, 176 p.

Komarov O. S. [*Enzyme Systems of lipid peroxidation requlation in preserved kidney*] (Russian). Dissertation. Moscow, 1975.

Komarov P., Morgunov A., Shdanov R. and Bilenko M. Anti-ischemic action of N-hydroxy spin label: a novel antioxidant agent. Abstract. *J. Mol. Cell. Cardiol.*, 1988, v. 20, Suppl. V, No. 23, p. S. 26.

Komarov P. G., Morgunov A. A., Zhdanov R. I. and Bilenko M. V. Anti-ischemic effect of 1-Hydroxy Derivatives of Nitroxyl Bioantioxidants. In: Zhdanov R. ed. *Bioactive Spin Labels.* Springer Verlag, 1992, pp. 491-507.

Komarov P. G., Taskaeva O. N. and Zhdanov R. I. [N-hydroxytetramethylpiperidines as bioantioxidants]. (Russian). *Dokl. Acad. Nauk SSSR,* 1987, v. 297, No. 3, pp. 734-737. English transl. see[6*], pp. 384-387.

Komarov P. G., Velikhanova D. M. and Bilenko M. V. Tolerance of intact and induced liver systems of metabolism of xenobiotics to ischemic and reperfused damaging effect. (Russian). In: *Cytochrome P-450 structure and function. All Union Symposium.* Abstracts. Minsk, 1982, p. 80.

Konda A., Baba S., Iwaki T. et al. Hyperbaric oxygenation prevents delayed neuronal death following transient ischaemia in the gerbil hippocampus. *Neuropathol. Appl. Neurobiol.,* 1996, v. 22, No. 4, pp. 350-360.

Kondakova I. V., Borunob E. V. and Antipova E. V. [Activation of phospholipase hydrolysis during oxidative cell damage]. (Russian). *Biul. Eksp. Biol. Med.* 1991, v. 112, No. 9, pp. 258-287. English transl. see[1*], pp. 1290-1301.

Kondo S. Altruistic cell suicide in relation to radiation hormesis. *Int. J. Radiat. Biol.,* 1988, v. 53, No. 1, pp. 95-102.

Kondo T., Reaume A. G., Huang T. T., et al Reduction of CuZn-superoxide dismutase activity exacerbates neuronal cell injury and edema formation after transient focal cerebral ischemia. *J. Neurosci.*, 1997, V. 17, No. 11, pp. 4180-4189.

Kondrashova M. U. and Maevsky E. I. [Variable application carbones and lipids as form of regulation of physiological state]. (Russian). In: *Regulation of power exchange and physiological state of organism.* Moscow, Nauka, 1978, pp. 5-14.

Kondrashova M. N., Mayevsky E. I., Balayan G. V. et al. [Hypoxia adaptation by means of metabolic switch for amber acid transformation]. (Russian). *Mitochondria. Biochemistry and Ultrastructure.* Moscow, Nauka, 1973, pp. 112-129.

Kono Y., Ozawa K., Tanaka J. et al. Significance of mitochondrial enhancement in restoring hepatic energy charge after revascularization of isolated ischemic liver. *Transplantation*, 1982, v. 33, No. 2, pp. 150-155.

Konorev E. A., Polumiskov V. Yu., Avilova O. A. and Golikov A. P. [Emoxypine during reperfusion of the ischemic dog myocardium: effect of infarct size and plasma creatine kinase activity]. (Russian). *Biul. Eksp. Biol. Med.*, 1990, v. 110, No. 9, pp. 252-254. English transl. see[1*], pp. 1182-1185.

Konorev A. E., Saks V. A., Sharov V. G. and Makhotina L. A. [Oxidative myocardial damage: protective action of exogenous phosphocreatine]. (Russian). *Biul. Eksp. Biol. Med.*, 1990, v. 109, No. 3, pp. 224-226. English transl. see[1*], pp. 281-284.

Konovalova G. G., Cherpachenko N. M., Lankin V. Z. et al. [Limitation of the zone of coronaro-occlusion myocardial infarction in rats by antioxidant therapy]. (Russian). *Biul. Eksp. Biol. Med.*, 1984, v. 98, No. 8, pp. 153-156. English transl. see[1*], pp. 1030-1033.

Kontorshchikova K. N. and Solov'eva T. I. Effect of osone on lipid peroxidation in rat hearts isolated after clinical death. *Biul. Eksp. Biol. Med.*, 1996, v. 121, No. 2, pp. 161-163. English transl. see[1*], pp. 148-150.

Kontos H. A., Wei E. P., Povlishock J. T. et al. Oxygen radicals mediate the cerebral arteriolar dilation from arachidonate and bradykinin in cats. *Circ. Res.*, 1984, v. 55, pp. 295-303.

Konvai V. D. [Effect of mechanical asphyxia on lipid peroxidation processes in the rat brain]. (Russian). *Patol. Fiziol Eksp. Ter.,* 1982, No. 5, pp 30-32.

Konvai V. D., Lukoshkin A. V. and Smirnova V. B. [Possible mechanisms of lipid peroxidation in rat liver tissue within the recovery period after mechanical asphyxia]. (Russian). *Vopr. Med. Khim.,* 1982, v. 28, No. 4, pp. 42-46.

Kornbrust D. J. and Mavis R. D. Microsomal lipid peroxidation. Characterization of the role of iron and NADPH. *Mol. Pharmacol.*, 1980, No. 17, pp. 400-407.

Korneev A. A., Komissarova I. A. and Nartsissov Ya. R. [The possibility of using glutathione as a protector during exposure to hypoxia]. (Russian). *Biul. Eksp. Biol. Med.*, 1993, v. 116, No. 9, pp. 261-263. English transl. see[1*], pp. 1089-1092.

Korochkin I. M. and Keiyan A. V. [Hyperbaric oxygenation in coronary heart disease]. (Russian). *Sov. Med.,* 1986, No. 9, pp. 8-11.

Korochkin I. M., Klebanov G. I. Chukaeva I. I. et al. [Chemiluminescence of leukocytes in the diagnosis of acute myocardial infarction]. (Russian). *Ter. Arhiv,* 1984, v. 8, pp. 29-31.

Korthuis R. J. and Gute D. C. Postischemic leukocyte/endothelial cell interactions and microvascular barrier dysfunction in skeletal muscle: cellular mechanisms and effect of Daflon 500 mg. *Int. J. Microcirc. Clin. Exp.*, 1997, v. 17, Suppl. 1, pp. 11-17.

Korthuis R. J., Smith J. K. and Carden D. L. Hypoxic reperfusion attenuates postischemic microvascular injury. *Am. J. Physiol.*, 1989, v. 256, No. 1, Pt. 2, pp. H315-H319.

Koshkin V. V. [Generation of the superoxide radical and lipid peroxidation in skeletal muscles]. (Russian). *Biokhimiya*, 1985, v. 50, No. 9, pp. 1406-1410. English transl. see[3*], pp. 1199-1203.

Kostyuk V. A. and Potapovich A. I. [Estimation of lipid peroxidation products using TBA-test under anaerobic conditions]. (Russian). *Vopr. Med. Khim.*, 1987, v. 33, No. 3, pp. 115-118.

Kotaka K., Miyzaki Y. Ogawa K. et al. Reversal of ischemia-induced mitochondrial dysfunction after coronary reperfusion. *J. Mol. Cell Cardiol.,* 1982, v. 14, pp. 223-231.

Kovacheva-Ivanova S., Bakalova R., Kagan V. and Georgiev G. [Activation of lipid peroxidation and changes in vitamin E levels in the lungs during oxidative stress]. (Russian.) *Biul. Eksp. Biol. Med.*, 1992, v. 113, No. 2, pp. 132-134. English transl. see[1*], pp. 164-166.

Kovalev I.E., Piruzian L. A., Shaternikov V. A. et al. [Participation of cytochrome P-450 system in immediate allergic disorder]. (Russian). *Dokl. Acad. Nauk SSSR*, 1982, v. 265, No. 7, pp. 247-249.

Kovanov V. V., Bilenko M. V., Amosov I. S. et al. [Haemodynamics and microcirculation of ischaemic and preserved kidneys during extracorporal perfusion]. (Russian). *Eksp. Khir. Anesteziol.,* 1983, No. 2, pp. 3-11.

Kovanov V. V., Bilenko M. V. and Shelenkova L. N. [Intensity of lipid peroxidation in the kidneys in the early postischemic period as a function of pO_2 in inspired air]. (Russian.) *Acute ischemia of organs and early postischemic disorders. 2nd All-Union Symposium*. Abstracts. Moscow, 1978, pp. 51-52.

Koyama I., Bulkley G. B., Williams G. H. and Im M. J. The role of oxygen free radicals in mediating the reperfusion injury of cold-preserved ischemic kidneys. *Transplantation*, 1985, v. 40, pp. 590-595.

Kozlov Yu. P., Danilov V. S. and Kagan V. E. [*Free radical lipid oxidation in biological membranes*]. (Russian). Moscow, MGU, 1972, 87 p.

Kozlov Yu. P., Kagan V. E. and Arkhipenko Yu. V. [*Molecular mechanisms of oxigenic injury of calcium transfer system in muscle sarcoplasmic reticulum*]. (Russian). Irkutsk. 1983, 134 p.

Kozlov A. V., Meschanov A. Yu., Nikolayeva I. G. et al. [Comparative study of the anti-oxidative properties of polyhydroxamic acids, desferal and other iron chelators. (Russian). *Biofizika*, 1993, v. 38, No. 4, pp. 708-713. English transl. see[2*], pp. 733-738.

Kozlov A. V., Shinkarenko L. I., Vladimirov Yu. A. and Azizova O. A. [Role of endogenous free iron in activation of lipid peroxidation during ischemia]. (Russian). *Biul. Eksp. Biol. Med.*, 1985, v. 99, No. 1, pp. 38-41. English transl. see[1*], pp. 47-50.

Kozlov A. V., Vdovin A. V., Azizova O. A. and Vladimirov Yu. A. [Role of free Fe^{2+} in lipid peroxidation during ischemia and reoxygenation of the liver]. (Russian). *Biul. Exsp. Biol. Med.*, 1987, v. 104, No. 8, pp. 165-167. English transl. see[1*], pp. 1069-1071.

Kozlov A. V., Yegorov D. Yu., Vladimirov Yu. A. and Azizova O. A. Intracellular free iron in liver tissue and liver homogenate: studies with electron paramagnetic resonance on the formation of paramagnetic complexes with desferal and nitric oxide. *Free Rad. Biol. Med.*, 1992, v. 13, No. 1, pp. 9-16.

Kramer J. H., Dickens B. F., Misik V. and Weglicki W. B. Phospholipid hydroperoxides are precursors of lipid alkoxyl radicals produced from anoxia/reoxygenated endothelial cells. *J. Mol. Cell. Cardiol.*, 1995, v. 27, No. 1, pp. 371-381.

Kramer J. H., Mak I. T. and Weglicki W. B. Differential sensitivity of canine cardiac sarcolemma and microsomol enzymes to inhibition by free radical-induced lipid peroxidation. *Circ. Res.*, 1984, Vol. 55, No. 1, pp. 120-124.

Krasnova E. I., Minin P. P. and Zavodskaya I. S. [Changes in creatine kinase and its mB isoenzyme activity after complete and partial occlusion of the coronary artery in rats.]. (Russian). *Biul. Eksp. Biol. Med.*, 1986, v. 102, No. 11. pp. 554-556. English transl. see[1*], pp. 1529-1530.

Krause G. S., Joyce K. M., Nayini N. R. et al. Cardiac arrest and resuscitation: brain iron delocalization during reperfusion. *Ann. Emerg. Med.*, 1985, v. 14, pp. 1037-1043.

Krause S. M. and Rozanski D. Effects of an increase in intracellular free $[Mg^{2+}]$ after myocardial stunning on sarcoplasmic reticulum Ca^{2+} transport. *Circulation*, 1991, v. 84, No. 3, pp. 1378-1383.

Kreuzer F. Facilitated diffusion of oxygen and its possible significance. *Respir. Physiol.*, 1970, v. 9, No. 1, pp. 1-30.

Krokan H., Grafstrom R. C., Sundqvist K. et al. Cytotoxicity, thiol depletion and inhibition of O^6-methylguanine-DNA methyltransferase by various aldehydes in cultured human bronchial fibroblasts. *Carcinogenesis*, 1985, v. 6, No. 12, pp. 1755-1759.

Kubo K., Yoshitake I., Kumada Y. et al. Radical scavenging action of flunarizine in rat brain in vitro. *Arch. Int. Pharmacodyn.*, 1984, v. 272, pp. 283-295.

Kubota M, Nakane M., Narita K. et al. Mild hypotermia reduces the rate of metabolism of arachidonic acid following postischemic reperfusion. *Brain Res.*, 1998, v. 779, No. 1-2, pp. 297-300.

Kucukcelebi A. and Ozcan M. The beneficial effect of cyclosporin A on the no-reflow phenomenon in rat skin island flaps. *Br. J. Plastic Surg.*, 1992, v. 45, No. 7, pp. 512-514.

Kudrin A. N., Kogan A. Kh., Korolev V. V. et al. [Free radical oxidation of lipids in pathogenesis of myocardium infarction and medico-prophylactic role of antioxidants - sodium selenite, and its combination with vitamine E]. (Russian). *Kardilogiya*, 1978, v. 18, No. 2, pp. 115-118.

Kulagin B. R. [*Pathogogical physiology of trauma and shock*]. (Russian) Leningrad, Meditsina, 1978, 296 p.

Kulikov V. Yu., Semenyuk A. V. and Kolesnikova L. I. [*Lipid peroxidation and cold temperature factor*]. (Russian). Novosibirsk, Nauka, 1988, 191 p.

Kulitskaya V. I. and Murzin G. V [Effect of post-traumatic modulations in oxygen tension in the arterial blood on the rate of lipid peroxidation in the lungs, liver and myocardium of rats with pulmonary contusion]. (Russian). *Biul. Eksp. Biol. Med.*, 1984, v. 98, No. 9, pp. 276-278. English transl. see[1*], pp. 1173-1176.

Kumar P. S., Georg T., Bai N. J. et al. Effect of lipid antioxidants on rat erythrocyte hemolysis. *Int. J. Vit. Nutr. Res.*, 1979, v. 49, pp. 352-358.

Kunitsa N. I., Kuz'menko I. V., Alekseev S. M. et al. [Effects of tocopherol and its analogs on *in vivo* lipid peroxidation and electron transport in rat liver mitochondria]. (Russian). *Biokhimiya*, 1993, v. 58, No. 11, pp. 1709-1713. English transl. see[3*], pp. 1256-1259.

Kupatt C., Habazettl H., Goedecke A. et al. Tumor necrosis factor-alpha contributes to ischemia- and reperfusion- induced endothelial activation is isolated hearts. *Circ. Res.*, 1999, v. 84, No. 4, pp. 392-400.

Kupatt C., Habazettl H., Zahler S. et al. ACE- inhibition prevents postischemic coronary leukocyte adhension and leukocyte-dependent reperfusion injury. *Cardiovasc. Res.*, 1997, v. 36, No. 3, pp. 386-395.

Kuppusamy P. and Zweier J. L. Characterization of free radical generation by xanthine oxidase. *J. Biol. Chem.*, 1989, v. 264, No. 17, pp. 9880-9884.

Kuthan H. and Ulrich V. Oxidase and oxygenase function of the microsomal cytochrom P-450 monooxygenase system. *Eur. J. Biochem.*, 1982, v. 126, pp. 583-588.

Kuwashima J., Nakamura K., Fujitani B. et al. Relationship between cerebral energy failure and free fatty acid accumulation following prolonged brain ischemia. *Jpn. J. Pharmacol.,* 1978, v. 28, pp. 277-287.

Kuzin M. I., Vornovitsky E. G., Zenkova N. A. et al. [Electromechanical coupling disturbances in myocardial cells in the compression syndrome]. (Russian). *Biul. Eksp. Biol. Med.*, 1983, v. 95, No. 2. pp. 14-16. English transl. see[1*], pp. 175-178.

Kuzmin A. I., Shulzhenko V. S. and Medvedev O. S. [Association of adenine nucleotide degradation and catecholamine loss in myocardial ischemia with contractility recovery in reperfusion]. (Russian). *Biul. VKNC Akad. Med. Nauk SSSR*, 1987, No. 1, pp. 75-82.

La Brecque D. R. and Pesch L. A. Preparation and partial characterisation of hepatic regenerative stimulator substance (SS) from rat liver. *J. Physiol. (London)* 1975, v. 248, No. 2, pp. 273-284.

Lakatta E. G., Nayler W. G. and Poole-Wilson P. A. Calcium overload and mechanical function in posthypoxic myocardium: biphasic effect of pH during-hypoxia. *Eur. J. Cardiol.,* 1979, v. 10, pp. 77-87.

Landi L., Cabrini L., Tadolini B. et al. Incorporation of ubiquinones into lipid vesicles and inhibition of lipid peroxidation. *Ital. J. Biochem.,* 1985, v. 34, No. 5, pp. 356-363.

Lankin V. Z., Kogan A. Kh., Kovalevskaya A. L. et al. [Enzymes that detoxify active oxygen and lipoperoxide forms in experimental ischemia and myocardial infarction]. (Russian). *Biul. Eksp Biol. Med.,* 1982, v. 93, No. 5. pp. 58-60. English. transl. see[1*], pp. 613-615.

Lankin V. Z., Tikhaze A. K., Lemeschko V. V. et al. Superoxiddismutase and glutathionperoxidase in cytosol and mitochondria of rat liverduring aging. *Biul. Exp. Biol. Med.*, 1981, v. 92, No. 9, pp. 310-311.

Lankin V. Z., Tikhaze A. K., Voronina N. V. et al. [Possibility of studying lipid peroxidation in surviving tissues]. (Russian). *Biul. Eksp. Biol. Med.*, 1981, v. 91, No. 3, pp. 327-329. English transl. see[1*], pp. 339-341.

Lankin V. Z., Zakirova A. N., Akhmetova B. Kh. et al. [Lipid peroxides and atherosclerosis. Free-radical peroxidation of polyene lipids in blood of patients with ischemic heart disease]. (Russian). *Kardiologiya*, 1980, v. 20, No. 7, pp. 96-99.

Lan'shina O. E. Loginov V. A. and Kovalenko E. A. [Effect of hyperbaric hyperoxia on blood levels of partial pressure of oxygen and water-soluble lipid peroxidation products]. (Russian.) *Biul. Eksp. Biol. Med.*, 1992, v. 114, No. 9, pp. 254-255. English transl. see[1*], pp. 1258-1260.

Lantos J., RÖth E. and Temes G. Effects of lidocaine on cerebral lipid peroxidation and neutrophil activation following complete compression ischemia. *Arch. Int. Pharmacodyn.*, 1996, v. 331, pp. 179-188.

Laster S. M., Wood J. G and Gooding L. R. Tumor necrosis factor can induces both apoptotic and necrotic forms of cell lysis. *J. Immunol.* 1988, v. 141, No. 8, pp. 2629-2635.

Lavallee M. and Vatner S. F. Regional myocardial blood flow and necrosis in primates following coronary occlusion. *Am. J. Physiol.*, 1984, v. 246, pp. H635-H639.

Lawrence G. D. and Cohen G. Concentrating ethane from breath to monitor lipid peroxidation in vivo. In: *Methods in Enzymology*, 1984, v. 105. *Oxygen Radicals in Biological Systems*. Packer L., ed. N. Y. Acad. Press Inc., pp. 305-311.

Lawrence J. H., Loomis W. F., Tobias C. A. and Turpin F. H. Preliminary observations on the narcotic effects of xenon with a review of values for solubilities of gases in water and oils. *J. Physiol.,* 1946, v. 105, pp. 197-204.

Lawson D. L., Mehta J. L. and Nichols W. W. Coronary reperfusion in dogs inhibits endothelium-dependent relaxations: role of superoxide radicals. *Free Rad. Biol. Med.*, 1990, v. 8, No. 4, pp. 373-380.

Lawson D. L., Mehta J. L., Nichols W. W. et al. Superoxide radical-mediated endothelial injury and vasoconstriction of rat thoracic aortic rings. *J. Lab. Clin. Med.,* 1990, v. 115, No. 5, pp. 541-548.

Lawson T. and Stohes S. Changes in endogenous DNA damage in ageing mice in response to butylated hydroxyanisole and oltipraz. *Mech. Ageing Dev.,* 1985, v. 30, pp. 179-185.

Lea M. A. and Weber G. Role of enzymes in homeostasis. *J. Biol. Chem.*, 1968, v. 243, No. 6. pp. 1096-1102.

Lebedev A. V., Levitsky D. O. and Loginov V. A. [Oxygen as an inductor of calcium ion tranfer through bilayer lipid membranes]. (Russian). *Dokl. Acad. Nauk SSSR*, 1980, v. 252, No. 6, pp. 1494-1497. English transl. see[8*], pp. 106-109.

Ledenev A. N. and Runge E. K. [Superoxide radical production by ischemic heart mitochondria]. (Russian). *Biul. Eksp. Biol. Med.*, 1985, v. 100, No. 9. pp. 303-305. English transl. see[1*], pp. 1204-1206.

Lee C. O. Ionic activities in cardiac muscle cells and application of ion-selective microelectrodes. *Am. J. Physiol.*, 1981, v. 241, No. 4. pp. H459-H478.

Lee J. W., Miyawaki H., Bobst E. V., et al. Improved functional recovery of ischemic rat hearts due to singlet oxygen scavengers histidine and carnosine. *J. Mol. Cell. Cardiol.* 1999, v. 31, No. 1, pp. 113-121.

Lee S. L., Balasabramanian V. and Dhalla W. S. Excitation-contraction coupling in heart. XIX. Effect of hypoxia on calcium transport by subcellular particles in the isolated perfused rat heart. *Can. J. Physiol. Pharmac.*, 1976, v. 54, No. 7, pp. 49-58.

Lee K. T., Tsai L. Y. and Sheen P. C. Effect of vitamin E, topical hypothermia and steroid on ischemic liver in rats. *Kao Hsiung I Hsueh Ko Hsueh Tsa Chih*, 1998, v. 14, No. 1, pp. 6-12.

Lefer A. M. Eicosanoids as mediators of ischemia and shock. *Fed. Proceed.*, 1985, v. 44, No. 2. pp. 275-280.

Lefer A. M. Interaction between myocardial depressant factor and vasoactive mediators with ischemia and shock. *Am. J. Physiol.*, 1987, v. 252, No. 2, Pt. 2, pp. 193-205.

Lefer A. M. Thromboxane A2 and leukotrienes are eicosanoid mediators of shock and ischemic disorders. *Review. Progr. Clin. Biol. Res.*, 1988, v. 264, pp. 101-114.

Lefer A. M. and Lefer D. J. Endothelial dysfunction in myocardial ischemia and reperfusion: role of oxygen-derived free radicals. *Basic Res. Cardiol.*, 1991, v. 86, Suppl. 2, pp. 109-116.

Lefer A. M. and Lefer D. J. Pharmacology of the endothelium in ischemia-reperfusion and circulatory shock. Review. *Ann. Rev. Pharmacol. Toxicol.*, 1993, v. 33, pp. 71-90.

Lefer A. M. and Ma X. L. Cytokines and growth factors in endothelial dysfunction. *Crit. Care Med.*, 1993, v. 21, No. 2, Suppl., pp. S9-S14.

Lefer A. M., Ma X. L., Weyrich A. S. et al. Mechanism of the cardioprotective effect of transforming growth factor beta1 in feline myocardial ischemia and reperfusion. *Proc. Natl. Acad. Sci. USA*, 1993, v. 90, No. 3. pp. 1018-1022.

Lefer A.M., Tsao P., Aoki N. et al. Mediation of cardioprotection by transforming growth factor-beta. *Science*, 1990, v. 249, No. 4964, pp. 61-64.

Lefer D. J., Scalia R., Campbell B. et al. Peroxynitrite inhibits leukocyte-endothelial cell interactions and protects against ischemia-reperfusion injury in rats. *J. Clin. Invest,* 1997, v. 99, No. 4, pp. 684-691.

Lehninger A. [Principles of biochemistry]. (Russian transl. from English). Moscow, Mir, 1985, v. 2, 368 p.

Lehninger A. L., Reynafarje B., Vercesi A. et al. Transport and accumulation of calcium in mitochondria. *Ann. N. Y. Acad. Sci.,* 1978, v. 307, pp. 160-176.

Lenaz G. The role of lipids in the structure and function of membrane. In: Roodyn D. B., ed., *Subcell. Biochem.,* 1979, v. 6, pp. 233-343.

Lennon S. V., Martin S. J. and Cotter T. G. Dose-dependent induction of apoptosis in human tumour cell lines by widely diverging stimuli. *Cell. Prolif.*, 1991, v. 24, No. 2, pp. 203-214.

Leo R., Pratico D., Iuliano. L. et al. Platelet activation by superoxide anion and hydroxyl radicals intrinsically generated by platelets that had undergone anoxia and then reoxygenated. *Circulation*, 1997, v. 95, No. 4, pp. 885-891.

Leoni S., Luly P, Mangiantini M. T. et al. Effect of free fatty acids and cholesterol in vitro on liver plasma membrane - bound enzymes. *Experientia*, 1982, v. 38, pp. 102-104.

Lerman A., Hildebrand F. L., Marqulies K. B. et al. Endothelium: a new cardiovascular regulatory peptide. *Mayo Clin. Proc.,* 1990, v. 65, pp. 1441-1455.

Leslie S. W., Gad S. C. and Acosta D. Cytotoxicity of butylated hydroxytoluene and butylated hydroxyanisole in cultured heart cells. *Toxicology*, 1978, v. 10, pp. 281-289.

Lesnefsky E. J., Van Benthuysen K. M., McMurtry I. F. et al. Lidocaine reduces canine infarct size and deacreses release of lipid peroxidation product. *J. Cardiovasc. Pharmacol.*, 1989, v. 13, pp. 895-901.

Lesnefsky E. J., Repine I.E. and Horwitz L. D. Oxidation and release of glutathione from myocardium during early reperfusion. *Free Radic. Biol. Med.* 1989, v. 7, pp. 31-35.

Letts L. G. and Piper P. J. The action of leukotrienes C_4 and D_4 on guinea-pig isolated hearts. *Br. J. Pharmacol.,* 1982, v. 76, pp. 169-176.

Levandovsky I. V. and Epishin Y. N. [The treatment of the early stages of homeostasis upset with the aid of cytochrome c and SPS (sodium persulfate) contrykal with the tourniquet shock]. *Biul. VKNC Acad. Med. Nauk SSSR,* 1981, v. 3. pp. 60-63.

Levandovsky I. V. and Shalnev A. N. [The role of sensitive receptor structures in the pathogenesis of disorders in the blood coagulation system and fibrinolysis in tourniquet shock]. (Russian). *Probl. Hematol.*, 1982, v. 27, No. 5, pp. 47-50.

Levin G. S. and Kamenetskaya Z. L. [*Lipid metabolism in bleeding and shock*]. (Russian). Tashkent, Meditsina, 1982, 168 p.

Levin R. M., Leggett R., Whitbeck C. and Horan P. Effect of calcium and calcium chelator on the response of the bladder to in vitro ischaemia. *Br. J. Urol.,* 1998, v. 82, No. 6, pp. 882-887.

Levitsky D. O., Benevolensky D. S. and Levchenko T. C. [The quantitaive estimation of calcium transfer ability of sarcoplasmatic reticulum in heart]. (Russian). In: *Myocardium metabolism.* Moscow, *Meditsina*, 1981, pp. 35-47.

Levitsky D. O., Levchenko T. S., Saks V. A. et al. [Functional coupling between Ca^{2+}-ATPase and creatine phosphokinase in sarcoplasmic reticulum of myocardium]. (Russian). *Biokhimiya*, 1977, v. 42, No. 10, pp. 1766-1773. English transl. see[3*], pp. 1389-1394.

Lewis M. J., Grey A. C. and Henderson A. H. Determinants of hypoxic contracture in isolated heart muscle preparations. *Cardiol. Res.,* 1978, v. 13, No. 2, pp. 86-94.

Lewis R. A. and Austen K. F. The biologically active leukotrienes. *J. Clin. Invest.*, 1984, v. 73, No. 4, pp. 889-897.

Li R. K., Shaikh N., Weisel R. et al. Cultured vascular endothelial cell susceptibility to extracellularly generated oxidant injury. *J. Mol. Cell. Cardiol.*, 1992, v. 24, No. 6, pp. 595-604.

Li X. Y., McCay P. B., Zughaib M. et al. Demonstration of free radical generation in the "stuned" myocardium in the conscious dog and identification of major differences between conscious and open-chest dogs. *J. Clin. Invest.*, 1993, v. 92, No. 2, pp. 1025-1041.

Liao H. J., Chen X. M. and Cui S. W. [Morphologic and molecular evidence of apoptosis during reperfusion phase after brief period of renal ischemia]. (Chinese). *Chinese Med. J.,* 1994, v. 74, No. 10, pp. 629-631.

Liashenko S. M. [*Effect of antioxidants and iron chelators on ischemic and regenerating liver*]. (Russian). Dissertation. Kupavna, 1994.

Liashenko S. M., Bilenko M. V. and Gefter Ju. V. [Comparative evaluation of influence of antioxidants on regenerative rate and function of ischemic liver]. (Russian). *Bioantioxidants. IVth All-Union Symposium.* Abstracts. Moscow, 1992, pp. 154-155.

Liedtke A. J., Nellis S. H. and Whitesell L. F. Effects of carnitine isomers on fatty acid metabolism in ischemic swine hearts. *Circ. Res.*, 1981, v. 48, pp. 859-866.

Lindert K. A., Caldwell-Kenkel J. C., Nukina S. et al. Activation of Kupffer cells on reperfusion following hypoxia: particle phagocytosis in a low-flow, reflow model. *Am. J. Physiol.*, 1992, v. 262, pp. G345-G350.

Lindroos P. H., Zarnegar R. and Michalopoulos G. K. Hepatocyte growth factor (hepatopotetin A) rapidly increases in plasma before DNA synthesis and liver regeneration stimulated by partial hepatectomy and carbon tetrachloride administration. *Hepatology*, 1991, v. 13, No. 4, pp. 743-750.

Lindsey J., Zhang H., Kaseki H. et al. Fatty acid metabolism and cell proliferation. VII. Antioxidant effects of tocopherols and their quinones. *Lipids*, 1985, v. 20, pp. 151-157.

Lindsay T. F., Liauw S., Romaschin A. D. and Walker P. M. The effect of ischemia/reperfusion on adenine nucleotide metabolism and xanthine oxidase production in skeletal muscle. *J. Vasc. Surg.*, 1990, v. 12, No. 1, pp. 8-15.

Lindsay T., Walker P. M., Mickle D. A. and Romaschin A. D. Measurement of hydroxy-conjugated diens after ischemia-reperfusion in canine skeletal muscle. *Am. J. Physiol.*, 1988, v. 254, No. 3, Pt. 2, pp. H578-H583.

Linnik M. D., Zobrist R. H. and Hatfield M. D. Evidence supporting a role for programmed cell death in focal cerebral ischemia in rats. *Stroke*, 1993, v. 24, pp. 2002-2009.

Liochev S. I. and Fridovich I. Vanadate-stimulated oxidation of NAD(P)H in the presence of biological membranes and other sources of O. *Arch. Biochem. Biophys.*, 1990, v. 279, No. 1, pp. 1-7.

Lipin'ski S., Lipetska K., Donets Y. et al. [Action of ionizing radiation of the activity of superoxidesmutase as measured by the chemiluminescent methods]. (Russian). *Radiobiologiya*, 1976, v. 16, No. 5, pp. 665-670.

Litvitsky P. F., Kogan A. Kh., Kudrin A. N. et al. [Pathogenetic role of activation of lipid peroxidation and protective effect of sodium selenite during ischemia and reperfusion of the myocardium]. (Russian). *Biul. Eksp. Biol. Med.*, 1981, v. 91, No. 3, pp. 271-274. English transl. see[1*], pp. 270-273.

Litvitsky P. F., Yakunin G. A., Lavrova L. A. et al. [Dinamics of some parameters of hemostasis in transistent coronary insufficiency depending on the duration of myocardial ischemia]. (Russian). *Biul. Eksp. Biol. Med.*, 1983, v. 95, No. 1, pp. 24-27. English transl. see[1*], pp. 31-34.

Liu P., Vonderfecht S. L., Fisher M. A. et al. Priming of phagocytes for reactive oxygen production during hepatic ischemia-reperfusion potentiates the susceptibility for endotoxin-induced liver injury. *Circ. Shock,* 1994, v. 43, No. 1, pp. 9-17.

Liu S., Beckman J. S. and Ku O. D. Peroxynitrite, a product of superoxide and nitric oxide, produces coronary vasorelaxation in dogs. *J. Pharmacol. Exp. Ther.*, 1999, v. 268, pp. 1114-1121.

Lizenko E. I. [Lipid composition and the lipolytic enzymes of the lysosomes. Review]. (Russian). *Usp. Sovr. Biol.,* 1982, v. 94, No. 1, pp. 94-110.

LoBalsamo L., Bergsland J., Lajos P. and Feldman M. J. Prevention of reperfusion injury in ischemic-reperfused hearts by oxypurinol and allopurinol. *Transpl. Int.,* 1989, v. 2, No. 4, pp. 218-222.

Lochner A., Kotze J. C. N., Benade A. J. S. et al. Mitochondrial oxidative phosphorylation in low-flow hypoxia: role of free fatty acids. *J. Mol. Cell. Cardiol.*, 1978, v. 10, pp. 857-875.

Lochner A., Van Niekerk J. and Kotze J. C. N. Mitochondrial acyl-CoA, adenine nucleotide translocase activity and oxidative phosphorylation in myocardial ischemia. *J. Molec. Cell. Cardiol.,* 1981, v. 13, pp. 991-997.

Lochner A., Van Niekerk I. and Kotze J. C. N. Normothermic ischemic cardiac arrest of the isolated perfused and lysolecithin on mechanical metabolic recovery. *Basic Res. Cardiol.*, 1985, v. 80, pp. 363-376.

Loeper J., Goy J., Rozensztajn L. et al. Lipid peroxidation and protective enzymes during myocardial infarction. *Clin. Chim. Acta,* 1991, v. 196, No. 2-3, pp. 119-125.

Lohse W., Otto G., Pahlig H. et al. Prognostic value of glutamyl dehydrogenase in clinical liver *Transplantation. Transplantation*, 1984, v. 38, No. 5, p. 559.

Longmuir I. S. and Pashko L. The induction of cytochrome P-450 by hypoxia. *Adv. Exp. Med. Biol.,* 1976, v. 75, pp. 121-125.

Longmuir I. S., Young A. and Mailman K. Induction by hypoxia of a new haemoglobin-line pigment. *Adv. Exp. Med. Biol.,* 1978, v. 94, pp. 297-300.

Lu Y. T., Liu S. F., Mitchell J. A. et al. The role of endogenous nitric oxide in modulating ischemia-reperfusion injury in the isolated, blood-perfused rat lung. *Am. J. Respir. Crit. Care Med.,* 1998, v. 157, No. 1, pp. 273-279.

Lucy A. J. Structural interactions between vitamin E and polyunsaturated phospholipids. In: de Duve O. and Hayaishi O., eds. *Tocopherol, oxygen and biomembranes.* N. Y., Elsevier, 1978, pp. 109-120.

Lucy J. A. The fusion of biological membranes. *Nature*, 1970, v. 227, pp. 815-817.

Lui X. K., Prasad M. R., Engelman R. M. et al. Role of iron on membrane phospolipid breakdown in ischemic reperfused rat heart. *Am. J. Physiol.* б 1990, v. 259, No. 4, Pt. 2, pp. H1101-H1107.

Lukacova N., Gottlieb M. and Marsala J. Lipid peroxidation and phospholipid composition in rat brain regions after ischemia and in early perfusion periods. *Arch. Ital. Biol.,* 1998, v. 136, No. 3, pp. 167-180.

Lukyanova L. D., Balmukhanov B. S. and Ugolev A. T. [*Oxygen-depended processes in cell and its functional state*]. (Russian). Moscow, Nauka, 1982, 301 p.

Lukyanova L. O., Kudrin A. N. and Kogan A. Kh. [Inhibition with obsidan of free radical lipid peroxidation in myocardial infarction induced by coronary occlusion]. (Russian). *Farmakol. Toksikol.*, 1983, v. 46, No. 1, pp. 45-48.

Lukyanova L. D., Popova O. A., Romanova B. E. et al. [The agents influencing the regulation of processes of biological oxidation in tissue preparation]. (Russian). In: Condrashova M. N., ed. *Energy exchange regulation and physiological condition of organism*. Moscow, Nauka, 1978, pp. 90-101.

Lunets E. F., Kuchuro S. V. and Kostyuk V. A. [Effect of lipid peroxidation on activity of Mg^{2+}-ATPase in brain and liver tissues]. (Russian). *Vopr. Med. Khim.,* 1986, v. 32, No. 4, pp. 32-35.

Luo Z., Diaco M., Murohara T. et al. Vascular endothelial growth factor attenuates myocardial ischemia-reperfusion injury. *Ann. Thorac. Surg.*, 1997, v. 64, No. 4, pp. 993-998.

Lupinetti F. M., Starnes V. A., Laws K. A. et al. Prostacyclin reduction of regional ischemic injury in the canine myocardium. *J. Surg. Res.*, 1986, v. 41, No. 2, pp. 146-157.

Lyakhovich V. V. and Tsyrlov I. B. [Structural aspects of monooxigenases biochemistry]. (Russian). *Novosibirsk, Nauka*, 1978. 238 p.

Lyzlova S. N., Stefanov B. E. and Taame N. [Possibility of participation of creatinkinase in cell metabolism regulation]. (Russian). In: *Myocardial Metabolism*. Moscow, Meditsina, 1981, pp. 107-116.

Ma X. L., Lopez B. L., Liu G. L. et al. Peroxynitrite aggravates myocardial reperfusion injury in the isolated perfused rat heart. *Cardivasc. Res.,* 1997, v. 36, No. 2, pp. 195-204.

Mackenzie R. J., Furnival C. M., Wood C. B. et al. The effect of prolonged hepatic ischemia before 70 percent partial hepatectomy in the dog. *Br. J. Surg.,* 1977, v. 64, No. 1, pp. 66-69.

Maclouf J., Sala A., Rossoni G. et al. Consequences of transcellular biosynthesis of leukotriene C4 on organ function. *Haemostasis*, 1996, suppl. 4, pp. 28-36.

MacManus J. P., Buchan A. M., Hill I.E. et al. Global ischemia can cause DNA fragmentation indicative of apoptosis in rat brain. *Neurosc. Lett.,* 1993, v. 164, pp. 89-92.

MacManus J. P., Hill I.E., Huang Z. -Gao et al. DNA damage consistent with apoptosis in transient focal ischemic neocortex. *Neuroreport,* 1994, v. 5, pp. 493-496.

MacMillan V. Cerebral Na^+, K^+ ATPase activity during exposure to and recovery from acute ischemia. *J. Cerebr. Blood Flow Metab.,* 1982, v. 2, pp. 457-465.

Madden M. C., Vender R. L. and Friedman M. Effect of hypoxia on prostacyclin production in cultured pulmonary artery endothelium. *Prostaglandins*, 1986, v. 3, No. 6, pp. 1049-1062.

Maessen J. G., van der Vusse G. J., Vork M. et al. Determination of warm ischemia time at donor nephrectomy. *Transplantation*, 1988, v. 45, №1, pp. 147-152.

Maezawa H., Manaka K., Yamakawa K., et al. Decreased sulfhydryl groups in the reperfused myocardial tissue of a rat model of myocardial infarction. *Jpn. Circ. J.* 1997 v. 61 No. 2 pp. 151-160.

Mahony Ch. Effect of aspirin on myocardial ischemia. *Am. J. Cardiol.,* 1989, v. 64, pp. 387-389.

Maiorino M., Thomas J. P., Girotti A. W. and Ursini F. Reactivity of phospholipid hydroperoxide glutathione peroxidase with membrane and lipoprotein lipid hydroperoxides. *Free Radic. Res. Commun.* 1991, v. 12-13, No. 1, pp. 131-135.

Majewska M. D., Strosznajder J. and Lazarewicz J. Effect of ischemic anoxia and barbiturate anesthesia on free radical oxidation of mitochondrial phospholipids. *Brain Res.,* 1978, v. 158, pp. 423-434.

Mak J. T. and Weglicki W. B. Comparative antioxidant activities of propranolol, nifedipine, verapamil and diltiazem against sarcolemmal membrane lipid peroxidation. *Circ. Res.*, 1990, v. 66, pp. 1449-1452.

Maksimova I. A., Maksimov V. M. and Piruzian L. A. [Quantitative estimation of kinetics of the free-radical processes in the organs of animals under hypoxia]. (Russian). *Fiziol. Zh. SSSR,* 1972, v. 63, No. 5, pp. 773-778.

Mal'tseva E. L., Kurnakova N. V., Burlakova E. B. and Pal'mina N. P. [Kinetic cooperativeness in the inhibition of the activity of protein kinase with diene conjugates form phospholipids and arachidonic acid *in vitro*]. (Russian). *Biokhimiya*, 1990, v. 55, No. 3, Pt. 1, pp. 471-479. English transl. see[3*], pp. 348-354.

Man R. Y. K., Wong T. and Choy P. C. Effect of lysophosphoglycerides on cardiac arrythmias. *Life Sci.*, 1983, v. 32, pp. 1325-1330.

Mandel L. J., Tanako T., Soltoff S. P. and Murdaugh S. Mechanisms whereby exogenous adenine nucleotides improve rabbit renal proximal function during and after anoxia. *J. Clin. Invest.*, 1988, v. 81, №4, pp. 1255-1264.

Mangino M. J., Mangino J. E., Kotadia B. and Sielczak M. Effects of the 5-lipoxygenase inhibitor A-64077 on intestinal hypothermic organ preservation injury. *J. Pharmacol. Exp. Ther.*, 1997, v. 281, №2, pp. 950-956.

Marion J. and Wolfe L. S. Origin of the arachidonic acid released post-mortem in rat forebrain. *Biochim. Biophys. Acta*, 1979, v. 574, pp. 25-32.

Markey B. A., Phan S. H., Varani J. et al. Inhibition of cytotoxicity by intracellular superoxide dismutase supplementation. *Free Rad. Biol. Med.*, 1990, v. 9, No. 4, pp. 307-314.

Marklund S. L. Oxygen toxicity and protective systems. *J. Clin. Toxicol.,* 1985, v. 23, No. 6, pp. 289-298.

Markov Kh. M., Kozlov A. V. and Pinelis V. G. [Effect of prostacycline and prostaglandins E_2 and $F_{2\alpha}$ on the tonus and reactivity of renal and limb vessels]. (Russian). *Kardiologiya*, 1981, v. 21, No. 3, pp. 75-79.

Marshall R. and Parratt J. The early consequences of myocardial ischemia and their modification. *J. Phisiol. (Paris),* 1980, v. 76, No. 7, pp. 699-715.

Marshall R. S., Koch C. J. and Rauth A. M. Measurement of low levels of oxygen and their effect on respiration in cell suspensions maintained in an open system. *Radiat. Res.*, 1986, v. 108, No. 1, pp. 91-101.

Marshansky V. N., Novgorodov S. A. and Yaguzhinsky L. S. [Effect of specific inhibitors of the respiratory chain enzymes and ATP-synthetase on ion transport in the mitochondria induced by non-enzymatic peroxide reactions]. (Russian). *Biofizika*, 1983, v. 28, No. 5, pp. 830-834. English transl. see[2*], pp. 879-884.

Martin L. F., Peter A. O., Fehr D. M. et al. ^{31}P-NMR evaluation of postischemia renal ATP and pH levels after ATP-$MgCl_2$ treatment in rabbits. *Am. J. Surg.*, 1992, v. 164, No. 2, pp. 132-139.

Marubayashi S., Dohi R., Ezaki H. et al. Preservation of ischemia rat liver mitochondrial functions and liver viability with CoQ_{10}. *Surgery*, 1982, v. 91, No. 6, pp. 631-637.

Marubayashi S., Dohi K., Ezaki H. et al. Preservation of ischemic liver cell prevention of damage by coenzyme Q_{10}. *Transplant. Proc.*, 1983, v. 15, No. 7, pp. 1297-1229.

Marubayashi S., Dohi K. and Kawasaki T. Role of free radical in ischemic rat liver cell injury-prevention of damages by vitamin E., coenzyme Q_{10} or reduced glutathione administration. *Surg. Forum*, 1985, v. 36, p. 136-138.

Maruyama H., Harada A., Kurokawa T. et al. Duration of liver ischemia and hepatic regeneration after hepatectomy in rats. *J. Surg. Res.*, 1995, v. 58, pp. 290-294.

Mason J., Forhorst J. and Welsch J. Role of the medullary perfusion defect in the pathogenesis of ischemic renal failure. *Kidney Int.*, 1984, v. 26, No. 3, pp. 283-293.

Mason J., Welsch J. and Takabatake T. Disparity between surface and deep nephron function early after renal ischemia. *Kidney Int.*, 1983, v. 24, No. 1, pp. 27-26.

Masson S. and Quistorff B. The 31P NMR visibility of ATP in perfused rat liver remains about 90%, unaffected by changes of metabolic state. *Biochemistry*, 1992, v. 31, №33, pp. 7488-7493.

Matoba S., Tatsumi T., Keira N. et al. Cardioprotective effect of angiotensin-converting enzyme inhibition against hypoxia/reoxygenation injury in cultured rat cardiac myocytes. *Circulation*, 1999, v. 99, No. 6, pp. 817-822.

Matsumoto F., Sakai H., Yamaguchi M. et al. Allopurinol reduced hepatic ischemia-reperfusion injury exacerbated by inhalation of high-concentration oxygen in rats. *Eur. Surg. Res.*, 1997, v. 29, No. 6, pp. 429-437.

Matsumoto K. and Nakamura T. Hepatocyte growth factor: molecular structure and implications for a central role in liver regeneration. *J. Gastroenterol. Hepatol.*, 1991, v. 6, No. 5, pp. 509-519.

Matsumoto K. and Nakamura T. Hepatocyte growth factor: molecular structure, roles in liver regeneration and other biological functions. *Crit. Rev. Oncog.*, 1992, v. 3, No. 1-2, pp. 27-54.

Matthes G. H., Hackensellner U. A. and Dahlke A. Successful cryo preservation of auricle fragments from rat hearts at -196°C. *Experientia*, 1978, v. 34, pp. 1024-1025.

Matthews N., Neale M. L., Jackson S. K. and Stark J. M. Tumor cell killing by fumour necrosis factor: inhibition by anaerobic condition free-radical scavengers, and inhibitors of arachidonate metabolism. *Immunology*, 1987, v. 62, pp. 151-155.

Maudal A. K. and Miller J. Protection against ischemic acute renal failure by prostaglandin infusion. *Prostaglandins, Leukotrienes and Medicine*, 1982, v. 8, pp. 361-373.

Maulik M., Seth S. D., Maulik S. K. and Manchanda S. C. Effects of supplementation with iloprost on the cardioprotection by BW755C (a dual inhibitor of cyclooxygenase and lipoxygenase enzymes) in myocardial reperfusion injury. *Pharmacol. Res.*, 1997, v. 36, No. 2, pp. 103-108.

Mayansky A. N. and Mayansky D. N. [*Essay about neutrocyte and macrophage*]. (Russian). Novosibirsk, Nauka, 1983, 256 p.

Mayansky D. N. [Kupfer cells and pathology of the liver]. (Russian). *Patol. Fisiol. Exper. Ter.*, 1985, No. 4, pp. 80-86.

McCluskey E. R., Corr P. B., Lee B. J., et al. The arachidonic acid metabolic capacity of canine myocardium is increased during heating of acute myocardial infarction. *Circ. Res.*, 1982. v. 57, pp. 743-750.

McCord J. M. Oxygen-derived free radicals in post-ischemic tissue injury. *New Engl. J. Med.*, 1985, Vol. 312, No. 3, pp. 159-163.

McCord I. M. and Fridovich I. Superoxide dismutase. *J. Biol. Chem.*, 1969, v. 244, No. 2, pp. 6049-6053.

McCord J. M. and Roy R. S. The pathophysiology of superoxide: role in inflammation and ischemia. *Can. J. Physiol. Pharmacol.*, 1982, v. 60, No. 11, pp. 1346-1352.

McCutchan H. J., Schwappach J. R., Enquist E. G. and Walden D. L. Xanthine oxidase-derived H_2O_2 contributes to reperfusion injury of ischemic skeletal muscle. *Am. J. Physiol.*, 1990, v. 258, No. 5, Pt. 2, pp. H1415-H1419.

McDonagh P. F. Coronary microvascular injury from myocardial ischemia reperfusion. *Int J. Microcirs. Clin. Exp.*, 1984, v. 3, No. 3/4, p. 441.

McKelvey T. G., Hollwarth M. E., Granger D. N. et al. Mechanisms of conversion of xanthine dehydrogenase to xanthine oxidase in ischemic rat liver and kidney. *Am. J. Physiol.*, 1988, v. 254, pp. G753-G760.

McLeod L. L. and Sevanian A. Lipid peroxidation and modification of lipid composition in an endothelial cell model of ischemia and reperfusion. *Free Radic. Biol. Med.*, 1997, v. 23, №4, pp. 680-694.

McNeil C. J., Banford J. C., Drown D. M. and Smith W. F. A relationship between thiols and the superoxide ion. *FEBS Lett.*, 1982, v. 133, No. 1, p. 175-177.

Meerson F. Z. [*Pathogenesis and prevention of stress and ischemic injuries of the heart*]. (Russian). Moscow, Meditsina, 1984, 272 p.

Meerson F. Z. and Vasil'ev V. K. [Damage and repair of myocardial DNA during emotional-painful stress]. (Russian). *Biul. Eksp. Biol. Med.*, 1981, v. 92, No. 9, pp. 297-299. English transl. see[1*], pp. 1187-1190.

Meerson F. Z., Didenko V. V., Savov V. M. et al. [Lipid peroxidation in experimental myocardial infarction: action of hyperbaric oxygenation]. (Russian). *Biul. Eksp. Biol. Med.*, 1984, v. 98, No. 10, pp. 398-400. English trans. see[1*], pp. 1316-1318.

Meerson F. Z., Kagan V. E., Azkhipenko Yu. V. et al. [Prevention of the lipids peroxidation activation and the myocardial antioxidant systems damage in stress and experimental infarction]. (Russian). *Kardiologiya*, 1981, v. 21, No. 12, pp. 55-60.

Meerson F. Z., Saltykova Y. A., Didenko V. V. et al. [The role of lipid peroxidation in the pathogenesis of arrhythmias and the antiarrhythmic effect of antioxidants]. (Russian). *Kardiologiya*, 1984, v. 24, No. 5, pp. 61-68.

Meerson F. Z., Vasil'ev K. V. and Dosmagambetova R. S. [DNA damage and its prevention in the nonischemic area of heart muscle in rats with experimental myocardial infarction]. (Russian). *Biul. Eksp. Biol. Med.*, 1982, v. 93, No. 6, pp. 57-60. English transl. see[1*], pp. 761-764.

Meerson F. Z., Vavich M. P., Rozhitskaya I. I. and Golubeva L. Yu. [Abolishing stress-induced activation of myocardial reparative DNA synthesis by an increased load on the heart]. (Russian). *Biul. Eksp. Biol. Med.*, 1988, v. 106, No. 12, pp. 655-657. English transl. see[1*], pp. 1678-1680.

Meerson F. Z., Vavich M. P. and Rozhitskaya I. I. [Effect of emotional-painful stress on the rate of DNA synthesis in heart and liver cells]. (Russian). *Vopr. Med. Khim.*, 1988, v. 34, No. 5, pp. 59-62.

Mehta J. L. and Conti C. R. Aspirin in myocardial ischemia: why, when, and how much? *Clin. Cardiol.*, 1989, v. 12, pp. 179-184.

Mekhfi H., Veksler V., Mateo P., et al. Creatine kinase is the main target of reactive oxygen species in cardiac myofibrils. *Circ. Res.*, 1996, v. 78, No. 6, pp. 1016-1027.

Meldrum D. R. Tumor necrosis factor in the heart. *Am. J. Physiol.*, 1998, v. 274, No. 3, Pt. 2, pp. R577-R595.

Meldrum D. G., Stephenson L. L. and Zamboni W. A. Effects of L-NAME and L-arginine on ischemia-reperfusion injury in rat skeletal muscle. *Plast Reconstr. Surg.* 1999, v. 103, No. 3, pp. 935-940.

Melikyants A. G. and Kucheryanu V. G. [Protective effect of ionol in ischemic injury to the small intestine]. (Russian). *Biul. Eksp. Biol. Med.*, 1983, v. 95, No. 6, pp. 43-45. English transl. see[1*], pp. 767-769.

Mellors A. and Tappel A. L. Hydrolysis of phospholipids by a lysosomal enzyme. *J. Lipid. Res.*, 1967, v. 8, No. 5, pp. 479-485.

Menger M. D., Pelikan S., Steiner D. and Messmer K. Microvascular ischemia-reperfusion in striated muscle: significance of "reflow paradox". *Am. J. Physiol.*, 1992, v. 263, No. 5, Pt. 2, pp. H1901-H1906.

Menger M. D., Lehr H. A. and Messmer K. Role of oxygen radicals in the microcirculatory manifestations of postischemic injury. (Review). *Klinische Wochenschrift*, 1991, v. 69, No. 21-23, pp. 1050-1055.

Menger M. D., Steiner D. and Messmer K. Microvascular ischemia-reperfusion injury in striated muscle: significance of "no-reflow". *Am. J. Physiol.*, 1992, v. 263, No. 6, Pt. 2, pp. H1892-H1900.

Mentz P., Opitz H. and Forster W. Action of unsaturated fatty acids and prostaglandins on parameters of exitation and contraction in the guinea pig myocardium. *Acta Biol. Med. Germ.,* 1976, v. 35, No. 8-9, pp. 1165-1166.

Merck Index. *An encyclopedia of chemicals and drugs*. Ed. Stecher P. G., Rahway, N. Y., 1960, p. 775.

Meredith M. J. and Reed D. J. Status of the mitochondrial pool of glutatione in the isolated hepatocyte. *J. Biol. Chem.*, 1982, v. 257, No. 7, pp. 3747-3753.

Meredith M. J. and Reed D. J. Depletion in vitro of mitochondrial glutathione in rat hepatocytes and enhancement of lipid peroxidation by adriamycin and 1, 3-bis(2-chloroethyl)-1-nitrosourea (BCNU). *Biochem. Pharmacol.,* 1983, v. 32, No. 8, pp. 1383-1388.

Metelitsa D. I. [*Enzyme systems in oxygen activation*]. (Russian). Moscow, Nauka, 1982, 256 p.

Michalopoulos G., Houck K. A., Dolan M. L. and Loutteke N. C. Control of hepatocyte replication by two serum factors. *Cancer Res.,* 1984, v. 44, No. 10, pp. 4414-4419.

Mihara M., Uchiyama M. and Fukuzawa K. Thiobarbituric acid value on fresh homogenate of rat as a parameter of lipid peroxidation in aging, CCl_4 intoxication and vitamin E deficiency. *Biochem. Med.*, 1980, v. 23, No. 3, pp. 302-311.

Michiels C., Arnould T., Houbion A. and Remacle J. Human umbilical vein endothelial cells submitted to hypoxia-reoxygenation in vitro: implication of free radicals, xanthine oxidase, and energy deficiency. *J. Cell. Physiol.*, 1992, v. 153, No. 1, pp. 53-61.

Mickle D. A., Li R. K., Weisel R. D. et al. Water-soluble antioxidant specificity against free radical injury using cultured human ventricular myocytes and fibroblasts and saphenous vein endothelial cells. *J. Mol. Cell. Cardiol.*, 1990, v. 22, No. 11, pp. 1297-1304.

Milhoan K. A., Lane T. A. and Bloor C. M. Hypoxia induces endothelial cells to increase their adherence for neutrophils: role of PAF. *Am. J. Physiol.*, 1992, v. 263, No. 8, Pt. 2, pp. H956-H962.

Milner S. M. Effects of the food additive butylated hydroxytoluene on monolayer cultures of primate cells. *Nature,* 1967, v. 216, No. 5115, pp. 557-560.

Milyutina N. P., Ananyan A. A., Sapozhnikov V. P. et al. [Effect of long-term exposure to hyperbaric conditions on lipid peroxidation and on erythrocyte structure and function]. (Russian). *Biul. Eksp. Biol. Med.*, 1992, v. 113, No. 5, pp. 474-476. English transl. see[1*], pp. 633-636.

Mimnaugh E. G., Trush M. A. and Gram T. E. Stimulation by adriamycin of rat heart and liver microsomal NADPH-dependent lipid peroxidation. *Biochem. Pharmac.,* 1981, v. 30, pp. 2797-2804.

Mink R. B., Dutka A. J., Kumaroo K. K. and Hallenbeck J. M. No conversion of xanthine dehydrogenase to oxidase in canine cerebral ischemia. *Am. J. Physiol.*, 1990, v. 259, No. 6, Pt. 2, pp. H1655-H1659.

Min'ko A. A., Petunin Yu. I. and Sushko B. S. [On the oxygen permeability of bilayer membranes]. (Russian). In: Berezovskiy V. A., ed. *Lung surfactants in norma and pathology*. Kiev, Naukova Dumka, 1983, pp. 85-90.

Minor T., Isselhard W., Yamamoto Y. et al. The effects of allopurinol and SOD on lipid peroxidation and energy metabolism in the liver after ischemia in an aerobic/anaerobic persufflation. *Surgery Today,* 1993, v. 23, No. 8, pp. 728-732.

Mitrev Z., Beyersdorf F., Hallmann R. et al. Reperfusion injury in skeletal muscle: controlled limb reperfusion reduces local and systemic complications after prolonged ischaemia. *Cardiovasc. Surg.*, 1994, v. 2, No. 6, pp. 737-748.

Miwa-Nishimura N., Kanaide H., Abe S. and Nakamura M. Does allopurinol prevent myocardial injury as a result of hypoxia-re-oxygenation in rats? *Int. J. Exp. Pathol.,* 1990, v. 71, No. 5, pp. 727-739.

Miyamae M., Camacho S. A., Weiner M. W. and Figueredo V. M. Attenuation of postischemic reperfusion injury is related to prevention of [Ca2+]m overload in rat hearts. *Am. J. Physiol.*, 1996, v. 271, №5, pt. 2, pp. H2145-H2153.

Mizukami M., Aono J., Sakai R. et al. Experimental myocardial infarction. Effects of lipid peroxide, 13-hydroperoxy linoleic acid on coronary circulation in rats. *Arzneim-Forsch. Drug Res.,* 1984, v. 34(1), No. 5, pp. 569-571.

Mkhitaryan V. G. and Badalyan G. E. [The influence of peroxidized and non-peroxidized unsaturated fatty acids on the superoxide dismutase activity]. (Russian). *Zh. Eksp. Klin. Med.*, 1978, No. 6, pp. 7-11.

Moak S. A. and Greenwald R. A. Enhancement of rat serum ceruloplasmin levels by exposure to hyperoxia. *Proc. Soc. Exp. Biol. Med.,* 1984, v. 177, No. 9, pp. 97-103.

Mogilevskii G. M., Sharov V. G., Severin V. V. and Semenovskii M. L. [Ultrastructural study of antiischemic protection of the normal and hypertrophied myocardium by creatine phosphate]. (Russian). *Biul. Eksp. Biol. Med.*, 1990, v. 109, No. 4, pp. 408-410. English transl. see[1*], pp. 538-541.

Mogil'nitskaya L. V., Prokofiev V. N. and Morgulis G. L. [Free-radical processes in the rat blood under hyperbarooxygenation and post-hyperoxic period]. (Russian). *Ukr. Biokhim. Zh.*, 1996, v. 68, No. 1, pp. 96-100.

Moknova E. N., Starkov A. A. and Bobylova V. A. [Uncoupling of oxidative phosphorylation by fatty acids in liver and muscle mitochondria]. (Russian). *Biokhimiya*, 1993, v. 58, №10, pp. 1513-1522. English transl. see [3*], pp. 1105-1112.

Molitoris B. A., Wilson P. D., Schrier R. W. and Simon F. R. Ischemia induces partial loss of surface membrane polarity and accumulation of putative calcium ionophores. *J. Clin. Invest.*, 1985, v. 76, No. 6, pp. 2097-2105.

Moncada S. and Vane J. R. Arachidonic acid metabolites and the interactions between platelets and blood-vessel walls. *N. Engl. J. Med.*, 1979, v. 300, pp. 1142-1149.

Montalescot G., Maclouf J., Drobinski G. et al. Eicosanoid biosynthesis in patients with stable angina: beneficial effects of very low dose aspirin. *J. Am. Cell Cardiol.*, 1994, v. 24, No. 1, pp. 33-38.

Moore K. H., Koen A. E. and Hull F. E. β-Hydroxy fatty acid production by ischemic rabbit heart. Distribution and chemical states. *J. Clin. Invest.*, 1982, v. 69, No. 2, pp. 377-383.

Morgan K. G. Calcium and vascular smooth muscle tone. *Am. J. Med.*, 1987, v. 82, No. 313, pp. 9-15.

Morgan J., Smith J. A., Wilkins G. M. and Leake D. S. Oxidation of low-density lipoprotein by bovine and porcine aortic endothelial cells and porcine endocardial cells in culture. *Atherosclerosis*, 1993, v. 102, No. 2, pp. 209-216.

Morgunov A. A. [*Protective effect of inhibitors of free-radical processes on the myocardium in total heart ischemia and ischemic shock*]. (Russian). Dissertation. Moscow, 1991.

Morgunov A. A. and Bulgakov V. G. [The influence of exogenous lipid peroxidation products and antioxidant ionol on contractile function of heart in postischemic period]. (Russian). In: *Bioantioxidant. 2nd All-Union Conference.* Abstracts. Chernogolovka, 1986, v. 2, pp. 28-29.

Morgunov A. A., Bulgakov V. G. and Bilenko M. V. [Effect of the antioxidant ionol as part of liposomes on the cardiac function of the isolated and ischemic heart]. (Russian). *Eksper. Farmakol.*, Moscow, 1992, No. 3, pp. 17-20.

Morgunov A. A., Komarov P. G., Bulgakov V. G. and Bilenko M. V. [Comparative analysis of anti-ischemic effect of water- and fat-soluble antioxidants on the cardiac function of an isolated ischemic rat heart]. (Russian). In: *Bioantioxidant. 3rd All-Union Conference.* Abstracts. Moscow, 1989, v. 2, pp. 118-119.

Mori S., Tanaka A., Kitai T. et al. Primary and reversible injury of H(+)-ATPase in warm ischemia and reperfusion of rat liver in relation to intramitochondrial adenine nucleotide. *J. Surg. Res.*, 1995, v. 58, №2, pp. 175-181.

Mori E., Haramaki N., Ikeda H. and Imaizumi T. Intra-coronary administration of L-arginine aggravates myocardial stunning through production of peroxynitrite in dogs. *Cardiovasc. Res.* 1998, v. 40, No. 1, pp. 113-123.

Mori T., Asano T., Matsui T. et al. Intraluminal increase of superoxide anion following transient focal cerebral ischemia in rats. *Brain. Res.* 1999, v. 816, No. 2, pp. 350-357.

Morrissey P. and O'Brien P. J. Selenium-independent glutathione peroxidase activity in rabbit liver. *Can. J. Biochem.*, 1980, v. 58, No. 10, pp. 1012-1017.

Moskaleva E. Yu. and Ilyushina N. O. [DNA injuries under ionizing radiation action and their reparation]. (Russian). In: *The results of science and technology. Series Radiational Biology. v. 9 DNA reparation and replication in irradiated cells*. Moscow, 1990, 207 p.

Mousa S. A., Ritger R. C. and Smith R. O. Efficacy and safety of deferoxamine conjugated to hydroxyethyl starch. *J. Cardiovasc. Pharmacol.*, 1992, v. 19, No. 3, pp. 425-429.

Mozhaitseva A. G. [The content of DNA disintegration products in the blood of patients with myocardial infarction]. (Russian). *Klin. Med.*, 1966, v. 44, No. 1, pp. 45-47.

Mukanov K. O., Buldygerova N. V., Kovaleva E. S. et al. [Antioxidant-induced calcium transport in biological membranes]. (Russian). *Biul. Eksp. Biol. Med.*, 1987, v. 104, No. 9, pp. 296-299. English transl. see[1*], pp. 1220-1223.

Muksimov F. A., Almatov K. T. and Tuichibaev M. U. [Effect of phospholipase A_2 of vespa orientalis poison on oxidative phosphorilation and activity of polyenzymatic systems of mitochondries]. (Russian). *Dokl. Akad. Nauk Uzb. SSR*, 1982, No. 12, pp. 37-39.

Mullane K. M., Read N., Salmon J. A. and Moncada S. Role of leukocytes in acute myocardial infarction in anesthetized dogs: relationship to myocardial salvage by anti-inflammatory drugs. *J. Pharmacol. Exp. Ther.*, 1984, v. 228, pp. 510-522.

Muller J. M., Vollmar B. and Menger M. D. Pentoxifylline reduces venular leukocyte adherence ("reflow paradox") but not microvascular "no reflow" in hepatic ischemia/reperfusion. *J. Surg. Res.*, 1997, v. 71, No. 1, pp. 1-6.

Mulligan M., Althaus B. and Linder M. C. Non-ferritin, non-heme iron pools in rat tissues. *Int. J. Biochem.*, 1986, v. 18, pp. 791-798.

Murota S., Nishida M. and Morita J. Hypoxia-reoxygenation inhibits gap junctional intercellular communication in vascular endothelial cells in culture. Abstract. *Hypoxia Medical J.,* 1998, v. 6, No. 2, p. 52.

Murrell G. A. C., Francis M. J. O. and Bromley L. Modulation of fibroblast proliferation by oxygen free radicals. *Biochem. J.*, 1990, v. 265, No. 3, pp. 659-665.

Musat S. and Dhalla N. S. Alteration in cardiac sarcolemmal ATP receptors by oxyradicals. *Ann. N. Y. Acad. Sci.*, 1996, v. 793, pp. 1-12.

Musil J. [*Principles of biochemistry of pathological processes*]. (Russian). Moscow, Meditsina, 1985, 430 p.

Myslovatyi B. S., Zairov D. K., Chebotarev A. A. et al. [Partial pressure of oxygen in different organs and tissues and some hemodynamic parameters in tourniquet shock]. (Russian). *Biul. Eksp. Biol. Med.*, 1982, v. 94, No. 7, pp. 19-22. English transl. see[1*], pp. 872-875.

Nadjaryan T. L., Pikovskaya G. N., Barsel V. A. et al. [The influence of dibunol on some physiological indexes of practically healthy people]. (Russian). *Izv. Akad. Nauk SSSR, Ser. Biol.*, 1985, No. 2, pp. 211-218.

Nagai S., Miazaki Y., Ogawa K. et al. The effect of coenzyme Q10 on reperfusion injury in canine myocardium. *J. Mol. Cell. Cardiol.*, 1985, v. 17, pp. 873-886.

Nagel E., Meyer zu Vilsendorf A., Bartels M. and Pichlmayr R. Antioxidative vitamins in prevention of ischemia/reperfusion injury. *Int. J. Vitam. Nutr. Res.*, 1997, v. 67, No. 5, pp. 298-306.

Nakamura H., Nemenoff R. A., Gronich J. H. and Bonventre J. V. Subcellular characteristics of phospholipase A2 activity in the rat kidney. Enhanced cytosolic, mitochondrial, and microsomal phospholipase A2 enzimatic activity after renal ischemia and reperfusion. *J. Clin. Invest.*, 1991, v. 87, No. 5, pp. 1810-1818.

Nakanishi T., Nishioka K. and Jarmakani J. M. Mechanism of tissue Ca^{2+} gain during reoxygenation after hypoxia in rabbit myocardium. *Am. J. Physiol.*, 1982, v. 242, pp. H437-H449.

Nakano M., Sugano M., Terasaki M. et al. Preserved mitochondrial function by allopurinol despite deteriorated hemodynamics in warm ischemia-damaged canine liver. *Res. Exper. Med.,* 1992, v. 192, No. 6, pp. 389-399

Nakashima M., Niwa M., Iwai T. and Uematsu T. Involvement of free radicals in cerebral vascular reperfusion injury evaluated in a transient focal cerebral ischemia model of rat. *Free Radic. Biol. Med.* 1999, v. 26, No. 5-6, pp. 722-729.

Naldini L., Weidner K. M., Vigna E. et al. Scatter factor and hepatocyte growth factor are indistinguishable ligands for the MET receptor. *EMBO J.,* 1991, v. 16, No. 10, pp. 2867-2878.

Nanri K., Montecot C., Springhetti V. et al. The selective inhibitor of neuronal nitric oxide synthase, 7-nitroindazole, reduces the delayed neuronal damage due to forebrain ischemia in rats. *Stroke*, 1998, v. 29, No. 6, pp. 1248-1253.

Näslung U., HäggmarkS., Johansson G. et al. Superoxide dismutase and catalase reduce infarct size in a porcine myocardial occlusion-reperfusion model. *J. Mol. Cell. Cardiol.*, 1986, v. 18, No. 10, pp. 1077-1084.

Nathan C.F. Secretion of oxygen intermediates: role in effector functions of activated macrophages. *Fed. Proc.,* 1982, v. 41, No. 6, pp. 2206-2211.

Nayini N. R., White B. C., Aust S. D. et al. Post resuscitation iron delocalization and malondialdehyde production in the brain following prolonged cardiac arrest. *Free Rad. Biol. Med.*, 1985, v. 1, No. 2, pp. 111-116.

Nayler W. G. and Elz J. S. Reperfusion injury: laboratory artifact or clinical dilemma? *Circulation*, 1986, v. 74, No. 2, pp. 215-227.

Nayler W. G. and Poole-Wilson P. A. Calcium antagonists: definition and mode of action. *Basic Res. Cardiol.*, 1981, v. 76, pp. 1-15.

Nayler W. G., Ferrari R., Poole-Wilson P. A. et al. A protective effect of a mild acidosis on hypoxic heart muscle. *J. Moll. Cell. Cardiol.,* 1979, v. 11, No. 10, pp. 1053-1071.

Nayler W. G., Poole-Wilson P. A. and Williams A. Hypoxia and calcium. *J. Mol. Cell. Cardiol.*, 1979, v. 11, No. 7, pp. 683-706.

Nazarova T. A. and Zhyetnisbaev B. A. [Induces of circulation dynamics of traumatic shock in albino rats according to data of integral rheography]. (Russian). *Krovoobrashchenie*, 1983, v. 3, pp. 58-60.

Neale M. L., Fiera R. A. and Matthews N. Involvement of phospholipase A_2 activation in tumor cell killing by tumor necrosis factor. *Immunology*, 1988, v. 64, pp. 81-85.

Neely J. R. and Grotyohann L. W. Role of glycolytic products in damage to ischemic myocardium. Dissociation of adenosine triphosphate levels and recovery of function of reperfused ischemic hearts. *Circ. Res.*, 1984, v. 55, No. 6, pp. 816-824.

Neely J. R. and Rovetto M. J. Techniques for perfusing isolated rat hearts. In: Hardman J. G. and O'Mallcy B. W, eds. *Methods in Enzymology*, N. Y., Acad. Press., 1975, v. 39; pp. 43-60.

Neely J. R., Rovetto M. J. and Whitmer J. T. Rate-limiting steps of carbohydrate and fatty acid metabolism in ischemic hearts. *Acta Med. Scand.*, 1976, v. 587, Suppl., pp. 9-15.

Nees S. and Gerlach E. Adenine nucleotide and adenosine metabolism in cultured coronary endothelial cells: formation and release of adenine compounds and possible functional implications. In: Berne R. M., Rall T. N. and Rubio R., eds. *Regulatory functions of adenosine*. 1983, pp. 347-355.

Nelson S. K., Bose S. K and McCord I. M. The toxicity of high-dose superoxide dismutase suggets that superoxide can both initiate and terminate lipid peroxidation in the reperflused heart. *Free Radic. Biol. Med.*, 1994, v. 16, No. 3, pp. 195-200.

Neubauer S., Ertl G., Pulzer F. et al. Effect of endothelin-1 in the isolated heart under ischemic and cardioplegic conditions. *J. Cardiovasc. Pharm.*, 1990, v. 16, pp. 804-811.

Newman G. C., Hospod F. E., Trowbridge S. D. et al. Restoring adenine nucleotides in a brain slice model of cerebral reperfusion. *J. Cereb. Blood Flow Metab.,* 1998, v. 18, No. 6, pp. 675-685.

Neyfakh E. A. and Kagan V. E. [Detection of lipid peroxides in the organs of normal animals *in vivo*]. (Russian). *Biokhimiya*, 1969, v. 34, No. 3, pp. 511-English transl. see[3*], pp. 406-411.

Nicolov R. Arachidonic acid cascade and anti-hypoxic drugs. *Meth. Find Exptl. Clin. Pharmacol.*, 1984, v. 6, No. 5, pp. 231-234.

Nielsen V. G., Tan S., Baird M. S. et al. Xanthine oxidase mediates myocardial injury after hepatoenteric ischemia-reperfusion. *Crit. Care Med.,* 1997, v. 25, No. 6, pp. 1044—1050.

Nigulyanu V. I., Elsky V. N., Krivoruchko B. I. et al. [*Crush syndrome*]. (Russian). Kishinyov, Shtiintsa, 1984, 222 p.

Nilsson U. A., Aberg J., Aneman A. and Lundgren O. Feline intestinal ischemia and reperfusion: relation between radical and tissue damage. *Eur. Surg. Res.,* 1993, v. 25, No. 1, pp. 20-29.

Nilson U. A., Haraldson G., Bratell S. et al. ESR-measurement of oxygen radicals in vivo after renal ischaemia in the rabbit. Effects of pre-treatment with superoxide dismutase and heparin. *Acta Physiol.* Scandinav., 1993, v. 147, No. 3, pp. 263-270.

Nishida K. [Measurement of free radical generation from endothelial cells and observation of cell injury exposed to anoxia-reoxygenation]. (Japan). *J. Jpn. Surg. Soc.*, 1992, v. 93, No. 4, pp. 369-376.

Nishida T., Koseki M., Kamiike W. et al. Levels of purine compounds in a perfusate as a biochemicol marker of ischemic injury of cold-preserved liver. *Transplantation*, 1987, v. 44, No. 1, pp. 16-21.

Nishida T., Morita S., Masuda M. et al. Lazaroid U74500A is superior to U74006F in preserving rabbit heart for 24 hours. *Ann. Thorac. Surg.* 1998, v. 66, No. 5, pp. 1647-1652.

Nishigaki I., Hagihara M., Tsunekawa H. et al. Lipid peroxide levels of serum lipoprotein fractions of diabetic patients. *Biochem. Med.,* 1981, v. 25, No. 3, pp. 373-378.

Nishigaya K., Yoshida Y., Sasuga M. et al. Effect of recirculation on exacerbation of ischemic vascular lesions in rat brain. *Stroke*, 1991, v. 22, No. 5, pp. 635-642.

Nishikawa T., Kirsch J. R., Koehler R. C. et al. Nitric oxide synthase inhibition reduces caudate injury following transient focal ischemia in cats. *Stroke*, 1994, v. 25, No. 4, pp. 877-885.

Nguyen V.D., Cieslinski D.A. and Humes H.D. Importance of adenosine triphosphate in phospholipase A2-induced rabbit renal proxiomal tubule cell injury. *J. Clin. Invest.*, 1988, v. 83, No. 3, pp. 1098-1105.

Nohl H. A novel superoxide radical generator in heart mitochondria. *FEBS Lett.*, 1987, v. 214, No. 2, pp. 269-273.

Nohl H. and Hegner D. Do mitochondria produce oxygen radicals in vivo? *Eur. J. Biochem.,* 1978, v. 82, pp. 563-567.

Noji S., Tashiro K., Koyama E. et al. Expression of hepatocyte growth factor gene in endothelial and Kupffer cells of damaged rat livers, as revealed by in situ hybridization. *Biochem. Biophys. Res. Commun.,* 1990, v. 173, No. 1, pp. 42-47.

Nördlinger B., Douvin D., Javaudin L. et al. An experimental study of survival after two hours of normothermic hepatic ischemia. *Surg. Gynec. Obstct.,* 1980, v. 150, pp. 859-864.

Nordström C. H. and Siesjö. Bo. K. Effect of phenobarbital in cerebral ischemia. Part 1: Cerebral energy metabolism during pronounced incomplete ischemia. *Stroke,* 1978, v. 9, No. 4, pp. 327-335.

Nose P. S. Cytokines and reperfusion injury. *J. Card. Surg.*, 1993, v. 8, No. 2, Suppl., pp. 305-308.

Nossuli T. O., Hayward R., Scalia R. et al. Peroxynitrite reduces myocardial infarct size and preserves coronary endothelium after ischemia and reperfusion in cats. *Circulation,* 1997, v. 96, pp. 2317-2324.

Novelli G. P., Martini G., Bordi G. et al. Fluidity of mitochondrial and microsomal membrane in traumatic shock. In: Workshop on: *Oxygen free-radicals in shock.* Florence, 1985, p. 18.

Nowak T. S., Fried R. L., Lust W. D. and Passonneau J. V. Changes in brain energy metabolism and protein synthesis following transient bilateral ischemia in the gerbil. *J. Neurochem.*, 1985, v. 44, pp. 487-494.

Nunnally R. L. and Hollis D. P. Adenosine triphosphate compartmentation in living hearts: a phosphorus nuclear magnetic resonance saturation transfer study. *Biochemistry,* 1979, v. 18, No. 16, pp. 3642-3646.

Obata M. [Investigation of lipid peroxidation in regenerating rat liver]. (Jpn.) *J. Jap. Surg. Soc.*, 1992, v. 93, No. 8, pp. 833-841.

Obeid L. H., Linardic C. M., Karolak L. A. et al. Programmed cell death induced by ceramide. *Science,* 1993, v. 259, No. 5102, pp. 1769-1771.

Oberley L. W., Bize J. B., Sahus K. et al. Superoxide dismutase activity of normal murine liver, regenerating liver, and H_6 hepatoma. *J. Nat. Cancer Inst.,* 1978, v. 61, No. 2, pp. 375-379.

Oda T. Normothermic anoxic arrest of the canine heart: estimation of the safe period. *Arch. Jpn. Chir.,* 1975, v. 44, pp. 313-329.

Ogilby J. D. and Apstein C. S. Reversal of reperfusion injury with post-ischemia EDTA pulse therapy. *J. Mol. Cell. Cardiol.*, 1983, v. 15, Suppl. 1, p. 330.

Ogletvee M. L., Lefer A. M., Smith G. B. and Nicolaou K. C. Studies on the protective effect of prostacyclin in acute myocardial ischemia. *Eur. J. Pharmacol.,* 1979, v. 56, pp. 95-103.

Oh S. M. and Betz A. L. Interaction between free radicals and excitatory amino acids in the formation of ischemic brain edema in rats. *Stroke,* 1991, v. 22, pp. 915-921.

Ohta H., Adachi T. and Hirano K. Internalization of human extracellular-superoxide dismutase by bovine aortic endothelial cells. *Free Rad. Biol. Med.*, 1994, v. 16, No. 4, pp. 501-507.

Ohta H., Sweeney E. A., Masamune A. et al. Induction of apoptosis by sphingosine in human leukemic HL-60 cells: a possible endogenous modulator of apoptotic DNA fragmentation occurring during phorbot ester-induced differentation. *Cancer Res.,* 1995, v. 55, pp. 691-697.

Okabe H. [The role of xanthine dehydrogenase (xanthine oxidase) in ischemia-reperfusion injury in rat kidney]. (Jpn.) *Nippon Jinzo Gakkai Shi,* 1996, v. 38, No. 12, pp. 577-584.

Okajima A., Hiyazawa K. and Kitamura N. Primary structure of rat hepatocyte growth factor and induction of its mRNA during liver regeneration following hepatic injury. *Eur. J. Biochem.,* 1990, v. 193, No. 2, pp. 375-381.

Okamoto H., Matsumoto M., Ohtsuki T. et al. Internucleosomal DNA cleavage involved in ischemia-induced neuronal death. Biochem. *Biophys. Res. Commun.,* 1993, v. 196, No. 3, pp. 1356-1362.

Okamura I., Sunamori M. and Suzuki A. Protective effect of lidocaine inreperfused ischemic myocardium. Evalution by hemodynamic and biochemical study. *Jap. Circulat. J.,* 1982, v. 46, pp. 657-662.

Okayasu T., Curtis M. T. and Farber J. L. Structural alterations of the inner mitochondrial membrane in ischemic liver cell injury. *Arch. Biochem. Biophys.*, 1985, v. 236, pp. 638-.

Oksman T. H., Dalin M. V. and Kovanov V. V. [Ishemic toxin]. (Russian). *Dokl. Acad. Nauk SSSR,* 1971, v. 199, No. 4, pp. 980-983. English transl. see[7*], pp. 528-531.

Okuma M., Takayama H. and Uchino H. Generation of prostacyclin - like substance and lipid peroxidation in vitamin E-deficient rats. *Prostaglandins,* 1980, v. 19, No. 4, pp. 527-536.

Olbinskaya L. I. and Litvitsky P. F. [*Coronary and myocardial insufficiency*]. (Russian). Moscow, Meditsyna, 1986, 272 p.

Omar B. A. and McCord J. M. The cardioprotective effect of Mn-superoxide dismutase in lost at high doses in the postischemic isolated rabbit heart. *Free Radic. Biol. Med.*, 1990, v. 9, pp. 473-478.

Omar R., Nomikos I., Piccorelli G. et al. Prevention of postischemic lipid peroxidation and liver cell injury by iron chelation. *Gut.*, 1989, v. 30, No. 4, pp. 510-514.

O'Neill M. J., Hicks C., Ward M. and Panetta J. A. Neuroprotective effects of the antioxidant LY231617 and NO synthase inhibitors in global verebral ischaemia. *Brain Res.,* 1997, v. 760, No. 1-2, pp. 170-178.

Onishchenko N. A., Kirpatovsky V. I., Artamonov S. D. et al. [Mechanism of antiischemic protection of the kidneys by agents of various pharmacological groups]. (Russian). *Patol. Fiziol. Eksp. Ter.,* 1980, No. 6, pp. 43-47.

Onufriev M. V., Stepanichev M. Yu., Lazareva N. A. et al. [Activity of NO-synthase and radical formation in rat brain compartments : age dependence]. (Russian). *Biul. Eksp. Biol. Med.*, 1995 v. 120, N7, pp. 54-56. English transl. See 1*., pp. 700-702.

Opie I. H. Role of carnitine in fatty acid metabolism of normal and ischemic myocardium. *Am. Heart J.,* 1979, v. 97, No. 3, pp. 375-388.

Opie L. H. Myocardial infarct size. Part I. Basic considerations. *Am. Heart. J.,* 1980, v. 100, No. 3, pp. 355-372.

Oredsson S., Arlock P., Plate G. and Qvarfordt P. Metabolic and electrophysiologocal changes in rabbit skeletal muscle during ischemia and reperfusion. *Eur. J. Surg.,* 1993, v. 159, No. 1, pp. 3-8.

O'Regan M. H., Smith-Barbour M., Perkins L. M. et al. The effect of amflutizole, a xanthine oxidase inhibitor, on ischemia-evoked purine release and free radical formation in the rat cerebral cortex. *Neuropharmacology*, 1994, v. 33, No. 10, pp. 1197-1201.

Orlov O. N. [Regulation mechanisms of intracellular calcium distribution. Review]. (Russian). *Usp. Sovr. Biol.*, 1981, v. 92, No. 1, pp. 19-34.

Orlov O. N. [Mechanisms of calcium metabolism regulation in cells]. (Russian). In: *Biological membranes.* Saransk, 1984, pp. 26-34.

Orrenius S. Oxidative stress studied in intact mammalian cells. *Phil. Trans. Royal Soc. Lond. (Biol.)* 1985, v. 311, No. 1152, pp. 673-677.

Oshiro Y., Marubayashi S., Maeda T. et al. Contribution of neutrophils and effect of monoclonal antibodies to adhesion molecules on ischemia and reperfusion injury in rat livers. *Transplantation. Proceedings.,* 1995, v. 27, No. 1, pp. 743-744.

Osipov A. N., Yakutova E. Sh. and Vladimirov Yu. A. [Formation of hydroxyl radicals on interaction of hypochlorite with ferrous ions]. (Russian). *Biofizika*, 1993, vol. 38, No. 3, pp. 390-396. English transl. see[2*], pp. 383-388.

Otani H., Engelman R. M., Ronson J. A. et al. Enhanced prostaglandin synthesis due to phospholipid breakdown in ischemic-reperfused myocardium. Control of its production by a phospholipase inhibitor or free radical scavengers. *J. Mol. Cell. Cardiol.*, 1986, v. 18, No. 9, pp. 953-961.

Otani H., Prasad M. R., Jones R. M. and Das D. K. Mechanism of membrane phospholipid degradation in ischemic-reperfused rat hearts. *Am. J. Physiol.*, 1989, v. 257, №1, Pt. 2, pp. H252-H258.

Otverchenko V. N. [*Ischemic injury of myocardium and some new methods of antiischemic protection in cardiosurgery*]. (Russian). Dissertation. Tbilisi, 1984.

Otverchenko V. N., Bulgakov V. G., Volkova I. A. and Bilenko M. V. [The time-course and pathogenesis of cardiac ischemie contracture]. (Russian). *Kardiologiya*, 1984, v. 24, No. 8, pp. 100-104.

Oubidar M., Boquillon M., Maril Ch. et al. Ischemia-induced brain iron delocalization: effect of iron chelators. *Free Rad. Biol. Med.*, 1994, v. 16, No. 6, pp. 861-867.

Ouriel K., Ginsburg M. E., Patti C. S. et al. Preservation of myocardial function with mannitol reperfusate. *Circulation*, 1985, v. 72, Suppl., pp. 11254-11258.

Ozaki M., Fuchinoue S., Teraoka S. and Ota K. Mobilization of low-molecular weight iron and peroxidative damage during ischemia and reoxygenation of the rat liver. *Transplant. Proceed.*, 1984, v. 26, No. 2, pp. 918-921.

Ozaki M., Fuchinoue S., Teraoka S. and Ota K. The in vivo cytoprotection of ascorbic acid against ishemia/reoxygenation injury of rat liver. *Arch. Biochem. Biophys.* 1995, v. 318, No. 2, pp. 439-445.

Ozawa K., Seta K., Araki H. and Handa H. Rapid liberation of potassium ions from brain mitochondria. *J. Biochem.,* 1967, v. 62, No. 5, pp. 584-590.

Pakhomov V. J. and Kostyushin V. N. [Partial oxygen pressure in tissues of the lower limbs in patients with obliterating vascular diseases under the effect of hyperbaric oxygenation]. *Klin. Med.,* 1985, v. 63, No. 10, pp. 73-77.

Palace G. P., Del-Vecchio P. J., Horgan M. J. et al. Release of tumor necrosis factor after pulmonary artery occlusion and reperfusion. *Am. Rev. Respir. Dis.,* 1993, v. 147, No. 1, pp. 143-147.

Palermartinez A., Panus P. C., Chumley P. H. et al. Endogenous xanthine oxidase does not significantly contribute to vascular endothelial production of reactive oxygen species. *Arch. Biochem. Biophys.,* 1994, v. 311, No. 1, pp. 79-85.

Paller M. S. Hydrogen peroxide and ischemic renal injury: effect of catalase inhibition. *Free Radic. Biol. Med.,* 1991, v. 10, No. 1, pp. 29-34.

Paller M. S. and Eaton J. W. Hazards of antioxidant combinations containing superoxide dismutase. *Free Rad. Biol. Med.,* 1995, v. 18, No. 5, pp. 883-890.

Paller M. S. and Hedlund B. E. Role of iron in postischemic renal injury in the rat. *Kidney Int.,* 1988, v. 34, pp. 474-480.

Palmer C., Vannucci R. C. and Towfighi J. Reduction of perinatal hypoxic-ischemic brain damage with allopurinol. *Pediatr. Res.,* 1990, v. 27, No. 4, Pt. 1, pp. 332-336.

Pal'mina N. P., Mal'tseva E. L., Kuznakova N. V. and Burlakova E. B. [Effect of α-tocopherol over a wide concentration range (10^{-17}-10^{-2} M) on protein kinase C activity: relationship to cell proliferation and tumor growth]. (Russian). *Biochimiya,* 1994, v. 59, No. 2, pp. 193-200. English transl. see[3*], pp. 135-139.

Palombo J. D., Bowers J. L., Clouse M. E. et al. Hepatic utilization of exogenous nucleotide precursors for restoration of ATP after cold ischemia in rats. *Am. J. Clin. Nutr.,* 1993, v. 57, No. 3, pp. 420-427.

Panchenko L. F., Gerasimov A. M., Nozdrachova L. I., Kariakina G. A. [Accumulation of malonyldialdehyde in isolated rat liver]. (Russian). *Vopr. Med. Khim.,* 1974, v. 20, No. 3, pp. 321-324.

Panchenko L. F., Gerasimov A. M. and Antonenkov V. D. [*Role of peroxisomes in cell pathology*]. (Russian). Moscow, Meditsina, 1981, 207 p.

Panchenko E. P., Smirnov I.E. and Zyabkina A. G. [Tromboxan and prostacyclin in patients with angine pectoris accompanied by induced and spontaneous ischaemia of the myocardium]. (Russian). *Biul. VKNC Akad. Med. Nauk SSSR,* 1986, v. 9, No. 1, pp. 73-79.

Pannen B. H., Al-Adili F., Bauer M., et al. Role of endothelins and the nitric oxide in hepatic reperfusion injury in the rat. *Hepatology* 1998, v. 27, No. 3, pp. 755-764.

Paramonova G. I., Gubsky Yu. I., Goryushko A. G. et al. [Stress effect on lipid peroxidation and physico-chemical state of membranes of endoplasmic reticulum of the liver of adult and old rats]. (Russian.) *Ukr. Biokhim. Zh.,* 1996, v. 68, No. 5, pp. 47-53.

Park T. S., Gonzales E. R. and Gidday J. M. Platelet-activating factor mediates ischemia-induced leukocyte-endothelial adherence in newborn pig brain. *J. Cereb. Blood. Flow Metab.,* 1999, v. 19, No. 4, pp. 417-424.

Park Y., Kaneka S. and Kehrer J. P. Oxidative changes in hypoxia rat heart tissue. *Am. J. Physiol.,* 1991, v. 260, pp. H1395-H1405.

Parker P. E., Bashour F. A., Downey H. F. et al. Coronary hemodynamics during reperfusion following acute coronary ligation in dog. *Amer. Heart. J.,* 1975, v. 90, pp. 593-599.

Parnev O. M., Naumov O. G., Ivkov N. N. et al. [The effect of phenothiazines on the viscosity and electrical stability of model phospholipid membranes]. (Russian). *Biul. Eksp. Biol. Med.,* 1987, v. 103, No. 5, pp. 555-557.

Parks D. A. and Granger D. N. Oxygen-derived radicals and ischemia-induced tissue injury. In: Greenwald R. A. and Chohen G., eds. *Oxy radicals and their scavenger systems.* 1983, v. 2, pp. 135-144.

Parks D. A., Bulkley G. B., Granger D. N. et al. Ischemic injury in the cat small intestine: role of superoxide radicals. *Gastroenterology,* 1982, v. 82, pp. 9-15.

Parks D. A., Williams T. K. and Beckman J. S. Conversion of xanthine dehydrogenase to oxidase in ischemic rat intestine: a reevaluation. *Am. J. Physiol.,* 1988, v. 254, pp. 6768-6774.

Parsadanian H. K., Ter-Tatevosian L. P., Aslanian J. G., Adunz G. T. [Heart muscle phosphoprotein phosphatase. Isolation of individual molecular forms and their properties]. (Russian). *Vopr. Med. Khim.,* 1985, v. 31, No. 5, pp. 45-49.

Parthasarathy S., Wieland E. and Steinberg D. A role for endothelial cell lipoxygenase in the oxidative modification of low density lipoprotein. *Proc. Natl. Acad. Sci. USA,* 1980, v. 86, pp. 1046-1050.

Partridge C. A., Horvath C. J., Delvecchio P. J. et al. Influence of extracellular matrix in tumor necrosis factor-induced increase in endothelial permeability. *Am. J. Physiol.*, 1992, v. 263, No. 6, Pt. 1, pp. 627-633.

Patton S. E., Rosen G. M. and Rauckman E. J. Superoxide production by purified hamster hepatic nuclei. *Molec. Pharmacol.*, 1980, v. 18, pp. 588-593.

Pavlov A. R., Revina A. A., Dupin A. M. et al. [Interaction of carnosine with superoxide radicals in aqueous solutions]. (Russian). *Biul. Eksp. Biol. Med.*, 1990, v. 110, No. 10, pp. 391-393. English transl. see[1*], pp. 1374-1378.

Pearson J., Carleton S. and Gordon J. L. Metabolism of adenine nucleotides by ectoenzymes of vascular endothelial and smooth-muscle cells in culture. *Biochem. J.*, 1980, v. 190, pp. 421-429.

Peng C. F., Murphy M. L. and Straub K. O. Effects of early reperfusion on the mechanical and biochemical characteristics of ischemic myocardium. *J. Surg. Res.*, 1986, v. 41, No. 5, pp. 493-502.

Perez N. G., Gao W. D. and Marban E. Novel myofilament Ca^{2+}-sensitizing property of xanthine oxidase inhibiotors. *Circ. Res.*, 1998, v. 83, No. 4, pp. 423-430.

Perry M. A. and Wadhwa S. S. Gradual reintroduction of oxygen reduces reperfusion injury in cat stomach. *Am. J. Physiol.*, 1988, v. 254, pp. G366-G372.

Persidskii Yu. V., Kudriavets Yu. I. and Barshtein Yu. A. [The action of tumor necrosis factor on the microvascular endothelium and its role in the morphological changes in the internal organs.] (Russian). *Biul. Eksp. Biol. Med.*, 1991, v. 111, No. 3, pp. 294-297. English transl. see[1*] pp. 369-372.

Peskin A. V., Zbarsky I. B. and Konstantinov A. A. [Investigation of electron transport systems in rat liver microsomal and nuclear membranes and in hepatoma 22a]. (Russian). *Biokhimiya*, 1981, v. 46, pp. 481-489. English transl. see[3*], pp. 579-589.

Peterson D. A., Kelly B. and Gerrard J. M. Allopurinol can act as an electron transfer agent. Is this relevant during reperfusion injury? *Biochem. Biophys. Res. Commun.*, 1986, v. 137, pp. 76-79.

Petrovich D. R., Finkelstein S., Waring A. et al. Liver ischemia increase the molecular order of microsomal membranes by increasing the cholesterol - to phospholipid ratio. *J. Biol. Chem.*, 1984, v. 10, pp. 13217-13223.

Petrovich Yu. A. and Gutkin D. V. [Free-radical oxidation and its role in the pathogenesis of inflammation, ischemia and stress. Review]. (Russian). *Patol. Fiziol. Eksp. Ter.*, 1986, No. 5, pp. 85-92.

Petzelbauer P., Bender J. R., Wilson J. and Pober J. S. Heterogeneity of dermal microvascular endothelial cell antigen expression and cytokine responsiveness in situ and in cell culture. *J. Immunol.*, 1993, v. 151, No. 9, pp. 5062-5072.

Phan S. H., Gannon D. E., Varani J. et al. Xanthine oxidase activity in rat pulmonary artery endothelial cells and its alteration by activated neutrophils. *Am. J. Pathol.*, 1989, v. 134, No. 6, pp. 1201-1211.

Phillis J. W. and O'Regan M. H. Mechanisms of glutamate and aspartate release in the ischemic rat cerebral cortex. *Brain Res.*, 1996, v. 730, No. 1-2, pp. 150-164.

Phillips S. J., Kungtahworn C., Zeff R. H. et al. Emergency coronary artery revascularization: a possible therapy for acute myocardial infarction. *Circulation*, 1979, v. 60, pp. 241-246.

Pincemail J., Detry O., Philippart C. et al. Diaspirin crosslinked hemoglobin (DCLHb): absence of increased free radical generation following administration in a rabbit model of renal ischemia and reperfusion. *Free Radic. Biol. Med.*, 1995, v. 19, No. 1, pp. 1-9.

Piper P. J., Letts L. G., Samhoun M. N. et al. The leukotrienes: their actions on gastrointestinal airway and vascular smooth muscle. *Allergologie*, 1982, v. 5, pp. 138-141.

Pisarenko O. I., Lakomkin V. L., Studneva I. M. et al. Allopurinol-enhanced postischemic recovery in the isolated rat heart involves repletion of high-energy phosphates. *Biochem. Med. Metab. Biol.*, 1994, v. 51, No. 1, pp. 16-26.

Pisarenko O. I., Studieva I. M., Shul'zhenko V. S. et al. [Relationship between the energy state of the myocardium and the release of products of anaerobic metabolism from it]. (Russian). *Biokhimiya*, 1990, v. 55, No. 1, pp. 114-125. English transl. see[3*], pp. 87-95.

Plotnikov M. B., Avdoshin A. D., Chernysheva G. A. and Saratikov A. S. [Dimethyl sulfoxide as a corrector of disturbances of the cerebral circulation and oxygen and carbohydrate metabolism and of lipid peroxidation soon agter intracerebral hemorrhage]. (Russian.) *Biul. Eksp. Biol. Med.*, 1990, v. 109, No. 4, pp. 357-359. English transl. see[1*], pp. 466-468.

Plotnikov M. B., Kobzeva E. A. and Plotnikova T. M. [Antioxidative effects of antioxidants in cerebral ischemia]. (Russian). *Biul. Eksp. Biol. Med.*, 1992, v. 113, No. 5, pp. 504-506. English transl. see[1*], pp. 679-681.

Plotnikov M. B., Vaizova O. E. and Suslov N. Y. [Effect of cerebrocrast on local cerebral blood flow and EEG of alert rats with brain ischemia.] (Russian). *Biul. Exp. Biol. Med.,* 1995, v. 119, No. 2, pp. 296-298. English transl. see[1]* pp. 288-290.

Poberezkina N. B. and Osinskaya L. F. [Biological role of superoxide dismutase. Review]. (Russian). *Ukr. Biokhim. Zh.*, 1989, v. 61, No. 2, pp. 14-27. English transl. See 5*., pp. 14-29.

Pogosian G. A., Dremina E. S., Sharov V. S. et al. [Kinetics of Fe^2-induced lipid peroxidation in suspension of mitochondrial and nuclear membrans]. (Russian). *Biophizika*, 1996, v. 41, No. 2, pp. 342-347.

Pohlman T. H. and Harlan J. M. Human endothelial cell response to lipopolysaccharide interleukin-1, and tumor necrosis factor is regulated by protein synthesis. *Cell. Immunology*, 1989, v. 119, pp. 41-52.

Pokrovsky A. A. [Enzymatic mechanisms of the functioning of lysosomes]. (Russian). *Dokl. Akad. Nauk SSSR*, 1975, v. 225, No. 3, pp. 689-692. English transl. see[7*], pp. 487-490.

Polunovsky V. A., Wendt C. H., Ingbar D. H. et al. Induction of endothelial cell apoptosis by TNFα: modulation by inhibitors of protein synthesis. *Exp. Cell Res.*, 1994, v. 214, pp. 584-594.

Poole-Wilson P. A. Measurement of myocardial intracellular pH in pathological states. *J. Mol. Cell. Cardiol.*, 1978, v. 10, No. 6, pp. 511-526.

Poole-Wilson P. A., Harding D. P., Bourdillon P. D. V. et al. Calcium out of control. *J. Mol. Cell. Cardiol.*, 1984, v. 16, No. 2, pp. 175-187.

Poot M. Lipid peroxidation and cell proliferation. In: Carmen Vigo-Pelfrey, ed. *Membrane Lipid Oxidation.* Boca Raton, CRC Press, 1990, pp. 121-133.

Popov V. G., Lazutin V. K., Khitrov N. K. et al. [Noradrenaline and adrenalin content in different sections of the heart in patients deseased due to myocardial infarction]. (Russian). *Kardiologiya*, 1975, v. 15, No. 10, pp. 102-107.

Portugal V., Alonso I., Bilbao J. et al. [Effect of antioxidative therapy on regeneration of ischemic liver]. (Spanish). *Revista Espanola de Enfermedades Digestivas,* 1993, v. 83, No. 6, pp. 434-438.

Prado R., Watson B. D. and Wester P. Effects of nitric oxide synthase inhibition on cerebral blood flow following bilateral carotid artery occlusion and recirculation in the rat. *J. Cerebral. Blood Metab.,* 1993, v. 13, No. 4, pp. 720-723.

Pratt R. M. and Greene R. M. Inhibition of palatal epithelial cell death by altered protein synthesis. *Develop. Biol.,* 1976, v. 54, No. 1, pp. 135-145.

Prehn J. H., Backhauss C. and Krieglstein J. Transforming growth factor - beta1 prevents glutamate neurotoxicity in rat neocortical cultures and protects mouse neocortex from ischemic injury in vivo. *J. Cereb. Blood Flow Metab.*, 1993, v. 13, No. 3, pp. 521-525.

Preobrazhensky A.N., Dzhavadov S.A. and Saks V.A. [Study of hypothetical mechanism of protective effect of phosphocreatine on ischemic myocardium]. (Russian). *Biokhimiya*, 1986, v. 51, No. 4, pp. 675-683. English transl. see[3*], pp. 583-590.

Preusse C. S., Gebhard M. M. and Bretschneider B. H. Interstital pH value in the myocardium as indicator of ischemic stress of cardioplegically arrested hearts. *Bas. Res. Cardiol.,* 1982, v. 77, No. 4, pp. 372-387.

Prilipko L. L., Orlov O. N., Ivanova S. M. et al. [Lipid peroxidation activation during the stress in the man estimated from the pentane content in expirated air]. (Russian). *Dokl. Acad. Nauk SSSR*, 1982, v. 265, No. 4, pp. 1010-1013. English transl. see[7*], pp. 450-452.

Pronchuk N. F., Gurevich M. I. and Meerson F. Z. [Depression of the contractile activity of isolated cardiomyocytes under the impact of high concentrations of catecholamines and the prevention of this phenomenon with an antioxidant]. (Russian). *Kardiologiya*, 1984, v. 24, No. 5, pp. 107-109.

Prosdocimi M., Finesso M., Gorio A. et al. Coronary and systemic 6-ketoprostaglandin F_1 alpha and tromboxane B_2 during myocardial ischemia in dog. *Am. J. Physiol.*, 1985, v. 148, No. 4, pp. 493-499.

Proskuryakov S. Ya., Gabai V. L., Mosin A. F. et al. [DNA degradation and changes in permeability and morphology of cells of the Ehrlich ascite tumor under aerobic incubation without glucose]. (Russian). *Tsitologiya*, 1989, v. 31, No. 6, pp. 690-695.

Przyklenk K. and Kloner R. A. Superoxide dismutase plus catalase improve contractile function in the canine model of the "stunned myocardium". *Circ. Res.*, 1986, v. 58, No. 1, pp. 148-156.

Przyklenk K., Vivaldi M. T., Schoen F. J. et al. Solvage of ischaemic myocardium by reperfusion: importance of collateral blood flow and myocardial oxygen demand during occlusion. *Cardiavasc. Res.*, 1986, v. 20, No. 6, pp. 403-414.

Pucheu S., Boucher F., Malfroy B. and De Leiris J. Protective effect of the superoxide scavenger EUK8 against ultrastructural alterations induced by ischemia and reperfusion in isolated rat hearts. *Nutrition.* 1995, V. 11, No. 5, (Suppl.), pp. 582-584.

Pucheu S., Coudray C., Tresallet N. et al. Effect of iron overload in the isolated ischemic and reperfused rat heart. *Cardiovasc. Drugs. and Therapy*, 1993, v. 7, No. 4, pp. 701-711.

Pulatova M. K., Todokov I. N., Kosaganova N. Yu. et al. [Ionol-dependent activation of protein and DNA biosynthesis in the regenerating mouse liver]. (Russian). *Khim. Farm. Zh.*, 1985, v. 19, No. 1, pp. 23-26.

Punch J., Rees R., Cashmer B. et al. Xanthine oxidase: its role in the no-reflow phenomenon. *Surgery*, 1992, v. 111, No. 2, pp. 169-176.

Punz A., Nanobashvili J., Fuegl A., et al. Effect of alpha-tocopherol pretreatment on high energy metabolites in rabbit skeletal muscle after ischemia-reperfusion. *Clin. Nutr.* 1998, v. 17, No. 2, pp. 85-87.

Qayumi A. K., Jamieson W. R., Godin D. V. et al. Response to allopurinol pretreatment in a swine model of heart-lung transplantation. *J. Invest. Surg.*, 1990, v. 3, No. 4, pp. 331-340.

Quinn P. J. The fluidity of cell membranes and its regulation. *Prog. Biophys. Mol. Biol.*, 1981, v. 38, pp. 1-104.

Racay P., Bezakova G., Kaplan P. et al. Alteration in rabbit brain endoplasmic reticulum Ca^{2+} transport by free oxygen radicals *in vitro*. *Biochem. Biophys. Res. Commun.*, 1994, v. 199, No. 1, pp. 63-69.

Radi R., Beckman J. S., Bush K. M. and Freeman B. A. Peroxynitrite-induced membrane lipid peroxidation: The cytotoxic potential of superoxide and nitric oxide. *Arch. Biochem. Biophys.*, 1991, v. 288, pp. 481-487.

Radkevich L. A., Kolotygina I. M. and Kabakov A. Yu. [Stability of the membrane potential of mouse liver cells on exposure to damaging agents]. (Russian). *Biofizika*, 1985, v. 30, No. 4, pp. 646-649. English transl. see[2*], pp. 703-707.

Raff M. C., Barres B. A., Burne J. F. et al. Programmed cell death and the control of cell survival: lessons from the nervous system. *Science*, 1993, v. 262, pp. 695-700.

Raguzin A. V. Anoxia of the brain of newborns and its sequelae in various methods of experimental reoxygenation. In: *Constituent Congress Intern. Soc. Pathophysiol.* Abstracts. Moscow, 1991, No. 15. 3. 15, p. 325.

Ramazanova L. I. and Bardakhchyan E. A. [Ultrastructure and biochemistry of ischemic myocardium after reperfusions]. (Russian). *Kardiologiya*, 1985, v. 25, No. 2, pp. 82-86.

Randall H. M. and Cohen J. J. Anaerobic CO_2 production by dog kidney in vitro. *Am. J. Physiol.*, 1966, v. 211, pp. 493-505.

Rao P. S., Cohen M. V. and Mueller H. S. Production of free radicals and lipid peroxides in early experimental myocardial ischemia. *J. Mol. Cell. Cardiol.*, 1983, v. 15, No. 10, pp. 713-716.

Rao P. S., Walsh T. R., Makowka L. et al. Purine nucleoside phosphorylase: a new marker for free oxygen radical injury to the endothelial cell. *Hepatology*, 1990, v. 11, No. 2, pp. 193-198.

Rapundalo S. T., Briggs F. N. and Feher J. J. Effects of ischemia on the isolation and function of canine cardial sarcoplasmic reticulum. *J. Mol. Cell. Cardiol.*, 1986, v. 188, pp. 837-851.

Rashba Yu. E., Chernikov V. A., Baider L. M. and Vartanyan L. S. The use of hydroxylamines of the piperidine and imidazoline series for the quantitative determination of superoxide radicals in microsomal membranes by the EPR method. *Biological membranes*, 1986, v. 3, No. 8, pp. 838-845.

Rashba Yu. E., Nagler L. G., Vartanian L. S. et al., [Is xanthine oxidase a universal source of superoxide radical in liver damage induced by ischemia and reperfusion?]. (Russian). *Biul. Eksp. Biol. Med.*, 1990, v. 109, N6, pp. 548-550. English transl. see1*, pp. 727-730.

Rashba Yu. E., Vartanyan L. S., Seregina L. A. et al. [Superoxide radical - superoxide dismutase system dysfunction in the rat liver during ischemia]. (Russian). *Biul. Eksp. Biol. Med.*, 1986, v. 102, No. 11, pp. 559-560. English transl. see[1*], pp. 1535-1537.

Rashba Yu. E., Vartanyan L. S., Seregina L. A. et al. [Uncompensated changes in the O2/SOD system as a possible factor of intensification of lipid peroxidation during hepatic ischemia]. (Russian). In: *Bioantioxidant. 2 nd All-Union Conference*. Abstracts. Chernogolovka, 1986, v. 2, pp. 235-236.

Rashba Yu. E., Chernikov V. A., Baider L. M. and Vartanyan L. S. The use of hydroxylamines of the piperidine and imidazoline series for the quantitative determination of superoxide radicals in microsomal membranes by the EPR method. *Biological membranes*, 1986, v. 3, No. 8, pp. 838-845.

Rashba-Step J., Tatoyan A., Duncan R. et al. Phospholipid peroxidation induces cytosolic phospholipase A_2 activity: membrane effects versus enzyme phosphorylation. *Arch. Biochem. Biophys.*, 1997, v. 343, No. 1, pp. 44-54.

Rasmussen H. and Goodman D. B. P. Relationships between calcium and cyclic nucleotides in cell activation. *Physiol. Reviews*, 1977, v. 57, No. 3, pp. 421-509.

Rasool N., Faroqui M. and Rubinstein E. H. Lidocaine accelerates neuroclectrical recovery after in complete global ischemia in rabbits. *Stroke*, 1990, v. 21, pp. 929-935.

Ratych R. E., Chuknyiska R. S. and Bulkly G. B. The primary localization of free radical generation after anoxia/reoxygenation in isolated endothelial cells. *Surgery*, 1987, v. 102, No. 2, pp. 122-131.

Rauckman E. J., Rosen G. M., Post S. E. et al. Effect of model traumatic injury on hepatic drug-metabolizing enzymes. *J. Trauma*, 1980, pp. 884-886.

Razack S., D'Agnillo F. and Chang T. M. Crosslinked hemoglobin-superoxide dismutase-catalase scavenges free radicals in a rat model of intestinal ischemia-reperfusion injury. *Artif. Cells. Blood Substit. Immobil. Biotechnol.*, 1997, v. 25, No. 1-2, pp. 181-192.

Redl H., Gasser H., Schlag G. and Marzi I. Involvement of oxygen radicals in shock related cell injury. Review. *Br. Med. Bull.*, 1993, v. 49, No. 3, pp. 556-565.

Reeves W. B. and Shah S. V. Activation of potassium channels contributes to hypoxic injury in proximal tubules. *J. Clin. Invest.*, 1994, v. 94, No. 6, pp. 2289-2294.

Rehncrona S., Folbergrova J., Smith D. and Siesjö B. K. Influence of complete and pronounced incomplete cerebral ischemia and subsequent recirculation on cortical concentration of oxidized and reduced glutathione in the rat. *J. Neurochem.*, 1980, v. 34, pp. 84-93.

Rehncrona S., Mela L., Siesjљ B.K. et al. Recovery of brain mitochondrial function in the rat after complete and incomplete cerebrae ischemia. *Stroke*, 1979, v. 10, pp. 437-446.

Rehncrona S., Smith D. S., Akesson B. et al. Peroxidative changes in brain cortical fatty acids and phospholipids, as characterized during Fe^{++}- and ascorbic acid-stimulated lipid peroxidation in vitro. *J. Neurochem.*, 1980, v. 34, pp. 1630-1638.

Reid T. R., Torti F. M. and Ringold G. M. Evidence for two mechanisms by which tumor necrosis' factor kills cells. *J. Biol. Chem.*, 1989, v. 264, No. 8, pp. 4583-4589.

Reimer K. A., Jennings R. B. and Hill M. L. Total ishemia in dog hearts in vitro. 2. High energy phosphate depletion and associated defects in energy metabolism, cell volume regulation and sarcolemmal integrity. *Circ. Res.*, 1981, v. 49, No. 4, pp. 901-911.

Riabinin V. E. and Lifshits R. I. [Characteristics and correlation between processes of mitochondrial and microsomal oxidation at various intensities of lipid peroxidation]. (Russian). *Biokhimiya*, 1991, v. 56, No. 11, pp. 1991-1998. English transl. see[3*], pp. 1406-1411.

Riley A. L. Effect of ischemia on renal blood flow in the rat. *Nephron*, 1978, v. 21, pp. 107-113.

Ritov V. B. [Molecular structure and mechanism of functioning of calcium-dependent ATPase of sarcoplasmatic reticulum]. (Russian). In: *Itogi Nauki i Tekhniki. Ser. Biol. Chem.*, 1977, v. 11, pp. 5-77.

Ritov V. B., Murzakhmetova H. K. and Kozlov Yu. P. [*Molecular structure and functional properties of transport calcium-dependent ATPase of sarcoplasmatic reticulum*]. (Russian). Irkutsk, Izdatelstvo Irkutsk. State Univ., 1983, 133 p.

Roberts A. B., Roche N. S., Winokur T. S. et al. Role of transforming growth factor-beta in maintenance of function of cultured neonatal cardiac myocytes. *J. Clin. Invest.*, 1992, v. 90, No. 5, pp. 2056-2062.

Roberts G. P., Perry M. O., Hariri R. G. and Shires G. T. Induced skeletal muscle ischemia in man. *Circ. Shock.*, 1985, v. 17, pp. 253-258.

Robinson E. and Hedlund B. Role of iron/ischemia and reperfusion. *Circ. Shock.*, 1989, v. 27, p. 367.

Romanova L. K. [*Experimental and clinical regeneration of lung*]. (Russian). Moscow, Meditsina, 1971, 199 p.

Romanova L. K. and Grushevskaya O. O. [Regularities of hepatocyte proliferation under liver regeneration after subtotal resection]. (Russian). In: Liozner L. D., ed. *Methods of regeneration and cell-division.* Moscow, Nauka, 1979, 121 p.

Romaschin A. D., Rebeyka I., Wilson G. I. et al. Conjugated diene in ischemic and reperfused myocardium: an in vivo chemical signature of oxygen free radical mediated injury. *J. Mol. Cell. Cardiol.*, 1987, v. 19, No. 3, pp. 289-302.

Romaschin A. D., Wilson G. J., Thomac. U. et al. Subcellular distribution of peroxidized lipids in myocardial reperfusion injury. *Am. J. Physiol.*, 1990, v. 259, No. 1, pp. H116-H123.

Romson J., Hook B., Kunkel S. et al. Reduction in the extent of myocardial ischemic injury by neutrophil depletion in the dog. *Circulation*, 1983, v. 67, pp. 1016-1023.

Rose I. C., Johnson M., Ramwell P. W. et al. Effect of arachidonic acid on systemic arterial pressure, myocardial contractility and platelets in the dog. *Proc. Soc. Exp. Biol. Med.*, 1974, v. 147, No. 3, pp. 652-655.

Rosenberg A. A. Cerebral blood flow and O_2 metabolism after asphyxia in neonatal lambs. *Pediatr. Res*, 1986, v. 20, No. 8, pp. 778-782.

Rosenblum W. I. Unsaturated fatty acids and cyclooxygenase inhibitors: effect on pial arterioles. *Am. J. Physiol.*, 1982, v. 242, No. 4. pp. H629-H142.

Rosenblum W. I. Effect of free radical generation on mouse pial arterioles: probable role of hydroxyl radicals. *Am. J. Physiol.*, 1983, v. 245, pp. H139-H142.

Rossi F. The O_2-forming HADPH oxidase of the phagocytes: nature, mechanisms of activation and function. *Biochim. Biophys. Acta*, 1986, v. 853, No. 1, pp. 65-89.

Röth E., Lantos J., Temes G. et al. Cardioprotection during heart ischemia-reperfusion. *Acta Chir. Hung.*, 1997, v. 36, No. 1-4, pp. 306-309.

Röth E., Török B., Zsoldos T. and Matkovics B. Lipid peroxidation and scavenger mechanisms in experimentally induced heart infarcts. *Basic Res. Cardiol.*, 1985, v. 80, pp. 530-536.

Rothman R. J., Serroni A. and Farber J. L. Cellular pool of transient ferric iron, chelatable by deferoxamine and distinct from ferritin, that is involved in oxidative cell injury. *Mol. Pharmacol.*, 1992, v. 42, pp. 703-710.

Rouslin W. Mitochondrial complexes I, II, III, IV and V in miocardial ischemia and autolysis. *Am. J. Physiol.*, 1983, v. 244, No. 6, pp. H743-H748.

Rovetto M. J., Lamberton W. F. and Neely J. R. Mechanism of glycolytic inhibition in ischemic rat hearts. *Circ. Res.*, 1975, v. 37, No. 6, pp. 742-751.

Roy R. S. and McCord J. M. Superoxide and ischemia: conversion of xanthine dehydrogenase to xanthine oxidase. In: Grenwald R. A. and Cohen G., eds. *Oxy radicals and their scavenger systems. Cellular and Medical Aspects.*, N. Y., Elsevier Sci., 1983, v. 2 pp. 145-153.

Rubanyi G. M. and Vanhoutte P. M. Superoxide anions and hypoxia inactivate endothelium derived relaxing factor. *Am. J. Physiol.*, 1986, v. 250, pp. H822-H827.

Rubin B. B., Liauw S., Tittley J. et al. Prolonged adenine nucleotide resynthesis and reperfusion injury in postischemic skeletal muscle. *Am. J. Physiol.*, 1992, v. 262, №5, Pt 2, pp. H1538-H1547.

Ruf W., McNamara J. J., Saehiro A. et al. Platelet trapping in myocardial infarct in baboons: therapeutic effect of aspirin. *Am. J. Cardiol.*, 1980, v. 46, pp. 405-405.

Runeberg L. and Pakarinen A. Anaerobic decarboxylation of α-ketoglutarate by rat mitochondria with fumorate as hydrogen acceptor. *Ann. Med. Exp. Biol. Fenn.*, 1967, v. 45, pp. 427-433.

Rusakov S. A., Filippova G. N., Pushkareva M. Yu. et al. [Effect of the tumor necrosis factor on the free schingosine content and sphingomyelinase activity in mouse liver cells and nuclei]. (Russian). *Biokhimiya*, 1993, v. 58, No. 5, pp. 724-732. English transl. see[3*], pp. 476-482.

Ryan J. W. and Ryan U. S. Endothelial metabolism. In: Effros R. M., Schmid-Schönbein H. and Ditzel J., eds. *Microcirculation. Current physiological, medical and surgical concepts.* N. Y., Academic Press, 1981, pp. 147-169.

Ryan J. W., Ryan U. S., Schultz D. R. et al. Subcellular localization of pulmonary angiotensin converting enzyme (kinase II). *Biochem. J.* 1975, v. 146, pp. 497-499.

Ryasina T. V., Korshunova T. S., Polyanskii N. B. and Muranov K. O. [Anti-ischemic effect of a new oxynicotinic acid derivative]. (Russian). *Biul. Eksp. Biol. Med.*, 1994, v. 118, No. 7, pp. 49-51.

Rymsa B., Wang I. F. and de Groot H. $O_2^{\bullet}$-release by activated Kupffer cells upon hypoxia-reoxygenation. *Am. J. Physiol.*, 1991, v. 261, No. 4, pp. G602-G607.

Rysmendiev A. Zh., Tukeshova V., Grabarova I. P. et al. [Free-radical lipid oxidation in acute myocardial infarction complicated by cardiogenic shock]. (Russian). *Kardiologiya*, 1986, v. 26, No. 5, pp. 37-40.

Saburova L. M. and Berezov T. T. [Ways and possibilities of biochemical evaluation of vitality of organs in acute ischemia]. (Russian). *Vestn. Akad. Med. Nauk SSSR*, 1975, No. 7, pp. 46-51.

Sacks T., Moldow C. F., Craddock P. R. et al. Oxygen radicals mediate endothelial cell damage by complement-stimulated granulocytes: an in vitro model of immune vascular damage. *J. Clin. Invest.*, 1978, v. 61, pp. 1161-1167.

Saez J. C., Cifuentes F., Ward P. H., et al. Tourniquet shock in rats: effects of allopurinol on biochemical changes of the gastrocnemius muscle subjected to ischemia followed by reperfusion. *Biochem. Med. Metabol. Biol.*, 1986, v. 35, No. 2, pp. 199-209.

Saffitz J. E., Corr P. B., Lee B. I. et al. Pathophysiological concentrations of lysophosphatides quantified by electron microscopic autoradiography. *Am. J. Cardiol.*, 1982, v. 49, pp. 970.

Safirstein R., Price P. M., Saggi S. J. and Harris R. C. Changes in gene expression after temporary renal ischemia. *Kidney Int.*, 1990, v. 37, No. 6, pp. 1515-1521.

Sagach V. F. [Leucotriens and the cardiovascular system. Review]. (Russian). *Patol. Fiziol. Eksp. Ter.*, 1986, No. 1, pp. 84-89.

Sagach V. F. and Dmitrieva A. V. [The role of the thrombocyte-activating factor in the development of circulatory disorders in the postischemic shock reaction]. (Russian.) *Fiziol. Zh.*, 1990, v. 36, No. 4, pp. 8-14.

Sailer N. Enzymes. In: Lajtha A., ed. *Handbook of Neurochemistry.*, N. Y., Plenum Press, 1969, v. 1, pp. 325-468.

Sairanen T., Ristimaki A., Karjlainen-Lindsberg M. L. et al. Cyclooxygenase-2 is induced globally in infarcted human brain. *Ann. Neurol.* 1998, v. 43, No. 6, pp. 738-747.

Sakay K., Ito T. and Ogawa K. Roles of endogenous prostacyclin and thromboxane A_2 in the ischemie canine heart. *J. Cardiovasc. Pharmacol.*, 1982, v. 4, pp. 129-135.

Sakr M. F., McClain C. J., Gavaler J. S. et al. FK 506 pretreatment is associated with reduced levels of tumor necrosis factor and interleukin 6 following hepatic ischemia reperfusion. *J. Hepatol.*, 1993, v. 17, No. 3, pp. 301-307.

Sakr M., Zetti G., McClaio C. et al. The protective effect of FK 506 pretreatment against renal ischemia/reperfusion injury in rats. *Transplantation*, 1992, v. 53, No. 5, pp. 987-991.

Saks V. A. [Intracellular energy transfer phosphocreatinic way]. (Russian). *Usp. Biol. Khim.*, 1983, v. 24, No. 1, pp. 40-64.

Saks V. A. [Metabolism of the myocardium results of 10-year studies in the All-Union Cardiology Research Centre of the SSSR Academy of Medical Science]. (Russian). *Biul. VKNS Akad. Med. Nauk SSSR*, 1985, v. 8, No. 2, pp. 31-38.

Saks V. A., Dzhaliashvili I. V., Konorev E. A. and Strumia E. [Molecular and cellular aspects of the cardioprotective effect of phosphocreatine]. (Russian). *Biokhimiya*, 1992, v. 57, No. 12, pp. 1763-1784. English transl. see[3]*, pp. 1223-1239.

Salaris S. C. and Babbs C. F. Effect of oxygen concentration on the formation of malondialdehyde-like material in a model of tissue ischemia and reoxygenation. *Free Radic. Biol. Med.*, 1989, v. 7, No. 6, pp. 603-609.

Sairanen T., Ristimaki A., Karjlainen-Lindsberg M. L. et al. Cyclooxygenase-2 is induced globally in infarcted human brain. *Ann. Neurol.* 1998, v. 43, No. 6, pp. 738-747.

Salganik R. I., Shabalina I. G., Kolosova N. G. et al. [Physicochemical properties of membranes and functional status of liver mitochondria in rats with an inherited capacity for increased radical formation]. (Russian). *Biul. Eksp. Biol. Med.*, 1995, v. 119, No. 6, pp. 628-631. English transl. see[1]*, pp. 605-607.

Saluja I., Song D., O'Regan M. H. and Phillis J. W. Role of phospholipase A2 in the release of free fatty acids during ischemia-reperfusion in the rat cerebral cortex. *Neurosci. Lett.*, 1997, v. 233, No. 2-3, pp. 97-100.

Salvati P., Whittle B. J. R. Investigation of the vascular actions of arachidonate lipoxygenase and cyclo-oxygenase products on the isolated perfused stomach of rat and rabbit. *Prostaglandins*, 1981, v. 22, No. 1, pp. 141-156.

Samuelsson B., Hammarstrom S., Murphy R. S. and Borgeat P. Leukotrienes and slow reacting substance of anaphylaxis (SRS-A). *Allergy*, 1980, v. 35, pp. 375-381.

Sandermann H. Regulation of membrane enzymes by lipids. *Biochim. Biophys. Acta.*, 1978, v. 515, No. 23, pp. 209-237.

Sanders S. P., Zweier J. L., Kuppusa M. Y. and Harrison S. J. Hyperoxic sheep pulmonary microvascular endothelial cells generate free radicals via mitochondrial electron transport. *J. Clin. Invest.*, 1993, v. 91, No. 1, pp. 46-52.

Sandstrom P. A. and Buttke T. M. Autocrine production of extracellular catalase prevents apoptosis of the human CEM T-cell line in serum-free medium. *Proc. Natl. Acad. Sci. USA*, 1993, v. 90, pp. 4708-4712.

Sandstrom P. A., Tebbey P. W., Cleave S. V. et al. Lipid hydroperoxides induce apoptosis in T-cells displaying a HIV-associated glutathione peroxidase deficiency. *J. Biol. Chem.*, 1994, v. 269, No. 2, pp. 798-801.

Sanina O. L. and Berdinskykh N. K. [Biological role of ceruloplasmin and its clinical potential. Review]. (Russian). *Vopr. Med. Khim.*, 1986, v. 32, No. 5. pp. 7-14.

Sarbash V. I., Rymarchuk V. I. and Malenkov A. G. [Effect hemodynamic disturbances on the rat liver transepithelial potential in vivo]. (Russian). *Biul. Eksp. Biol. Med.*, 1987, v. 104, No. 10, pp. 44-445, English transl. see[1*], pp. 1401-1403.

Sasaki R., Watanabe Y., Morishita T. and Yamagata S. Estimation of the cell number of heart muscles in normal rats. *Tohoku J. Exp. Med.*, 1968, v. 95, pp. 177-184.

Sasaki T., Wakai S. and Asano T. The effect of lipid hydroperoxide of arachidonic acid on the canine basilar artery. An experimental study on cerebral vasospasm. *J. Neurosurg.*, 1981, v. 54, No. 3, pp. 357-365.

Sato Y., Hotta N., Sakamoto N. et al. Lipid peroxide level in plasma of diabetic patients. *Biochem. Med.*, 1979, v. 21, No. 1, pp. 104-107.

Satomi N., Sakurai A., Haranaka R. et al. Preventive effects of several chemicals against lethality of recombinant human tumor necrosis factor. *J. Biol. Response Mod.*, 1988, v. 7, pp. 54-63.

Savage M. P., Goldberg S., Bove A. A. et al. Effect of thromboxane A2 blockade on clinical outcome and restenosis after successful coronary angioplasty. Milti-Hospital Eastern Atlantic Restenosis trial (M-Heart II). *Circulation*, 1995, v. 92, No. 11, pp. 3194-3200.

Savel'ev V. S., Chekareva G. A., Mishnev O. D. et al. [Morphological changes in limb skeletal muscle tissue during experimental postischemic recirculation]. (Russian). *Biul. Eksp. Biol. Med.*, 1985, v. 100, No. 9, pp. 351-355. English transl. see[1*], pp. 1266-1269.

Savov V. M., Babijaev M. A. and Kagan V. E. [Mechanisms underlying Ca^{2+} effects on lipid peroxidation.]. (Russian). *Biul. Eksp. Biol. Med.*, 1986, v. 101, No. 6, pp. 693-695. English transl. see[1*], pp. 771-773.

Sawczuk I. S., Hoke G., Olsson C. A. et al. Gene expression in response to acute unilateral obstruction. *Kidney Int.*, 1989, v. 35, pp. 1315-1319.

Sazontova T. G., Arkhipenko Yu. V. and Meerson F. Z. [Activation rat brain Na-, K-ATPase by stress]. (Russian). *Biul. Eksp. Biol. Med*, 1984, v. 97, No. 5, pp. 556-558. English transl. see[1*], pp. 590-592.

Scales W. E., Campbell D. A. Jr., Green M. E. and Remick D. G. Hepatic ischemia reperfusion injury: importance of oxidant tumor necrosis factor interactions. *Am. J. Physiol.*, 1984, v. 267, No. 6, Pt. 1, pp. G1122-G1127.

Schaper J., Mulch J., Winkler B. et al. Ultrastructural, function and biochemical criteria for estimation of reversibility of ischemic injury: a study on the effects of global ischemia on the isolated dog heart. *J. Mol. Cell. Cardiol.*, 1979, v. 11, pp. 521-541.

Schepetkin I. A. [Tumor necrosis factor as polypeptid growth factor]. (Russian). *Usp. Sovr. Biol.*, 1993, v. 113, No. 5, pp. 617-625.

Scherphof G., Roerdink F., Hoekstra D. et al. [Liposome stability in the presence of blood constituents. The succession of liposomal lipids and their contents inclusion into rat liver cells]. (Russian). In: Gregoriadic G. and Allison A., eds. *Liposomes in biological systems*. Moscow, Meditsina, 1983, pp. 183-210.

Schimizu A. and Yamanaka N. Apoptosis and cell desquamation in repair process of ischemic tubular necrosis. Virchows Arch. B. *Cell Pathol.*, 1993, v. 64, pp. 171-180.

Schinetti M. L., Sbarbati R. and Scarlattini M. Superoxide production by human umbilical vein endothelial cells in an anoxia-reoxygenation model. *Cardiovasc. Res.*, 1989, v. 23, No. 1, pp. 76-80.

Schlag G. and Redl H. Oxygen radicals in hypovolemic-traumatic shock. In: Workshop on: *Oxygen free-radicals in shock.* Florence, 1985, pp. 32-33.

Schlag M. G., Clarke S., Carson M. W. et al. The effect of mannitol versus dimethyl thiourea at attenuating ischemia/reperfusion-induced injury to skeletal muscle. *J. Vasc. Surg.* 1999, v. 29, No. 3, pp. 511-521.

Schmid-Shonbein G. W. The damaging potential of leukocyte activation in the microcirculation. *Angiology*, 1993, v. 44, No. 1, pp. 45-56.

Schoenberg M. H., Yones M., Muhl E. et al. Free radical involvement in ischemic damage of the small intestine. Oxy radicals and their scavenger systems. *Cell. Med. Aspects.*, 1983, v. 2, pp. 154-158.

Schömig A., Dart A. M., Dietz R. et al. Release of endogenous catecholamines in the ischemic myocardium of the rat. Part A: Locally mediated release. *Circ. Res.*, 1984, v. 55, No. 5, pp. 689-701.

Schrijvers A. H., De Groot M. J., Heijnen V. V. et al. Ischemia and reperfusion induced multilamellar vesicles in isolated rabbit hearts: time correlation between morphometric data and metabolic alterations. *J. Mol. Cell. Cardiol.*, 1990, v. 22, No. 6, pp. 653-665.

Schrör K., Smith E. F., Bickerton M. et al. Preservation of ischemic myocardium by pinane thromboxane A_2. *Am. J. Physiol.*, 1980, v. 238, pp. H87-H92.

Schulze-Osthoff K., Bakker A. C., Vanhaesbroek B. et al. Cytotoxic activity of tumor necrosis factor is mediated by early damage of mitochondrial functions. Evidence for the involvement of mitochondrial radical generation. *J. Biol. Chem.*, 1992, v. 267, pp. 5317-5323.

Schumer M., Colombel M. C., Sawczuk I. S. et al. Morphologic, biochemical, and molecular evidence of apoptosis during the reperfusion phase after brief periods of renal ischemia. *Am. J. Pathol.,* 1992, v. 140, pp. 831-838.

Scommotau S., Uhlmann D., Loffler B. M. et al. Involvement of endothelin/nitric oxide balance in hepatic ischemia/reperfusion injury. *Langenbecks Arch. Surg.*, 1999, v. 384, No. 1, pp. 65-70.

Seeger W., Wolf H., Stahler G. et al. Influence of tocopherol, its chromane compound, phytyl chains and superoxide dismutase on increased vascular resistance and permeability due to arachidonate metabolism in isolated rabbit lung. *Prostaglandins*, 1982, v. 23, No. 2, pp. 175-184.

Seekamp A., Jochum M., Ziegler M. et al. Cytokines and adhesion molecules in elective and accidental trauma-related ischemia/reperfusion. *J. Trauma,* 1998, v. 44, No. 5, pp. 874-882.

Seekamp A., Warren J. S., Remick D. G. et al. Requirements for tumor necrosis factor-alpha and interleukin-1 in limb ischemia/reperfusion injury and associated lung injury. *Am. J. Pathol.,* 1993, v. 143, No. 2, pp. 453-463.

Seelye R. N., Nevalainen T. J., Gavin J. B. et al. Physical and biochemical changes in rigor mortis of cardiac muscle. *Biochem. Med.,* 1979, v. 21, No. 3, pp. 323-332.

Segawa D., Hatori N., Yoshizu H. et al. The effect of nitric oxide synthase inhibitor on reperfusion injury of the brain under hypothermic circulatory arrest. *J. Thorac. Cardiovasc. Surg.*, 1998, v. 115, No. 4, pp. 925-930.

Seifulla R. D., Onishchenko N. A., Krasova E. I. et al. [Molecular mechanisms of antiischemic protection of the myocardium with pharmacological agents. Review.]. (Russian). *Farmakol. Toksikol.,* 1980, v. 43, No. 4, pp. 433-439.

Sekamova S. M. [Morphology and some aspects of crush syndrome]. (Russian). *Arkh. Patol.,* 1987, v. 49, No. 2, pp. 3-12.

Seleznev S. A. [*Liver in the dynamics of traumatic shock*]. (Russian). Moscow, Meditsina, 1971, p. 118.

Selivanova L. P. and Saburova L. M. [Glycogen metabolism in ischemized organs]. (Russian). *Biul. Eksp. Biol. Med.*, 1978, v. 86, No. 10, pp. 414-416. English transl. see[1*], pp. 1299-1301.

Semenyuk A. V., Voronin A. Yu., Kulikov V. Yu. et al. [Lipid peroxidation in the rat liver during stimulation of macrophages]. (Russian). *Biul. Eksp. Biol. Med.,* 1986, v. 101, No. 4, pp. 453-455. English transl. see[1*], pp. 484-487.

Semchenko V. V., Poluektov L. V. and Konvai V. D. [Role of lipid peroxidation in injuries to brain neurons during ischemia and postischemic period]. (Russian). *Biul. Eksp. Biol. Med.*, 1983, v. 96, No. 7, pp. 12-14. English transl. see[1*], pp. 892-899.

Serbinova E., Kharfus M., Ukhin L. Yu. et al. [Mechanisms of the antioxidative action of screened phenols in biomembranes: effects of ionol and its derivatives]. (Russian). *Biul. Eksp. Biol. Med.*, 1990, v. 110, No. 11, pp. 486-489. English transl. see[1*], pp. 1499-1502.

Seregina L. A. [Protective effect of antioxidants in total ischemia of rat liver]. (Russian). *Bioantioxidant. 2nd All-Union Conference.* Abstracts. Chernogolovka, 1986, v. 2, pp. 247.

Sergeev P. V. [*Steroid hormones*]. (Russian). Moscow, Nauka, 1984, 240 p.

Sergeeva V. S., Altuckova L. K. and Kremlin S. M. [Early postischemic liver disorder and its therapy with medical supplies]. (Russian). *2nd Moscow Med. Inst. Proceedings*., 1979, v. 129, No. 3, pp. 81-83.

Serhan C., Anderson P., Goodman E. et al. Phosphatidate and oxidized fatty acids are calcium ionophores: studies employing arsenazo III in liposomes. *J. Biol. Chem.*, 1981, v. 256, No. 6, pp. 2736-2741.

Serov R. A., Terekhova T. V., Makarova L. D. et al. [Prostaglandins and cyclic nucleotides in ventricular myocardium of patients with ventricular noradrinaline-induced heart damage]. (Russian). *Kardiologiya*, 1985, v. 25, No. 2, pp. 86-90.

Seth S. D., Maulik M., Manchada S. C. and Maulik S. K. Role of aspirin in modulating myocardial ischemic reperfusion injury. *Agents Actions,* 1994, v. 41, No. 3-4, pp. 151-155.

Sevanian A. Lipid peroxidation, membrane damage, and phospholipase A_2 action. In: Chow C. K., ed. *Cellular antioxidant defense mechanisms.* Florida, Boca Raton CRC Press. Inc., 1988, v. 2, pp. 78-95.

Sevanian A. and Hochstein P. Mechanisms and consequences of lipid peroxidation in biological systems. *Ann. Rev. Nutr.,* 1985, v. 5, pp. 365-390.

Sevanian A. and Kim E. Phospholipase A_2 dependent release of fatty acids from peroxidized membranes. J. *Free Rad. Biol. Med.*, 1986, v. 1, pp. 263-271.

Sevanian A., Muakkassah-Kelly S. and Montestruque S. The influence of phospholipase A_2 and glutathion peroxidase in the elimination of membrane lipid peroxides. *Arch. Biochem. Biophys.*, 1983, v. 233, No. 2, pp. 441-452.

Sevanian A., Nordenbrand K., Kim E. et al. Microsomal lipid peroxidation: the role of NADPN-cytochrome P450 reductase and cytochrome P450. *Free Rad. Biol. Med.*, 1990, v. 8, pp. 145-152.

Severina E. P. and Evtodienko Yu. V. [Investigation of the localization of phospholipase A_2 in the mitochondria]. (Russian). *Biokhimiya,* 1981, v. 46, No. 7, pp. 1199-1201. English transl. see[3*], pp. 964-966.

Seyama A. The role of oxygen-derived free radicals and the effect of free radical scavengers on skeletal muscle ischemia/reperfusion injury. *Surgery Today,* 1993, v. 23, No. 12, pp. 1060-1067.

Shackelford D. A., Tobaru T., Zhang S. and Zivin J. A. Changes in expression of the DNA repair protein complex DNA-dependent protein kinase after ischemia and reperfusion. *J. Neurosci.,* 1999, v. 19, No. 12, pp. 4727-4738.

Shadid M., Moison R., Steendijk P. et al. The effect of antioxidative combination therapy on post hypoxic-ischemic perfusion, metabolism, and electrical activity of the newborn brain. *Pediatr. Res.*, 1998, v. 44, No. 1, pp. 119-124.

Shadid M., Van Bel F., Steendijk P. et al. Effect of deferoxamine on post-hypoxic-ischemic reperfusion injury of the newborn lamb heart. *Biol. Neonate.,* 1999, v. 75, No. 4, pp. 239-249.

Shafranova V. P. and Ryasina T. V. [On the anatomical prerequisites of ischemic stroke modelling]. (Russian). *Laboratory animals in medical research.* Moscow, 1974, pp. 208-210.

Shaikh N. A. and Downar E. Time course of changes in porcine myocardial phospolipid levels during ischemia. *Circ. Res.*, 1981, v. 49, No. 2, pp. 316-325.

Shaikh N. A. and Downar E. Activation of phospholipases during myocardial ischemia and their probable role in arrhythmogenesis. *Develop. Cardiovasc. Med.,* 1985, v. 49, pp. 298-316.

Shakhman E. V., Donchenko G. V. and Datsenko Z. M. [Superoxide dismutase activity of microsomes from functionally different tissues of rats in the case of E-vitamin dificiency]. (Russian). *Ukr. Biokhim. Zh.*, 1995, v. 67, No. 6, pp. 88-93.

Shal'nev A. N., Levandovsky I. V. and Oksman T. M. [Effect of the duration of preceding ischemia and the mass of ischemized tissue on the state of the clotting and anticlotting systems of the blood in tourniquet shock]. (Russian). *Biul. Eksp. Biol. Med.*, 1979, v. 87, No. 6, pp. 436-539. English transl. see[1*], pp. 558-561.

Shanskaya A.I., Kochetygov W.I., Pozdnyakov P.K. and Puchkova S.H. [Parameters of systemic hemodynamics and microcirculation under intravenous infusion of fat emulsions]. (Russian). *Probl. Hemat.,* 1982, v. 27, No. 6, pp. 30-32.

Shappell S. B., Taylor, A. A., Hughes H. et al. Comparison of antioxidant and nonantioxidant lipoxygenase inhibitors on neutrophil function, implications for pathogenesis of myocardial reperfusion injury. *J. Pharmacol. Exp. Ther.*, 1990, v. 252, No. 2, pp. 531-538.

Sharapov V. I., Grek O. R., Dolgov A. V. [Mechanisms of lesion of liver monooxygenase systems in thermal ischemia]. (Russian). *Farmakol. Toksikol.,* 1986, v. 49, No. 2, pp. 64-68.

Sharapov V. I., Sokirchenko I. A., Greek O. R. et al. [Structural and functional aspects of postischemic regeneration of the rat hepatocyte endoplasmic reticulum]. (Russian). *Biul. Eksp. Biol. Med.*, 1986, v. 101, No. 3, pp. 277-279. English transl. see[1*], pp. 269-271.

Shargorodsky B. M., Rastorguev B. P. and Raiskina M. E. [Changes in redox processes in the heart in the acute stage of miocardium experimental infarction]. (Russian). *Biofizika*, 1967, v. 12, No. 1, pp. 114-119. English transl. see[2*], pp. 131-137.

Sharma H. S., Wuensch M., Brand T. et al. Molecular biology of the coronary vascular and myocardial responses to ischemia. *J. Cardiovasc. Pharmacol.*, 1992, v. 20, suppl. 1, pp. S23-S31.

Sharov V. G., Jennings R. B. and Hawkins H. K. [Comparative study of defects of cardiomyocyte membrane permeability in deep ischemia and after treatment with isoproterenol using colloidal lanthanum]. (Russian). *Arkh. Patol.,* 1980, v. 42, No. 10, pp. 35-44.

Sharov V. G., Severin V. V. and Mogilevsky G. M. [The ultrastructure of myocardial calcium necroses]. (Russian). *Biul. VKNC Akad. Med. Nauk SSSR,* 1985, v. 8, No. 1, pp. 63-77.

Shatos M. A., Doherty J. M. and Hoak J. C. Alteration in human vascular endothelial cell function by oxygen free radicals. Platelet adherence and prostacyclin release. *Arterioscl. Thromb.,* 1991, v. 11, No. 3, pp. 594-601.

Shedlofsky S. I., Bonkowsky M. L., Sinclair P. R. et al. Iron loading of cultured hepatocytes: effect of iron on 5' aminolaevulinate synthase is independent of lipid peroxidation. *Biochem. J.,* 1983, v. 212, No. 2, p. 321-330.

Sheker A. and Singh M. Role of eicosanoid inhibition of ischemia reperfusion injury: intact and isolated rat heart studies. *Methods Find. Exp. Clin. Pharmacol.,* 1997, v. 19, No. 4, pp. 223-229.

Shelenkova L. N. *Lipid peroxidation in ischemic organs and use of antioxidants for antiischemic protection.* [Dissertation]. (Russian). Moscow, 1982.

Shelenkova L. N. and Bilenko M. V. [Stabilization of membrane lipids by antioxidants as a method for preventing ischemic injuries to organs]. (Russian). In: *Structure, biosynthesis and metabolism of lipids in animal and human organisms. 3rd All-Union Symposium.* Abstracts. Leningrad, Nauka, 1978, p. 148.

Sherstnev M. P., Klebanov G. I. and Vladimirov Yu. A. [Antioxidant activity of synthetic glicerophospholipids]. (Russian). *Zh. Fiz. Khim.,* 1985, v. 59, No. 9, pp. 2283-2286.

Sheve T., Khalanchik V., Khibish C. and Rapaport S. M. [Mitochondrial lipids peroxidation and mitochondrial lysic by protein factor from reticulocytes]. (Russian). In: *Mitochondria accumulation of energy and regulation of enzymatic processes.* Moscow, Nauka, 1977, pp. 40-45.

Shigeno T., Yamasaki Y., Kato G. et al. Reduction of delayed neuronal death by inhibition of protein synthesis. *Neurosci. Lett.,* 1990, v. 120, pp. 117-119.

Shilina M. K., Ivashchenko S. P., Shrainberg A. I. et al. [Peroxidation of unsuturated lipids under experimental acute hypoxia of liver]. (Russian). In: *Liver insufficiency and issues of liver transplantation.* Moscow, 1972, pp. 70-73.

Shimamura T., Zhu Y., Zhang S. et al. Protective role of nitric oxide in ischemia and reperfusion injury of the liver. *J. Am. Coll. Surg.* 1999, v. 188, No. 1, pp. 43-52.

Shimkovich M. V. and Petkov O. [Prevention of adrenaline-induced heart lesions by the antioxidant ionol]. (Russian). *Biul. Eksp. Biol. Med.,* 1984, v. 98, No. 7, pp. 16-18. English transl. see[1*], pp. 869-870.

Shine K. I. Ionic events in ichemia and anoxia. Am. J. Path., 1981, v. 102, No. 2, pp. 256-261.

Shinkarenko V. S. and Morozov S. E. [Microhemodynamics in experimental ischemia]. (Russian). *Vestn. Acad. Med. Nauk SSSR,* 1988, No. 2, pp. 21-26.

Shipp J. C., Opie L. H. and Challoner D. Fatty acid and glucose metabolism in the perfused heart. *Nature,* 1961, v. 189, pp. 1018-1019.

Shiratori Y., Kiriyama H., Fukushi Y. et al. Modulation of ischemia-reperfusion-induced hepatic injury by Kupffer cells. *Dig. Dis. Sci.,* 1994, v. 39, No. 6, pp. 1265-1272.

Shkolovoy V. V. [Effect of myocardial ischemia on the ATPase systems]. (Russian). *Vopr. Med. Khim.,* 1984, v. 30, No. 5, pp. 41-43.

Shleikin A. G., Shulutko B. I., Oksman T. M. and Orlov A. V. [Effect of protectors on morphofunctional changes in rat kidneys in tourniquet shock]. (Russian.) *Patol. Fiziol. Eksper. Ter.,* 1992, No. 5-6, pp. 36-37.

Shlyapnikov V. N. and Bogdanova A. A. [Morphological and histochemical myocardium caused by adrenaline]. (Russian). *Arkh. Patol.,* 1979, v. 41, No. 9, pp. 22-28.

Shneivas V. B., Amilov K. S. and Levin G. S. [Role of lipid peroxidation in liver damage in visceral-ischemic shock]. (Russian.) *Patol. Fiziol. Eksper. Ter.,* 1994, No. 1, pp. 27-30.

Shoskes D. A., Xie Y. and Gonzalez-Cadavid N. F. Nitric oxide synthase activity in renal ischemia-reperfusion injury in the rat: implications for renal transplantation. *Transplantation,* 1997, v. 63, No. 4, pp. 495-500.

Shug A. L., Shrago E., Bittas N. et al. Acyl-CoA inhibition of adenine nucleotide translocation in ischaemic myocardium. *Am. J. Physiol.,* 1975, v. 228, pp. 689-692.

Shug A. L., Thomsen J. H. and Folts J. D. Changes in tissue level of carnitine and other metabolites during myocardial ischemia and anoxia. *Arch. Biochem. Biophys.,* 1978, v. 187, pp. 25-33.

Shuliakovskaya T. S., Arshinov V. Yu, Pakhomov V. Ju. et al. [Antioxidants as protectors of toxic action of diethylnitrosoamine. Study of antitoxic properties and mechanism of action]. (Russian). *Dokl. Acad. Nauk SSSR,* 1980, v. 254, No. 1, pp. 242-246. English transl. see[6*], pp. 278-281.

Shul'zhenko V. S., Kuz'min A. I. and Kapel'ko V. I. [Restoration of myocardial contractility during gradual reperfusion after total ischemia]. (Russian.) *Biul. Eksp. Biol. Med.*, 1989, v. 107, No. 4, pp. 407-409. English transl. see[1*], pp. 450-453.

Shumakov V. I. and Onishchenko N. A., eds. [*Actual questions of organs conservation.* Review]. (Russian). 1979, v. 2, Moscow, 107 p

Shumakov V. I., Galperin E. I. and Nekhludova E. A. [*Transplantation of liver*]. (Russian). Moscow, Meditsina, 1981, 288 p.

Shumakov V. I., Onishchenko N. A. and Shtengol'd E. S. [*Conservation of organs*]. (Russian). Moscow, Meditsina, 1975, 250 p.

Shumakov V. I., Onishchenko N. A. and Kirpatovsky V. I. [*Pharmocological protection of transplant*]. (Russian). Moscow, Meditsina, 1983, 203 p.

Siems W., Dubiel W., Dumdey R. et al. Accounting for the ATP-consuming processes in rabbit reticulocytes. *Eur. J. Biochem.*, 1984, v. 139, pp. 101-107.

Siems W., Kowalewski J., Werner A. et al. Radical formation in the rat small intestine during and following ischemia. *Free Rad. Res. Commun.*, 1989, v. 7, No. 3-6, pp. 347-353.

Siems W., Schmidt H., Müller M. et al. H_2O_2 formation during nucleotide degradation in the hypoxic rat liver: a quantitative approach. *Free Rad. Res. Commun.*, 1986, v. 1, No. 5, pp. 289-295.

Sies H. Oxidative Stress: introductory remarks. In: Sies H., ed. *Oxidative stress.* London, Academic Press. 1985, pp. 1-8.

Sies H. and Akerboom P. M. T., Glutathione disulfide efflux from cells and tissues. *Methods Enzymol.*, 1984, v. 105, pp. 445-451

Sies H., Roch O. R., Martin O. E. and Boveris A. Increased biliary glutathione disulfide release in chronically ethanol-treated rats. *FEBS Lett.*, 1979, v. 103, pp. 287-290.

Silver E. M. and Szabo S. Role of lipid peroxidation in tissue injury after hepatic ischemia. *Exp. Mol. Pathol.*, 1983, v. 38, No. 1, pp. 69-76.

287 Simidzhev I., Kagan V. E. and Minkov I. [Effect of alpha-tocopherol and its derivatives on ATPase activity and oxidative phosphorylation in rat liver mitochondria]. (Russian). *Biul. Eskp. Biol. Med.*, 1987, v. 104, No. 9, pp. 299-301. English transl. See[1*], pp. 1224-1226.

Simonsen S. G., Zhang J., Canada A. T. et al. Hydrogen peroxide production by monoamine oxidase during ischemia-reperfusion in the rat brain. *J. Cereb. Blood Flow Metab.*, 1993, v. 13, pp. 125-134.

Sinet R. M. and Garber P. Inactivation of the human CuZn superoxide dismutase during exposure to O_2 and H_2O_2. *Arch. Biochem. Biophys.*, 1981, v. 212, No. 2, pp. 411-416.

Singal P. K., Kapur N., Dhillon K. S. et al. Role of free radicals in catecholamine-induced cardiomyopathy. *Can. J. Phisiol. Pharmacol.*, 1982, v. 60, No. 11, pp. 1390-1397.

Skulachev V. P. [Energy, metabolite oxygen and electron transport along biological membranes. Review]. (Russian). *Usp. Sovr. Biol.*, 1989, v. 88, No. 2, pp. 163-180.

Slater T. F. Overview of methods used for detecting lipid peroxidation. *Methods in Enzymology*, 1984, v. 105. *Oxygen Radicals in Biological Systems.* Packer L. red. N. Y. Acad. Press. Inc. pp. 283-293.

Smirnov L. D. and Dyumaev K. N. [β-Hydroxyderivates of six membered nitrogenous heterocycles. Synthesis inhibiting effects and biological propeties. Review]. (Russian). *Khim. Farm. Zh.*, 1982, v. 16, No. 4, pp. 412-429.

Smirnova S. G., Synzynys B. I., Zamulaeva I. A. et al. [DNA replication in mammalian cells after the action of physical, chemical or biological agents. VI. DNA synthesis restoration in the cultured irradiated cells]. (Russian). *Tsitologiya*, 1991, v. 33, No. 8, pp. 75-81.

Smith D. S., Rehncrona S. and Siesjö. Bo. K. Barbiturates as protective agents in brain ischemia and as free radical scavengers in vitro. *Acta Physiol. Scand.*, 1980, Suppl. 492, pp. 129-134.

Smith M. W., Collan Y., Kahng M. W. and Trump B. F. Changes in mitochondrial lipids of rat kidney during ischemia. *Biochim. Biophys. Acta*, 1980, v. 618, pp. 192-201.

Smith W. L., De Witt D. L. and Day J. S. Purification, quantitation and localization of PgI_2 synthase using monoclonal antibodies. In: Samuelsson B. et al., eds. *Adv. Prostaglandin, Thromboxane, and Leukotriene Res.* N. Y., Raven Press, 1983, v. 11, pp. 87-92.

Snyder D. W., Grafford W. A., Glashow J. L. et al. Lysophosphoglycerides in ischemic myocardium effluents and potentiation of their arrhythmogenic effects. *Am. Physiol.*, 1981, v. 241, pp. H700-H707.

Sobel B. E., Corr P. B., Robinson A. K. et al. Accumulation of lysophosphoglycerides with arrhythmogenic properties in ischemic myocardium. *J. Clin. Invest.*, 1978, v. 62, pp. 546-553.

Sokirchenko I. A. [Changes in Kupffer cells after reversible ischemia in rat liver]. (Russian). *Biul. Eksp. Biol. Med.*, 1993, v. 116, No. 11, pp. 548-550. English transl. see[1*], pp. 1440-1442.

Sokirchenko I. A. and Shkurupii V. A. [Subcellular manifestations of postischemic changes in rat liver parenchyma under the correcting effect of α-tocopherol]. (Russian). *Biul. Eksp. Biol. Med.*, 1987, v. 103, No. 2, pp. 226-229. English transl. see[1*], pp. 260-263.

Sokolov V. S., Churakova T. D., Bulgakov V. G. et al. [Study of the mechanizms of action of the lipid peroxide oxidation products on the permeability of bilayer lipid membranes]. *Biofizika*, 1981, v. 26, No. 1, pp. 147-149. English transl. see[2*], pp. 152-155.

Solez K., Idbura T., Silva C. B. et al. Clonidine after renal ischemia to lessen acute renal failure and microvascular damage. *Kidney Int.*, 1980, v. 18, pp. 309-322.

Soloviev A. I. [Stimulation mechanisms of contractile activity of vascular smooth muscle at their hyperoxygenation]. (Russian). *Fiziol. Zh. SSSR*, 1984, v. 30, No. 4, p. 444-449.

Soloviev A. I. and Bershtein S. A. [Influence of glycolysis blokade in the vascular wall on phasic activity of the smooth muscles]. (Russian). *Biul. Eksp. Biol. Med.*, 1982, v. 94, No. 8, pp. 11-13. English transl. see[1*], pp. 1022-1024.

Sonnino R. E., Pigatt L., Schrama A. et al. Phospholipase A2 secretion during intestinal graft ischemia. *Dig. Dis. Sci.*, 1997, v. 42, No. 5, pp. 972-981.

Sorensen V., Nilsson U., Pettersson S. et al. Effects of pretreatment with an indeno-indole compound on lipid peroxidation in the cortex and medulla of rabbit kidneys after ischaemia-reperfusion. *Acta Physiol. Scand.*, 1997, v. 161, No. 3, pp. 403-409.

Soriano M. A., Ferre I., Rodriguez-Farre E. and Planas A. M. Apoptosis and c-Jun in the thalamus of the rat following cortical infarction. *Neuroreport.*, 1996, v. 7, No. 2, pp. 425-428.

Soriano S. G., Coxon A., Wang Y. F. et al. Mice deficient in Mac-1(CD11b/CD18) are less susceptible to cerebral ischemia/reperfusion injury. *Stroke*, 1999, v. 30, No. 1, pp. 134-139.

Sorokovoy V. I. and Vladimirov Yu. A. [Mitochondrial injury under anoxia]. (Russian). *Itogi Nauki i Tekhniki. Ser. Biophysics*. v. 5. *Molecular pathology of membrane structures*. Moscow, 1975, pp. 12-55.

Sorokovoy V. I., Petrenko Yu. U., Mochenova G. M. et al. [Changes in some lipid peroxidation parameters of albino rat liver mitochondria during anoxic injury simulated in vitro]. (Russian). *Biul. Eksp. Biol. Med.*, 1984, v. 98, No. 11, pp. 543-546. English transl. see[1*], pp. 1492-1495.

Soussi B., Idström J. P., Schersten T. and Bylund-Fellenius A. C. Cytochrome c oxidase and cardiolipin alterations in response to skeletal muscle ischaemia and reperfusion. *Acta Physiol. Scand.*, 1990, v. 138, pp. 107-114.

Southard J. H., Ametani M. S., Lutz M. F. and Belser F. O. Effects of hypothermic perfusion of kidneys on tissue and mitochondrial phospholipids. *Cryobiology*, 1984, v. 21, No. 1, pp. 20-24.

Southard J. H., Senzig K. A., Hoffmann R. et al. Toxicity of oxygen to mitochondrial respiration activity in hypothermically perfused canine kidneys. *Transplantation*, 1980, v. 29, pp. 459-461.

Spector A. A. and Yorek M. A. Membrane lipid composition and cellular function. *J. Lipid Res.*, 1985, v. 269, pp. 1015-1035.

Spector W. S. *Handbook of Biological Data*. Philadelphia, W. B. Saunders Co., 1966.

Spencer K. T., Lindower P. D., Buettner G. R. and Kerber R. E. Transition metal chelators reduce directly measured myocardial free radical production during reperfusion. *Cardiovasc. Pharmacol.*, 1998, v. 32, No. 3, pp. 343-348.

Spirichev V. B. and Kon' I. Ya. [Fat-soluble vitamins and membranes]. (Russian). *Zh. All-Union Khim., Society*, 1978, v. 23, No. 4, pp. 425-434.

Spiricheva N. V., Aksenova M. V., Orlov O. N. et al. [Damage to rat brain synaptosomal creatine phosphokinase through activation of molecular oxygen]. (Russian). *Biul. Eksp. Biol. Med.*, 1989, v. 107, No. 6, pp. 680-682. English transl. see[1*], pp. 791-793.

Spyridopoulos I., Brogi E., Kearney M. et al. Vascular endothelial growth factor inhibits endothelial cell apoptosis induced by tumor necrosis factor-α: balance between growth and death signals. *J. Mol. Cell. Cardiol.*, 1997, No. 29, pp. 1321-1330.

Spragg R. G. DNA strand break formation following exposure of bovine pulmonary artery and aortic endothelial cells to reactive oxygen products. *Am. J. Respir. Cell. Mol. Biol.*, 1991, v. 4, No. 1, pp. 4-10.

Squadrito F., Altavilla D., Zingarelli B. et al. Tumor necrosis factor involvement in myocardial ischaemia/reperfusion injury. *Eur. J. Pharmacol.*, 1993, v. 237, No. 2-3, pp. 223-230.

Stamler J. S., Singel D. J. and Loscalzo J. Biochemistry of nitric oxide and redox-activated forms. *Science*, 1992, v. 258, pp. 1898-1902.

Starzl T. E., Terblanche J., Porter K. A. et al. Growth stimulating factor in the regenerating canine liver. *Lancet*, 1979, v. 1, No. 8108, pp. 127-130.

Steen E. M., Noronha-Dutra A. M. and Rump S. The effects of lipid peroxidation on isolated heart muscle cells. *J. Clin. Pathol.,* 1982, v. 137, No. 1, p. 84.

Steenbergen C., Murphy E., Levy L. and London R. E. Elevation in cytosolic free calcium concentration early in myocardial ischemia in perfused rat heart. *Circ. Res.*, 1987, v. 60, No. 5, pp. 700-707.

Stefanov A. V. and Soloviev A. I. [Influence of liposomes on electrical and contractile activity of vascular smooth muscles]. (Russian). *Dokl. Acad. Nauk SSSR*, 1983, v. 273, No. 1, pp. 227-231. English transl. see[8*], pp. 238-241.

Stefanov A. V., Kreines V. M., Dmitrieva A. V. et al. [Use of liposomes as a nonspecific detoxicator the crush syndrome]. (Russian). *Biul. Eksp. Biol. Med.*, 1984, v. 97, No. 5, pp. 569-572. English transl. see[1*], pp. 607-610.

Stefanov A. V., Khromov A. S., Zhukova A. V. et al. [Prevention of postischemic myocardial damage by means of liposomes]. (Russian). *Biul. Eksp. Biol. Med.*, 1992, v. 113, No. 6, pp. 590-593. English transl. see[1*], pp. 787-790.

Steinbrecher U. P. Role of superoxide in endothelial-cell modification of low-density lipoproteins. Biochem. *Biophys. Acta,* 1988, v. 959, No. 1, pp. 20-30.

Stirpe F. and Della Corte E. The regulation of rat liver xantine oxidase. *J. Biol. Chem.*, 1969, v. 244, pp. 3855-3863.

Stohes S. J., Laxson T. A., Anderson L. and Bulding E. Effects of oltipraz, BHA, ADT and cabbage on glutathione metabolism, DNA damage and lipid peroxidation in old mice. *Mech. Aging Dev.,* 1986, v. 37, pp. 137-145.

Stohes S. J. and Bagchi D. Oxidative mechanisms in the toxicity of metal ions. *Free Rad. Biol. Med.*, 1995, v. 18, No. 2, pp. 321-336.

Strelkov R. B., Kucherenko N. G. and Kozlov V. M. [Modification of intestinal syndrome using gas hypoxic mixture in different irradiation conditions]. (Russian). *Radiobiologiya*, 1983, v. 23, No. 3, pp. 392-395.

Strenbergh W. C. 3rd, Tuttle T. M., Makhoul R. G. et al. Postischemic extremities exhibit immediate release of tumor necrosis factor. *J. Vasc. Surgery,* 1994, v. 20, No. 3, pp. 474-481.

Strosznaider J. and Dabrowitecki Z. Peroxidation of fatty acids in microsomes prepared from normal and ischemic guinea pig brains. *Bull. Acad. Pol. Sci., Ser. Sci. Biol.,* 1975, v. 23, No. 9, pp. 647-653.

Strubelt O., Younes M. and Li Y. Protection by albumin against ischaemia- and hypoxia-induced hepatic injury. *Pharmacol. Toxicol.*, 1994, v. 75, No. 5, pp. 280-284.

Struchkov V. A. and Strazhevskaya N. B. [DNA-bound lipids: composition and possible functions]. (Russian). *Biochimiya*, 1993; v. 58, No. 8, pp. 1154-1175. English transl. see[3*], pp. 815-830.

Suffys P., Beyoerk R., Van Roy F. and Fiers W. Reduced tumor necrosis factor-induced cytotoxicity by inhibitors of the arachidonic acid metabolism. *Biochem. Biophys. Res. Commun.,* 1987, v. 149, pp. 735-743.

Sugiyama S., Miyazaki Y., Kotaka K. and Ozawa T. On the mechanism of ischemia-induced mitochondrial disfunction. *Jpn. Circ. J.,* 1982, v. 46, No. 3, pp. 296-302.

Sukolinsky V. N., Konoplya E. F., Morozkina T. S. et al. [The influence of an antioxidant complex on macromolecule biosynthesis in hepatocytes of acutely exposed organisms]. (Russian). *Radiobiologiya*, 1988, v. 28, No. 2, pp. 238-240.

Sunde R. A. The biochemistry of selenoproteins. *J. Amer. Oil. Chem. Soc.,* 1984, v. 61, No. 12, pp. 1891-1900.

Sutton H. C. Efficiency of chelated iron compounds as catalysts for the Haber-Weiss reaction. *Free Rad. Biol. Med.*, 1985, v. 1, No. 3, pp. 195-202.

Suttorp N., Toepten W. and Roka L. Antioxidant defense mechanisms endothelial cells: glutathione redox cycle versus catalase. *Am. J. Physiol.*, 1986, v. 251, pp. C671-C680.

Suzuki Y., Lamkawa T. and Yamazaki N. Effects of L-carnitine on ventricular arrhythmias in dogs with acute myocardial ischemia and a supplement of excess fatty acids. *Jpn. Circ. J.,* 1981, v. 45, pp. 552-559.

Swain J. L., Sabina R. L., McHall Ph. A. et al. Prolonged myocardial nucleotide depletion after brief ischemia in the open-chest dog. *Am. J. Physiol.*, 1982, v. 242, pp. H818-H826.

Swoboda G., Fritzsche J. and Hasselbach W. Effect of phospholipase A_2 and albumin on the calcium-dependent ATPase and lipid composition of sarcoplasmic membranes. *Eur. J. Biochem.*, 1979, v. 95, pp. 77-88.

Takacs L. and Scolnik J. H. Effect of tourniquet shock and acute hemorrhage on the circulation on various organs in the rat. *Circulat. Res.*, 1962, v. 10, pp. 753-757.

Takahashi K. and Kako K. J. Ishemia-induced changes in sarcolemmal (Na^+, K^+)-ATPase, K^+-pNPPase, sialic acid, and phospholipid in the dog, and effects of the nisoldipine and chlorpromazine treatment. *Biochem. Med.*, 1984, v. 31, pp. 271-286.

Takahashi T., Takeyoshi I., Hasegawa Y. et al. Cardioprotective effects of Lazaroid U-74389G on ischemia-reperfusion injury in canine hearts. *J. Heart Lung Transplant.*, 1999, v. 18, No. 4, pp. 285-291.

Takenaka M., Tatsukawa Y., Dohi K. et al. Protective effects of α-tocopherol and coenzyme Q10 on warm ischemic damages of the rat kidney. *Transplantation*, 1981, v. 32, No. 2, pp. 137-141.

Takeo S. and Sakanashi M. Possible mechanisms for reoxygenation-induced recovery of myocardial high-energy phosphates after hypoxia. *J. Mol. Cell. Cardiol.*, 1983, v. 15, pp. 577-594.

Takeshiga H., Takayanagi R. and Manakami S. Lipid peroxidation and the reduction of ADP-Fe^{3+} chelate by NADH-ubiquinone reductase preperation from bovine heart mitochondria. *Biochem. J.*, 1980, v. 192, No. 3, pp. 861-866.

Takumura A., Fukuzuwa K., Akamatsu Y. et al. Identification of vasopressor phospolipid in crude soubean lecithin. *Lipids*, 1978, v. 13, No. 7, pp. 468-472.

Tan S., Gelman S., Wheat J. K. et al. Circulating xanthin oxidase in human ischemia-reperfusion. *South Med. J.*, 1995, v. 88, pp. 479-482.

Tan S., Yokoyama Y., Wang Z. et al. Hypoxia-reoxygenation is an damaging as ischemia-reperfusion in the rat liver. *Crit. Care Med.*, 1998, v. 26, No. 6, pp. 1089-1095.

Tan S., Yokoyama Y., Dickens E. et al. Xanthine oxidase activity in the circulation of rats following hemorrhagic shock. *Free Rad. Biol. Med.*, 1993, v. 15, No. 4, pp. 407-414.

Tanguy S., Boucher F., Besse S. et al. Trace elements and cardioprotection: increasing endogenous glutathione peroxidase activity by oral selenium supplementation in rats limits reperfusion-induced arrhythmias. *J. Trace Elem. Med. Biol.* 1998, v. 12, No. 1, pp. 28-38.

Tappel A. L. Lipid peroxidation damage to cell components. Fed. Proc., 1973, v. 32, pp. 1870-1874.

Tarasov V. A. [*Molecular mechanisms of reparation and mutagenesis*]. (Russian). Moscow, Nauka, 1982, 226 p.

Tarasova N. I., Kovalenko O. A. and Vanin A. F. [Mechanism of iron incorporation into animal tissues]. (Russian). *Biofizika*, 1981, v. 26, No. 4, pp. 678-782. English transl. see[2*], pp. 688-693.

Tatsukawa Y., Dohi J., Yamada K. et al. The role of coenzyme Q10 for preservation of the rat kidney: a model experiment for kidney transplantation. *Life Sci.*, 1979, v. 24, pp. 1309-1314.

Tavazzi B., Di Pierro D., Bartolini M. et al. Lipid peroxidation, tissue necrosis, and metabolic and mechanical recovery of isolated reperfused rat heart as a function of increasing ischemia. *Free Radic. Res.*, 1998, v. 28, No. 1, pp. 25-37.

Tel'pukhov V. I., Bilenko M. V., Tikhomirova A. I. et al. [Anti-ischemic protection of the brain by acelizine, a water-soluble aspirin]. (Russian). *Biul. Eksp. Biol. Med.* 1992, v. 113, No. 2, pp. 156-159. English transl. see[1*], pp. 202-206.

Tel'pukhov V. I., Bilenko M. V., Khokhlov A. V. and Komarov P. G. [Changes in EEG and higher nervous activity of rats receiving acelizine as anti-ischemic protection of the brain]. (Russian). *Biul. Eksp. Biol. Med.* 1992, v. 113, No. 3, pp. 245-247. English transl. see[1*], pp. 312-315.

Tel'pukhov V. I., Trenin S. O., Knyazeva G. D. and Lapkina T. I. [Prolongation of safe times of brain ischemia]. (Russian). *Vestn. Akad. Med. Nauk SSSR*, 1985, No. 4, pp. 65-71.

Terada L. S. Suppression of rabbit myocardial xanthine dehydrogenase activity by an endogenous compound. *J. Mol. Cell Cardiol.*, 1994, v. 26, pp. 125-132.

Terada L. S., Guidot D. M., Leff J. A. et al. Hypoxia injuries endothelial cells by increasing endogenous xanthine oxidase activity. *Nat. Acad. Sci. USA*, 1992, v. 89, No. 8, pp. 3362-3366.

Terada L. S., Willinghaw I. R., Rosandich M. E. et al. Generation of superoxide anion by brain endothelial cell xanthine oxidase. *J. Cell. Physiol.*, 1991, v. 148, No. 2, pp. 191-196.

Terao Y., Shibata O., Goto S. et al. Phospholipase A2 is activated in the kidney, but not in the liver during ischemia-reperfusion. *Res. Commun. Mol. Pathol. Pharmacol.*, 1997, v. 96, No. 3, pp. 277-289.

Thilman P., Xie Y., Kleihues P. and Kiessling M. Persistent inhibition of protein synthesis precedes delayed neuronal death in postischemic gerbil hippocampus. *Acta Neuropathol.,* 1986, v. 71, pp. 88-93.

Thomas C. E. and Jackson R. L. Lipid hydroperoxide involvement in copper-dependent and independent oxidation of low-density lipoproteins. *J. Pharmacol. Exp. Ther.*, 1991, v. 256, pp. 1182-1188.

Thomas G. R. and Thibodaux H. Transforming growth factor-beta 1 inhibits postischemic increases in splanchnic vascular resistance. *Biotechnol. Ther.,* 1992, v. 3, No. 3-4, pp. 91-100.

Thompson-Gorman S. L. and Zweier J. L. Evaluation of the role of xanthine oxidase in myocardial reperfusion injury. *J. Biol. Chem.*, 1990, v. 265, No. 12, pp. 6656-6663.

Thornton M. A., Winn R., Alpers C. E. and Zager R. A. An evaluation of the neutrophil as a mediator of *in vivo* renal ischemic-reperfusion injury. *Am. J. Pathol.,* 1989, v. 135, pp. 509-515.

Tien M., Svingen B. A. and Aust S. D. Superoxide dependent lipid peroxidation. *Fed. Proc.,* 1981, v. 40, No. 2, pp. 179-182.

Timoshin A. A., Tskitishvili O. V., Serebryakova L. I. et al. [Formation hydroxyl radicals in local ischemia of the canine heart]. (Russian). *Biofizika*, 1994, No. 3, pp. 502-506. English transl. see[2*], pp. 495-499.

Tkachenko S. S., Tikhilov R. M., Belousov A. E. et al. [Oxygenobarotherapy after emergency and planned microvascular operations in traumatological patients]. (Russian). *Khirurgia*, 1985, No. 11, pp. 14-17.

Toba R., Katagiri T. and Takeyama U. Studies on the cardiac sarcoplasmic reticulum in myocardial infarction. *Jpn. Circ. J.,* 1978, v. 42, No. 4, pp. 447-453.

Togashi H., Shinzawa H., Yong H. et al. Ascorbic acid radical, superoxide, and hydroxyl radical are detected in reperfusion injury of rat liver using electron spin resonance spectroscopy. *Arch. Biochem. Biophys.*, 1994, v. 308, No. 1, pp. 1-7.

Tokumura A., Kume T., Fukusawa K. et al. Cardiovascular effects of lysophosphatidic acid and its structural agents in rats. *J. Pharmacol. Exp. Ther.*, 1981, v. 219, pp. 219-224.

Toleikis A. I., Vavzhinkova Kh. and Tutterova M. [Role of carnitine in energy metabolism of mitochondria and myocardial cell in ischemia and postischemic reperfusion]. (Russian). *Vopr. Med. Khim.*, 1986, v. 32, No. 6, pp. 106-110.

Tominaga T., Kure S., Narisawa K. and Yoshimoto T. Endonuclease activation following focal ischemic injury in the rat brain. *Brain Res.,* 1993, v. 608, No. 1, pp. 21-26.

Tominaga T., Sato S., Ohnishi T. and Ohnishi S. T. Potentiation of nitric oxide formation following bilateral carotid occlusion and focal cerebral ischemia in the rat - *in vivo* detection of the nitric oxide radical by electron paramagnetic resonance spin trapping. *Brain Res.,* 1993, v. 614, No. 1-2, pp. 342-346.

Topham R. W., Walker M. C. and Calisch M. P. Liver xanthine dehydrogenase and iron mobilization. *Biochem. Biophys. Res. Commun.,* 1982, v. 109, No. 4, pp. 1240-1246.

Torchinsky Yu. M. [*Serum in proteins*]. (Russian). Moscow, Nauka, 1977, 302 p.

Török B., Röth E., Zsoldos T. et al. Lipid peroxidation in lung of rats exposed to hyperoxic, hypoxic and ischemic effects. *Exp. Pathol.,* Hungary, 1986, v. 29, No. 4, pp. 221-226.

Torres M., Jarvisalo J., Hakim J. The liver - protective enzymes against reduced forms of oxygen in phenobarbital-treated rats. *Enzyme*, 1981, v. 26, pp. 129-135.

Tosaki A., Blasig I.E., Pali T. and Ebert B. Heart protection in isolated working rat hearts. *Free Rad. Biol. Med.*, 1990, v. 8, No. 4, pp. 363-372.

Tovey M. G., Gugenheim J., Guymarho J. and Blanchard B. Genes for interleukin-1, interleukin 6, and tumor necrosis factor are expressed at markedly reduced levels in the livers of patients with severe liver disease. *Autoimmunity*, 1991, v. 10, No. 4, pp. 297-310.

Trachte G. J., Lefer A. M., Aharony D. and Smith J. B. Potent constriction of cat coronary arteria by hydroperoxides of arachidonic acid and its blockade by anti-inflammatory agents. *Prostaglandins*, 1979, v. 18, pp. 909-914.

Tranum-Jensen J., Janse M. I., Fidett I. W. et al. Tissue osmolility cell swelling and reperfusion in acute regional myocardial ischemia in the isolated porcine heart. *Circulat. Res.,* 1981, v. 49, No. 2, pp. 364-381.

Treshchinsky A. I., Levenets V. N. and Shlapak I. P. [The main principles and peculiarities of intensive therapy of shock caused by trauma]. (Russian). Klin. *Khirurgiya*, 1987, v. 4, pp. 21-24.

Trifillis A. L., Kahug M. W., Cowley Kh. et al. Metabolic studies of postischemic acute renal failure in the rat. *Exp. Mol. Pathol.,* 1984, v. 40., pp. 155-168.

Trinus F.P., Pisarev A.A., Chubenko A.V. et al. [Experimental-morphological studies of the effect of liposomes in CC_{14} poisoning]. (Russian). *Biul. Eksp. Biol. Med.*, 1985, v. 100, No. 12, pp. 714-715. English transl. see[1*], pp. 1703-1706.

Tripathy Y. and Hegde B. M. Effect of alpha-tocopherol pretreatment on infarct size following 90 minutes of ischemia and 4 hours of reperfusion in dogs. *Indian J. Physiol. Pharmacol.* 1997, v. 41, No. 3, pp. 241-247.

Tripathy Y., Hegde B. M. and Raghuveer C. V. Effect of superoxide dismutase and acidified sodium nitrite on infarct size following ischemia and reperfusion in dogs. *Indian J. Physiol. Pharmacol.* 1997, v. 41, No. 3, pp. 248-256.

Tritto I., Duilio C., Santoro G., et al. A short burst of oxygen radicals at reflow induces sustained release of oxidized glutathione from postischemic hearts. *Free Radic. Biol. Med.*, 1998, v. 24, No. 2, pp. 290-297

Trubetskoy A. V. and Lysenko L. T. [Features specific for the reaction of coronary vessels and metabolism of the myocardium in changing over from high- to low-frequency electric stimulation]. (Russian). *Kardiologiya*, 1975, v. 15, No. 3, pp. 76-81.

Tsao P. S. and Lefer A. M. Time course and mechanism of endothelial dysfunction in isolated ischemic- and hypoxic-perfused rat hearts. *Am. J. Physiol.*, 1990, v. 259, No. 6, Pt. 2, pp. H1660-H1666.

Tsutsumi E. and Takenaka F. Inhibition of piruvate kinase by free fatty acids in rat heart muscle. *BBA*, 1969, v. 171, pp. 355-357.

Tsyomik V. A. [Lipid peroxidation in myocardial mitochondria and sarcoplasmic reticulum in short-term deterioration of coronary circulation]. (Russian). *Vopr. Med. Khim.*, 1985, v. 31, No. 1, pp. 109-111.

Ugryumov A. [Morphological and hystoenzymic study of kidneys preserved by hypothermic perfusion]. (Russian). Dissertation, Moscow, 1976.

UK TIA Study Group: United Kingdom transient ischemic attack (UK-TIA) aspirin trial: Interim results. *Br. Med. J.*, 1988, v. 296, pp. 316-320.

Umemura A. [Regional difference in free fatty acids release and the action of phospholipase during ischemia in rat brain]. (Japanese). *No-To-Shinkei,* 1990, v. 42, №10, pp. 979-986.

Uraizee A., Reimer R. A., Murry Ch. E. and Jennings R. B. Failure of superoxide dismutase to limit size of myocardial infarction after 40 minutes of ischemia and 4 days of reperfusion in dogs. *Circulation*, 1987, v. 75, No. 6, pp. 1237-1248.

Vaage J., Ytrehus K., Myklebust R. et al. Oxygen free radicals and cardiac injury. Workshop on: *Oxygen free-radicals in shock.* Florence, 1985, pp. 66-67.

Vale M. G., Osorio V. R. and Carvalho A. P. Synthesis of adenosine triphosphate during release of intravesicular and membrane bound calcium ions from passively loaded sarcoplasmic reticulum. *Biochem. J.*, 1976, v. 156, pp. 239-244.

Val'dman A. V., Voronina T. A., Smirnov L. D. et al. [Effect of 3-hydroxypyridin derivatives on the central nervous system]. (Russian). *Biul. Eksp. Biol. Med.*, 1985, v. 99, No. 1, pp. 60-62. English transl. see[1*], pp. 73-75.

Van Belle H., Goossens F. and Wynantsi J. Formation and release of purine catabolites during hypoperfusion, anoxia and ischemia. *Am. J. Physiol.*, 1987, v. 252, No. 5, Pt. 2, pp. H886-H893.

van den Hoek T. L., Shao Z., Li C. et al. Mitochondrial electron transport can become a significant source of oxidative injury in cardiomyocytes. *J. Mol. Cell. Cardiol.*, 1997, v. 29, No. 9, pp. 2441-2450.

Vander Kraaij A. M., van Eijk H. G. and Koster J. F. Prevention of postischemic cardiac injury by the orally active iron chelator 1, 2-dimethyl-3-hydroxy-4-pyridone (L1) and the antioxidant (+)-cyanidanol-3. *Circulation*, 1989, v. 80, No. 1, pp. 158-164.

Vandeplassche G., Hermans C., Thone F. and Borgers M. Mitochondrial hydrogen peroxide generation by NADH-oxydase activity following regional myocardial ischemia in the dog. *J. Mol. Cell. Cardiol.*, 1989, v. 21, No. 4, pp. 383-392.

Van der Vusse G. J., Reneman R. S. and van Bilsen M. Accumulation of arachidonic acid in ischemic reperfusion cardiac tissue: possible causes and consequences. *Prostagland. Leucot. Essent. Fatty Acids,* 1997, v. 57, №1, pp. 85-93.

Van der Vusse G. I., Roemen Th. A., Prinzen F. W. et al. Uptake and tissue content of fatty acids in dog myocardium under normoxic and ischemic conditions. *Circulat. Res.*, 1982, v. 50, No. 4, pp. 538-546.

van der Worp H. B., Thomas C. E., Kappelle L. J. et al. Inhibition of iron-dependent and ischemia-induced brain damage by the alpha-tocopherol analogue MDL 74, 722. *Exp. Neurol.* 1999, v. 155, No. 1, pp. 103-108.

Van Diest M. I., Verbeuren T. J. and Herman A. G. Cyclo-oxygenase blockers influence the effects of 15-lipoxygenase metabolites of arachidonic acid in isolated canine blood vessels. *Prostaglandins*, 1986, v. 32, No. 1, pp. 97-100.

Vanin A. F., Kubrina L. N., Kiladze S. V. et al. [On factors influencing the formation of dinitrozyl complexes of non-haem iron in liver samples of mice and yeasts in vitro]. (Russian). *Biofizika*, 1977, v. 22, No. 4, pp. 646-650.

Van Jaarsveld H., Potgieter G. M., Kuyl J. M. et al. The effect of desferal on rat heart mitochondrial function, iron content, and xanthine dehydrogenase oxidase conversion during ischemia/reperfusion. *Clin. Biochem.*, 1990, v. 23, No. 6, pp. 509-513.

Van-Thiel D. H., Sakr M., Zetti G. and McClain C. FK 506 reduces the injury experienced following renal ischemia and reperfusion. *Renal Failure,* 1992, v. 14, No. 3, pp. 285-288.

van Jaarsveld H., Kuyl J. M., van Zyl G. F. and Barnard H. C. Salicylate in the perfusate during ischemia/reperfusion prevented mitochondrial injury. *Res. Comm. Mol. Path. Pharm.* 1994, v. 86, No. 3, pp. 287-295.

Van Waarde A. Avison M. J. Thulin G. et al. Role of nucleoside uptake in renal postischemic ATP synthesis. *Am. J. Physiol.*, 1992, v. 262, №6, pt. 2, pp. F1092-F1099.

Van Ye T. M., Roza A. M., Pieper G. M. et al. Inhibition of intestinal lipid peroxidation does not minimize morphologic damage. *J. Surg. Res.*, 1993, v. 55, No. 5, pp. 553-558.

Varani J., Phan S. H., Gibbs D. F. et al. H_2O_2-mediated cytotoxicity of rat pulmonary endothelial cells. Changes in adenosine triphosphate and purine products and effects of protective interventions. *Lab. Invest.,* 1990, v. 63, No. 5, pp. 683-689.

Varfolomeev S. D. and Mezch A. T. [*Prostaglandines as molecular bioregulators*]. (Russian). Moscow, MGU, 1985, 303 p.

Varskeviciene Z. Z., Cerniauskiene R. C. and Grybauskas P. S. Effect of α-tocopherol on the production of malondialdehyde in rat tissue homogenates after hypobaric exposure. *Gen. Physiol. Biophys*., 1984, v. 3, pp. 47-53.

Vartanyan L. S., Rashba Yu. E., Nagler L. G. et al. [Uncompensated (irreversible) changes in SOD activity, and superoxide radicall formation rate in different subcellular organelles during ischemia of the liver]. (Russian). In: *Bioantioxidant. 3rd All-Union Conference.* Abstracts. Moscow, 1989, v. 1, . pp. 38-39.

Vartanyan L. S., Rashba Yu. E., Nayler L. G. et al. [Membranes of subcellular organelles as a sourse of superoxide radicals in ischemia of the liver]. (Russian). *Biul. Eksp. Biol. Med.*, 1990, v. 55, No. 6, pp. 550-552. English transl. see[1*], pp. 731-734.

Vartanyan L. S., Sadovnikova I. P., Gurevich S. M. and Sokolova I. S. [Superoxide radical generation in membranes of subcellular organelles of regenerating rat liver]. (Russian). *Biokhimiya*, 1992, v. 57, No. 5, pp. 671-678. English transl. see[3*], pp. 454-459.

Vasdev S. C., Biro G. P., Narbaiz R. and Kako K. J. Membrane changes induced by early myocardial ischemia in the dog. *Can. J. Biochem.,* 1980, v. 58, pp. 1112-1119.

Vashchenko V. I., Esyrev O. V. and Uspanova Zh. K. [Effect of fatty acids on Ca^{2+} channels of sarcoplasmic reticulum skeletal muscles]. (Russian). *Ukr. Biokhim. Zh.,* 1995, v. 67, №2, pp. 41-45.

Vasilets L. A., Vornovitsky E. G. and Chodorov B. I. [Change in rabbit myocardial contractile function after burn shock different duration]. (Russia). *Biul. Eksp. Biol. Med.*, 1979, v. 87, No. 5, pp. 402-405.

Vasiliev R. F. and Vishutinskii A. A. [Chemiluminescent method of measuring the relations between elementary constants in liquid-phase oxidation of hydrocarbons]. (Russian). *Dokl. Acad. Nauk SSSR*, 1962, v. 145, No. 6, pp. 1301-1304.

Vejar M., Hackett D., Brunelli C. et al. Comparison of low-dose aspirin and coronary vasodilators in acute unstable angina. *Circulation*, 1990, v. 81, Suppl. 1, pp. I-4–I-11.

Veksler V. I., Khatkerich A. N. and Kapel'ko V. I. [High myocardial mitochondrial creatine kinase activity during ischemia is combined with impaired recovery of contractility after reperfusion]. (Russian). *Biul. Eksp. Biol. Med.*, 1991, v. 111, No. 6, pp. 572-574. English transl. see[1*], pp. 742-745.

Velikhanova D. M. [*Use of inhibitors of lipid peroxidation and inductors of microsomal monooxidase in ischemic and reperfusion injury of the liver*]. Dissertation. (Russian). Moscow, 1983.

Velikhanova D. M., Bilenko M. V. and Kagan V. E. [Peroxidation of lipids and injury to mixed function oxygenases in rat liver endoplasmic reticulum membranes during ischemia]. (Russian). *Biul. Eksp. Biol. Med.*, 1981, v. 92, No. 7, pp. 50-52. English transl. see[1*], pp. 913-915.

Verbolovich V. P., Teplova L. L., Petrenko E. P. et al. [Lipid peroxidation in the pathogenesis of postperfusion syndrom]. (Russian). *Izv. Akad. Nauk. KazSSR, Seriya Biol.,* 1982, No. 1, pp. 76-84.

Ver Donck L., Van Reempts J., Vandeplasche G. and Bogers M. A. A new method to study activated oxygen species induced damage in cardiomyocytes and protection by Ca^{2+}-antagonists. *J. Mol. Cell. Cardiol.*, 1988, v. 20, pp. 811-823.

Verkhatskii A. N. and Pidoplichkov I. [Ion currents in the membrane of isolated heart cells. Review]. (Russian). *Usp. Fiziol. Nauk,* 1987, v. 18, No. 1, pp. 23-42.

Vermylen J. and Deckmyn H. Reorientation of prostaglandin endoperoxide metabolism by a thromboxane synthetase inhibitor: in vitro and clinical observations. Br. *J. Clin. Pharmacol.*, 1983, v. 15, Suppl. 1, pp. 17S-22S.

Victor T., La Cock C. and Zochner H. Myocardial tissue free fatty acids. *J. Mol. Cell. Cardiol.*, 1984, v. 16, No. 8, pp. 709-721.

Villicara A., Kempski O. and Spatz M. Arachidonic acid and cerebromicrovascular endothelial permeability. *Adv. Neurol.,* 1990, v. 52, pp. 195-201.

Viner R. I., Novikov K. N., Kozlov Yu. P. et al. [Does alpha-tocopherol interact with the active site of cytochrome p-450 in liver microsomes]. (Russian). *Biul. Eksp. Biol. Med.*, 1987, v. 103, No. 3, pp. 304-306. English transl. see[1*], pp. 336-339.

Vitola J. V., Forman M. B., Holsinger J. P. et al. Role of endothelin in a rabbit model of acute myocardial infarction: effects of receptor antagonists. *J. Cardiovasc. Pharmacol.*, 1996, v. 28, No. 6, pp. 774-783.

Vladimirov Y. A. [Cholesterol and physical properties of membrane lipid bilayer]. (English transl. from Russian). In: Lopukhin Yu. M., Archakov A. I., Vladimirov Yu. A. and Kogan E. M. Cholesterenosis. *Membrane Cholesterol: Theoretical and Clinical Aspects.* Moscow, Meditsina, Harwood Acad. Publ., London, 1984, pp. 27-79.

Vladimirov Yu. A. [Physico-chemical mechanismus of barrier function breach in biological membrane lipid layers]. Biological membranes and membrane-active compounds. *Tashkent, FAN,* 1985, pp. 14-28.

Vladimirov Yu. A. and Archakov A. I. [*Lipid peroxidation in biological membranes*]. (Russian). Moscow, Nauka, 1972, 252 p.

Vladimirov Yu. A., Olenev V. I., Suslova T. B. et al. [Mechanism of lipid peroxidation and its effect on biological membranes]. (Russian). *Itogi Nauki i Tekhniki. Ser. Biophys.*, Molecular pathology of membrane structures. Moscow, 1975, v. 5, pp. 56-117.

Vladimirov Yu. A., Parfenov E. A., Epanchintseva O. M. et al. [Antiradical activity of copper(II) coumarin chelators]. (Russian). *Biul. Eksp. Biol. Med.*, 1992, v. 113, No. 5, pp. 479-481. English transl. see[1*], pp. 641-644.

Vladimirov Yu. A., Suslova T. B. and Olenev V. I. [Regulatorary role of Fe-ions in lipid peroxidation in mithochondries]. (Russian). *Mithochondrias. Electron-transport and transformation of energy.* Moscow, Nauka, 1976, pp. 109-125.

Vogt M. T. and Farber E. On the molecular pathology of ischemic renal cell death: reversible and irreversible cellular and mitochondrial metabolic alterations. *Am. J. Pathol.*, 1968, v. 53, pp. 1-26.

Volkova I. A. [*Changes in adenylic nucleotide content, the functions of the heart and kidneys following their total ischemia, as well as some ways of treatment these organs in the early postischemic period*]. Dissertation. (Russian). Moscow, 1985.

Volkova L. P., Bondarenko M. I. and Severin N. F. [The dynamics of lipoperoxides and free fatty acids in patients with acute myocardial infarction]. (Russian). *Vrach. Delo,* 1981, v. 12, pp. 35-38.

Von Asmuth E. J., Leeuwenberg J. F., Vander-Linden C. J. and Buurman W. A. Tumor necrosis factor-alpha induces neutrophil-mediated injury of cultured human endothelial cells. *Scand. J. Immunol.*, 1991, v. 34, No. 2, pp. 197-206.

Voogd A., Sluiter W., Van Eijk H. G. and Koster J. F. Low molecular weight iron and the oxygen paradox in isolated rat hearts. *J. Clin. Invest.*, 1992, v. 90, pp. 2050-2055.

Vornovitsky E. G., Len'kova N. A. and Vasilets L. A. [Changes in contractive activity of the rabbit myocardium in burn shock]. (Russian). *Biul. Eksp. Biol. Med.*, 1979, v. 87, No. 1, pp. 6-8. English transl. see[1*], pp. 5-7.

Voronov G. G. and Lukienko P. I. [Activity of microsomal oxidation enzymes in ischemized and intact lobes of the rat liver after resumption of liver circulation]. (Russian). *Farmakol. Toksikol.*, 1982, v. 45, No. 4, pp. 55-58.

Voronov G. G. and Lukienko P. I. [Stabilizing action of α-tocopherol in postischemic damage to the membrane hydroxylating system of the endoplasmic reticulum of rat liver]. (Russian). *Biul. Eksp. Biol. Med.*, 1983, v. 95, No. 4, pp. 33-34. English transl. see[1*], pp. 441-444.

Voronov G. G., Lukienko P. I. and Bushma M. I. [Variation in the activity of the enzymes of the rat liver endoplasmic reticulum hydroxylation system during ischemia]. (Russian). *Farmakol. Toksikol.*, 1982, v. 45, No. 1, pp. 54-58.

Wada K., Ito M., Miyazawa T. et al. Repeated hyperbaric oxygen induces ischemic tolerance in gerbil hippocampus. *Brain Res.*, 1996, v. 740, No. 1-2, pp. 15-20.

Wahlländer A., Soboll S. and Sies H. Hepatic mitochondrial and cytosolic glutathione content and the subcellular distribution of GSH-S-transferases. *FEBS Lett.*, 1979, v. 97, No. 1, pp. 138-140.

Waite M. Approaches to the study of mammalian cellular phospholipases. Review. *J. Lipid Res.*, 1985, v. 26, No. 12, pp. 1379-88.

Walker P. M., Lindsay T. F., Labbe R. et al. Salvage of skeletal muscle with free radical scavengers. *J. Vasc. Surg.*, 1987, v. 5, pp. 68-75.

Walsh T. R., Rao P. N., Makowka L. and Starzl T. E. Lipid peroxidation is a nonparenchymal cell event with reperfusion after prolonged liver ischemia. *J. Surg. Res.*, 1990, v. 49, No. 1, pp. 18-22.

Wang H. G., Shibamoto T. and Koyama S. Effect of platelet-activating factor on hepatic capillary pressure in isolated dog liver. *Prostagland. Leukot. Essent. Fatty Acids*, 1997, v. 57, No. 3, pp. 293-298.

Wang D., Yang X. P., Liu Y. H. et al. Reduction of myocardial ifarct size by inhibition of inducible nitric oxide synthase. *Am. J. Hypertens.* 1999, v. 12, No. 2, Pt. 1, pp. 174-182.

Wang P., Chen H., Qin H., et al. Overexpression of human copper, zinc-superoxide dismutase (SOD1) prevents postischemic injury. *Procced. Natl. Acad. Sci. USA*, 1998, v. 95, No. 8, pp. 4556-4560.

Wang W. Z., Anderson G., Fleming J. T. et al. Lack of nitric oxide contributes to vasospasm during ischemia/reperfusion injury. *Plast. Reconstr. Surg.*, 1997, v. 99, No. 4, pp. 1099-1108.

Wang X., Yang J. and Liu F. [Superoxide dismutase encapsulated erythrocytes used in the study of cerebral ischemia-reperfusion]. (Jpn). *Chung Kuo I Hsueh Ko Hsueh Yuan Hsueh Pao*, 1996, v. 18, No. 5, pp. 392-396.

Wang X., Zhou H. and Yang J. [GSH. Px or SOD encapsulated erythrocytes in the study of cerebral ischemia-reperfusion]. (Chinese). *Chung. Hua. I. Hsueh. Tsa. Chih.*, 1997, v. 77, No. 1, pp. 43-46.

Wanna F. S., Obayashi D. Y., Young J. N. and DeCampil W. M. Simultaneous manipulation of the nitric oxide and prostanoid pathways reduces myocardial reperfusion injury. *J. Thorac Cardiovasc. Surg.*, 1995, V. 110, No. 4, pt1, pp. 1054-1062.

Wanner G. A., Muller P., Ertel W. et al. Differential effect of cyclooxygenase metabolites on proinflammatory cytokine release by Kupffer cells after liver ischemia and reperfusion. *Am. J. Surg.* 1998, v. 175, No. 2, pp. 146-151.

Ward P. A. Mechanisms of endothelial cell killing by H_2O_2 or products of activated neutrophils. *Am. J. Med.*, 1991, v. 91, No. 3C, pp. 89S-94S.

Ward P. H., Maldonado M. and Vivaldi E. Oxygen-derived free radicals mediate liver damage in rats subjected to tourniquet shock. *Free Rad. Res. Commun.*, 1992, v. 17, No. 5, pp. 313-325.

Warnick C. T. and Lazarus H. M. Recovery of nucleotide levels after cell injury. *Can. J. Biochem.*, 1981, v. 59, pp. 116-121.

Watanabe N., Kuriyama H., Sone et al. Signalling pathway of TNF in normal and tumor cells. *J. Biol. Chem.*, 1988, v. 263, pp. 10262-10266.

Watanabe N., Onda M., Genga A. and Asano G. [The role of Ca^{2+}-ATPase and oxygen radical in reperfusion injury of rat liver]. (Jpn). *J. Jpn. Surg. Soc.*, 1993, v. 94, No. 12, pp. 1269-1276.

Watanabe T., Nishiyama M., Hori T. et al. Ebselen (DR3305) ameliorates delayed cerebral vasospasm in a canine two-hemorrhage model. *Neurol. Res.*, 1997, v. 19, No. 5, pp. 563-565.

Watts J. A., Maiorano L. J and Maiorano P. C. Protection by verapamil of globally ischemic rat hearts: energy preservation, a partial explanation. *J. Mol. Cell. Cardiol.*, 1985, v. 17, No. 8, pp. 797-804.

Wawra E., Zollner H., Schaur R. J. et al. The inhibitory effect of 4-hydroxy-nonenal on DNA polymerases alpha and beta from rat liver and rapidly dividing Yoshida ascites hepatoma. Cell. *Biochem. Function,* 1986, v. 4, pp. 31-36.

Weaver, J. and Pollack, S., Low-M_r iron isolated from guinea pig reticulocytes as AMP-Fe and ATP-Fe complexes. *Biochem. J.*, 1989, v. 261, pp. 787-792.

Weglicki W. B., Dickens B. F. and Tong-Mak I. Enhanced liposomal phospholipid degradation and lipophospholipid produced due to free radicals. *Biochem. Biophys. Res. Commun.* 1984, v. 124, №, pp. 229-235.

Weglicki W. B., Mak I. T. and Simic M. G. Mechanisms of cardiovascular drugs as antioxidants. *J. Mol. Cell. Cardiol.*, 1990, v. 22, pp. 1199-1208.

Weglicki W. B., Owens K., Urchel C. W. et al. Hydrolysis of myocardial lipids during acidosis and ischemia. *Recent Adv. Stud. Cardiac. Struct. Metab.*, 1973, v. 3, pp. 781-793.

Weinberg J. M. Oxygen deprivation induced injury to isolated rabbit kidney tubules. *J. Clin. Invest.*, 1985, v. 76, No. 3, pp. 1193-1208.

Wei J. and Quast M. J. Effect of nitric oxide synthase inhibitor on a hyperglycemic rat model of reversible focal ischemia: detection of excitatory amino acids release and hydroxyl radical formation *Brain Res.*, 1998, v. 791, No. 1-2, pp. 146-156.

Weisbrot-Lefkowitz M., Reuhl K., Perry B., et al. Overexpression of human glutathione peroxidase protects transgenic mice against focal cerebral ischemia/reperfusion damage. *Brain Res. Mol. Brain Res.*, 1998, v. 53, No. 1-2, pp. 333-338

Weisiger R. A. Oxygen radicals and ischemic tissue injury. *Gastroenterology*, 1986, v. 90, No. 2, pp. 494-496.

Weiss E. B. and Mullick P. C. Leukotriene effect in airways smooth muscle: calcium dependency and verapamil inhibitions. *Prostaglandins, Leukotrienes and Medicine, 1983*, v. 12, pp. 53-66.

Weiss S. J. Oxygen, ischemia and inflammation. *Acta Physiol. Scand.*, 1986, Suppl. 548, pp. 9-37.

Weixiong H., Aneman A., Nilsson U. and Lundgren O. Quantification of tissue damage in the feline small intestine during ischaemia-reperfusion: the importance of free radicals. *Acta Physiol. Scand.*, 1994, v. 150, No. 3, pp. 241-250.

Werns S. W., Grum C. M., Ventura A et al. and Lucochesi B. R. Xanthine oxidase inhibition does not limit canine infarct size. *Circulation*, 1991, v. 83, No. 3, pp. 995-1005.

Wexler L. F., Weinberg E. O., Ingwall J. S. et al. Acute alterations in diastolic left ventricular chamber distensibility: mechanistic differences between hypoxemia and ischemia in isolated perfused rabbit and rat hearts. *Circ. Res.*, 1986, v. 59, No. 5, pp. 515-528.

White B. C., De Gracia D. J., Krause G. S. et al. Brain nuclear DNA survives cardiac arrest and reperfusion. *Free Rad. Biol. Med.*, 1991, v. 10, pp. 125-135.

White F. C. and Bloor C. M. Coronary collateral circulation in the pig: correlation of collateral flow with coronary bed size. *Basic Res. Cardiol.*, 1981, v. 76, pp. 189-196.

Whitman G. J., Niibori K., Yokoyama H. et al. The mechanisms of coenzyme Q10 as therapy for myocardial ischemia/reperfusion injury. *Mol. Aspects Med.*, 1997, v. 18 (Suppl.), pp. S195-S203.

Whorton A. R., Montgomery M. E. and Kent R. S. Effect of hydrogen peroxide on prostaglandin production and cellular integrity in cultured porcine aortic endothelial cells. *J. Clin. Invest.*, 1985, v. 76, pp. 295-302.

Wildhirt S. M., Suzuki H., Horstman D. et al. Selective modulation in inducible nitric oxide synthase isozyme in myocardial infarction. *Circulation*, 1997, v. 96, No. 5, pp. 1616-1623.

Williams R. E., Zweier J. L. and Flaherty J. T. Treatment with deferoxamine during ischemia improves functional and metabolic recovery and reduces reperfusion-induced oxygen radical generation in rabbit hearts. *Circulation*, 1991, v. 83, No. 3, pp. 1006-1014.

Wills E. D. Effects of lipid peroxidation on membrane-bound enzymes of the endoplasmic reticulum. *Biochem. J.*, 1971, v. 123, No. 5, pp. 983-991.

Willson R. L. Free radical protection: why vitamin E, not vitamin b-carotene or glutathione? *Ciba Found. Symp.*, 1989, v. 101, pp. 19-44.

Wilson D. R., Arnold P. E., Burke Th. J. and Schrier R. W. Mitochondrial calcium accumulation and respiration in ischemic acute renal failure in the rat. *Kidney Int.*, 1984, v. 25, No. 3, pp. 519-526.

Wimalasena J., Davis M. and Kitabchi A. E. Characterization and solubilization of the specific binding sites for d-α-tocopherol from human erythrocyte membranes. *Biochem. Pharmacol.*, 1982, v. 31, No. 21, pp. 3455-3461.

Witschi H. and Saheb W. Stimulation of DNA synthesis in mouse lung following intraperithoneal injections of butilated hydroxytoluene. *Proc. Soc. Exp. Biol. Med.*, 1974, v. 147, No. 3, pp. 690-693.

Wolbarsht M. L. and Fridovich I. Hyperoxia during reperfusion is a factor in reperfusion injury. (Hypothesis Paper). *Free Rad. Biol. Med.*, 1989, v. 6, pp. 61-62.

Wolf H. K., Zarnegar R. and Michalopulos G. K. Localization of hepatocyte growth factor in human and rat tissues: an immunochistochemical study. *Hepatology*, 1991, v. 14, pp. 488-494.

Wolf R. A. and Gross R. W. Identification of neutral active phospholipase C which hydrolyzes choline glycerophospholipids and plasmalogen selective phospholipase A_2 in canine myocardium. *J. Biol. Chem.*, 1985, v. 260, pp. 7295-7303.

Wolfe J. T., Ross D., Cohen G. M. A role for metals and free radicals in the induction of apoptosis in thymocytes. *FEBS Lett.*, 1994, v. 352, No. 1, pp. 58-62.

Wolgast M., Karlberg L., Källskog Ö. et al. Hemodynamic alterations in ischemic acute renal failure. *Nephron*, 1982, v. 31, pp. 301-303.

Wolin M. S. Activated oxygen metabolites as regulators of vascular tone. (Review). *Klinische Wochenschrift.* 1991, v. 69, No. 21-23, pp. 1046-1049.

Wong G. H. W. and Goeddel D. V. Fas antigen and P55 TNF receptor signal apoptosis through distinct pathways. *J. Immunol.*, 1994, v. 152, No. 14, pp. 1751-1755.

Wong G. H. W., Elwell J. H., Oberley L. W. and Goeddel D. V. Manganous superoxide dismutase is essential for cellular resistance to cytotoxicity of tumor necrosis factor. *Cell,* 1989, v. 58, No. 5. pp. 923-931.

Wong J. K., Pruett T. L. and Jones R. S. Short-term effect of hepatic arterial versus portal venous reperfusion on energy levels of liver tissue. *Dig. Dis. Sci.,* 1990, v. 35, No. 11, pp. 1397-1402.

Wong P. Y. K. and Cheung W. Y. Calmodulin stimulates human platelet phosholipase A_2. *Biochem. Biophys. Res. Commun.,* 1979, v. 90, pp. 473-480.

Wood J. M., Rush B. and Pitts B. J. Inhibition of bovine heart Na^+ K^+ ATPase by palmitoylcarnitine and palmitoyl-CoA. *Biochem. Biophys. Res. Commun.*, 1977, v. 74, pp. 677-684.

Workshop on: *Oxygen free-radicals in shock.* Florence, 1985.

Wright J. G., Araki C. T., Belkin M. and Hobson R. W. 2nd. Postischemic hypothermia diminishes skeletal muscle reperfusion edema. *J. Surg. Res.*, 1989, v. 47, No. 5, pp. 389-396.

Wu S., Nagashima T., Ikeda K. et al. The mechanism of free radical generation in brain capillary endothelial cells after anoxia and reoxygenation. *Acta. Neurochir. Suppl. (Wien),* 1997, v. 70 pp. 37-39.

Wu T. W., Hasmihoto N., Au J. X. et al. Trolox protects rat hepatocytes against oxyradical damage and the ischemic rat liver from reperfusion injury. *Hepatology*, 1991, v. 13, iss. 3, pp. 575-580.

Wu W. and Qiu F. Experimental study on ischemia and reperfusion injury of rat liver and effects of ligustrazine and slavia compound. *Chinese Med. Sci. J.,* 1994, v. 9, No. 3, pp. 162-166.

Yadav S. S., Howell D. N., Steeber D. A. et al. P-Selectin mediates reperfusion injury through neutrophil and platelet sequestration in the warm ischemic mouse liver. *Hepatology*, 1999, v. 29, No. 5, pp. 1494-1502.

Yakhnina D. N. [Possible mechanism of a protective effect of α-tocopherol in experimental hypoxia]. (Russian). *Vopr. Med. Khim.,* 1980, v. 26, No. 1, pp. 88-92.

Yakubenko E. D., Mareeva T. E. and Manankov A. K. Correction of lipid peroxidation disturbances under experimental traumatic shock by the ionol antioxidant]. (Russian). *Ukr. Biokhim. Zh.,* 1985, v. 57, No. 1, pp. 79-82.

Yakushev V. S. and Lifshits R. I. [Effect of mistidine on content of malone dialdehyde in tissues under experimental myocardial infarction]. (Russian). *Vopr. Med. Khim.*, 1976, v. 22, No. 4, pp. 476-478.

Yakutova E. Sh., Osipov A. N., Kostenko O. V. et al. [Interaction of hypochlorite with oxyhemoglobin leads to liberation of iron in a catalytically active form]. (Russian). *Biofizika*, 1992, v. 37, No. 6, pp. 1021-1028. English transl. see[2*], pp. 919-926.

Yamada M., Hearse D. J. and Curtis M. J. Reperfusion and readmission of oxygen. Pathophysiological relevance of oxygen-derived free radicals to arrhythmogenesis. *Circ. Res.*, 1990, v. 67, No. 5, pp. 1211-1224.

Yamamoto M., Shima T., Uozumi T. et al. A possible role of lipid peroxidation in cellular damages caused by cerebral ischemia and protective effect of α-tocopherol administration. *Stroke*, 1983, v. 14, No. 6, pp. 977-982.

Yamamoto T., Cohen A. M., Kakar N. R. et al. Productions of prostanoids and nitric oxide by infarcted heart in situ and the effect of aspirin. *Biochem. Biophys. Res. Commun.* 1999, v. 257, No. 2, pp. 488-493.

Yamauchi N., Kuriyama H., Watanabe N. et al. Infracellular hydroxyl radical production induced by recombinant human tumor necrosis factor and its implication in the killing of tumor cells in vitro. *Cancer Res.,* 1989, v. 49, pp. 1671-1675.

Yanagisawa M., Kurihara H., Kimura S. et al. Endothelin a novel potent vasconstrictor peptide produced by vascular endothelial cells. *Nature*, 1988, v. 322, pp. 411-415.

Yanagishita T., Tomita M., Itoh S. et al. Protective effect of captopril on ischemic myocardium. *Jpn. Circ.* J., 1997, v. 61, No. 2, pp. 161-169.

Yang B. C. and Mehta J. L. Inhibition of nitric oxide does not affect reperfusion-induced myocardial injury, but it prevents lipid peroxidation in the isolated rat heart. *Life Sci.*, 1997, v. 61, No. 3, pp. 229-236.

Yano J., Riggs Th. R., Milam D. F. and Alexander J. C. Calcium-accentuated ischemic damage during reperfusion: the time course of the reperfusion injury in the isolated working rat heart model. *J. Surg. Res.*, 1987, v. 42, No. 1, pp. 51-55.

Yao Y. M., Bahrami S., Redl H. et al. IL-6 release after intestinal ischemia/reperfusion in rats is under partial control of TNF. *J. Surg. Res.*, 1997, v. 70, No. 1, pp. 21-26.

Yarmonenko S. P., Veinson A. A. and Magdon O. [*Oxygen action and tumour X-ray therapy*]. (Russian). Moscow, Meditsyna, 1980, 247 p.

Yavich M. P., Rozhitskaya I. I., Golubeva L. Yu. et al. [Effects of surgical stress on DNA synthesis in liver and brain cells]. (Russian). *Vopr. Med. Khim.*, 1990, v. 36, No. 6, pp. 8-11.

Yokoyama K., Nakamura K., Nakamura K. et al. Effect of coenzyme Q10 on superoxide production in rats with reperfusion injuries. *Scand. J. Reconstr. Surg. Hand. Surg.* 1999, v. 33, No. 1, pp. 1-5.

Yokoyama Y., Beckman J. S., Beckman T. K. et al. Circulating xanthine oxidase: potential mediator of ischemic injury. *Am. J. Physiol.*, 1990, v. 258, No. 4, Pt. 1, pp. G564-G570.

Yoshida S., Abe K., Busto R. et al. Influence of transient ischemia on lipid-soluble antioxidants. Free fatty acids and energy metabolites in rat brain. *Brain Res.,* 1982, v. 245, No. 2, pp. 307-316.

Yoshida S., Inoh S., Asano T. et al. Effect of transient ischemia on free fatty acids and phospholipids in the gerbil brain. *J. Neurosurg.,* 1980, v. 53, No. 3, pp. 323-331.

Yoshida T., Maulik N., Engelman R. M. et al. Glutathione peroxidase knockout are susceptible to myocardial ischemia reperfusion injury. *Circulation*, 1997, v. 96, No. 9, Suppl: II, pp. 216-220.

Yoshida T., Tanaka M., Sotomatsu A. and Hirai S. Activated microglia cause superoxide-mediated release of iron from ferritin. *Neurosci. Lett.,* 1995, v. 190, No. 1, pp. 21-24.

Yoshikawa T., Furukawa Y., Wakamatsu Y. et al. Experimental hypoxia and lipid peroxide in rats. *Biochem. Med.,* 1982, v. 27, No. 2, pp. 297-213.

Yoshikawa T., Ueda S., Naito Y. et al. Role of oxygen-derived free radicals in gastric mucosal injury induced by ischemia or ischemia-reperfusion in rats. *Free Rad. Res. Commun.*, 1989, v. 7, No. 3-6, pp. 285-291.

Yoshitatsu S., Von Lubitz D. K. J. E., Basile A. S. et al. Internucleosomal DNA fragmentation in gerbil hippocampus following forebrain ischemia. *Neurosci. Letters.,* 1994, v. 171, pp. 179-182.

Ytrehus K., Gunnes S., Myklebust R. et al. Protection by superoxide dismutase and catalase in the isolationed rat heart reperfused after prolonged cardioplegia: a combined study of metabolic functional and morphometric ultrastructural viriables. *Cardiovasc. Res.*, 1987, v. 21, No. 7, pp. 492-499.

Ytrehus K., Myklebust R. and Mjösi O. D. Influence of oxygen radical generated by xanthine oxidase in the isolated perfused rat heart. *Cardiovasc. Res.*, 1986, v. 20, No. 8, pp. 597-603.

Yu Z. F., Bruce-Keller A. J., Goodman Y. and Mattson M. P. Uric acid protects neurons against excitotoxic and metabolic insults in cell culture, and against focal ischemic brain injury *in vivo. J. Neurosci. Res.,* 1998, v. 53, No. 5, pp. 613-625.

Yusa K. and Shikama K. Oxidation of oxymyoglobin to metmyoglobin with hydrogen peroxide: involvement of ferryl intermediate. *Biochemistry*, 1987, v. 26, pp. 6684-6688.

Xia Y. and Zweier J. L. Substrate control of free radical generation from xanthine oxidase in the postischemic heart. *J. Biol. Chem.*, 1995, v. 270, No. 32, pp. 18797-18803.

Xu K. Y., Zweier J. L. and Becker L. C. Hydroxyl radical inhibits sarcoplasmic reticulum Ca(2+)-ATPase function by direct attack on the ATP binding site. *Circ. Res.*, 1997, v. 80, №1, pp. 76-81.

Zager R. A., Iwata M., Conrad D. S. et al. Altered ceramide and sphingosine expression during the induction phase of ischemic acute renal failure. *Kidney Int.,* 1997, v. 52, No. 1, pp. 60-70.

Zager R. A. and Gmur D. J. Effects of xanthine oxidase inhibition on ischemic acute renal failure in the rat. *Am. J. Physiol.*, 1989, v. 257, No. 6, Pt. 2, pp. F953-F958.

Zager R. A., Mahan J. and Merola A. J. Effects of manintol on the postishemic kidney. Biochemical, functional, and morphologic assessments. *Laboratory Investigation* v. 53, No. 4, pp. 433-442.

Zager R. A., Fuerstenberg S. M., Baehr P. H. et al. *An evaluation of antioxidant effects on recovery from postischemic acute renal failure.*

Zairov D. K., Bukov V. A., Sorokina M. I. et al. [Pathogenetic significance of central hemodynamic disorders and their association with myocardial ultrastructural changes in the early period of protracted crush syndrome in experiment]. (Russian). *Patol. Fiziol. Eksp. Ter.,* 1982, No. 2, pp. 14-19.

Zakharova M. V., Shkurupii V. A. and Grek O. R. [Effects of preadministration of phenobarbital and zyxorin on repair of rat liver parenchyma after ischemic injury]. (Russian). *Biul. Eksp. Biol. Med.*, 1994, v. 117, No. 2, pp. 185-188. English transl. see[1*], pp. 188-191.

Zakim D. The role of membrane lipids in the regulation of membrane bound enzymes. *Prog. Liver Dis.*, 1986, v. 8, pp. 65-80.

Zaks I. O. and Gabrielyan N. I. [Toxic activity of blood plasma in the early postresuscitation period]. (Russian). *Biul. Eksp. Biol. Med.*, 1984, v. 97, No. 5, pp. 545-547. English transl. see[1*], pp. 573-575.

Zamuruev O. N. [Restoration of blood supply and intensity of phospholipid metabolism in various parts of the postischemic rat brain]. (Russian). *Biul. Eksp. Biol. Med.*, 1979, v. 87, No. 2, pp. 139-142. English Transl. see[1*], pp. 118-120.

Zamuruev O. N. [Phospholipids and malonyl dialdehyde content in rat cerebral cortex during incomplete ischemia and in the postischemic period]. (Russian). *Biul. Eksp. Biol. Med.*, 1984, v. 98, No. 11, pp. 523-525. English transl. see[1*], pp. 1466-1468.

Zaslavsky Yu. S. [*Role of ubiquinone as antioxidant in membrane structure and function in norma and underadiation*]. (Russian). Dissertation, Puschino, 1984.

Zar H. A., Tanigawa K., Kim Y. M. and Lancaster J. R. Jr. Mild therapeutic hypothermia for postischemic vasoconstriction in the perfused rat liver. *Anesthesiology* 1999, v. 90, No. 4, pp. 1103-1111.

Zarnegar R., Muga S., Rahija R. and Michalopoulos G. Tissue distribution of hepatopoietin-A: a heparin-binding polypeptide growth factor for hepatocytes. *Proc. Natl. Acad. Sci. USA.*, 1990, v. 87, pp. 1252-1256.

Zaugg C. E., Hornstein P. S., Zhu P. et al Endothelin-1-induced release of thromboxane A_2 increases the vasoconstrictor effect of endothelin-1 in postischemic reperfused rat hearts. *Circulation*, 1996, v. 94, No. 4, pp. 742-747.

Zeeb I., Pfenninger E. and Grunert A. [The bowel as an ischemic organ]. (German.) *Anaesthesist*, 1990, v. 39, No. 7, pp. 343-352.

Zhang Y. H., Takahashi K., Zjang G. et al. In vivo induction of apoptosis (programmed cell death) in mouse thymus by administration of lipopolysaccharide. *Infection and Immunity*, 1989, v. 61, No. 12, pp. 5044-5048.

Zhang W., Naomoto Y., Tanaka N. et al. Sinergistic effect of tumor necrosis factor-alpha and interferon-alpha on the induction of apoptosis detected by BM-1 /Jimro - a new marker of apoptosis. *Acta Medica Okayama*, 1994, v. 48, No. 1, pp. 51-55.

Zhang Z., Blake D. R., Stevens C. R. et al. A reappraisal of xanthine dehydrogenase and oxidase in hypoxic reperfusion injury: the role of NADH as an electron donor. *Free Rad. Res.*, 1998, v. 28, No. 2, pp. 151-164.

Zhao Z. S., Khan S. and O'Brien P. J. Catecholic iron complexes as cytoprotective superoxide scavengers against hypoxia: reoxygenation injury in isolated hepartocytes. *Biochem. Pharmacol.*, 1998, v. 56, No. 7, pp. 825-830.

Zhdanov R. I. and Komarov P. G. Sterically-hindered hydroxylamines as bioactive spin labels. *Free Rad. Res. Commun.*, 1990, v. 9, No. 3-6, pp. 367-377.

Zherdev V. P., Sariev A. K., Dvoryaninov A. A. et al. [Pharmacokinetics of a water-soluble antioxidants of 3 hydroxypyridine type]. (Russian). *Biul. Eksp. Biol. Med.*, 1986, v. 101, No. 3, pp. 325-327. English transl. see[1*], pp. 333-335.

Zhestyannikov V. D. [*DNA reparation and its biological meaning*]. (Russian). Leningrad, Nauka, 1979, 285 p.

Zhizhina G. P., Karobko V. G. and Alessenko A. V. [Correlation of DNA degradation induced by the tumour necrosis factor alpha and with nuclear sphingosin content and DNA peroxide accumulation]. (Russian). *Biokhimiya*, 1994, v. 59, No. 11, pp. 1307-1313. English transl. see[3*], pp. 1756-1765.

Zhuravlev A. I. [Development of B. N. Tarusov's concepts on the role of chain processes in biology]. (Russian). In: Zhuravlev A. I., ed. *Bioantioxidants in regulation of metabolism in norma and pathology*. Moscow, Nauka, 1982, pp. 3-36.

Zieve L., Anderson W. R. and Lindblad S. Course of hepatic regeneration after 80-90 percent resection of normal rat liver. Comparison with two-lobe and one-lobe hepatectomy. *J. Lab. Clin. Med.*, 1985, v. 105, No. 3, pp. 331-336.

Zimmerman B. J., Parks D. A., Grisham M. B et al. Allopurinol does not enhance antioxidant properties of extracellular fluid. *Am. J. Physiol.*, 1988. v. 255, No. 1, Pt. 2, pp. H202-H206.

Zimmermann R., Schaper J., Munkel B. et al. In situ hybridization studies of acidic fibroblast growth factor (aFGF) in ischemic porcine myocardium. *J. Mol. Cell. Cardiol.*, 1993, v. 25, Suppl. 1, p. S. 84.

Zoli M., Ferraguti F., Zini I. et al. Increases in sulphated glycoprotein-2 mRNA levels in the rat brain after transient forebrain ischemia or partial mesodiencephalic hemitransection. *Brain. Res. Mol. Brain. Res.*, 1993, v. 18, No. 1-2, pp. 163-177.

Zorin A. V., Isanin N. A. and Yakovlev A. Yu. [The kinetic analysis of the cells starting DNA replication in systems stimulated to proliferate. III. The regenerating liver]. (Russian). *Tsitologiya*, 1975, v. 17, No. 7, pp. 783-789.

Zweier J. L., Duke S. S., Kuppusamy P. et al. Electron paramagnetic resonance evidence that cellular oxygen toxicity is caused by the generation of superoxide and hydroxyl free radicals. *FEBS Lett.*, 1989, v. 252, No. 1-2, pp. 12-16.

Zweier J. L., Kuppusamy P. and Lutty G. A. Measurement of endothelial cell free radical generation: evidence for a central mechanism of free radical injury in postischemic tissues. *Proc. Nat. Acad. Sci. USA*, 1988, v. 85, No. 1, pp. 4046-4050.

Zweier J. L., Samouilov. and Kuppusamy P. Non-enzymatic nitric oxide synthesis in biological systems. *Biochim. Biophys. Acta.*, 1999, v. 1411, No. 2-3, pp. 250-262.

INDEX

1

2

5

8

A

B

C

D

E

F

G

I

J

K

L

M

N

O

P

Q

R

S

V

W

X

α

β